Atomic Astrophysics and Spectroscopy

Spectroscopy allows the precise study of astronomical objects and phenomena. Bridging the gap between physics and astronomy, this is the first integrated graduate-level textbook on atomic astrophysics. It covers the basics of atomic physics and astrophysics, including state-of-the-art research applications, methods and tools.

The content is evenly balanced between the physical foundations of spectroscopy and their applications to astronomical objects and cosmology. An undergraduate knowledge of physics is assumed, and relevant basic material is summarised at the beginning of each chapter.

The material is completely self-contained and contains sufficient background information for self-study. Advanced users will find it useful for spectroscopic studies. Websites hosted by the authors contain updates, corrections, exercises and solutions, and news items from physics and astronomy related to spectroscopy. Links to these can be found at www.cambridge.org/9780521825368.

ANIL K. PRADHAN is a Professor in the Department of Astronomy, the Chemical Physics Program, and the Biophysics Graduate Program at The Ohio State University. He has taught a course on astrophysical spectroscopy for the past 20 years at The Ohio State University.

SULTANA N. NAHAR is a Senior Research Scientist in the Department of Astronomy at The Ohio State University. She has an extensive background in atomic physics and astrophysics and is one of the foremost researchers on atomic processes.

Atomic Astrophysics and Spectroscopy

Anil K. Pradhan
and
Sultana N. Nahar
Department of Astronomy
The Ohio State University

CAMBRIDGE
UNIVERSITY PRESS

CAMBRIDGE
UNIVERSITY PRESS

University Printing House, Cambridge CB2 8BS, United Kingdom

Cambridge University Press is part of the University of Cambridge.

It furthers the University's mission by disseminating knowledge in the pursuit of education, learning and research at the highest international levels of excellence.

www.cambridge.org
Information on this title: www.cambridge.org/9781107483583

First published 2011
First paperback edition 2015

A catalogue record for this publication is available from the British Library

Library of Congress Cataloguing in Publication data
Pradhan, Anil K.
Atomic astrophysics and spectroscopy / Anil K. Pradhan and Sultana N. Nahar.
p. cm.
Includes bibliographical references and index.
ISBN 978-0-521-82536-8
1. Atomic spectroscopy. 2. Astronomical spectroscopy. I. Nahar, Sultana N. II. Title.
QC454.A8P73 2011
523.01′97–dc22

2010036196

ISBN 978-0-521-82536-8 Hardback
ISBN 978-1-107-48358-3 Paperback

Contents

Preface

This text is aimed at students and researchers in both astronomy and physics. Spectroscopy links the two disciplines; one as the point of application and the other as the basis. However, it is not only students but also advanced researchers engaged in astronomical observations and analysis who often find themselves rather at a loss to interpret the vast array of spectral information that routinely confronts them. It is not readily feasible to reach all the way back into the fundamentals of spectroscopy, while one is involved in detailed and painstaking analysis of an individual spectrum of a given astrophysical object. At the same time (and from the other end of the spectrum, so to speak) physics graduate students are not often exposed to basic astronomy and astrophysics at a level that they are quite capable of understanding, and, indeed, that they may contribute to if so enabled.

Therefore, we feel the need for a textbook that lays out steps that link the mature field of atomic physics, established and developed for well over a century, to the latest areas of research in astronomy. *The challenge is recurring and persistent: high-resolution observations made with great effort and cost require high-precision analytical tools, verified and validated theoretically and experimentally.*

Historically, the flow of information has been both ways: astrophysics played a leading role in the development of atomic physics, and as one of the first great applications of quantum physics. As such, it is with basic quantum mechanics that we begin the study of astrophysical spectroscopy. The atomic physics and the astrophysics content are intended to be complementary, and attempt to provide a working knowledge in the two areas, as necessary for spectral analysis and modelling. The emphasis is on the introductory theoretical basics, leading up to a practical framework for applications of atomic spectroscopy. While we limit ourselves to atomic physics, we have attempted to highlight and delineate its reach into the main areas of astronomy.

The link between basic-to-advanced atomic physics and spectral analysis is increasingly important in ever more sophisticated astrophysical models. But the challenge of writing a book such as this one has been to find a balance between basic physics treatment that is not superficial, and state-of-the-art astrophysical applications that are not too technical. Though that defined and delimited the scope, it was still clear from the outset that the material should encompass a wide variety of topics. But what is essential and what is superfluous is, to some extent, a matter of subjective judgement. The level of depth and breadth of each topic is subject to these constraints. However, owing to the objective needs before us, we have tried to be as comprehensive as possible (limited by our own expertise, of course).

The text is evenly divided into atomic physics and astrophysics. The first seven chapters form the foundational elements of atomic processes and spectroscopy. The next seven chapters deal with astrophysical applications to specific objects and physical conditions. Each chapter follows the same plan. We begin with the essentials that all readers should be able to follow easily. However, towards the end of each chapter we outline some of the more advanced or specialized areas. The subject matter is broadly divided into 'basic' material in both areas, and 'advanced' material that incorporates state-of-the-art methods and results. The underlying atomic physics is intended as an introduction to more specialized areas, such as spectral diagnostics, astrophysical models, radiative transfer, plasma opacities, etc.

Emphasizing the unifying and connecting themes, the text is planned as follows. Following the Introduction, the next six chapters cover 'basic' collisional and radiative atomic structure and processes. The second part of the text, the other seven chapters, are the 'applications' of the physical framework developed in the first part. Chapters 8 and 9 describe the interaction of radiation with matter and spectral formation. The remainder of the text, Chapters 10–14, deals with descriptions of astronomical sources: stars, nebulae, active galactic nuclei and cosmology. A special chapter is devoted to a description of the largest single application of atomic physics to astronomy: stellar opacities (Chapter 11). However, the content of these chapters is *not* designed to be exhaustive, but mainly to exemplify spectral formation in astrophysical environments. Each of Chapters 10–14 contains tables and sample spectra characteristic of the particular astrophysical source(s). The appendices provide some of the tools, and some of the atomic data, needed in spectral modelling. However, they are not comprehensive and readers are advised to consult the websites described below.

Supplementary to the present text are the authors' websites.[1] They will provide continual updates and revisions related to atomic data and developments in atomic astrophysics. Eventually, this facility is designed to be user-interactive, with features such as on-line calculation of spectral line intensities and ratios, model calculations of ionization fractions, etc., using up-to-date atomic data.

1 www.astronomy.ohio-state.edu/ ~pradhan and www.astronomy.ohio-state.edu/ ~nahar.

Acknowledgements

The material in this book is partially based on several courses that Anil Pradhan has taught over the years. First of all, it is from a course on Theoretical Spectroscopy taught to astronomy graduate students at the Ohio State University every alternate year for nearly 20 years. Some of the material is also derived from graduate courses taught on atomic structure at the University of Windsor, scattering theory at the University of Colorado, and advanced undergraduate causes on stellar astrophysics at Ohio State. In addition, teaching introductory astronomy courses to non-science majors at Ohio State has been a valuable exercise in learning that, in addition to the discovery of wondrous new objects, some of the most basic and common phenomena in the Universe remain poorly understood (and *that* is the fun and raison d'etre for doing astronomy!).

But it is in the inspiration derived from our teachers and mentors wherein lies the foundation. The first acknowledgement – indeed a debt of gratitude – is due to Mike Seaton, advisor and mentor for over three decades. Mike was among the foremost pioneers who developed atomic astrophysics into the discipline it is today. Although he was not aware of this effort, and, regrettably, would not see it, Mike's monumental contributions are self-evident throughout the text. Nearly a decade ago, Dimitri Mihalas first suggested to Anil Pradhan the need for a book such as this. Dimitri has since then encouraged and advised on several aspects of the presentation, so well exemplified in his classic *Stellar Atmospheres*. From the observational side, Don Osterbrock continually pointed out over many years the specific needs for astrophysical diagnostics that could be fulfilled by the state-of-the-art atomic physics he appreciated so well. We also regret that Don is no more to see the fruit of his inspiration, howsoever imperfect this may be.

A number of our colleagues have read parts of the material and made numerous suggestions. We are especially grateful to our long-time collaborator, Werner Eissner, for revisions of the chapter on atomic structure. Special thanks are due to former student, postdoc, and now a valued colleague, Max Montenegro, for expertly and patiently (re-)drawing most of the figures in the book. Among the several colleagues who reviewed the material, Gerry Newsom and Bob Wing made particularly valuable comments on the chapters on stars, emission lines and nebulae. We also thank Dirk Grupe, Hong Lin Zhang, Prajaval Shastri, Bob Williams, Belinda Wilkes and David Branch, who read and suggested many corrections or improvements. But although all the material in the text has been reviewed by at least one of them, any errors, omissions and inaccuracies are entirely our responsibility. All we can say is that, fortunately, we have the electronic means to correct or revise any material, and would greatly appreciate readers pointing those things out to us. We shall endeavour to post all updates on the special website meant for this book (with due acknowledgement).

Finally, we would like to acknowledge the immense support and motivation derived from our families, to whom this work is dedicated. Anil Pradhan would like to thank his wife, Indira, and children, Alka and Vivek, his mother, Sarojini (who is no more), grandparents, Laksman Swarup and Phoolmati Pradhan, and parents Mahesh Chandra and Kunj Bala Pradhan. Sultana Nahar is grateful for inspiration from parents, Abdul Razzaq and Shamsun Nahar, teachers and family, especially her son, Alburuj R. Rahman.

1 Introduction

1.1 Atomic astrophysics and spectroscopy

Spectroscopy is the science of light–matter interaction. It is one of the most powerful scientific tools for studying nature. Spectroscopy is dependent on, and therefore reveals, the inherent as well as the extrinsic properties of matter. Confining ourselves to the present context, it forms the link that connects astronomy with fundamental physics at atomic and molecular levels. In the broadest sense, spectroscopy explains all that we see. It underlies vision itself, such as the distinction between colours. It enables the study of matter and light through the wavelengths of radiation ('colours') emitted or absorbed uniquely by each element. Atomic astrophysics is atomic physics and plasma physics applied to astronomy, and it underpins astrophysical spectroscopy. Historically, astrophysical spectroscopy is older than modern astrophysics itself. One may recall Newton's experiments in the seventeenth century on the dispersion of sunlight by a prism into the natural rainbow colours as an identification of the visible band of radiation. More specifically, we may trace the beginning of astrophysical spectroscopy in the early nineteenth century to the discovery of dark lines in the solar spectrum by Wollaston in 1802 and Fraunhofer in 1815. The dark lines at discrete wavelengths arise from removal or absorption of energy by atoms or ions in the solar atmosphere. Fraunhofer observed hundreds of such features that we now associate with several constituent elements in the Sun, such as the *sodium D lines*.

Figure 1.1 shows the *Fraunhofer lines*. Fraunhofer himself did not associate the lines with specific elements; that had to await several other crucial developments, including laboratory experiments, and eventually quantum theory. He labelled the lines alphabetically, starting from A in the far red towards shorter wavelengths. It is instructive to revisit the proper identification of these historic lines. Going from right to left in Fig. 1.1, the first two lines A (7594 Å) and B (6867 Å) do not originate in the Sun but are due to absorption by oxygen in the terrestrial atmosphere. The line C at 6563 Å is due to absorption by hydrogen (the same transition in *emission* is a bright red line). The three lines A, B and C lie towards the red end of the visible spectrum. In the middle region of the spectrum are the two orange lines D1 and D2 (5896, 5890 Å, respectively) that are the characteristic 'doublet' lines of sodium (sodium lamps have an orange hue, owing to emission in the very same transitions). Towards the blue end we have the strong line E at 5270 Å, due to absorption by neutral iron, and another line, F (4861 Å), due to hydrogen. The molecular G band of CH lies around 4300 Å. Farther into the blue, there are the H and K lines (3968, 3934 Å, respectively) from singly ionized calcium, which are among the strongest absorption lines in the solar spectrum. Although the letters have no physical meaning, this historic notation is carried through to the present day. Much of early astrophysics consisted of the identification of spectral lines, according to the presence of various atomic species in stars and nebulae.

The lightest and most abundant element in the Universe is hydrogen, chemical symbol H. The abundances and line intensities of other elements are expressed relative to H, which has the most common spectroscopic features in most astronomical sources. Observed line wavelengths led to an early grasp of specific spectra, but it needed the advent of quantum mechanics to understand the underlying structure. The pioneering exploration of the hydrogen spectrum and of alkali atoms by Rydberg was the first systematic attempt to analyze the pattern of spectral lines. We shall see later how useful simple variants of the empirical Rydberg formula can be in the analysis of astrophysical spectra.

Spectroscopy also predates quantum mechanics. In spite of the empirical work and analysis, a quantitative understanding of spectroscopy had to await the quantum

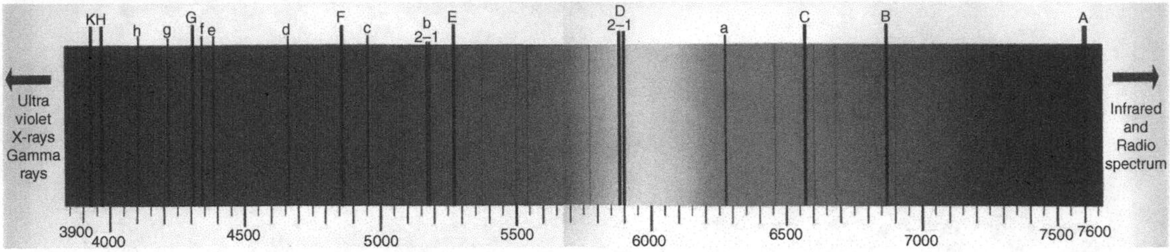

FIGURE 1.1 The Fraunhofer lines (Courtesy *Institute for Astronomy*, University of Hawaii, www.harmsy.freeuk.com).

theory of the atom. Schrödinger's success in finding the right equation that would reproduce the observed hydrogenic energy levels according to the Bohr model and the Rydberg formula was the crucial development. Mathematically, Schrödinger's equation is a rather straightforward second-order differential equation, well-known in mathematical analysis as Whittaker's equation [1]. It was the connection of its eigenvalues with the energy levels of the hydrogen atom that established basic quantum theory. In the next chapter, we shall retrace the derivation that leads to the quantization of apparently continuous variables such as energy. However, with the exception of the hydrogen atom, the main problem was (and to a significant extent still is!) that atomic physics gets complicated very fast as soon as one moves on to non-hydrogenic systems, starting with the very next element, helium. This is not unexpected, since only hydrogen (or the hydrogenic system) is a two-body problem amenable to an exact mathematical solution. All others are three-body or many-body problems that mainly have numerical solutions obtained on solving generalized forms of the Schrödinger's equation. With modern-day supercomputers, however, non-hydrogenic systems, particularly those of interest in astronomy, are being studied with increasing accuracy. A discussion of the methods and results is one of the main topics of this book.

Nearly all astronomy papers in the literature identify atomic transitions by wavelengths, and not by the spectral states involved in the transitions. The reason for neglecting basic spectroscopic information is because it is thought to be either too tedious or irrelevant to empirical analysis of spectra. Neither is quite true. But whereas the lines of hydrogen are well-known from undergraduate quantum mechanics, lines of more complicated species require more detailed knowledge. Strict rules, most notably the Pauli exclusion principle, govern the formation of atomic states. But their application is not straightforward, and the full algebraic scheme must be followed, in order to derive and understand which states are allowed by nature to exist, and which are not. Moreover, spectroscopic information for a given atom can be immensely valuable in correlating with other similar atomic species.

While we shall explore atomic structure in detail in the next chapter, even a brief historical sketch of atomic astrophysics would be incomplete without the noteworthy connection to stellar spectroscopy. In a classic paper in 1925 [2], Russell and Saunders implemented the then new science of quantum mechanics, in one of its first major applications, to derive the algebraic rules for recoupling total spin and angular momenta S and L of all electrons in an atom. The so-called *Russell–Saunders coupling* or *LS coupling* scheme thereby laid the basis for spectral identification of the states of an atom – and hence the foundation of much of atomic physics itself. Hertzsprung and Russell then went on to develop an extremely useful phenomenological description of stellar spectra based on spectral type (defined by atomic lines) vs. temperature or colour. The so-called *Hertzsprung–Russell (HR) diagram* that plots luminosity versus spectral type or temperature is the starting point for the classification of all stars (Chapter 10).

In this introductory chapter, we lay out certain salient properties and features of astrophysical sources.

1.2 Chemical and physical properties of elements

There are similarities and distinctions between the *chemical* and the *physical* properties of elements in the periodic table (Appendix 1). Both are based on the electronic arrangements in shells in atoms, divided in *rows* with increasing atomic number Z. The electrons, with principal quantum number n and orbital angular momentum ℓ, are arranged in *configurations* according to shells (n) and subshells (nl), denoted as 1s, 2s, 2p, 3s, 3p, 3d... (the number of electrons in each subshell is designated as the exponent). The chemical properties of elements are well-known. Noble gases, such as helium, neon and argon, have low chemical reactivity owing to the tightly bound closed shell electronic structure: $1s^2$ (He, $Z = 2$) $1s^2 2s^2 2p^6$ (Ne, $Z = 10$) and $1s^2 2s^2 2p^6 3s^2 3p^6$ (argon, $Z = 18$). The alkalis, lithium (Li, $Z = 3$), sodium (Na, $Z = 11$), potassium (K, $Z = 19$), etc., have relatively high chemical reactivity owing to the single valence electron ns outside a closed shell configuration, e.g., $1s^22s^1$ (Li) (see Chapter 2 for a detailed discussion). Chemical reactivity is responsible for molecular formation and processes. Sodium or potassium atoms combine easily with chlorine, a halogen with a vacancy in the 3p electronic orbit ($1s^22s^22p^63s^23p^5$), to form NaCl or KCl (common salts); the pairing is through an ionic bond, reflecting the fact that the Na atom 'donates' an electron, while the chlorine atom gains an electron to fill the 'vacancy' to close the outer shell. The chemical properties involving valence electrons and the reactivity of an element are determined by the *electron affinity*, the energy required to remove valence electrons. Atoms with more than one valence electron in an open shell form molecular bonds in a similar manner. The carbon atom has two valence electrons in the 2p shell, which can accommodate six electrons as a closed shell. The four vacancies can be filled by single electrons from four H atoms to form one of the most common molecular compounds in nature, CH_4 (methane), which, for instance, is probably the predominant constituent of 'oceans' on Saturn's moon Titan. Carbon monoxide, CO, is one of the most abundant molecular species in astronomical sources. Its stability lies in the match between the two valence electrons in the carbon atom and the two vacancies in the oxygen atom, which has four electrons in the 2p shell. In general, the chemical properties of elements are concerned with valence electrons and shells of atoms.

On the other hand, by physical properties of elements, we refer largely to spectroscopic and atomic processes, such as energy level structure, radiative transitions, excitations, ionization and more. Of course, these are also based on the electronic structure of atoms and ions but in a different manner than those of chemical processes. To begin with, the physical and chemical properties are expected to be similar for elements along the *columns* of the periodic table, since the electronic structures are similar (discussed in detail in Chapter 2). For example, boron (B) and aluminium (Al) both have a single valence electron in the p shell, preceded by an inner two-electron filled s shell: $1s^22s^22p^1$ (B) and $1s^22s^22p^63s^23p^1$ (Al). Therefore, the energy-level structure and processes involving those levels are usually similar. Both boron and aluminium display two-level fine structure splitting of the ground state $np\left({}^2P^o_{1/2} - {}^2P^o_{3/2}\right)$. Transitions between these two levels generate a weak 'forbidden' spectroscopic line in both the elements. Likewise, the atoms of flourine and chlorine in the halogen column have energy-level structures and spectral features similar to B and Al, owing to the fact that a single-vacancy p shell has the same spectral composition as a single-valence p electron: $1s^22s^22p^5$ (F) and $1s^22s^22p^63s^23p^5$ (Cl); both atoms also have the same ground state as B and Al, ${}^2P^o$, and the same type of forbidden transition.

From the point of view of atomic and astrophysical spectroscopy one of the most important manifestations of physical properties is for ions along an *isoelectronic sequence*: ions of different elements and atomic number Z, but with the same number N of electrons. For example, the helium isolectronic sequence consists of the ions of all elements of the periodic table stripped down to two electrons: $1s^2$ in the ground state (*He-like ions*. The columns of the periodic table already provide a guide to similarity of physical properties, for if similar electronic structure leads to similar properties, then the same electronic structure should do so also. For example, the singly charged carbon ion (expressed by C^+ or C II) has five electrons, isoelectronic with boron; similarly the nitrogen ion N III, the oxygen ion O IV and each ion of an element ($Z > 5$) with five electrons belongs to the boron sequence. However, there is a crucial physical difference with neutral elements along a column in the periodic table: not only is the atomic number Z different, but also the charge on each ion $+z = Z - N$ in the isoelectronic sequence is different. Therefore, the atomic physics, which depends basically on the electromagnetic potential in the atom or ion, is different for each ion. As Z increases, the attractive electron–nucleus Coulomb potential increases, resulting in higher-speed electrons. When the velocities are sufficiently high, relativistic effects become important. The energy-level splittings and processes dependent on relativistic and inter-electron interactions lead to significant differences in spectral formation for ions within the same isoelectronic sequence. We shall discuss a number of aspects of isoelectronic sequences in much more detail in later chapters.

Physical properties of elements also refer to interaction of radiation with matter on the atomic scale, which brings forth some physical processes not usually within the realm of chemistry, such as excitation and ionization of electrons.[1] Finally, physical phenomena are dependent

[1] To some extent the distinction between physical and chemical processes, as we have drawn here, is superficial from a fundamental viewpoint. But we do so purposefully to emphasize the physical nature of elemental species as they lead to atomic and astrophysical spectroscopy.

on environmental properties, primarily the temperature and density of the ambient plasma medium. The diversity of astrophysical environments makes it necessary to consider the intrinsic physical properties of atoms in conjunction with extrinsic plasma parameters. This book is concerned with the physical properties of elements in various ionization stages, particularly from an astrophysical perspective.

1.3 Electromagnetic spectrum and observatories

Astrophysical observations using state-of-the-art spectrometers on-board space missions and on ground-based telescopes are revealing spectral features at very high resolution in all wavelength ranges. Indeed, one may view astronomical sources as 'astrophysical laboratories' for atomic physics – a reversal of roles that greatly enhances the reach of both disciplines, atomic physics and astronomy.

Radiation emission from astronomical objects ranges over the whole electromagnetic spectrum from radio waves to gamma rays. The photon energy $h\nu$ and wavelength λ corresponding to each type of radiation are related inversely as

$$\nu = c/\lambda. \tag{1.1}$$

The least energetic radio wave photons have the longest wavelength $\lambda > 1$ m, and the most energetic gamma rays have wavelengths more than ten orders of magnitude smaller, $\lambda < 0.1$ Å. (Note that $1\,\text{Å} = 10^{-10}\,\text{m} = 10^{-4}\,\mu\text{m} = 0.1\,\text{nm}$, where a μm is also referred to as *micron* and 'nm' refers to nano-metre.)

Figure 1.2 is a schematic representation of the different regions of the electromagnetic spectrum of solar radiation transmitting through the terrestrial atmosphere. The atmosphere blocks out most regions of the spectrum (shaded area), except the optical or visible (vis), the near infrared (NIR), and the radio waves. The visible band is, in fact, a very narrow range in wavelength, but of course the one most accessible. The shaded regions are opaque to an observer on the ground, owing to higher atmospheric opacity. For example, water vapour in the atmosphere is very effective in blocking out IR radiation, owing to absorption by H_2O molecules, except in a few 'windows' or bands around 100–1000 nm or 1–10 μm (discussed later). This atmospheric 'blanketing' is also beneficial to us since it not only retains the re-radiated energy from the Earth (the greenhouse effect), but also absorbs the more energetic radiation from the Sun. Even a little more of the Sun's ultraviolet (UV) radiation could be biologically disastrous, not to mention the effect of high energy particles and other cosmic radiation of shorter wavelengths, which, although some do get through, are largely blocked out by the atmosphere. The use of ground-based telescopes is, therefore, confined to the wavelength ranges accessible from the Earth, after propagation of radiation through the atmosphere. For all other wavelengths we need to go into Outer Space.

Figure 1.2 also shows the general division of the electromagnetic spectrum for the Earth-based and space-based telescopes. Satellite-based space observatories make observations in the opaque regions. Some recent space observatories are the Compton Gamma-Ray Observatory (GRO), the X-ray Multi-Mirror Mission-Newton (XMM-Newton), the Chandra X-ray Observatory (CXO), Hubble Space Telescope (HST), Spitzer Infra-red Observatory, etc., respectively named after famous scientists: Arthur Compton, Isaac Newton, Subrahmanyan Chandrasekhar, Edwin Hubble and Lyman Spitzer. Another current mission includes the multi-wavelength X-ray–γ-ray NASA satellite Swift, to study gamma-ray bursts that are found to occur all across the sky, and X-ray observations from active galactic nuclei and other sources.

There is significant overlap in the approximate wavelength ranges given, depending on the detectors and instrumentation. Ground-based telescopes have sensitive spectrometers that can range somewhat outside the range

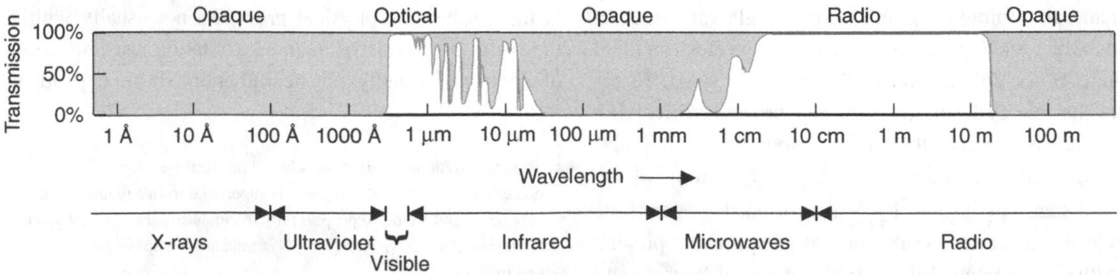

FIGURE 1.2 The electromagnetic spectrum of transmitted radiation through the Earth's atmosphere (http://imagine.gsfc.nasa.gov).

visible to the human eye, 4000–7000 Å. Optical CCDs (charge-coupled devices) can detect radiation from near UV to near IR, 3000–10 000 Å. The detector capability, measured in terms of the percentage of photons detected, called the *quantum efficiency*, deteriorates rapidly near the edges of some spectral windows. Subsequent chapters on astronomical objects will describe the prominent atomic species and spectral features. Atomic and molecular processes play the dominant role at all wavelengths except gamma-rays due to nuclear processes and electron–positron annihilation or synchrotron radiation.

Exercise 1.1 *Compile a list of current major ground and space observatories with spectroscopic instruments and corresponding wavelength ranges.*

1.4 Astrophysical and laboratory plasmas

Ionized materials in astrophysical plasmas constitute over 99% of the observed matter in the Universe – that is, all the matter in stars, nebulae and interstellar matter, which comprise observable galaxies.[2] As we mentioned, the analysis of characteristic light is the science of spectroscopy, and nearly all information on observable matter is derived from spectroscopy. This is how we really see the Universe in all its glory. Observable matter spans a huge range in density–temperature parameter space. Whereas the interstellar medium may be cold and thin, down to a few K and to less than one particle per cm^3, highly energetic plasmas in the vicinity of black holes at centres of galaxies may approach a thousand million K and immense (as yet unknown) densities. An important set of temperature–density combinations is the one in stellar cores: exceeding ten million K and 100 g cm^{-3}, conditions required for hydrogen nuclear fusion that provides most of stellar energy.

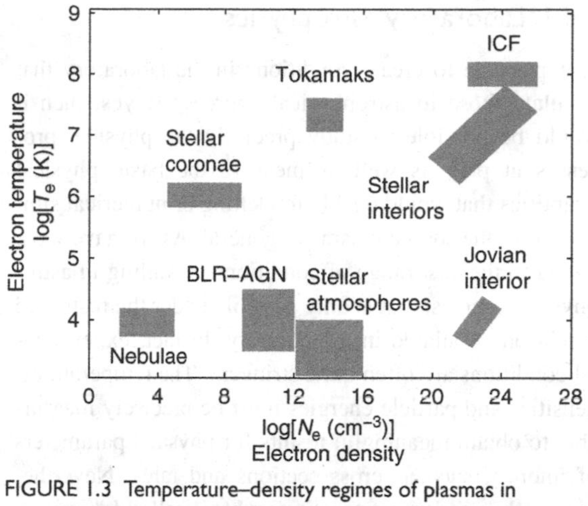

FIGURE 1.3 Temperature–density regimes of plasmas in astrophysical objects, compared with laboratory plasmas in magnetic confinement fusion devices, such as tokamaks, and inertial confinement fusion (ICF) devices, such as Z-pinch machines and high-powered laser facilities. BLR-AGN refers to 'broad-line regions in active galactic nuclei', where many spectral features associated with the central massive black hole activity manifest themselves.

Figure 1.3 shows astrophysical and laboratory plasma sources and their approximate temperatures and densities. As one might see, the astrophysical objects correspond to several regimes of electron temperature T_e and density n_e. Often, only some parts of a source are observed. Ordinary stars, for instance, range from a temperature $T \sim 2000$–3000 K in their outer atmospheres to $>10^7$ K in the core, where thermonuclear fusion creates their energy. The directly observable parts of a star are its *photosphere*, from which most of its radiation is emitted, and the hot highly ionized gas in the *corona*, a tenuous but extended region surrounding the main body of the star. In an extreme manifestation of temperature ranges, a stellar condition called supernova begins with an explosive plasma ball of some thousand million degrees, to less than 10^3 K after a few years of expansion into a 'nebular' remnant of the diffuse ionized plasma. It contains mainly H II and the material ejected from the progenitor star, as well as matter swept up from the interstellar medium. The detailed temperature–density–abundance–ionization structure of objects is revealed by spectral analysis of the observable regions of each type of object in different wavelength ranges, as discussed in individual chapters.

[2] It is worth mentioning how astronomers currently view matter and energy. There is considerable evidence that observable matter comprises only 4% of the Universe. About 22% is so-called 'dark matter' that apparently does not interact with electromagnetic radiation to emit or absorb light, and is therefore not observed. The existence of dark matter may be inferred by its gravitational influence on objects. For example, the rotation rate of matter within galaxies is observed *not* to decrease with increasing distance from the centre as expected, but rather remains roughly constant to very large distances. This implies that there is unseen matter in and beyond the observable halo of galaxies. Some of the matter may also be hidden in hot and highly ionized gas in the intergalactic medium, which is indicated by X-ray spectroscopy. The remaining 74% constituent is called *dark energy*, if one interprets the observed acceleration in the expansion rate of the Universe as part of the gravitational mass-energy balance. We discuss these topics in detail in Chapter 14.

1.4.1 Laboratory astrophysics

Is it possible to create conditions in the laboratory that simulate those in astrophysical sources? If yes, then it would be possible to study precisely the physical processes at play as well as measure the basic physical quantities that would enable modelling or numerical simulation of the source plasma in general. As we have seen, owing to the vast range of conditions prevailing in astrophysical sources, that is only possible under the restricted conditions available in a laboratory. In fact, experimental conditions are often quite stringent. The temperatures, densities and particle energies must be precisely measurable to obtain meaningful results for physical parameters of interest, such as cross sections and rates. Nevertheless, such experiments constitute what is called *laboratory astrophysics*. Laboratory devices need to create and maintain a plasma for sufficiently long periods of time for measurements to be carried out. This is quite difficult, especially for measuring *absolute* cross sections with high accuracy, since the results need to be calibrated relative to some independent criterion.

Essentially, the experimental techniques are designed to enable electrons and photons to interact with atoms or ions. This is accomplished either by *colliding* or *merging* beams of interacting particles, or in devices that confine an (electron + ion) plasma. Among these instruments for high-resolution measurements are *electron beam ion traps* (EBIT), which create a trapped plasma of ions interacting with an electron beam of controlled energy within some beam width, *ion storage rings*, where ions are magnetically trapped for long periods of time in a ring-like structure, enabling electron-ion experiments, and *advanced light sources* (ALS) of photons mounted on synchrotron accelerators for targeting the 'stored' ions and measuring photoionization cross sections. We will describe these experiments later, while discussing benchmarking and validation of theoretical results.

Laboratory plasma sources, while quite different from astrophysical sources in spatial and temporal behaviour, also span a wide range of temperature–density regimes. Most of the spectral diagnostics and atomic data we describe in this text also apply to laboratory plasmas. In particular, two classes of device for controlled thermonuclear fusion are shown in Fig. 1.3: (i) magnetically confined plasma reactors, called *tokamaks*, and (ii) *inertial confinement fusion* (ICF) devices. The ICF machines are essentially of two types. The first kind is based on laser-induced fusion, wherein an arrangement of symmetrically directed and powerful lasers is fired at a small deuterium-tritium pellet (heavy isotopes of hydrogen containing either one or two neutrons respectively), causing it to implode. The second kind are the so-called Z-pinch[3] machines, wherein a very high electrical discharge passes through wires of a heavy element, arranged cylindrically, which explode and emit X-rays that are directed towards the pellet. The fusion pellet is placed inside the cylindrical formation or cavity, which is called the *hohlraum*. At high temperatures, heavy elements exist in many ionization stages and emit copious amount of radiation; gold ($Z = 79$) is the common choice for hohlraum wires.

1.4.2 Astrophysical plasma composition and abundances

Astronomical objects are generally electrically neutral, i.e., an equal number of negative (electrons) and positive (protons and other ions) charges exists. The electrons are the dominant and most 'active' constituents, since their velocities compared with those of protons are $v_e/v_p = \sqrt{m_p/m_e} = 42.85$. In astronomical plasmas, typical proton densities are $\sim 80\%$ of n_e, and other heavier ions, such as helium nuclei (α particles), and partially or fully stripped ions of heavier elements constitute the rest of the positively charged particles. In astrophysical nomenclature, all elements heavier than helium are called 'metals'. Cosmic plasma compositions denote the H-abundance as 'X', He-abundance as 'Y', and all other metals combined as 'Z'. For instance, the solar elemental composition by mass is X = 0.70, Y = 0.28, Z = 0.02. Although metals constitute only 2% of the plasma, they are responsible for most of the spectral features, and they crucially determine properties, such as the plasma opacity that governs the transfer of radiation through the source (Chapter 10). Further study of plasmas in various situations requires us to consider the fundamental bulk properties associated with this most prevalent state of matter.

1.5 Particle distributions

A plasma of charged particles and a radiation field of photons can be treated with certain distribution functions.

[3] The 'Z' here refers *not* to the atomic number but the fact that a current passing along the z-axis through a wire creates a surrounding magnetic field, which acts naturally to constrain or 'pinch' the exploding plasma; the wires are indeed made of high-Z material!

1.5.1 Fermions and bosons

Concepts of indistinguishability and symmetry play a fundamental role in quantum mechanics. All particles of a given kind; electrons, protons, photons, etc., have the same observational property of being indistinguishable from other particles of the same kind. This universal fact is of profound importance and is known as the *principle of indistinguishability* [3]. Quantum mechanically, all observable quantities are expressed in terms of probabilities derived from a wavefunction formed for each kind of particle in terms of its spatial and spin coordinates. But the probabilities are related to the *squares* of the wavefunctions. That introduces an ambiguity in the actual sign of the wavefunction, which can be '+' or '−'. The total wavefunction of an ensemble of identical particles is therefore fixed by nature into two kinds. The first kind, *bosons*, refers to a *symmetric* total wavefunction, corresponding to the fact that interchange of coordinates of any two particles leaves the sign of the wavefunction unchanged. The second kind are called *fermions*, which correspond to an *antisymmetric* total wavefunction that changes sign upon interchange of coordinates.[4] The spin is a special 'coordinate', and has a value that is either integral (or zero), or half-integral. Bosons are particles of zero or integral spin, and fermions possess half-integral intrinsic angular momentum. Bosons and fermions obey different statistical mechanics: *Bose–Einstein* statistics in the case of bosons, and *Fermi–Dirac statistics* for fermions, both discussed in the next section.

1.5.2 Temperature: Maxwellian and Planck functions

The concept of 'temperature', which gives a measure of hot and cold in general sense, needs a more precise description in astronomy. For a given system of particles, say photons or electrons, the temperature has a meaning if and only if it corresponds to a distinct radiation (photon) or particle (electron) energy *distribution*. In the ordinary sense, the 'temperature' of a photon or an electron, or even a photon or electron beam, is meaningless. But the root-mean-square (rms) particle energy may be simply related to the *kinetic* temperature according to

$$E = h\nu \sim kT, \text{ and } E = 1/2\, mv^2. \tag{1.2}$$

With three-dimensional compoments of the velocity,

$$\frac{1}{2}mv^2 = \frac{3}{2}kT, \; k = 8.6171 \times 10^{-5}\,\mathrm{eV\,K^{-1}} = 1.380\,62 \times 10^{-16}\,\mathrm{erg\,K^{-1}} \tag{1.3}$$

where k is the Boltzmann constant. Consider a star ionizing a molecular cloud into a gaseous nebula. *Nebulae* are a class of so-called H II regions where the principal ionic species is ionized hydrogen (protons). The two distinct objects, the star and the nebula, have different temperatures; one refers to the energy of the radiation emitted by the star, and the other to the energy of electrons in the surrounding ionized gas heated by the star.

1.5.2.1 *Black-body radiation and the Sun*

The total energy emitted by an object per unit area per unit time is related to its temperature by the *Stefan–Boltzmann Law*

$$E = \sigma T^4, \;\text{ where } \sigma = 5.67 \times 10^{-8}\,\mathrm{W\,(m^{-2}\,K^{-4})} \tag{1.4}$$

is known as the Stefan constant. The Stefan–Boltzmann relation holds for a body in *thermal equilibrium*. The term *black body* expresses black colour or rather the lack of any preferred colour, absorbing radiation most efficiently at *all* wavelengths. *Kirchhoff's law* states that the emissivity of a black body is related to its absorptivity; a black body is also the most efficient radiator (emitter) at all wavelengths (discussed further in Chapter 10). At any temperature, a black body emits energy in the form of electromagnetic radiation at *all* wavelengths or frequencies. However, the distribution of emitted radiation changes with temperature such that the peak value and form of the distribution function defines a unique temperature for the object (black body), as discussed in the next section. The total *luminosity* L of a spherical black body of radius R, such as a star, integrated over all frequencies, is called the *bolometric luminosity*,

$$L = 4\pi R^2 \sigma T^4. \tag{1.5}$$

The radiation field of a star, considered to be a black body, is given by the Planck distribution function,[5] which defines the energy–frequency relationship at a given temperature:

$$B_\nu(T_*) = \frac{2h\nu^3}{c^2}\frac{1}{\exp(h\nu/kT_*) - 1}, \tag{1.6}$$

where T_* is the *radiation temperature* of the star and ν is the frequency of the photons. In terms of wavelengths it reads

[4] A simple and elegant 'proof' is given in the classic textbook by E. U. Condon and G. H. Shortley [3].

[5] This is the underlying radiation field, which is attenuated by spectral features, such as lines and bands particular to the star.

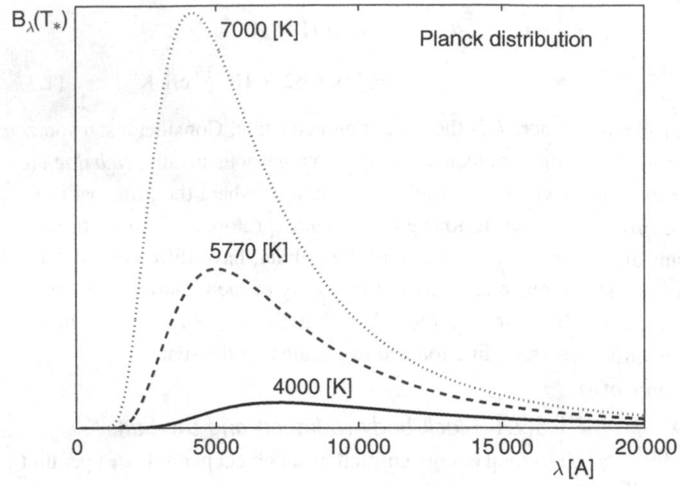

FIGURE 1.4 Planck distributions of photon intensity vs. wavelength at radiation temperatures T_* of various stars. Light from the Sun corresponds to $T_* = 5770$ K, which peaks at wavelengths around yellow; stars with higher(lower) temperature are bluer (redder). The Planck function B_λ is discussed in the text.

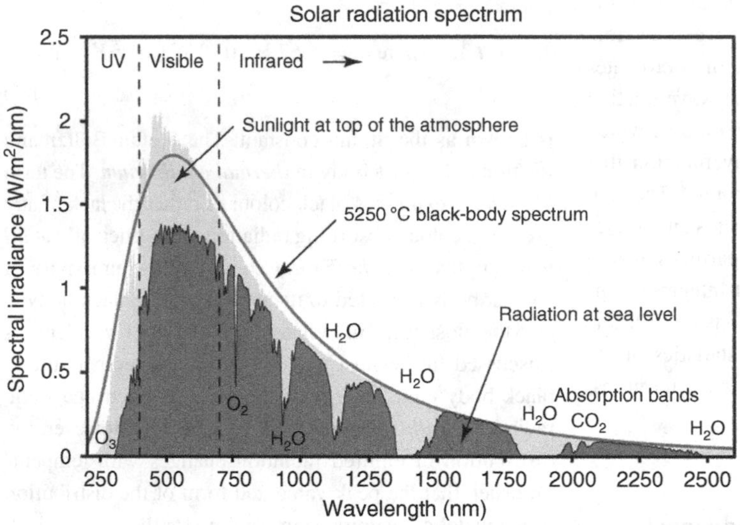

FIGURE 1.5 Sunlight as received at the top of the atmosphere and at sea level after attenuation by atmospheric constituents, primarily ozone, oxygen, water and carbon dioxide. (Courtesy Robert A. Rhode, http://globalwarmingart.com/wiki/Image:Solar_Spectrum.png).

$$B_\lambda(T_*) = \frac{2hc^2}{\lambda^5} \frac{1}{\exp(hc/\lambda k T_*) - 1}. \tag{1.7}$$

A surface temperature of $T_* = 5770$ K corresponds to the peak emission of a black body in the characteristic colour of the Sun – yellow – around 5500 Å. Hotter stars radiate more in the blue or ultraviolet and are 'bluer'; cooler stars radiate a greater fraction in the infrared and are 'redder' than the Sun. Figure 1.4 shows the black-body curves for several temperatures T_* representative of stars. *Wien's law* states that the black body distribution $B_\lambda(T)$ peaks at

$$\lambda_p = \frac{2.8978 \times 10^7 \text{ Å}}{T/\text{K}}, \tag{1.8}$$

thus peaking at $\dfrac{2.898 \times 10^3 \text{ Å}}{T/10\,000\text{ K}}$ or 2900 Å at 10 000 K.

We have already noted the historical relevance of the spectrum of the Sun. But, of course, the Sun is of great importance otherwise. It is therefore instructive to introduce a few salient features of solar spectra, some of which we shall deal with in later chapters.

Figure 1.5 illustrates several aspects of 'sunlight' as received on the Earth. First, following the discussion above about black-body curves associated with the radiation of stars, Fig. 1.5 is fitted to a black body at a slightly lower temperature, 5250 K, than at the surface of the Sun. The best fitting Planck function corresponds to a somewhat lower temperature than the actual spectrum observed *above* the atmosphere (light grey). The spectrum at *sea level* (dark grey) is seen to be significantly attenuated by absorption by the constituents of the atmosphere, primarily molecular bands due to water, oxygen and ozone. Figure 1.5 also shows that, although the peak of Sun's radiation is in the visible, there is a long tail indicating significant flux in the infrared. Water vapour in the atmosphere absorbs much of the longer wavelength ($\lambda > 1000$ nm $= 10\,000$ Å $= 1\,\mu$m) IR radiation via molecular transitions in H_2O.[6] But water also allows a considerable amount of solar radiation to be transmitted through the atmosphere through what are referred to as 'windows' in certain wavelength bands where H_2O has inefficient absorption (weak molecular transitions). Three of these windows are of particular importance, since they enable astronomical observations to be made from ground level in these IR bands, referred to as the *J, H and K bands*[7] centred around 1.2 μm, 1.6 μm and 2.2 μm, respectively.

Another interesting feature of Fig. 1.5 is the difference in radiation above and below the atmosphere on the UV side where, unlike the IR, the solar UV flux drops off rapidly. This is the *ozone effect*, as O_3 prevents the harmful UV radiation from reaching the Earth and thereby makes life as we know it possible.

1.5.2.2 *Maxwellian particle distribution*

Using old quantum theory (before the invention of wave mechanics), Einstein proposed an explanation of the *photoelectric effect* that relates Planck's quantum of energy $h\nu$ to absorption by an atom with the ejection of an electron. For instance, if the atom is surrounded by other atoms as in a metal, then a certain amount of energy is needed for the electron to escape. Hence the kinetic energy of the photoelectron is obtained as

$$\frac{1}{2}mv^2 = h\nu - W, \tag{1.9}$$

where W is called a *work function*. In the process of photoionization, where an atom or ion is ionized by absorbing a photon, W may be thought of as the *ionization energy* E_I of a bound electron.

The charged particles in the plasma ionized by a star in an H II region have an *electron temperature* T_e associated with the mean kinetic energy of the electrons given by Eq. 1.2. But it makes little sense to refer to the temperature of a single particle. Hence an averaged kinetic energy over a specified distribution of particle velocities is

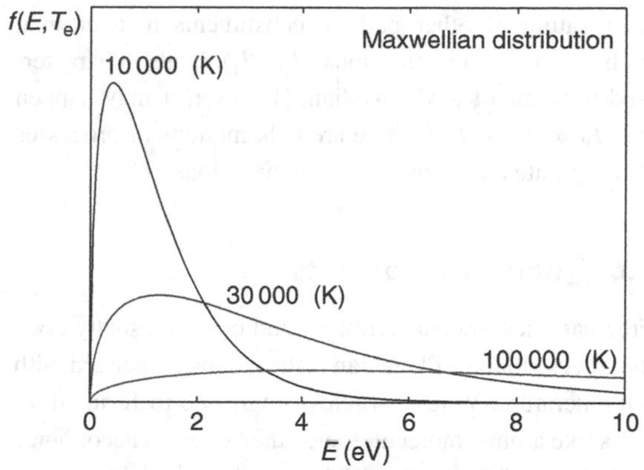

FIGURE 1.6 Maxwellian distributions, $f(E, T_e)$ of free electron energies at three bulk plasma kinetic temperatures T_e.

defined. In most astrophysical sources the fractional probability of electrons as a function of velocity or energy is characterized by a Maxwellian distribution of electrons at temperature T_e as

$$f(v) = \frac{4}{\sqrt{\pi}}\left(\frac{m}{2kT}\right)^{3/2} v^2 \exp\left(-\frac{mv^2}{2kT}\right). \tag{1.10}$$

Figure 1.6 shows the general form of the Maxwellian distribution functions at a few characteristic temperatures T_e. An example of the distinction, as well as the physical connection, between the Planck function and the Maxwellian function is found in H II regions. They are ionized by the hottest stars with black-body temperatures of $T_* \approx 30\,000$–$40\,000$ K. The resulting ionization of hydrogen in a molecular cloud (see Chapter 12) creates a plasma with electron kinetic energies that can be described by a Maxwellian distribution at $T_e \approx 10\,000$–$20\,000$ K. Since 1 eV $\equiv 11\,600$ K (see Eq. 1.3), the electron temperature is of the order of 1 eV in H II regions.

Plasmas need not always have a Maxwellian distribution; electron velocity (energy) distributions may not be given by Eq. 1.10. For example, in the expanding ejecta of supernovae, solar flares or laboratory fusion devices, some electrons may be accelerated to very high velocities. Such non-Maxwellian components or high-energy 'tails' may co-exist in a source otherwise characterized by a Maxwellian plasma that defines the bulk kinetic temperature. Another example is that of mono-energetic beams used in laboratory experiments; the beam widths may be described by the well-known Gaussian distribution centred around a given energy. It should be noted that often the subscripts on temperature T are omitted, and it is the context that determines whether the reference is to radiation or the electron temperature, T_* or T_e. The kinetic

[6] The basic composition of the Earth's atmospheric gases by volume at sea level is $N_2 \sim 78\%$, $O_2 \sim 21\%$. Note that these two gases alone comprise $\sim 99\%$ of the dry atmosphere. But there are variations, allowing for $H_2O < 4\%$ and $CO_2 \sim 0.036\%$, and some other trace gases. Both H_2O and CO_2 are greenhouse gases that regulate the greenhouse effect on Earth. Although the CO_2 concentration is usually only about one hundredth of that of H_2O, it can be pivotal in global warming, since it is directly affected by life on Earth and carbon-based fossil fuels.

[7] Not to be confused with the H and K lines of ionized calcium.

temperature of other particle constituents in a plasma, such as protons or other ions (T_p, T_i), is also characterized in terms of a Maxwellian. However, it may happen that $T_e \neq T_p$ or T_i, if there are bulk motions or processes that separate electrons from protons or ions.

1.6 Quantum statistics

Free particles, such as fermions and bosons, usually obey the Maxwellian or Planckian distributions associated with a temperature. When particles congregate to form structures like atoms, molecules, etc., they do so in accordance with laws of quantum mechanics described by energy levels quantized in energy and other variables such as momentum. The statistical mechanics of quantum distribution of particles among those levels is *quantum statistics*. There are three statistical distributions that relate to plasma sources. Once again, temperature is the crucial variable that determines the energies of particles and the levels they can occupy, subject to the principle of indistinguishability (and hence their fundamental classification as fermions or bosons) and quantum mechanical rules, such as the *Pauli exclusion principle*.

1.6.1 Maxwell–Boltzmann statistics

In thermal equilibrium, temperature determines the energy available for particles to be excited to higher levels, and the population distribution among them. Assuming a temperature T and a given excited energy level E_i, the distribution of the number of particles in level i relative to the total is

$$\frac{N_i}{N} = \frac{g_i \, e^{-E_i/kT}}{\sum_j g_j \, e^{-E_j/kT}}. \tag{1.11}$$

Here, g_i is the statistical weight for level i or its maximum possible occupancy number. The Maxwell–Boltzmann distribution is the one most frequently used to evaluate the number of electrons in excited levels of an atom or ion. The denominator in Eq. 1.11 is referred to as the *partition function*,

$$U = \sum_i g_i \, e^{-E_i/kT}. \tag{1.12}$$

It is related to what is known as the *equation-of-state* for a plasma and is discussed in detail in Chapter 10.

1.6.2 Fermi–Dirac statistics

What happens as $T \to 0$? In that limit, the distribution of particles depends on their basic nature; fermions or bosons. Since fermions are particles with half-integral spin they must occupy discrete states in accordance with the Pauli exclusion principle, which states that no two fermions can occupy the same quantum mechanical state. This basic fact leads to atomic structure, corresponding to the states of the atom defined by the couplings of angular and spin momenta of all electrons (Chapter 2). As the temperature approaches absolute zero, the electrons have no energy to be excited into higher levels. But not all atomic electrons can occupy the same quantum mechanical state, in particular the ground state, since that would violate the exclusion principle. So they occupy the next available higher levels, until a highest level, called the *Fermi level*, with energy E_F. The Fermi–Dirac probability distribution is given by

$$f(E_i, T) = \frac{1}{\exp[(E_i - E_F)/kT] + 1}. \tag{1.13}$$

At $T = 0$, we have probabilities $f(E, T) = 1$ if $E \leq E_F$ and $f(E, T) = 0$ otherwise. We may visualize the situation as in Fig. 1.7. All levels up to the Fermi level are filled at absolute temperature zero, constituting an ensemble of fermions called the *Fermi sea*. As T increases, particles get excited to higher levels, out of the Fermi sea. Eventually, for sufficiently high temperature and $kT \gg E$, the Fermi–Dirac distribution approaches the Maxwell–Boltzmann distribution characterized by the exponentially decaying probability as $\exp(-E/kT)$ in Eq. 1.11. The probability (Eq. 1.13) is related to the actual number of particles in an energy level i as

$$N_i(\text{FD}) = \frac{g_i}{\exp[(E_i - E_F)/kT] + 1}. \tag{1.14}$$

Thus far we have considered only the temperature as the primary physical quantity. But in fact the density of the plasma plays an equally important role. Intuitively one can see that for sufficiently high densities, at any temperature, particles may be forced together so that the exclusion principle applies. In such a situation one can think of a 'quantum degeneracy pressure' owing to the fact that no two electrons with all the same quantum numbers can be forced into the same state. When that happens, all accessible levels would again be occupied at the given temperature *and* density. The foremost example of Fermi–Dirac distribution in astrophysics is that of *white dwarfs* Chapter 10). These are stellar remnants of ordinary stars, like the Sun, but at the end of stellar evolution after the nuclear fuel (fusion of H, He, etc.) that powers the star runs out. The white dwarfs have extremely high densities, about a million times that of the Sun. The electrons in white dwarfs experience degeneracy pressure, which in fact prevents their gravitational collapse by forcing the

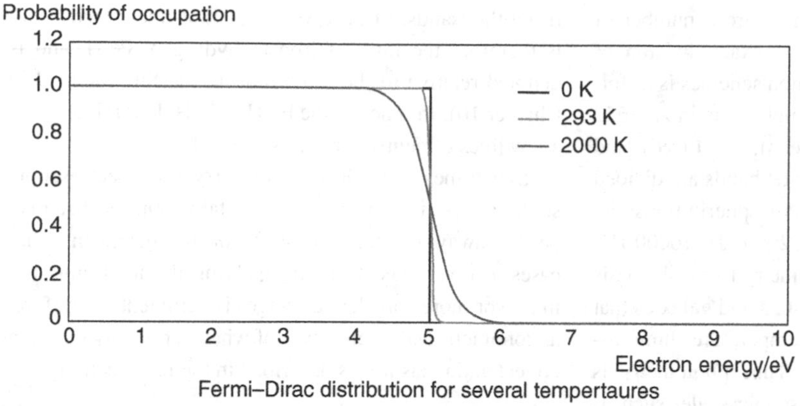

FIGURE 1.7 Fermi–Dirac distribution.

electrons to remain apart (up to a certain limit, as we shall see in Chapter 10).

1.6.3 Bose–Einstein statistics

Zero-spin or integral-spin particles are unaffected by the Pauli exclusion principle and any number may occupy any energy level. They follow the Bose–Einstein probability distribution for the number of particles

$$N_i(BE) = \frac{g_i}{\exp E_i/kT - 1}. \tag{1.15}$$

All bosons at absolute zero tend to congregate in the same quantum mechanical state, in what must be the ground state. The Bose–Einstein distribution also approaches the Maxwell–Boltzmann distribution for $kT \gg E_i$.

Bosons might be not only single particles, such as photons (spin 0), but also a system of atomic particles with the total spin of all electrons, protons and neutrons adding up to an integral value or zero in each atom. Atoms of alkali elements, such as rubidium, are examples of bosons, which have been experimentally shown to undergo condensation into the same structure-less state. As an alkali, rubidium atoms have an unpaired electron and an odd-atomic numbered nucleus, both of which have a spin quantum number of 1/2 that, in the lowest state, corresponds to a total spin of 0. The so-called Bose–Einstein condensation (BEC) is achieved by lowering the kinetic temperature to practically zero by slowing down the atomic velocities through laser impact (*laser cooling*). Recall the simple kinetic theory expression (Eq. 1.2), which relates velocity to temperature; bringing the atoms in a gas to a virtual standstill occurs in the μK range. At such a temperature the atoms coalesce into a *Bose–Einstein condensate*. There is very little hard scientific evidence on any astrophysical entity that would be a boson condensate. But the hypothesis of a 'boson star', perhaps following gravitational collapse, has been contemplated. That, if observed, would be complementary to known objects, such as neutron stars, made of fermions.

1.7 Spectroscopy and photometry

Spectroscopy and imaging of astrophysical sources, i.e., spectra and 'pictures', complement each other in astrophysical studies. In between the two lies *photometry*, or the calibrated measurement of brightness in a given wavelength band (or 'colour'). The division between spectroscopy and photometry rests essentially on the study of energy *at* a given wavelength of spectroscopy or *in* a given wavelength region of photometry. Finer divisions between the two rest on resolution, techniques and instruments.

1.7.1 Photometry and imaging

Photometry involves measurement and calibration of brightness in certain wavelength ranges or bands, e.g. in the optical. As already mentioned, the general division between spectroscopy and photometry is that, while the former refers to the study of energy emitted in the continuum and lines, the latter concerns total emission across a region in the electromagnetic spectrum. However, the difference may be thought of simply in terms of resolution: photometry measures spectral energy with low resolution, and spectroscopy determines the division of energy with high resolution at specific wavelengths (usually associated with atomic and molecular transitions).

Photometric observations correspond to integrated energy (brightness) in a wavelength band weighted by the response function of the filter or the detector. Around the visible region of the spectrum, transmitted through the atmosphere and observable from the ground, the main wavelength bands are denoted as ultraviolet, violet, blue, green, red and infrared regions. There are a number of systems in observational use for the exact division of wavelengths. One of the more common schemes is as follows. The approximate peak wavelengths are, in Å: 3650 (ultraviolet), 4800 (blue), 5500 (green), 7000 (red), and 8800 (infrared). In addition the near-IR bands are divided into three parts, corresponding to atmospheric transmission windows (Figs 1.2 and 1.5): 12000 (J), 16000 (H) and 22000 (K). Astronomical photometry forms the basis of 'colour-magnitude' diagrams of stars and galaxies that relate the bulk energy emitted to temperature, luminosity and other physical properties. Thus photometry is useful for information on a macroscopic scale, such as size, location or surroundings of an astronomical object, whereas spectroscopy yields more detailed information on microscopic physical processes.

1.7.2 Spectroscopy

The formation of the spectrum from an astrophysical plasma depends on atomic processes that emit or absorb radiation. The astrophysical plasma constituents are in general electrons, protons and trace elements in various ionization stages. Also, there is often an external radiation field, for example, from another star or galactic nucleus, interacting with the plasma. The radiative and collisional interactions, in turn, depend on the prevailing density, temperature and radiation source. A variety of atomic interactions, mainly between the electrons and ions, determine the observed spectral features that are divided into the primary components of a typical spectrum: (i) a continuum defining the background radiation, if present, and (ii) a superposition of lines that add or subtract energy to or from the continuum, characterized by emission or absorption, respectively. The relative magnitudes of intensities of the continuum and the lines is a function of the densities, temperatures, radiation field and abundances of elements in the source. Atomic astrophysics seeks to study the qualitative and quantitative nature of the microscopic atomic interactions and the observed spectra.

Among the first quantities to be obtained from spectroscopy is the temperature, which generally determines the wavelength range of the resulting spectra, as well as a measure of the total energy output of the source (such as a star). Another useful parameter derived from spectroscopy of an astronomical object is its gross composition in terms of the amount of 'metals' present or its *metallicity*. But the metallicity is generally measured not with respect to all the metals in the source but to iron, which is one of the most abundant elements. Iron often provides hundreds, if not thousands, of observable spectral lines. Metallicity is therefore the ratio of iron to hydrogen, Fe/H, and is denoted relative to the same ratio in the Sun (defined in Chapter 10). In practice the Fe/H ratio is determined from a few lines of neutral or singly ionized iron.

Sometimes it is difficult to carry out spectroscopic studies, especially in the case of faint objects that may be far away (at high redshift for instance). In other cases a broad classification involving the total emission in two or more wavelength ranges is sufficient. Therefore, astronomers avail themselves of whatever energy they can collect and measure, as described in the next section.

1.7.3 Spectrophotometry

Spectroscopy and photometry may be combined as *spectrophotometry*, which refers to calibrated spectral energy distribution. It is also useful to carry out narrow-band or single-line imaging of a given source, say at a specific wavelength, e.g., the well-known 5007 green line from O III (Chapter 9). The advantage of such a combination of observations is that one can ascertain the spatial distribution, as well as the emission, from plasma in an extended source such as a nebula. For instance, the λ 5007 line may indicate the temperature distribution in the source, e.g., a supernova remnant in a late phase that resembles a gaseous nebula.

An example of measurements that lie in between photometry and spectroscopy is that of *photometric redshifts* of distant objects, now being derived observationally. The redshift of a spectral line, usually the strongest line Lyα, indicates the distance of the source at the present epoch due to the cosmological expansion of the Universe (Chapter 14). A similar redshift also occurs in the entire wavelength region, since all photons from the receding source undergo the same redshift. If spectroscopic observations are not possible or difficult owing to the large distance of an object, photometric redshifts may be derived from the much larger photon flux that can be detected in a wavelength region as opposed to a single wavelength.

1.8 Spectroscopic notation

A Roman numeral or a numerical superscript after the chemical symbol of the element denotes its ionization state: I or 0 for neutral, II or + for singly ionized, and so on; e.g., Li I or Li^0, Li II or Li^+, and Li III or Li^{2+}. The last ionization state, the fully stripped bare ion, has a numeral equal to Z, the atomic (proton) number in the nucleus. These are the common notations in atomic physics and spectroscopy. However, the astronomy usage can vary according to context in a sometimes confusing manner. While the spectroscopic notation with the numeral refers to an ionization state, the superscript notation may refer to the *abundance* of an element. For example, Fe II refers to singly ionized iron, but Fe^+ refers to its abundance when written as, e.g., O^+/Fe^+, which means the abundance by number of O II ions relative to Fe II.

1.9 Units and dimensions

Until one turns to routine calculations, in particular to computers, which have no concept of physical quantities and, apart from logical and bookkeeping operations, are good only for *number* crunching, one stays with *physical* equations. Being invariant to choice and change of 'yardsticks', the outcome of Eq. 1.7, for instance, remains unaltered if one measures the wavelength λ in units of Å, km or, if one prefers, units bigger than parsecs or smaller than fm (femtometres, 10^{-15} metres).[8] The electric and magnetic interactions are controlled by the dimensionless *electromagnetic coupling parameter*

$$\alpha = \frac{e^2}{\hbar c} = 1\,/\,137.0360\,, \tag{1.16}$$

often referred to as the *fine-structure constant*. In the atomic shell environment, energies are most naturally measured in units of

$$1\,\text{Ry} = \frac{\alpha^2}{2} mc^2 = 13.6\,\text{eV}\,, \tag{1.17}$$

the ionization energy of a hydrogen atom out of its ground state.

Strict observance of *phase invariance* for canonical pairs of observables fixes the unit of time:

$$\tau_0 = \hbar/\text{Ry} = 4.8378 \times 10^{-17}\ \text{s}\,, \tag{1.18}$$

the 'Rydberg' time of around 50 as (attosecond, 10^{-18}s, lasts as long as it takes a hydrogen electron in its ground state with velocity $c \cdot \alpha$ to traverse the diameter of the atom, whose radius a_0 of about 0.5 Å we pick as the unit of length:

$$a_0 = \frac{\hbar}{mc} \Big/ \alpha = \lambda^{\text{C}}/(2\pi\,\alpha) = 0.529177 \times 10^{-8}\ \text{cm}. \tag{1.19}$$

Again it is a 'mechanical' property of the electron, now its Compton wavelength, that leads to the Bohr radius a_0; momentum mc rather than $\lambda^{\text{C}} = h/(mc)$. This fixes the unit of linear momentum $p = \hbar k$ or wavenumber k, such that

$$k \to k \cdot a_0 \tag{1.20}$$

secures an invariant phase (kr) if $r \to r/a_0$.

The third canonical pair of observables in atomic physics is a familiar affair, which in a sense started atomic physics and spectroscopy; the uncertainty relation for angular momentum d,

$$\Delta d \cdot \Delta\varphi = h\,, \tag{1.21}$$

yields $h/(2\pi) = \hbar$ as the natural unit of angular momentum, because the angle φ is uncertain by 2π in a closed orbit.

For convenience, one uses less natural units, such as for energy the Hartree (H) and Rydberg units (also referred to as *atomic units* (au)

$$1\,\text{H} = 2\,\text{Ry} = 27.21\,\text{eV}, \tag{1.22}$$

which implies $\tau^{\text{H}} = \tau_0/2\,,$

The atomic units are arrived at technically on *dropping* the quantities $\hbar$, m and e (or equivalently setting them equal to unity), while the physical units are *derived* from the equation of motion. This is akin to saying that in the MKS (metre–kilogram–second) system a day lasts 86 400 MKS or the equator spans 4×10^7 MKS. We rather focus on (conjugate pairs of) observables. Notably the elementary electric charge e is not an observable: it enters atomic structure by way of coupling with an external electromagnetic field.[9] The assignments in Eqs 1.17 and 1.19 conceptually define its role via

$$e^2 = 2\,a_0\,\text{Ry}\,. \tag{1.23}$$

[8] Physical quantities are the product of a 'quality' (a yardstick) and an 'intensitiy' (the measure taken with it), and as one factor in the product increases, the other decreases (e.g., in Eq. 2.40). Turning from *scale invariance* to calculations involving the atomic shell, 'qualities' that keep the intensities within a convenient range are as described herein.

[9] As W. Eissner points out, only via approaches such as Millikan's experiment, followed by the long [hi]story of QED. Historically, Sommerfeld named α the fine-structure constant before its primary role in QED could be appreciated. The square e^2 does appear in the equations of motion, but only as the electromagnetic coupling parameter α with the two other *universal structure constants* (Eq. 1.16). In Chapter 14, we address the issue of *variation* of fundamental natural constants, which of course would alter phase space and physical relations such as the Heisenberg uncertainty principle and, in fact, lead to a different Universe or an *evolution* thereof.

Rydberg's constant began in spectroscopy as

$$\mathcal{R}_\infty = \frac{\mathrm{Ry}}{\hbar c} = 109\,737.32\,\mathrm{cm}^{-1}, \tag{1.24}$$

or rather in a form corrected along

$$1\,\mathcal{R}_M = \frac{M}{M + m_\mathrm{e}} \mathcal{R}_\infty \tag{1.25}$$

for the finite mass $M \approx 1820\,m_\mathrm{e}$ of the hydrogen nucleus:

$$\mathcal{R}_\mathrm{H} = 109,677.576\,/\,\mathrm{cm}\,. \tag{1.26}$$

The wavenumber $\mathcal{R}$ readily translates into the (vacuum) wave length λ of a photon needed to ionize the ground state 1s ^{2}S:

$$\lambda_\mathrm{H}^\mathrm{PI} = 1/\mathcal{R} = 911.76\,\text{Å}\,. \tag{1.27}$$

The equivalent temperature of 1 Ry follows from Eqs 1.17 and 1.3 as

$$1\,\mathrm{Ry} \equiv 157\,885\,\mathrm{K}\,. \tag{1.28}$$

It is interesting that atomic sizes of all elements are remarkably similar. Given the Bohr radius (Eq. 1.19) of the electron orbit in the hydrogen atom the size (diameter) of the H atom is ~1 Å. One might think that heavier atoms would increase in size according to atomic number along the periodic table (see Appendix A). But this is not so when one examines the calculated radii of atoms of various elements. All atoms lie in the narrow range ~1–3 Å. This is because the inner electrons are pulled in closer to the nucleus as Z increases. Nonetheless, it is remarkable that atomic size is constant to within a factor of three for all elements, though Z varies by nearly a factor of 100. Of course, the size of atomic *ions* varies significantly from these values since for ions the size depends on both Z and the number of electrons N in the ion, i.e., the ion charge $z \equiv Z - N + 1$.

A table of physical constants useful in atomic physics and astronomy is given in Appendix A.

Exercise 1.2 *Write a program to compute and plot the Maxwellian and Planck functions corresponding to a range of* T_e *and* T_*, *respectively.*

Exercise 1.3 *Plot the black-body function (a) at the effective temperature of the Sun,* $T_* = 5700\,\mathrm{K}$, *and (b) at* $T_* = 2.73\,\mathrm{K}$, *the microwave background temperature of the Universe; compare the latter with data obtained by space satellites, such as the Cosmic Background Explorer (COBE). The Universe would be a perfect black body, provided one ignored all the matter in it!*

2 Atomic structure

As mentioned in the first chapter, astrophysical applications played a crucial role in the development of atomic physics. In their 1925 paper, Russell and Saunders [2] derived the rules for spectroscopic designations of various atomic states based on the coupling of orbital angular momenta of all electrons into a total L, and the coupling of all spin momenta into a total S, called the LS coupling scheme. Each atomic state is thus labelled according to the total L and S.

Atomic structure refers to the organization of electrons in various shells and subshells. Theoretically it means the determinations of electron energies and wavefunctions of bound (and quasi-bound) states of all electrons in the atom, ion or atomic system (such as electron–ion). As fermions, unlike bosons, electrons form *structured* arrangements bound by the attractive potential of the nucleus. Different atomic states arise from quantization of motion, orbital and spin angular momenta of all electrons. Transitions among those states involve photons, and are seen as lines in observed spectra.

This chapter first describes the quantization of individual electron orbital and spin angular momenta as quantum numbers l and s, and the principal quantum number n, related to the total energy E of the hydrogen atom. The dynamic state of an atom or ion is described by the Schrödinger equation. For hydrogen, the total energy is the sum of electron kinetic energy and the potential energy in the electric field of the proton.

For multi-electron atoms the combination of individual l and s follows strict coupling rules for the total angular momenta, which define the *symmetry of atomic levels*. In a given atom or ion, the rules constitute the angular algebra for all possible atomic states to be determined *independently* of dynamical variables in any effective atomic potential. The orbital spin and the dynamical parts are separately quantized and therefore separable in the Hamiltonian. With a given spin-orbital nl, the dynamical quantities determine the *stationary states* and expectation values, such as the mean radius of each orbital $\langle r_{nl} \rangle$. These concepts are introduced here through the simplest atomic system, hydrogen. It is the most abundant element in the Universe (90% by number and 70% by mass). The series of spectral lines due to absorption or emission of photons by hydrogen lie in the ultraviolet (UV), optical and infrared (IR) wavelength ranges in the spectra of nearly all astrophysical objects.

Subsequent sections discuss the atomic structure for multi-electron atoms, beginning with the two electrons atom, helium. For a multi-electron atomic system, electron–electron correlation interactions are to be added, introducing complexity in determining the energies and wavefunctions. An approximate treatment of a multi-electron atom, in analogy with the central potential field in an H-atom, comprises the *central-field approximation*. The most common and complete treatment is the generalization of the Schrödinger equation into the *Hartree–Fock equations*.

In addition, if the velocity of the electrons in the atom is significant compared with c, such as in heavy elements or highly charged ions, relativistic effects come into play. The primary effect is the explicit consideration of fine structure, in addition to the total LS scheme, and consequent splitting of LS states or terms into fine-structure levels J. The atomic levels are then designated as LSJ. The fully relativistic version of the equation of motion of an electron is described by the Dirac equation. However, relativistic effects may be incorporated in successively complex approximations, depending on the nuclear charge or *atomic number Z*, to varying extent, that are *intermediate* between the non-relativistic Schrödinger equation and the Dirac equation. The intermediate methods form a class of *Breit–Pauli approximations* appropriate for most atomic systems of astrophysical importance up to the iron group elements.

Finally, the behaviour of energy levels along *isoelectronic sequences*, that is, ions with the same number of electrons but different number Z of protons in the nucleus, illustrates a number of practical useful features of atomic spectroscopy.

2.1 The hydrogen atom

The study of the hydrogen atom underpins the basic concepts of atomic spectroscopy. Therefore the quantum mechanical treatment for this atom is discussed in some detail, leading up to the Rydberg series of levels that define the series of spectral lines.

The classical equation of motion of an electron with mass m moving in the central field of a heavy nucleus with electric charge number Z is

$$\frac{\mathbf{p}^2}{2m} - \frac{Ze^2}{r} = \frac{p_r^2 + p_\perp^2}{2m} - \frac{Ze^2}{r} = E, \tag{2.1}$$

with $\boldsymbol{p}$ split as indicated by subscripts. The quantum mechanical analogue is obtained on replacing the momentum and energy differential operators $\boldsymbol{p} \to -i\hbar\nabla$ and $\mathrm{E} \to i\hbar\partial/\partial t$ to obtain the Schrödinger form,

$$\left[-\frac{\hbar^2}{2m}\left(\nabla^2\right) + V(r)\right]\Psi = E\,\Psi \tag{2.2}$$

$$\left[-\frac{\hbar^2}{2m}\left(\nabla_r^2 + \nabla_\perp^2\right) + V(r)\right]\Psi = E\,\Psi, \tag{2.3}$$

$$\text{specifically}\quad V(r) = -\frac{Ze^2}{r} = -\frac{2Z}{r/a_0}\,\mathrm{Ry}. \tag{2.4}$$

In standard notation for spherical coordinates we have

$$\nabla_r^2 = \frac{1}{r^2}\frac{\partial}{\partial r}\left(r^2\frac{\partial}{\partial r}\right) \tag{2.5}$$

$$\nabla_\perp^2 = \frac{1}{r^2\sin\vartheta}\frac{\partial}{\partial\vartheta}\left(\sin\vartheta\frac{\partial}{\partial\vartheta}\right) + \frac{1}{r^2\sin^2\vartheta}\frac{\partial^2}{\partial\varphi^2} \tag{2.6}$$

$$\Psi(r,\vartheta,\varphi) = R(r)\,\mathrm{Y}(\vartheta,\varphi),$$

as the wavefunction factorizes accordingly. Substitution into the Schrödinger equation gives

$$\frac{1}{R}\frac{\mathrm{d}}{\mathrm{d}r}\left(r^2\frac{\mathrm{d}R}{\mathrm{d}r}\right) + \frac{2mr^2}{\hbar^2}[E - V(r)] \tag{2.7}$$

$$= -\frac{1}{\mathrm{Y}}\left[\frac{1}{\sin\vartheta}\frac{\partial}{\partial\vartheta}\left(\sin\vartheta\frac{\partial\mathrm{Y}}{\partial\vartheta}\right) + \frac{1}{\sin^2\vartheta}\frac{\partial^2\mathrm{Y}}{\partial\phi^2}\right]. \tag{2.8}$$

Since the left-hand side depends only on r and the right only on the two spherical angles, both sides must equal some constant λ. Dealing with the angular equation first conveniently leads to the radial problem, as in the following subsections.

2.1.1 Angular equation

The expression 2.8 leads to the angular equation

$$\frac{1}{\sin\vartheta}\frac{\partial}{\partial\vartheta}\left(\sin\vartheta\frac{\partial\mathrm{Y}}{\partial\vartheta}\right) + \frac{1}{\sin^2\vartheta}\frac{\partial^2\mathrm{Y}}{\partial\phi^2} + \lambda\mathrm{Y} = 0, \tag{2.9}$$

with solutions $\mathrm{Y}(\vartheta,\varphi)$, known as spherical harmonics.[1] The equation can be expressed in the convenient form

$$L^2\mathrm{Y}(\vartheta,\varphi) = \lambda\mathrm{Y}(\vartheta,\varphi), \tag{2.10}$$

with an angular momentum operator L. Writing the solution as

$$\mathrm{Y}(\vartheta,\varphi) = \Theta(\vartheta)\,\Phi(\varphi) \tag{2.11}$$

and substituting in Eq. 2.9, the equation separates to the form

$$\frac{\mathrm{d}^2\Phi}{\mathrm{d}\varphi^2} + \nu\Phi = 0, \tag{2.12}$$

$$\frac{1}{\sin\vartheta}\frac{\mathrm{d}}{\mathrm{d}\vartheta}\left(\sin\vartheta\frac{\mathrm{d}\Theta}{\mathrm{d}\vartheta}\right) + \left(\lambda - \frac{\nu}{\sin^2\vartheta}\right)\Theta = 0, \tag{2.13}$$

where ν is another constant. If ν is the square of an integer, i.e., $\nu = m^2$, Φ and its derivative $\mathrm{d}\Phi/\mathrm{d}\varphi$ are finite and continuous in the domain 0 to 2π:

$$\Phi(\varphi) = (2\pi)^{-1/2}\mathrm{e}^{im\varphi}; \tag{2.14}$$

m is called the *magnetic angular quantum number* and equals $0, \pm1, \pm2, \ldots$ On replacing ϑ by $w = \cos\vartheta$ the equation for Θ reads

$$\frac{\mathrm{d}}{\mathrm{d}w}\left[(1-w^2)\frac{\mathrm{d}\Theta}{\mathrm{d}w}\right] + \left[\lambda - \frac{m^2}{1-w^2}\right]\Theta(w) = 0. \tag{2.15}$$

A finite solution Θ requires

$$\lambda = l(l+1), \tag{2.16}$$

with positive integers $l = 0, 1, 2\ldots$ The solutions are associated Legendre polynomials of order l and m,

$$\mathrm{P}_l^m(w) = (1-w^2)^{|m|/2}\frac{\mathrm{d}^{|m|}}{\mathrm{d}w^{|m|}}\mathrm{P}_l(w), \tag{2.17}$$

[1] Like sin, exp and other standard mathematical functions, Y is set in *roman* type since it is taken for a filter or operator that creates a *value* from arguments or *variables*, which appear in *italic* type in scientific notation. In this sense, non-standard functions are taken for variables as a whole. Thus the Legendre polynomials P(cos ϑ) appear in roman type, so there is no notation clash with radial functions $P(r)$ in the next sections.

where $m = l, l-1, \ldots, -l$. For $m = 0$ the function $P_l(w)$ is a Legendre polynomial of order l. The angular solution of normalized spherical harmonic is (e.g. [4])

$$Y_{lm}(\vartheta, \varphi) = N_{lm}\, P_l^m(\cos\vartheta)\, e^{im\varphi}, \tag{2.18}$$

where

$$N_{lm} = \epsilon \left[\frac{2l+1}{4\pi}\frac{(l-|m|)!}{(l+|m|)!}\right]^{1/2}, \tag{2.19}$$

with $\epsilon = (-1)^m$ for $m > 0$ and $\epsilon = 1$ for $m \leq 0$. Spherical harmonics satisfy the orthogonality condition,

$$\int_{\varphi=0}^{2\pi}\int_{\vartheta=0}^{\pi} Y^*_{l_1 m_1}(\vartheta,\varphi) Y_{l_2 m_2}(\vartheta,\varphi)\ \sin\vartheta\, d\vartheta\, d\varphi = \delta_{l_1,l_2}\,\delta_{m_1,m_2}. \tag{2.20}$$

The equation with angular momentum operator can now be written as

$$L^2 Y_l^m(\vartheta,\varphi) = l(l+1)\,\hbar^2\, Y_l^m(\vartheta,\varphi) \tag{2.21}$$

and

$$m = l, l-1, \ldots, -l. \tag{2.22}$$

With angular momentum $L = mvr = mwr^2$ the angular frequency $w = L/mr^2$, the centripetal force is $mw^2 r = L^2/mr^3$ and the corresponding potential energy is

$$\frac{1}{2}mw^2r^2 = \frac{L^2}{2mr^2}. \tag{2.23}$$

This is similar to the second potential term of hydrogen, provided

$$L^2 = l(l+1)\hbar^2. \tag{2.24}$$

2.1.2 Radial equation

We now turn to the radial coordinate representing the dynamical motion of the electron in the atom. Equation 2.7 leads to the radial equation

$$\left[\frac{1}{r^2}\frac{d}{dr}\left(r^2\frac{d}{dr}\right) + \frac{2m}{\hbar^2}(E - V(r)) - \frac{\lambda}{r^2}\right] R(r) = 0, \tag{2.25}$$

λ being established in Eq. 2.24. It simplifies on substituting $R(r) = P(r)/r$:

$$\left[\frac{\hbar^2}{2m}\frac{d^2}{dr^2} - V(r) - \frac{l(l+1)\hbar^2}{2mr^2} + E\right] P(r) = 0. \tag{2.26}$$

Using atomic units we write ($e = m_e = a_0 = \hbar = 1$)[2]

$$\left[\frac{d^2}{dr^2} - V(r) - \frac{l(l+1)}{r^2} + E\right] P(r) = 0, \tag{2.27}$$

2.1.3 Rydberg states and hydrogenic energy levels

It may appear that it is easier to express the radial equation than its angular counterpart, but its solution is not only more difficult: it is *always* approximate, with the outstanding exception of the single electron. The angular algebra embodied in the angular equation can be evaluated *exactly* for an atomic transition matrix element, but the solution of the radial equation entails the use of an effective potential, constructed in various approximations, as described in Chapter 4. For hydrogenic systems this is the well-known Coulomb potential Ze^2/r, as discussed below.

Equation 2.27 can be solved on specifying boundary conditions. The bound electron moves in the attractive potential of the nucleus, which behaves as $\lim_{r\to\infty} V(r) = 0$. Let us look for solutions at two limiting cases of the electron motion: (i) r at infinity and (ii) r near $r = 0$.

For case (i) with $r\to\infty$ the radial *number* equation reduces to

$$\left[\frac{d^2}{dr^2} + E\right] P(r) = 0, \tag{2.28}$$

which has solutions

$$P(r) = e^{\pm ar}, \qquad a = \sqrt{-E}. \tag{2.29}$$

Taking $E < 0$, implying bound states, a runaway solution $e^{ar} \to \infty$ for $r\to\infty$ is not acceptable. On the other hand, $\lim_{r\to\infty} e^{-ar} = 0$ is a possible solution, and is also valid for $E > 0$ when a becomes imaginary, implying free spherical waves. We concentrate on $E < 0$.

[2] The radial motion displayed in this equation is reminiscent of the one-dimensional motion of a particle in a potential, namely

$$V(r) + \frac{l(l+1)\hbar^2}{2mr^2},$$

where the last term is a centrifugal potential. Moving away from physics for a moment to computers, which know nothing about physics but can deal superbly with mere numbers, we divide Eq. 2.26 with $V(r)$ from Eq. (1.23) by 1 Ry as expressed in Eq. 1.17:

$$\left[\frac{d^2}{d(r/a_0)^2} + \frac{2Z}{r/a_0} - \frac{l(l+1)}{(r/a_0)^2} + E/\text{Ry}\right] P(r) = 0$$

This is a pure *number equation*, having exploited a_0 from Eq. 1.19.

The asymptotic behaviour suggests that the solution $P(r)$ should have the form

$$P(r) = e^{-ar} f(r) \tag{2.30}$$

subject to the condition $\lim_{r \to 0} f(r) = 0$. On substitution, the radial number Eq. 2.27 leads to

$$\frac{d^2 f}{dr^2} - 2a\frac{df}{dr} + \left[\frac{2Z}{r} - \frac{l(l+1)}{r^2}\right] f(r) = 0 \tag{2.31}$$

for a hydrogen-like ion with one electron and a positively charged nucleus with Z protons. If $r \ll 1$, the solution $f(r)$ may be expressed as a power series

$$f(r) = r^s[A_0 + A_1 r + A_2 r^2 + \ldots]. \tag{2.32}$$

For f to be finite as $r \to 0$, consistent with the behaviour of an orbital nl 'bound' at the nucleus, requires $s > 0$ for the exponent.

Exercise 2.1 *Use a power series expansion in the radial equation to show that*

$$s = l + 1 > 0, \tag{2.33}$$

i.e., $\lim_{r \to 0} P(r) \sim r^{l+1}$. *Prove that the coefficients A obey the recursion relation*

$$\frac{A_k}{A_{k-1}} = \frac{2[(l+k)a - Z]}{k^2 + (2l+1)k}, \tag{2.34}$$

$$\lim_{r \to \infty} \frac{A_k}{A_{k-1}} = \frac{2a}{k}. \tag{2.35}$$

We note that the exponential e^{2ar} has the following expansion:

$$e^{2ar} = 1 + 2ar + \frac{(2ar)^2}{2!} + \cdots + \frac{(2ar)^k}{k!} + \cdots, \tag{2.36}$$

$$\frac{(2a)^k/k!}{(2a)^{k-1}/(k-1)!} = \frac{2a}{k}. \tag{2.37}$$

Equation 2.32 indicates that the radial solution f behaves as $r^s e^{2ar}$ for large k. Therefore,

$$P(r) = f(r)e^{-ar} \approx r^{l+1} e^{ar} \tag{2.38}$$

at large distances r.

The above solution diverges at infinity, i.e., $P(r) \to \infty$ for $r \to \infty$ unless the series terminates at some finite values of k. Eq. 2.34, along with a from Eq. 2.29, shows that the coefficient A_k vanishes if the following condition is met:

$$(l+k)\sqrt{-E} - Z = 0 \tag{2.39}$$

or, reverting E to energies from shorthand for numbers E/Ry,

$$E = -\frac{Z^2}{n^2} \times \text{Ry}; \tag{2.40}$$

One may also replace Ry with 'Hartrees/2' or 'au/2' in atomic units. The boundary conditions on the radial wavefunctions have forced the bound states to be discrete with integer n. The equation gives an infinite number of discrete energy levels $-Z^2/n^2$ asymptotically approaching zero for any finite charge number Z. It also shows that the energy is degenerate with respect to l and m. *Degeneracy* in energy or state is defined as the number of eigenfunctions associated with a particular energy.

2.1.4 *Hydrogenic wavefunctions*

The full series solution for the hydrogen radial function $P(r)$ may be expressed in terms of Laguerre polynomials (e.g., [4])

$$L_{n+l}^{2l+1}(r) = C_0 + C_1 r + C_2 r^2 + \cdots + C_{n-l-1}\, Eq.r^{n-l-1}, \tag{2.41}$$

where

$$C_{n-l-1} = (-1)^{n-l} \frac{(n+l)!}{(n-l-1)!}. \tag{2.42}$$

For C_{n-l-1} to remain finite, $n - l - 1$ must be *zero* or a positive integer (note that $0! = 1$ and $n! = \pm\infty$ for a negative integer value of n). Hence

$$n = l+1, l+2, \ldots \tag{2.43}$$

The radial function then becomes

$$P_{nl}(r) = \sqrt{\frac{(n-l-1)!Z}{n^2[(n+l)!]^3 a_0}} \left[\frac{2Zr}{na_0}\right]^{l+1} \times e^{-Zr/na_0} L_{n+l}^{2l+1}\left(\frac{2Zr}{na_0}\right), \tag{2.44}$$

where the Laguerre polynomial is given by

$$L_{n+l}^{2l+1}(\rho) = \sum_{k=0}^{n-l-1} (-1)^{k+2l+1} \times \frac{[(n+l)!]^2 \rho^k}{(n-l-1-k)!(2l+1+k)!k!}. \tag{2.45}$$

The orthogonality condition of the radial function is

$$\int_0^\infty P_{nl}(r)\, P_{n'l}(r)\, dr = \delta_{nn'}, \tag{2.46}$$

$\sqrt{1/a_0}$ of Eq. 2.44 securing scale invariance. In *bra-ket* notation, the complete solution for the bound states of hydrogen may now be written as

$$\langle r|nlm\rangle \equiv \psi_{nlm}(r,\vartheta,\varphi) = R_{nl}(r)Y_{lm}(\vartheta,\varphi) = \frac{1}{r}P_{nl}(r)Y_{lm}(\vartheta,\varphi), \quad (2.47)$$

which satisfies the orthogonality condition

$$\langle nlm|n'l'm'\rangle = \int \psi^*_{nlm}(r,\vartheta,\varphi)\,\psi_{n'l'm'}(r,\vartheta,\varphi)\,d\tau = \delta_{nn'}\,\delta_{ll'}\,\delta_{mm'}. \quad (2.48)$$

where $d\tau = r^2 dr\ \sin\vartheta d\vartheta\ d\varphi$.

2.1.5 Charge density and expectation values

$R^2_{nl}(r)$ is the *radial charge density* describing the distribution of electrons of different symmetries (ℓ values) at a distance r from the nucleus. One can compute the charge density, that is the probability of finding an electron in volume element $d\tau$ as

$$\psi^*\psi\ d\tau = \frac{1}{r^2}P^2_{nl}(r)\,Y^*_{lm}(\vartheta,\varphi)\,Y_{lm}(\vartheta,\varphi)\,d\tau. \quad (2.49)$$

One may obtain the *expectation values* $\langle nl|r^k|nl\rangle$ to moments of order k:

$$\langle r^k\rangle = \int_0^\infty P^2_{nl}(r)\,r^k dr = \int_0^\infty R^2_{nl}\,r^{k+2}\,dr, \quad (2.50)$$

for example

$$\langle r\rangle = \frac{a_0}{2Z}[3n^2 - l(l+1)],$$

$$\begin{aligned}\langle r^2\rangle &= \frac{a_0^2}{Z^2}\frac{n^2}{2}[5n^2+1-3l(l+1)],\\ \left\langle\frac{1}{r}\right\rangle_{nl} &= \frac{Z}{n^2 a_0},\\ \left\langle\frac{1}{r^2}\right\rangle_{nl} &= \frac{Z^2}{n^2\left(l+\frac{1}{2}\right)a_0^2},\\ \left\langle\frac{1}{r^3}\right\rangle_{nl} &= \frac{Z^3}{n^3 l\left(l+\frac{1}{2}\right)(l+1)a_0^3}.\end{aligned} \quad (2.51)$$

These relations are useful in atomic structure calculations of matrix elements comprising radial integrals over wavefunctions.

States with $E > 0$, in contrast to $E < 0$, form a continuum instead of a discrete spectrum, because their orbits are not closed and thus not quantized. A continuum state is a free state, except that it is designated with an angular momentum (and either box or flux normalization).

Exercise 2.2 Obtain from the full expression for $P_{nl}(r)$ in terms of Laguerre polynomials, the radial functions for the 1s and the 2p orbitals.

2.2 Quantum numbers and parity

Atomic structure depends on quantization of continuous variables (r, E), ϑ, and φ. They are associated with discrete quantum numbers as

$$\begin{aligned} r, E &\rightarrow n \text{ (principal quantum number)}\\ &= 1, 2, 3, \ldots\infty\\ \vartheta &\rightarrow l \text{ (orbital quantum number)}\\ &= 0, 1, 2, \ldots\ (n-1)\\ \varphi &\rightarrow m_\ell \text{ (magnetic quantum number)}\\ &= 0, \pm1, \pm2, \ldots \pm l,\end{aligned} \quad (2.52)$$

where n represents a shell consisting of ℓ number of subshells, or $n\ell$ orbitals, which further subdivide into m_ℓ suborbitals.

The shells with n = 1, 2, 3, 4, 5, 6, ... are referred to as K, L, M, N, O, P, ... — as *values* set in *roman* type. Each shell can accommodate a maximum number of $2n^2$ electrons. A shell is *closed* when full, i.e., all $n\ell m$ orbitals are fully occupied, and *open* when there are vacancies. By long-standing convention, angular momenta l are represented by alphabetic characters s, p, d, f, g, h, i, k, ... for l = 0, 1, 2, 3, 4, 5, 6, 7, ... (note that there is no value j; scientific notation aims to avoid confusion with *variables* like spin momentum s, linear momentum p, angular momentum l or oscillator strength f). Thus, an electron in an orbital of nl = 1s is in the first or K-shell (n = 1) and in an orbit with l = 0 (s orbital). The total angular momentum L for more than one electron follows the same alphabetic character notation, but in the upper case. For example, $L = 0$ is denoted as S, and the higher values are L = 1, 2, 3, 4, 5, 6, 7, etc., are P, D, F, G, H, I, K, etc. (note again the absence of 'J'). The orbital magnetic quantum number m depends on l and is written as m_l. For ions with more than one electron, the total orbital magnetic angular momentum can be obtained as $M_L = \sum_i m_{li}$ where L is the total orbital angular momentum. There are $2L + 1$ possible values of M_l for the same L and this is called the angular momentum multiplicity of L.

These quantum numbers reflect the shape and symmetry of the density distribution through the angular function $P_l^m(\vartheta,\varphi)$ and the radial function $R_{nl}(r)$. The latter exhibits nodes (intersecting zeros along the radius vector), the former exhibits nodes at well-defined angles. The higher the value of n, the looser the binding and the greater the number of nodes in R_{nl} for a given value of l. There are $n - l - 1$ nodes in the wavefunction of an electron labelled nl, counting the sloping one far out and the one at the (pointlike) nucleus as one. Hence for a 2s orbital, the number of nodes is $2 - 0 - 1 = 1$; for a 3d orbital, it is $3 - 2 - 1 = 0$, etc.

The intrinsic angular momentum s of the electron manifests itself via the associated magnetic *Bohr* moment μ_B: as orbitals $l > 0$ create a magnetic field, this moment aligns in quantized positions of s, which leads to term splitting and Pauli's ad-hoc theory. It was overtaken by the Dirac equation, where both s and μ_B rather miraculously appear (it took a while to see why). The spin is separately quantized in non-relativistic quantum mechanics. The associated spin quantum number S is defined such that S^2 commutes with all dynamical variables and, similar to L^2, the eigenvalue S^2 is $S(S+1)\hbar^2$, that is,

$$S^2\psi_s = \hbar^2 S(S+1)\psi_s. \tag{2.53}$$

However, spin s can be an integer or half an odd integer. For a single electron, $s = 1/2$. While s refers to spin angular momentum of a single electron, S refers to the total or net spin angular momentum. As m is related to l, the spin magnetic quantum number m_s is related to S such that its values vary from $-S$ to $+S$, differing by unity. Hence for a particle with $S = 1/2$, m_s has two values, $-1/2$, $1/2$, describing spin down and spin up. The spin multiplicity of an LS state is given by $2S + 1$, and is labelled singlet, doublet, triplet, quartet, quintet, sextet, septet, octet, etc., for $2S+1 = 1, 2, 3, 4, 5, 6, 7, 8$, etc.

We noted that non-relativistic hydrogenic energies depend only on the principal quantum number n and are degenerate with respect to both l and m_ℓ. For a given n the value of l can vary from 0 to $n - 1$, and for each l, m_ℓ between $-l$ to l, the eigenfunctions are $(2\ell + 1)$ degenerate in energy. The total degeneracy of the energy level E_n is

$$\sum_{l=0}^{n-1}(2l+1) = n^2. \tag{2.54}$$

This degeneracy for a one-electron atom is said to be 'accidental', and is a consequence of the form Z/r of the Coulomb potential. Because it depends only on radial distance, the hydrogenic Hamiltonian is not affected by angular factors, rendering it invariant under rotations. Including the two-spin states that nature distinguishes along some axis, the total number of degenerate levels for a given n is $2n^2$.

Finally, we define the *parity* π of an atomic state. It refers to the symmetry of the state in spatial coordinates. It expresses the phase factor that describes the behaviour of the wavefunction, either positive or negative, with respect to its mirror image or flipping of the distance coordinate. Considering the sum of the integer values l, one speaks of 'even' parity π is $+1$, of 'odd' parity otherwise:

$$\pi = (-1)^{\sum_i l_i} = \begin{cases} +1, \text{ even} \\ -1, \text{ odd}, \end{cases} \tag{2.55}$$

where i is the index of (valence) electrons. Typically odd parity is expressed as superscript 'o' (in roman type since a value or label, not a variable), while even parity is either not marked or denoted by a superscript 'e'. Parity change is a crucial criterion for dipole allowed transitions between two atomic states (Chapter 9).

2.3 Spectral lines and the Rydberg formula

Photons are emitted or absorbed as electrons jump down or up between two energy levels and produce spectral lines. The energy *difference* between two levels is also expressed in terms of frequencies or wavelengths of the spectral lines. For a hydrogen atom, the wavenumber of the spectral line is given by

$$\Delta\mathcal{E}_{n,n'} = \mathcal{R}_H\left[\frac{1}{n'^2} - \frac{1}{n^2}\right] \quad (n' > n), \tag{2.56}$$

where $\mathcal{R}_H$ is the Rydberg constant of Eq. 1.26; finite atomic masses of elements often introduce very small but spectroscopically significant corrections in wavenumbers and lengths.

The Rydberg formula (Eq. 2.56) yields series of spectral lines, each corresponding to a fixed initial n and final $n < n' \leq \infty$, as seen in Fig. 2.1. The first five series are

(i) $\Delta\mathcal{E}_{n,n'} = \mathcal{R}_H\left[1 - \frac{1}{n'^2}\right]$, $n' = 2, 3, 4, \ldots$ Lyman (Ly),

(ii) $\Delta\mathcal{E}_{n,n'} = \mathcal{R}_H\left[\frac{1}{2^2} - \frac{1}{n'^2}\right]$, $n' = 3, 4, 5, \ldots$ Balmer (Ba),

(iii) $\Delta\mathcal{E}_{n,n'} = \mathcal{R}_H\left[\frac{1}{3^2} - \frac{1}{n'^2}\right]$, $n' = 4, 5, 6, \ldots$ Paschen (Pa),

(iv) $\Delta\mathcal{E}_{n,n'} = \mathcal{R}_H\left[\frac{1}{4^2} - \frac{1}{n'^2}\right]$, $n' = 5, 6, 7, \ldots$ Brackett (Br),

(v) $\Delta\mathcal{E}_{n,n'} = \mathcal{R}_{\mathrm{H}}\left[\frac{1}{5^2} - \frac{1}{n'^2}\right]$, $n' = 6, 7, 8, \ldots$ Pfund (Pf).

The computer delivers numbers for energies, typically in Rydberg units, whereas an observer measures (Fabry–Perot) wavenumbers $\mathcal{E} = E/(\hbar c)$, in particular Eq. 1.26 for ionizing hydrogen out of its ground state, i.e.,

$$\mathcal{R}_{\mathrm{H}} = 109,677.576\ \mathrm{cm}^{-1} = \frac{1}{911.76\,\text{Å}},$$

or its inverse, namely wavelengths in angstroms:

$$\lambda = \frac{911.76\,\text{Å}}{\Delta E/\mathrm{Ry}}. \tag{2.57}$$

The Lyman, Balmer, Paschen, and other series of H-lines are found to lie in distinct bands of the *electromagnetic spectrum*, as shown in Fig. 2.1. In particular, the Lyman series from 1215–912 Å lies in the far ultraviolet (FUV), the Balmer series from 6564–3646 Å in the optical and near ultraviolet regions, and the Paschen series from 18 751–8204 Å in the infrared (IR). The sequence of transitions in each series is denoted as α, β, γ, δ, etc., such that the first transition ($\Delta n = 1$) is α, the second ($\Delta n = 2$) is β, and so on. The wavelengths in the Lyman series are (Fig. 2.1): Ly α (1215.67 Å), Ly β (1025.72 Å), Ly γ (972.537 Å), ..., Ly$_\infty$ (911.76 Å). The Lyα line is the resonance line in hydrogen, i.e., it corresponds to a (2p–1s) transition, that is, from the first excited level to the ground level. Historically, the Balmer series in the visible (optical) range, readily accessible to ground-based telescopes, has been associated explicitly with hydrogen, and labelled as Hα (6562.8 Å), Hβ (4861.33 Å), Hγ (4340.48 Å), Hδ (4101.73 Å), and so on, towards shorter wavelengths.

Spectral lines of H and other elements had been identified in astronomical objects long before their quantum mechanical interpretation, such as the Fraunhofer absorption lines from the Sun, which have been observed since 1814.

Exercise 2.3 *(a) Use the formulae above to show for which series the hydrogen spectral lines overlap. (b) Calculate the Ly α transitions in H-like ions of all elements from C to Fe. [Hint: all wavelengths should lie in the X-ray range $\lambda < 40$ Å.] (c) Give examples of transitions in H-like ions that may lie in the extreme ultraviolet (EUV) wavelength range 100 Å$< \lambda <$ 600 Å.*

2.4 Spectroscopic designation

Before we describe the details of atomic calculations to determine the energy levels of a multi-electron atom, it is useful to describe its angular momenta as a guide to multi-electron structures.

A multi-electron system is described by its configuration and a defined spectroscopic state. The *electronic configuration* of an atomic system describes the arrangement of electrons in shells and orbitals, and is expressed as nl^q. In the case of a helium atom, the ground configuration is $1s^2$, where the superscript gives the occupancy number or the number of electrons in orbital 1s. For carbon with six electrons the configuration is $1s^2 2s^2 2p^2$, that is, two electrons in the 1s shell, two in 2s and two in 2p orbitals (when both s shells are full). The angular momenta of an atom depend on its electronic configuration.

The spectroscopic state of the atom is described by the total orbital angular momentum L, which is the vector sum of the individual angular momenta of all electrons. Likewise, the total spin angular momentum S is the vector sum over spin quantum numbers of all electrons. However, the state is not unique, and depends on physical factors, such as the number of electrons in the atom and its nuclear charge. The spectroscopic identification is based on the coupling of angular and spin quantum numbers of all electrons in the atom. The basic scheme is known as LS coupling or Russell–Saunders coupling, mentioned earlier. The main point is that in LS coupling the orbital motion of the electron is not strongly coupled to the spin momentum. Therefore, the orbital momenta of all electrons can be added together separately to yield a total L for the whole atom, and the spin momenta can likewise be added together to give total S. More precisely, both L and S are treated as separate constants of motion. The Hamiltonian is then *diagonal* in L^2 and S^2 operators, as both angular quantities commute with H:

$$[H, L^2] = 0 = [H, S^2]. \tag{2.58}$$

This secures *simultaneous* eigenfunctions $|LSM_L M_S\rangle$ of the operators L^2 and S^2, and of the component L_z and S_z.

Vector addition of angular momenta means that the total is a set of all possible positive numbers with a difference of unity ranging from the simple addition and subtraction of the component momenta. Hence, vector addition of L_1 and L_2 is the set of positive values from $|L_2 - L_1|, |L_2 - L_1 + 1|, \ldots, |L_2 + L_1|$. Similar addition holds for spin S. These can be added for the total angular momentum, $J = L + S$. These sums, along with the Bohr atomic model and the Pauli exclusion principle, which states that no two electrons in an atom can be in the same level, determine the total final number of possible states of the atomic system.

The total *symmetry of an atomic state* is specified by L, S and the parity. The LS coupling designation of

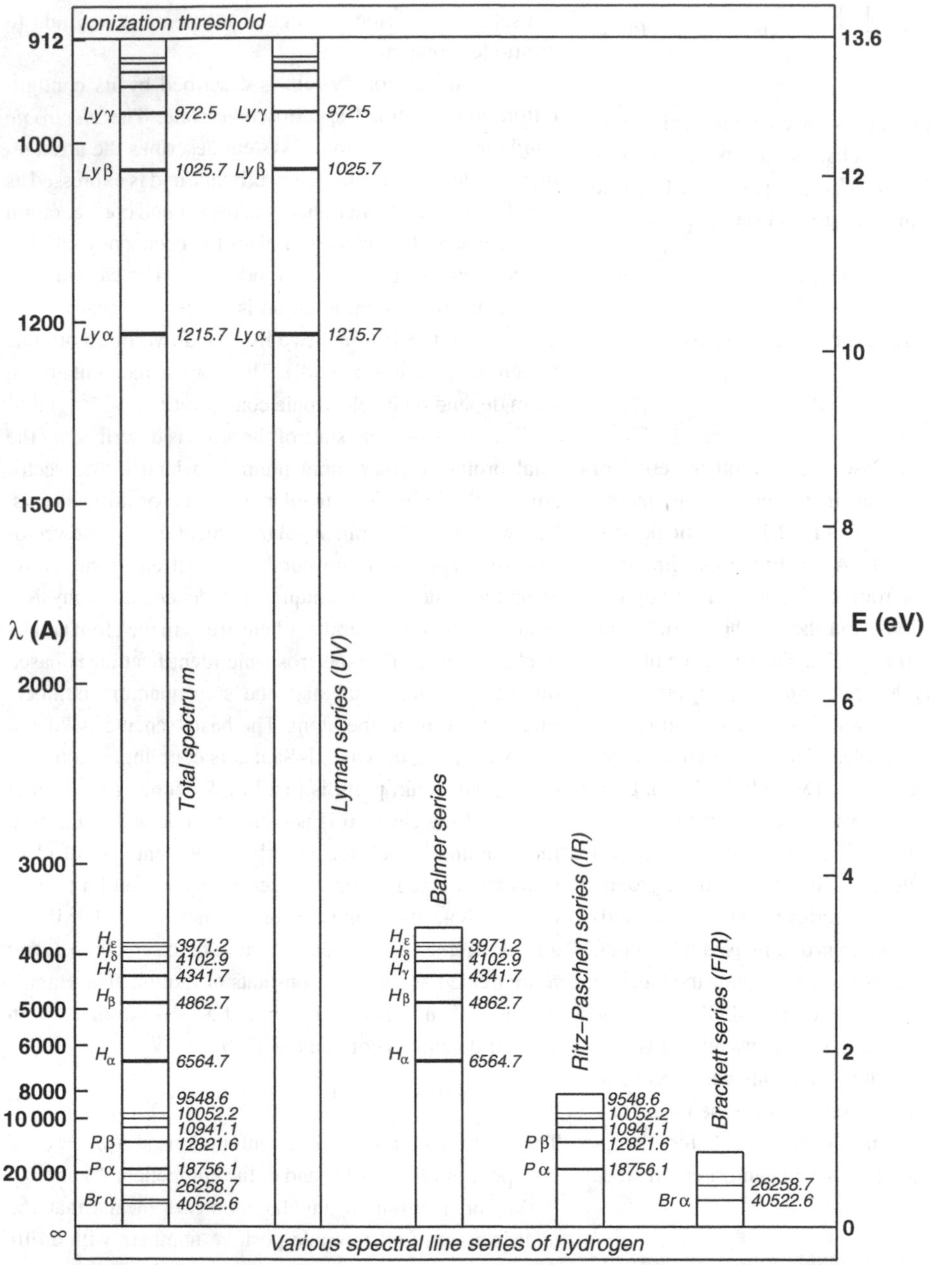

FIGURE 2.1 The hydrogen spectrum and energy levels. The Lyman series lies in the UV, the Balmer series in the optical, the Paschen series in the near IR, and the Brackett series in the far IR.

an atomic state is conventionally expressed as $^{(2S+1)}L^{\pi}$ and is called LS term. For a single-electron system, i.e., hydrogen, total $L = \ell$ and total $S = s = 1/2$. Since the spin multiplicity $(2S + 1) = 2$, the 1s ground state term with $L = 0$ and even parity is designated as ^{2}S. Similarly the excited state of 2p electron with its odd parity $[(-1)^1]$ is ^{2}P^o, of 3d is ^{2}D, and so on. Thus, all LS terms of hydrogen have doublet spin multiplicity and are denoted $^2L^{\pi}$.

Now consider two electrons with $l_1 = 1$ and $l_2 = 2$. Vector addition of these two gives three total $\boldsymbol{L}$; 1, 2, 3, from $|l_2 - l_1|$ to $l_1 + l_2$. The spin quantum number s is always 1/2. So the vector addition gives two possible total spin S; $s_1 - s_2 = 0$ and $s_1 + s_2 = 1$. Therefore, all two-electron states, such as of helium, are either singlets or triplets since $(2S + 1) = 1$ and 3, respectively. Generally, the LS coupling designation of an atomic state is referred to as an *LS term*. To emphasize: *LS* coupling is applicable when relativistic effects are not important enough to couple individual orbital and spin angular momenta together (this is discussed later).

2.5 The ground state of multi-electron systems

As mentioned above, the Bohr atomic model and the Pauli exclusion principle play crucial roles in structuring electrons in multi-electron systems. The $n = 1$ or K-shell has two levels, since $m_l = 0$ ($l = 0$) and $m_s = \pm 1/2$ (spin up and down). Hence, with occupancy number 2 of the K-shell, we can write

$$\text{K-shell}: \; n = 1, \; \ell = 0 \; m_\ell = 0, \; m_s = \pm\frac{1}{2}.$$

The ground configuration of the two electrons in helium is $1s^2$ with opposite spins, that is total sum $S = 0$, and the spin multiplicity is $(2S + 1) = 1$. Since both are s-electrons with $\ell = 0$, therefore $L = 0$ and parity is even. Hence, the helium ground state is ^1S. With both $L = 0 = S$ for the filled K-shell $1s^2$ (^1S), the helium 'core' will not add to the total L and S of the electronic configurations of elements with more than two electrons.

The situation gets a bit complicated with the next L-shell for which

$$\text{L-shell}: \; n = 2, \; \ell = 0, 1, \; m_\ell = 0, \pm 1, \; m_s = \pm\frac{1}{2}. \tag{2.59}$$

The electrons can fill up the orbitals, giving electronic configurations of various elements as

$$\begin{aligned} &[\ell = 0, \; m_\ell = 0] &&\rightarrow 1s^2 2s^1 \text{ (Li)}, \; 1s^2 2s^2 \text{ (Be)}, \\ &[\ell = 1, \; m_\ell = 0, \pm 1] &&\rightarrow (1s^2 2s^2) + 2p^1 \text{ (B)}, \\ & && +2p^2 \text{ (C)}, \; +2p^3 \text{ (N)}, \\ & && +2p^4 \text{ (O)}, \; +2p^5 \text{ (F)}, \\ & && +2p^6 \text{ (Ne)}. \end{aligned}$$

The lithium ground state *LS* depends only on the single 2s-electron, i.e., in analogy with hydrogen, the Li ground terms must be $1s^2 2s$ (^2S). With core $1s^2$ all other excited terms of helium must be of the form ^2L, where $L = \ell$, the orbital angular momentum of the outer valence electron. The beryllium ground state is $1s^2 2s^2$ (^1S) since the 2s subshell is also filled (paired spins and orbital momenta), in analogy with helium.

Moving on to the $\ell = 1$ open subshell, the ground *LS* term for boron is simple: $1s^2 2s^2 2p^1$ ($^2P^o$); again, the analogy with hydrogen may be invoked since the closed Be-like electronic core $1s^2 2s^2$ does not contribute to total L or S. But the *LS* assignment by inspection breaks down for carbon, and all other open L-shell elements, since we now have more than one electron in the p-shell and it is no longer obvious how the exclusion principle allows the *LS* designation of possible atomic states. Furthermore, if we consider not just the ground configuration but also excited configurations then we have (with the exception of hydrogen) a myriad of couplings of spin and orbital angular momenta of two or more electrons.

The general question then is: what are the spectroscopic *LS* terms for a given electronic configuration with open-shell non-equivalent electrons (single electron in outer orbit), e.g., $n\,\ell\,n'\,\ell'$ ($n \neq n'$), and equivalent electrons $n\ell^q$ with the same n and ℓ in a configuration with occupancy number q? We need not consider the closed shells or subshells, since their total $L = 0 = S$, do not affect the *LS* states of open-shells.

The exclusion principle states that no two electrons in an atom may have the same four quantum numbers $(n, \; \ell, \; m_{\ell i}, \; m_{si})$. At first sight, it appears straightforward to apply this rule to construct a list of allowed *LS* terms. However, it turns out to be rather involved in terms of bookkeeping, related to combinations of total M_L, M_S, consistent with the four quantum numbers of all electrons in a configuration. They are illustrated next.

2.5.1 Non-equivalent electron states

The *LS* coupling is simple for non-equivalent electrons since the n are different and the exclusion principle is not invoked; that is, terms of all possible L and S values are allowed. The possible values are simply vectorial sums of the individual ℓ and s values. The easiest example is that of two s-electrons, i.e., $ns\,n's$ of an excited configuration of helium. The total $L = 0$, since both electrons have $\ell = 0$. But the spins $\pm\frac{1}{2}$ can now add up to $S = 1$ or 0, i.e., the multiplicity $(2S+1)$ can be 3 or 1, respectively. Therefore, we have two $ns\,n's$ (^1S, ^3S) *LS* terms, e.g., the first two excited terms in helium; similarly the next two higher terms are 1s2p (^3P, ^1P). We ignore the parity for the time being, since it is easy to determine even or odd parity from summed l (Eq. 2.55). For more than two electrons we can couple L and S in a straightforward manner; say for three electrons,

$$\begin{aligned} &ns\,n'p\,(^1P)\,n''d \rightarrow {}^2P, {}^2D, {}^2F, \\ &ns\,n'p\,(^3P)\,n''d \rightarrow {}^{(2,4)}(P, D, F). \end{aligned}$$

The couplings are the same for any three electrons spd. The first two electrons s and p give sp (^1P, ^3P) terms, which couple to the d electron as above. The singlet ^1P *parent term* ($S = 0$) yields only the doublet (S=1/2) terms, but the triplet ^3P term ($S = 1$) gives both doublets (S=1/2)

TABLE 2.1 Six possible combinations for a p-electron.

m_l m_s	1 1/2	0 1/2	−1 1/2	1 −1/2	0 −1/2	−1 −1/2
Notation	1^+	0^+	-1^+	1^-	0^-	-1^-

TABLE 2.2 Twenty possible distributions for the np^3-electrons.

M_L / M_S	2	1	0	−1	−2
3/2			$1^+ 0^+ -1^+$		
1/2	$1^+ 0^+ 1^-$	$1^+ 0^+ 0^-$	$1^+ 0^+ -1^-$	$1^+ -1^+ -1^-$	$0^+ -1^+ -1^-$
1/2		$1^+ -1^+ 1^-$	$1^+ -1^+ 0^-$	$0^+ -1^+ 0^-$	
1/2			$0^+ -1^+ 1^-$		
−1/2	$1^+ 1^- 0^-$	$1^+ -1^- 1^-$	$1^+ -1^- 0^-$	$0^+ 0^- -1^-$	$-1^+ 0^- -1^-$
−1/2		$0^+ 1^- 0^-$	$0^+ 1^- -1^-$	$-1^+ 1^- -1^-$	
−1/2			$-1^+ 1^- 0^-$		
−3/2			$1^- 0^- -1^-$		

and quartets (S=3/2) when coupled to the third d-electron. It is clear that one obtains the same coupled LS terms regardless of the order in which the terms are coupled, i.e., same for pds, dsp, etc.

Similarly, for three non-equivalent p-electrons we can write down the LS terms as follows. Dropping the n prefix, assuming that $n \neq n' \neq n''$, we have pp′ (^{1}S, ^{1}P, ^{1}D, ^{3}S, ^{3}P, ^{3}D) as the parent terms, which yield

(^{1}S) p″ → ^{2}P,
(^{1}P) p″ → ^{2}(S, P, D),
(^{1}D) p″ → ^{2}(P, D, F),
(^{3}S) p″ → $^{(2,4)}$P,
(^{3}P) p″ → $^{(2,4)}$(S, P, D),
(^{3}D) p″ → $^{(2,4)}$(P, D, F).

2.5.2 Equivalent electron states

For equivalent electron configurations nl^q, both the vector addition of angular and spin momenta and the Pauli exclusion principle are to be considered. The exclusion principle disallows certain LS terms, and requires an explicit evaluation of all possible combinations of (m_{li}, m_{si}) for the q equivalent electrons to form the *allowed* values of total (M_L, M_S).

Consider the equivalent electron configuration np^3. As seen above for non-equivalent electrons not subject to the exclusion principle, the six possible LS states are: ^{2,4}S, ^{2,4}P, ^{2,4}D and ^{2,4}F. But many of these states are eliminated by the exclusion principle. To wit: no L = 3 or F terms can be allowed since two of the three electrons will have the same (m_{li}, m_{si}). But we must do the book-keeping systematically as follows. The p subshell can have

$$m_l = 1,\ 0,\ -1;\quad m_s = 1/2, -1/2, \tag{2.60}$$

Hence, with common values of n and l(=1) but differing in m_l and m_s, there are six possible combinations of m_l and m_s, or cells, as given in Table 2.1.

Now the combination of three 2p electrons can be expressed as all possible distributions of the type $(1^+0^+1^+)$, $(1^+1^-1^+)$, etc. (Do we detect a problem with these combinations?) A distribution in the cell $(1^+\ 0^+\ 1^-)$ is associated with values $M_L = 2$ and $M_S = \frac{1}{2}$. Since electrons are indistinguishable, they may be permuted without affecting the distribution; thus $(1^+\ 0^+\ 1^-)$, $(0^+\ 1^+\ 1^-)$ are the same. These electrons are grouped according to their respective values of M_L and M_S. Since

$$M_L = \sum_i m_l;\quad M_S = \sum_i m_s, \tag{2.61}$$

each LS term must have a cell with the highest $M_L = L$ or $M_S = S$. The rule is that *the highest* M_L *or* M_S *must have a cell with* $M_L - 1$, $M_L - 2, \ldots, -M_L$ *and* $M_S - 1, M_S - 2, \ldots, -M_S$. The number of independent distributions of three indistinguishable electrons in six orbitals is $(6 \times 5 \times 4)/3! = 20$. The combined electronic cells are now grouped according to M_L and M_S in Table 2.2. A distribution of $(1^+\ 0^+\ 1^-)$ has the values $M_L = 2$ and $M_S = 1/2$. Table 2.2 shows all 20 possible distributions of (m_l, m_s), following the exclusion principle.

TABLE 2.3 Summary of electron distribution.

M_L	M_S	(m_l, m_s) cells	Term (^{2S+1}L)
$2, 1, 0-1, -2$	$\pm\frac{1}{2}$	10	^2D
$1, 0, -1$	$\pm\frac{1}{2}$	6	^2P
0	$\pm\frac{3}{2}, \pm\frac{1}{2}$	4	^4S

With reference to Table 2.3, starting with highest value of M_L, it is apparent that the entries for $M_L = \pm 2$ and $M_S = \pm 1/2$, belong to the ^{2}D term, which also includes entries $M_L = \pm 1$, 0. We remove all these entries (ten in total) for ^{2}D. From the remaining entries, the highest M_L is 1 with $M_S = 1/2$ belonging to ^{2}P. ^{2}P will also include $M_L = 0, -1$ and $M_S = -1/2$, and hence will take six entries out. The remaining four entries, with $M_L = 0$ and $M_S = \pm(3/2, 1/2)$, give ^{4}S.

An important fact is that *the number of allowed LS terms from an open shell configuration nl^q, with q electrons or occupancy, is the same as with q electron vacancies*. For example, the ground configuration of oxygen, $2p^4$, has two vacancies in the 2p shell, and gives the same three LS terms as $2p^2$. A simpler example is that of configuration np or np^5, both of which give a single LS term ^2P.

Exercise 2.4 *(a) Show by direct evaluation that the configurations np^2 and np^4 give rise to the same LS terms. (b) List all possible combinations of $(m_{\ell i}, m_{si})$ for three equivalent electrons in a 'd' open subshell, i.e., nd^3. Note that the available number of combinations give two ^2D LS terms.*

While we have determined the physical LS states using the Pauli exclusion principle, we still do not know their energies and the *order* in which they exist and are measured in the laboratory. But before we describe the theoretical framework, the Hartree–Fock method, it is useful to note an empirical rule: the lowest state of a given configuration is the LS term with the highest spin multiplicity (2S+1). Further discussion is given in the section on *Hund's rules*.

2.6 Empirical rules for electronic configurations

Multi-electron elements may be divided into two groups, 'light' and 'heavy', depending on their Z numbers. However, this division is imprecise. Two particular criteria play for the elements, (i) nuclear charge and (ii) the number of electrons. The former, causing Coulomb force, determines whether relativistic effects are important or not. The latter criterion is related to electron correlation or interaction that comes into play in establishing multiplet structure and levels. We will consider relativistic effects later.

On the basis of the electronic structure we may somewhat arbitrarily categorize elements as 'light' for $Z \leq 18$ and 'heavy' for $Z > 18$. This division rests partially on a complexity in the ground configuration electronic shell structure that occurs in argon ($Z = 18$). The subshell structure of elements up to argon is filled up in a naturally straightforward manner, first according to n and then according to ℓ. For example, the outermost open electronic configuration of chlorine ($Z = 17$) is $3p^5$ with one vacancy in the 3p subshell (chlorine is designated as a halogen in the periodic table in Appendix A). The 3p subshell is all occupied in argon (a noble gas) with a closed subshell $3p^6$. However, the next element with $Z = 19$, potassium (K), begins by filling in the 4s subshell, instead of 3d, with the outermost subshells of the ground configuration as $3p^6 4s$; i.e., the 19th electron goes into the 4s subshell. Furthermore, the ground configuration of calcium (Ca) with $Z = 20$ is $3p^6 4s^2$; again the 4s, not the 3d, subshell is occupied. This is a manifestation of a more general rule that *subshells with lower $(n + \ell)$ are filled in first*. For both K and Ca $(n + \ell) = 4$ for the 4s subshell, and 5 for the 3d; hence the former fills up first. Moreover, if $(n + \ell)$ is the same, then the higher ℓ is filled up first. For instance, in the iron group elements from Ca to Zn ($Z = 20$–30), we have the 3d subshell filling in after the 4s rather than the 4p, although both have $(n + \ell) = 5$. But the picture becomes further complicated, since for elements heavier than nickel ($Z = 28$) there are deviations from the $(n + \ell)$-first rule.

2.7 Intermediate coupling and jj coupling

In heavier (large Z) atoms the electron–nuclear force becomes strong enough to cause the breakdown of the LS coupling scheme. The velocity of an electron increases to relativistic level, its electrostatic interaction with other electrons weakens, and the total angular momentum of individual electrons needs to be considered. Hence, for high-Z atoms this leads to *jj coupling* where j is sum of the individual total electron angular and spin momenta, that is,

$$j_i = l_i + s_i, \quad J = \sum_i j_i, \tag{2.62}$$

where J is the total angular momentum of all electrons in the atom with multiplicity or degeneracy $2J+1$. The total J follows the vector sum: for two electrons, the values of J range from $|j_1+j_2|$ to $|j_1-j_2|$. The states are denoted as $(j_i\,j_2)_J$. For example, for a (pd) configuration $j_1(1\pm 1/2) = 1/2,\ 3/2$, and $j_2(2\pm 1/2) = 3/2,\ 5/2$; the states are designated as $(1/2\ 3/2)_{2,1}$, $(1/2\ 5/2)_{3,2}$, $(3/2\ 3/2)_{3,2,1,0}$, $(3/2\ 5/2)_{4,3,2,1}$ (note that the total discrete $J = 0-4$). The J-state also includes the parity and is expressed as $J\pi$ or, more completely, as ${}^{(2S+1)}L^{\pi}_{J}$. The latter designation relates the fine-structure level to the parent LS term. For each LS term there can be several fine-structure levels. The total angular magnetic quantum number J_m runs from $-J$ to J.

Fine-structure levels can be further split into hyperfine structure when nuclear spin $\boldsymbol{I}$ is added vectorially to $\boldsymbol{J}$ to yield the quantum state $\boldsymbol{J}+\boldsymbol{I} = \boldsymbol{F}$. Figure 2.2 shows the schematics of energy levels beginning with a given electronic configuration.

For cases where LS coupling is increasingly invalid because of the importance of relativistic effects, but the departure from pure LS coupling is not too severe and full consideration of relativistic effects is not necessary, an *intermediate coupling* scheme designated as LSJ is employed. (We discuss later the physical approximations associated with relativistic effects and appropriate coupling schemes.) In intermediate coupling notation, the angular momenta l and s of an interacting electron are added to the total orbital and spin angular momenta, J_1 of all other electrons in the following manner,

$$\boldsymbol{J}_1 = \sum_i \boldsymbol{l}_i + \sum_i \boldsymbol{s}_i, \quad \boldsymbol{K} = \boldsymbol{J}_1 + \boldsymbol{l}, \quad \boldsymbol{J} = \boldsymbol{K} + \boldsymbol{s}, \tag{2.63}$$

where $s = 1/2$. The multiplicity is again $2J+1$, and the total angular magnetic quantum number J_m runs from $-J$ to J.

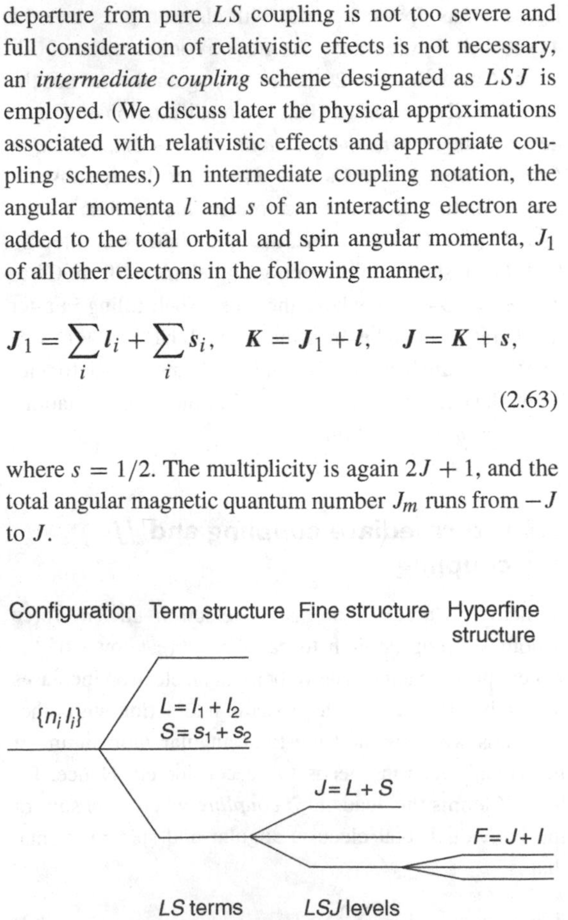

FIGURE 2.2 Electronic configuration and energy level splittings.

An important point to note is that the physical existence of atomic energy states, as given by the number of total J-states, must remain the same, regardless of the coupling schemes. Therefore, the total number of J-levels $\left(\sum_i j_i = J \text{ or } L+S = J\right)$ is the same in intermediate coupling or jj-coupling. A general discussion of the relativistic effects and fine structure is given in Section 2.13

2.8 Hund's rules

The physical reason for the variations in subshell structure of the ground configuration of an atom is the electron–electron interaction. It determines the energies of the ground and excited states. We need to consider both the direct and the exchange potentials in calculating these energies. Before we describe the atomic theory to ascertain these energies, it is useful to state some empirical rules. The most common is the *Hund's rules* that governs the spin multiplicity $(2S+1)$, and orbital L and total J angular momenta, in that order.

The *S-rule* states that an LS term with the highest spin multiplicity $(2S+1)$ is the lowest in energy. This rule is related to the exchange effect, whereby electrons with like spin spatially avoid one another, and therefore see less electron–electron repulsion (the exchange potential, like the attractive nuclear potential, has a negative sign in the Hamiltonian relative to the direct electron–electron potential, which is positive). For example, atoms and ions with open subshell np^3 ground configuration (N I, O II, P I, S II) have the ground state $^4S^o$, lower than the other terms $^2D^o$, $^2P^o$ of the ground configuration.

The *L-rule* states that for states of the same spin multiplicity the one with the larger total L lies lower, again owing to less electron repulsion for higher orbital angular momentum electrons that are farther away from the nucleus. Hence, in the example of np^3 above, the $^2D^o$ term lies lower than the $^2P^o$. Another example is the ground configuration of O III, which is C-like $2p^2$ with the three LS terms ^3P, ^1D, ^1S in that energy order. A more complex example is Fe II, with the ground $3p^6 3d^6 4s$ and the first excited $3p^6 3d^7$ configurations. The LS terms in energy order *within* each configuration are $3d^6 4s$ (^6D, ^4D) and $3d^7$ (^4F, ^4P). But the two configurations overlap and the actual observed energies of these four terms lie in the order ^6D, ^4F, ^4D, ^4P.

The *J-rule* refers to fine-structure levels $\boldsymbol{L}+\boldsymbol{S} = \boldsymbol{J}$. For less than half-filled subshells, the lowest J-level lies lowest, but for more than half-filled subshells it is the reverse, that is, the highest J-level lies lowest in energy.

For example, both O III (C-like) and Ne III (O-like) have the ground state ^{3}P, but the ground configuration open subshells are $2p^2$ and $2p^4$, respectively (recall that both configurations have the same LS term structure). Hence, the fine-structure energy levels are $J = 0, 1, 2$ for O III (and C I), and $J = 2, 1, 0$ for Ne III (and O I). Again, a more interesting application of the J-rule is the Fe ions with ground configurations containing the 3d open shell: Fe I ($3d^64s^2$), Fe II ($3d^64s$) and Fe III ($3d^6$) have more than half of the 3d subshell occupied, and the resulting fine-structure energy levels are in descending order of J-values. On the other hand, Fe V ($3d^4$), Fe VI ($3d^3$), Fe VII ($3d^2$) and Fe VIII (3d) have less than half of the 3d subshell unoccupied and the levels are in ascending order of J-values. For example, the fine-structure levels of the ground term of Fe II are $3d^64s\ ^6D_{9/2,7/2,5/2,3/2,1/2}$, and for Fe VII, $3d^2\ ^3F_{2,3,4}$.

Hund's rules are useful but show increasing deviations with higher nuclear charge. They may apply to excited configurations of an ion, albeit with exceptions, owing to configuration interaction and the fact that more than one subshell is open. These empirical rules are particularly useful in large-scale theoretical calculations of energy levels of complex atomic systems where only small number of energy levels have been experimentally observed.

Exercise 2.5 *From the observed energy levels of Fe ions with the 3d open shell, available from the NIST website www.nist.gov/physlab/data/asd.cfm, construct a table for the LSJ levels of the ground and the first few excited configurations. Plot the energy levels on a Grotrian diagram sorted out by symmetry LSJ. Show the transitions responsible for some prominent IR and optical forbidden lines. For what excited configurations is the J-rule violated?*

We emphasize that while these empirical rules illustrate both the simplicity and the complexity of the atomic energy level structure, a complete understanding is based on actual computations and observations. This is particularly so for neutral elements beyond argon, where two valence subshells are open. Appendix A lists the ground state and configuration of all natural elements, and their ionization energies obtained from the National Institute of Standards and Technology (NIST), www.nist.gov.

Before we describe the basic theory that underpins atomic structure calculations, it is helpful to consider approximate (and partly semi-empirical) determination of level energies of outermost electrons in the valence subshell. The basic idea is to treat that electron separately from the 'core' with all other inner electrons, as described in the next section.

2.9 Rydberg formula with quantum defect

Based on empirical work *before* the advent of quantum mechanics, Rydberg obtained an expression for the energy levels not only of H-like ions, as described earlier, but also of atoms with one valence electron outside a closed shell, such as the alkali elements Li, Na, K. The general Rydberg formula is modified from that for H-like ions (Eq. 2.40), to account for the screening effect on the valence electron by the core electrons of the closed shell. The outer electron sees an *effective charge* $z = Z - N + 1$, where N is the number of electrons. The effective Coulomb potential is similar to that of H, but screening modifies it. This departure from a pure Coulomb form effectively *reduces* the principal quantum numbers n by an amount called the *quantum defect* $\mu \geq 0$, which modifies the Rydberg formula to

$$E(n\ell) = \frac{z^2}{(n-\mu)^2}. \tag{2.64}$$

Excited energy levels described by the Rydberg formula are often labelled as 'Rydberg levels'. This is generally true for all atoms and ions, provided the outer electron is in sufficiently high-n state, i.e., sufficiently far away from all the inner electrons so as to experience only the residual charge z. However, the amount of screening depends on the orbital symmetry via the orbital angular momentum ℓ. Assuming that there are no other potentials involved, μ is a unique positive constant for each ℓ. We can write,

$$E(n\ell) = \frac{z^2}{(n-\mu_\ell)^2}. \tag{2.65}$$

This relation is very useful in estimating the energies of successively higher n levels, where the quantum defect μ_ℓ may be obtained empirically from fits to observed spectra. The above equation may then be used to obtain energy levels of an arbitrarily large number of levels up to the series limit at $n = \infty$ for any given ℓ. It should be noted that the formula may not be accurate at all for low-n levels, but is progressively more so with increasing n of the outer electron. Also, it does not take account of relativistic fine structure.

We can now define an *effective quantum number* ν for a given bound level as

$$\nu \equiv n - \mu; \qquad E_n = -\frac{z^2}{\nu^2}. \tag{2.66}$$

As $n \to \infty$, $E \to 0$; the bound electron reaches the series limit or *threshold* where it becomes free at $E_{n=\infty} = 0$. Effective quantum numbers are similar to n and become n when the quantum defect of a higher orbital becomes zero;

TABLE 2.4 Binding energies, effective quantum numbers and quantum defects of C I.

Configuration	LS	$E_\infty - E(LS)$	ν	μ
$2s^22p^2$	^3P	0.8277 Ry	1.0992	0.9008
$2s^22p3s$	$^1P^o$	0.2631 Ry	1.9493	1.0507
$2s^22p4s$	$^1P^o$	0.1141 Ry	2.9603	1.0397
$2s^22p5s$	$^1P^o$	0.0637 Ry	3.9635	1.0365
$2s^22p3p$	^1P	0.2005 Ry	2.2331	0.7669
$2s^22p4p$	^1P	0.0939 Ry	3.2641	0.7359
$2s^22p5p$	^1P	0.0548 Ry	4.2727	0.7273
$2s^22p3d$	$^1D^o$	0.1201 Ry	2.8855	0.1145
$2s^22p4d$	$^1D^o$	0.0671 Ry	3.8605	0.1395
$2s^22p5d$	$^1D^o$	0.0426 Ry	4.8450	0.1550

ν increases approximately by unity. However, it is often a decimal number, especially for s, p, and d orbitals where μ is relatively large. A Rydberg series of states is usually represented by νl. It may be proved that the nature of the Rydberg series arises from the property of an attractive Coulomb potential to support (bind) an infinite number of atomic states.

We shall employ the simple quantum defect formula in a variety of ways to analyze the spectrum of bound states, and states known as *quasi-bound* or *resonant states*. In Chapter 3 we will describe *quantum defect theory* that will enable an approximate analysis of *both* bound and continuum states of atomic systems. For the present we illustrate various properties of the quantum defect, and demonstrate the utility of the Rybderg formula for a *non-hydrogenic* ion.

Table 2.4 gives the observed energy levels, effective quantum numbers and quantum defects for neutral carbon. The values given correspond to the first few members of Rydberg series $n\ell$ for a given ℓ and $n = 2 - 5$. For consistency, we consider the same LS term for each series.

Several properties of the quantum defect are evident from Table 2.4: (i) $\mu_\ell(n)$ is a slowly varying quantity nearly independent of n, (ii) it approaches a constant value with increasing n, $\mu_\ell(n) \to$ constant as $n \to \infty$, and (iii) $\mu_s > \mu_p > \mu_d > \mu_f > \ldots$ For light elements, such as carbon, $\mu_\ell \approx 0$ for $\ell > 3$ (f-electron). This is a consequence of the fact that with increasing angular momentum the valence (Rydberg) electron sees a constant Coulomb potential; there is practically no departure from it from an f-electron onwards. For heavy elements, where f-orbitals may be occupied, the same trend continues but for still higher ℓ values.

Since the Rydberg formula was initially derived from fits to alkali spectra, it describes the energy levels of the valence electron quite accurately. Consider a valence electron ns in sodium (Na):

ns :	3s	4s	5s	6s	7s
ν :	1.626	2.643	3.647	4.649	5.650
μ :	1.374	1.357	1.353	1.351	1.350

Whereas the change in effective quantum number $\nu(ns)$ is nearly unity, the quantum defect μ_{ns} in Na approaches the limiting value of 1.35 for high n. Additional properties of the quantum defect may be obtained readily by an investigation of the energy levels, as in the following examples [5].

Exercise 2.6 *(a) Show that the series limits in Na for $\ell > 0$ are: $\mu_p \approx 0.86$, $\mu_d \approx 0.01$, $\mu_f \approx 0.00$. The progression of μ_l demonstrates a decrease in electron screening with ℓ. (b) The ν_n and μ_n for alkalis vary systematically with Z along the periodic table. For the simple alkali-like atomic structure, ν_n changes little, as also reflected in the ionization potential $E_{IP} = -1/\nu^2$. On the other hand, the quantum defect μ_{ns} increases by roughly unity for each additional inner s-shell, which results in enhanced screening of s-electrons (similar consideration applies to other ℓ-series). Using experimental values of ionization energies of the ns-electron in Li, Na, K, Rb and Cs, show that the $\nu_n \approx 1.6 - 1.9$, and μ_{ns} increases by $\approx$ 1 for each successive element from Li onwards. (c) The μ_l decreases with z along an isoelectronic sequence, as the perturbing effect of the core on the outer electron decreases and the pure Coulomb potential dominates. The variation of μ_ℓ vs. z may be seen by considering the Na isoelectronic sequence, Na I, Mg II, Al III, Si IV, P V, Si VI, etc. Show that the μ_{3s} decreases from ~1.4 for neutral Na to ~0.6 for S VI and so on (compute precise values using experimental ionization energies).*

To see how useful the simple Rydberg formula with non-vanishing quantum defect can be, let us apply some of the points described above in the following example. Suppose a C IV line is observed at 64555.6 cm^{-1}. What spectral parameters can be determined quantitatively? The ionization energy E_{IP} of C IV is 64.476 eV = 64.476 eV / (13.6 eV/Ry) = 4.74 Ry. C V is a Li-like ion of carbon with three electrons; therefore, the ion charge seen by an electron in a Rydberg level is $z = Z - N + 1 = 4$. The effective quantum number of the ground state $1s^2 2s$ (^2S) is

$$E_{\mathrm{IP}}(1s^2 2s)/\mathrm{Ry} = \frac{z^2}{\nu^2}, \tag{2.67}$$

$$\nu = \frac{z}{\sqrt{E_{\mathrm{IP}}/\mathrm{Ry}}} = \frac{4}{\sqrt{4.74}} = 1.837, \tag{2.68}$$

hence, $\mu_s = n - \nu = 2 - 1.837 = 0.163$ (2.69)

is the quantum defect of a $1s^2 ns$ Rydberg series of states in C IV. (In general, the quantum defect derived from the *ground* state energy is not good enough since the valence electron penetrates the core too deeply.) Assuming the observed line to be from the first spectral transition above the ground state, $1s^2 2s \leftarrow 1s^2 2p$, with transition energy $\Delta E(2s - 2p) = 64\,555.6\,\mathrm{cm}^{-1}/(109\,737.3\,\mathrm{cm}^{-1}\,\mathrm{Ry}^{-1} = 0.5882\,\mathrm{Ry}$ and wavelength $\lambda = 1/(64555.6\,\mathrm{cm}^{-1})$= 1549 Å, the ionization energy $E_{\mathrm{IP}}(1s^2 2p)$ of the first excited state $1s^2 2p$ is 4.74 Ry−0.5882 Ry = 4.15 Ry. Since

$$\nu_{2p} = \frac{4}{\sqrt{4.15}} = 1.964, \tag{2.70}$$

the quantum defect of the $1s^2 np$ series is then

$$\mu_p = 2 - 1.964 = 0.036. \tag{2.71}$$

Now let us see how accurately we can determine the transition wavelengths from the ground state to the excited levels of the $1s^2 np$ series. Taking μ_p to be constant for all levels, we have

$$\nu(1s^2 3p) = 3 - 0.036 = 2.964, \tag{2.72}$$

$$E(1s^2 3p)/\mathrm{Ry} = \frac{16}{(2.964)^2} = 1.82, \tag{2.73}$$

$$\Delta E(2s - 3p)/\mathrm{Ry} = 4.74 - 1.82 = 2.92, \tag{2.74}$$

$$\lambda(1s^2 2s\text{–}1s^2 3p) = \frac{912\text{Å}}{2.92} = 312.3\,\text{Å}. \tag{2.75}$$

The actual observed value for this second excited transition in C IV is 312.43 Å. Thus the quantum defect analysis yields an accurate estimate. However, let us remind ourselves that C IV is a Li-like ion with H-like configurations, with a closed He-like core and an outer valence electron, where the Rydberg formula is expected to be accurate. This is generally not so for low-lying levels of more complex atoms. We can similarly estimate the remaining transitions in the ns–np Rydberg series, up to the series limit at 912 Å/4.74 = 192.4 Å. C IV lines are prominent UV features in the spectra of AGN, quasars, Lyα clouds and other sources.

Rydberg series of levels play a crucial role in atomic processes. They also determine the positions for autoionizations and resonances, as will be discussed later. However, we define the states of an electron in or with an atom to be bound when the ground or excited discrete energies are negative, that is, lie below the first ionization threshold or energy of the ground state of the ionized core.

Exercise 2.7 *Suppose the following absorption lines are observed in the spectrum of a neutral gas, with wavenumbers 1.301, 2.471, 2.900, 3.107 in units of $10^4\mathrm{cm}^{-1}$. Fit the wavenumbers to the Rydberg formula to determine (a) the quantum defect, (b) the Rydberg series, (c) the ionization energy with respect to the ground state and (d) the atomic species.*

As mentioned before, semi-empirical studies of atomic structure described above can be useful and illustrative. But now we must turn to proper theoretical methods that deal with the full quantum mechanical complexity of multi-electron atoms.

2.10 Multi-electron atomic systems

Treatment of a multi-electron atomic system is much more complex than of hydrogen. The one-electron Hamiltonian has only one potential term, the electron–nucleus Coulomb attraction potential $V(r)$. Extension to a many-electron system requires one (i) to sum over all one-electron operators, that is the kinetic energy and the attractive electron–nucleus potential, and (ii) to sum over two-electron Coulomb repulsion operators. Helium, for example, or helium-like ions, lead us to the two-electron Hamiltonian

$$H = \frac{p_1^2}{2m} + \frac{p_2^2}{2m} - \frac{e^2 Z}{r_1} - \frac{e^2 Z}{r_2} + \frac{e^2}{|\boldsymbol{r}_1 - \boldsymbol{r}_2|}, \tag{2.76}$$

with the two electrons in positions $\boldsymbol{r}_1$ and $\hat{\boldsymbol{r}}_2$.

The multi-electron Hamiltonian, reduced to *number* equations with atomic units (Ry), and $H = H_0 + H_1$,

$$H_0 = \sum_{i=1}^{N} \left[-\nabla_i^2 - \frac{2Z}{r_i} \right], \tag{2.77}$$

$$H_1 = \sum_{j<i} \frac{2}{r_{ij}}, \tag{2.78}$$

which accounts for the two-electron Coulomb interaction. Beyond the single electron case, a solution Ψ from the Schrödinger equation no longer exists. One has to start with a trial function Ψ^t in some parametric form. The best-known example in single-particle expansions is the Slater-type orbital, $P_{nl}^{\rm STO}(r)$. A trial function would be of no use without conditions for the parameters to satisfy. That prescription is known as a *variational principle*. The standard *Rayleigh–Ritz variational principle* places an upper bound on the eigenvalue obtained from a trial function in the Schrödinger equation. Particularizing it to the many-electron Hamiltonian discussed above, as embodied in the Hartree–Fock scheme discussed in the next section, yields the *Hartree–Fock variational principle*:

$$\delta \langle \Psi | H | \Psi \rangle = 0 \tag{2.79}$$

with solution $E = E_{\rm min}$,

because the spectrum of the non-relativistic Hamiltonian is bounded from below.[3]

2.11 The Hartree–Fock method

There is no analytical solution to the non-relativistic Hamiltonian once it deals with more than one electron, and the variational equation (Eq. 2.79) replacing the Schrödinger equation. The most widely employed basic approach, which underlies most treatments, is the *Hartree–Fock method.* Starting with a complete set of integro-differential equations, its trial functions iterate very closely towards the 'true' solution, albeit within the constraints given by the finite number of configurations specified for the functions. The numerical approach follows the prescription first laid down by Hartree for constructing a *self-consistent* iterative procedure for the inter-electron potentials and the electronic wavefunctions [6]. However, the Hartree method did not account for electron exchange. That was accomplished by Fock [7], who extended the treatment by introducing antisymmetrization of wavefunctions, which includes exchange in an *ab-initio* manner. It is an iterative scheme [6] referred to as the self-consistent Hartree–Fock method.

We begin with the exact Hamiltonian as a Rydberg–Bohr scaled number equation (Eqs 2.77 and 2.78):

$$H = \sum_{i=1}^{N} \left[-\nabla_i^2 - \frac{2Z}{r_i} \right] + \sum_{j \neq i} \frac{2}{r_{ij}}. \tag{2.80}$$

For future use, we also define the short-hand notation for the one-electron operator,

$$f_i \equiv -\nabla_i^2 - \frac{2Z}{r_i}, \tag{2.81}$$

representing the kinetic energy $p^2/2m$ and the nuclear potential, and the two-electron operator,

$$g_{ij} \equiv \frac{2}{r_{ij}}. \tag{2.82}$$

Then the Rydberg–Bohr scaled Hamiltonian reads

$$H = \sum_i f_i + \sum_{j \neq i} g_{ij} \equiv F + G, \tag{2.83}$$

where F and G are interaction operators for all electrons in the atom. It is the electron–electron interaction that, together with the electron–nuclear interaction, makes the non-hydrogenic atom a many-body problem not amenable to exact solutions. The Coulomb potential $2Z/r$ in the Schrödinger equation for hydrogen needs to be replaced by a potential $V(r)$ that yields the individual electron wavefunctions $\psi(r, \vartheta, \varphi, m_s)$. The wavefunction ψ is written as a *spin-orbital*, the product of a function of spatial coordinates $\phi(r)$ and spin function ζ_{ms} with binary components $\pm 1/2$.

$$\psi_{n,l,m_l,m_s}(r, \vartheta, \varphi, m_s) = \varphi(\boldsymbol{r}) \zeta_{m_s} \tag{2.84}$$

We can also denote the spin-orbital by replacing the spatial coordinates with quantum numbers (Eq. 2.52) as $\psi_{n,l,m_l,m_s} = \phi(nlm_\ell)\zeta(m_s)$. Since each electron moves in a potential created by all other electrons, the crucial problem is to construct the potential from the set of all ψ_i, where i refers to all atomic electrons. In other words, the potential $V(r_i)$ for the ith electron is obtained *self-consistently*. That is the aim of the self-consistent iterative scheme employed to solve the Hartree–Fock equations. Moreover, the individual electronic states ψ_i are used to construct the total atomic wavefunction for a given state of the multi-electron atom. Assuming the atom to be an N-electron system, the total atomic wavefunction is a product of the one-electron spin-orbitals

[3] For states ψ^{LS} one will safely obtain the one with lowest energy – and higher ones must be orthogonal. A well-known application is the Hartree–Fock method with its specific type of trial function, to be dealt with in some detail in Section 2.11. In between, we will unravel the coupling to LS and to other schemes from spectroscopy. Technically, two particles can be dealt with on the elementary level of Clebsch–Gordan coefficients. Equivalent electrons have, in principle, been addressed in the classic book by Condon and Shortley [3] as reflected in the following sections.

$$\psi_{n,l,m_l,m_s}(r,\theta,\phi,m_s) = \prod_{i=1}^{N} \psi_{n_i,\ell_i,m_{\ell_i},m_{s_i}}$$
$$= (\psi_{n_1,l_1,m_{l_1},m_{s_1}})(\psi_{n_2,l_2,m_{l_2},m_{s_2}})\cdots(\psi_{n_N,\ell_N,m_{\ell_N},m_{s_N}}). \quad (2.85)$$

But a simple product form of the total wavefunction does *not* incorporate electron exchange, which requires interchange of electronic coordinates to satisfy the general antisymmetry postulate. The wavefunction must change sign upon interchange of the coordinates of any two electrons, i.e., the interchange of any of their spatial or spin coordinates. The Hartree–Fock representation of the multi-electron system incorporates the antisymmetrization in the wavefuction representation. For illustration, consider the helium atom. The mathematical form of a two-particle antisymmetric wavefunction can be written as

$$\Psi(1,2) = \frac{1}{\sqrt{2}}\left[\psi_1(1)\psi_2(2) - \psi_1(2)\psi_2(1)\right]. \quad (2.86)$$

This is, in fact, the usual expansion of the *determinant* of a 2×2 matrix with elements that correspond to the two electrons and their coordinates, i.e.,

$$\Psi = \frac{1}{\sqrt{2}}\begin{vmatrix} \psi_1(1) & \psi_1(2) \\ \psi_2(1) & \psi_2(2) \end{vmatrix}. \quad (2.87)$$

The antisymmetry is now clear. Interchange of coordinates 1 and 2 changes the sign of the determinant (Eq. 2.87). But if both coordinates of the two electrons are the same, then the state cannot exist, since the antisymmetric wavefunction (Eq. 2.87) would have two identical rows or columns; ergo, the determinant would be zero.

For helium-like two-electron systems, the average potential energy of electron 1 in the field of electron 2 is

$$U_1(r_1) = \int \psi^*(\boldsymbol{r}_2)\,\frac{2}{r_{12}}\,\psi(\boldsymbol{r}_2)\,\mathrm{d}\boldsymbol{r}_2. \quad (2.88)$$

Similarly, we have the potential energy of electron 2 in the field of electron 1. We may now define an effective one-electron Hamiltonian operator as

$$H_i = \frac{p_i^2}{2m} - \frac{e^2 Z}{r_i} + U_i(r_i). \quad (2.89)$$

Analogous to the Schrödinger equation for each electron, we have

$$H_1(\boldsymbol{r}_1)\psi(\boldsymbol{r}_1) = \epsilon_1\psi(\boldsymbol{r}_1)$$
$$H_2(\boldsymbol{r}_2)\psi(\boldsymbol{r}_2) = \epsilon_2\psi(\boldsymbol{r}_2). \quad (2.90)$$

These are the Hartree–Fock equations for the two electrons. They are coupled through r-dependence (as well as spin dependence related to the exchange effect, but not explicitly expressed above). $H_1(\mathbf{r}_1)$ depends on $\psi(\boldsymbol{r}_2)$, implying that $\psi(\boldsymbol{r}_2)$ must be known before solving $H_1(\boldsymbol{r}_1)$. Hence a trial $\psi(\boldsymbol{r}_2)$ is adopted and used to obtain $\psi(\boldsymbol{r}_1)$, according to the variational criterion (Eq. 2.79). Since the forms of $\psi(\boldsymbol{r}_1)$ and $\psi(\boldsymbol{r}_2)$ are identical, the new $\psi(\boldsymbol{r}_2)$ is used again to obtain $\psi(\boldsymbol{r}_1)$. This continues until the desired accuracy is attained. The scheme is often referred to as the *Hartree–Fock self-consistent field* method (HF-SCF).

In analogy with helium, we can write the N-electron wavefunction in the determinantal representation as

$$\Psi = \frac{1}{\sqrt{N}}\begin{vmatrix} \psi_1(1) & \psi_1(2) & \ldots & \psi_1(N) \\ \psi_2(1) & \psi_2(2) & \ldots & \psi_2(N) \\ \ldots & \ldots & \ldots & \ldots \\ \psi_N(1) & \psi_N(2) & \ldots & \psi_N(N) \end{vmatrix}. \quad (2.91)$$

Equation (2.91) is called the *Slater determinant.* Like the two-electron determinant (Eq. 2.87), if all coordinates of any two electrons are the same, then two rows or two columns would be identical and the determinant vanishes. The Pauli exclusion principle immediately follows: no two electrons can have all spatial and spin quantum numbers the same. Each subscript in ψ_a represents a set of four quantum numbers (n, l, m_l, m_s), and each variable i represents spatial coordinates $\boldsymbol{r}$ and spin-coordinates in position i of electron a. Since a spin-orbital has a parity $(-1)^l$, the Slater determinant has the well-defined parity $(-1)^{\sum_i l_i}$ and can therefore be even or odd under the inversion transformation $\boldsymbol{r}_i \to -\boldsymbol{r}_i$ depending on whether the sum $\sum_i l_i$ is even or odd.

Calculations of atomic structure follow the prescription first laid down by Hartree for constructing a *self-consistent* iterative procedure for the inter-electron potentials and the electronic wavefunctions [6]. However, the Hartree method did not account for electron exchange effect. The self-consistent Hartree–Fock method also relies on an iterative scheme [6–8]. Given an electronic configuration characterized by a set of (nl), the atomic wavefunction is composed of individual spin-orbital wavefunctions, that is, $\psi_{nl} = \phi(n_\ell m_\ell)\zeta(m_s)$. The $\psi_a(j)$ are subject to the orthonormality condition

$$\langle\psi_a(j)\psi_b(j)\rangle = \delta_{ab}. \quad (2.92)$$

We note that the Hartree–Fock variational principle implies for the chosen wavefunction that the ground state energy

$$E_0 \leq E[\Psi] = \langle\Psi|H|\Psi\rangle. \tag{2.93}$$

In the Hartree–Fock method the first trial wavefunction is a Slater determinant. Substitution of the wavefunction determinant introduces one-operator (involving one-electron function) and two-operator integrals, as in the helium case. The expectation value of the one-electron or one-body term is readily evaluated as

$$\langle\Psi|H_0|\Psi\rangle = \sum_k \langle\psi_k(i)|H_0|\psi_k(i)\rangle = \sum_k I_k. \tag{2.94}$$

H_1 is the sum of two-electron or two-body operators, for which we can write

$$\langle\Psi|H_1|\Psi\rangle = \sum_{k,l\neq k} \left[\left\langle\psi_k(i)\psi_l(j)\left|\frac{2}{r_{ij}}\right|\psi_k(i)\psi_l(j)\right\rangle - \left\langle\psi_k(i)\psi_l(j)\left|\frac{1}{r_{ij}}\right|\psi_l(i)\psi_k(j)\right\rangle\right], \tag{2.95}$$

summing over all $N(N-1)/2$ pairs of orbitals. We can also write it as

$$\langle\Psi|H_1|\Psi\rangle = \frac{1}{2}\sum_k\sum_l \left[\left\langle\psi_k(i)\psi_l(j)\left|\frac{2}{r_{ij}}\right|\psi_k(i)\psi_l(j)\right\rangle - \left\langle\psi_k(i)\psi_l(j)\left|\frac{1}{r_{ij}}\right|\psi_l(i)\psi_k(j)\right\rangle\right]. \tag{2.96}$$

The first term is called the *direct term*,

$$J_{kl} = \left\langle\psi_k(i)\psi_l(j)\left|\frac{1}{r_{ij}}\right|\psi_k(i)\psi_l(j)\right\rangle, \tag{2.97}$$

which is the average value of the interaction $1/r_{ij}$ relative to $\psi_k(i)\psi_l(j)$. The second term is called the *exchange term*,

$$K_{kl} = \left\langle\psi_k(i)\psi_l(j)\left|\frac{1}{r_{ij}}\right|\psi_l(i)\psi_k(j)\right\rangle, \tag{2.98}$$

which is the matrix element of the interaction $1/r_{ij}$ between two states $\psi_k(i)\psi_l(j)$ and $\psi_l(i)\psi_k(j)$, obtained by interchange of the electrons. Hence, the total energy is given by

$$E[\Psi] = \sum_i I_i + \frac{1}{2}\sum_i\sum_j [J_{ij} - K_{ij}]. \tag{2.99}$$

E should be stationary with respect to the variations of the spin-orbitals; ψ_i subject to N^2 orthonormality conditions. Hence the variational principle introduces N^2 Lagrange multipliers (or variational parameters) λ_{kl}, such that (incorporating the orthonormal conditions)

$$\delta E - \sum_k\sum_l \lambda_{kl}\,\delta\langle\psi_k|\psi_l\rangle = 0. \tag{2.100}$$

From the above equation it is seen that $\lambda_{kl} = \lambda^*_{kl}$ and hence N^2 Lagrange multipliers may be considered as the elements of a Hermitian matrix. Any Hermitian matrix can be diagonalized by a unitary transformation. Hence, we can assume that the matrix of Lagrange multipler λ_{kl} is diagonal with elements $E_k\delta_{kl}$, that is,

$$\delta E - \sum_k E_k \delta\langle\psi_k|\psi_k\rangle = 0. \tag{2.101}$$

Varying the Schrödinger equation with respect to spin-orbitals ψ_i and using the above relations, we can find, for the N spin-orbitals, the set of integro-differential equations

$$\left[-\nabla_i^2 - \frac{2Z}{r_i}\right]\psi_k(i) + \left[\sum_l \int \psi_l^*(j)\frac{2}{r_{ij}}\psi_l(j)\mathrm{d}j\right]\psi_k(i) - \sum_l\left[\int \psi_l^*(j)\frac{2}{r_{ij}}\psi_k(j)\mathrm{d}j\right]\psi_l(i) = E_k\psi_k(i), \tag{2.102}$$

where the summation over k extends over the N occupied spin-orbitals, and the integral $\int \ldots \mathrm{d}j$ implies an integration over the spatial coordinates $\boldsymbol{r}$ and a summation over the spin-coordinate of electron j. These equations are the *Hartree–Fock equations* of a multi-electron system. We can separate the spin functions from the spin-orbital by writing $\psi_k(i) = u_k(\boldsymbol{r}_i)\chi_{1/2,m_l^k}$ and using the orthonormality condition $\langle\chi_{1/2,m_l^k}\chi_{1/2,m_l^j}\rangle = \delta_{m_L^k,m_L^j}$. Then the Hartree–Fock equations are in a form that involves only the spatial part,

$$\left[-\nabla_i^2 - \frac{2Z}{r_i}\right]u_k(\boldsymbol{r}_i) + \left[\sum_l \int u_l^*(\boldsymbol{r}_j)\frac{2}{r_{ij}}u_l(\boldsymbol{r}_j)\mathrm{d}\boldsymbol{r}_j\right]u_k(\boldsymbol{r}_i) - \sum_l \delta_{m_L^k,m_L^l}\left[\int u_l^*(\boldsymbol{r}_j)\frac{2}{r_{ij}}u_k(\boldsymbol{r}_j)\mathrm{d}\boldsymbol{r}_j\right]u_l(\boldsymbol{r}_i) = E_k u_k(\boldsymbol{r}_i). \tag{2.103}$$

The integrals are commonly expressed as *direct* and *exchange operators*. The direct operator V_l^{d} is

$$V_l^{\mathrm{d}}(\boldsymbol{r}_i) = \int u_l^*(\boldsymbol{r}_j)\frac{1}{r_{ij}}u_l(\boldsymbol{r}_j)\mathrm{d}\boldsymbol{r}_j, \quad (2.104)$$

which is the electrostatic repulsion potential due to electron j when averaged over the orbital u_l. The non-local exchange operator is defined as

$$\begin{aligned} V_l^{\mathrm{ex}}(\boldsymbol{r}_i)\psi(i) &= \delta_{m_L^k, m_L^l}\left[\int u_l^*(\boldsymbol{r}_j)\frac{1}{r_{ij}}u_k(\boldsymbol{r}_j)d\boldsymbol{r}_j\right] \\ &\quad \times u_l(\boldsymbol{r}_i)\chi_{1/2,m_L^l} \\ &= \delta_{m_L^k, m_L^l} V_l^{\mathrm{ex}}(\boldsymbol{r}_i)\chi_{1/2,m_L^l}; \end{aligned} \quad (2.105)$$

$V_l^{\mathrm{ex}}(\boldsymbol{r}_i)$ acts on the spatial coordinates only.

The Hartree-Fock equation (Eq. 2.104) for a given spin-orbital for electron i may be written in terms of the one- and two-electron operators f_i and g_{ij} defined in Eqs 2.81 and 2.82.

$$\begin{aligned} & f_i\psi_i(1) + \left[\sum_{j\neq i}\int \psi_j^*(2)\, g_{12}\, \psi_j(2)\, \mathrm{d}V\right]\psi_i(1) \\ & - \sum_{j\neq i}\left[\int \psi_j^*(2)\, g_{12}\, \psi_i(2)\, \mathrm{d}V\right]\psi_j(1) = \epsilon_i\psi_i(1). \end{aligned} \quad (2.106)$$

Several physical aspects of these equations are important to note. First, the two-electron integration variable dV refers to all space and spin coordinates written simply as (1) and (2). Second, the summation for each electron i necessarily involves interaction with all other electrons (except itself of course). Third, whereas the second term on the left is simply the repulsive Coulomb interaction between electron i and j, the third term on the left is the so-called exchange term, which has no classical analogue. But the exchange effect is quite physical and related to the spin of the electron: two electrons with the same spin quantum number $\zeta(m_s)$ may not occupy the same spatial position $\boldsymbol{r}$. The two integrals on the left are the direct integral J (Eq. 2.97) and the exchange integral K (Eq. 2.98), which may also be expressed as

$$J_j(1)\psi_i(1) = \left[\int \psi_j^*(2)\, g_{12}\, \psi_j(2)\, \mathrm{d}V\right]\psi_i(1) \quad (2.107)$$

and $$K_j(1)\psi_j(1) = \left[\int \psi_j^*(2)\, g_{12}\, \psi_i(2)\, \mathrm{d}V\right]\psi_j(1). \quad (2.108)$$

Note the interchange of i and j in the K-integral with respect to the J-integral. In the next chapter, we shall make use of these integrals to express the matrix elements for energies of atomic states and radiative transitions among them.

We shall now need to evaluate matrix elements with interaction operators and the one-electron orbitals. Using a Legendre expansion in Eq. 2.78,

$$\begin{aligned} 1/r_{ij} &= \sum^{\infty} \mathrm{P}_k(\boldsymbol{r}_i\cdot\boldsymbol{r}_j)\, r_<^k / r_>^{k+1} \\ \hat{\boldsymbol{r}}_i\cdot\hat{\boldsymbol{r}}_j &= \cos\vartheta, \end{aligned} \quad (2.109)$$

where $r_<$ and $r_>$ refer to the lesser or greater of r_i, r_j. one could then exploit the *addition theorem* for spherical harmonics defined in Section 2.1.1:

$$\frac{1}{r_{12}} = \sum_{l=0}^{\infty}\frac{r_<^l}{r_>^{l+1}}\left[\frac{4\pi}{2l-1}\sum_{m=-l}^{l}\mathrm{Y}_{lm}^*(\vartheta_1,\varphi_1)\,\mathrm{Y}_{lm}(\vartheta_2,\varphi_2)\right] \quad (2.110)$$

in H matrix elements. The radial part of the matrix elements is evaluated in terms of the so-called *Slater integrals* (see Eq. 2.26),

$$R^k(abcd) = \int_0^{\infty}\mathrm{d}r\int_0^{\infty}\mathrm{d}s\, P_a(r)P_b(s)\frac{r_<^k}{r_>^{k+1}}P_c(r)P_d(s). \quad (2.111)$$

The four-orbital *Slater integrals* R^k display a high degree of symmetry; for a start one may swap (a, c) and (b, d). The evaluation of interaction matrix elements is discussed in more detail in Chapter 4.

2.11.1 Configuration mixing

The topic of this section is crucial to practical atomic structure calculations. The N-electron wavefunction for different states of the atom within a given configuration $(nl)^q$ may be represented by a linear combination of Slater determinants (Eq. 2.91). But in general all states of the same angular and spin symmetry interact with one another. They are eigenvectors of the same Hamiltonian, specified by the total angular and spin quantum numbers and parity. Therefore, an accurate representation of the wavefunction of a given state must generally consider CI in atomic structure calculations. A single configuration often does not represent the atomic state accurately, as other configurations can give rise to the same LS state (assuming LS coupling to be valid, for simplicity). For example, the ground state of S IV is $[1s^22s^22p^6]3s^23p\,(^2P^o)$. However, the $^2P^o$ state can also be formed from $3p^3$ ($^4S^o$, $^2D^o$, $^2P^o$), $3s^24p$ ($^2P^o$) 3s3p3d ($\ldots$,$^2P^o$) and so on. Let us say that these four configurations contribute with different amplitudes to form the four state vectors $^2\mathrm{P^o}$ of a 4×4 Hamiltonian matrix. Hence, to obtain the optimized energy and wavefunction for each $^2\mathrm{P^o}$ state, all four configurations should be included;

$$\Psi(^2\mathrm{P^o}) = \sum_{i=1}^{4} a_i \psi[C_i(^2\mathrm{P^o})]$$
$$= \Big[a_1\psi(3\mathrm{s}^2 3\mathrm{p}) + a_2\psi(3\mathrm{p}^3) + a_3\psi(3\mathrm{s}^2 4\mathrm{p}) + a_3\psi(3\mathrm{s}3\mathrm{p}4\mathrm{p})\Big], \quad (2.112)$$

where the mixing coefficients a_i are amplitudes to different configurations C_i for each $^2\mathrm{P^o}$ state.

It is often useful to consider isoelectronic sequences as a whole between the neutral atom and ions with the same number of electrons but different Z. In a geminal paper of 1959, D. Layzer [9] demonstrated that along with the configuration that usually labels a spectroscopic term, for example the ground term $[1\mathrm{s}^2]2\mathrm{s}^2\ ^1\mathrm{S}$ in the Be-sequence, one *must* allow for a contribution from the two electrons of [closed shell]$2\mathrm{p}^2$, which is easily understood, as it accounts at a low level for mutual 'polarization'. Layzer's theory puts configuration interaction (or CI, although it is Coulomb interaction all the same) on an analytic footing based on scaling laws. Along with [ ]$2\mathrm{s}^2$ all other orbital combinations *within a principal shell denoted by n (and the same parity)* make up a complete wavefunction; such a set, comprising two elements in this example of Be-like ions, is called a *complex of configurations* associated with a given n-value.

While the concept of a configuration C serves as a suitable label in designating spectroscopic terms of ions of astrophysical interest up to and somewhat beyond the iron group, a closer look at an isoelectronic sequence reveals a more subtle picture. Layzer [9] unravelled the erratic trends in LS-term separation along sequences. The radial coordinate scales as $1/Z$; it follows straight from the Schrödinger equation (Eq. 2.27) on dividing by Z^2 that,

$$\left[\frac{\mathrm{d}^2}{\mathrm{d}\rho^2} + \frac{2}{\rho} - \frac{l(l+1)}{\rho^2}\right] = \frac{1}{n^2}, \quad (2.113)$$

as $\rho = Z \cdot r$ no longer depends upon Z. This Z-invariant length appeared earlier in the rigorous formulation (Eq. 2.44) for the hydrogenic radial solution.

As Z goes up, the radial functions of a many-electron system increasingly resemble hydogenlike orbitals, whose energy eigenvalues are degenerate in their orbital quantum number l. As a function of Z the non-relativistic Hamiltonian yields eigenvalues[4]

$$H'_{\Gamma SL} = W_2 Z^2 + W_1 Z + W_0 + \mathcal{O}(1/Z) \quad (2.114)$$

$$W_2 = \sum_{i=1}^{N} -\frac{1}{n_i^2}\ \mathrm{Ry}$$

$$W_1 = \big\langle \Gamma \big| C \big\rangle \Big\langle CSL \Big| \sum_{j<i} \frac{e^2}{r_{ij}} \Big| C'SL \Big\rangle \big\langle C' \big| \Gamma \big\rangle \quad (2.115)$$

for a given value SL (Γ refers to a set of n_i), say the ground state $C_1\ ^3\mathrm{P}$ = $1\mathrm{s}^2 2\mathrm{s}^2 2\mathrm{p}^2(^3\mathrm{P})$ of the carbon isoelectronic sequence with its $N = 6$ electrons (and $W_2 = -3$ Ry). What is less obvious from Eq. 2.80 is the structure of W_1 required for a non-relativistic answer (Eq. 2.114) that exhibits the correct quadratic and linear dependence in Z: to be an eigenvalue ΓSL of the $2/r_{ij}$-matrix comprising *all* configurations C, C' associated with the same set of principal quantum numbers n; this adds $C_3 = 1\mathrm{s}^2 2\mathrm{p}^4$ in the carbon-like case (the configuration $C_2 = 1\mathrm{s}^2 2\mathrm{s} 2\mathrm{p}^3$ in between is of opposite parity). If the Slater integrals (Eq. 2.111) for the matrix elements $CSL - C'SL$ are to be evaluated with *hydrogenic* ($Z = 1$) wavefunctions, then diagonalizing this matrix, which ensures stationary solutions H', is our first application of the Hartree–Fock principle.

Table 2.5 summarizes results for the ground configuration. Ratios like 0.105 : 0.994 or even 0.200 : 0.980 from the components of eigenvectors — as a measure of admixture – do not look dramatic until one compares the consequences in applications such as dipole oscillator strengths (see Chapter 4) or collision strengths (Chapter 5).

If the concept of a single configuration has been supplanted by the *complex* of configurations quasi-degenerate in the high-Z limit, it still plays a role of its own near the neutral end of sequences, notably for 'heavy' elements, whose subshells may not be built up in hydrogenic order,

[4] The full explanation is as follows: single orbital functions approach (scaled) hydrogenic form and thus l-degeneracy in energy, and the associated configurations mix strongly, with coefficients at the high Z limit equal to the result with Z-scaled hydrogenic radial functions. Omitting the other members of a complex leads to poor results near the neutral end of a sequence for such quantities as term energies, radiative transition probabilities or electron excitation rates, and the answer may be outright wrong towards high z limits, notably when plotting an oscillator strength f (see next chapter) vs. $1/Z$. Layzer's scaling laws are readily read off the hydrogenic equation of motion in its Z-scaled form on absorbing Z into r. The length r scales as $1/Z$, since $\rho = Zr$ is invariant, so energies due to the potential Z/r scale as Z^2, as confirmed by the Rydberg formula. And energies of a term LS within a complex follow a linear plot along the Z axis, as they arise from $1/r_{ij}$ and length scales as Z. We shall come across all these applications throughout the book, especially when discussing radiative transition quantities and their thresholds and high-Z limits.

TABLE 2.5 [Layzer-coefficients W_1 for the ground complex of the C-sequence] Coefficients W_1/Ry as in Eq. 2.115 (n.b. 1.23: $e^2 = 2a_0$ Ry), and eigenvectors $\langle \Gamma SL|C'SL\rangle$ of the ground complex in the carbon isoelectronic sequence computed with hydrogenic wavefunctions ('i' refers to LS term indices also including configuration other than C_1 and C_3).

	$C_1 = 1s^22s^22p^2$			$C_3 = 1s^22p^4$		
	^{3}P	^{1}D	^{1}S	^{3}P	^{1}D	^{1}S
No CI	6.5239	6.5661	6.6294	7.0724	7.1146	7.1779
Full W_1	6.5177	6.5599	6.6054	7.0786	7.1208	7.2019
i	1	2	3	10	11	12
$\langle i\|C_1\rangle$	0.9945	0.9945	0.9797	0.1050	0.1050	0.2005
$\langle i\|C_3\rangle$	−0.1050	−0.1050	−0.2005	0.9945	0.9945	0.9797

1s-2s-2p-3s-3p-3d, and so on. Up to argon ($Z = 18$), the subshells of elements are filled in a straightforward manner: first according to n and then according to l, and then as explained in Section 2.6.

Within the context of Hartree–Fock formulation, the consideration of more than one configuration or CI is known as the *multi configuration Hartree–Fock (MCHF)* approximation. The accuracy of energies and wavefunctions naturally increases with inclusion of more configurations, as the wavefunction expansion converges. However, this can lead to very large numerical calculations. Although a limited number of configurations can provide sufficiently accurate results for practical applications, very extensive CI calculations are necessary for high precision. Powerful methods and computer codes have been developed for computational treatment for fully self-consistent MCHF of CI calculations, including relativistic effects in varying approximations (discussed later), e.g., CIV3, SUPERSTRUCTURE [10], MCDF [8], GRASP [11].

2.12 Central-field approximation

The simplest treatment for complex atoms is in analogy with hydrogen. The many-electron Hamiltonian would be much easier to handle without the repulsive electron–electron Coulomb term, which is too large to be treated as a perturbation. H_1 in Eq. 2.78 consists of non-central forces between electrons due to non-radial motions. However, the inter-electron repulsion contains a large spherically symmetric component. This entails the construction of an effective potential that simulates a *central force* under the assumption of spherical symmetry, as included in H_0 in Eq. 2.77.

We assume that each electron is acted on by the averaged charge distribution of all the other electrons and construct a potential energy function $V(r_i)$ with a one-electron operator, and is a good approximation to the actual potential of the ith electron in the field of the nucleus and the other $N-1$ electrons. When summed over all electrons in the atom this charge distribution is spherically symmetric.

The *effective potential* experienced by each electron due to electron–nuclear attraction and electron–electron repulsion, in H_0 and H_1 respectively, consists of a radial and a non-radial part. The central-field approximation involves neglecting the non-radial part, while retaining the radial part assumed to be dominant. Unlike the Hartree–Fock method, where we explicitly account for the electron–electron interaction, we now seek an effective potential $U(r)$, which combines the radial electron–nuclear term Ze^2/r_i with an averaged *radial component* of the electron–electron term. The approximate N-electron Hamiltonian then becomes

$$H = -\sum_{i=1}^{N} \frac{h^2}{2m}\nabla_i^2 + U(r), \tag{2.116}$$

where

$$U(r) = -\sum_{i=1}^{N} \frac{e^2 Z}{r_i} + \left\langle \sum_{i\neq j}^{N} \frac{e^2}{r_{ij}} \right\rangle \tag{2.117}$$

is designated as the *central-field potential*, and does not have any explicit dependence on the two-electron operator. While this may appear to be a drastic approximation at first sight, it should be noticed that $U(r)$ contains the electron–nuclear potential term Z/r, which increases in magnitude with Z. Therefore, for a given number of electrons N the central-field approximation improves in accuracy with increasing charge $z \equiv (Z-N+1)$ of an ion, or along an isoelectronic sequence. The wavefunction of each electron may now be computed using the radial equation Eq. 2.27, provided $U(r)$ is known. First, we note that the boundary conditions on $U(r)$ are

$$U(r)/(2\text{Ry}) = \begin{cases} -\dfrac{Z}{r} & \text{if } r\rightarrow 0 \\ -\dfrac{z}{r} & \text{if } r\rightarrow\infty \end{cases} \tag{2.118}$$

in atomic units (Hartrees or 2Ry).

2.12.1 Thomas–Fermi–Dirac approximation

Consistent with the limits in Eq. 2.118 there are several ways to choose the central potential $U(r)$. A particularly useful procedure is the *Thomas–Fermi–Dirac–Amaldi* (TFDA) model [12], where the charge distribution is assumed to be spherically symmetric. Its precursor, the *Thomas–Fermi model*, treated the atomic electrons as a degenerate Fermion gas obeying the relation between electron density and the maximum momentum, the Fermi momentum p_{F}. Electrons are said to occupy cells in phase space of volume h^3 with two electrons in each cell, one with spin up and the other with spin down. These cells are all occupied up to a maximum Fermi momentum p_{F}.[5] The spatial density of electrons is then

$$\rho = \frac{(4/3)\pi p_{\text{F}}^3}{h^3/2}. \tag{2.119}$$

An improvement over the Thomas–Fermi model is to take account of the exchange effect in a simplified manner. With electron–electron exchange effect, any two electrons with the same spin spatially avoid each other – a necessary condition from the antisymmetry postulate for the system to exist. Based on quantum statistics, the TFDA model gives a continuous function $\phi(x)$ such that

$$U(r) = \frac{Z_{\text{eff}}(\lambda_{nl}, r)}{r} = -\frac{Z}{r}\phi(x), \tag{2.120}$$

where

$$\phi(x) = \mathrm{e}^{-Zr/2} + \lambda_{nl}(1 - \mathrm{e}^{-Zr/2}), \quad x = \frac{r}{\mu}, \tag{2.121}$$

and μ is a constant:

$$\mu = 0.8853\left(\frac{N}{N-1}\right)^{2/3} Z^{-1/3}. \tag{2.122}$$

[5] The concept of Fermi momentum may be utilized to understand extremely high density objects such as the *white dwarf* stars (Chapter 9). An arrangement of electrons constrained by the Pauli exclusion principle is called the 'Fermi sea', with electrons filling in cells up to a highest Fermi level at zero temperature. As the temperature rises, energies of electrons excited out of the Fermi sea close to the 'surface' levels at $p = p_{\text{F}}$ approach a Maxwellian distribution (Chapter 1). Owing to very high densities and temperatures in white dwarfs, the electron energies and velocities are in the relativistic range.

The function $\phi(x)$ is a solution of the potential equation

$$\frac{\mathrm{d}^2\phi(x)}{\mathrm{d}x^2} = \frac{1}{\sqrt{x}}\phi(x)^{\frac{3}{2}}. \tag{2.123}$$

From Eq. 2.118 the boundary conditions on $\phi(x)$ are

$$\phi(0) = 1, \quad \phi(\infty) = -\frac{Z-N+1}{Z}. \tag{2.124}$$

Having determined a central potential $U(r)$, for example as in the TFDA approximation above, we compute the one-electron orbitals $P_{nl}(r)$ by solving the wave equation

$$\left[\frac{\mathrm{d}^2}{\mathrm{d}r^2} - \frac{l(l+1)}{r^2} + 2U(r) + \epsilon_{nl}\right] P_{nl}(r) = 0. \tag{2.125}$$

This is similar to the radial equation (2.27) for the hydrogenic case, with the same boundary conditions on $P_{nl}(r)$ as $r\rightarrow 0$ and $r\rightarrow\infty$, and $(n-l+1)$ nodes. The second-order radial differential equation is solved numerically since, unlike the hydrogenic case, there is no general analytic solution. Equation 2.125 may be solved by both inward and outward integration, matching the two solutions at a suitable point. As $r\rightarrow 0$, the outward solution is given by the first few points of a power series expansion. The inward solution begins from the asymptotic region $r\rightarrow\infty$, using an exponentially decaying function appropriate for a bound state, such as the normalized Whittaker function,

$$W(r) = \mathrm{e}^{-zr/\nu}\left(\frac{2zr}{\nu}\right)\left(1 + \sum_{k=1}^{\infty}\frac{a_k}{r^k}\right)\mathcal{N}, \tag{2.126}$$

where $\nu = z/\sqrt{\epsilon}$ is the effective quantum number (and as such not necessarily an integer) and ϵ is the eigenvalue. The coefficients are

$$a_1 = \nu\,\{l(l+1) - \nu(\nu-1)\}\,\frac{1}{2z}, \tag{2.127}$$

$$a_k = a_{k-1}\,\nu\,\{l(l+1) - (\nu-k)(\nu-k+1)\}\,\frac{1}{2kz} \tag{2.128}$$

and the normalization factor is

$$\mathcal{N} = \left\{\frac{\nu^2}{z}\Gamma(\nu+l+1)\,\Gamma(\nu-1)\right\}^{-1/2}. \tag{2.129}$$

The one-electron spin-orbital functions then assume the familiar hydrogenic form

$$\begin{aligned}\psi_{n,\ell,m_\ell,m_s}(r,\theta,\phi,m_s) &= \phi(r,\theta,\phi)\zeta_{m_s} \\ &= R_{n\ell}(r)\mathrm{Y}_{\ell,m_\ell}(\theta,\phi)\zeta_{m_s} \\ &= \frac{P_{n\ell}(r)}{r}\mathrm{Y}_{\ell,m_\ell}\zeta_{m_s}.\end{aligned} \tag{2.130}$$

Once the set of $P_{nl}(r)$ has been obtained numerically, the eigenfunctions and eigenenergies for the spectroscopic states can be obtained. However, we do note the distinction between the strict Hartree–Fock orbitals as solutions of Eq. 2.104, as opposed to orbitals computed in the TFDA potential, or any central-field potential generally. The one-electron TFDA orbitals are derived using a statistical treatment of the free-electron gas, which therefore neglects the shell-structure that is inherent in the Hartree–Fock method. To wit: a TFDA $1s$ orbital remains invariant whether in the $1s^2$ configuration or a $1s2p$ configuration. But the Hartree–Fock $1s$ orbitals are different for each configuration. Although in principle this seems rather a serious limitation of the TFDA potential, in practice configuration interaction accounts for much of the discrepancy that might otherwise result. Taking account of configuration interaction, as well as the flexibility afforded by scaling the TFDA orbitals, yields results that are comparable in accuracy to the full MCHF treatment.

2.13 Relativistic fine structure

In earlier sections we had considered the non-relativistic Hamiltonian, Eq. 2.80. But fine structure and the dependent coupling schemes were introduced algebraically in Eqs. 2.62 and 2.63 without consideration of relativistic effects. Physically, however, magnetic coupling between the orbital angular momentum and spin angular momentum splits LS term energies into a number of components or fine structure levels. Relativistically, the electron and the nuclear motion each represents a moving charge observed at the position of the other: ergo, a magnetic field and resulting magnetic moment, as follows from Ampère's law. This magnetic moment due to orbital motion, in turn, interacts with the *intrinsic spin*.

The observed energy splitting of a state could not be explained by the Schrödinger equation until the idea of electron spin or intrinsic angular momentum was introduced by Goudsmit and Uhlenbeck in 1925. Pauli implemented the idea in the equations of motion. But it was Dirac whose equation not merely satisfied special relativity requirements, but could account for spin and magnetic moment.[6] Before we delve into the basics of the Dirac equation, we attempt to understand the spin–orbit interaction at an elementary level.

[6] The Klein–Gordon equation, a straightforward relativistic extension of the Schrödinger equation, could not.

2.13.1 Spin–orbit interaction

Since magnetic monopoles are not known to exist like the electric charge e, we consider the Bohr magneton

$$\mu_B = \frac{e\hbar}{2m} \tag{2.131}$$

of an electron, postulated empirically before it emerged from the Dirac equation. The magnetic dipole moment is vectorially associated with the spin: $\vec{\mu} = -g_s \mu_B \vec{s}/\hbar$, $g_s = 2$. If the electron orbits in angular momentum state l around a nucleus with charge number Z, such a steady electric current $e\vec{l}$ creates a magnetic field according to the Biot–Savart law, the magnetic counterpart of Coulomb's law. The potential energy of the magnetic moment in this field reads

$$H^{so} = \alpha^2 Z \frac{s \cdot l}{\hbar^2 \, (r/a_0)^3} \text{ Ry}, \tag{2.132}$$

if one exploits identities as in the Coulomb case. Most remarkable is the numerical remainder once one has extracted the values for H, r and angular momenta: in this equation, the dimensionless quantity $e^2/(\hbar c) = \alpha$ makes its debut in the history of physics, then called the 'fine-structure constant' – long before quantum electrodynamics. The laws of quantum mechanics for angular momenta lead to discrete eigenvalues of H^{so}, two for a single electron with $l > 0$; there are a minimum of $(2S+1, 2L+1)$ in the multi-electron case (see Section 2.8), hence the name *multiplicity* for $2S+1$.

It is instructive to state a few expressions, which we describe later in this section. The energy difference due to the *spin–orbit interaction* in a one-electron atom of charge Z is

$$E_{so} = \mu \cdot H^{so} = \frac{Ze^2\hbar^2}{2m^2c^2r^3} \mathbf{l} \cdot \mathbf{s}, \tag{2.133}$$

where μ is the magnetic moment and $\mathbf{l} \cdot \mathbf{s}$ is the spin–orbit operator in the Hamiltonian

$$\mathbf{l} \cdot \mathbf{s} = \frac{1}{2}[j(j+1) - l(l+1) - s(s+1)]. \tag{2.134}$$

Therefore, the spin–orbit energy shift is

$$E_{so} = \frac{Z^4\alpha^2}{n^3} \left[\frac{j(j+1) - l(l+1) - \frac{3}{4}}{1\left(l+\frac{1}{2}\right)(l+1)} \right] Ry. \tag{2.135}$$

Scaling with such a high power in Z quickly overcomes Coulombic term separation, which scales as Z for terms within a complex and Z^2 if n changes. A relativistic approach accounts for the particle velocities involved. As the atomic number Z increases along the periodic table, the velocity of the inner-shell electrons (particularly the

1s subshell closest to the nucleus) also increases, since the energy balance of the electron in an orbit is expressed as $mv_e^2/2 \sim Z/r_e$. For elements $Z > 10$ the velocity of an electron in the first Bohr orbit is already a good fraction of the the speed of light c. Balancing Coulombic attraction with the inertial centrifugal force yields proportionality $v_e \approx \sqrt{Z}$, while for $Z = 1$ we have $v_e = \alpha \cdot c$. A rigorous inclusion of relativistic effects entails the transition from the Schrödinger to the Dirac equation (Section 2.13.2). The relativistic Hamiltonian includes 'correction' terms that shift a LS term energy into a number of fine-structure levels, as formulated below. Consequently, the angular momenta coupling LS changes to jj or intermediate LSJ coupling.

The hydrogen energy levels at extremely high resolution show evidence of some other small effects on the energy, collectively treated in *quantum electrodynamics* (QED). The 2p level splits into a pair of level by the spin–orbit effect. The level $2p_{1/2}$ is degenerate with $2s_{1/2}$, but drops below it by a small amount in what is called the *Lamb shift*, caused by the emission and re-absorption of a virtual photon by the *same* electron. Even the 1s ground state splits by the interaction between the magnetic moments of electron and proton in what is called *hyperfine structure*. Quantum electrodynamics effects break the j-degeneracy of H-like levels, weakening the binding energy mainly of low-n states ns as a result of the emission and re-absorption of a virtual photon by the same electron. Using microwave techniques, Lamb and Retherford measured the excitation energy of the dipole transition from $2p_{1/2}$ to $2s_{1/2}$ in H as $E/h \approx$ 11.06 GHz.

For non-hydrogenic systems, in the first instance, alkali atoms, we may write (Bohr–Rydberg scaled)

$$H^{so} = S \cdot L\,\zeta \tag{2.136}$$

$$\zeta_{nl,n'l} = \alpha^2 \int_0^\infty \mathrm{d}r\, P_{nl}(r) \frac{1}{r^3} P_{n'l}(r) \tag{2.137}$$

with a *spin–orbit parameter* ζ, which at this stage is definite only for hydrogen-like ions as Eq. 2.137; yet if a valence shell with a single type nl of electrons is involved, the value ζ_{nl} can be obtained from observation. Computing it involves magnetic interaction with the other $N-1$ electrons. In Eq. 2.132, retardation in particular is ignored and will have to wait for the discussion on Dirac formulation.

A striking feature that emanates from Eq. 2.132 is that H^{so} breaks the LS symmetry of the non-relativistic Hamiltonian, which separately preserves orbital and spin angular momentum, having eigenstates $|SM_S\rangle$ and $|LM_L\rangle$ (and thus of total angular momentum $|JM_J\rangle$ as well). Because both vectors making up the scalar product

$$S \cdot L = \sum_{\kappa=-1}^{+1} (-1)^\kappa\, S_{-\kappa} L_\kappa \tag{2.138}$$

act as stepping operators, it preserves only the total magnetic quantum number $m_j = m_s + m_l$ or $M_J = M_S + M_L$, and eigenstates

$$|SLJM_J\rangle = |SM_S\, LM_L\rangle\, \langle SM_S\, LM_L|SLJM_J\rangle \tag{2.139}$$

of *intermediate coupling* SLJ are readily built with plain Clebsch–Gordan coefficients (Appendix C). A semi-relativistic multi-electron tretment of intermediate coupling is based on the Breit–Pauli approximation, which is discussed following the fully relativistic one-electron Dirac theory.

2.13.2 The Dirac equation

The Lorentz-invariant single-particle Dirac equation is the relativistic equation of motion for a fermion without restrictions on v/c, and weak magnetic fields are readily included. In its time-independent form for a free particle it reads [7]

$$(\boldsymbol{\alpha} \cdot \boldsymbol{p}\, c + \beta mc^2)\psi = E\psi, \tag{2.140}$$

where $\boldsymbol{\alpha}$ and β are four elements of a non-commutative algebra over the field of complex numbers that *can* be represented by 4×4 matrices; then the wavefunction ψ becomes a four-component 'spinor':

$$\alpha_x = \begin{pmatrix} 0 & 0 & 0 & 1 \\ 0 & 0 & 1 & 0 \\ 0 & 1 & 0 & 0 \\ 1 & 0 & 0 & 0 \end{pmatrix}, \quad \alpha_y = \begin{pmatrix} 0 & 0 & 0 & -\mathrm{i} \\ 0 & 0 & \mathrm{i} & 0 \\ 0 & \mathrm{i} & 0 & 0 \\ \mathrm{i} & 0 & 0 & 0 \end{pmatrix},$$

$$\alpha_z = \begin{pmatrix} 0 & 0 & 1 & 0 \\ 0 & 0 & 0 & 1 \\ 1 & 0 & 0 & 0 \\ 0 & 1 & 0 & 0 \end{pmatrix},$$

$$\beta = \begin{pmatrix} 1 & 0 & 0 & 0 \\ 0 & 1 & 0 & 0 \\ 0 & 0 & -1 & 0 \\ 0 & 0 & 0 & -1 \end{pmatrix}, \quad \psi = \begin{pmatrix} \psi_1 \\ \psi_2 \\ \psi_3 \\ \psi_4 \end{pmatrix}.$$

7 The discussion in this section is based in part on the treatment by M. Weissbluth in *Atoms and Molecules* [13].

The four coupled equations reduce to two separate sets of two coupled equations each, owing to the matrix choice for β:

$$c\,\boldsymbol{\sigma}\cdot\boldsymbol{p}\,\psi_v + (mc^2 - E)\psi_u = 0,$$
$$c\,\boldsymbol{\sigma}\cdot\boldsymbol{p}\,\psi_u - (mc^2 + E)\psi_v = 0, \tag{2.141}$$

where the 2×2 Pauli spin matrices $(\sigma_{x,y,z})$ and the identity matrix I are

$$\sigma_x = \begin{pmatrix} 0 & 1 \\ 1 & 0 \end{pmatrix},\ \sigma_y = \begin{pmatrix} 0 & -\mathrm{i} \\ \mathrm{i} & 0 \end{pmatrix},$$

$$\sigma_z = \begin{pmatrix} 1 & 0 \\ 0 & -1 \end{pmatrix},\ \mathrm{I} = \begin{pmatrix} 1 & 0 \\ 0 & 1 \end{pmatrix}.$$

Writing $\boldsymbol{\sigma} = (\sigma_x, \sigma_y, \sigma_z)$, we get

$$\boldsymbol{\alpha} = \begin{pmatrix} 0 & \boldsymbol{\sigma} \\ \boldsymbol{\sigma} & 0 \end{pmatrix},\ \beta = \begin{pmatrix} \mathrm{I} & 0 \\ 0 & -\mathrm{I} \end{pmatrix},$$

when

$$\psi_u = \begin{pmatrix} \psi_1 \\ \psi_2 \end{pmatrix},\ \psi_v = \begin{pmatrix} \psi_3 \\ \psi_4 \end{pmatrix}.$$

From the free particle Dirac equation,

$$\psi_u = \frac{c\,\boldsymbol{\sigma}\cdot\boldsymbol{p}}{E - mc^2}\psi_v. \tag{2.142}$$

Approximating $\boldsymbol{\sigma}\cdot\boldsymbol{p} \approx mv\,\mathrm{I}$, $E - mc^2 \approx (1/2)mv^2$,

$$\psi_u/\psi_v \approx 2c/v. \tag{2.143}$$

Hence, for $v/c \ll 1$, the component ψ_u is known as the 'large' component and ψ_v as the 'small' component. For a central-field potential, the Dirac equation can be reduced to the form similar to the Schrödinger equation, as will be seen later.

The free-particle Dirac equation may be modified to include the effects of external fields. In the presence of an external electromagnetic field the Hamiltonian can be

$$H = \frac{1}{2}mv^2 + \frac{q}{c}\boldsymbol{v}\cdot\boldsymbol{A} - q\phi, \tag{2.144}$$

where vector and scalar potential $\boldsymbol{A}$ and ϕ relate to the Lorentz force,

$$\boldsymbol{F} = q\left(\boldsymbol{E} + \frac{1}{c}\frac{\partial\boldsymbol{A}}{\partial t} - \nabla\phi\right), \tag{2.145}$$

and to the field $\boldsymbol{B} = \nabla\times\boldsymbol{A}$. The Dirac equation can be written as

$$\left[\boldsymbol{\alpha}\cdot(c\boldsymbol{p} - q\boldsymbol{A}) + \beta mc^2 + q\phi\right]\psi = E\psi. \tag{2.146}$$

The Dirac equation of a particle in an electromagnetic field can be reduced to the coupled equations of large and small components of the spinor. By replacing q by $-e$, it can be written as [13]

$$(E + e\phi)\psi = \left[\frac{1}{2m}\left(\boldsymbol{p} + \frac{e}{c}\boldsymbol{A}\right)^2 + \frac{e\hbar}{2mc}\boldsymbol{\sigma}\cdot\nabla\times\boldsymbol{A}\right.$$
$$\left. - \frac{p^4}{8m^3c^2} - \frac{e\hbar^2}{8m^2c^2}\nabla\cdot\nabla\phi - \frac{e\hbar}{4m^2c^2}\,\boldsymbol{\sigma}\cdot(\nabla\phi\times\boldsymbol{p})\right]\psi. \tag{2.147}$$

The terms in the above equation can be interpreted as follows. The first term on the right-hand side,

$$\frac{1}{2m}\left(\boldsymbol{p} + \frac{e}{c}\boldsymbol{A}\right)^2, \tag{2.148}$$

contains the kinetic energy and interaction terms with a vector potential field. The interaction terms contribute to numerous physical processes, such as absorption, emission, scattering of electromagnetic waves, diamagnetism and the Zeeman effect. The second term,

$$\frac{e\hbar}{2mc}\boldsymbol{\sigma}\cdot\nabla\times\boldsymbol{A},$$

represents the interaction of the spin magnetic moment $\boldsymbol{\sigma}$ with a magnetic field $\boldsymbol{B}$. The third term,

$$H^{\mathrm{mass}} = -\frac{p^4}{8m^3c^2} \tag{2.149}$$

$$= -\frac{\alpha^2}{4}\left(\frac{p\,a_0}{\hbar}\right)^4 \mathrm{Ry} = -\frac{\alpha^2}{4}\left(\frac{\mathrm{d}^2}{\mathrm{d}(r/a_0)^2}\right)^2 \mathrm{Ry}, \tag{2.150}$$

expressed in Rydberg energy units and Bohr radii on exploiting Eqs 1.17 and 1.19, is the relativistic mass–velocity correction, and we see that it scales as $\alpha^2 Z^4$ Ry. It can be derived from

$$E' = E - mc^2 = (c^2p^2 + m^2c^4)^{1/2} - mc^2$$
$$= \frac{p^2}{2m} - \frac{p^4}{8m^3c^2}. \tag{2.151}$$

The fourth term,

$$H^{\mathrm{Dar}} = \frac{e\hbar^2}{8m^2c^2}\nabla\cdot\nabla\phi, \tag{2.152}$$

is known as the Darwin term. If ϕ depends only on r, as $\phi(r) = Ze^2/r$, then

$$H^{\mathrm{Dar}} = -\frac{Ze^2\hbar^2}{8m^2c^2}\nabla^2\left(\frac{1}{r}\right). \tag{2.153}$$

Using $\nabla^2(1/r) = -4\pi\delta(r)$, and assuming that only the radial part of the wavefunction will be in the matrix element of the H^{Dar},

$$\langle H^{\mathrm{Dar}}\rangle = \frac{Z\pi e^2\hbar^2}{2m^2c^2}\,|\psi(0)|^2 = E^{\mathrm{Dar}}. \tag{2.154}$$

The matrix elements are non-zero only for s-states. Using $|\psi(0)|^2 = Z^3/(\pi n^3)a_0^3$,

$$\langle H^{\rm Dar}\rangle = \frac{Z^4 e^2\hbar^2}{2m^2c^2n^3a_0^3} = \frac{Z^4\alpha^2}{n^3}{\rm Ry} \qquad (l = 0 \text{ only}). \tag{2.155}$$

The fifth term,

$$H^{\rm so} = \frac{e\hbar}{4m^2c^2}\,\boldsymbol{\sigma}\cdot(\nabla\phi\times\boldsymbol{p}), \tag{2.156}$$

describes spin–orbit coupling of $\boldsymbol{\sigma} = \mathbf{2s}$ with the magnetic field generated by $\boldsymbol{r}\times\boldsymbol{p} = \hbar\boldsymbol{l}$. With $\phi(r) = Ze^2/r$,

$$H^{\rm so} = -\frac{e\hbar^2}{2m^2c^2}s\cdot\left(\frac{1}{r}\frac{{\rm d}\phi}{{\rm d}r}\,\boldsymbol{r}\times\boldsymbol{p}\right)$$

$$= \frac{Ze^2\hbar^2}{2m^2c^2\,r^3}\,\boldsymbol{l}\cdot\boldsymbol{s} \tag{2.157}$$

$$= Z\alpha^2\frac{\boldsymbol{l}\cdot\boldsymbol{s}}{(r/a_0)^3}\,{\rm Ry}. \tag{2.158}$$

The spin–orbit interaction term was first included by Pauli on empirical grounds, but it follows directly from Dirac's linearization of the relativistic equation of motion. Substituting the terms in the Dirac equation, one gets

$$(E+e\phi)\psi = \left[\frac{1}{2m}\left(\boldsymbol{p}+\frac{e}{c}\boldsymbol{A}\right)^2 + \frac{e\hbar}{2mc}\boldsymbol{\sigma}.\nabla\times\boldsymbol{A} - \frac{\alpha^2}{4}\sum_i p_i^4 - \frac{Z\alpha^2}{4}\sum_i\nabla^2\left(\frac{1}{r_i}\right) + Z\alpha^2\sum_i\frac{1}{r_i^3}\,\boldsymbol{l}_i\cdot\boldsymbol{s}_i\right]\psi. \tag{2.159}$$

The total angular momentum $\boldsymbol{J} = \boldsymbol{L}+\boldsymbol{S}$ can be defined as

$$2\mathbf{L}\cdot\mathbf{S} = \mathbf{J}^2 - \mathbf{L}^2 - \mathbf{S}^2. \tag{2.160}$$

Since $\mathbf{J}^2$, $\mathbf{L}^2$, $\mathbf{S}^2$ and J_z commute with each other, there are stationary states, which are simultaneous eigenstates of these operators and of $\mathbf{L}\cdot\mathbf{S}$. For such a stationary state, we can replace the angular momentum operator by their eigenvalues:

$$\mathbf{L}^2 \to l(l+1),\ \mathbf{J}^2 \to j(j+1),$$
$$\mathbf{S}^2 \to s(s+1),\ s = 1/2,$$
$$2\mathbf{L}\cdot\mathbf{S} \to j(j+1) - l(l+1) - s(s+1). \tag{2.161}$$

It can be shown that $j = l+1/2$ or $j = l-1/2$ and the possible values of j are $(-l+1/2, -l+3/2, \ldots, l-1/2)$.

If we consider a hydrogenic system of a single electron, then

$$\left\langle\frac{1}{r^3}\right\rangle nl = \frac{Z^3}{n^3 l\left(l+\frac{1}{2}\right)(l+1)a_0^3}.$$

Using this expression, we can determine the shift of an energy level (Eq. 2.135) due to spin–orbit interaction as

$$H_{nl}^{\rm so} = \alpha^2\frac{j(j+1) - l(l+1) - s(s+1)Z^4}{2l\left(l+\frac{1}{2}\right)(l+1)}\frac{Z^4}{n^3},$$

in Rydberg units.

2.13.3 Pauli approximation for a central field

The Pauli approximation can be made at non-relativistic energies when $v/c \ll 1$, and the wavefunction is replaced by the large component. Assuming an electron in a central electric field and that the vector potential $\boldsymbol{A}$ and the magnetic field $\boldsymbol{B} = \nabla\times\boldsymbol{A}$ are zero and the electric field ϕ is a function of radial distance only, $\boldsymbol{E} = -(\boldsymbol{r}/r){\rm d}\phi/{\rm d}r$, the Dirac equation reduces to

$$(E+e\phi)\psi = \left[\frac{\boldsymbol{p}^2}{2m} - \frac{p^4}{8m^3c^2} - \frac{e\hbar^2}{8m^2c^2}\nabla\cdot\nabla\phi - \frac{e\hbar}{4m^2c^2}\boldsymbol{\sigma}\cdot(\nabla\phi\times\boldsymbol{p})\right]\psi, \tag{2.162}$$

which includes mass correction and Darwin and spin–orbit terms. This is known as the approximate Pauli equation.

2.13.4 Dirac equation in a central field

Consider a projectile of rest mass m_0 traveling with a velocity v in a central field $V(r)$, which can be real or complex. Defining $\gamma = 1/\sqrt{1-v^2/c^2}$, the total energy $E = mc^2 = m_0\gamma c^2$ can be expressed in terms of momentum $p = mv = m_0\gamma v$ as

$$E = \sqrt{\frac{m_0^2c^4}{1-v^2/c^2}} = \sqrt{\frac{m_0^2v^2c^2 + m_0^2c^4 - m_0^2v^2c^2}{1-v^2/c^2}}$$
$$= \left(p^2c^2 + m_0^2c^4\right)^{1/2} = E' + m_0c^2, \tag{2.163}$$

where E' is the kinetic energy. The Dirac equation[8]

$$[\boldsymbol{\alpha}\cdot\boldsymbol{p} + \beta m_0c + V(r)]\,\psi = E\psi,$$
$$\text{or}\qquad [\boldsymbol{\alpha}\cdot\boldsymbol{p} + \beta(m_0c - p_0)]\,\psi = 0, \tag{2.164}$$

[8] The discussion in this section is based on C. G. Darwin [14], H. A. Bethe and E. E. Salpeter [15], and S. N. Nahar and J. M. Wadehra [16].

on defining $\quad p_0 = (E - V)/c, \qquad (2.165)$

can be written as a set of coupled equations for the four spinor components:

$$-\frac{\mathrm{i}}{\hbar}(p_0 - m_0 c)\psi_1 + \left[\frac{\partial}{\partial x} - \mathrm{i}\frac{\partial}{\partial y}\right]\psi_4 + \frac{\partial}{\partial z}\psi_3 = 0, \qquad (2.166)$$

$$-\frac{\mathrm{i}}{\hbar}(p_0 - m_0 c)\psi_2 + \left[\frac{\partial}{\partial x} + \mathrm{i}\frac{\partial}{\partial y}\right]\psi_3 - \frac{\partial}{\partial z}\psi_4 = 0, \qquad (2.167)$$

$$-\frac{\mathrm{i}}{\hbar}(p_0 - m_0 c)\psi_3 + \left[\frac{\partial}{\partial x} - \mathrm{i}\frac{\partial}{\partial y}\right]\psi_2 + \frac{\partial}{\partial z}\psi_1 = 0, \qquad (2.168)$$

$$-\frac{\mathrm{i}}{\hbar}(p_0 - m_0 c)\psi_4 + \left[\frac{\partial}{\partial x} + \mathrm{i}\frac{\partial}{\partial y}\right]\psi_1 + \frac{\partial}{\partial z}\psi_2 = 0. \qquad (2.169)$$

Since the potential $V(r)$ is a function of r only, each component $\psi_i (i = 1, 2, 3, 4)$ of ψ can be expressed as a product of a spherical harmonic Y_{lm} and some radial function $R(r)$, as $R(r)\mathrm{Y}_{lm}(\vartheta, \varphi)$.

Using the standard recursion relation and properties of the spherical harmonics, one can choose:

$$\psi_1 = a_1\, G(r)\, \mathrm{Y}_{lm}, \qquad \psi_2 = a_2\, G(r)\, \mathrm{Y}_{l,m+1}, \qquad (2.170)$$

$$\psi_3 = -\mathrm{i}a_3\, F(r)\, \mathrm{Y}_{l+1,m}, \quad \psi_4 = -\mathrm{i}a_4\, F(r)\, \mathrm{Y}_{l+1,m+1}. \qquad (2.171)$$

The factor $-\mathrm{i}$ in ψ_3 and ψ_4 makes the radial function $F(r)$ real. The values of the coefficients a are adjusted such that all four simultaneous equations are satisfied. Substituting these into the first and third equations and using angular properties, the spinors can be written as

$$\psi_1 = \sqrt{l+m+1}\, G_l(r)\, \mathrm{Y}_{lm}(\hat{r}),$$
$$\psi_2 = -\sqrt{l-m}\, G_l(r)\, \mathrm{Y}_{l,m+1}, \qquad (2.172)$$
$$\psi_3 = -\mathrm{i}\sqrt{l-m+1}\, F_l(r)\, \mathrm{Y}_{l+1,m},$$
$$\psi_4 = -\mathrm{i}\sqrt{l+m+2}\, F_l(r)\, \mathrm{Y}_{l+1,m+1}, \qquad (2.173)$$

where F and G, the radial parts of the small and the large components of ψ, have been subscripted with l. The set of coupled equations is now reduced to two equations satisfied by F_l and G_l:

$$\frac{1}{\hbar}(p_0 - m_0 c)G_l + \left[\frac{2l+3}{2l+1}\right]^{1/2}\left[\frac{\mathrm{d}F_l}{\mathrm{d}r} + \frac{l+2}{r}F_l\right] = 0, \qquad (2.174)$$

$$-\frac{1}{\hbar}(p_0 + m_0 c)F_l + \left[\frac{2l+1}{2l+3}\right]^{1/2}\left[\frac{\mathrm{d}G_l}{\mathrm{d}r} - \frac{l}{r}G_l\right] = 0. \qquad (2.175)$$

These two equations can be combined for the large component G_l as

$$G_l'' + \left[\frac{2}{r} - \frac{\eta'}{\eta}\right]G_l + \left[\eta\delta + \frac{l\eta'}{r\eta} - \frac{l(l+1)}{r^2}\right]G_l = 0, \qquad (2.176)$$

where prime and double prime stand for the first- and second-order spatial derivatives, and

$$\eta = \frac{p_0 + m_0 c}{\hbar}, \qquad \delta = \frac{p_0 - m_0 c}{\hbar},$$
$$\eta\delta = K^2 - \frac{2EV - V^2}{\hbar^2 c^2}, \qquad K^2 = \frac{E^2 - m_0^2 c^4}{\hbar^2 c^2}. \qquad (2.177)$$

For an electrostatic potential $V(r) = Ze^2/r = \frac{2Z}{(r/a_0)} \times \mathrm{Ry}$, and thus $V(r) = \alpha^2\ V \times m_0 c^2/2 = \alpha^2\ Z/(r/a_0) \times m_0 c^2$ (see Eqs 1.17 and 2.4), the Compton-like reciprocal length η reads $\eta = (m_0 c/\hbar)\left(1 + \gamma - \alpha^2 V\right)$; and γ from the identity $E = \gamma m_0 c^2$ may be written with the quantity K, contributing the factor α^2 over the Compton cross section when expressed as an inverse square of a_0: $\gamma = \left(1 + \alpha^2 K^2\, a_0^4\right)^{1/2}$. This paves the way to pure number equations along the lines that lead to Eq. 2.27. Taking $G_l = \sqrt{\eta}g_l(r)/r$, the equation for the large component can be rewritten as

$$g_l''(r) + \left[K^2 - \frac{l(l+1)}{r^2} - U_l^+(r)\right]g_l(r) = 0, \qquad (2.178)$$

where

$$-U_l^+(r) = -2\gamma V + \alpha^2 V^2 - \frac{3}{4}\frac{\eta'^2}{\eta^2} + \frac{1}{2}\frac{\eta''}{\eta} + \frac{l+1}{r}\frac{\eta'}{\eta}. \qquad (2.179)$$

On the other, hand a choice of

$$\psi_1 = a_1\, G(r)\, \mathrm{Y}_{lm}, \qquad \psi_2 = a_2\, G(r)\, \mathrm{Y}_{l,m+1} \qquad (2.180)$$

$$\psi_3 = -\mathrm{i}a_3\, F(r)\, \mathrm{Y}_{l-1,m}, \psi_4 = -\mathrm{i}a_4\, F(r)\, \mathrm{Y}_{l-1,m+1} \qquad (2.181)$$

will show that the simultaneous equations are satisfied for

$$a_1\sqrt{l+m+1} = a_2\sqrt{l-m},$$
$$a_4\sqrt{l+m} = -a_3\sqrt{l-m-1}, \qquad (2.182)$$

and the corresponding spinor components are

$$\psi_1 = \sqrt{l-m}\, G_l(r)\, \mathrm{Y}_{lm},$$
$$\psi_2 = -\sqrt{l+m+1}\, G_l(r)\, \mathrm{Y}_{l,m+1}, \qquad (2.183)$$

$$\psi_3 = -\mathrm{i}\sqrt{l+m}\, F_l(r)\, \mathrm{Y}_{l-1,m},$$
$$\psi_4 = \mathrm{i}\sqrt{l-m-1}\, F_l(r)\, \mathrm{Y}_{l-1,m+1}. \tag{2.184}$$

A similar procedure will give, for the larger component,

$$g_l''(r) + \left[K^2 - \frac{l(l+1)}{r^2} - U_l^-(r)\right] g_l(r) = 0, \tag{2.185}$$

where

$$-U_l^-(r) = -2\gamma V + \alpha^2 V^2 - \frac{3}{4}\frac{\eta'^2}{\eta^2} + \frac{1}{2}\frac{\eta''}{\eta} - \frac{l}{r}\frac{\eta'}{\eta}. \tag{2.186}$$

Combining the two equation for $g_l(r)$,

$$g_l^{\pm''}(r) + \left[K^2 - \frac{l(l+1)}{r^2} - U_l^{\pm}(r)\right] g_l^{\pm}(r) = 0. \tag{2.187}$$

This is the Schrödinger equivalent of the Dirac equation, where the potentials $U_l^{\pm}$ are the effective Dirac potentials due to spin up and spin down, respectively, eigenvalues of the well-known spin–orbit interaction,

$$H^{\mathrm{so}} = \frac{1}{4m_0^2c^2}\frac{1}{r}\frac{\mathrm{d}V(r)}{\mathrm{d}r}\,\boldsymbol{\sigma}\cdot\boldsymbol{L}, \tag{2.188}$$

where $\boldsymbol{\sigma}$ is related to spin $\boldsymbol{S}$ as $\boldsymbol{\sigma} = 2\boldsymbol{S}$ and the value of $(\boldsymbol{\sigma}\cdot\boldsymbol{L})$ equals l for $j = l+1/2$ and $-(l+1)$ for $j = l-1/2$.

2.13.5 Multi-electron systems and the Breit equation

In three classic papers [17], G. Breit introduced the two-electron effects not included in the Dirac equation. The terms were written ad hoc, to order α^2Ry. Bethe and Salpeter [15] showed that the interaction can be derived by making simplifications in the photon exchange between the electrons. A Lorentz-invariant theory for a multi-electron system cannot be written in closed form. Yet relativistic effects can be treated as perturbative correction terms.[9] We may 'add' a relativistic term H_2 to the non-relativistic Hamiltonian and treat it as a perturbation:

$$H = H_0 + H_1 + H_2, \tag{2.189}$$

where H_0 represents the one-electron terms and H_1 is the electron–electron interaction. The relative magnitudes of Coulombic H_1 terms and H_2 of order α^2 Ry depend on Z and the residual or ion charge $z = Z - N$. For an isoelectronic sequence, i.e., a given value z, H_1 dominates H_2 at low Z but H_2 typically scales Z^2-fold. Energy eigenvalues can be obtained accurately in powers of the electromagnetic coupling parameter α. For small charge numbers Z, the energy can also be obtained in powers of $Z\alpha$. For large values of Z the energy can be expanded in powers of $1/Z$ or of α.

The generalized perturbation theory treatment to include relativistic effects is accomplished using the Breit–Pauli approximation, which incorporates both the electron correlation and relativistic effects via the so-called *Breit interaction*. Consider scattering of two identical fermions. The Møller interaction between them represents current and dipole terms. The Møller term in the Fourier transform of the potential gives the repulsive Coulomb potential $e^2/(4\pi r)$. It also gives rise to current–current, current–dipole, and dipole–dipole interactions. The dipole–dipole term is identified as the energy between the magnetic moments of two electrons. The current–dipole interaction leads to spin–orbit coupling between the two electrons. However, prior to Møller, Breit derived these corrections in a Coulomb potential using classical arguments and applied them to the He atom. Hence, they are collectively known as the Breit interaction between two electrons.

The Breit equation is a differential equation, similar to the Dirac equation but for a relativistic wavefunction involving a second electron. Hence, it includes additional terms for two electrons interacting with each other and with an external electromagnetic field. But unlike the Dirac equation it is not fully Lorentz invariant and thus an approximation. The stationary equation for two electrons reads

$$\left[E - \sum_{i=1}^{2}\left\{\boldsymbol{\alpha}_i\cdot\left(c\boldsymbol{p}_i - q\boldsymbol{A}(\boldsymbol{r}_i)\right) + \beta_i mc^2 + q\phi(\boldsymbol{r}_i)\right\} + \frac{e^2}{r_{12}}\right]\psi = \frac{e^2}{r_{12}}\left[\boldsymbol{\alpha}_1\cdot\boldsymbol{\alpha}_2 + \frac{(\boldsymbol{\alpha}_1\cdot\boldsymbol{r}_{12})\boldsymbol{\alpha}_2\cdot\boldsymbol{r}_{12})}{r_{12}^2}\right]\psi, \tag{2.190}$$

where the wavefunction ψ now depends on the positions $\boldsymbol{r}_1$ and $\boldsymbol{r}_2$ of two electrons and has 16 spinor components, four for each electron, 1 and 2. The two terms on the right-hand side, the Gaunt term and the retarded interaction terms, are together known as the *Breit interaction* or *Breit operator*. It gives the leading contribution for the relativistic corrections of the interaction between the two electrons and is also of order $\alpha(Z\alpha)(Z^2)$ Ry.

The Breit equation is conveniently converted to momentum space and expressed in terms of the two-component Pauli spinors. This reduces the 16-component

[9] The references on the Breit equation used in this section are: Bethe and Salpeter [15] and Sakurai [18].

two-particle wavefunction to a four-component wavefunction. For a weak external field the Breit equation can be expressed in terms of spin-up spinors (Bethe and Salpeter). Then for a weak external electromagnetic field, such that the total energy of the two electrons is close to $2mc^2$, the Breit equation can be solved.

Keeping terms in the lowest order of the coupling parameter α and assuming an external field that is weak compared to the binding energy, the Breit equation reduces to

$$H_{\rm BP}\psi = [H_{\rm NR} + H_{\rm mass} + H_{\rm Dar} + H_{\rm so} + \frac{1}{2}\sum_{i\neq j}^{N}[g_{ij}(so+so') + g_{ij}(ss') + g_{ij}(css') + g_{ij}(\mathrm{d}) + g_{ij}(oo')]]\,\psi\,; \tag{2.191}$$

the advantage of the formulation $\frac{1}{2}\sum_{j\neq i}$ over $\sum_{j<i}$ will become clear when discussing $g_{ij}(so+so')$. Written as number equations, $H_{\rm NR}$ is the non-relativistic Hamiltonian

$$H_{\rm NR} = \sum_{i=1}^{N}\left\{-\nabla_i^2 - \frac{2Z}{r_i} + \sum_{j>i}^{N}\frac{2}{r_{ij}}\right\}, \tag{2.192}$$

and

$$H_{\rm mass} = -\frac{\alpha^2}{4}\sum_i p_i^4,\quad H_{\rm Dar} = -\frac{\alpha^2}{4}\sum_i \nabla^2\left(\frac{Z}{r_i}\right),$$

$$H_{\rm so} = \alpha^2\sum_{i=1}\frac{Z}{r_i^3}\,\boldsymbol{l}(i)\cdot\boldsymbol{s}(i) \tag{2.193}$$

are the relativistic mass-velocity correction, Darwin and spin–orbit terms. The rest are two-body terms with notation c for contraction, Dar or d for Darwin, o for orbit, s for spin and a prime indicating 'other'.

The full Breit interaction term includes all magnetic effects among the electrons, i.e., mutual spin–orbit cum spin–other-orbit coupling $g_{ij}(so+so')$ and spin–spin coupling (ss') as

$$H^{\rm B} = \sum_{i>j}[g_{ij}(so+so') + g_{ij}(ss')], \tag{2.194}$$

where

$$g_{ij}(so+so') = -\alpha^2\left[\left(\frac{\boldsymbol{r}_{ij}}{r_{ij}^3}\times\boldsymbol{p}_i\right)\cdot(\boldsymbol{s}_i + 2\boldsymbol{s}_j) + \left(\frac{\boldsymbol{r}_{ij}}{r_{ij}^3}\times\boldsymbol{p}_j\right)\cdot(\boldsymbol{s}_j + 2\boldsymbol{s}_i)\right],$$

$$g_{ij}(ss') = 2\alpha^2\left[\frac{\boldsymbol{s}_i\cdot\boldsymbol{s}_j}{r_{ij}^3} - 3\frac{(\boldsymbol{s}_i\cdot\boldsymbol{r}_{ij})(\boldsymbol{s}_j\cdot\boldsymbol{r}_{ij})}{r_{ij}^5}\right]. \tag{2.195}$$

The two-body non-fine-structure terms are

$$g_{ij}(\mathrm{d}) = \frac{\alpha^2}{2}\nabla_i^2\left(\frac{1}{r_{ij}}\right),$$

$$g_{ij}(css') = -\frac{16\pi\alpha^2}{3}\,\boldsymbol{s}_i\cdot\boldsymbol{s}_j\delta^3(\boldsymbol{r}_{ij}),$$

$$g_{ij}(oo') = -\frac{\alpha^2}{r_{ij}}\left(\boldsymbol{p}_i\cdot\boldsymbol{p}_j + \frac{(\boldsymbol{r}_{ij}\cdot\boldsymbol{p}_i)(\boldsymbol{r}_{ij}\cdot\boldsymbol{p}_j)}{r_{ij}^3}\right); \tag{2.196}$$

$g_{ij}(\mathrm{d})$ is similar to the Darwin term of the nuclear Coulomb field in the Dirac equation and represents a single electron in an electric field of another electron; $g_{ij}(ss')$ and the contact term $g_{ij}(css')$ contribute to the ordinary interaction between two Bohr magnetons (cf. Section 2.13), while $g_{ij}(oo')$ is the classical relativistic correction to electron–electron interaction due to the retardation of the electromagnetic field produced by an electron. There can be additional terms caused by magnetic fields.

There are good reasons for keeping mutual spin–orbit and spin–other-orbit combined in the shape of $g_{ij}(so+so')$, i.e., symmetric in the two-particle indices. The kinematics of the first coupling term in Eq. 2.195, that is, before '$+2\boldsymbol{s}_j$', are clear enough: particle i with momentum $\boldsymbol{p}_i$ is orbiting electron j. In a discussion, one would continue with what one calls 'spin–other-orbit' coupling, namely that j is itself orbiting unlike an infinitely heavy nucleus in ordinary spin–orbit coupling, say in $H_{\rm so}$ of Eq. 2.193 (or more to the point: if its orbital quantum number $l_j > 0$). Strangely enough for classical mechanics, such a term appears in the fourth or last position in $g_{ij}(so+so')$, unmistakable since the s-index must be i. But this is not classical mechanics as testified by the Thomas-factor of 2! Thus all arrangements in $g_{ij}(so+so')$ are falling into place.

The electron–electron correlations contribute dominantly through $1/Z$ dependence, while the relativistic corrections play an important role in the distribution of fine-structure components for weaker transitions. The one-body terms with weight $\alpha^2 Z^4$Ry contribute, in general, more than the two-body terms, which scale only with the third power of Z.

2.13.6 Dirac–Fock approximation

Next in importance to Breit–Pauli terms, which stop at order $\alpha^2 Z^4$ Ry $= (\alpha Z)^4 mc^2/2$, follow QED effects:

self-energy (corrected for finite size of a structured nucleus) scales with another factor α, screened self-energy as $\alpha(Z\alpha)^3 mc^2$ and vacuum polarization. The self-energy behaves as $1/n^3$ in an interacting outer electron nl. Vacuum polarization is usually appreciable for s orbitals at large Z.

When the Breit–Pauli approximation or inclusion of other higher-order contributions are not valid, a fully relativistic (not perturbative) treatment based on the Dirac theory must be employed. This relativistic approach is known as the Dirac–Fock approximation. The Dirac–Coulomb Hamiltonian in this approximation reads as

$$H_{\rm DC} = \sum_{i=1}^{N} H_{\rm D}(i) + \sum_{i=1}^{N-1}\sum_{j=i+1}^{N} \frac{e^2}{|\boldsymbol{r}_i - \boldsymbol{r}_j|} \qquad (2.197)$$
$$= \sum_{i=1}^{N}\left(H_{\rm D}(i) + \alpha^2 mc^2 \sum_{j<i} \frac{a_0}{|\boldsymbol{r}_i - \boldsymbol{r}_j|}\right),$$

where the first term is the one-body contribution for an electron due to kinetic energy and interaction with the electric charge of the nucleus, as

$$H_{\rm D} = c\,\boldsymbol{\alpha}\cdot\boldsymbol{p} + (\beta - 1)mc^2 - Ze^2/r \qquad (2.198)$$
$$= c\,\boldsymbol{\alpha}\cdot\boldsymbol{p} + \left(\beta - 1 - \alpha^2\frac{Za_0}{r}\right)mc^2,$$

employing the identity (Eq. 1.23) for e^2 (along with the definition, (Eq. 1.17)). This can be seen as the relativistic version of the Hartree–Fock equations. They are solved in a similar manner for more accurate energies and wavefunctions. Numerical difficulties, though, limit the amount of configurations one can include.

2.13.7 Z-scaling of fine structure

We end by returning to a more heuristic description of the fine-structure effects by noting another striking feature of Eq. 2.132: scaling with $\alpha^2 Z^4$ Ry. As long as $\alpha^2 Z^4 \ll 1$ it looks all set for perturbation theory to obtain hydrogen-like doublet levels $nl_j \equiv nlj$: find eigenvalues E_{nlj} of matrices $H^{\rm NR} + H^{\rm so}$ with Hamiltonian constituents (Eqs. 2.80 plus 2.132) in the representation (Eq. 2.139), using ordinary hydrogen-like radial wavefunctions. The result is disappointing: j-degeneracy of $\left(nl_{l+1/2},\, n(l+1)_{l-1/2}\right)$, observed as well as predicted for a Dirac electron, is missed by a wide margin. Dirac theory by itself delivers the remedy for perturbative treatment: there are two more terms of order $\alpha^2 Z^4$ Ry in Section 2.13.2, though not breaking LS symmetry. The first arises when applying Einstein's $m/\sqrt{(1 - v^2/c^2)} \approx m + \frac{1}{2}mv^2/c^2 + \mathcal{O}(v^4/c^4)$ to the kinetic energy in the non-relativistic Hamiltonian. With $v^4 \approx (p/m)^4$ (if $v \ll c$) one arrives at the purely mechanical mass–velocity operator $H^{\rm mass}$ of Eq. 2.150, the electromagnetic coupling parameter α entering by plain arithmetic via the quantities Ry and a_0, see Eqs. 1.17 and 1.19. If $H^{\rm mass}$ is carried alongside $H^{\rm so}$, everything falls into place except s-levels, which coincide with $p_{1/2}$ only when including the non-classical Darwin term (Eq. 2.152), the third and last of order $\alpha^2 Z^4$ Ry in what is called the 'low-Z Breit–Pauli Hamiltonian'. Following Eq. 2.113, electron velocities behave as $v \approx \alpha Zc$, which should be kept small compared with the speed of light, say not exceeding a fifth. This translates into stopping at iron ions in isoelectronic sequences. Within principal shells n of H-like ions, the following splitting pattern emerges for successive levels E_{nj}:

$$E_{n,l+1/2} - E_{n,l-1/2} = \alpha^2 Z^4 \frac{1}{n^3}\frac{1}{l(l+1)}\,\text{Ry}. \qquad (2.199)$$

As we move along a multi-electron sequence, let us focus on the Coulombic term separation, which goes linear as Z Ry within a complex (otherwise as Z^2 Ry). Level splitting is tiny at the neutral end, but begins to outstrip separation once the effective Z reaches $Z \approx \alpha^2 Z^4$: again soon after ions of iron. Thus a perturbative approach serves most ions of astrophysical interest, as outlined for the H-sequence. Now with more than one electron participating we can choose how to combine the angular momenta. In line with the low-Z approach is *intermediate coupling SLJ*, where the assignment ${}^{2S+1}L_J$ points towards a dominant term ΓSL. This ceases to be true once Z has increased enough for the various multiplets to overlap.

Because the nucleus is 'replaced' by one electron in two-body terms, they scale with a relative factor $-1/Z$. The kinematics for fine structure will be discussed in Section 2.13.5, Eqs. 2.195, are transparent enough, but less so their shape in reduced tensorial form, a topic outside the scope of this text – unlike the simple Coulombic (Eq. 2.109), which involves merely two Racah tensors $C^{[\lambda]}$) (a limited discussion of the algebra involved is given in Chapters 3 and 4). This makes one wonder how it shows up in the simplest multi-electron case, *He-like ions*, neutral He in particular, when the effective charge is small, somewhere between 1 and 2. And indeed, observation yields triplet energies in J-order 1s 2p, ${}^3\mathrm{P}^{\rm o}_{2,1,0}$ at splitting distances of 0.0675 cm and 0.0879 cm, and similar but even smaller for the M-shell triplets of 1s3p and 1s3d: thus effects among the two electrons are winning over the ordinary spin–orbit coupling in the field of the nucleus.

The '$-1/Z$-effect' in 1snl triplets weakens immediately past Li II. Yet full $J = l-1; l; l+1$ order prevails only past C V, because $J = l$ is pushed down by multiplet mixing with the associated singlet due to ordinary spin–orbit coupling. Naturally, Hund's criteria of Section 2.8 are not quite valid for this sequence, and show their limitations even for a simple case.

Two-body magnetic interaction comes decisively into play when evaluating spin–orbit parameters ζ. In a key paper of 1962, M. Blume and R. E. Watson [19] broke the deadlock about magnetic closed shell effects on valence electrons. Here one cannot construct a potential that mimicks the effects of exchange, while it proved so successful in the Coulomb case outlined in Section 4.6. It has to be done explicitly with properly antisymmetrized wavefunctions. But the effects can be absorbed in a relatively simple procedure into ζ as *effective spin–orbit parameters*, employing the two types of magnetic integral, the magnetic counterpart of the four-orbital Slater integrals (Eq. 2.111). Moreover, angular factors, based on Racah or $6j$-coefficients (cf. Appendix C) in the general case (Eq. 2.195), simplify to square roots over simple ratios of angular quantum numbers. Thus the two parameters ζ_{3p} and ζ_{3d} of valence electrons in an M-shell will be reduced by the closed shell electrons 1s, 2s and 2p, and by a sizable fraction by $1s^2$ alone. An interesting aside concerns the exchange portion, which unlike the Coulomb case (e.g., in Section 2.11) has the same sign as the direct contribution, making ζ even smaller. At high Z, two-body magnetic interaction among valence electrons is often dominated by second-order ordinary spin–orbit coupling, an effect automatically included through matrix diagonalizations in the course of the calculation.

Significant for radiative probabilities and, because of flux redistribution, in collision rates is multiplet mixing, where terms interact in first order magnetically, because $S' \neq S$ precludes Coulomb interaction. To start with He-like ions: it is spin–orbit interaction (cf. Eq. 2.132) within 1snl between triplet and singlet $J = l$ that moves the former level down and the latter up, in the triplet case closer to the unchanged $J = l - 1$ below. For instance, level $1s2p\,^3P^o_1$ of Fe-like He lies 0.1484 Ry above $^3P^o_0$ (1.0847 Ry below $^3P^o_2$ from observation), whereas statistical weights without admixture from $1s2p\,^1P^o$ would place it 0.4111 Ry above $^3P^o_0$. A close look reveals a ratio 0.283 : 0.959 for the *term coupling coefficients* (cf. Ch. 4) $\left\langle ^{1,3}P^o \middle| ^3P^o_1 \right\rangle$, and there is 8% of $^1P^o$ in $1s2p\,^3P^o_1$.

We mention in passing that the nucleus of isotopes with odd atomic weight number always has, and others can have, non-vanishing spin I and an associated magnetic moment (and an electric quadrupole moment if $I \geq 1$). Conservation of the vectorial sum F with J marks the background of hyperfine structure, in magnitude three orders smaller than fine structure following Eq. 2.131: see Fig. 2.2, which schematically shows successive term splitting for a given electronic configuration.

3 Atomic processes

Spectral formation depends on a variety of intrinsic atom–photon interactions. In addition, external physical conditions, such as temperature, density and abundances of elements determine the observed spectrum. As described in later chapters, spectral analysis is therefore often complicated and it is difficult to ascertain physical effects individually (and even more so collectively). The main aim of this chapter is to provide a unified picture of basic atomic processes that are naturally inter-related, and may be so considered using state-of-the-art methods in atomic physics. A quantum mechanical treatment needs to take the relevant factors into account. An understanding of these is essential, in order to decide the range and validity of various theoretical approximations employed, and the interpretation of astrophysical observations. From a practical standpoint, it is necessary to determine when and to what extent a given effect or process will affect spectral lines under expected or specified physical conditions.

For example, at low temperatures and densities we may expect only the low-lying atomic levels to be excited, which often give rise to infrared (IR) and optical forbidden emission lines. But the presence of a background ultraviolet (UV) radiation field from massive young stars in star-forming regions of molecular clouds (e.g., the Orion Nebula discussed in Chapter 12), may excite low-lying levels via UV absorption to higher levels and subsequent radiative cascade of emission lines that would appear not only in the UV but also contribute to the intensities of the IR/optical lines. Therefore, the intensities of observed forbidden lines may involve a contribution from allowed transitions via muliple-level cascades. That, in turn, requires a more extensive atomic model with all relevant excited levels.

Fundamental atomic parameters associated with atomic processes are mostly computed theoretically and, to a much lesser extent, measured experimentally. Our treatment is aimed at a theoretical description of atomic processes. That basically requires a knowledge of the wavefunctions of quantum mechanical states of the atom, and electron–atom–photon interactions with an appropriate Hamiltonian in a generalized Schrödinger formulation. The task is to compute fundamental quantities, such as cross sections for ionization, recombination and scattering, and transition probabilities for radiative transitions.

The atomic parameters are obtained from transition matrix elements whose general form is

$$\langle \psi_f | H_{\rm int} | \psi_i \rangle \,. \tag{3.1}$$

The ψ_i and ψ_f are the initial and final quantum mechanical wavefunctions representing the system, and $H_{\rm int}$ is the interaction operator associated with a given process. The processes with a free electron involve continuum states of the (e + ion) system. If there are several initial and final states then a *transition matrix* can also be obtained from appropriate wavefunctions. Once the cross sections are calculated as functions of energy, we can obtain the *rate coefficient* at a given temperature by convolving over a characteristic temperature distribution in the source, as described in a later chapters on emission and absorption lines (Chapters 6 and 7, respectively). As we emphasize in this chapter and throughout the text, resonances in cross sections play an important role in most atomic processes. But delineation of detailed resonance structures often requires the calculation of cross sections at a large number of energies.

While we will study individual atomic processes in detail in the following chapters, we can understand their overall nature and role in astrophysical plasmas in an elementary manner, introducing the basic concepts that we shall use later in a more advanced description of scattering and radiative theory. In this chapter we describe the general quantum mechanical considerations of the atomic processes and methods: multi-electron variants of the Schrödinger equation, wavefunction expansions for a multi-level formulation of a given process, and the Hamiltonian terms including (or excluding) important atomic interactions. Subsequent chapters deal with specific processes and the properties of atomic parameters for each process, such as cross sections and rate coefficients. We begin with the physical description of the states of an atom.

3.1 Bound, continuum and resonance states

Excited bound states decay via spontaneous or stimulated emission according the Einstein relations (Eq. 4.18). Discrete bound state energies are given by the Rydberg formula $E_n = (-z^2/\nu_n^2)$ Ry < 0, and the series of excited states or levels converge on to the series limit $E = 0$ as $n \to \infty$. We may denote the ground state ionization energy as E_g, equal to the energy difference between the ground state and the series limit. Excited state ionization energies E_i are also measured relative to the ionization threshold $E = 0$ when the electron has zero energy and becomes free. Ionization energies of the outermost electron(s) in excited states are *lower* than those of the ground state. On the other hand, ionization energies of inner electronic shells are higher, in increasing order from the outside inwards up to the most tightly bound electrons in the K-shell, and consequently the highest binding or ionization energy.

Excitation of a bound state to another bound state may occur due to impact by other particles or photons. Figure 3.1 shows free electrons and the energy levels of an ion. The bound electron(s) in the ion may be excited, or ionized, by the free electrons. Since the ionized electron may have any energy, there is an infinite *continuum of positive (kinetic) energies* $E > 0$ above the ionization threshold at $E = 0$, where no 'pure' bound states can exist. Here, the meaning of the term 'continuum state' of an (e + ion) system, with a free electron and an ion, needs to be understood. The state refers to a continuum of kinetic energies, which the free electron may have, but the total (e + ion) energy is relative

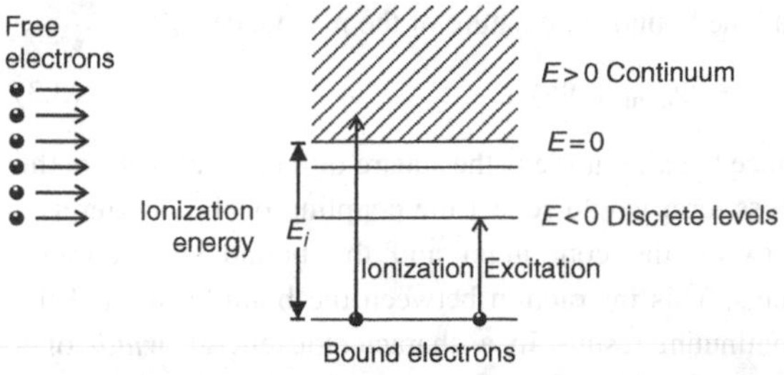

FIGURE 3.1 Excitation of bound states and ionization into the continuum.

to a specific bound state of the ion, usually the ground state.

Figure 3.2 is an illustration of a bound electron excitation by impact of a free or continuum electron. The excited electron decays radiatively, emitting a photon $h\nu$ in an emission line. *Electron impact excitation* (EIE), and (e + ion) recombination into an excited level, are both followed by the emission of a photon and are the two most important processes in the formation of *emission spectra* in astrophysical sources.

While that seems relatively straightforward, there are also *quasi-bound states* formed by two (or more) excited electrons in an atomic system that play an important role. A free electron with sufficiently low kinetic energy and in close proximity to an ion can form quasi-bound states. If the total energy is below the ionization threshold, that is, less than the ionization energy, the (e + ion) system is a pure bound state. However, if the total (e + ion) energy lies above the ionization threshold, it could be in an unstable *autoionization state*, which ionizes spontaneously, i.e., breaks up into an ion and a free electron. While in an autoionizing state, an electron is in excited state loosely bound to the ion, which is also in excited state. Hence, these are also referred to as *doubly excited states*. The 'active' electrons are thus both in excited states in an electronic configuration whereby they repel each other; the combined state is thus unstable and breaks up after a lifetime that is orders of magnitude shorter than that of singly excited bound states. For instance, the lifetime of an autoionizing state is about 10^{-14} s compared with the typical lifetime of about 10^{-9} s for the low-lying excited bound states of neutral atoms.

An autoionizing state generally decays into the continuum, with the outermost electron going free, while the excited ion core electron drops back to the ground state. Figure 3.3 illustrates the autoionization process. In contrast to radiative decay, which involves the emission of a photon, autoionization is a *radiation-less* decay of a compound unstable system, where the ejected electron carries away the energy released when the inner electron makes a downward transition, say to the ground state. While in a doubly excited state, we have two or more electrons in excited states, and therefore electron–electron repulsion drives its break-up or *autoionization*. It is of great importance since the process of formation and break-up gives rise to *resonances* that manfiest themselves in the cross sections of atomic processes.[1] Resonances, in turn,

[1] All three expressions are equivalent, and interchangeably used, to describe autoionizing, quasi-bound, or resonance states of an (e + ion) system with two or more electrons.

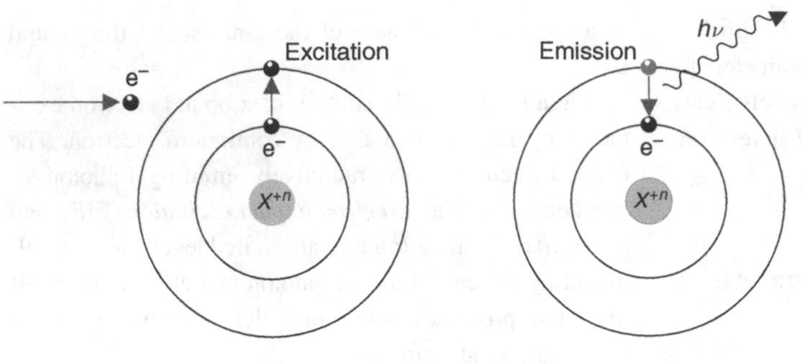

FIGURE 3.2 Electron impact excitation of an electron from a lower bound level to an upper bound level, followed by downward radiative transition of the excited electron.

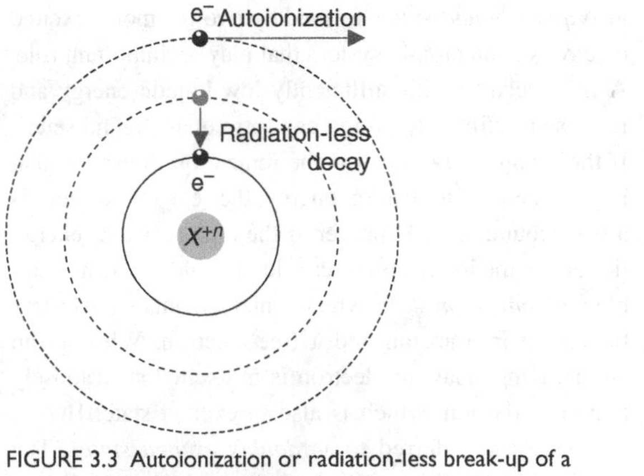

FIGURE 3.3 Autoionization or radiation-less break-up of a resonant state. The inner electron undergoes a downward transition and releases energy, which ionizes the outer electron, carrying away the energy that could otherwise be emitted as a photon.

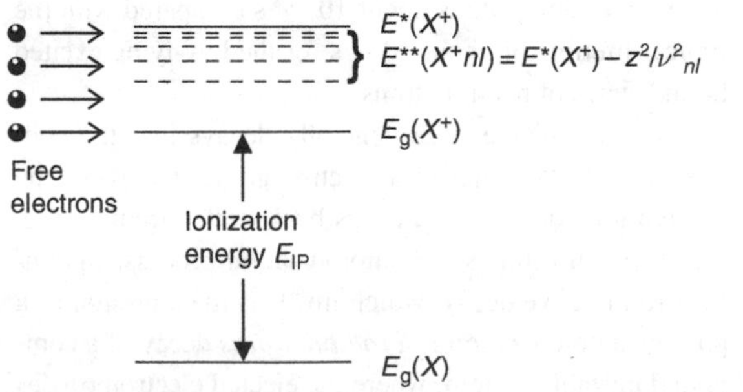

FIGURE 3.4 Excitation and autoionization of a doubly excited resonance state.

generally enhance the cross section of a given process, such as collisional excitation or photoionization, by large factors, as we shall see in later chapters.

Figure 3.4 shows an energy diagram of bound states and autoionizing resonance states of an (e + ion) system. We have an atom X with ground state at energy $E_g(X)$. A continuum of states then exists with the ion X^+ with ground-state energy $E_g(X^+)$, and a free electron with positive energy. An excited state of the ion is denoted with energy $E^*(X^+)$. Now a series of doubly excited states (dashed lines) could form with energy $E^{**}(X^+)n\ell$, where both the ion and the Rydberg electron are in excited states. Owing to the infinite range of the attractive Coulomb potential, $V(r) \sim -2z/(r/a_0)$ Ry, it can support an infinite number of discrete bound states. Furthermore, each excited bound state of the ion can, in principle, serve as a core for an entire Rydberg series of infinite resonances given by the Rydberg formula with energies converging on to the core level energy $E^*(X^+)$

$$E^{**}(X^+n\ell) = E^*(X^+) - \frac{z^2}{\nu_{nl}^2}\,\text{Ry}, \tag{3.2}$$

where ν_{nl} is the effective quantum number relative to the excited core level. In general, resonances can also be formed by ionization (viewed as excitation into the continuum) of any inner-shell electron. The resulting (e + ion) system stabilizes by decay to the ground state, releasing energy either as photons or ejected electrons. This is referred to as the *Auger process* and is discussed in Section 5.9.

To account for resonances in atomic processes, the wavefunction must include not only the bound state components but also the (e + ion) continuum. It is the quantum mechanical interference between the two that gives rise to resonances. While a rigorous theory of resonance phenomena is quite involved (and will be discussed later in this chapter and in other chapters), we can obtain a physical understanding by writing the wavefunction as a sum of two *independent* components, referring to the continuum and the bound state nature of the resonance,

$$\Psi_{\text{res}} = \Psi_{\text{cont}} + \Psi_{\text{bbd}}. \tag{3.3}$$

Since the amplitude is the square of the wavefunction, the cross term would reflect the coupling or the interference between the continuum and the bound state components. This interaction between the bound state and the continuum results in a characterstic energy *width* of a resonance, in contrast to a bound state, which, in principle, has an infinitesimally small width. The width of a resonance is related to the lifetime in accordance with the uncertainty principle $\Delta E\ \Delta t \geq \hbar$: a larger energy width implies a shorter lifetime of the resonance, compared with a bound state.

If we neglect the interference term then we may carry out a purely atomic structure calculation to determine the position and the strength of the resonance, treating it (albeit unphysically) as a bound state of the atom X with no associated width. But a proper description of the resonance, including all terms in the wavefunction expansion above, should entail not only the position but also the *shape* of the resonance in the cross section.

In the following sections, we will describe the theoretical approximations usually employed, including or excluding resonance phenomena. But first, we give an overview of the primary physical processes leading to spectral formation.

3.2 Collisional and radiative atomic processes

The electromagnetic spectrum is formed as a result of the interaction of matter with light. In addition, characteristics of a spectrum depend on the physical and chemical conditions in the source. For example, at low temperatures and densities only low-lying levels are excited, leading to emission of IR and optial lines. The atomic processes in astrophysical and laboratory plasmas involve electron–ion–photon interactions. While we describe various aspects of the basic atomic processes in this and later chapters, it is useful to introduce their definition in a heuristic manner at this stage. The dominant atomic processes involving electrons, ions and photons in astronomical plasmas are schematically described in Fig. 3.5.

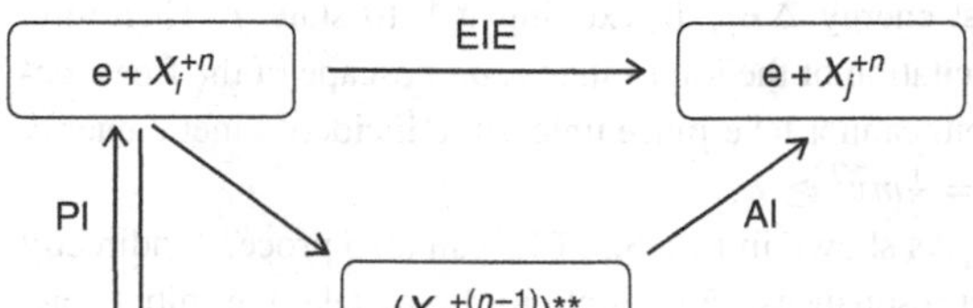

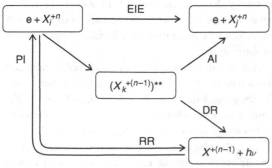

FIGURE 3.5 Unified picture of dominant atomic processes in plasmas: electron impact excitation (EIE), photoionzation (PI), autoionization (AI), dielectronic recombination (DR) and radiative recombination (RR). Note the often important role of resonance states in the centre, mediating atomic processes.

These atomic processes dominate spectral formation in most plasma sources, and form the bulk of the discussion in this text. However, there are other processes, which are important in many special circumstances and need to be considered accordingly. We mention a few of these additional atomic processes. One common theme among these less dominant processes is that they are heavy-particle collisions and are often treated as 'molecular' problems.

3.2.1 Detailed balance

As evident from the unified picture of atomic processes in Fig. 3.5, the principle of detailed balance plays a fundamental role in the determination of atomic rates for inverse processes. We apply it later to a variety of collisional and radiative processes in subsequent chapters on electron impact excitation (Chapter 5), photoionization (Chapter 6), and recombination (Chapter 7).

Consider a collision process where $\mathbf{k}_i$ and $\mathbf{k}_f$ are the incident and scattered wave vectors and m_i and m_f are the masses in the initial and final states or 'channels' (readers should refer to Figs. 3.7 and 5.2 for illustrations). We choose the incident beam along z-axis, i.e., $k_i = k_{iz}$. Let Ψ_i and Ψ_f be the incident and final wavefunctions. Then the total scattered wavefunction (Chapter 5) can be expressed as

$$\Psi = \Psi_i + \Psi_f = A\left[e^{i\mathbf{k}_i.\mathbf{r}} + f(\vartheta)\sqrt{\frac{m_f}{m_i}}\frac{e^{ik_f r}}{r}\right], \quad (3.4)$$

where A is a normalization constant and $f(\vartheta)$ is the amplitude of the scattered wave. The current density of a beam is defined as

$$J = \frac{\hbar}{2im}[\Psi^*\nabla\Psi - \Psi\nabla\Psi^*], \quad (3.5)$$

with a nabla operator $\mathbf{e}_r\frac{\partial}{\partial r} + \mathbf{e}_\vartheta\frac{1}{r}\frac{\partial}{\partial\vartheta} + \mathbf{e}_\varphi\frac{1}{r\sin\vartheta}\frac{\partial}{\partial\varphi}$ in spherical coordinates. Then the current density of the incident beam of wavefunction Ψ_i reads

$$J_i = A\frac{\hbar k_i}{m_i} = Av_i; \quad (3.6)$$

similarly,

$$J_f = A\frac{m_f}{m_i}\frac{|f(\vartheta)|^2}{r^2}v_f \quad (3.7)$$

for the scattered beam Ψ_f, neglecting terms of the order of $1/r^3$.

The number dn/dt of particles detected per unit time, in solid angle element $d\Omega$ is related to both J_i and J_f separately. We can write

$$\frac{dn}{dt} \propto J_i d\Omega, \text{ i.e., } \frac{dn}{dt} = \sigma(\vartheta)J_i d\Omega, \quad (3.8)$$

where σ is the proportionality constant, here called the *cross section*. Writing $d\Omega = dS/r^2$, where dS is the area subtended by the solid angle $d\Omega$ at radius r, we have

$$\frac{dn}{dt} = \sigma(\vartheta) A v_i \frac{dS}{r^2} \quad (3.9)$$

Since $J_f = (dn/dt)/dS$, we have

$$\frac{\sigma(\vartheta) A v_i}{r^2} = A \frac{m_f}{m_i} \frac{|f(\vartheta)|^2}{r^2} v_f, \text{ i.e.,}$$

$$\sigma_D(\vartheta) = \frac{p_f}{p_i} |f(\vartheta)|^2 \quad (3.10)$$

where the subscript D has been introduced to denote the direct process. If we reverse the process to go to state i from state f, we obtain

$$\sigma_R(\vartheta) = \frac{p_i}{p_f} |f(\vartheta)|^2. \quad (3.11)$$

The two relations of the direct and reverse processes can be combined to obtain

$$\frac{\sigma_D}{p_f^2} = \frac{\sigma_R}{p_i^2}. \quad (3.12)$$

If the initial state is degenerate with g_i states corresponding to the initial energy for the direct process, and g_f states to the reverse process, then the cross sections are obtained as averages over all processes, that is, $\sigma_D \equiv (1/g_i) \sum_i \sigma_D$ and $\sigma_R \equiv (1/g_f) \sum_f \sigma_R$. Thus we have

$$\frac{g_i \sigma_D}{p_f^2} = \frac{g_f \sigma_R}{p_i^2} \quad (3.13)$$

This relation is a formal statement of the *principle of detailed balance* for collisional processes (see also Chapters 5, 6 and 7).

3.2.2 Radiative transitions – photo-excitation

An ion with positive charge $+n$ may be excited radiatively from an initial state i to another state j by absorption of an incident photon energy $h\nu_{ij}$ characteristic to the ion:

$$X_i^{+n} + h\nu_{ij} \rightleftharpoons X_j^{+n}. \quad (3.14)$$

Excited states have finite lifetimes relative to the ground state (which supposedly has 'infinite' lifetime). The reverse arrow refers to spontaneous radiative decay after a characterstic lifetime by emission of the same photon. The radiative decay may also take place with the emission of several photons, i.e., *fluorescence*, consistent with conservation of energy. The computed parameters for these atomic processes are the physically equivalent quantities: the oscillator strength f for photo-excitation, the radiative decay rate or transition probability A, and the line strength S, which is symmetric with respect to degeneracies of the initial and final states. The oscillator strength f, known from damped and enforced oscillations in classical physics, is dimensionless, S is usually given in units of a_0^2 for electric dipole transitions, and A as decays per second.

3.2.3 Electron impact excitation

Electron–ion collision in a plasma with free electrons and positive ions could excite the ion in initial state i to a higher state j

$$e(\epsilon) + X_i^{+n} \to e(\epsilon') + X_j^{+n}. \quad (3.15)$$

The excited state j decays by emission of a photon $h\nu_{ij}$ producing an emission line. This mechanism is the dominant contributor to emission of *forbidden lines* (Chapter 4) due to transitions among low-lying energy levels with small energy differences (in contrast, (e + ion) recombination lines seen in emission spectra generally occur via *allowed transitions*).

Shown at the top of Fig. 3.5, the first reaction, electron impact excitation (EIE), occurs with transfer of kinetic energy $\frac{1}{2}mv^2 = \Delta E_{ij}$ from a free electron to the bound electron(s) in the ion X^+. Suppose the incident free electron has kinetic energy $\frac{1}{2}mv^2 = (ka_0)^2$ Ry when measured in wavenumbers k at distance $r \to -\infty$ from the ion X^+. As it approaches the ion, the free electron gains kinetic energy from 'falling' into the attractive positive Coulomb potential. Similarly, it loses kinetic energy when receding from the ion back to $r \to +\infty$. But if the incident electron loses energy in exciting the ion, it must have sufficient residual kinetic energy to 'climb' out of the attractive potential well. This energy must be at least equal to that gained while falling towards the ion, in order for the electron to escape *after* having lost energy ΔE_{ij} in exciting X^+ to state j. Therefore, excitation of the ion to state i, *and* escape of the free electron, cannot take place unless the incident kinetic energy $E = \frac{1}{2}mv^2 \geq E_{ij}$.

As shown in Fig. 3.5, EIE can also proceed indirectly via resonances. For simplicity let us take the initial state to be the ground state, i.e., $E_i \equiv E_g = 0$, and $\Delta E_{ij} = E_j$. Consider the case where initially at $r \to -\infty$ the electron has *insufficient* energy to excite level j, i.e., $E < E_j$. However, the electron may gain sufficient energy to do so while approaching the ion. If the distance of closest approach is up to an *impact parameter* $r = b$ from the ion, then the electron kinetic energy on impact is

$$\frac{1}{2}mv_b^2 = \frac{1}{2}mv_\infty^2 + \frac{2z}{b/a_0} \text{Ry}, \quad (3.16)$$

where z is the charge on the ion (the effective constant charge number outside the radius b). Let us assume $z = 1$ for ion X^+. If $\frac{1}{2}mv_b^2 > E_j$ then the electron may excite the ion to level j from the ground state on impact, but then would lack the energy $2za_0/b$ Rydbergs required to escape back to $r \to +\infty$. As in Fig. 3.4, the electron could therefore become 'bound' to the ion in a doubly excited resonance state whose energy E_r is given by the Rydberg formula $E_r = E_j - \left(z^2/\nu_j^2\right)$ Ry. Resonances, denoted as $X^{+(n-1)**}$, are centrally located in Fig. 3.5 to emphasize their role in all the inter-related processes, and the fact that the *same* resonances manifest themselves in the cross sections for all processes – (e + ion) scattering, photoionization and (e + ion) recombination (DR).

The atomic parameter to compute is the cross section $\sigma_{EIE}(E)$ as a function of continuum electron energy E, an area expressed in cm^2 or πa_0^2, as discussed further in Chapter 5.

3.2.4 Photoionization and radiative recombination

If the incident photon energy exceeds the binding energy of an electron in the target ion, then the bound electron may be knocked out leaving an ion with one charge higher; this *photoionization* (PI) process is

$$X + h\nu \rightleftharpoons X^+ + e(\epsilon). \tag{3.17}$$

The free electron, often termed a *photoelectron*, is said to be in the continuum with positive energy ϵ that is equal to the difference between the incident photon energy $h\nu$ and the electron binding energy or the ionization potential IP, that is $h\nu-$ IP $= \frac{1}{2}mv^2$. The inverse process occurs when a free electron combines with an ion, followed by the emission of a photon hν. This is called *radiative recombination* (RR). Beginning at the diagonally opposite end from EIE in Fig. 3.5, we have the ionization of an ion $X^{+(n-1)}$ by an incident photon $h\nu$ leading back to the free electron and the ion.

The parameters needed for these atomic processes are the cross section for photoionization (σ_{PI}) and the recombination cross section (σ_{RC}), described in Chapter 6.

3.2.5 Autoionization and dielectronic recombination

Photoionization and elelctron–ion recombination can also proceed through autoionizing resonances. The resonance states may be formed provided that hν corresponds exactly to the energies of the compound (e + ion) system given by the Rydberg formula (Eq. 3.2). The PI cross section thus has the same resonances as the cross section for EIE, and the energy of a free electron is such that an autoionizing state of the (e + ion) system is formed, followed by break-up, as follows

$$e+X^{+n} \leftrightarrow (X^{+n-1})^{**} \leftrightarrow \begin{cases} e+X^{+n} & \text{AI} \\ X^{+n-1}+h\nu & \text{DR.} \end{cases} \tag{3.18}$$

There are two pathways for the autoionizing or resonant state, as shown in Fig. 3.6. The first branch is autoionization (AI) or radiation-less break-up of the intermediate resonance back to a free electron and the ion X^+. The second, the radiative branch, is dielectronic recombination (DR) that leads to recombination and radiative stablization into an (e + ion) bound state and an emitted photon. The DR process, i.e., electron recombination via resonances, is also referred to as 'inverse autoionization', but note that it is accompanied by the emission of a photon, which carries away the excess energy from the doubly excited resonant state, leading to radiative stabilisation of the (e + ion) system into a bound state of the recombined ion ($X^{+(n-1)}$ in Fig. 3.5). Since the DR process is essentially electron impact excitation of an ion and capture into a doubly excited autoionizing state, followed by radiative decay, the DR probability is coupled to that of the primary EIE process. In fact, in many situations the resonant DR process dominates RR, since the EIE rate is generally much higher than the rate for direct RR. Therefore, the total (e + ion) recombination rate may be enhanced by orders of magnitude when DR is also considered in addition to RR.[2]

Now if we consider the inverse of the photoionization process *including resonances*, then both the non-resonant RR and the resonant DR processes may be treated together. Whereas the RR and the DR recombination processes may be considered individually, there is no such separation in nature between the two. All observations or experiments of the (e + ion) recombination always measure both processes together. It is therefore desirable to treat (RR $\oplus$ DR) in a unified manner. The unified treatment of electron–ion recombination is discussed in Chapter 7.

3.2.6 Electron impact ionization

In addition to photoionization, ionization can also occur by impact of a free electron and an ion, producing another ion with one higher charge and the release of another

2 In Chapter 7 we shall see that the RR rate, in turn, is related to the PI rate which is much smaller than EIE rates.

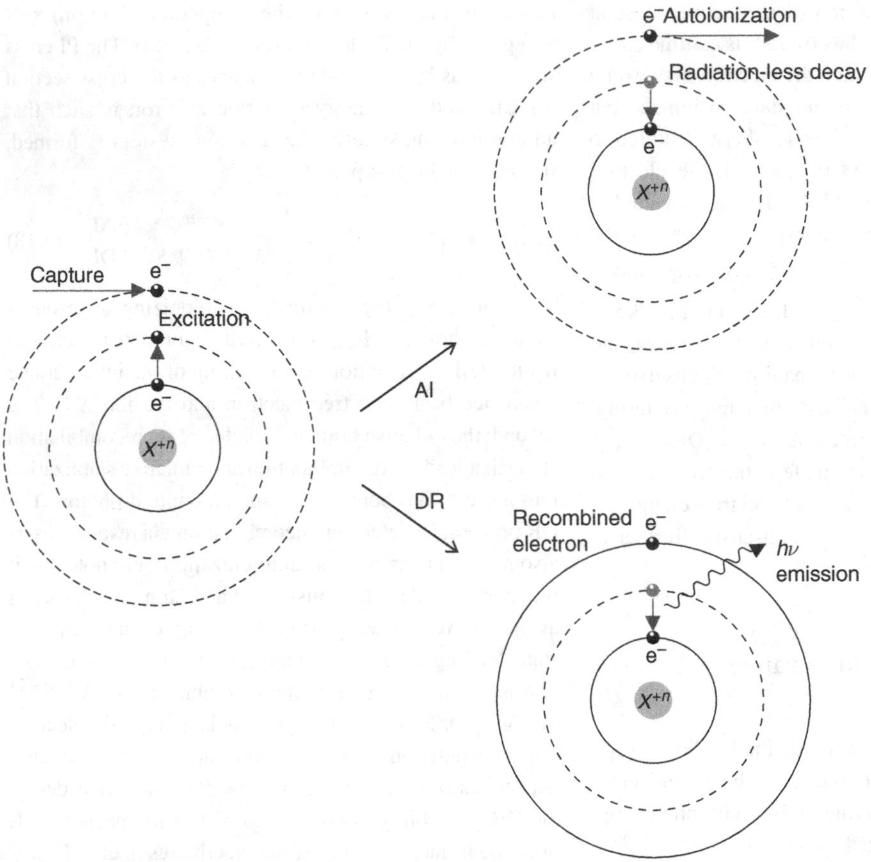

FIGURE 3.6 Autoionization and radiative decay – dielectronic recombination.

bound electron. This is referred to as *electron impact ionization* (EII):

$$e(\epsilon) + X^{+n} \leftrightarrow e(\epsilon') + e(\epsilon'') + X^{+n+1}. \quad (3.19)$$

It is a three-body process, and rather more involved theoretically since the final state has two free electrons moving with different energies ϵ' and ϵ'' in the potential of the residual ion. Provided the energies ϵ, ϵ' are sufficiently low, the two post-collision electrons move in a *correlated* manner. Such two-electron correlations are difficult to treat theoretically. Experimentally, however, the situation is more tractable and a large number of measurements have been carried out for most of the astrophysically abundant atoms and ions. The EII process also has different dynamics and resonant behaviour from EIE or photoionization and (e + ion) recombination, as described in Chapter 5.

The inverse of the EII process is the collision and recombination of two free electrons and an ion, referred to as *three-body recombination*. It has a low rate, except at high plasma densities, where it scales as n_e^2.

3.2.7 Charge exchange recombination and ionization

In circumstances where free charged particles in a plasma source come into contact with neutral gas, ions and neutral atoms (or molecules) may undergo a *charge exchange (CEX)* process through the transfer of an electron. The most common CEX reaction is with neutral hydrogen:

$$X^{+n} + \mathrm{H\,I} \longrightarrow X^{n-1} + \mathrm{p}^{+}. \quad (3.20)$$

An extremely important example is CEX of neutral oxygen and charged particles in the ionosphere of the Earth, the atmospheric layer that lies above the ozone layer. Above the ozone layer, the atmosphere is subject to solar UV radiation, which ionizes the otherwise neutral constituents into ions and electrons. Since oxygen is a prime constituent of Earth's atmosphere, the CEX reaction

$$\mathrm{O}^{+} + \mathrm{H\,I} \leftrightarrow \mathrm{O\,I} + \mathrm{p}^{+} \quad (3.21)$$

proceeds with a fast rate both ways, i.e., recombination and ionization via charge exchange between oxygen ions and hydrogen, and vice versa. In fact the O–H CEX reaction is referred to as *resonant charge exchange*, since both the O I and the H I ionization potentials are nearly the same, 13.6 eV. This implies that each ion reacts with the neutral atom of the other at that resonant energy. The rates of non-resonant CEX reactions are naturally lower. In general, CEX reactions can considerably affect ionization balance (Chapter 7), i.e., the distribution of elements in different ionic states in the plasma. A variety of methods, including elaborate molecular close coupling calculations, has been carried out to compute CEX cross sections (see [20] and references therein).

Charge exchange recombination can sometimes be level-specific, and affect level populations. This is because electron transfer into excited states could subsequently decay to lower states, resulting in specific emission lines. Examples of such reactions are found in laboratory fusion plasmas, where highly charged ions recombine with neutral atoms in the colder peripheral regions of an inertial or magnetic confinement device (such as a tokamak). Since recombination usually occurs into excited states, and spectral formation due to recombination is always seen as emission lines, CEX reactions lead to specific emission spectra, characteristic of the plasma conditions in the laboratory device. A prominent example is the observation of lines of helium-like ions from excited states, caused by CEX between hydrogen-like ions and neutral hydrogen [21, 22].

3.2.8 Proton impact excitation

At first sight, it may appear that positively charged protons are not likely to impact and excite positive ions in the plasma. That is, in fact, generally true, owing to Coulomb repulsion. Also the proton is 1836 times more massive than the electron, and the proton velocity at a given plasma temperature is about 43 times slower than that of the electron. So protons do not effectively compete with electrons in exciting ions. But at sufficiently high temperatures, and consequently high proton velocities, the proton thermal energy kT_p may not only exceed the Coulomb repulsion potential but may also be able to excite low-lying atomic levels with small energy separation ΔE. Such a situation occurs with closely spaced fine-structure levels. In plasmas with $kT_p \gg \Delta E$, proton impact excitation needs to be considered. There are two other points to bear in mind. First, the density of protons in astrophysical plasmas is usually not much less than that of electrons, i.e., $n_p \approx 0.8 n_e$, and second, the proton's much heavier mass ensures that a large number of orbital angular momenta (ℓ) or partial waves contribute to the scattering cross section (Chapter 5). For these reasons proton impact rates at appropriate temperatures may be comparable to, or even exceed, electron impact excitation rates under certain conditions ([23, 24]; see also reviews [25, 26] and recommended data [27]).

3.2.9 Ion–atom collisions

More generally, ions or atoms can collide with another ion or atom and excite low-lying levels, albeit in uncommon circumstances. For instance, excitation of fine structure levels due to hydrogen atom impact is observed in mostly neutral sources, such as cooler regions of nebulae and stars. The resulting spectra are usually in the infrared. An example is the excitation of low-lying C or O atomic energy levels by H I in partially ionized or neutral regions of nebulae with low H I densities comparable to or exceeding electron densities.[3]

One common theme among the last three processes – proton impact excitation, charge exchange, and ion–atom collisions – is that they are generally treated as molecular collision problems, which we do not consider in this text.

3.3 Theoretical approximations

In the past few decades, considerable progress has been made in precise theoretical calculations that may be carried out using sophisticated computational programs. However, large-scale calculations involving many atomic states need extensive computing resources and effort. In this section we describe some of the main theoretical methods that enable the calculation of parameters for atomic processes needed to model atomic spectra in plasma sources. Quantum mechanically, the basic problem is to describe the electron–ion–photon system with (i) a wavefunction expansion and (ii) a Hamiltonian that includes the dominant interactions. The wavefunction expansion is a quantum superpostion of the possible pathways or channels for a given process or reaction. In general, the wavefunction expansion in theoretical calculations is not exact and far from complete. Practical calculations aim at limited though sufficient completness and convergence according to specified criteria. The most powerful methods were developed from atomic collision theory and have been widely employed in a number of computer codes used to generate vast quantities of atomic data. However, the quality of the parameters is not of

[3] Recommended data for many atomic and molecular processes are available from the Research Report Series from the National Institute for Fusion Science of Japan [27].

uniform accuracy. It is important to understand the limitations and validity of different methods theoretically as well as in practice.

Atomic calculations are typically named after the approximation used in obtaining the wavefunction representation of the atomic system. The theoretical methods are described in decreasing order of complexity as follows. First, we elaborate the general form of the wavefunction expansion in atomic collision theory that forms the basis of the so-called *coupled-channel approximation*, usually referred to as the *close coupling (CC) approximation*. That is followed by a detailed description of the powerful *R-matrix method* [28], which is highly efficient in implementing the CC approximation. We also discuss the distorted wave approximation, which does not include channel coupling but is otherwise of considerable utility. Other approximations of more limited validity are the Coulomb–Born and the Coulomb–Bethe approximations used in (e + ion) scattering.

We also outline in some detail an extremely useful theoretical tool, the quantum defect theory, used for the analysis of computed energy levels and cross sections and extension to energies not directly calculated. In addition, the central-field model (cf. Chapters 2 and 4) entails a radial potential without channel coupling in the (e + ion) continuum. But it is often used owing to its simplicity, compared with the Hartree–Fock or the coupled-channel approximations.

3.3.1 Channels in atomic processes

The processes outlined previously show how atomic collisions and radiation are responsible for spectral formation. Before describing some of the methods that deal with these processes, we discuss an essential concept in scattering theory: *channels* that refer to possible pathways in which an atomic process can occur. A particular channel specifies the quantum mechanical states of interacting particles. For example, (e + ion) channels specify a target (or core) ion in some state and an interacting electron with a given energy and angular momentum. Figure 3.7 is a schematic illustration of incoming and outgoing channels into and from an interaction region.[4] For instance, an electron incident on the ion in the channels on the left-hand side may exit via channels on the right-hand side.

[4] In this text we confine ourselves to electron–ion and electron–photon processes. But multi-channel scattering refers to *all* collision processes. A general treatment of scattering theory is given in several excellent textbooks, such as J. R. Taylor [29], R. G. Newton [30], U. Fano and A. R. P. Rau [31] and, of course, Mott and Massey [32].

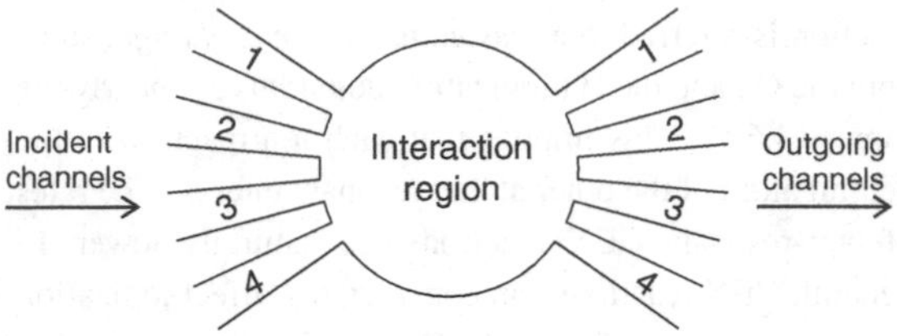

FIGURE 3.7 Collision channels and the interaction region (modelled after [29]).

However, the energy of the incident electron is the factor that determines whether it will scatter from the ion elastically without exchange of energy, or inelastically with transfer of energy to the ion, (de-)exciting it from one state to the other.

For (e + ion) processes, each channel corresponds to the total spin and angular symmetry of the reactants, the electron and the ion. A channel satisfies conservation of (i) energy and (ii) total angular momentum and parity. Consider an atomic process with the channel symmetry $SL\pi$, representing the vector sum of the spin and angular momemtum of the incident electron s, ℓ and the state of the 'target' ion $S_i L_i$. In LS coupling:

$$\mathbf{S} = \mathbf{S_i} + \mathbf{s}, \quad \mathbf{L} = \mathbf{L_i} + \mathbf{l}, \quad \pi = (-1)^{\pi_i + \ell}. \tag{3.22}$$

The parity π of the (e + ion) system is the *scalar* sum of the parity of the ion state ($\pi_i = -1$ for odd, and +1 or 0 for even) and the incident electron angular momentum ℓ. Hence, we label a channel i by $S_i L_i (J_i) \pi_i \, k_i \ell_i (SL(J)\pi)$. We may specify another qunatum number $\mathbf{K}$ in the algebraic pair coupling; representation as: $\boldsymbol{L_i} + \boldsymbol{l} = \boldsymbol{K}$ and $\boldsymbol{K} + \boldsymbol{s} = \boldsymbol{J}$. Conservation of total energy E requires, for all channels,

$$E = E_i + k_i^2, \tag{3.23}$$

loosely writing k_i^2 for channel energies $(k_i a_0)^2$ Ry. In LS coupling, the Hamiltonian is invariant with respect to the total (e + ion) symmetry $SL\pi$. Calculations are carried out for all $SL(J)\pi$ symmetries that might contribute to the problem at hand, such as excitation, photoionization or reccombination (discussed in Chapters 5, 6 and 7, respectively).

While the angular and spin quantum numbers specify the symmetry, the channel wavefunction is the product of the two independent wavefunctions: Φ_i for the core or target ion in state i, and θ_i for a free electron interacting with the target ion. The total (e + ion) wavefunction expansion is a sum over all channels contributing to any particular symmetry. Given an N-electron target ion, the total wavefunction for the $(N + 1)$ electron (e + ion) system in the coupled channel approximation is

$$\Psi_E(\mathrm{e}+\mathrm{ion}) = \mathcal{A}\sum_i^{n_f} \Phi_i(\mathrm{ion})\theta_i\,, \tag{3.24}$$

for n_f channels. A is the antisymmetrization operator for the $(N+1)$-electron system, explicitly accounted for as in the Hartree–Fock approximation (Chapter 2). All wavefunction components, each defining a channel, are coupled because the square of the wavefunction includes not only the probabilities associated with individual channels but also the quantum mechanical interference among them.

We illustrate these concepts for electron scattering with the simplest ion, He^+ or He II. Let us assume that we are interested in only the first three states of the hydrogen-like He II: 1s (^{2}S), 2s (^{2}S) and 2p (^{2}P$^\mathrm{o}$). This problem could refer to excitation of 2s or the 2p level from the ground state 1s. The target ion angular momenta are then $L_i = 0$ and 1. Let us also restrict the orbital angular momentum of the free electron to only three values; $\ell = 0, 1, 2$ (s, p, d orbitals continuum waves). The total (e + ion) spin quantum numbers are $S = 0$ and 1, or $(2S+1) = 1$ and 3 (singlet and triplet), and total angular momenta $L = 0, 1, 2, 3$. Each total $SL\pi$ symmetry for the (e + He$^+$) system has a set of associated channels. For example, the channel wavefunction expansion for ^{3}S (triplet S even) symmetry is

$$\Psi(^3\mathrm{S}, k^2) = \Phi(1\mathrm{s})\theta(k_1\mathrm{s}) + \Phi(2\mathrm{s})\theta(k_2\mathrm{s}) + \Phi(2\mathrm{p})\theta(k_3\mathrm{p}). \tag{3.25}$$

So we have a three-channel expansion for (e + He II) with ^{3}S symmetry. Note the last term on the right where both the target ion and the free electron have odd parities, and hence even (e + ion) symmetry. The total energy is related to the ion energy in each state and the *relative* kinetic energy of the electron as

$$E = k^2 = k_1^2 + E_1 = k_2^2 + E_2 = k_3^2 + E_3, \tag{3.26}$$

where the subscripts 1, 2, 3 refer to the three target states 1s (^{2}S), 2s (^{2}S) and 2p (^{2}P$^\mathrm{o}$). In general, there is a much larger number of target states, many more orbital angular momenta of the free electron (partial waves), and all possible total $SL\pi$ symmetries. In calculations, it is common practice to specify k_1^2 (which is a measure for the 'far-away' kinetic energy of the electron relative to the ground state) rather than E.

The simple example above refers to the three free channels ($n_\mathrm{f} = 3$). Levels 2 and 3 are degenerate in LS coupling: $E(2\mathrm{s}) = E(2\mathrm{p})$. Now consider the case $k_1^2 < E_2 = E_3$: the incident electron has insufficient energy to excite levels 2 and 3. In that case, channel 1 said to be 'open' if $k_1^2 > 0 = E_1$, whereas the other two are 'closed', since $k_1^2 < E_2 = E_3$, and the electron cannot excite the ion to levels 2 or 3 and still escape the attractive Coulomb field of the ion. In other words, a channel is open if the incident electron kinetic energy is equal to or higher than the target level associated with the channel; otherwise it is closed. As discussed above, there are infinite Rydberg series of resonances with levels 2 and 3 as series limits (Fig. 3.4).

3.3.2 Scattering phase shift

Before proceeding into details of theoretical formulations, we introduce the concept of phase shift in the asymptotic form of a wavefunction by an interacting potential. Phase shift will later be generalized to the *scattering matrix* for the many-particle system when we discuss electron–ion scattering.

The radial wave equation for an electron moving freely with angular momentum l and wavenumber k,

$$\left[\frac{\mathrm{d}^2}{\mathrm{d}r^2} - \frac{l(l+1)}{r^2} + k^2\right] f(r) = 0\,, \tag{3.27}$$

is scale-invariant in the absence of any interacting potential. The solutions of this second-order linear ordinary differential equation are the Riccati–Bessel functions $j_l(kr)$:

$$j_l(kr) = \left[\frac{\pi kr}{2}\right]^{1/2} J_{l+1/2}(kr), \tag{3.28}$$

$$\lim_{kr\to\infty} j_l(kr) = C\sin\left(kr - \frac{1}{2}\ell\pi\right)$$
$$= C\left(\mathrm{e}^{-\mathrm{i}\left(kr-\frac{1}{2}\ell\pi\right)} + \mathrm{e}^{+\mathrm{i}\left(kr-\frac{1}{2}\ell\pi\right)}\right), \tag{3.29}$$

where $J_{l+1/2}$ are the ordinary Bessel functions. They have the desired behaviour $\lim_{r\to 0} j_l(kr) \sim r^l$, and asymptotically they are harmonic functions as the centrifugal potential goes to zero, neglecting terms of the order $\mathrm{O}(1/kr)$. The spherical Bessel functions are part and parcel of the plane-wave expansion (with current or flux normalization C),

$$C\,\mathrm{e}^{\mathrm{i}kz} = C\sum_{l=0}^{\infty} \mathrm{i}^l j_l(kr)\mathrm{P}_l(\cos(\vartheta)),$$
$$z = r\cos(\vartheta),\ C = \sqrt{1/(ka_0)}\,, \tag{3.30}$$

which is the starting point when scattering a particle on a target with spherical symmetry. Thus a partial wave phase shift $\pi\ell$ between incoming wave $\exp(-\mathrm{i}kr)$ and outgoing spherical wave indicates no interaction.

3.3.3 Coulomb phase shift

The presence of an interaction potential alters the wavefunction of the free particle by a 'shift' in its phase after, and due to, the interaction. The *phase shift* is introduced with orbital angular momentum l, as the wave moves away towards infinity from the interaction region. In general for any potential that decays *faster* than the Coulomb potential $V(r) \sim 1/r$, as $r \to \infty$, the asymptotic behaviour of the radial function is

$$f(r) \sim \sin\left(kr - \frac{1}{2}l\pi + \delta_l\right), \tag{3.31}$$

where δ_l is the phase shift associated with angular momemtum l.

The radial wave equation for a free electron in a scaled Coulomb potential $V(r) = 2z/r$, and hence itself Ryberg–Bohr-scaled, reads

$$\left[\frac{d^2}{dr^2} - \frac{l(l+1)}{r^2} + \frac{2z}{r} + k^2\right] f(r) = 0, \tag{3.32}$$

and k^2 must now be read as the (asymptotic) energy in Rydbergs, following Section 2.1.2, in particular Eq. 2.27. The asymptotic form of the solutions $f(r)$ involves a phase shift, owing to its long-range behaviour and a logarithmic term as

$$\lim_{r\to\infty} f(r) = \sin\left(kr - \frac{1}{2}l\pi - \frac{z}{k}\ln(2kr) + \sigma_l\right), \tag{3.33}$$

where the *Coulomb phase shift* is

$$\sigma_l = \arg\Gamma\left(l + 1 - i\frac{z}{k}\right). \tag{3.34}$$

The functional form of Eq. 3.33 can be represented by a cosine function as well. Considering both ingoing and outgoing spherical waves in the total wavefunction, the Coulomb functions are defined by

$$\phi^{\pm} = c \pm i s, \tag{3.35}$$

where the functions $s_i(r)$ and $c_i(r)$ have the asymptotic form [33]

$$\begin{aligned} s(\epsilon, l; r) &\sim (\pi k)^{-1/2}\sin(\zeta), \\ c(\epsilon, l; r) &\sim (\pi k)^{-1/2}\cos(\zeta), \end{aligned} \tag{3.36}$$

and the Coulomb phase in Eq. 3.33 reads

$$\zeta = kr - \frac{1}{2}l\pi - \frac{z}{k}\ln(2kr) + \arg\Gamma\left(l + 1 - i\frac{z}{k}\right). \tag{3.37}$$

Hence asymptotically the Coulomb functions are written as

$$\phi^{\pm} \sim (\pi k)^{-1/2}\exp(\pm i\zeta). \tag{3.38}$$

3.4 The close coupling approximation

In the close coupling (CC) approximation, the (e + ion) atomic system is described by a system of $(N+1)$ electrons. The ion core, also termed as the *target or the core*, consists of N-electron states as in Chapter 2: *frozen-cores approximation*; the $(N+1)$th electron is the 'free' electron.[5] The $(N+1)$-electron system may also represent a bound state, either the ground or an excited state, as well as the (e + ion) continuum, such as in the scattering process or after photoionization. We write the continuum solutions in a general form with the formal notation of the CC method as implemented in atomic collision theory. For simplicity, we restrict ourselves to the LS coupling approximation. With full quantum number designations, the target states of the N-electron ion are denoted as

$$\Phi\left(S_i, L_i, M_{S_i} M_{L_i} \middle| r_{N+1}^{-1}\right), \tag{3.39}$$

dependent on all coordinates *except* r_{N+1}. The target state is specified in terms of the angular momenta, spin and parity as S_i, L_i, $M_{S_i} M_{L_i}$ and π_i. The target functions Φ_i are vector-coupled products of the angular and spin variables. For the $(N+1)$-electron system, the quantum numbers

$$SLM_S M_L \pi \tag{3.40}$$

are the total angular, spin and parity of the (e + ion) system; these are conserved during the collision process. The channel index i refers collectively to the quantum numbers

$$S_i L_i \ell_i s_i \longrightarrow SL\pi, \tag{3.41}$$

where ℓ_i is the orbital angular quantum number of the free electron (also referred to as a *partial wave*). The complete CC wavefunction expansion for the $(N+1)$ electrons is

$$\Psi = \mathcal{A}\left[\sum_i^{n_f} \Phi_i(x_1, \ldots x_N)\frac{1}{r}F_i(r)\right] + \sum_j^{n_b} \chi_j(x_1, \ldots x_{N+1})c_j, , \tag{3.42}$$

where the target function Φ_i involves all spatial and spin coordinates x_i of the N target electrons, and $F_i(r)$ is the radial wavefunction θ_i of the colliding electron in Eq. 3.24. The first term on the right-hand side is similar to the one we have already discussed, but where we have introduced the radial variables for the target ion and radial functions F_i(r) for the free electron in

[5] The 'free' refers to freely varying wavefunction according to the Kohn variational principle (Eq. 3.119 and Section 5.1.3). The collisional electron may be in a physical continuum or bound (e + ion) state depending upon the energy boundary condition we impose.

channel i; n_f is the number of free electron channels included in the CC wavefunction expansion. The second term refers to n_b bound electron channel functions that compensate for the restrictions on the total (e + ion) bound state wavefunction. These $(N+1)$-electron bound channel functions may be composed of the *same* one-electron bound orbitals as used to construct the target ion wavefunction Φ. However, in general the χ_j may be any square-integrable functions. Therefore, the χ_j may also be utilized for representing short-range interactions that are always important for the accuracy of the total wavefunction in the inner strong interaction region.

The energies and wavefunctions are obtained from variational solutions

$$H_{N+1}\Psi_E = E\Psi_E \tag{3.43}$$

of the $(N+1)$-electron Hamiltonian

$$H_{N+1} = \sum_{i=1}^{N+1}\left\{-\nabla_i^2 - \frac{2Z}{r_i} + \sum_{j>i}^{N+1}\frac{2}{r_{ij}}\right\}; \tag{3.44}$$

E is the total energy, with E_i the target energy in state i and ϵ_i the energy of the added electron, $E = E_i + \epsilon_i$. The energy $E = \langle\Psi|H|\Psi\rangle$ and, as part of the trial function Ψ_E, the coefficients c_j and free-electron radial functions $F_i(r)$ are variationally determined (as discussed for the distorted wave case with Eq. 3.119). In the R-matrix method implementing the CC approximation, Eq. 3.42 (also Eq. 3.45), $F_i(r)$ is composed of basis functions $u_{i,j}(r)$ (Section 3.5). At positive energies $\epsilon_i = k_i^2 > 0$ the channel is open, and at negative energies $\epsilon_i = -z^2/\nu_i^2 < 0$ it is closed. In the latter case, the radial functions $F_i(r)$ must decay exponentially in the limit $r \to \infty$. Bound-state solutions Ψ_B are normalized to unity. A continuun wavefunction Ψ_F, which is flux-normalized, describes the collisional process with the free electron interacting with the target at positive values of ϵ.

Substituting the expansion (Eq. 3.42) into the $(N+1)$-election trial function Eq. 3.43 yields a set of coupled equations for the radial components $F_i(r)$:

$$\left[\frac{d^2}{dr^2} - \frac{l(l+1)}{r^2} + \frac{2Z}{r} + k_i^2\right]F_i(r) \tag{3.45}$$

$$= 2\sum_j\left[V_{ij}(r)F_j(r) + \int_0^\infty dr' W_{ij}(r,r')F_j(r')\right]$$

$$+ \sum_{nl}\lambda_{i,nl}P_{nl}(r)\delta_{l,l_i},$$

where V_{ij} is the direct potential representing multipole potentials and W_{ij} is the exchange potential. The Lagrange multipliers $\lambda_{i,nl}$ are a result of the orthogonality condition imposed on the continuum functions F_i, such that they are orthogonal to the spectroscopic bound orbitals $P_{nl}(r)$ of the target ion

$$\langle F_i(r)|P_i(nl)\rangle = 0. \tag{3.46}$$

Each channel i is coupled to all other channels j summed over on the right. Note that the integral in the exchange potentials W_{ij} implies that the CC equations are *coupled integro-differential* equations [34]. However, the exchange potential $\int WF$ between the free electron and the bound electrons in the ion is evaluated such that it is negligible outside the envelope of bound electrons. This is one of the essential features of the CC approximation. In the R-matrix method (Section 3.5), configuration space is divided into the so-called inner and outer regions, as shown in Fig. 3.8. In the outer region, i.e., beyond the effective radial extent of the bound electrons in the set $[P_{nl}(r)]$ exchange is neglected, but in the inner region the total (N+1)-electron wavefunction, Eq. 3.42, is explicitly antisymmetrized. The coupled equations (Eq. 3.45) are formally equivalent to Hartree–Fock equations we discussed earlier in Chapter 2, except that there is one electron in the continuum.[6]

3.4.1 Scattering matrices and cross section

Let us first investigate the asymptotic forms of the coupled channel continuum wavefunctions. The coupled differential equations (Eq. 3.45) can be integrated outward subject to the R-matrix boundary conditions at $r = a$ and then fit to an asymptotic expansion. We denote the free electron energy k_i^2 as ϵ_i, and let $n_f = n_o + n_c$, where n_o is the number of open channels and n_c is the number of closed channels. Then

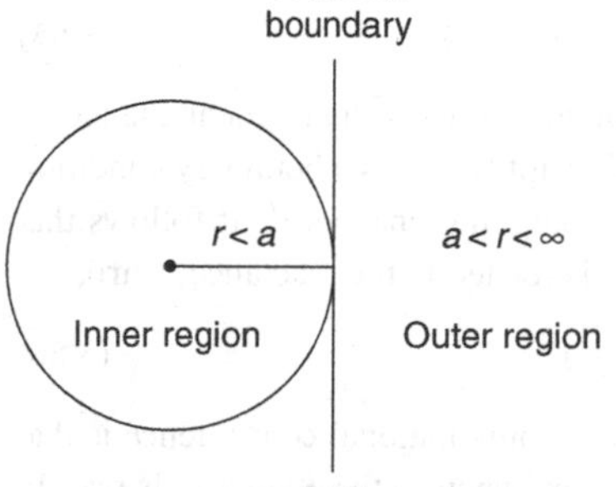

FIGURE 3.8 Configuration space in the coupled channel R-matrix method.

[6] The earliest reference to the close coupling method is by M. J. Seaton, who named it the 'continuum-state Hartree–Fock method' and carried out the first CC calculation for electron impact excitation of O I in 1953 [35].

$$k_i^2 > 0 \qquad i = 1, n_o, \tag{3.47}$$

$$k_i^2 < 0 \qquad i = (n_o + 1), n_f. \tag{3.48}$$

For open channels, the continuum radial functions must refer to both the incident and the outgoing channels, say i and i'. The free electron radial functions are then $F_{ii'}(r)$. As discussed above, the continuum waves have an asymptotic form that may be described in terms of functions related to sine and cosine functions, with Coulomb phase ζ from Eq. 3.37:

$$F_{ii'}(r) \sim s_i(r)A_{ii'} + c_i(r)B_{ii'} \quad (i, i' = 1, n_o), \tag{3.49}$$

$$F_{ii'}(r) \quad \sim \exp(-\zeta r) \sim 0 \quad (i = n_o + 1, n_f), \tag{3.50}$$

where A and B are $n_o \times n_o$ square matrices and

$$s(\epsilon, \ell; r) \sim (\pi k)^{-1/2} \sin(\zeta), \tag{3.51}$$

$$c(\epsilon, \ell; r) \sim (\pi k)^{-1/2} \cos(\zeta). \tag{3.52}$$

Note that for the multi-channel case we now have matrices that define the amplitudes of the continuum waves; these are related to the flux (as in Eq. 3.4), as we shall discuss later in the chapter on electron–ion scattering. For the time being, it is useful to define these matrices in general. First, the *reactance matrix*

$$\boldsymbol{K} = \boldsymbol{B}\boldsymbol{A}^{-1}. \tag{3.53}$$

With Coulomb functions $\phi^{\pm} = c \pm \mathrm{i}s$ and their asymptotic form given in Eqs 3.35 and 3.38 we now define the most important quantity in the collision process:[7] the *scattering matrix* $\boldsymbol{S}$. The $\boldsymbol{S}$-matrix quantifies the flux in the outgoing continuum waves asymptotically as

$$F_{ii'}(r) \sim \frac{1}{2}\left[\phi^{-}(r)\delta(i, i') - \phi^{+}(r)S(i, i')\right]$$

$$\text{for } i, i' = 1, n_o, \tag{3.54}$$

$$F_{ii'} \sim 0 \qquad \text{for} \quad i = (n_o + 1), n_f. \tag{3.55}$$

The first subscript of $F_{ii'}$ denotes the incident channel i, while the second subscript specifies a boundary condition corresponding to the outgoing channel i'. It follows that the scattering matrix is related to the reactance matrix as

$$\boldsymbol{S} = (1 + \mathrm{i}\boldsymbol{K})(1 - \mathrm{i}\boldsymbol{K})^{-1}. \tag{3.56}$$

A fact of considerable computational convenience is that the $\boldsymbol{S}$-matrix is complex, whereas the K-matrix is real. In practice, it is often easier to compute the K-matrix. The matrix $\boldsymbol{S}$ satisfies the *unitarity condition*

$$\boldsymbol{S}^{\dagger}\boldsymbol{S} = 1 \rightarrow \boldsymbol{S} = \boldsymbol{S}^{\mathsf{T}}, \; \boldsymbol{S}^{*}\boldsymbol{S} = 1, \tag{3.57}$$

where the supercript T denotes the transposed matrix. The mathematical condition of unitarity (see also Chapter 5) is related to the physical requirement of conservation of incident and outgoing fluxes in the scattering process. For the single channel case, it also follows that

$$\boldsymbol{S} = \frac{1 + \mathrm{i}\boldsymbol{K}}{1 - \mathrm{i}\boldsymbol{K}} = \mathrm{e}^{2\mathrm{i}\delta}, \tag{3.58}$$

and the real part of the $\boldsymbol{S}$-matrix or the K-matrix is related to the quantum defect μ as

$$K = \tan(\pi\mu), \qquad \delta = \pi\mu. \tag{3.59}$$

The elastic scattering cross section σ in terms of the phase shift is given by the well-known formula (discussed in Chapter 5)

$$\sigma(\vartheta) = \left| \frac{1}{2ik} \sum_{l=0}^{\infty} (2l+1)(\mathrm{e}^{2\mathrm{i}\delta_l} - 1)\mathrm{P}_l(\cos\vartheta) \right|^2, \tag{3.60}$$

where k is the *wavenumber* which we employed in Eq. 3.32 for dimensionless (ka_0), the *number of waves* along a distance of one Bohr radius.[8]

To summarize some of the salient features of the CC approximation:

- The coupled channel approach is, in principle, equivalent to quantum superposition of (e + ion) pathways for electron and photon interactions. The total wavefunction for the (e + ion) system is a sum over all individual channel wavefunctions; the square of the wavefunction then yields the coupling.
- The *open* and *closed* channels are characterized according to the kinetic energy of the free electron relative to the target ion.
- Resonances are a particular manifestation of channel coupling: they arise from coupling among the open channels with continuum wavefunctions, and closed channels with exponentially decaying wavefunctions.
- The presence of resonances in cross sections makes them highly energy dependent, necessitating CC calculations for a large number of energies. An important feature of CC calculations is that resonant structures in the atomic processes are included naturally. The complex resonant structures in collisional excitation, photoionization and recombination result from couplings between continuum channels that are open ($k_i^2 > 0$)

[7] The physical description of these quantities is given in detail in Chapter 5 in the context of electron–ion collisions.

[8] It also means full circle in physics: to re-composition from de-composition, when the colliding electron – a plane wave (Eq. 3.30), later modified for the long-range Coulomb potential – was expanded in spherical partial waves via Legendre polynomials. It implies summing over the various expansions, the finite range of partial waves ℓ in particular.

and bound channels that are closed ($k_i^2 < 0$) and form at electron energies k_i^2 corresponding to the Rydberg series of states converging onto the target thresholds.

- At negative electron energies all channels are closed. Bound states of the (e + ion) $(N + 1)$-electron system occur at eigenvalues of the $(N + 1)$-electron Hamiltonian. For a given symmetry $SL\pi$ all closed channels represent bound states, and are exactly equivalent to electronic configurations. In principle, the CC method for bound states is equivalent to the multi-configuration Hartree–Fock method with configuration interaction included via coupled closed channel functions. Since generally there are many closed channels for each $SL\pi$, a CC calculation amounts to a large configuration interaction atomic structure calculation for all bound states of each symmetry. This powerful technique enables the calculation of a large number of bound state wavefunctions, transition proabilities and photoionization cross sections for arbitrarily high states of excitation along a Rydberg series, as done in the work on the Opacity Project [36, 37] and the Iron Project [38].

3.5 The R-matrix method

Based on the general concepts of the CC approximation, we outline the *R-matrix method* as developed by P. G. Burke and collaborators [39, 28]. The R-matrix method forms the basis for a powerful set of computer codes for a comprehensive treatment of the atomic processes in Fig. 3.5. Although the method was first employed in nuclear physics [40, 41, 42, 43], its physical aspects render it highly suitable for atomic (and molecular) problems. The basic idea is to consider the projectile–target interaction space (or configuration space) as two distinct regions, as in Fig. 3.8: an *inner region*, where the projectile is close to the target and interactions are strong, as in the CC approximation already described. In the *outer region*, the interactions are weak (particularly exchanges) or long-range, and may be treated via asymptotic approximations. Whereas in the inner region all close interactions must be fully accounted for, in the outer region the potentials, and hence the wavefunctions, have simpler and generally well-known forms, such as in terms of the Coulomb functions. The (e + ion) continuum wavefunction can be expanded in the inner region at any energy in terms of a basis set of square-integrable eigenfunctions. The R-matrix is defined as the inverse of the logarithmic derivative of the wavefunction at the boundary between the inner and outer regions. Then the (e + ion) scattering wavefunctions can be computed by matching suitably asymptotic functions at the R-matrix boundary (or the channel radius). In addition, bound state wavefunctions can be obtained by matching to exponentially decaying functions asymptotically in all channels.

3.5.1 Single-channel problem

Prior to multi-channel generalization, a simple sketch of the R-matrix method can be drawn for a single-channel scattering [44]. First, consider s-wave ($\ell = 0$) scattering with a short-range central potential described by

$$\left(\frac{d^2}{dr^2} + V(r) + k^2\right) u(r) = 0. \tag{3.61}$$

Suppose $V(r) = 0$ for $r > a$. Then the radius $r \leq a$ defines the aforementioned inner region and $r > a$ defines the outer region. The radial functions then have asymptotic form

$$u(r) = \sin(kr) + \boldsymbol{K}\cos(kr), \qquad (r \geq a). \tag{3.62}$$

Now we invoke one of the main points of the R-matrix method: the wavefunction $u(r)$ in the inner region $r < a$ is expanded in terms of a complete orthogonal basis set of eigenfunctions $[u_j(r)]$ for a given value ℓ,

$$u_l(r) = \sum_j^{\infty} a_j u_j(r), \qquad (0 \leq r \leq a) \tag{3.63}$$

$$\int_0^a u_j(r) u_{j'}(r) \mathrm{d}r = \delta_{jj'}, \tag{3.64}$$

obtained in a suitably chosen central-field potential $V(\ell, r)$

$$\left(\frac{d^2}{dr^2} + V(lr) + k_j^2\right) u_j(r) = 0\,. \tag{3.65}$$

As we shall see later, it is necessary in practice to constrain the infinite expansion in Eq. 3.63 to a finite number $j \leq N_c$. The number of nodes in $u_j(r)$ increases with j. We need functions $u_l(r)$ regular at $r = 0$, i.e., $\lim_{r\to 0} = r^{\ell+1}$; at $r = a$ we impose the logarithmic boundary condition

$$a\,\frac{u'(r=a)}{u(r=a)} = b, \tag{3.66}$$

for the slope $u' = du/dr$, and b is a constant usually taken to be simply zero. The expansion Eq. 3.63 converges uniformly for all values of the logarithmic derivative b except at $r = a$. N_c must be large enough to support $u(r)$, hence increasing roughly linearly with wavenumber k, which relates to the collision energy $(ka_0)^2$ Ry, or k^2 in atomic units (see Chapter 4). The expansion coefficients can be derived as in the following exercise.

Exercise 3.1 *Show that the expansion coefficients a_j are*

$$a_j = \frac{1}{a}\frac{u_j(a)}{\left(k_j^2 - k^2\right)}[au'(a) - bu(a)]. \tag{3.67}$$

Hint: premultiply Eq. 3.61 by $u_j(r)$ and Eq. 3.65 by $u(r)$, and integrate both equations in the range $r = [0, a]$. Subtraction of the two equations yields an equation that may be evaluated using Green's theorem and the specified boundary conditions on $u_j(r)$.

Once the expansion coefficients a_j are determined, we may substitute in Eq. 3.63 to evaluate the radial functions $u(r)$ everywhere in the inner region $0 \leq r \leq a$. The R-matrix can now be defined in terms of the functional values at the boundary $r = a$ as

$$R = \frac{1}{a}\sum_{j=1}^{\infty}\frac{[u_j(a)]^2}{k_j^2 - k^2}. \tag{3.68}$$

The R-matrix relates the amplitude of $u(r)$ to its derivative on the boundary such that

$$R = \frac{u(a)}{[au'(a) - bu(a)]}. \tag{3.69}$$

Note that with the choice $b = 0$, we have the R-matrix as the simple logarithmic derivative on the boundary. But computationally the most significant point is that in the R-matrix the energy enters as k^2 in Eq. 3.68 *independently* of the inner region expansion. Therefore, once the R-matrix is evaluated, other energy-dependent quantities may be obtained in the outer region at any energy provided their functional forms are known. The outer region functions, of course, have known asymptotic wavefunctions dependent explcitly on the energy, as in Eq. 3.62.

It follows that we can obtain quantities such as the K-matrix in terms of the R-matrix by *matching* the inner region expansion to the asymptotic form Eq. 3.62 at the boundary. For the single channel case this is

$$K = \frac{-sinka + R(kacoska - bsinka)}{coska + R(kasinka + bcoska)}. \tag{3.70}$$

This relation demonstrates, without loss of generality, how the physical reactance matrix K can be obtained in terms of the R-matrix basis functions. The single channel model can be generalized in a straightforward manner to non-zero angular momenta by adding the centrifugal potential $-\ell(\ell+1)/r^2$ to Eq. 3.61. Recall that the radial functions are then spherical Bessel functions with simple asymptotic forms

$$f_\ell(kr) = j_{\ell+1/2}(kr) \sim_{r\to\infty} \sin(kr - \ell\pi/2), \tag{3.71}$$

$$g_l(kr) = j_{-\ell-1/2}(kr) \sim_{r\to\infty} \cos(kr - \ell\pi/2). \tag{3.72}$$

The model can be further extended for a long-range Coulomb potential in the external region, where the solutions are the regular and irregular Coulomb functions discussed earlier (Eq. 3.35). To complete the single-channel scattering picture we can write the radial function in terms of the asymptotic ingoing and outgoing spherical waves

$$f(r) \sim \exp(-ikr) - S_\ell \exp(ikr), \tag{3.73}$$

for particle scattering with angular momemtum ℓ in a short-range potential. The single-channel scattering matrix is

$$S_\ell = (1 + iK_\ell)(1 - iK_\ell)^{-1}. \tag{3.74}$$

3.5.2 Multi-channel problem

The generalization to the coupled multi-channel case entails more than a few modifications of the single-channel problem. First, we have the target, which is an ion with internal structure that comprises of several, if not many, excited states. The general form of the potential is, therefore, not given by a central field, but by explicit computation of the potentials in all states of the ion of interest, or included in the (e + ion) problem. The target ion must therefore be described accurately with suitably high precision. Second, the R-matrix, and hence the physical parameters such as the K-matrix, cannot be obtained in closed analytic form as in the single-channel case. The expansion coefficients for the channel radial functions in the inner region must be obtained on solving the coupled integro-differential equations (Eq. 3.45) for the (e + ion) system, with an $(N+1)$-electron Hamiltonian containing all necessary interaction terms. Third, the matching procedure at the R-matrix boundary involves inward integration of asymptotic functions computed in the outer region.

We begin with the division of configuration space described above in Fig. 3.8: an inner region circumscribed by the radial extent of the ion $r < r_a$. In the outer region, exchange is neglected; the interaction potentials $V_{ii'}(r)$ are the Coulomb and other long-range multipole potentials. The R-matrix boundary a is set such that *all* target orbitals have decayed to $|P_{nl}(a)|/\max(|P_{nl}|) < \delta$, to some pre-specified small value δ of say 10^{-3}, i.e., the radial functions decay down to one-thousandth of their peak value.

The $(N+1)$-electron system behaves in a similar way to a bound state inside the boundary and is represented by a configuration interaction expansion analogous to that used in bound state calculations. The coupled integro-differential equations (Eq. 3.45) are solved in this region. In the outer region, the electron moves in the long-range multipole potentials of the target and the wavefunction is represented in terms of Coulomb functions. The target or the core state wavefunctions are expanded in terms of configurations $\Phi_i = \sum_i a_i v_i$ in Eq. 3.42. The configurations are built up from one-electron orbitals coupled together to give an antisymmetric function (Slater determinant), and configuration interaction is taken into account as in the standard Hartree–Fock method (Chapter 2). Each orbital is a product of a radial function, a spherical harmonic and a spin function

$$v = \chi_{m_s}(\sigma) \mathrm{Y}_{\ell m_\ell}(\vartheta, \varphi)(1/r) P_{n\ell}(r). \tag{3.75}$$

Atomic structure codes such as CIV3 [45] or SUPERSTRUCTURE [10] can be used to obtain the set of one-electron orbital wavefunctions $P_{n\ell}$ that represent the target states Φ employed in Eq. 3.42.

3.5.3 Inner region

As for the single-channel case, the basic idea is that in the inner region the total wavefunction can be expanded in terms of basis functions, which now involve the target ion and a set of continuum-type orbitals representing the free electron [46]. The $(N+1)$-electron wavefunction expansion is

$$\begin{aligned} \Psi_k(x_1, \ldots, x_N, x) &= \mathcal{A} \sum_{ij} c_{ijk} \Phi_i(x_1, \ldots, x_N) \frac{1}{r} u_{ij}(x) \\ &+ \sum_j d_{jk} \chi_j(x_1, \ldots, x_{N+1}). \end{aligned} \tag{3.76}$$

The N-electron target eigenfuctions Φ_i are expanded in terms of configuration interaction (CI) basis functions ϕ_k (cf. Chapter 2) as

$$\Phi_i(x_1, \ldots, x_N) = \sum_k b_{ik} \phi_k(x_1, \ldots, x_N), \tag{3.77}$$

where $x_i \equiv \mathbf{r}_i s_i$ represents the spatial coordinate and spin of the ith electron. Generalizing the single-channel case (Eq. 3.63), the u_{ij} are the free-channel radial functions, expanded in terms of continuum basis functions of a (computationally mangeable) set of angular momenta. Comparing with the general form of the CC wavefunction, Eq. 3.42, the coefficients c_{ijk} and d_{jk} are obtained on diagonalization of the basis functions Ψ_k with respect to the $(N+1)$-electron Hamiltonian

$$\left(\Psi_k | H(N+1) | \Psi'_k\right) = E_k \delta_{kk'}, \tag{3.78}$$

where the parentheses replace bras and kets so as to indicate indicate spatial integration over the inner region only, in the range from $r = 0$ up to the R-matrix boundary, $r = a$. The continuum orbitals $u_{ij}(r)$ in Eq. 3.76, for each angular momentum l_i in the basis set, are obtained in a suitable model central potential $V(r)$, i.e.,

$$\begin{aligned} &\left[\frac{\mathrm{d}^2}{\mathrm{d}r^2} - \frac{l_i(l_i+1)}{r^2} - V(r) + k_{ij}^2\right] u_{ij}(r) \\ &\quad = \sum_n \lambda_{ijn} P_{nl_i}(r). \end{aligned} \tag{3.79}$$

The R-matrix boundary conditions are

$$u_{ij}(0) = 0, \qquad \frac{u'}{u} = b \quad \left(u' = \frac{\mathrm{d}u}{\mathrm{d}r}\right). \tag{3.80}$$

The role of the Lagrange multipliers on the right-hand side of Eq. 3.79 is to ensure that the continuum orbitals of a given angular momentum ℓ_i are orthogonal to the bound one-electron orbitals $P_{n\ell_i}$,

$$(u_{ij} | P_{n\ell_i}) = 0, \tag{3.81}$$

integrated over the finite range up to $r = a$. Since the bound one-electron functions $P_{n\ell_i}$ are also orthogonal to one another, all of the bound and continuum orbitals taken together

$$\left[P_{n_{\min}\ell_i}, \ldots P_{n_{\max}\ell_i};\; u_{i,1}, u_{i,2}, \ldots u_{i,n}\right]$$

form a complete set in the range $r = [0,a]$ for each ℓ_i; n is the total number of continuum orbitals. This basis set, and hence the set ψ_k, is *independent* of the total energy E of the $(N+1)$-electron system. The energy dependence arises through the coefficients A_{E_k}, which are obtained as follows. The total wavefunction expansion in the inner region is

$$\Psi_E = \sum_k A_{Ek} \Psi_k. \tag{3.82}$$

Though themselves not physical, the functions Ψ_k provide a basis for the expansion of the physical multi-channel wavefunction Ψ_E in terms of the target eigenfunctions Φ_i and radial channel functions $F_{ik}(r)$. In the inner region (sometimes denoted by subscript 'I') the total wavefunctions Ψ_k are diagonalized and normalized, such that

$$\left(\Psi_k | H^{N+1} | \Psi'_k\right)_{\mathrm{I}} = E_k \delta_{kk'}. \tag{3.83}$$

In the inner region we have

$$(\Psi_E | \Psi_E)_{\mathrm{I}} = \sum_k |A_{kE}|^2, \tag{3.84}$$

and from Eq. 3.82 the coefficients

$$A_{kE} = (\Psi_E | \Psi_k). \tag{3.85}$$

The radial functions at energy E are expanded in terms of their basis functions as

$$F_{iE} = \sum_k u_{ik}(r) A_{kE}. \tag{3.86}$$

It follows that

$$(H - E_k)(\Psi_k|\Psi_E) - (\Psi_k|(H - E)|\Psi_E) = 0, \tag{3.87}$$

which on integration by parts gives

$$(E - E_k)(\Psi_k|\Psi_E) = \sum \left(u'_{ik}(a) F_{iE}(a) - u_{ik} F' i E(a)\right), \tag{3.88}$$

as before scaled in Rydberg units (minus a good number of additional factors of two from the identity 2 Ry = H, Eq. 1.22, if one measures energy in 'atomic' units such as in [46]). From the equation above [33] we have

$$A_{kE} = (E_k - E)^{-1} \sum_i u_{ik} \left(F'_{iE}(a) - b F_{i'E}(a)\right) \tag{3.89}$$

and, using Eq. 3.86,

$$F_{iE}(a) = \sum_{i'} R_{ii'}(E) \left(F'_{i'E}(a) - \frac{b}{a} F_{i'E}(a)\right), \tag{3.90}$$

where the R-matrix has been introduced with elements

$$R_{ii'}(E) = \sum_k u_{ik}(a)(E_k - E)^{-1} u_{i'k}(a). \tag{3.91}$$

There is another practical difficulty not thus far addressed: how to alleviate the restriction on the basis functions $u(r)$, which end at the boundary $r = a$ with the same slope, usually horizonal – while their true physical compositions $F(r)$ do so only at discrete collision energies in quasi-periodical sequence. Hence, computed quantities such as cross sections, while correct at those discrete energies (within the trial function expansion), show sinusoidal behaviour in between in cases expected to be smooth, with amplitudes growing more wildly as one reduces the number N_c of basis functions for a given value of ℓ, as expected from Fourier experience. To choose N_c much bigger than required to support the physical solutions, say with about twice as many nodes, is not economical and can be numerically detrimental. But one can pursue the idea as follows, devised by Buttle [47]. Since the contribution to the R-matrix from N_c to ∞ is from high-lying terms in energy, they may be evaluated assuming a model problem that neglects multipole potentials (including exchange) with *uncoupled channels*. These additional terms are then important only for the diagonal terms in the R-matrix. The correction to the diagonal elements R^c_{ii} (with channel index i) is given by the *Buttle correction*

$$\begin{aligned} R^c_{ii}\left(N_c, k_i^2\right) &\approx \frac{1}{a} \sum_{k=N_c+1}^{\infty} \frac{|u_{ik}(a)|^2}{E_{ik} - E_i} \\ &= \left[\frac{a}{F_i^0(a)} \left(\frac{dF_i^0}{dr}\right)_{r=a} - b\right]^{-1} - \frac{1}{a} \sum_{k=1}^{N_c} \frac{|u_{ik}(a)|^2}{E_{ik} - E_i}. \end{aligned} \tag{3.92}$$

Hence, the Buttle corrected R-matrix is given by

$$R_{ii'} = \frac{1}{a} \sum_k \frac{u_i(a) u_{i'}(a)}{E_k - E} + R^c_{ii'}\left(N_c, k_i^2\right) \delta_{ii'}. \tag{3.93}$$

We again note that the R-matrix is obtained by diagonalizing H^{N+1} *once* for each symmetry, specified by a set of conserved quantum numbers and parity of the electron–target system (Eq. 3.41). Thereafter the wavefunctions in the outer region and all needed physical quantities may be obtained. Generalizing for all channels inside the sphere we can write the radial wavefunction (Eq. 3.90) in matrix form, introducing the compact notation

$$\boldsymbol{F} = a\, \boldsymbol{R} \cdot \boldsymbol{F'} - b\, \boldsymbol{R} \cdot \boldsymbol{F}, \qquad (r \leq a), \tag{3.94}$$

where each boldface variable is a matrix whose size is given by the number of channels; the 'dot' indicates matrix multiplication. The logarithmic derivative of the radial wavefunction of the collisional electron on the boundary is given by $F(a)$, and is to be matched across the boundary to the external or outer region. The 'surface' amplitudes $u_{ik}(a)$ and the poles E_k of the basis functions Ψ_k are obtained directly.

3.5.4 Outer region

By choice, the outer region (Fig. 3.8) is outside the charge distribution of the ion, and exchange between the free electron and the target ion is neglected. In the outer region the $(N + 1)$-electron-coupled integro-differential equations (Eq. 3.45) reduce to coupled differential equations, without the W-integral that represents exchange terms. Therefore, the channel radial functions $F_i(r)$ in the outer region are given by

$$\left[\frac{d^2}{dr^2} - \frac{l_i(l_i + 1)}{r^2} + \frac{2z}{r} + k_i^2\right] F_i(r) - \sum_j V_{ij} F_j(r) = 0,$$

$$i = 1, n_f, \; (r \geq a), \tag{3.95}$$

where n_f is the number of free channels retained in the expansion and $V_{ij}(r)$ are multipole potentials due to electron–electron interaction of the colliding electron with the target electrons. The elements V_{ij} are given in terms of the target states Φ_i and Φ_j as

$$V_{ij}(r) = \left\langle \Phi_i \left| \sum_{m=1}^{N} \frac{1}{r_{m,N+1}} \right| \Phi_j \right\rangle . \quad (3.96)$$

Expanding the two-electron operator in terms of spherical harmonics,

$$V_{ij}(r) = \left\langle \Phi \left| \sum_{i=1} \sum_{\lambda\mu} \frac{4\pi}{2\lambda+1} \mathrm{Y}^{\mu}_{\lambda}(\hat{r}_i) \mathrm{Y}^{\mu *}_{\lambda}(\hat{r}_{N+1}) \frac{1}{r^{\lambda+1}} \right| \Phi \right\rangle$$

$$= \sum_{\lambda} \frac{C^{\lambda}_{ij}}{r^{\lambda+1}} . \quad (3.97)$$

The long-range potential coefficients C^{λ}_{ij} are defined in terms of the Legendre polynomials as

$$C^{\lambda}_{ij} = \left\langle \Phi_i \left| \sum_{m=1}^{N} r^{\lambda}_{m} \mathrm{P}_{\lambda}(\cos\vartheta_{m,N+1}) \right| \Phi_j \right\rangle . \quad (3.98)$$

In practice, we include multipole contributions of $\lambda = 1$ and 2 (dipole and quadrupole). From the fact that the target wavefunctions Φ_i (Eq. 3.77) are negligibly small for $r \geq a$, it follows that

$$|V_{ij}(r) \ll \frac{2z}{r}, \qquad \text{for } r \geq a . \quad (3.99)$$

The multipole potentials are, therefore, small perturbations. The outer-region-coupled differential equations then become

$$\left(\frac{d^2}{dr^2} - \frac{l_i(l_i+1)}{r^2} + \frac{2z}{r} + k_i^2 \right) F_i(r)$$

$$= \sum_{\lambda} \sum_{j=1}^{n_f} \frac{C^{\lambda}_{ij}}{r^{\lambda+1}} F_j(r). \quad (3.100)$$

These solutions from the external region are matched to the inner region at the R-matrix boundary $r = a$, and propogated outward to fit to asymptotic forms (Eqs. 3.49 and 3.50), where we have expressed the open-channel solutions in terms of the reactance matrix $\boldsymbol{K}$ with indices running over all open channels n_o, and closed-channel solutions in terms of an exponentially decaying factor.

3.5.5 Open channels

The open-channel solutions are required for electron-ion scattering (Chapter 5), and for photoionization and (e + ion) recombination (Chapters 6 and 7). To derive formal expressions, it is useful to adopt the compact matrix notation introduced earlier (Eq. 3.94). The open-channel wavefunction in the outer region is

$$\boldsymbol{F} = s + c\boldsymbol{K}, \quad (r \geq a). \quad (3.101)$$

Substituting it into the R-matrix boundary condition at $r = a$ and eliminating $\boldsymbol{F}$, we get

$$s + c\boldsymbol{K} = a\boldsymbol{R}(s' + c'\boldsymbol{K}) - b\boldsymbol{R}(s + c\boldsymbol{K}). \quad (3.102)$$

Exercise 3.2 *Derive Eqs 3.101 and 3.102 and matrices* **A, B** *that represent the asymptotic form of the reactance matrix* $\mathbf{K} = \mathbf{B}\mathbf{A}^{-1}$ *in Eq. 3.53.*

Solving for $\boldsymbol{K}$ (symmetric matrix),

$$\boldsymbol{K} = \boldsymbol{A}^{-1}\boldsymbol{B} = [c - \boldsymbol{R}(ac' - bc)]^{-1}[-s + \boldsymbol{R}(as' - bs)], \quad (3.103)$$

where

$$\mathbf{A} = c - \mathbf{R}(ac' - bc), \quad \mathbf{B} = -s + \mathbf{R}(as' - bs). \quad (3.104)$$

As noted before for the single-channel case (Eq. 3.70), once the multi-channel R has been calculated, K can be obtained directly. And from K we obtain the $\boldsymbol{S}$-matrix and hence the cross sections (Chapter 5). The R-matrix is determined by a single diagonalization in the inner region, and the outer region wavefunctions are matched at the R-matrix boundary, to yield cross sections at all energies. The multi-channel scattering problem in general has some channels open and some closed. The interaction between the open and closed channels gives rise to resonances, which we shall study in later chapters.

3.5.6 Bound states: all channels closed

In the CC approximation, the (e + ion) wavefunction may correspond to bound states if the total $E < 0$, and therefore all n_f free channels are closed. Bound states occur at negative eigenenrgies of the $(N+1)$-electron Hamiltonian (Eq. 3.43)

$$(H_{N+1} - E)\Psi_{N+1}(\mathrm{e} + \mathrm{ion}) = 0. \quad (3.105)$$

We expand F_i in terms of closed channel functions (Eqs. 3.76 and 3.77)

$$c_{ij}(r) \sum_{r \to \infty} \exp(-\phi_i)\delta_{ij}, \qquad (i, j = 1, n_o + 1, n_f). \quad (3.106)$$

Putting $x_j \equiv \exp(-\phi_j)$,

$$F_i(r) = \sum_{j=1}^{n} c_{ij} x_j, \quad (r \geq a), \quad (3.107)$$

which in matrix notation simply appears as $\boldsymbol{F} = \mathbf{cx}$. As for open channels, the matching involves equating the coefficients of the outer region expansion above to the inner region radial functions in terms of the R-matrix basis. Adopting matrix notation, and putting F_i in the R-matrix boundary condition at $r = a$, we get

$$\boldsymbol{F} = \mathbf{cx} = a\mathbf{Rc}'\mathbf{x} - b\mathbf{Rcx}, \tag{3.108}$$

which is a set of n_f homogenous equations for determining x_j. Writing it as

$$\mathbf{Bx} = \left[\mathbf{c} - a\mathbf{Rc}' - b\mathbf{Rc}\right]\mathbf{x} = 0, \tag{3.109}$$

the components x_j of the matrix columns of $\mathbf{x}$ are obtained from $\sum_{j=1}^{n_f} B_{ij}x_j = 0$. These equations have only non-trivial solutions at the negative energy eigenvalues corresponding to bound states of the electron–target system. In matrix form: $\mathbf{Bx} = \mathbf{0}$, and therefore the non-trivial solutions occur at eigenvalues that satisfy the determinantal condition

$$|\mathbf{B}| = \mathbf{0}. \tag{3.110}$$

This equation is solved iteratively and the bound state energies and wavefunctions of the $(N+1)$-electron system are obtained [48, 33]. To treat radiative processes with the R-matrix method we require a large number of excited bound state energies and wavefunctions to compute bound–bound radiative transition probabilities, bound–free photoionization cross sections, and free–bound (e + ion) recombination cross sections and rate coefficients, as discussed in Chapters 4, 6 and 7, respectively.

3.5.7 The R-matrix codes

The R-matrix framework outlined in the preceding sections enables us to compute the bound and free atomic wavefunctions and the basic quantities, (i) the scattering matrix and (ii) the dipole (or multipole) matrix elements. In subsequent chapters, we discuss their usage in the treatment of atomic interactions, such as electron impact excitation (Chapter 4), bound–bound radiative transitions (Chapter 5), and bound–free photoionization (Chapter 6) together with the inverse process free–bound recombination (Chapter 7). But prior to that, in the next section, we briefly describe the computational methodology schematically illustrated in Fig. 3.9.

Although other computer program packages were developed to implement the CC method, most notably the IMPACT package from University College London [49], the R-matrix method proved to be the most efficient for large-scale calculations. As mentioned earlier, The R-matrix package of codes has been used for the Opacity Project [37] and the Iron Project [38] for radiative and collisional CC calculations for nearly all atomic systems of astrophysical interest.[9] Although not apparent from the simplified discussion in preceding sections, the R-matrix codes in fact comprise at least three sets of codes that implement the CC method to account for physical effects with approximations of varying complexity. In addition, technical improvements, such as massive parallelization on high-performance supercomputing platforms, constitute different branches of the large suite of computer programs (e.g. [50]). Also, a variety of schemes exist for simplifying or approximating the full R-matrix calculations through *frame transformations*, such as the *intermediate coupling frame transformation* (ICFT) applied to highly charged ions (e.g. [51]).

The various branches of the R-matrix computer packages (Fig 3.9) entail (i) pure LS coupling, (ii) LS coupling with algebraic transformation and limited account of relativistic effects, (iii) relativistic effects in the Breit–Pauli R-matrix (BPRM) approximation including (a) only the one-body terms in the (e + ion) Hamiltonian (Chapter 2), and (b) 'full' Breit–Pauli calculations, including nearly all terms of the Breit interaction (Eq. 3.111), (iv) a new version of R-matrix II codes with an enhanced algebraic treatment of *all* terms LS in the target configurations [52], (v) Dirac R-matrix (DARC) methodology with four-component wavefunctions [53, 54].

The matrix diagonalization of the $(N + 1)$-electron Hamiltonian that yields the R-matrix is related to atomic parameters, i.e., scattering matrices and dipole matrix elements, from which the excitation and photoionization cross sections and radiative transition probabilities may be obtained. The (e + ion) wavefunctions $\Psi_F(SL\pi; E)$ and $\Psi_B(SL\pi; E')$ may be calculated for both $E > 0$ and $E < 0$; the former case corresponds to electron scattering, i.e., *continuum states* and the latter to *bound states* of the system. The functions χ_j form a set of $\mathcal{L}^2$-integrable $(N+1)$-electron correlation functions required by orthogonality constraints (the radial functions Eq. 3.86 of the colliding electrons); they may also represent a pseudostate expansion. For moderate charge number, say $Z < 30$, the BPRM method entails intermediate coupling [55], with a pair-coupling representation $S_iL_l(J_i)l_i(K_i)s_i(J\pi)$. As the individual states S_iL_i split into the fine structure levels J_i, the number of channels becomes several times larger than the corresponding LS coupling case, but the computational problem may be more than an order of magnitude larger.

9 A detailed description of these two projects, databases and their astrophysical applications is given on the authors' websites: www.astronomy.ohio-state.edu/~pradhan and www.astronomy.ohio-state.edu/~nahar.

In Fig. 3.9 the LS or the BP (intermediate coupling) calculations are based on accurate configuration interaction representation of the N-electron target states by two atomic structure codes, SUPERSTRUCTURE [10]) and CIV3 [45]. The first two R-matrix codes, STG1 and STG2, are then employed to generate multipole integrals and algebraic coefficients and set up the $(N+1)$-electron Hamiltonian corresponding to the coupled integro-differential equations. The Hamiltonian is diagonalized in STGH; in the BP calculations the diagonalization is preceded by LSJ recoupling in RECUPD, as shown in the left branch of Fig. 3.9. The R-matrix basis set of functions and the dipole matrix elements so produced are then input into STGB for bound state wavefunctions, STGF for continuum wavefunctions, STGBB for radiative transition probabilities, and STGBF for photoionization cross sections. In addition, STGF[J] is used to obtain collision strengths for electron impact excitation in LS or intermediate coupling and fine structure transitions.

3.5.7.1 Full Breit–Pauli R-matrix (BPRM) method

Until recently the Breit–Pauli version of R-matrix codes include only the one-body operators – mass–velocity, Darwin, and spin–orbit – had been incorporated [46, 56]. This is inadequate even for some low-Z ions, and the two-body 'magnetic' terms must be included for high accuracy, as discussed in Section 2.13.2. Further work has now been concluded to include all eight operators in the full Breit interaction [52], as in the BP Hamiltonian (Section 2.13.2),

$$H^{\mathrm{BP}}_{N+1} = H_{N+1} + H^{\mathrm{mass}}_{N+1} + H^{\mathrm{Dar}}_{N+1} + H^{\mathrm{so}}_{N+1} + H_{oo} + H_{ssc} + H_{ss} + H_{soo}. \tag{3.111}$$

3.5.7.2 R-matrix II method

To improve the electron-correlation treatment within the R-matrix formulation, Burke and collaborators have developed new algebraic algorithms to enable complete sets of N- and $(N+1)$-electron configurations in Eq. 3.42 to be generated, with one- or few-electron excitations. The Rmatrix II codes are efficiently parallelized into a package called PRMAT [57]. Recent calculations for astrophysically important iron-group systems, such as Fe II and N III include several hundred terms LS [58]. The Rmatrix II codes (the middle branch of Fig. 3.9) employ the two-dimensional R-matrix propagator technique through sub-divisions of the atomic radius into several internal regions. These sub-regions are circumscribed by separate R-matrix boundaries. The R-matrix proper is subdivided into block propagator matrices. Schematically, if s and $s-1$ denote two adjacent regions, then the corresponding R-matrices are related as

$$\mathbf{R_s} = \mathbf{r_{ss}} - \mathbf{r_{ss-1}}[\mathbf{r_{s-1s-1}}\mathbf{R_{s-1}}]^{-1}\mathbf{r_{s-1s}}. \tag{3.112}$$

The middle branch in Fig. 3.9 illustrates the correspodence with the earlier version of codes in the left branch: STG2 → ANG, RECUPD → FINE, STGH → HAM and STGF → FARM. Parallel versions of asymptotic codes are denoted, for example, as PFARM.

3.5.7.3 Dirac R-matrix (DARC) method

For heavy atomic systems it is essential to use the fully relativistic Dirac treatment in jj-coupling (the right branch of Fig. 3.9), rather than the intermediate coupling BPRM approach [59, 60]. The R-matrix methodology using the Dirac Hamiltonian paved the way for the development of the Dirac R-matrix (DARC) codes [53, 61]. The four-component spinors are used to describe orbital wavefunctions:

$$\psi_{n\kappa m_j} = \begin{pmatrix} \psi_{\mathrm{L}} \\ \psi_{\mathrm{S}} \end{pmatrix} = \frac{1}{r}\begin{pmatrix} \chi_{\kappa m_j} P_{n\kappa}(r) \\ \mathrm{i}\chi_{-\kappa m_j} Q_{n\kappa}(r) \end{pmatrix}. \tag{3.113}$$

The large radial component of the continuum electron, P_i at the boundary can be expressed in terms of the R-matrix (Eq. 3.91) as [59],

$$P_i(a) = \sum_j R_{ij}\left[2acQ_j(a) - (b+\kappa_j)P_j(a)\right]. \tag{3.114}$$

To construct the target ion eigenfunctions for the fully relativistic DARC calculations, a relativistic atomic structure code with the Dirac formulation called GRASP2 is employed [11]. To facilitate interface with the asymptotic BPRM codes, DARC calculations sometimes employ four-component spinors in the inner region but two-component spinors using the large component $P_{n\kappa}$ in the outer region. This may lead to some inaccuracy for very highly charged ions or heavy elements. The effects of small radial component $Q_{n\kappa}$ may be significant for certain transitions and energy regions [62].

3.6 Approximate methods

The CC approximation incorporates most of the atomic effects necessary for a comprehesive and accurate treatement of (e + ion) processes. The importance of resonance and coupling effects, inherent in the CC approximation, is well-established. Although the CC calculations are complex and intensive in terms of effort and computational resources, these are now commonplace, especially using the R-matrix method, and have been carried out for most

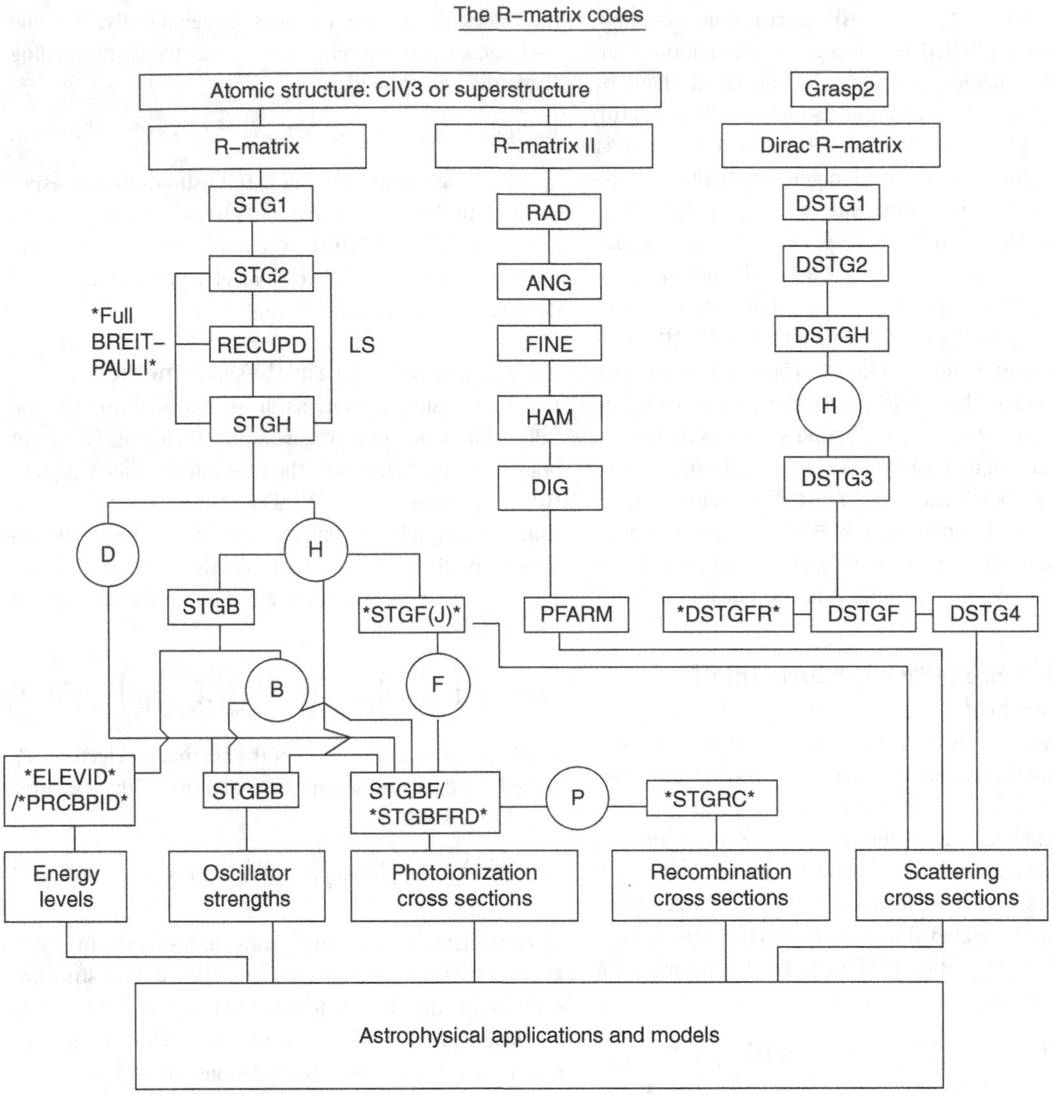

FIGURE 3.9 The three main branches represent different versions of the R-matrix package of close-coupling codes as described in the text.

astrophysically important atomic systems under the Opacity Project and the Iron Project [37, 38]. Nevertheless, for large-scale production of atomic data, simpler methods remain in common usage, particularly for extensive modelling of plasmas in laboratory fusion devices employing magnetic confinement (e.g., tokamaks) or inertial confinement fusion (ICF). These approximations, which we describe next, form the basis for several existing codes.

3.6.1 Distorted wave method

The distorted wave (DW) method does not include coupling among channels; only the initial and final channels are considered in the scattering or the continuum problem. It follows that resonance effects, which arise due to coupling among open and closed channels, are not considered in the DW approximation in an *ab-inito* manner. But they may be included perturbatively using a variety of schemes. The DW approximation is valid when coupling to channels other than the initial and the final states is weak. Such a situation does not occur in general in (e + ion) collision processes, but may be assumed to be the case for strong (say dipole) transitions. In highly charged ions, the Coulomb potential dominates the electron–electron interaction, and coupling among *all* channels is weak. Therefore, an example of where the DW approximaion may be valid would be strong dipole transitions in highly charged ions. Furthermore, the DW approximation is conveniently employed in constructing collisional–radiative models that couple not only a large number of levels within an ion, but also different ionization states of an element.[10]

The main feature of the DW method may be illustrated simply by neglecting the summation in the total (e + ion) wavefunction expansion in CC approximation (Eq. 3.24). More precisely, we may write the uncoupled Eq. 3.45 with a single index i in the potential term

$$\left[\frac{d^2}{dr^2} - V_i(r) - \frac{l(l+1)}{r^2} + k^2\right] F_i(r) = 0, \tag{3.115}$$

where V_i assume the asymptotic form

$$\lim_{r \gg R} V_i(r) = -2z/r \tag{3.116}$$

far enough outside the target. Its solution is the radial function in channel i, the continuum electron moving in the potential of the ion in state i. The name of the DW method stems from the idea that the radial functions are thought to be 'distorted' from a free plane wave form in the presence of the ionic potential of the initial and the final state.

Given the asymptotic form

$$F_{ii'}(r) \sim \sin(\zeta_i)\delta_{ii'} + \cos(\zeta_i)K_{ii'} \tag{3.117}$$

of the radial functions of the colliding electron in the total wavefunction Ψ, a solution of the *Kohn variational equation* (see Section 5.1.3)

$$\delta\left[\langle \Psi_i | H - E | \Psi_i' \rangle - K_{ii'}\rangle\right] = 0 \tag{3.118}$$

secures stability for small variations in Ψ_i, Ψ_i'. With $\boldsymbol{K}^{\text{trial}}$ from trial functions Ψ, the Kohn variational principle (cf. Section 5.1.3) yields

$$K_{ii'}^{\text{Kohn}} = K_{ii'}^{\text{trial}} - \left(\Psi_i | H - E | \Psi_i'\right)^{\text{trial}}, \tag{3.119}$$

which differs from the exact matrix $\boldsymbol{K}$ in second order of the error in the trial functions. Using the Kohn variational principle with $K^{\text{trial}} = 0$ (and appropriate choice of phases) we can write the K-matrix in the DW formulation as [34]

$$K_{ii'}^{\text{DW}} = -\left(\Psi_i | H - E | \Psi_i'\right). \tag{3.120}$$

We recall that in its simplest form the total (e + ion) wavefunction Ψ (Eq. 3.24) is the product $\Phi\theta$, where Φ is the target function and θ is the free electron function with radial functions $F(r)$. However, we can formulate the DW approximation such that the total Ψ also includes the functions χ (Eq. 3.42), which, in addition to providing short-range correlation, would also lead to resonances at eigenvalues of these functions. Since resonances are of crucial importance in atomic processes, DW calculations sometimes include their effect in a perturbative or ad-hoc manner. The $(N+1)$-electron bound state (e + ion) wavefunctions may be explicitly included in the collision problem [64]. These would then give rise to resonances in cross sections at the eigenvalues of the $(N+1)$-electron configurations [65, 66]. However, only a limited number of resonances can be included individually in this way, and Rydberg series of resonances converging on to target thresholds may not be considered (see, however, the section on quantum defect theory). Note that we are able to write the K-matrix elements directly in Eq. 3.120, because the channels are uncoupled, except of course for initial and final channels.

There are a number of computer codes based on the DW method, or variants thereof, such as the University College London DW code [67], which uses the atomic structure code SUPERSTRUCTURE [10] to construct the target ion eigenfunctions Φ (Eq. 3.24). There are also several relativistic DW (RDW) codes in pure *jj*-coupling. A fully relativistic approach for calculating atomic data for a variety of atomic processes, and a RDW code for highly charged ions, has been developed by D. H. Sampson, H. L. Zhang, and C. Fontes (e.g., [68]). The atomic structure data necessary for electron–ion collision calculations are either obtained by the Dirac–Fock–Slater code or by a multi-configuration Dirac–Fock (MCDF) code [11, 54]. In the former case, the same potential is used for treating both bound and free electrons so that all electron orbitals are automatically orthogonal and the exchange integrals are calculated in a consistent manner. In the latter case, since a potential is constructed from bound wavefunctions to solve for the free-electron orbitals, a different potential from the one in the MCDF code can be used, such as a semi-classical exchange potential.

In the RDW approach [68], the radial wavefunction containing a large and small component P and Q (cf. Eq. 3.113), for all orbitals, bound and free, that are solutions of the single-electron Dirac equation,

$$\left[\frac{d}{dr} + \frac{\kappa}{r}\right] P_i(r) = \frac{\alpha}{2}\left[\epsilon_i - V + \frac{4}{\alpha^2}\right] Q_i(r) \tag{3.121}$$

and

$$\left[\frac{d}{dr} - \frac{\kappa}{r}\right] Q_i(r) = \frac{\alpha}{2}(V - \epsilon_i) P_i(r), \tag{3.122}$$

[10] A discussion of the related physics and the widely used code for semi-empirical atomic calculations for plasma applications is given by R. D. Cowan [63].

where α is the fine-structure constant, $V(r)$ is a central potential and the relativistic quantum number κ has the values

$$\kappa = l, \quad j = l - \tfrac{1}{2};$$
$$\kappa = -(l+1), \quad j = l + \tfrac{1}{2}. \tag{3.123}$$

These single electron wavefunctions are used to form a basis set of multi-electron wavefunctions for structure calculations or for e + ion collision calculations. In doing this, a standard jj-coupling scheme, the natural scheme for a fully relativistic treatment, is followed. The electron impact excitation collision strength is then given by the K-matrix or reactance-matrix elements according to

$$\Omega_{ij} = 2\sum_J (2J+1) \sum_{\substack{\ell,\ell' \\ j,j'}} |K(J_i\ell j, J_j\ell' j'; J)|^2, \tag{3.124}$$

where J represents the total angular momentum of the (e + ion) system. The inner summations are performed over all channels associated with J, and that correspond to initial and final levels i and j.

The RDW approach has been extended to treat other atomic processes: electron–impact ionization, photoionization, electron capture and autoionization, and dielectronic recombination [68]. When coupling with other channels, except between the relevant initial and final channels, is negligible, this RDW approach is expected to yield accurate results, especially for high-Z highly charged ions or for sufficiently high energies. The former is especially true, since it treats relativistic effects non-perturbatively. The latter is practically useful for light elements of astrophysical interest, since the data calculated for high energies can be used to complement low-energy R-matrix results, as done for photoionization in [69]. However, for light elements and for low energies, the RDW approach suffers from two deficiencies. First, the RDW approach generally overestimates the background contribution to the cross sections (or collision strengths) since this approach neglects channel coupling and often employs the weak-coupling approximation, Eq. 3.120, which does not satisfy the principle of unitarity, Eq. 3.57. We note that a fully unitarized DW calculation can always be performed by calculating T- or S-matrix elements from the K-matrix elements in precisely the same manner that is employed in the R-matrix approach, i.e., via the exact relation, Eq. 3.57. Thus, it is possible to eliminate a lack of unitarity as a cause of discrepancies between DW and R-matrix results, in favor of a lack of channel coupling in the former calculations. Second, the RDW approach does not include resonances in an *ab-initio* manner. This latter concern can be remedied by using Cowan's two-step method [63], also known as the independent-process and isolated-resonance approximation (IRA; cf. Chapter 7), to include resonance contribution to the relevant processes. For example, the resonance contribution to electron impact excitation can be considered as the two-step process of electron capture followed by autoionization [68].

It should be mentioned that, when using DW data to solve rate equations associated with collisional-radiative modelling, autoionization levels are treated explicitly, on a par with bound levels. With this type of explicit treatment, the resonance contribution to all processes, including the appropriate radiation damping, is automatically taken into account for all processes and their inverses within the IRA approximation (see also Chapter 7).

3.6.2 Coulomb–Born approximation

A simpler variant of the DW approximation is the Coulomb–Born (CB) approximation which results if we replace the potential term $V_i(r)$ by the Coulomb potential $-2z/r$ in Eq. 3.115. The radial function solutions are then the Coulomb functions, known analytically (Eqs. 3.51, 3.52). Physically, the CB approximation is valid when the free electron is relatively far from the ion, moving in the Coulomb potential of the ionic charge on the whole, and not interacting with the target electrons closely. Such situations occur for large angular momenta of the incident electron, where the CB radial waves do not differ much from the DW functions. Therefore, the CB method may be used for large ℓ-waves, but not for low-ℓ scattering, for which the CC, or at least the DW, approximation should be employed.

An important difference between the CC and the DW methods on the one hand, and the CB approximation on the other, is that exchange is not included in the CB method (although another variant called the Coulomb–Born–Oppenheimer approximation does allow for exchange). The CB approximation, therefore, cannot be used for spin-change transitions, which require close low-ℓ interactions between the electron and the ion.

We may write the CB K-matrix elements, using Eq. 3.24 as

$$K_{ii'}^{\mathrm{CB}} = -\left(\Phi_i\theta_i \left| \sum_{n=1}^{N} \frac{1}{r_{n,(N+1)}} \right| \Phi_i'\theta_i'\right), \tag{3.125}$$

where the operator is the direct two-electron interaction potential between the N target electron(s) and the free colliding electron '$N + 1$'; exchange is neglected. Note the difference with the similar expression in the DW approximation, where the Hamiltonian may include other interactions in addition to the direct Coulomb potential.

3.6.3 Coulomb–Bethe approximation

We may treat the collisional transition as an induced radiative process. Provided the incident electron is sufficiently far from the ion, and close interactions, such as electron exchange, are not important, the approximate effect of the collision is to induce a 'radiative' dipole transition in the target ion. This is called the Coulomb–Bethe (CBe) approximation.

The CBe approximation is valid only for sufficiently large angular momenta, which may be determined approximately as follows [70]. Since the Coulomb potential is the only one considered, we can write

$$\frac{2z}{r_0} - \frac{\ell(\ell+1)}{r_0^2} + k^2 = 0. \tag{3.126}$$

The solution of the equation yields the conditionality

$$\ell > \ell_0 = \left[k^2 r_0^2 + z r_0 + \frac{1}{4}\right]^{1/2} - \frac{1}{4}, \tag{3.127}$$

where we may take r_0 to be the mean radius of the target orbitals. This condition should then be satisfied for both the incident and the outgoing channel momenta ℓ, ℓ' in order for the CBe method to be valid. The condition also provides a useful marker for the value of angular momenta sufficiently large for close interactions to be unimportant. Conversely, for partial waves less than l_0, close encounters should be considered such as in the CC method.

Although the CBe method is not accurate for low partial waves, it is often useful to ensure convergence of the partial wave expansion. The CC method is emoloyed for low-ℓ wave scattering, and the CBe method is used to complete the summation to high partial waves ($\ell > \ell_0$) for dipole allowed transitions whose cross sections converge slowly with ℓ. Since Coulomb functions and related integrals may be evaluated analytically, efficient procedures have been developed to compute CBe cross sections up to arbitrarily high ℓ [70, 71].

3.6.4 Central-field approximation

As discussed in Chapter 2, bound or continuum state wavefunctions of an atomic system may be computed in a suitably chosen central potential $V(r)$, such as the Thomas–Fermi–Dirac potential described in Chapter 2. The radial wave equation with a central potential yields the one-electron orbitals $P_{n\ell}(r)$ for an ion with effective charge $z = (Z-N+1)$, where Z is the nuclear charge and N is the number of electrons. Past calculations of many astrophysical parameters have been made with no further refinement, i.e., with no multiplet structure. For example, photoionization cross sections for individual $n\ell$-subshells may be obtained using the central-field model, but without the further division according to the LS terms that constitute atomic term structure. Obviously, effects such as channel coupling or resonances cannot be considered using a simple central potential. We discuss the physical effects in the central-field method for photoionization in Chapter 6, in comparison with more elaborate CC work using the R-matrix method.

3.6.5 Quantum defect theory

The quantum defect theory (QDT) relates (e + ion) bound states and the continuum. As the level of excitation of a bound electron increases in energy, the energy levels become less and less negative until they approach zero energy, when the electron becomes free and is moving in the potential of the ion. We have already seen that the bound energy levels in a Rydberg series of a given orbital angular momemtum $n\ell$ approach a constant quantum defect μ_ℓ. Moreover, a free electron under the influence of a potential undergoes a phase shift that is also dependent on ℓ and energy k^2. It is therefore physically reasonable that the quantum defect and the phase shift should be related at zero energy, the Rydberg series limit

$$\lim_{n\to\infty} \pi\mu_{n\ell} = \lim_{k^2\to 0} \delta_\ell(k^2). \tag{3.128}$$

This fundamental relationship of QDT is *Seaton's theorem*.[11] It is a powerful tool, since the (e + ion) scattering cross section is directly related to the phase shift (or the multi-channel generalization thereof discussed below). The phase shift may be obtained from quantum defects by analytic continuation from negative bound state energies according to the modified Rydberg formula (Eq. 2.65), and extrapolation to positive energies of the continuum.

11 Note the formal resemblance to Levinson's theorem: given a radial potential, the phase shift $\delta_l(k)$ satisfies $\lim_{k\to 0}\delta_l(k) = n_\ell\pi$, where n_ℓ is the number of bound states of angular momentum ℓ supported by the potential.

Now consider the reverse process. Recall that in the single-channel case $K_\ell = \tan\delta_\ell$ (Eq. 3.59). It follows that

$$|K_\ell - \tan(\pi\mu_\ell)| = 0, \tag{3.129}$$

which again relates the continuum phase shift to the quantum defect of bound energy levels. Therefore, if the phase shift is known then we may obtain the energy levels of bound states by extrapolation from positive energies to negative energies, from the continuum to the bound state region.

Quantum defect theory is often useful in the multi-channel case. The S- and the K-matrices are generalized multi-channel phase shift matrices. Perhaps the most important use of multi-channel QDT, or MCQDT, is in the analysis of resonance structures, which involve coupled closed and open channels. In scaled form, the energy balance for closed channels reads

$$k_i^2 = -z^2/\nu_i^2, \qquad i > n_\text{o}, \tag{3.130}$$

where ν_i is the effective quantum number in the closed channel i. As we have seen, the coupled radial solutions in the open and closed channels have the asymptotic form

$$F_{ii'} \sim \sin(\zeta_i)\,\delta_{ii'} + \cos(\zeta_i)\,K_{ii'}, i = 1, n_\text{o},$$
$$F_{ii'} \sim C\,\exp(-zr/(n-\mu_\ell)), \quad i > (n_\text{o}+1), n_f, \tag{3.131}$$

respectively. The K-matrix (and the **S**-matrix) is defined only for energies above threshold(s) where all channels are open. However, the multi-channel K-matrix elements may be fitted to analytic expressions in energy and extrapolated to energies *below* the threshold where resonances occur. This analytic continuation of the *K-matrix* to energies where some channels are open and some closed, is denoted as K and partitioned as $\mathcal{K}_\text{oo}, \mathcal{K}_\text{oc}, \mathcal{K}_\text{co}\mathcal{K}_\text{cc}$, where the sub-matrices stand for open–open, open–closed, closed–open and closed–closed parts of the K-matrix. The main result of the MCQDT is then

$$\boldsymbol{K} = \mathcal{K}_\text{oo} - \mathcal{K}_\text{oc}\,[\mathcal{K}_\text{cc} + \tan(\pi\nu_c)]^{-1}\,\mathcal{K}_\text{co}, \tag{3.132}$$

where ν_c are the effective quantum numbers in closed channels. The denominator on the right-hand side gives rise to poles in the K-matrix when

$$|\mathcal{K}_{cc} + \tan(\pi\nu_c)| = 0. \tag{3.133}$$

The pole position and the residues reflect energies and widths of the resonance. Thus MCQDT gives the Rydberg series of resonance structures from a scattering calculation above threshold, where there are no resonances, to energies below threshold, where they occur. Expressions may be readily obtained for the S-matrix and its analytic continuation S.[12] We will apply MCQDT expressions to problems in (e + ion) scattering, photoionization and recombination in Chapters 5, 6 and 7.

In summary: following up from Chapter 2, where we revisited basic quatum mechanics of atomic structure, we have described a theoretical and computational framework that underlies modern calculations for most of the atomic processes in astrophysics.

Basically, the atomic processes and spectra in astrophysical and laboratory plasmas involve electron–ion–photon interactions including a variety of processes (not all of which are considered herein). Figure 3.5 shows the main reactions that can take place leading to excitation and ionization with one free electron and photon, that are the main subject of this chapter. As might be seen from the above sketch of the dominant atomic processes in a plasma, a theoretical description requires a knowledge of the wavefunctions of the quantum mechanical states of the ion, and electron–ion–photon interactions that must be included in the appropriate Hamiltonian in a generalized Schrödinger equation. In the past few decades, considerable progress has been made and precise theoretical calculations may be carried out using sophisticated computational programs. However, large-scale calculations involving many atomic states need extensive computing resources and effort.

Most of the fundamental atomic parameters associated with primary atomic processes are either computed theoretically or, to a much lesser degree measured experimentally. The sections above describe the individual processes with a view to providing a unified and self-consistent theoretical framework. The atomic theory outlined in this chapter describes the main features of interest in astrophysical calculations. The close coupling or the coupled channel method, particularly as implemented through the R-matrix method, is discussed in detail, since it is the most general and the most advanced formulation to treat electron–photon–ion reactions in a unified manner. Other methods, such as the distorted wave method, may be regarded as approximations to the general theory; they are nevertheless useful in computations of data for atomic systems where coupled channel calculations

[12] Further extensions of the matrix expressions (Eq. 3.132) may be made by combining it (i) with matrix operators for transformation from LS coupling to a pair-coupling scheme, taking account of fine structure algebraically, and (ii) also incorporating limited relativistic effects in the target ion via *term coupling coefficients* (TCC) transformation. Both (i) and (ii) operations are performed by codes such as in computer programs JAJOM and STGFJ [38, 72], shown in Fig. 3.9. Further discussion is given in the next chapter.

are computationally demanding or may not yield significantly higher accuracy (such as for highly charged ions). Although considerable work has been done, the atomic physics methods are yet to be generalized to treat several important problems, such as external electric or magnetic field effects in all plasma environments. Particular applications of the methods discussed in this chapter to specific atomic processes are described in the next four chapters on radiative transitions, electron impact excitation, photoionization and recombination, respectively.[13]

[13] In this chapter we have limited outselves to a fairly concise treatment of atomic processes in astrophysics. A more general exposition is given in the classic treatise by N. F. Mott and H. S. W. Massey [32]. More recent and extensive essays on a variety of topics in atomic–molecular-optical (AMO) physics are given in the compendium edited by G. W. F. Drake [73].

4 Radiative transitions

A radiative transition between the bound states of an atom occurs when an electron jumps to an upper or excited level by absorption of a photon (photo-excitation), or jumps to a lower level by emission of a photon (de-excitation or radiative decay). For an atom X, these processes can be expressed as

$$X + h\nu \rightleftharpoons X^*, \tag{4.1}$$

where the superscript $*$ denotes an excited state. Excited states have finite lifetimes relative to the ground state, which is supposed to be infinitely long-lived. Radiative processes introduce absorption and emission lines at particular transition energies in a spectrum (discussed in Chapters 8 and 9). The strength of an observed emission or absorption line depends on both the internal properties of the atom, and the external environmental conditions where the atomic system exists. To analyze a spectrum, we need to understand the qualitative nature of the observed features as well as their measured quantitative strength. The physics of the former determines the latter, when coupled with external factors, such as the temperature or density. Spectral formation under laboratory conditions is often different from that encountered in astrophysical conditions. In this chapter, we describe the internal atomic physics of radiative transitions and outline the rules that govern spectral formation.

In common astrophysical or spectroscopic parlance, transitions are often classified as 'allowed', 'forbidden' or 'intersystem'. For example, in many astrophysical sources, some of the most important observed lines are classified as 'forbidden'. But what does that mean exactly? The answer to this question requires an understanding of the physical basis, as well as the quantum mechanical rules, that determine the strength of each type of transition. 'Forbidden' clearly does not mean that the associated transitions cannot take place; they obviously do since the lines are observed. Rather, it means that their transition rates are orders of magnitude smaller than allowed lines, so that the forbidden lines are extremely weak when observed in laboratory experiments compared with those resulting from allowed transitions. But in many astrophysical sources, especially H II regions (gaseous nebulae, supernova remnants, the interstellar medium), the external physical conditions are such that, despite small intrinsic probabilities, the forbidden transitions are the dominant mode excitation and radiative decay. In such sources, the electron temperatures are low, about 10 000 K ($\sim$1 eV) or lower, and the densities are often lower than those attained in the laboratory, typically 10^3–10^6 cm^{-3}. We shall see why forbidden lines can form just as easily as allowed lines, depending on physical conditions.

The classification of radiative transition rests on transition rates described by Einstein's well-known coefficients A and B, which depend only on *intrinsic* atomic properties. They are computed quantum mechanically or measured in the laboratory. However, it is important to keep in mind that the intensities of lines depend on external physical conditions. The Einstein transition probabilities or rates are independent of *extrinsic factors*, such as temperature or density in the source. We first outline the fundamentals of atomic transitions related to the formation of emission and absorption lines in terms of the Einstein relations. This is followed by a first-order quantum mechanical treatment usually valid for astrophysical applications. More elaborate treatments, such as including quantum electrodynamic (QED) effects, are necessary for exceptional accuracy. But we limit our exposition to an outline of the basic formulation employed in most practical calculations.

The incident photons from the ambient radiation field may induce various transitions in the atom. Quantum mechanically, their probability is computed from a transition matrix element that, in turn, depends on the wavefunctions of the initial and final states, and an operator corresponding to the appropriate moment of the radiation field. These moments are generally characterized as the dipole moments that correspond to allowed transitions, or higher-order multipoles that correspond to non-dipole or forbidden transitions. In addition, radiative transitions are governed by the symmetries of the atomic states involved, as specified by the quantized angular and spin momenta of the initial and final states. The specification of these symmetries with respect to a given type of transition is referred to as *selection rules*. An outline of the computational framework for the calculation of transition matrix elements, and hence the probabilities and rates for radiative transitions, forms the bulk of this chapter.

4.1 Einstein *A* and *B* coefficients

Interaction of electromagnetic radiation with atoms may be treated in what is known as the *first quantization*: the energy levels are quantized (discrete), but the radiation field is continuous, i.e., forms a continuum of photon energies. This assumption simplifies to a perturbative treatment suitable for piece-wise consideration of transition operators. *Second quantization*, where the radiation field is also quantized, constitutes QED, which enables high precision but entails far more elaborate calculations. Accuracy at QED level is not generally required for most astrophysical situations.[1] Here, we confine ourselves to the former case, with stationary atomic levels under perturbation by an external radiation field.

We begin with a simple picture of two atomic levels, such that $E_2 > E_1$; the transition energy is given by a discrete quantum of energy

$$E_{21} = E_2 - E_1 = h\nu \,. \tag{4.2}$$

In accordance with the old quantum theory – which entails Planck's definition of light quanta – Einstein postulated three distinct radiative processes that connect the two levels shown in Fig. 4.1. The intrinsic probability coefficients are: (i) *spontaneous decay* from 2 to 1 with probability coefficient A_{21}, (ii) *absorption* of a photon from the radiation field of density ρ and transition from 1 to 2 with the probability coefficient B_{12}, and (iii) the inverse of (ii), namely *stimulated emission* with a probability coefficient B_{21}, resulting in a transition from 2 to 1 induced by some other photon. We assume that the levels are degenerate with weights g_1 and g_2, that is, there are g_i states with the same energy E_i.

Let the level populations denoted as N_i, $\rho(\nu)$ be the radiation density. Then the rates for these processes

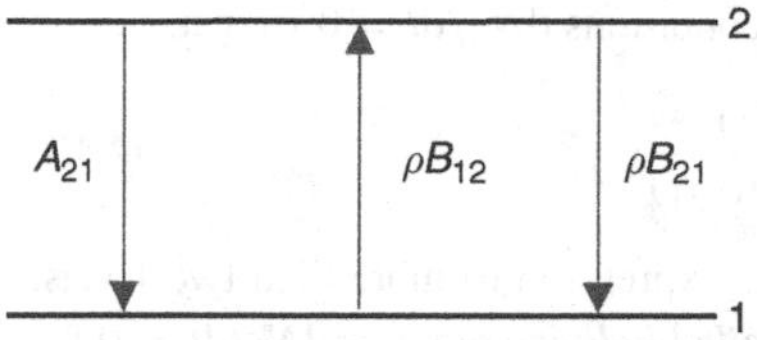

FIGURE 4.1 The three radiative processes connecting two levels 1 and 2.

depend on their respective coefficients A and B. In detail, the rate of spontaneous decay depends on N_2, the rate of absorption depends on both N_1 and the photon number $\rho(\nu_{12})$ with the right energy $h\nu = E_2 - E_1$. Likewise, from detailed balance, the rate of stimulated or induced emission depends on N_2 and $\rho(\nu_{21})$. In equilibrium, the time-dependent level populations satisfy the rate equations

$$-\frac{dN_2}{dt} = \frac{dN_1}{dt} = A_{21}N_2 - B_{12}\rho_{12}N_1 + B_{21}\rho_{21}N_2, \tag{4.3}$$

that is, the population loss in level 2 equals the population gain in level 1. For a steady state $dN_2/dt = dN_1/dt = 0$ we get

$$\frac{N_2}{N_1} = \frac{B_{12}\rho(\nu)}{A_{21} + B_{21}\rho(\nu)}. \tag{4.4}$$

In the absence of an external radiation field, i.e., $\rho(\nu) = 0$, the equation simplifies to

$$\frac{dN_2}{N_2} = -A_{21}dt. \tag{4.5}$$

It follows that

$$N_2(t) = N_2(0)\, e^{-A_{21}t}, \tag{4.6}$$

which implies that an excited-level population decays exponentially with time at an e-folding rate given by A_{21}. This rate determines the *lifetime* τ_2 of level 2 as

$$A_{21} \equiv \frac{1}{\tau_2}\,. \tag{4.7}$$

Hence, the Einstein coefficient A is an inverse time, for the hydrogenic Lyα line in the order of 10^9 transitions per second, namely $A_{2p,1s} = 6.25 \cdot 10^8 s^{-1}$. Inserting this value into Eq. 4.7 results in a lifetime $\tau_{2p} \sim 1.6$ ns.

Next, we define the probabilities P_{21} and P_{12} for emission and absorption per atom per unit time:

$$-\frac{dN_2}{dt} = P_{21}N_2 - P_{12}N_1, \tag{4.8}$$

and then

$$P_{emi} = A_{21} + B_{21}\rho(\nu) \text{ and } P_{abs} = B_{12}\rho(\nu). \tag{4.9}$$

[1] Hydrogen and helium are outstanding exceptions, for which the most advanced radiative calculations have been carried out (e.g. [73] and references therein).

Under steady-state conditions $dN_2/dt = 0$ we get

$$\frac{N_2}{N_1} = \frac{B_{12}\rho(\nu_{21})}{A_{21} + B_{21}\rho(\nu_{21})}. \quad (4.10)$$

Generalizing this expression to more than two levels, the *principle of detailed balance* requires $dN_i/dt = 0$ for each level i, taking account of all transitions in and out of level i to and from all other levels j. In fact, detailed balance further requires that the steady-state condition applies to each pair of levels (i, j) for *each* process and its inverse separately. Otherwise, the system can deviate from the equilibrium steady state via any particular process that remains 'unbalanced'. Thus we have

$$\frac{N_j}{N_i} = \frac{B_{ij}\rho(\nu_{ij})}{A_{ji} + B_{ji}\rho(\nu_{ij})}. \quad (4.11)$$

If there is no radiation field, Eqs 4.5 and 4.6 generalize to

$$-\frac{dN_j}{dt} = \sum_{i<j} A_{ji}N_j, \quad (4.12)$$

$$N_j(t) = N_j(0)\,e^{-\sum_i A_{ji}t} \quad (4.13)$$

for all levels $i < j$. Now the lifetime of level j against radiative decay to any level i becomes

$$\tau_j = \left(\sum_i A_{ji}\right)^{-1}. \quad (4.14)$$

For hydrogen, it may be shown that the lifetime of a state nl varies as $\tau_{nl} \propto n^3$ and that the l-avaraged lifetime is given by

$$\tau_n = \left(\frac{1}{n^2}\sum_l \frac{2l+1}{\tau_{nl}}\right)^{-1} \propto n^5. \quad (4.15)$$

It is instructive to examine the relationship between the coefficients A and B by working through the following example.

Exercise 4.1 *In steady state and thermal equilibrium at temperature T, the level populations are expressed by the* Boltzmann equation

$$\frac{N_j}{N_i} = \frac{g_j}{g_i}\exp(-h\nu_{ij}/kT), \quad (4.16)$$

with statistical weights g of the respective levels. Assuming a black-body radiation field given by the Planck function

$$\rho_\nu = \frac{8\pi\nu^2}{c^3}\frac{h\nu}{\exp(h\nu/kT) - 1}, \quad (4.17)$$

derive the Einstein relations

$$A_{ji} = B_{ji}\frac{8\pi\nu_{ij}^2}{c^3}h\nu_{ij} = \frac{g_i}{g_j}B_{ij}\frac{8\pi\nu_{ij}^2}{c^3}h\nu_{ij}. \quad (4.18)$$

Hint: detailed balance implies $g_iB_{ij} = g_jB_{ji}$.

From Exercise 4.1, at a specific frequency ν_{ij} for transition $i \to j$ we obtain the coefficient

$$B_{ji} = \left(\frac{c^3}{8\pi h\nu^3}\right)A_{ji} \quad (4.19)$$

as stimulated emission. With the transition wavelength $c/\nu = \lambda$ and plain arithmetic with atomic shell identities from Section 1.9 such as $h = 2\pi\cdot \mathrm{Ry}\cdot\tau_0 = 3.0397 \times 10^{-16}$ Ry s, and ΔE_{ij} in Ry, one arrives at

$$B_{ji} = \frac{\lambda_{ij}^3}{8\pi h}A_{ji} = \frac{8.833 \times 10^{38}a_0^3}{\mathrm{Ry}\cdot\mathrm{s}}\cdot\left(\frac{\lambda_{ij}}{\mathrm{cm}}\right)^3 A_{ij} \quad (4.20)$$

$$= 6.0048 \times 10^{24}A_{ji}\,\lambda_{ij}^3 \quad (4.21)$$

$$= 4.54161 \times 10^9\frac{A_{ji}}{\Delta E_{ij}^3}. \quad (4.22)$$

With λ in angstroms, $B = 6.0048\ \lambda^3\ A$, which shows that B increases rapidly with λ. Because it depends on energy $\propto \lambda^{-3}$, Einstein's A is extremely small at radio frequencies. Therefore, longer-wavelength transitions, say in the cm range of microwaves, are more readily subject to stimulated emission than shorter-wavelength transitions. That is because longer-wavelength transitions occur between more closely spaced levels than shorter wavelength; the associated coefficients A are smaller than B, and therefore radiative depopulation via spontaneous decay is relatively slow. Stimulated emission in the microwave region is called *microwave amplification of stimulated emission of radiation*, or maser. More generally it is a particular case of the well-known *light amplification of stimulated emission of radiation* and its acronym, laser. Astrophysical sources with stimulated emission in the microwave region occur, for example, in the molecular torus around black holes. The reason for molecular emission in the microwave region is that molecular energy levels are closely spaced, and the transition energies $E_{ij} = hc/\lambda_{ij}$ are very small. Of course, for sufficient intensity to build up in a maser or laser, we must have a large population of atoms in the upper level. This usually (but not always) requires *population inversion*. It can be achieved if A is very small, which is often true for forbidden transitions among closely spaced levels in the context of selection rules discussed later.

In thermal equilibrium, according to the Boltzmann equation, there are far fewer atoms in the upper level than in the lower level. Therefore, although the coefficients for stimulated absorption and emission are equal, i.e., $B_{ij} = B_{ji}$ (modulo the statistical weight ratio g_j/g_i), the actual *number* or rate of upward excitations is more than downward stimulated radiative decays. The balance of upward and downward transitions under equilibrium conditions in the rate equations is achieved by the addition of downward spontaneous decays, governed by the A coefficient, to the stimulated emission rate. Population inversion generally means that the ratio of the upper to lower level population *exceeds* that given by the Boltzmann equation, that is,

$$\frac{N_j}{N_i} > \frac{g_j}{g_i} \mathrm{e}^{-E_{ij}/kT}. \tag{4.23}$$

A quantitative examination of the numerial values of A and B may be made as in the following example.

Exercise 4.2 *The ratio of the stimulated emission rate and the spontaneous emission rate under thermal equilibrium at temperature T is given in terms of the black-body Planck function for the radiation density of photons ρ as*

$$\frac{A_{ji}}{B_{ji}\rho(\nu_{ij})} = \mathrm{e}^{h\nu_{ij}/kT} - 1. \tag{4.24}$$

Compare this ratio for different regions of the electromagnetic spectrum in thermal equilibrium. Show in particular that, as discussed above, stimulated emission dominates in the microwave region by a factor of about 1000. In the ultraviolet or X-ray region, the reverse is the case. Of course, the actual rate depends on the external radiation source and photon density, and therefore for a non-thermal external source this ratio can assume arbitrary values, particularly in tunable and high-intensity laboratory sources.

The coefficients A and B are independent of level populations and external parameters, such as the temperature, density and radiation field (which was assumed to be a black body *only* in order to derive the relations in Exercise 4.1). As stressed before, the Einstein coefficients depend only on intrinsic atomic properties, such as nuclear charge, number of atomic electrons and electronic structure, and level of excitation. Evaluation of A and B requires us to describe the interaction of the electromagnetic radiation field with atomic levels quantum mechanically.

4.2 Electron motion in an electromagnetic field

A semi-classical perturbative treatment of the interaction of a radiation field with matter, embodied in first quantization, follows from the time-dependent Schrödinger equation

$$\mathrm{i}\hbar\frac{\partial\Psi}{\partial t} = H\Psi. \tag{4.25}$$

As long as time and spatial dependences are separable, the wavefunction can be written in terms of *stationary states* or time-independent eigenstates of the atom, since we are interested in changes in states of the atom with time rather than the state of the field. We expand

$$\Psi = \sum_n a_n(t)\psi_n \mathrm{e}^{-\mathrm{i}E_n t/\hbar} \tag{4.26}$$

with time-dependent coefficients $a_n(t)$ over space-dependent atomic wavefunctions satisfying

$$\left[\frac{\boldsymbol{p}^2}{2m} + V\right]\psi_n = E_n\psi_n, \quad \langle\psi_j|\psi_i\rangle = \delta_{ij}, \tag{4.27}$$

and select $a_n(0) = 1$ for a particular state i at time $t = 0$, while $a_{n\neq i}(0) = 0$ for all the other states. Then $|a_j(t)|^2$ is the probability to find the atom in an excited state j after a time t.

Our interaction Hamiltonian involves the electron charge $-e$ moving with velocity $\boldsymbol{v}$ in some applied radiation or electromagnetic field $(\boldsymbol{B}, \boldsymbol{E})$:

$$\boldsymbol{E} = -\nabla\phi - \frac{1}{c}\frac{\partial\boldsymbol{A}}{\partial t}, \quad \boldsymbol{B} = \nabla\times\boldsymbol{A}, \tag{4.28}$$

where $\boldsymbol{A}$ is a vector potential and ϕ is a scalar potential. The Lorentz force experienced by the electron is

$$\boldsymbol{F} = -e\left[\boldsymbol{E} + \frac{1}{c}\boldsymbol{v}\times\boldsymbol{B}\right]. \tag{4.29}$$

The radiation field can be regarded as a superposition of plane waves periodic in time:

$$\boldsymbol{A} = \boldsymbol{A}_0\left[\mathrm{e}^{\mathrm{i}(\omega t - \boldsymbol{k}\cdot\boldsymbol{r})} + \mathrm{e}^{-\mathrm{i}(\omega t - \boldsymbol{k}\cdot\boldsymbol{r})}\right], \tag{4.30}$$

where ω is the angular frequency and $\boldsymbol{k}$ is the wave vector; hence,

$$\boldsymbol{E} = -\nabla\phi + \frac{2\omega}{c}\boldsymbol{A}_0\sin(\omega t - \boldsymbol{k}\cdot\boldsymbol{r}). \tag{4.31}$$

The magnetic field $\boldsymbol{B}$ is usually much weaker than the electric field $\boldsymbol{E}$. Then we can write the Lorentz force (Eq. 4.29) as

$$\boldsymbol{F} = -e\boldsymbol{E} = -e\nabla\phi - \frac{e}{c}\frac{\partial\boldsymbol{A}}{\partial t} = -e\nabla\phi - \frac{\partial\boldsymbol{p}_E}{\partial t}, \tag{4.32}$$

where $p_E = \frac{e}{c}A$ is a momentum affecting or 'retarding' the internal motion or momentum of the atomic electron due to the electric force of the varying external field; p_E can be treated as the retardation correction to the unperturbed momentum p. The radiatively *perturbed* Hamiltonian is then written as

$$H = \frac{1}{2m}\left(p + \frac{eA}{c}\right)^2 + e\phi + V. \tag{4.33}$$

Substituting in the Schrödinger equation (Eq. 4.25),

$$i\hbar\frac{\partial\Psi}{\partial t} = \left[-\frac{\hbar^2}{2m}\nabla^2 - \frac{e}{2mc}(p\cdot A + A\cdot p) + \frac{e^2}{2mc^2}A^2 + e\phi + V\right]\Psi. \tag{4.34}$$

Using $[f(x), p_x]\psi = i\hbar\frac{\partial f}{\partial x}\psi$, it can be shown that $A\cdot p - p\cdot A = i\hbar\nabla.A$. With the choice $\nabla.A = 0$,[2] the above equation reduces to

$$i\hbar\frac{\partial\Psi}{\partial t} = \left[-\frac{\hbar^2}{2m}\nabla^2 + \frac{i\hbar e}{mc}A\cdot\nabla + \frac{e^2}{2mc^2}A^2 + e\phi + V\right]\Psi. \tag{4.35}$$

Assuming no additional charge, $\phi = 0$. The square term A^2 describes two-photon processes, but for the current task we consider weak external time-dependent fields that induce the atom to emit or absorb a single photon, which is an adequate assumption in astrophysical environments. Compared with other terms, the contribution A^2 may therefore be neglected in the first-order perturbation approach. Then the Schrödinger equation (Eq. 4.25) in an electromagnetic field can be rewritten as

$$i\hbar\frac{\partial\Psi}{\partial t} = \left[-\frac{\hbar^2}{2m}\nabla^2 - \frac{e}{mc}A\cdot p + V\right]\Psi. \tag{4.36}$$

However, we do note that A^2 is important in cases where the external electromagntic field is strong, i.e., comparable to internal atomic fields, such as in laser interactions with atoms or multi-photon processes.

4.2.1 Radiative transition probability

Dividing the current Hamiltonian,

$$H = H_0 + H' \tag{4.37}$$

$$H_0 = \frac{p^2}{2m} + V \tag{4.38}$$

$$H' = \frac{e}{mc}(A\cdot p), \tag{4.39}$$

with a first-order perturbation Hamiltonian H' and substituting the wavefunction expansion $\Psi = \sum_n a_n(t)\psi_n e^{-iE_nt/\hbar}$ in the Schrödinger equation (Eq. 4.25) with the total Hamiltonian gives

$$\sum_n\left[i\hbar\frac{da_n}{dt} + E_na_n\right]\psi_n\exp\left(-\frac{iE_nt}{\hbar}\right) = (H_0 + H')\sum_n a_n\psi_n\exp\left(-\frac{iE_nt}{\hbar}\right). \tag{4.40}$$

Since the unperturbed Hamiltonian satisfies $H_0\psi_n = E_n\psi_n$,

$$\sum_n\left(i\hbar\frac{da_n}{dt}\right)\psi_n\exp\left(-\frac{iE_nt}{\hbar}\right) = \sum_n a_nH'\psi_n\exp\left(-\frac{iE_nt}{\hbar}\right). \tag{4.41}$$

We assume that H' is so small that the initial $a_n(0) = 1$ does not change much with time, so that, on the right-hand side, we can approximate a_n ($a_n(0) = 1, a_n(\neq 0) = 0$):

$$\sum_n i\hbar\frac{da_n}{dt}\psi_n\exp\left(-\frac{iE_nt}{\hbar}\right) = H'\psi_i\exp\left(-\frac{iE_it}{\hbar}\right), \tag{4.42}$$

in which $a_i(0) = 1$. Multiplying on the left by ψ_j^* and integrating over spatial coordinates, we obtain

$$i\hbar\frac{da_j}{dt} = \langle j|H'|i\rangle\exp\left[-\frac{i(E_j - E_i)t}{\hbar}\right]. \tag{4.43}$$

Writing $(E_j - E_i) = \hbar\omega_{ji}$ and substituting for A from Eq. 4.30, we have

$$i\hbar\frac{da_j}{dt} = \left\langle j\left|\frac{e}{mc}A_0\cdot p\,e^{-ik\cdot r}\right|i\right\rangle e^{-i(\omega_{ji}+\omega)t} + \left\langle j\left|\frac{e}{mc}A_0\cdot p\,e^{ik\cdot r}\right|i\right\rangle e^{i(\omega_{ji}-\omega)t}, \tag{4.44}$$

which on integration over t from $t = 0$ gives

$$a_j(t) = \left\langle j\left|\frac{e}{mc}A_0\cdot p\,e^{-ik\cdot r}\right|i\right\rangle\left[\frac{1 - e^{-i(\omega_{ji}+\omega)t}}{\hbar(\omega_{ji}+\omega)}\right] + \left\langle j\left|\frac{e}{mc}A_0\cdot p\,e^{ik\cdot r}\right|i\right\rangle\left[\frac{1 - e^{i(\omega_{ji}-\omega)t}}{\hbar(\omega_{ji}-\omega)}\right]. \tag{4.45}$$

Consider $E_j > E_i$ so that the transition $i \to j$ is an absorption process. The first term can then be ignored

2 This choice – the transverse gauge – corresponds to the reasonable assumption that radiation propagates perpendicular to the direction of the electric and magnetic field vectors, and that the induced motion of the electron is along the electric field given by A.

compared with the second because of its large value when $\omega_{ji} \sim \omega$. Then the transition probability at time t is

$$|a_j(t)|^2 = \left|\left\langle j \left| \frac{e}{mc} A_0 \cdot p\, e^{ik\cdot r} \right| i \right\rangle\right|^2 \frac{\sin^2[\{(\omega_{ji}-\omega)/2\}t]}{\hbar^2\{(\omega_{ji}-\omega)/2\}^2}. \tag{4.46}$$

Note that

$$\begin{aligned}&\{1-e^{i(\omega_{ji}-\omega)t}\}\{1-e^{-i(\omega_{ji}-\omega)t}\}\\ &\quad = 2-2\cos\{2(\omega_{ji}-\omega)t/2\}\\ &\quad = 2[1-\cos^2\{(\omega_{ji}-\omega)t/2\}+\sin^2\{(\omega_{ji}-\omega)t/2\}]\\ &\quad = 4\sin^2\{(\omega_{ji}-\omega)t/2\}. \end{aligned}\tag{4.47}$$

But $|a_j(t)|^2$ increases with t^2 for small t, a behaviour that does not look physically realistic before recognizing the distribution function

$$\frac{\sin^2\{(\omega_{ji}-\omega)/2\}t}{\{(\omega_{ji}-\omega)/2\}^2} \to 2\pi\hbar t\delta(E_{ji}-\hbar\omega). \tag{4.48}$$

The function on the left peaks at $(\omega_{ji}-\omega)=0$ but decreases rapidly with decaying oscillations. It is similar to the delta function, namely

$$\delta(x) = \lim_{\epsilon\to 0} \frac{\epsilon}{\pi}\frac{\sin^2(x/\epsilon)}{x^2} \tag{4.49}$$

and therefore approaches $2\pi t\delta(\omega_{ji}-\omega)$ for large t using $\delta(ax)=\delta(x)/|a|$, and hence the right-hand side of Eq. 4.48.

With the unit vector $\hat{e}$ of polarization along the direction of E or A_0 in $A_0 = \hat{e}\, A_0$, the transition probability per unit time can be written as

$$\frac{|a_j(t)|^2}{t} = \frac{2\pi e^2}{\hbar} A_0^2 \left|\left\langle j \left| \frac{\hat{e}\cdot p}{mc}\, e^{ik\cdot r} \right| i \right\rangle\right|^2 \delta(E_{ij}-\hbar\omega) \tag{4.50}$$

$$= \frac{2\pi}{\hbar}\, \left|\langle j|H'|i\rangle\right|^2 \delta(E_j - E_i), \tag{4.51}$$

where $\langle\ldots\rangle$ in (Eq. 4.50) is dimensionless, and the expression (Eq. 4.51) is explicitly an inverse time observing the identity $\delta(ax) = \delta(x)/|a|$. In first order, this is the probability per unit time for a transition from an arbitrary state $|i\rangle$ to a state $\langle j|$ under the perturbing potential H'.

It is worthwhile to extend the formulation of the discrete bound–bound transitions to the case where the final states may encompass a range of continuous energies, such as in the vicinity of a resonance in the bound–free process. The well-known *Fermi's golden rule* gives the *radiative transition rate* (as opposed to the probability Eq. 4.51)

$$T_{ij} = \frac{2\pi}{\hbar}|\langle j|H'|i\rangle|^2 n_j, \tag{4.52}$$

where n_j is the density of final states j, or the number of states per unit energy, with lifetime $\tau_j = 1/T_{ji}$. Following the uncertainty principle, the energy 'width' of the state is then

$$\Gamma_j = \hbar T_{ji}, \tag{4.53}$$

identified as the *autoionization width* and considered in detail in the three subsequent chapters.

4.3 Transition matrix elements

The derivation for transition probability in the previous section has been carried through at a single frequency, and should be integrated over the frequency distribution of the incident waves. The field energy is given by

$$W = \frac{1}{8\pi}\int_V (E^*\cdot E + B^*\cdot B)\, dV, \tag{4.54}$$

which on substitution of A and using some identities gives

$$W = (V/2\pi c^2)\sum_k\sum_j \omega_k^2 |A_{kj}|^2, \tag{4.55}$$

where the sum over k is over all frequencies, and the sum over j reflects possible polarizations of the radiation. This gives radiation energy density $\rho = W/V = (1/2\pi c^2)\omega^2|A|^2$. Introducing ρ in the probability expression, Eq. 4.50

$$\frac{|a_j(t)|^2}{t} = \frac{4\pi^2c^2e^2}{\hbar\omega^2}\left|\left\langle j \left| \frac{\hat{e}\cdot p}{mc}\, e^{ik\cdot r} \right| i \right\rangle\right|^2 \times \rho(\nu)\,\delta(E_{ij}-\hbar\omega). \tag{4.56}$$

The total transition probability over a distribution of frequencies is obtained by integrating as

$$\begin{aligned} P_{ij} &= \int \frac{|a_j(t)|^2}{t} d\nu \\ &= \frac{4\pi^2c^2e^2}{\hbar^2}\left|\left\langle j \left| \frac{\hat{e}\cdot p}{mc}\, e^{ik\cdot r} \right| i \right\rangle\right|^2 \\ &\quad \times \int \frac{\rho(\nu)}{\omega^2}\delta(\omega_{ij}-\omega)\frac{d\omega}{2\pi}, \end{aligned}\tag{4.57}$$

again exploiting the identity $\delta(ax) = \delta(x)/|a|$. The major contributions to the integral come from a narrow region where $\omega = \omega_{ji}$ during absorption. We therefore put

$\omega^2 = \omega_{ji}^2$ and assume that the radiation density does not vary much in this region, that is $\rho(\nu) = \rho(\nu_{ji})$. Then

$$P_{ij} = 2\pi \frac{c^2 e^2}{h^2 \nu_{ji}^2} \left| \left\langle j \left| \frac{\hat{e} \cdot \boldsymbol{p}}{mc} \mathrm{e}^{\mathrm{i}\boldsymbol{k}\cdot\boldsymbol{r}} \right| i \right\rangle \right|^2 \rho(\nu_{ji}) \tag{4.58}$$

is the transition probability between two levels i and j per atom per unit time. Note that the transition probability is independent of time. A similar expression for $|a_j(t)|^2$ can be obtained from emission, where the indices (i, j) will interchange and $\mathrm{e}^{\mathrm{i}\boldsymbol{k}\cdot\boldsymbol{r}}$ will be replaced by $\mathrm{e}^{-\mathrm{i}\boldsymbol{k}\cdot\boldsymbol{r}}$.

Using Eq. 4.9 the absorption coefficient B_{ij} can now be obtained as

$$B_{ij} = P_{ij}/\rho(\nu) = 2\pi \frac{c^2 e^2}{h^2 \nu_{ji}^2} \left| \left\langle j \left| \frac{\hat{e} \cdot \boldsymbol{p}}{mc} \mathrm{e}^{\mathrm{i}\boldsymbol{k}\cdot\boldsymbol{r}} \right| i \right\rangle \right|^2. \tag{4.59}$$

For non-degenerate levels we have $B_{ij} = B_{ji}$. This result is important, as it is an example of microscopic reversibility, the quantum mechanical basis for the principle of detailed balance. In a central-field potential $V(\boldsymbol{r}) = V(r)$, the states i, j are products of radial functions and spherical harmonics Y_{lm} discussed in Chapter 2. In that case, or for random relative orientation between $\hat{\mathbf{e}}$ and $\boldsymbol{p}$, the average of $\cos^2\vartheta$ from the dot product contributes a factor of $\frac{1}{3}$. The above equation for unpolarized radiation is, therefore,

$$B_{ij} = \frac{2\pi}{3} \frac{c^2 e^2}{h^2 \nu_{ji}^2} \left| \left\langle j \left| \frac{p}{mc} \mathrm{e}^{\mathrm{i}\boldsymbol{k}\cdot\boldsymbol{r}} \right| i \right\rangle \right|^2. \tag{4.60}$$

The coefficients A and B are related (Eq. 4.18, Exercise 4.1), so if one coefficient is determined from the transition matrix element as above, then the other is as well.

4.4 Multipole expansion

Evaluating the dimensionless transition matrix element $\langle \ldots \rangle$ in Eq. 4.60 is not trivial, because it involves a matrix or exponential operator and several commutation relations. This is readily anticipated from the familiar Legendre expansion

$$\begin{aligned} \mathrm{e}^{\mathrm{i}\boldsymbol{k}\cdot\boldsymbol{r}} &= 1 + \mathrm{i}\boldsymbol{k}\cdot\boldsymbol{r} + (\mathrm{i}\boldsymbol{k}\cdot\boldsymbol{r})^2/2! + \cdots \\ &= \sum_{l=0}^{\infty} \mathrm{i}^l\,(2l+1)\,j_l(kr)\,\mathrm{P}_l(\cos\vartheta) \end{aligned} \tag{4.61}$$

with spherical Bessel functions $j_l(kr)$ (and ϑ in the Legendre polynomial as the angle between the vector $\boldsymbol{k}$ describing the radiation field and the particle in position $\boldsymbol{r}$). Consider an electron at a distance of 10^{-8} cm, twice a Bohr radius from the nucleus, and visible light, when the wavenumber k is of order $10^5\,\mathrm{cm}^{-1}$. In this long-wavelength scenario, Eq. 4.61 does not deviate much from the identity 1 over the volume of an ion, and a good answer can be expected – if the angular momenta L associated with i and j are not both zero and differ by one unit, at most (case of circular polarization $\hat{e}_{\pm 1}$). This leads to the *electric dipole approximation* $\mathrm{e}^{\mathrm{i}\boldsymbol{k}\cdot\boldsymbol{r}} \approx 1$. Successive interaction terms $\boldsymbol{k}\cdot\boldsymbol{r}$ between photon $\boldsymbol{k}$ and electron $\boldsymbol{r}$ contribute one more weakening factor α to the magnitude and correspond to higher-order components in radiative transitions. We discuss the effect of incident electromagnetic radiation for the first two terms in the expansion and of related multipole moments in the following sections.

4.5 Electric dipole approximation

Dipole and non-dipole transitions are generally referred to as allowed and forbidden transitions, respectively. The practical difference lies in transition rates that differ by orders of magnitude, as determined by intrinsic atomic properties. Evaluation of the particular transition matrix elements involves the use of appropriate moment operators. We first carry through the formulation for the dipole moment, and later generalize to non-dipole transitions.

The electric dipole transition matrix element has been identified as the first term of the exponential factor $\mathrm{e}^{\mathrm{i}\boldsymbol{k}\cdot\boldsymbol{r}}$. The physical basis of this approximation becomes evident through simple considerations. Incident radiation wavelengths in the optical–UV region of $\lambda \sim 10^3$–10^4 Å are large compared with the size a of an atom, which is of order a_0/Z. With the magnitude $k = 2\pi/\lambda$ of the radiation wave vector one has

$$\begin{aligned} ka &\sim \frac{2\pi}{\lambda}\frac{a_0}{Z} = \frac{a_0}{Zhc}(E_j - E_i) < \frac{a_0}{Zhc}\frac{Z^2 e^2}{a_0} \\ &= Z\alpha \sim \frac{Z}{137}, \end{aligned} \tag{4.62}$$

where Z^2e^2/a_0 is the binding energy of a hydrogen-like ground state electron. The condition means that the amplitude of the radiation wave varies little over the much smaller size of the atom. Such a situation obviously prevails when the wavelength of the incident radiation $\lambda \gg a_0 = 0.529$ Å. That is indeed the case for electromagnetic radiation of the strongest spectral lines from most astrophysical sources. For example the IR–radio range begins with $\lambda > 10\,000$ Å, the optical (visible) radiation lies in the 3000–7000 Å band (Fig. 1.1). So $a_0/\lambda \approx 10^{-3}$ at $\lambda_{\mathrm{opt}} = 5000$ Å. The physical consequence is that the atoms feel a nearly uniform external electric field, which thereby induces an electric dipole on the atomic charges.

The B-coefficient for absorption transition probability per unit time (Eq. 4.9) can now be written quantum mechanically as

$$P_{ij} = \frac{2\pi c^2 e^2}{h^2 \nu_{ji}^2} \left| \left\langle j \left| \frac{\hat{e} \cdot p}{mc} \right| i \right\rangle \right|^2 \rho(\nu_{ji}). \tag{4.63}$$

Let us assume that the electric vector is polarized along the x-axis. The momentum operator of an unperturbed atom can be written in terms of the Hamiltonian H_0 as

$$\langle j | p_x | i \rangle = \left\langle j \left| m \frac{dx}{dt} \right| i \right\rangle = \frac{im}{\hbar} \langle j | H_0 x - x H_0 | i \rangle \tag{4.64}$$

for the unperturnbed atom. With the commutation relation $[x_i, p_j]_\psi = x_i p_j \psi - p_j (x_i \psi) = x_i p_j \psi - (p_j x_i)\psi - (p_j \psi) x_i = -i\hbar \delta_{ij} \psi$ one obtains

$$[\boldsymbol{r}, \boldsymbol{p}^2] = \boldsymbol{p} \cdot [\boldsymbol{r}, \boldsymbol{p}] + [\boldsymbol{r}, \boldsymbol{p}] \cdot \boldsymbol{p} = 2i\hbar \boldsymbol{p}. \tag{4.65}$$

Replacing $\boldsymbol{p}^2 = 2m(H_0 - V)$, $[\boldsymbol{r}, H_0] = i\hbar \frac{\boldsymbol{p}}{m}$: since H_0 is Hermitian,

$$\left\langle j \left| \frac{p_x}{mc} \right| i \right\rangle = \frac{i}{\hbar c}(E_j - E_i) \langle j | x | i \rangle = \frac{2\pi i}{c} \nu_{ji} \langle j | x | i \rangle, \tag{4.66}$$

where $h\nu_{ji} = E_j - E_i$. Loosely writing $\boldsymbol{r}$ for (x, y, z):

$$\frac{e}{mc} \langle j | \boldsymbol{p} | i \rangle = \frac{2\pi i e}{c} \nu_{ji} \langle j | \boldsymbol{r} | i \rangle = \frac{2\pi i}{c} \nu_{ji} \langle j | \boldsymbol{D} | i \rangle, \tag{4.67}$$

where $\boldsymbol{D} = e\boldsymbol{r}$, charge e times the mean distance between opposite charges, is the dipole moment operator.[3] The *transition matrix element* with operator $\boldsymbol{r}$ in Eq. 4.67 will lead to radiative results in *length* form, and $\boldsymbol{p}$ to the *velocity* formulation: see equations like (Eq. 4.80) further along. The integration in Eq. 4.67 is over all space with respect to the initial and final states i and j. The associated dipole allowed transition is denoted as an *E1 transition*.

We rewrite the E1 absorption transition probability per unit time (Eq. 4.63),

$$\begin{aligned} P_{ij} &= \frac{2\pi c^2}{h^2 \nu_{ji}^2} \left| \left\langle j \left| \frac{e}{mc} \hat{e} \cdot \boldsymbol{p} \right| i \right\rangle \right|^2 \rho(\nu_{ji}) \\ &= \frac{2\pi}{\hbar^2} |\langle j | \hat{e} \cdot \boldsymbol{r} | i \rangle|^2 \rho(\nu_{ji}), \end{aligned} \tag{4.68}$$

and the absorption B coefficient as

$$B_{ij} = \frac{2\pi}{\hbar^2} |\langle j | \hat{e} \cdot \boldsymbol{r} | i \rangle|^2 = \frac{2\pi e^2}{h^2 m^2 \nu_{ji}^2} |\langle j | \hat{e} \cdot \boldsymbol{p} | i \rangle|^2. \tag{4.69}$$

While E1 transitions are referred to as allowed transitions, it really means that electric dipole transitions are the strongest types of electromagnetic transitions corresponding to the dominant first term in the multipole expansion of the radiation field (Eq. 4.61). However, a very important caveat is that a transition is allowed if and only if the initial and final states can be connected by an electric dipole moment. That in turn depends on the angular and spin symmetries of the states involved, and corresponding selection rules discussed later.

The transition matrix element yields several quantities that are physically equivalent but useful to define for different applications. In the dipole approximation, the transition probability from state i to j per unit time, by absorption of electric dipole radiation of energy density $\rho(\nu)$ per unit frequency, polarized in the x-direction, is

$$P_{ij} = \frac{2\pi e^2}{\hbar^2} |\langle j | x | i \rangle|^2 \rho(\nu_{ij}), \tag{4.70}$$

and for unpolarized radiation, averaging out the effects of $\hat{e}$,

$$P_{ij} = \frac{2\pi e^2}{3\hbar^2} |\langle j | \boldsymbol{r} | i \rangle|^2 \rho(\nu_{ij}), \tag{4.71}$$

where

$$|\langle j | \mathbf{r} | i \rangle|^2 = |\langle j | x | i \rangle|^2 + |\langle j | y | i \rangle|^2 + |\langle j | z | i \rangle|^2. \tag{4.72}$$

This expression applies to induced emission when the levels are non-degenerate. In the semi-classical treatment, the transition probability A_{ji} per unit time for spontaneous emission by electric dipole radiation is obtained as

$$B_{ij} = \frac{P_{ij}}{\rho(\nu)} = \frac{8\pi^3 e^2}{3h^2} |\langle j | \boldsymbol{r} | i \rangle|^2. \tag{4.73}$$

Then Einstein's A coefficient, the radiative decay rate, reads

$$A_{ji} = \frac{8\pi \nu_{ji}^2}{c^3} h\nu_{ji} B_{ij} \frac{g_i}{g_j} = \frac{64\pi^4 e^2}{3hc^3} \nu_{ji}^3 \frac{g_i}{g_j} |\langle j | \boldsymbol{r} | i \rangle|^2 \tag{4.74}$$

$$= \frac{4e^2}{3hc^3} \omega_{ji}^3 \frac{g_i}{g_j} |\langle j | \boldsymbol{r} | i \rangle|^2, \tag{4.75}$$

with $\omega_{ji} = 2\pi \nu_{ji}$ for unpolarized radiation. A_{ji} is also referred to as the *transition probability* (although 'transition rate' per unit time might be more appropriate).

To get an order-of-magnitude estimate for A at optical wavelengths we can put $|\langle j | \boldsymbol{r} | i \rangle|^2 \sim a_0^2$ for a transition allowed by electric dipole radiation. To within an order of magnitude (i.e., factors of $\frac{1}{4}$ or 2π), the quantity A becomes

$$A \approx \frac{\omega^3}{\hbar c^3} (e a_0)^2 \approx \alpha \omega^3 \left(\frac{a_0}{c}\right) \propto 10^8 \text{s}^{-1} \tag{4.76}$$

$$\approx \alpha^3 \omega^3 \tau_0^2 \propto 10^8 \text{s}^{-1}. \tag{4.77}$$

[3] **D** is rather a superfluous quantity: as in atomic structure work the charge enters radiative results only as the $e^2 = 2a_0$ Ry – see Eq. 1.23 – and one can do without the third expression in Eq. 4.67; but it is often employed in literature.

The travel time of light is $2a_0/c$ across the ground state of H atom, whereas the Rydberg time $\tau_0 \approx 50$ attoseconds for an electron to traverse H with velocity αc (see Section 1.9). Conversion factors from atomic to cgs units are:

$$A_{ji} = A_{ji}(\mathrm{au})/\tau_0 \ \mathrm{second}^{-1},$$
$$\tau_0 = 2.419 \times 10^{-17} \ \mathrm{second}, \tag{4.78}$$

where τ_0 is also called the atomic unit of time. The transition matrix element depends on the orientation and the direction of polarization, and hence on the magnetic quantum numbers m_i, m_j of the initial and final states. Suppose for the general case that the lower level is g_i-fold degenerate in m_i and the upper level is g_j-fold degenerate in m_j; then the decay rate from each state j is the same, meaning

$$\sum_{m_i} |\langle jm_j|\boldsymbol{r}|im_i\rangle|^2 = \sum_{m_i} \left|\langle j|r|i\rangle \left\langle jm_j \left|\mathsf{C}^{[1]}_{mj-m_i}\right| im_i\right\rangle\right|^2 \tag{4.79}$$

is independent of m_j, and $\mathsf{C}^{[1]}$ is an algebraic Clebsch–Gordon coefficient, defined in Appendix C. For each j the coefficient A implies a summation over degenerate final levels of i. Hence

$$A_{ji} = \frac{4}{3}\frac{\omega_{ji}^3 e^2}{\hbar c^3} \sum_{m_i} |\langle jm_j|\boldsymbol{r}|im_i\rangle|^2 = \frac{4e^2}{3m^2}\frac{\omega_{ji}}{\hbar c^3} \sum_{m_i} |\langle jm_j|\boldsymbol{p}|im_i\rangle|^2 : \tag{4.80}$$

pick one observable of the canonical pair $(\boldsymbol{r}, \boldsymbol{p})$ (and approximate the canonical by the unperturbed mechanical momentum $\boldsymbol{p} \approx m\boldsymbol{v}$ of the electron), and one gets A_{ji} either in the length form or the velocity form.

4.5.1 Line strengths

Consider a lower degenerate level with g_i values of m_i and an upper degenerate level with g_j values m_j. Recalling the footnote past Eq. 4.67 we introduce the line strength as the effective area[4] or 'cross section'

$$S_{ji} = \sum_{m_j}\sum_{m_i} |\langle jm_j|\boldsymbol{r}|im_i\rangle|^2 \tag{4.81}$$
$$= \frac{1}{m^2\omega_{ji}^2}\sum_{m_j}\sum_{m_i} |\langle jm_j|\boldsymbol{p}|im_i\rangle|^2 \tag{4.82}$$

facing radiation at excitation energy E_{ji}, i.e., without a charge factor e^2, which cancels in *all* applications. The line strength remains unchanged with interchange of indices, i.e., it is symmtrical: $S_{ji} = S_{ij}$. As in Eq. 4.67, length and velocity form are related by the operator identity $\boldsymbol{p} = \mathrm{i}\hbar\nabla$. For the radial component is $p = \mathrm{i}\hbar\frac{\partial}{\partial r}$, while the angular part, involving just the Racah tensor, remains unchanged (see Appendix C). In line strength terms, the A coeffcient (with $e^2 = \alpha\hbar c$ outside S) reads

$$A_{ji} = \frac{4}{3}\frac{\omega_{ji}^3 e^2}{\hbar c^3}\frac{S_{ji}}{g_j} = \frac{2\alpha\,\omega_{ji}^3}{3\pi c^2}\frac{S_{ji}}{g_j}. \tag{4.83}$$

4.5.2 Oscillator strengths

The *oscillator strength* is a useful dimensionless response function, in the first place applied to electric dipole transitions [75]. It offers a measure of the *intrinsic strength* of an atomic transition and the intensity of a spectral line. It usually serves to describe the radiative *absorption* strength (see Chapter 8), which is, of course, related to the *emission* strength via detailed balance as

$$f_{ij} = -\frac{g_j}{g_i} f_{ji} \quad \left(\text{or as '}gf\text{-values'} \quad g_i f_{ij} = -g_j f_{ji}\right). \tag{4.84}$$

The name derives from the analogy with an ensemble of classical oscillators with dipole moments in shifted phase with the incoming electromagnetic wave. The definition of absorption oscillator strength is such that f_{ij} is a fraction, so normalized as to obey the sum rule

$$\sum_j f_{ij} = 1, \tag{4.85}$$

if i is the ground level of an atomic system with *one* active electron. For an excited state j, emission to lower levels i, as well as absorption to higher levels k, occurs, leading to the one-electron sum rule,

$$\sum_{i,k} (f_{ji} + f_{jk}) = 1, \tag{4.86}$$

where f_{ji} is a negative fraction. Generally, if N electrons participate in photo-excitation then the sum equals N, and one gets

$$\sum_{i,k} (f_{ji} + f_{jk}) = N. \tag{4.87}$$

We can express an averaged oscillator strength as the average over all inital degenerate states and summed over

4 This is consistent with applications such as for E1-type electron impact collision strength at the high-energy limit [74].

all final states,

$$\bar{f}_{j,i} = \frac{1}{2l_i+1}\sum_{m_i}\sum_{m_j} f_{ij}(m_i, m_j)$$
$$= \frac{2m}{3\hbar}\omega_{ij}\sum_{m_j}|\langle j|\mathbf{r}|i\rangle|^2, \qquad (4.88)$$

as the dimensionless atomic quantity common to all three Einstein coefficients with a norm compatible with the sum rules. It is used in general by omitting the *bar* over f. Expressed in terms of the line strength it becomes

$$f_{ij} = \frac{2m}{\hbar}\frac{\omega_{ij}}{3g_i}S_{ij} = \frac{1}{3g_i}\frac{E_j - E_i}{\mathrm{Ry}}\frac{S_{ij}}{a_0^2}, \qquad (4.89)$$

exploiting $\mathrm{Ry} = e^2/(2a_0)$, $a_0 = \hbar^2/(me^2)$. With this manifestly dimensionless quantity f, the B coefficient for absorption becomes

$$B_{ij} = \frac{8\pi^3}{3h^3}|\langle j|\mathbf{D}|i\rangle|^2 = \frac{4\pi^2 e^2}{h\nu_{ji}m}f_{ij}. \qquad (4.90)$$

Similarly the coefficient A for absorption from atomic state i to j reads

$$A_{ji} = \frac{\alpha^3}{\tau_0}\frac{g_i}{g_j}\frac{(E_j - E_i)^2}{\mathrm{Ry}^2}f_{ij}$$
$$= 8.032\times 10^9\,\frac{g_i}{g_j}\left(\frac{E_{ji}}{\mathrm{Ry}}\right)^2 f_{ij}\,\mathrm{s}^{-1}, \qquad (4.91)$$

where again the 'travel' time $\tau_0 = \hbar/\mathrm{Ry} \approx 50$ attoseconds across the Rydberg atom appears.[5]

Dipole oscillator strengths are also related to another useful quantity called *dipole polarizability* α_i^{d} of a level i [3],

$$\alpha_i^{\mathrm{d}} = \sum_j f_{ij}\alpha(j,i), \qquad (4.92)$$

where i is usually the ground state. The quantity $\alpha(j,i)$ is the *classical polarizability of an oscillator* defined as

$$\alpha(j,i) = \frac{e^2}{4\pi^2 m}\left(\frac{1}{\nu_{ij}^2 - \nu^2}\right), \qquad (4.93)$$

where ν is the frequency of the incident photon.

Exercise 4.3 *Using hydrogenic wavefunctions from Chapter 2, calculate A(Lyα: 1s–2p). Hint: the initial and final (sub)levels $|nlm>$ are $|100>$ and $|210>$, $|211>$, $|21-1>$. It is helpful to prove the following identity*

$$\frac{1}{2}| < 21\pm 1|x\pm y|100 > |^2 = | < 210|z|100 > |^2$$
$$= \frac{2^{15}}{3^{10}}a_0^2. \qquad (4.94)$$

A general example is described in the next section using the central field approximation.

4.6 Central-field approximation

From the relations derived above for photoabsorption or photo-emission, we note that the line strength $S_{ij} = \sum_{m_i}\sum_{m_j}|\langle jm_j|\mathbf{r}|im_i\rangle|^2$ is independent of any kinematical factors, and is common to both A- and f-values. Later we will see that the same quantity S_{ij} can be generalized for bound–free radiative transition probabilities as well (Chapter 6).

In this section, we employ the central-field approximation to evaluate the wavefunctions and expressions for the S_{ij}. Whereas state-of-the-art calculations using advanced methods, such as the multi-configuration Hartree–Fock method or the R-matrix method are rather complex in nature, it is useful to work through an example to show the essential details of the calculations with a convenient potential. Like the hydrogenic problem studied in Chapter 2, in the central-field approximation the potential depends only on the radial coordinate. Hence we can express an atomic state by independent radial and angular functions as $R_{nl}(r)\mathrm{Y}_{lm}(\hat{r})$. Let us define a quantity

$$T_{ij} = \sum_{m_j}| < jm_j|\mathbf{D}|im_i > |^2, \qquad (4.95)$$

and begin with the transition matrix element

$$< j|\mathbf{r}|i > = \int_0^\infty R^*_{n_jl_j}(r)rR_{n_il_i}(r)r^2\mathrm{d}r$$
$$\times\int \mathrm{d}\Omega\mathrm{Y}^*_{l_jm_j}(\hat{r})\hat{r}\mathrm{Y}_{l_im_i}(\hat{r}). \qquad (4.96)$$

Expand the radial operator as

$$\mathbf{r} = \frac{\hat{\mathbf{x}}}{2}[(x+\mathrm{i}y)+(x-\mathrm{i}y)] + \frac{\hat{\mathbf{y}}}{2\mathrm{i}}[(x+\mathrm{i}y)-(x-\mathrm{i}y)] + \hat{\mathbf{z}}z, \qquad (4.97)$$

which in spherical coordinates is

$$r\hat{\mathbf{r}} = \frac{\hat{\mathbf{x}}r}{2}[\sin\theta\mathrm{e}^{\mathrm{i}\phi}+\sin\theta\mathrm{e}^{-\mathrm{i}\phi}] + \frac{\hat{\mathbf{y}}r}{2\mathrm{i}}[\sin\theta\mathrm{e}^{\mathrm{i}\phi}-\sin\theta\mathrm{e}^{-\mathrm{i}\phi}]$$
$$+\hat{\mathbf{z}}r\cos\theta. \qquad (4.98)$$

[5] Take an experimenter's dial reading of $E = 10.2$ eV at some resonance, perhaps observing $f = 0.416$ as an absorption feature. If $g_i = 2$ and $g_j = 8$, then Eq. 4.91 yields $A = 4.70\times 10^8\,\mathrm{s}^{-1}$. It goes without saying that $\frac{3}{4} = 1 - \frac{1}{4}$; from the numerator, 10.2 eV, and the denominator, 13.59 eV, in the equation, the case is revealed as Ly α.

To carry out the angular integral we use the following three recursion relations:

$$\cos\theta \mathrm{Y}_{lm} = \left[\frac{(l-m+1)(l+m+1)}{(2l+1)(2l+3)}\right]^{1/2} \mathrm{Y}_{l+1,m} + \left[\frac{(l-m)(l+m)}{(2l-1)(2l+1)}\right]^{1/2} \mathrm{Y}_{l-1,m}$$

$$\mathrm{e}^{\mathrm{i}\phi}\sin\theta \mathrm{Y}_{lm} = \left[\frac{(l+m+1)(l+m+2)}{(2l+1)(2l+3)}\right]^{1/2} \mathrm{Y}_{l+1,m+1} - \left[\frac{(l-m)(l-m-1)}{(2l-1)(2l+1)}\right]^{1/2} \mathrm{Y}_{l-1,m+1}$$

$$\mathrm{e}^{-\mathrm{i}\phi}\sin\theta \mathrm{Y}_{lm} = -\left[\frac{(l-m+1)(l-m+2)}{(2l+1)(2l+3)}\right]^{1/2} \mathrm{Y}_{l+1,m-1} + \left[\frac{(l+m)(l+m-1)}{(2l-1)(2l+1)}\right]^{1/2} \mathrm{Y}_{l-1,m-1} \tag{4.99}$$

Let $< j|z|i > \equiv z_{n_i l_i m_i}^{n_j l_j m_j}$, and using the first recursion relation, the z-component of the transition matrix element is

$$\begin{aligned} z_{n_i l_i m_i}^{n_j l_j m_j} &= \int_0^\infty R^*_{n_j l_j}(r) R_{n_i l_i}(r) r^3 \mathrm{d}r \int \mathrm{d}\Omega \mathrm{Y}^*_{l_j m_j}(\hat{r}) \\ &\times \left\{ \left[\frac{(l_i - m_i + 1)(l_i + m_i + 1)}{(2l_i + 1)(2l_i + 3)}\right]^{1/2} \times \mathrm{Y}_{l_i+1,m_i} + \left[\frac{(l_i - m_i)(l_i + m_i)}{(2l_i - 1)(2l_i + 1)}\right]^{1/2} \mathrm{Y}_{l_i-1,m_i} \right\} \\ &= \int_0^\infty R^*_{n_j l_j}(r) R_{n_i l_i}(r) r^3 \mathrm{d}r \\ &\times \left\{ \left[\frac{(l_i - m_i + 1)(l_i + m_i + 1)}{(2l_i + 1)(2l_i + 3)}\right]^{1/2} \times \delta_{l_j,l_i+1}\delta_{m_j,m_i} + \left[\frac{(l_i - m_i)(l_i + m_i)}{(2l_i - 1)(2l_i + 1)}\right]^{1/2} \delta_{l_j,l_i-1}\delta_{m_j,m_i} \right\}. \end{aligned} \tag{4.100}$$

The selection rules manifest themselves in accordance with the symmetries of initial and final levels. For non-vanishing elements $\Delta m = m_j - m_i = 0$, $\Delta l = l_j - l_i = \pm 1$. Hence all compoments of $z_{n_i l_i m_i}^{n_j l_j m_j}$ vanish except

$$z_{n_i l_i m_i}^{n_j, l_i+1, m_j} = \left[\frac{(l_i+1)^2 - m_i^2}{(2l_i+1)(2l_i+3)}\right]^{1/2} R_{n_i l_i}^{n_j, l_i+1},$$
$$z_{n_i l_i m_i}^{n_j, l_i-1, m_j} = \left[\frac{l_i^2 - m_i^2}{(2l_i-1)(2l_i+1)}\right]^{1/2} R_{n_i l_i}^{n_j, l_i-1}, \tag{4.101}$$

where

$$R_{n_i l_i}^{n_j l_j} = \int_0^\infty R^*_{n_j l_j}(r) R_{n_i l_i}(r) r^3 \mathrm{d}r. \tag{4.102}$$

Similarly,

$$\begin{aligned} (x \pm \mathrm{i}y)_{n_i l_i m_i}^{n_j l_j m_j} &= \int_0^\infty R^*_{n_j l_j}(r) R_{n_i l_i}(r) r^3 \mathrm{d}r \\ &\left\{ \left[\frac{(l_i \pm m_i + 1)(l_i \pm m_i + 2)}{(2l_i + 1)(2l_i + 3)}\right]^{1/2} \times \mathrm{Y}_{l_i+1,m_i\pm1} \mp \left[\frac{(l_i \mp m_i)(l_i \mp m_i + 1)}{(2l_i - 1)(2l_i + 1)}\right]^{1/2} \times \mathrm{Y}_{l_i-1,m_i\pm1} \right\} \end{aligned} \tag{4.103}$$

vanish unless $\Delta m = m_j - m_i = \pm 1$ and $\Delta l = l_j - l_i = \pm 1$. Therefore, the surviving components are

$$(x + \mathrm{i}y)_{n_i l_i m_i}^{n_j, l_i+1, m_i+1} = \left[\frac{(l_i + m_i + 1)(l_i + m_i + 2)}{(2l_i + 1)(2l_i + 3)}\right]^{1/2} \times R_{n_i l_i}^{n_j, l_i+1},$$
$$(x + \mathrm{i}y)_{n_i l_i m_i}^{n_j, l_i-1, m_i+1} = -\left[\frac{(l_i - m_i)(l_i - m_i - 1)}{(2l_i - 1)(2l_i + 1)}\right]^{1/2} \times R_{n_i l_i}^{n_j, l_i-1},$$
$$(x - \mathrm{i}y)_{n_i l_i m_i}^{n_j, l_i+1, m_i-1} = -\left[\frac{(l_i - m_i + 1)(l_i - m_i + 2)}{(2l_i + 1)(2l_i + 3)}\right]^{1/2} \times R_{n_i l_i}^{n_j, l_i-1},$$
$$(x - \mathrm{i}y)_{n_i l_i m_i}^{n_j, l_i-1, m_i-1} = \left[\frac{(l_i + m_i)(l_i + m_i - 1)}{(2l_i - 1)(2l_i + 1)}\right]^{1/2} \times R_{n_i l_i}^{n_j, l_i-1}. \tag{4.104}$$

If all directions in space are equivalent, then the atom can be in any of the m_i states with equal probabilities. Adding all the intensities of the transitions from a certain state $n_i l_i m_i$ to all sublevels m_j of the level $n_j l_j$, without regard to the direction of polarization of the interacting photon, one can find that the sum is independent of m_j. Substituting $< i| \equiv < \psi_{n_i l_i m_i}|$ and $< j| \equiv < \psi_{n_j l_j m_j}|$, we get for transitions $l_j = l_i + 1$,

$$\begin{aligned} T_{ij} &= \sum_{m_j} | < \psi_{n_j, l_{i+1}, m_j} |\mathbf{r}| \psi_{n_i l_i m_i} > |^2 \\ &= \sum_{m_j} | < \psi_{n_j, l_{i+1}, m_j} | \frac{\hat{\mathbf{x}}}{2}[(x + \mathrm{i}y) + (x - \mathrm{i}y)] \\ &+ \frac{\hat{\mathbf{y}}}{2i}[(x + \mathrm{i}y) - (x - \mathrm{i}y)] + \hat{\mathbf{z}} z | \psi_{n_i l_i m_i} > |^2. \end{aligned} \tag{4.105}$$

Substituting from Eq. 4.104

$$T_{ij} = \sum_{m_j} \left\{ \frac{1}{4} \left| (x+\mathrm{i}y)^{n_j,l_i+1,m_j}_{n_i l_i m_i} + (x-\mathrm{i}y)^{n_j,l_i+1,m_j}_{n_i l_i m_i} \right|^2 + \frac{1}{4} \left| (x+\mathrm{i}y)^{n_j,l_i+1,m_j}_{n_i l_i m_i} - (x-\mathrm{i}y)^{n_j,l_i+1,m_j}_{n_i l_i m_i} \right|^2 + \left| z^{n_j,l_i+1,m_j}_{n_i l_i m_i} \right|^2 \right\}, \tag{4.106}$$

which reduces to

$$T_{ij} = \frac{1}{2} \left| (x+\mathrm{i}y)^{n_j,l_i+1,m_i+1}_{n_i l_i m_i} \right|^2 + \frac{1}{2} \left| (x-\mathrm{i}y)^{n_j,l_i+1,m_i+1}_{n_i l_i m_i} \right|^2 + \left| z^{n_j,l_i+1,m_i}_{n_i l_i m_i} \right|^2. \tag{4.107}$$

Exercise 4.4 *Show that*

$$T_{ij} = \sum_{m_j} | < \psi_{n_j,l_{i+1},m_j} | \mathbf{r} | \psi_{n_i l_i m_i} > |^2 = \frac{l_i+1}{2l_i+1} \left| R^{n_j,l_i+1}_{n_i l_i} \right|^2. \tag{4.108}$$

In a similar manner, for transitions with $l_j = l_i - 1$,

$$T_{ij} = \sum_{m_j} | < \psi_{n_j,l_{i-1},m_j} | \mathbf{r} | \psi_{n_i l_i m_i} > |^2 = \frac{l_i}{2l_i+1} \left| R^{n_j,l_i-1}_{n_i l_i} \right|^2 \tag{4.109}$$

Now the line strength for $l_j = l_i + 1$ transitions is

$$S_{ij} = \sum_{m_j} \sum_{m_i} | < jm_j | \mathbf{D} | im_i > |^2 = \sum_{m_i} e^2 \frac{l_i+1}{2l_i+1} \left| \int_0^\infty R^*_{n_j l_{i+1}}(r) R_{n_i l_i}(r) r^3 \mathrm{d}r \right|^2. \tag{4.110}$$

and for $l_j = l_i - 1$ transitions,

$$S_{ij} = \sum_{m_i} e^2 \frac{l_i}{2l_i+1} \left| \int_0^\infty R_{n_j l_{i-1}}(r) R_{n_i l_i}(r) r^3 \mathrm{d}r \right|^2. \tag{4.111}$$

$S_{i,j}$ does not depend on m_i and hence $\sum_{m_i} = 2l_i + 1 = g_i$. Since $l_j = l_i \pm 1$, the values $(l_i + 1)$ and l_i are often referred to as $\max(l_i, l_j)$. The corresponding f-value for any two transitions is then

$$f_{ij} = \frac{E_{ji}(Ry)}{3 g_i e^2} S_{ij} = \frac{E_{ij}}{3} \frac{\max(l_i, l_j)}{2l_i+1} \left| \int_0^\infty R_{n_j l_j}(r) R_{n_i l_i}(r) r^3 \mathrm{d}r \right|^2, \tag{4.112}$$

and the A-values,

$$A_{ji}(s^{-1}) = 0.8032 \times 10^9 \tfrac{g_i}{g_j} E^2_{ji} f_{ij} = 0.8032 \times 10^9 \tfrac{g_i}{g_j} \tfrac{E^3_{ij}}{3} \tfrac{\max(l_i,l_j)}{2l_i+1} \times \left| \int_0^\infty R_{n_j l_j}(r) R_{n_i l_i}(r) r^3 \mathrm{d}r \right|^2. \tag{4.113}$$

The expressions derived above correspond to the length formulation. The derivation of the velocity forms is similar but a little longer. Evaluation of the quantity $T_{ij} = \sum_{m_j} | < jm_j | \mathbf{p} | im_i > |^2$ in velocity form can be carried out by writing

$$< j | \mathbf{p} | i > = < j | \frac{\hat{\mathbf{x}}}{2} [(p_x + \mathrm{i}p_y) + (p_x - \mathrm{i}p_y)] + \frac{\hat{\mathbf{y}}}{2i} [(p_x + \mathrm{i}p_y) - (p_x - \mathrm{i}p_y)] + \hat{\mathbf{z}} p_z | i >, \tag{4.114}$$

where, for example, $p_x = -\mathrm{i}\hbar \frac{\partial}{\partial x}$. Since $\mathbf{p}$ is a Hermitian operator, $|<j|\mathbf{p}|i>|^2 = |<i|\mathbf{p}|j>|^2$. Then we can get expressions for $<i|p_z|j>$, $<i|p_x + \mathrm{i}p_y|j>$ and $<i|p_x - \mathrm{i}p_y|j>$ using standard factors for spherical harmonics (see Appendix C), obeying the same selection rules for quantum numbers l and m in dipole transitions as for the length formulation. The non-vanishing terms for $l_j = l_i \pm 1$ are

$$\begin{aligned}
(p_z)^{n_j,l_i+1,m_i}_{n_i l_i m_i} &= -\mathrm{i}\hbar \left[\frac{(l_i+1)^2 - m_i^2}{(2l_i+1)(2l_i+3)} \right]^{1/2} \times (R_+)^{n_j,l_i+1}_{n_i l_i}, \\
(p_z)^{n_j,l_i-1,m_i}_{n_i l_i m_i} &= -\mathrm{i}\hbar \left[\frac{l_i^2 - m_i^2}{(2l_i-1)(2l_i+1)} \right]^{1/2} \times (R_-)^{n_j,l_i-1}_{n_i l_i}, \\
(p_x \pm \mathrm{i}p_y)^{n_j l_i+1,m_i\mp 1}_{n_i l_i m_i} &= \pm \mathrm{i}\hbar \times \left[\frac{(l_i \mp m_i + 1)(l_i \mp m_i + 2)}{(2l_i+1)(2l_i+3)} \right]^{1/2} \times (R_+)^{n_j,l_i+1}_{n_i l_i}, \\
(p_x \pm \mathrm{i}p_y)^{n_j l_i-1,m_i\mp 1}_{n_i l_i m_i} &= \mp \mathrm{i}\hbar \left[\frac{(l_i \pm m_i)(l_i \pm m_i - 1)}{(2l_i+1)(2l_i-1)} \right]^{1/2} \times (R_-)^{n_j,l_i-1}_{n_i l_i},
\end{aligned} \tag{4.115}$$

where

$$\begin{aligned}
(R_+)^{n_j l_j}_{n_i l_i} &= \int_0^\infty R_{n_i l_i} \left[\frac{\mathrm{d}R_{n_j l_j}}{\mathrm{d}r} + (l_j + 1) \frac{R_{n_j l_j}}{r} \right] r^2 \mathrm{d}r, \\
(R_-)^{n_j l_j}_{n_i l_i} &= \int_0^\infty R_{n_i l_i} \left[\frac{\mathrm{d}R_{n_j l_j}}{\mathrm{d}r} - l_j \frac{R_{n_j l_j}}{r} \right] r^2 \mathrm{d}r.
\end{aligned} \tag{4.116}$$

Carrying out the summations, it can be shown that for $l_j = l_i + 1$,

$$T_{ij}^{l_i+1} = \sum_{m_j} || < jm_j|\mathbf{p}|im_i > ||^2$$
$$= \hbar^2 \frac{l_i+1}{2l_i+1} \left| \int_0^\infty R_{n_il_i} \times \left[\frac{dR_{n_jl_i+1}}{dr} + (l_i+2)\frac{R_{n_jl_i+1}}{r} \right] r^2 \mathrm{d}r \right|^2 . \quad (4.117)$$

With $P_{nl} = rR_{nl}$, the expression above reduces to

$$T_{ij}^{l_i+1} = \hbar^2 \frac{l_i+1}{2l_i+1} \left| \int_0^\infty P_{n_il_i} \times \left[\frac{\mathrm{d}P_{n_jl_i+1}}{\mathrm{d}r} + (l_i+1)\frac{P_{n_jl_i+1}}{r} \right] \mathrm{d}r \right|^2 . \quad (4.118)$$

The line strength for the $l_j = l_i + 1$ transition is then given by

$$S_{ji} = \frac{e^2}{m^2\omega_{ji}^2} \sum_{m_j} \sum_{m_i} | < jm_j|\mathbf{p}|im_i > |^2$$
$$= \frac{\hbar^2 e^2}{m^2\omega_{ji}^2} \sum_{m_i} \frac{l_i+1}{2l_i+1} \left| \int_0^\infty P_{n_il_i} \times \left[\frac{\mathrm{d}P_{n_jl_i+1}}{\mathrm{d}r} + (l_i+1)\frac{P_{n_jl_i+1}}{r} \right] \mathrm{d}r \right|^2 . \quad (4.119)$$

Similarly for $l_j = l_i - 1$,

$$T_{ij}^{l_i-1} = \sum_{m_j} || < jm_j|\mathbf{p}|im_i > ||^2 = \frac{\hbar^2 l_i}{2l_i+1} \left| \int_0^\infty P_{n_il_i} \times \left[\frac{\mathrm{d}P_{n_jl_i-1}}{\mathrm{d}r} - l_i\frac{P_{n_jl_i-1}}{r} \right] \mathrm{d}r \right|^2 , \quad (4.120)$$

and the corresponding line strength is

$$S_{ji} = \frac{\hbar^2 e^2}{m^2\omega_{ji}^2} \sum_{m_i} \frac{l_i}{2l_i+1} \left| \int_0^\infty P_{n_il_i} \times \left[\frac{\mathrm{d}P_{n_jl_i-1}}{\mathrm{d}r} - l_i\frac{P_{n_jl_i-1}}{r} \right] \mathrm{d}r \right|^2 . \quad (4.121)$$

The oscillator strengths and radiative decay rates can now be obtained from S_{ji} as before. A couple of useful formulae for oscillator strengths are as follows. The average oscillator strengths for transitions between two Rydberg states, n_il_i and $n_jl_i \pm 1$ where $\Delta n = n_j - n_i \ll n_i, n_j$, varies as $f(n_il_i \to n_jl_j) \sim n_i$ [76]. The average oscillator strengths for transition between two Rydberg states with principle quantum numbers n_i and n_j such that $\Delta n \gg 1$ is given by [76]

$$f(n_i \to n_j) = \frac{4n_i}{3^{1.5}\pi(\Delta n)^3} . \quad (4.122)$$

4.7 Length, velocity and acceleration

In Eq. 4.67 we introduced two equivalent forms of radiative transition matrix elements, involving either the length or the velocity (momentum) operator. As just outlined, all it needs when extending from 1 to N electrons is a summation over the particle index n appended on the respective observables. Then the oscillator strength reads

$$f_{ij} = \frac{2}{m\hbar\omega_{ij}} \left| \sum_{n=1}^{N} \int \Psi_j^* \left(\hat{e} \cdot \boldsymbol{p}_n \mathrm{e}^{\mathrm{i}\boldsymbol{k}\cdot\boldsymbol{r}_n} \right) \Psi_i \, \mathrm{d}\mathbf{r}_1 \ldots \mathrm{d}\mathbf{r}_N \right|^2 , \quad (4.123)$$

writing $\mathbf{dr}$ for the volume elements $r^2\mathrm{d}r\,\mathrm{d}\vartheta\,\cos\vartheta\mathrm{d}\varphi$ in the integral $\langle j|(\ldots)|i\rangle$. In the dipole approximation the operator exponential is cut down to the identity operator, when

$$f_{ij} = \frac{2}{m\hbar\omega_{ij}} \left| \hat{\mathbf{e}} \cdot \sum_{n=1}^{N} \int \Psi_j^* \, \boldsymbol{p}_n \, \Psi_i \, \mathrm{d}\mathbf{r}_1 \ldots \mathrm{d}\mathbf{r}_N \right|^2 \quad (4.124)$$

$$= \frac{2m\omega_{ij}}{\hbar} \left| \hat{\mathbf{e}} \cdot \sum_{n=1}^{N} \int \Psi_j^* \, \boldsymbol{r}_n \, \Psi_i \mathrm{d}\mathbf{r}_1 \ldots \mathrm{d}\mathbf{r}_N \right|^2 \quad (4.125)$$

are the velocity and length forms for N active electrons. Having derived the length from the momentum or velocity form with the help of the commutation relation $[\boldsymbol{r}, H] = \mathrm{i}\hbar\boldsymbol{p}/m$, one can again exploit the commutation property of the previous result with H. Next one arrives at

$$[\boldsymbol{p}, H] = [\boldsymbol{p}, V] = -\mathrm{i}\hbar\nabla V, \quad (4.126)$$

so that

$$\langle j|\,[\boldsymbol{p}, H]\,|i\rangle = (E_j - E_i)\langle j|\boldsymbol{p}|i\rangle = \langle j| - \mathrm{i}\hbar\nabla V|i\rangle, \quad (4.127)$$

which leads to the 'acceleration' form

$$\langle j|\boldsymbol{p}|i\rangle = \frac{-\mathrm{i}\hbar}{E_{ij}}\langle j|\nabla V|i\rangle = \frac{-\mathrm{i}\hbar e^2}{E_{ij}} \left\langle j \left| \frac{\boldsymbol{r}}{r^3} \right| i \right\rangle . \quad (4.128)$$

Exercise 4.5 *Derive the above relation by using $\nabla\frac{1}{r_{ij}} = -\nabla\frac{1}{r_{ji}}$ in $V = -\sum_n \frac{e^2}{r_n} + \sum_{n<m} \frac{e^2}{r_{mn}}$. Hence the oscillator strength has the following acceleration form,*

$$f_{ij} = \frac{2e^4\hbar}{m\omega_{ij}^2} \left| \hat{\mathbf{e}}.\sum_{n=1}^{N} \int \Psi_j^* \frac{\mathbf{r}_n}{r_n^3} \Psi_i d\mathbf{r_1}...d\mathbf{r_N} \right|^2 . \quad (4.129)$$

Chandrasekhar [77] verified that the three forms give the *same* value for the oscillator strength, if Ψ_i and Ψ_j are *exact* solutions of the Hamiltonian H. With approximate wavefunctions though, the three values differ. The extent of agreement among the three forms is an indicator of the consistency and accuracy of the approximation made in computing the wavefunctions. The basic physical fact is that the length, velocity and acceleration operators sample, or operate on, different parts of the wavefunction with different weights. The length operator in the integrand leads to predominant contributions from large distances r; the first derivative velocity operator $\partial/\partial r$ collects most of its contribution from shorter distances, and the acceleration operator with its second derivative $\partial^2/\partial r^2$ most sensitively probes the sections of large curvature in a wavefunction at even shorter distances. Thus small details in this range of the integrand carry an inordinate weight, and results from this form may be poor, while the other two, especially the length form, make good sense. Disagreement between length and velocity is a signal for caution, yet one may still be able to pick one as the better choice, often the length form, depending upon the nature of the observable.

4.8 Oscillator strengths for hydrogen

Transitions $nl \rightarrow n'\,l \pm 1$ in hydrogen are of obvious interest, since H I exists in nearly all astrophysical environments, from extremely cold interstellar and intergalactic media, to extremely hot stellar and galactic interiors (the latter are also the sites of massive black hole activity). For the Lyman series 1s–np an analytic expression can be used [78]:

$$g_1 f_{1n'} = \frac{2^9 n'^5 (n'-1)^{2n'-4}}{3(n'+1)^{2n'+4}}, \tag{4.130}$$

where $g_1 = 2$. For high values of n the oscillator strengths decrease rapidly,

$$g_1 f_{1n'} = \frac{2^9\left(1-\frac{1}{n'}\right)^{2n'-4}}{3n'^3\left(1+\frac{1}{n'}\right)^{2n'+4}} \approx \frac{2^9}{3n'^3}\frac{e^{-2}}{e^2} \approx 3.1\frac{1}{n^3}, \tag{4.131}$$

saying that the hydrogenic oscillator strengths decrease as n^{-3}.

A variety of approximations has been employed to overcome the difficulty with rapidly oscillating wavefunctions associated with high-(nl) orbitals (e.g., [79, 80]). We reproduce some of the key steps in a semi-classical calculation [80] using the well-known *Wentzel–Kramers–Brillouin* (WKB) method from basic quantum mechanics (viz. [4]), which gives [80]

$$f_{nl}^{n'l'} = \frac{\Delta(E_{n'l',nl})}{\text{Ry}}\frac{\max(l,l')}{3(2l+1)}\left|\frac{D_{nl}^{n'l'}}{a_0}\right|^2 Z^2 \tag{4.132}$$

at transition energies $\Delta(E_{n'l',nl})/\text{Ry} = (1/n^2 - 1/n'^2)$ between Rydberg levels and with radial integrals

$$D_{nl}^{n'l'} = \int R_{nl}(r)\,r\,R_{n'l'}(r)\mathrm{d}r. \tag{4.133}$$

For the 1s–2p transition this integral (of dimension length) becomes

$$\begin{aligned} D_{1s}^{2p}(\text{WKB}) &= \int R_{1s}\,r\,R_{2p}\mathrm{d}r \\ &= \frac{1}{\sqrt{6}}\left(\frac{4}{3}\right)^4\frac{a_0}{Z} = 1.2903\frac{a_0}{Z}, \end{aligned} \tag{4.134}$$

or f(1s–2p)= $4(2/3)^9$ = 0.4657, which differs considerably (~10%) from the exact value 0.4162 [81]. However, the accuracy of the WKB values improves with nl, as shown in Table 4.1, which gives a representative sample of f-values for relatively low-lying transitions.

We mentioned that the dipole integral becomes more involved with higher nl. To obtain a general form of the dipole integral for any transition using the WKB method [80], a spherically symmetrical system is represented by the sum of incoming and outgoing waves as

$$R_{nl}(r) = \frac{1}{2}\left[\Psi_{nl}^{(+)} + \Psi_{nl}^{(-)}\right], \tag{4.135}$$

where

$$R_{nl}^{(\pm)} = \frac{C_{nl}}{\sqrt{k(r)}}\exp(\pm \mathrm{i}[\phi(r) - \pi/4]). \tag{4.136}$$

The normalization coefficient of the wavefunction can be determined by the WKB normalization factor

$$|C_{nl}|^2 = \frac{2}{\int_{r_1}^{r_2}\frac{\mathrm{d}r}{k(r)}} = \frac{2m}{\pi\hbar^2}\frac{\mathrm{d}E_{nl}}{\mathrm{d}n}. \tag{4.137}$$

The function $k(r)$ is the wavenumber, proportional to the radial velocity,

$$k(r) = \sqrt{\frac{2m}{\hbar^2}}\left[E_{nl} + eV(r) - \frac{\hbar^2}{2m}\frac{\left(l+\frac{1}{2}\right)^2}{r^2}\right]^{1/2}, \tag{4.138}$$

where E_{nl} is the energy eigenvalue, $V(r)$ is the electrostatic potential, and ϕ is the phase integral

$$\phi^{\text{WKB}}(r) = \int_{r_1}^{r} q(r')\mathrm{d}r', \tag{4.139}$$

TABLE 4.1 Oscillator strengths for hydrogen.

Line	Transition nl–$n'l'$	$<\Delta E/\mathrm{Ry}>$ avg	$\left|D_{nl}^{n'l'}/a_0\right|^2$ exact	$\left|D_{nl}^{n'l'}/a_0\right|^2$ WKB	f_{abs} exact
$Ly\alpha$	1s–2p	0.75	0.166 479	1.565	0.416 2
$Ly\beta$	1s–3p	0.889	0.266 968	0.228 7	0.079 1
$Ly\gamma$	1s–4p	0.937 5	0.092 771	0.078 18	0.028 99
$Ly\delta$	1s–5p	0.96	0.043 56	0.036 48	0.013 9
$H\alpha$	2s–3p	0.138 9	9.393 1	9.884	0.434 9
$H\alpha$	2p–3s	0.138 9	0.880 6	1.004	0.013 59
$H\alpha$	2p–3d	0.138 9	22.543 4	21.273	0.695 8
$H\beta$	2s–4p	0.187 5	1.644 4		0.102 8
$H\beta$	2p–4s	0.187 5	0.146 2		0.003 045
$H\beta$	2p–4d	0.187 5	2.923 1		0.121 8
$Pa\alpha$	3s–4p	0.048 6	29.913	32.962 5	0.484 7
$Pa\alpha$	3p–4s	0.048 6	5.929 2	7.280 3	0.032 2
$Pa\alpha$	3p–4d	0.048 6	57.235 3	58.213 8	0.618 3
$Pa\alpha$	3d–4p	0.048 6	1.696 0	2.067 6	0.010 99
$Pa\alpha$	3d–4f	0.048 6	104.659	98.511	1.017 5

with $r_1 = r_1(n, l)$ as the inner turning point defined by $k_{nl}(r_1) = 0$. The hydrogenic case is characterized by

$$V(r) = Ze/r,\ E_{nl} = -\frac{Z^2e^2}{2a_0n^2},$$

$$C_{nl} = \sqrt{\frac{2Z^2}{\pi a_0^2 n^3}},\ r_1 = \frac{a_0 Z^2}{Z}(1-\epsilon), \tag{4.140}$$

where

$$\epsilon = \left[1 - \left(\left(l+\frac{1}{2}\right)^2 / n^2\right)\right]^{1/2} \tag{4.141}$$

is the 'eccentricity' of the elliptical orbit. Allowing the radius r to be complex, $k_{nl}(r)$ has a a branch-cut along $r_1 \le r \le r_2$; $k(r)$ is defined to be positive (real) above the real axis and negative (real) below. Then the WKB approximation yields

$$D_{nl}^{n'l'}(\mathrm{WKB}) = \frac{1}{2}\Re\left[R_{nl}^{(+)}(r_s) r_s R_{n'l'}^{(-)}(r_s\sqrt{\frac{2i\pi}{G'(r_s)}}\right], \tag{4.142}$$

where $\Re$ is the real part and r_s is a saddle point. Then

$$G(r) = k_{nl}(r) - k_{n'l'}(r) - \frac{i}{r} + \frac{1}{2}\left[\frac{k'_{nl}(r)}{k_{nl}} + \frac{k'_{n'l'}(r)}{k_{n'l'}}\right]. \tag{4.143}$$

The variables r_s, $G(r)$, $G'(r)$ are directly calculated from the central potential $V(r)$ (the 'prime' represents derivative with respect to r). In addition to the few low-n f-values in Table 4.1, a larger set of transitions can be computed (e.g., [82]).

The total inter-shell n–n' hydrogenic oscillator strengths obey the recurrence relation

$$f_{nn'} = \frac{1}{2n^2}\sum_{l=0}^{n-1} 2(2l+1)\left[f_{nl}^{n'l+1} + f_{nl}^{n'l-1}\right]. \tag{4.144}$$

For $l = 0$, a term $f_{nl}^{n'l-1}$ does not exist.

Finally, the spectroscopic quantities for hydrogenic ions of nuclear charge Z are related to those of hydrogen as

$$(\Delta E)_Z = Z^2 E_H,\quad S_Z = \frac{S_H}{Z^2},$$

$$A_Z = Z^4 A_H,\quad f_Z = f_H. \tag{4.145}$$

It may be noted that the *collective* oscillator strength $f_{nn'}$ remains the same regardless of Z. The generalized treatment of radiative transitions for non-hydrogenic systems is given in Appendix C.

4.9 Configuration interaction

For non-hydrogenic atoms and ions where analytic expressions are no longer possible, evaluation of transition matrix elements is subject to numerical accuracy

TABLE 4.2 Coefficients F_{-1} between the first two configurations of the carbon sequence: $F_0 = 0$ in the expansion $f^{abs} = F_0 + F_{-1}/Z + F_{-2}/Z^2 + \ldots$ for the oscillator strength in transitions within one principal shell.

$1s^22s^22p^2$	—	^3P			^1D		^1S
$1s^22s^12p^3$	$^5S^o$	$^3D^o$	$^3P^o$	$^3S^o$	$^1D^o$	$^1P^o$	$^1P^o$
W_1/Ry	6.6380	6.7599	6.8021	6.8724	6.8771	6.9193	
CI(+$1s^22p^4$)	—	1.180	0.834	1.394	2.799	1.060	3.479
Full CI	—	0.958	1.031	1.122	2.258	1.303	2.206

with which the wavefunctions can be computed. As discussed earlier in Chapters 2 and 3, the accuracy of the wavefunction representation of an atomic level depends on configuration mixing. We again refer to the example on neutral carbon discussed in Chapter 2 (see Table 2.5), with a simple three-configuration expansion C_1 :$1s^22s^22p^2$, C_2 :$1s^22s^12p^3$ and C_3 :$1s^22p^4$, belonging to the $n = 2$ complex. Employing Layzer's isoelectronic Z-expansion [8] referred to in Chapter 2, Table 4.2 illustrates the effect of configuration mixing for the high-Z absorption oscillator strengths from the even-parity ground configuration C_1 terms ^3P, ^1D, ^1S in transitions to odd-partiy terms associated with the first excited configuration C_2: $^5S^o$, $^3D^o$, $^3P^o$, $^3S^o$, $^1D^o$, $^1P^o$. Inclusion of the 'quasi-degenerate' C_3 configuration $1s^22p^4$ has no effect on terms arising from C_2 of opposite parity. But C_1 interacts strongly with C_3, and calculations ignoring this fact may be off by perhaps 50 % – see the last entry in the table with 2.206 : 3.479.

4.10 Fine structure

Relativistic effects couple the spin and angular momenta and give rise to fine structure. Recalling the discussion in Chapter 2, whereas LS coupling is a reasonable approximation for light elements $Z \sim 10$ when fine structure components are often blended, for higher Z they are readily resolved in practical observations. For most astrophysically abundant elements, such as the iron-group elements with $20 \leq Z \leq 30$, intermediate coupling suffices, i.e. $\mathbf{L} + \mathbf{S} = \mathbf{J}$ (however, for high-Z a pure jj-coupling scheme is required, see Section 2.13.2). The fine structure levels in intermediate coupling are designated as LSJ. Radiative transitions are governed by *selection rules* as discussed later (Table 4.4) for dipole and non-dipole transitions including fine structure (see Section 4.13). In LSJ-coupling, an additional selection rule for dipole fine structure transitions (with parity change as required) is

$$\Delta J = 0, \pm 1, \tag{4.146}$$

except that there is no $J = 0 \to 0$ transition, which is strictly forbidden. The physical reason is simple: a photon carries angular momentum, therefore its absorption or emission from a $J = 0$ level must involve a change in J-value. However, because of vectorial addition of angular momenta, one can readily have transitions from a non-zero J-level to another one with the same J but opposite parity, e.g., the C II: $1s^22s^22p$ ($^2P^o{}_{1/2}$)–$1s^22s2p^2$ ($^2P_{1/2}$). Since the spin remains the same, this is a dipole allowed E1 transtion.

However, the J-selection rule for LSJ coupling also allows for spin flip, giving rise to an intermediate class of semi-forbidden transitions. They are called *intercombination* transitions, and involve *intersystem* LS multiplets. Like dipole allowed transitions they entail parity change, but initial and final levels of different spin symmetries, i.e., $\Delta S \neq 0$. The intercombination transitions arise because of departure from LS coupling. They are not 'pure' transitions in the sense that they do not correspond to either of the two terms in the radiation operator (Eq. 4.61) considered thus far. They arise from relativistic mixing between allowed and forbidden transitions, implying that we must consider fine structure. While forbidden in LS coupling, intercombination transitions behave as E1 transitions, and become increasingly so with Z. An intercombination line is symbolically written with a right square bracket], e.g., the two fine structure lines C II] 2323.5 Å and C II] 2324.69 Å, corresponding to transitions $1s^22s^22p$ $^2P^o_{1/2}$ $\to 1s^22s2p^2$ $^4P_{3/2,1/2}$ in the $^2P^o_J$–$^4P_{J'}$ multiplet (note the spin-flip *and* the parity change).

A good number of astrophysically important transitions occur between atomic states that must be treated beyond Russell–Saunders coupling, because magnetic effects due to μ_B are too large at higher Z to be neglected. The case of states i in intermediate coupling has been expanded in Chapter 2 and is applied in Section 4.14 in particular. The initial and final states involve term coupling coefficients based on matrix diagonalization of the Breit–Pauli Hamiltonian – and discussed in Section 2.13

as 'significant for radiative transitions'. A simple example is as in Eq. 4.168; the radiative kernels of the matrix elements are as derived in this chapter. The Z-scaled behaviour of dipole allowed E1 vs. intercombination E1 transitions has been explored extensively in the literature for radiative transitions [83], as well as for collision strengths (Chapter 5, [74, 84, 85, 86]). In general, dipole allowed E1 transitions with the same-spin multiplicity are much stronger than the intercombination E1 transitions. The selection rules associated with different transitions are listed in Table 4.4.

4.11 R-matrix transition probabilities

In Section 4.6 we worked through the formulation using the simplest atomic model, the central-field approximation. However, in Chapter 3 we also sketched out the formulation of the R-matrix method (Section 3.5) which has generally enabled the most elaborate numerical procedure based on the close coupling approximation for the treatment of bound and continuum (e + ion) processes. A large volume of radiative atomic data for astronomy has been, and is, computed using the R-matrix method, such as from the Opacity Project and the Iron Project ([37, 38]).[6] It is therefore worthwhile to relate the earlier discussion to the final expressions required to obtain transition probabilties.

Let Ψ_i be the initial wavefunction of a bound state i of the target with decaying waves in all channels, and $\Psi_f^-(\hat{k})$ be the wavefunction of the final scattering state of the (e + ion) system corresponding to a plane wave in the direction of the ejected electron momentum $\hat{k}$ and incoming waves in all open channels. Both Ψ_i and $\Psi_f^-(\hat{k})$ are now expanded in terms of the R-matrix basis set in the inner region as described in Eq. 3.82 (note the resemblance of the coefficients of the R-matrix basis and the Einstein A coefficient),

$$\Psi_i(E_i) = \sum_k \Psi_k A_{E_i ki}; \quad \Psi_f^-(\hat{k})(E_f) = \sum_k \Psi_k A^-_{E_f kf}(\hat{k}). \tag{4.147}$$

The coefficients $A_{E_i ki}$ and $A^-_{E_f kf}(\hat{k})$ are determined by solving the coupled differential equations in the external region (Eq. 3.95) subject to the bound state and free state boundary conditions and matching to the R-matrix boundary condition at $r = a$ (Fig. 3.8). The energy dependence is carried out through the A_{Ek} coefficients; the energy independent R-matrix basis functions Ψ_k are the same in both the bound and the free state. The bound state wavefunctions are normalized, e.g., $\langle \Psi_b | \Psi_b \rangle^2 = 1$. When the final state lies in the continuum, as for photoionization or bound–free, the free state wavefunctions are normalized per unit energy (say Rydbergs) as $\langle \Psi\left(E''_f\right) | \Psi\left(E'_f\right)\rangle = \delta(E' - E'')$.

The angular algebra associated with generalized radiative transitions is sketched out in Appendix C. To compute dipole matrix elements for either the bound–bound or the bound–free transitions we consider the dipole length operator

$$\boldsymbol{D} = \boldsymbol{R} + \boldsymbol{r} \tag{4.148}$$

with $\boldsymbol{R} = \sum_{n=1}^{N} \boldsymbol{r}_n$ for N target electrons and a colliding or outer electron $\boldsymbol{r}$. As a vector in ordinary space, the photon interaction $\boldsymbol{D}$ is a tensor of rank $k = 1$, namely the vector $\boldsymbol{D}_\mathrm{L} = \sum_n \boldsymbol{r}_n$ in the length form of the dipole matrix and $\boldsymbol{D}_\mathrm{V} = -\sum_n \nabla_n$ in the velocity form.

The *reduced* dipole matrix[7] in the R-matrix formulation is the sum of two parts:

$$\langle a || \boldsymbol{D} || b \rangle = D^\mathrm{i}(a, b) + D^\mathrm{o}(a, b), \tag{4.149}$$

where D^i represents the contribution from the inner region ($r \leq a$), and D^o from the outer region, ($r \geq a$). In the outer region, antisymmetrization or exchange can be neglected. Then

$$D^\mathrm{o}(a, b) = \sum_{ii'} \alpha_{ii'} (F_{ia} | F_{i'b}) + \beta_{ii'} (F_{ia} | r | F_{i'b}), \tag{4.150}$$

where

$$\alpha_{ii'} = \langle \Phi || \boldsymbol{R} || \Phi \rangle, \quad \beta_{ii'} = \langle \Phi || \boldsymbol{r} || \Phi \rangle. \tag{4.151}$$

Φ is the usual target wavefunction in the close coupling expansion (Eq. 3.42). The coefficient $\alpha_{ii'}$ is non-zero only if there is an optically allowed transition between the target states belonging to channels i and i', and $l_i = l_{i'}$ The coefficient $\beta_{ii'}$ is non-zero if the channels i and i' belong to the same target and if $l_i = (l_{i'} \pm 1)$. The integrals in D^o can be evaluated as described in [87]. In this manner the complete dipole matrix elements are computed from the wavefunction expansions in the inner region and the outer region. The extension to transition elements with higher multipoles is straightforward in principle, though computationally more demanding, where bound state wavefunctions are normalized to unity.

[6] See also the on-line database NORAD: www.astronomy.ohio-state.edu/~nahar, for up-to-date extensions of the OP and IP work.

[7] This important concept refers to the separation of radial and angular components according to the basic *Wigner–Eckart theorem*, Eq. C.18 discussed in Appendix C.

The coupled channel R-matrix method has been generalized to a uniform consideration of radiative processes: bound–bound, bound–free, and free–free transitions in the (e + ion) system. The division between the inner and the outer region in the operator Eq. 4.148 directly enables the computation of the two separate terms involving **R** and **r**, as in Eq. 4.150: the radial matrix elements with the **R** operator involving the target ion in the inner region, and with the **r** operator. If the initial and the final (e + ion) states are both bound then the free channel radial functions F_{ia} and $F_{i'b}$ are exponentially decaying, representing bound–bound transitions. On the other hand, if the final state is in the continuum, with oscillating radial functions, then the radial matrix elements refer to the bound–free process. Likewise, when both the initial and the final states are in the continuum then the same radial matrix element represents the free–free process.

As in Section 4.5.2 in the bound–bound case $\langle b'|b\rangle^2 = \delta_{b',b}$ for the dimensionless oscillator strength

$$f(b, a) = \frac{1}{3g_a} \frac{E_b - E_a}{\mathrm{Ry}} \frac{S_{a,b}}{a_0^2} \tag{4.152}$$

(Eq. 4.89), now with weight factors $(2S_a + 1)(2L_a + 1)$ in LS or $(2J_a + 1)$ in intermediate coupling. Again the atomic details for a transition between two bound states at energies E_a and E_b are carried through the line strength

$$S_{\mathrm{L}}(b; a) = |\langle b||D_{\mathrm{L}}||a\rangle|^2,$$

$$S_{\mathrm{V}}(b; a) = \frac{\mathrm{Ry}^2}{(E_b - E_a)^2} |\langle b||D_{\mathrm{V}}||a\rangle|^2, \tag{4.153}$$

which is symmetric in a and b. In the current case of invariably approximate $(N + 1)$-particle functions the arguments at the end of Section 4.10 in favour of the length formulation S_{L} pertain, yet with modifications if the initial or the final state involves an oscillating radial wavefunction (an incoming or outgoing electron).

Large-scale calculations of transition probabilities are illustrated in Fig. 4.2. Dipole oscillator strengths are computed in the length and velocity formulations using the close coupling R-matrix method, and plotted against each other as an overall measure of accuracy. Figure 4.2 (a) compares the length and velocity f-values for dipole E1 transitions obtained for the relatively simple case of Li-like Fe XXIV. The agreement is consistently good over the entire range of magnitudes from Log $gf_L = 0$ to -6 (it is customary to compare the symmetric quantity gf, the statistically weighted average). That implies that even for extremely small f-values the length and velocity operators in the dipole matrix element (Eq. 4.149) yield the same numerical value, both of which are therefore accurate. A rather different picture emerges for a somewhat

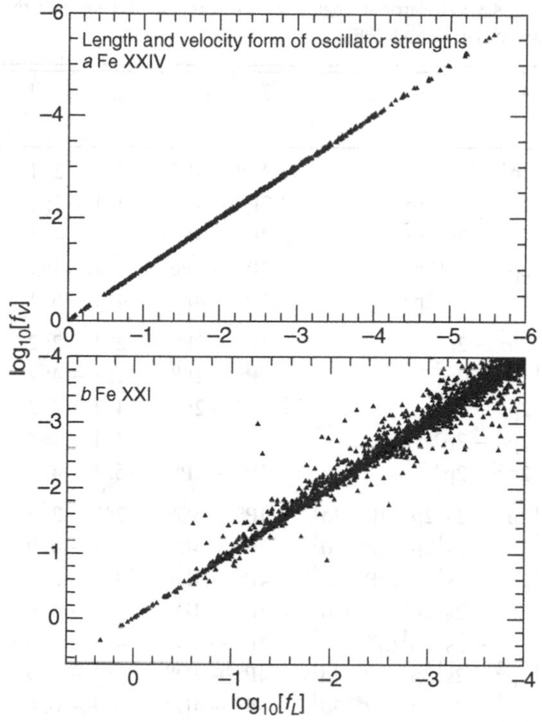

FIGURE 4.2 Dispersion in oscillator strengths computed in the length and velocity formulation as an indicator of overall accuracy of large-scale computations (the magnitude *decreases* to the right): (a) the relatively simple case of dipole transitions in He-like Fe; (b) the more complex case of B-like Fe shows much greater dispersion. The length values gf_L are used in practical applications. The calculations were performed in the close coupling approximation using the relativistic Breit–Pauli R-matrix method.

more complicated ion, B-like Fe XXI. Here the agreement between f_L and f_V is good for relatively strong transitions with large f-values, say $gf > 0.1$, but increasingly worse as they get smaller. The dispersion in the length and velocity f-values is not only significant, but could also range over an order of magnitude for very weak transitions. The behaviour in Fig. 4.2 (b) is quite typical of complex atomic systems. As explained earlier, we generally find that the length f-value gf_L is the more accurate.

A sample table of a large-scale calculation of transition probabilities computed using the Breit–Pauli R-matrix method is presented in Table 4.3. The table lists all three forms that are useful in practice: oscillator strength f, the line strength S, and the Einstein A coefficient. The first two sets of values correspond to those between the fine structure components of intercombination transitions, ${}^2\mathrm{P}^{\mathrm{o}}_J - {}^4\mathrm{P}^{\mathrm{e}}_{J'}$ and ${}^4\mathrm{P}^{\mathrm{e}}_J - {}^2\mathrm{P}^{\mathrm{o}}_{J'}$. The components are distinguished by their statistical weights $g = (2J + 1)$. The

TABLE 4.3 Oscillator strengths (*f*), line strengths (*S*) and radiative decay rates (*A* values) for intercombination and allowed E1 transitions in Fe XXII [88].

$C_i - C_k$	$T_i - T_k$	g_i:I$-g_j$:K	E_{ik} (Å)	f	S	A (s^{-1})
$2s^22p^1 - 2s^12p^2$	$2P^o - 4P^e$	2: 1 − 2: 1	247.63	7.58×10^{-4}	1.24×10^{-3}	8.24×10^{7}
$2s^22p^1 - 2s^12p^2$	$2P^o - 4P^e$	4: 1 − 2: 1	349.14	1.22×10^{-4}	5.61×10^{-4}	1.33×10^{7}
$2s^22p^1 - 2s^12p^2$	$2P^o - 4P^e$	2: 1 − 4: 1	217.49	2.70×10^{-5}	3.87×10^{-5}	1.91×10^{6}
$2s^22p^1 - 2s^12p^2$	$2P^o - 4Pe$	4: 1 − 4: 1	292.07	1.15×10^{-4}	4.41×10^{-4}	8.96×10^{6}
$2s^22p^1 - 2s^12p^2$	$2P^o - 4P^e$	4: 1 − 6: 1	253.13	1.04×10^{-3}	3.46×10^{-3}	7.21×10^{7}
$2s^12p^2 - 2p^3$	$4P^e - 2P^o$	2: 1 − 2: 2	85.81	2.31×10^{-4}	1.30×10^{-4}	2.09×10^{8}
$2s^12p^2 - 2p^3$	$4P^e - 2P^o$	2: 1 − 4: 4	81.73	1.77×10^{-5}	9.51×10^{-6}	8.82×10^{6}
$2s^12p^2 - 2p^3$	$4P^e - 2P^o$	4: 1 − 2: 2	90.14	3.07×10^{-5}	3.65×10^{-5}	5.04×10^{7}
$2s^12p^2 - 2p^3$	$4P^e - 2P^o$	4: 1 − 4: 4	85.65	3.30×10^{-4}	3.72×10^{-4}	3.00×10^{8}
$2s^12p^2 - 2p^3$	$4P^e - 2P^o$	6: 1 − 4: 4	89.69	1.18×10^{-4}	2.10×10^{-4}	1.47×10^{8}
$2s^12p^2 - 2s^12p^1(3P^*)3d^1$	$4P^e - 4D^o$	2: 1 − 2: 6	11.79	5.12×10^{-1}	3.97×10^{-2}	2.46×10^{13}
$2s^12p^2 - 2s^12p^1(3P^*)3d^1$	$4P^e - 4D^o$	2: 1 − 4: 9	11.80	6.72×10^{-1}	5.22×10^{-2}	1.61×10^{13}
$2s^12p^2 - 2s^12p^1(3P^*)3d^1$	$4P^e - 4D^o$	4: 1 − 2: 6	11.87	1.25×10^{-2}	1.95×10^{-3}	1.18×10^{12}
$2s^12p^2 - 2s^12p^1(3P^*)3d^1$	$4P^e - 4D^o$	4: 1 − 4: 9	11.87	1.08×10^{-1}	1.68×10^{-2}	5.10×10^{12}
$2s^12p^2 - 2s^12p^1(3P^*)3d^1$	$4P^e - 4D^o$	4: 1 − 6: 6	11.75	1.18×10^{-1}	1.83×10^{-2}	3.81×10^{12}
$2s^12p^2 - 2s^12p^1(3P^*)3d^1$	$4P^e - 4D^o$	6: 1 − 4: 9	11.95	3.03×10^{-3}	7.14×10^{-4}	2.12×10^{11}
$2s^12p^2 - 2s^12p^1(3P^*)3d^1$	$4P^e - 4D^o$	6: 1 − 6: 6	11.82	3.54×10^{-1}	8.26×10^{-2}	1.69×10^{13}
$2s^12p^2 - 2s^12p^1(3P^*)3d^1$	$4P^e - 4D^o$	6: 1 − 8: 2	11.84	6.73×10^{-1}	1.57×10^{-1}	2.40×10^{13}
LS	$4P^e - 4D^o$	12 − 20		7.92×10^{-1}	3.69×10^{-1}	2.27×10^{13}
$2s^12p^2 - 2s^12p^1(3P^*)3d^1$	$4P^e - 4P^o$	2: 1 − 2: 8	11.67	1.03×10^{-3}	7.89×10^{-5}	5.03×10^{10}
$2s^12p^2 - 2s^12p^1(3P^*)3d^1$	$4P^e - 4P^o$	2: 1 − 4: 12	11.67	4.97×10^{-3}	3.82×10^{-4}	1.22×10^{11}
$2s^12p^2 - 2s^12p^1(3P^*)3d^1$	$4P^e - 4P^o$	4: 1 − 2: 8	11.75	1.79×10^{-1}	2.76×10^{-2}	1.73×10^{13}
$2s^12p^2 - 2s^12p^1(3P^*)3d^1$	$4P^e - 4P^o$	4: 1 − 4: 12	11.75	2.34×10^{-1}	3.63×10^{-2}	1.13×10^{13}
$2s^12p^2 - 2s^12p^1(3P^*)3d^1$	$4P^e - 4P^o$	4: 1 − 6: 4	11.89	4.35×10^{-1}	6.81×10^{-2}	1.37×10^{13}
$2s^12p^2 - 2s^12p^1(3P^*)3d^1$	$4P^e - 4P^o$	6: 1 − 4: 12	11.82	1.12×10^{-1}	2.62×10^{-2}	8.03×10^{12}
$2s^12p^2 - 2s^12p^1(3P^*)3d^1$	$4P^e - 4P^o$	6: 1 − 6: 4	11.96	2.34×10^{-2}	5.52×10^{-3}	1.09×10^{12}
LS	$4P^e - 4P^o$	12 − 12		3.51×10^{-1}	1.64×10^{-1}	1.68×10^{13}
$2s^12p^2 - 2s^12p^1(3P^*)3d^1$	$4P^e - 2P^o$	2: 1 − 2: 11	11.68	5.36×10^{-3}	4.12×10^{-4}	2.62×10^{11}
$2s^12p^2 - 2s^12p^1(3P^*)3d^1$	$4P^e - 2P^o$	2: 1 − 4: 15	11.40	2.88×10^{-2}	2.16×10^{-3}	7.39×10^{11}
$2s^12p^2 - 2s^12p^1(3P^*)3d^1$	$4P^e - 2P^o$	4: 1 − 2: 11	11.76	1.24×10^{-4}	1.92×10^{-5}	1.20×10^{10}
$2s^12p^2 - 2s^12p^1(3P^*)3d^1$	$4P^e - 2P^o$	4: 1 − 4: 15	11.47	3.99×10^{-4}	6.03×10^{-5}	2.03×10^{10}
$2s^12p^2 - 2s^12p^1(3P^*)3d^1$	$4P^e - 2P^o$	6: 1 − 4: 15	11.54	1.53×10^{-4}	3.48×10^{-5}	1.15×10^{10}
$2s^22p^1 - 2s^12p^2$	$2P^o - 2P^e$	2: 1 − 2: 2	117.28	7.85×10^{-2}	6.06×10^{-2}	3.80×10^{10}
$2s^22p^1 - 2s^12p^2$	$2P^o - 2P^e$	4: 1 − 2: 2	136.01	1.87×10^{-4}	3.35×10^{-4}	1.35×10^{8}
$2s^22p^1 - 2s^12p^2$	$2P^o - 2P^e$	2: 1 − 4: 3	100.80	1.87×10^{-2}	1.24×10^{-2}	6.13×10^{9}
$2s^22p^1 - 2s^12p^2$	$2P^o - 2P^e$	4: 1 − 4: 3	114.34	8.59×10^{-2}	1.29×10^{-1}	4.38×10^{10}
LS	$2P^o - 2P^e$	6 − 6		8.98×10^{-2}	2.02×10^{-1}	4.60×10^{10}
$2s^12p^2 - 2p^3$	$2P^e - 2P^o$	2: 2 − 2: 2	139.55	8.33×10^{-3}	7.66×10^{-3}	2.85×10^{9}
$2s^12p^2 - 2p^3$	$2P^e - 2P^o$	2: 2 − 4: 4	129.07	1.69×10^{-2}	1.44×10^{-2}	3.38×10^{9}
$2s^12p^2 - 2p^3$	$2P^e - 2P^o$	4: 3 − 2: 2	173.24	4.58×10^{-3}	1.05×10^{-2}	2.04×10^{9}
$2s^12p^2 - 2p^3$	$2P^e - 2P^o$	4: 3 − 4: 4	157.39	6.87×10^{-2}	1.42×10^{-1}	1.85×10^{10}
LS	$2P^e - 2P^o$	6 − 6		5.73×10^{-2}	1.75×10^{-1}	1.60×10^{10}

next four sets of transitions are within several LS multiplets and their fine structure components. The total LS multiplet strengths are also given at the end of each set of dipole allowed E1 transitions with $\Delta S = 0$, as the sum over all fine structure components. However, for the intercombination E1 transitions $\Delta S \neq 0$, the total is omitted since it does not strictly refer to an LS multiplet. All transitions in Table 4.3 are between levels whose energies have been measured. Since the experimental energies are generally more accurate than theoretically computed ones, they are used in transforming line strengths into oscillator strengths and A-values, according to Eq. 4.89.

4.12 Higher-order multipole transitions

The radiative dipole moment has been discussed in detail deriving transition probabilities. Not every pair of states constitutes a dipole allowed transition leading to electric dipole radiation E1. Yet if the exponential operator is expanded high enough in powers of $\mathrm{i}\boldsymbol{k}\cdot\boldsymbol{r}$, one comes across *non*-vanishing radiative matrix elements of

$$\langle j|\mathrm{e}^{\mathrm{i}\boldsymbol{k}\cdot\boldsymbol{r}}(\hat{e}\cdot\boldsymbol{p})|i\rangle.$$

To study the higher-order multipole transitions of character $\lambda = l+1$ we recall the Legendre expansion (Eq. 4.61), namely

$$\mathrm{e}^{\mathrm{i}\boldsymbol{k}\cdot\boldsymbol{r}} = \sum_{l}^{\infty} \mathrm{i}^l\,(2l+1)\, j_l(kr)\mathrm{P}_l(\cos\vartheta),$$

once more emphasizing that ϑ is the angle between photon-$\boldsymbol{k}$ and electron-$\boldsymbol{r}$. This expression implies that the angular dependences can be expressed in terms of spherical harmonics. Successive terms in the transition matrix element can be interpreted as dipole, quadrupole, octupole, etc., involving correspondingly higher powers of ka. However, the numerical magnitude of the transition matrix element is reduced by a factor of kr in each term. Therefore, transitions due to higher multipoles are much weaker and called *forbidden*. With respect to Eq. 4.62, the next higher term is approximately lower by the factor $(ka)^2$ with respect to the dipole term. Typically $(ka)^2 \ll 1$, where a is the linear size of the particle or electron wavefunction. A forbidden transition is often symbolically written within square brackets [ ]. The far-IR [C II] $\lambda = 157\mu$m forbidden line for example is due to the transition $^2\mathrm{P}^o_{1/2}$–$^2\mathrm{P}^o_{3/2}$ between the two levels of the $1s^2 2s^2 2p$ ground term.

Up to the second-order contribution to the radiative interaction operator, the perturbation Hamiltonian reads

$$H' = \frac{e}{mc}(1+\mathrm{i}\boldsymbol{k}\cdot\boldsymbol{r})\boldsymbol{p}. \tag{4.154}$$

In transverse gauge, $\boldsymbol{k}\cdot\boldsymbol{p}=0$, each component (p_x, p_y, p_z) of the electron momentum $\boldsymbol{p}$ must be taken together with the components of the radiation field $\boldsymbol{k}\cdot\boldsymbol{r}$ perpendicular to it. In rectangular coordinates this leads to $(1+\mathrm{i}k_y y)p_z$ and two more cyclically equivalent components. Then it can be shown that in the expression

$$\left|\left\langle j\left|\frac{e}{mc}\hat{e}\cdot\boldsymbol{p}(1+\mathrm{i}\boldsymbol{k}\cdot\boldsymbol{r})\right|i\right\rangle\right|^2$$

only the first-order (electric dipole) and the second-order piece survive; the cross terms cancel out. If i and j are states of well-defined parity, the first-order operator connects only the states of opposite parity, while the second-order operator will connect states of the same parity. Looking at the second-order decomposition

$$\begin{aligned}(\boldsymbol{k}\cdot\boldsymbol{r})(\hat{e}\cdot\boldsymbol{p}) &= \frac{1}{2}\left[(\hat{e}\cdot\boldsymbol{p})(\boldsymbol{k}\cdot\boldsymbol{r})+(\hat{e}\cdot\boldsymbol{r})(\boldsymbol{p}\cdot\boldsymbol{k})\right]\\ &\quad+\frac{1}{2}\left[(\hat{e}\cdot\boldsymbol{p})(\boldsymbol{k}\cdot\boldsymbol{r})-(\hat{e}\cdot\boldsymbol{r})(\boldsymbol{p}\cdot\boldsymbol{k})\right]\\ &= \frac{1}{2}\left[(\hat{e}\cdot\boldsymbol{p})(\boldsymbol{k}\cdot\boldsymbol{r})+(\hat{e}\cdot\boldsymbol{r})(\boldsymbol{p}\cdot\boldsymbol{k})\right]\\ &\quad+\frac{1}{2}\left[(\boldsymbol{k}\times\hat{e})\cdot(\boldsymbol{r}\times\boldsymbol{p})\right]\end{aligned} \tag{4.155}$$

of the radiative operator, the first of the two terms represents the *electric quadrupole or E2* interaction. The cross products expose the second term as a *magnetic dipole or M1* interaction, because it is related to $\boldsymbol{B}\cdot\boldsymbol{L}$, where $\boldsymbol{L}=\boldsymbol{r}\times\boldsymbol{p}$ is the angular momentum component along the magnetic field $\boldsymbol{B}$. The magnetic dipole moment is given by $\vec{\mu} = e\hbar/(2mc)\vec{L}$. Since they physically arise from the same second-order interaction in the perturbation Hamiltonian due to radiation, the probability of M1 is of the same order of magnitude as that of E2. Moreover, the two operators are of even parity, hence *no parity change between states for E2 and M1 transitions.*

Expansion of the electric quadrupole term gives rise to components

$$\begin{aligned}\frac{m}{2}(xv_y+v_x y) &= \frac{m}{2}\frac{\mathrm{i}}{\hbar}(xH_0 y - xyH_0 + H_0 xy - xH_0 y)\\ &= \frac{m}{2}\frac{\mathrm{i}}{\hbar}(H_0 xy - xyH_0).\end{aligned} \tag{4.156}$$

Thus the contribution to the matrix element of electric quadrupole is

$$\frac{e^2}{m^2c^2}\frac{\omega^2}{c^2}\frac{1}{4}\frac{m^2}{\hbar^2}(E_j - E_i)^2|\langle j|xy|i\rangle|^2$$

$$= \frac{1}{4}\frac{\omega^4}{c^4}| < j|exy|i > |^2. \quad (4.157)$$

The operator in the matrix element has the appearance of a second-order spherical tensor, and is called the atomic electric quadrupole moment **Q**. Pure quadrupole transition arises when two parts of the charge distribution are oscillating like electric dipoles out of phase so that the dipole contribution vanishes. For this kind of radiation to be important, the wavelength of the incident wave must be comparable with the size of the charge distribution.

In analogy with the electric quadrupole and magnetic dipole transitions, E2 and M1, the transition rates via various higher-order multipole transitions in general can be expressed as the line strength [89],

$$S^{X\lambda}(ij) = \left|\left\langle \Psi_j \left\| O^{X\lambda} \right\| \Psi_i \right\rangle\right|^2, \quad (4.158)$$

where $X\lambda$ represents the electric or magnetic type and λ represents various multipoles with notations: electric dipole (E1), quadrupole (E2), octupole (E3), magnetic dipole (M1), quadrupole (M2), etc. The operator for *electric multipole transitions* in the length formulation is given by

$$O^{E\lambda} = b^{[\lambda]} \sum_{i=1}^{N+1} \mathrm{C}^{[\lambda]}(i) r_i^{\lambda}, \quad b^{[\lambda]} = \sqrt{\frac{2}{\lambda+1}}. \quad (4.159)$$

The magnetic multipole operators, which are slightly more complex,[8] are expressed to the lowest order as

$$O^{M\lambda} = b^{[\lambda]} \sum_i r_i^{\lambda-1} \left[\mathrm{C}^{[\lambda-1]}(i) \times \left\{\mathrm{l}(i) + (\lambda+1)\mathrm{s}(i)\right\}\right]^{[\lambda]}, \quad (4.160)$$

where the sum runs over electron coordinates and l and $\mathrm{s} = \frac{\sigma}{2}$ are the single electron orbital and spin operators.

For magnetic dipole transitions the radiative operator in the line strength Eq. 4.158 reads

$$O^{M1} = \mu_B \sum_i^N \left[(\mathrm{l}(i) + 2\mathrm{s}(i)) \right] \quad (4.161)$$

to first order. Including higher-order M1 contributions [90]

$$O^{M1} = \mu_B \sum_i^N \left[(\mathrm{l}(i) + 2\mathrm{s}(i)) \times \left\{1 + \frac{\alpha^2}{2}\left(\frac{\partial^2}{\partial r_i^2} - \frac{l(l+1)}{r_i^2} - \frac{\epsilon^2}{20}r_i^2\right)\right\} + \frac{\alpha^2}{4}\mathbf{p}_i \times (\mathbf{p}_i \times \sigma_i) + \frac{\alpha^2}{2}[\mathbf{r}_i \times (\mathbf{r}_i \times \sigma_i)] \times \left(\frac{\epsilon^2}{20} - \frac{Z}{r_i^3}\right) + \frac{\alpha^2}{8}\epsilon\sigma_i + \frac{\alpha^2}{2}\sum_{j<i} \times \left\{\frac{\mathbf{r}_{ij} \times [\mathbf{r}_{ij} \times (\sigma_i + \sigma_j)] + (\mathbf{r}_i \times \mathbf{r}_j)\mathbf{r}_i.(\mathbf{p}_i + \mathbf{p}_j)}{r_{ij}^3} - \frac{(\mathbf{r}_i \times \mathbf{p}_j) - (\mathbf{r}_j \times \mathbf{p}_i)}{r_{ij}}\right\}\right], \quad (4.162)$$

where $\epsilon = E_i - E_j$ is the photon energy for the transition. The higher-order terms were formulated [90] to explain the observed forbidden line due to the M1 transition $1s^1ns^1\,^3S_1 \rightarrow 1s^2\,^1S_0)$ in He-like ions (Section 4.14). Some of these higher contributing terms due to higher partial waves were later included in the M1 formulation [89] in the atomic structure code SUPERSTRUCTURE [10] (Section 3.5, Fig. 3.9) to calculate accurately the forbidden transition probabilities in the well-known fine structure doublet $\lambda\lambda$ 3626, 3629 lines of [O II], and similar $\lambda\lambda$ 6717, 6731 lines in [S II] due to the transitions np^3: $^4S^o_{3/2}-^2D^o_{3/2,5/2}$ between same-parity levels within the ground configuration $2p^3$ of O II and $3p^3$ of S II, as discussed later in Chapter 8.

Exercise 4.6 *The orbital angular momentum operator* $\mathbf{L} = 1/\hbar(\mathbf{r} \times \mathbf{p})$*, where* $\mathbf{p} = -i\hbar\nabla = -i\hbar(\partial/\partial x, \partial/\partial y, \partial/\partial z)$.

Derive the relations for the components of **L***:* $L_x = -i\,(y\partial/\partial z - z\partial/\partial y)$*,* $L_y = -i\,(z\partial/\partial z - x\partial/\partial z))$*,* $L_z = -i\,(x\partial/\partial y - y\partial/\partial x)$*. What are the equivalent expressions involving orbital and azimuthal angles* (θ, ϕ) *(e.g.,* $L_z = -i\partial/\partial\phi$*)?*

One of the points of this exercise is to note that the magnetic dipole moment is expressed in terms of the z-component of the orbital angular momentum: $\mu = (e\hbar/2mc)L_z$. It follows that the second-order radiation operator yields the *magnetic dipole* matrix element $(eh\nu_{jk}/2mc^3)|\langle j|\mu_{L_z}|k\rangle|^2$. Likewise, other second-order terms of the radiation operator are quadratic r^2 (*xy*, *yz*, etc.) and yield *electric quadrupole* matrix elements.

Exercise 4.7 *The radiative matrix element involves the square* $|\langle j|H'|k\rangle|^2$*. Show that the cross terms vanish and one is left with either the dipole term, when the states are of opposite parity, or the quadratic non-dipole term, when they have the same parity.*

[8] A brief introduction to the algebra related to the calculation of angular coefficients of mulitipole radiative transition matrix elements is given in Appendix C.

4.13 Selection rules and Z-scaling

Since both the M1 and E2 non-dipole transitions arise from the second-order term in the radiation field operator, it follows from Eqs 4.61 and 4.62 that their intrinsic strengths are lower than those of dipole E1 transitions by a factor of $\alpha^2 Z^2$ or about four to six orders of magnitude smaller for low-Z elements than high-Z ones, depending on the particulars of the two transitions. As Z increases, the higher-order transitions become more probable. The order of magnitude in difference between E1 and (M1, E2) transition strengths may also be obtained by considering the ratio of the dipole moment d_e on the one hand, and the magnetic moment μ_m and the quadrupole moment Q_e on the other hand [5], i.e., $d_e^2 : \mu_m^2 : Q_e^2 \sim 1 : (\alpha Z)^2 : (\alpha Z)^2$.

The individual transition rates vary widely not only according to Z but also the particulars of each transition. Whether or not a transition is probable, and its precise strength, depends on the angular and spin symmetry of the initial and final states. The primary reason for the algebraic derivations above is to highlight the relations that connect quantum numbers of different levels. These relations in turn are governed by selection rules for dipole and non-dipole transitions, summarized in Table 4.4.

Finally, as with Eq. 4.113 for E1 transitions, the A-coefficients for spontaneous decay by electric quadrupole E2 and magnetic dipole M1 transition (temporarily turning E and S dimensionless along the lines of (Eq. 2.27) are

$$g_j A_{ji}^{E2} = 2.6733 \times 10^3 \, s^{-1} \, (E_j - E_i)^5 S^{E2}(i, j), \tag{4.163}$$

$$g_j A_{ji}^{M1} = 3.5644 \times 10^4 \, s^{-1} \, (E_j - E_i)^3 S^{M1}(i, j); \tag{4.164}$$

and for electric octopole E3 and magnetic quadrupole M2 transitions

$$g_j A_{ji}^{E3} = 1.2050 \times 10^{-3} \, s^{-1} \, (E_j - E_i)^7 S^{E3}(i, j), \tag{4.165}$$

$$g_j A_{ji}^{M2} = 2.3727 \times 10^{-2} \, s^{-1} \, (E_j - E_i)^5 S^{M2}(i, j). \tag{4.166}$$

Like E2 and M1, the electric octupole and the magnetic quadrupole transitions arise from the same term in the multipole expansion, Eq. 4.61. However, approximations may be made in treating magnetic quadrupole ($\lambda = 2$) terms to lowest order from the general expression, Eq. 4.160. The transition probabilities also vary with ion charge z. Although allowed transitions are stronger than forbidden ones, the latter increase more rapidly with Z as $A^{E1} \propto Z^4$, A^{M1} eventually $\propto Z^{10}$, $A^{E2} \propto Z^8$. In accordance with these, while the relative strengths of E1 transitions are a factor of $(\alpha Z)^2$ or up to six orders of magnitude stronger than for E2 and M1, the E2 and M1 lines become increasingly more prominent with Z along an isoelectronic sequence.

4.14 Dipole and non-dipole transitions in He-like ions

The helium isoelectronic sequence provides the best example of a set of multipole transitions: allowed, forbidden and 'mixed' intercombination system. The He-like ions are of great importance in X-ray astronomy. The two-electron closed K-shell core requires high energies to excite or ionize. These energies therefore lie in the X-ray range, at wavelengths of the order of the size $\sim$ 1–20 Å of the atom for the astrophysically abundant ions from O to Fe. The He-like spectra are modelled in Chapter 8; here, we confine ourselves to a designation of the relevant radiative transitions that form the well-known emission lines. Their line ratios afford a range of diagnostics of physical parameters, such as the density, temperature and ionization state of the source or the medium. These lines are usually well separated and hence can be observed easily.

TABLE 4.4 Selection rules for radiative transitions of multipole order λ: note that the dipole allowed E1 and intercombination E1 transitions differ only in ΔS; the Δl, ΔL, ΔJ change by unity in both.

Type	$\Delta\pi$(Eλ,Mλ) $(-1)^\lambda, -(-1)^\lambda$	Δl	ΔS	ΔL(Eλ,Mλ) $\leq \pm\lambda, \leq \pm(\lambda-1)$	$L \neq L'$	$\Delta J(\Delta M)$ $\leq \pm\lambda(\leq \lambda)$	$J \neq J'$
Dipole allowed (E1)	Yes	±1	0	$\leq \pm1$	$0 \neq 0$	$\leq \pm1$	$0 \neq 0$
Intercombination (E1)	Yes	±1	1	$\leq \pm1$	$0 \neq 0$	$\leq \pm1$	$0 \neq 0$
Forbidden (M1)	No	0	0	0		$\leq \pm1$	$0 \neq 0$
Forbidden (E2)	No	$0, \pm2$	0	$\leq \pm2$	$0 \neq 0,1$	$\leq \pm2$	$0 \neq 0, 1, \frac{1}{2} \neq \frac{1}{2}$

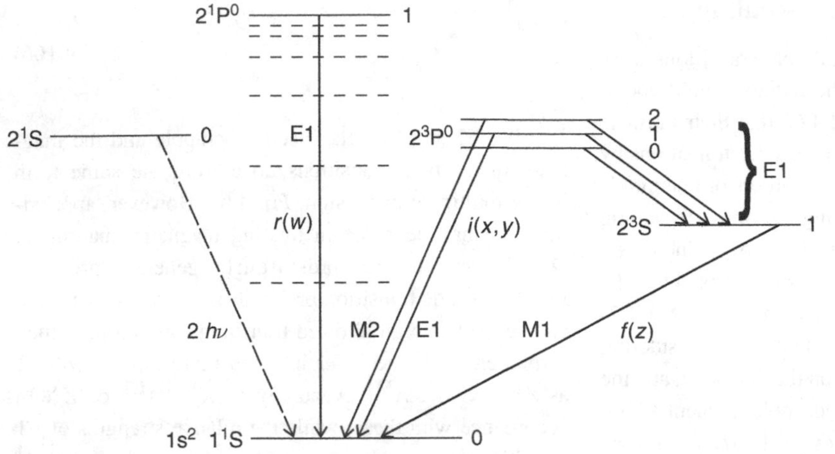

FIGURE 4.3 Multipole radiative transitions in He-like ions, showing the He Kα complex of fine structure levels (solid lines), with transitions (arrows) and autoionizing levels (dashed lines), that form the primary X-ray lines and dielectronic satellite lines (Chapter 7). The *LS* designation of each level *nSLJ* is shown on the left of each level and the value of *J* on the right.

Excitations from the ground to excited $n = 2$ levels are responsible for the formation of lines usually labelled as w, x, y and z arising from the following four transitions:

$$\begin{aligned} w &: 1s^2\,^1S_0 \longleftarrow 1s^1 2p^1\,^1P^o_1 \text{ dipole allowed (E1)}, \\ x &: 1s^2\,^1S_0 \longleftarrow 1s^1 2p^1\,^3P^o_2 \text{ forbidden (M2)}, \\ y &: 1s^2\,^1S_0 \longleftarrow 1s^1 2p^1\,^3P^o_1 \text{ intercombination (E1)}, \\ z &: 1s^2\,^1S_0 \longleftarrow 1s^1 2s^1\,^3S_1 \text{ forbidden (M1)}. \end{aligned} \quad (4.167)$$

In the *intersystem transition* $1s^2\,^1S$–$1s^1 2p^1, {}^3P^o$ spin–orbit interaction couples level $^3P^o_1$ with $1s2p\,(^1P^o)$, i.e.,

$$\Psi\left(1s^1 2p^1\,^3P^o_1\right) = c_1 \Psi\left(1s^1 2p^1\,^3P^o_1\right) + c_2 \Psi\left(1s^1 2p^1\,^1P^o_1\right). \quad (4.168)$$

For (near) neutral species the term coupling coefficient c_2 is small, a few percent. It implies that the transition $1\,^1S_0 - 2\,^3P^o_1$, which is spin-forbidden in LS coupling, might have a coefficient A about two orders of magnitude lower than in the allowed transition $1s^2\,^1S$–$1s^1 2p^1\,^1P^o$. Transition $1s^1 2s^1\,^1S_0 \rightarrow 1s^2\,^1S_0$ decays by a two-photon continuum (denoted by $2h\nu$ in Fig. 4.3); it does not give rise to a single line but nevertheless contributes to the level populations and line intensities (see Chapter 8).

Table 4.5 gives A-coefficients for the relevant radiative transitions of He-like ions.[9]

[9] They are taken from [91], which is an extensive compilation of accurate transition rates for E1, M1, E2 and M2 multipole transitions among levels up to $n \leq 6$ in He-like C, N, O, Ne, Si and Ar.

4.15 Angular algebra for radiative transitions

Appendix C provides an introduction to the angular algebra necessary for dealing with generalized radiative transition matrix elements, including multipole moments. In addition to the angular-spin decomposition of the relevant matrix elements computed from elaborate methods, it is often useful to obtain expressions for purely algebraic transformation from LS coupling to include fine structure ([36]) as described next.

4.15.1 Fine structure components of LS multiplets

The relative strength of observed lines is a central topic in astrophysics, in particular for transitions between pairs of multiplets to terms ^{2S+1}L. We introduced the concept of a *multiplet* in Section 2.4, supplemented in Section 2.13 with observed features in the shape of fine structure, an effect the electron's magnetic moment $\mu_B = e\hbar/2m$ and of relative magnitude $\alpha^2 Z^2$ over the equivalent terms of its electric charge e. Now we are going to deal with the *relative* strength of lines or transitions $J_j - J_i$ within a multiple $S_j L_j - S_i L_i$.

In this section we ignore multiplet mixing, explained in Section 2.13 as a high-Z effect. Then the line strength $S_{ij} = S_{ij}$, the *primary* quantity in atomic physics calculations, can be decomposed purely algebraically following Racah techniques for the respective angular operator $\mathrm{C}^{[\lambda]}$, so that

TABLE 4.5 *A*-coefficients for X-ray decay to the ground state of He-like ions.

Ion	M1 − $2\,^3S_1$ (z)		E1 − $2\,^3P^o_1$ (y)		M2 − $2\,^3P^o_2$ (x)		E1 − $2\,^1P^o_1$ (w)	
	λ/Å	$A\cdot$s	λ/Å	$A\cdot$s	λ/Å	$A\cdot$s	λ/Å	$A\cdot$s
C V	41.47	4.87×10^{1}	40.73	8.37×10^{7}	40.73	5.72×10^{7}	40.27	8.86×10^{11}
N VI	29.53	2.54×10^{3}	29.08	2.03×10^{8}	29.08	6.93×10^{7}	29.79	1.81×10^{12}
O VII	22.10	1.05×10^{3}	21.80	6.15×10^{8}	21.80	8.22×10^{7}	21.60	3.30×10^{12}
Ne IX	13.71	1.09×10^{4}	13.56	5.36×10^{9}	13.56	1.13×10^{8}	13.45	8.85×10^{12}
Si XIII	6.74	3.61×10^{5}	6.69	1.55×10^{11}	6.69	3.33×10^{8}	6.65	3.76×10^{13}
Ar XVII	3.995	4.80×10^{6}	3.970	1.78×10^{12}	3.967	6.70×10^{8}	3.950	1.07×10^{14}
Ca XIX	2.838	1.42×10^{7}	2.854	4.79×10^{12}	2.857	7.55×10^{8}	2.868	1.64×10^{14}
Fe XXV	1.8682	2.12×10^{8}	1.8595	4.42×10^{13}	1.8554	6.64×10^{9}	1.8504	4.57×10^{14}

$$S^{LSJ}_{ij} = C(L_i, L_j; J_i, J_j)\frac{S^{LS}_{ij}}{(2S+1)(2L_i+1)(2L_j+1)}, \quad (4.169)$$

in the Eλ case, $\lambda = 1$.[10] Contrary to the general case, not restricted by $S_j = S_i \equiv S$, notably in the He-transition 'y' of Section 4.14, the coupling coefficients C are rational numbers, explicitly displayed here as Racah products for f-values, which, along with Einstein's A, are *secondary* quantities.

Fine-structure transitions are restricted by the selection rule $\Delta J = 0, \pm1$. The values of the coefficients $C(L_i, L_j; J_i, J_j)$ depend on the difference $\Delta J = 0, \pm1$, $|\Delta L| = 0, 1$ and on total multiplicity factors (2S+1)$(2L_i+1)(2L_j+1)$. The quantities S^{LSJ}_{ij} satisfy the sum rule

$$S^{LS}_{ij} = \sum_{J_i, J_j} S^{LSJ}_{ij}, \quad (4.170)$$

where the indices run over all allowed transitions $J_i - J_j$.

Based on the work by E. U. Condon and G. H. Shortley [3], C. W. Allen [81] deduced $C(L_i, L_j; J_i, J_j)$ for all possible transitions $L_j - L_i$ up to 5–6, and spin multiplicities $(2S+1)$ up to 11. Table 4.6 gives the values of $C(L_i, L_j; J_i, J_j)$ for various transitions of normal multiplets ($|\Delta L| = 1$) and symmetrical multiplets ($|\Delta L| = 0$). The transformation coefficients are arranged in matrix form for the strongest or principal transitions: $\Delta J = -1$ for normal multiplets and and $\Delta J = 0$ for symmetrical multiplets. The associated coefficients $C(L_i, L_j; J_i, J_j)$ are denoted by x_1, x_2, etc. Similarly, the coefficients for the first satellites ($\Delta J = 0$ for normal and $\Delta J = \pm1$ for symmetrical multiplets) are denoted by y_1, y_2, etc., and for the second satellites ($\Delta J = 1$ for normal multiplets) by z_1, z_2, etc. Numerical values of these coefficients are given in Appendix C.

The calculated line strength in length form is independent of the transition energy E_{ij}. Line strengths are therefore useful to employ experimentally observed energies, wherever available, in place of calculated transition energies to provide more accurate fine-structure results for the oscillator strengths f. The fine-structure values f_{LSJ} may be computed directly from f_{LS} [36]:

$$\begin{aligned} f_{LSJ}(n_jSL_jJ_j, n_iSL_iJ_i) = {} & f_{LS}(n_jSL_j, n_iSL_i) \\ & \times (2J_j+1)(2L_i+1) \\ & \times \mathrm{W}^2(L_jL_iJ_jJ_i; 1S), \end{aligned} \quad (4.171)$$

where $\mathrm{W}(L_jL_iJ_jJ_i; 1S)$ is a Racah coefficient and is often expressed in the form of a Wigner 6-j symbol (see Appendix C):

$$\mathrm{W}(L_jL_iJ_jJ_i; 1S) = \begin{Bmatrix} L_j & S & J_i \\ J_j & 1 & L_i \end{Bmatrix}. \quad (4.172)$$

These components satisfy the sum rule

$$\sum_{J_iJ_j}(2J_i+1)f_{LSJ}(n_jSL_jJ_j, n_iSL_iJ_i) = f_{LS}(n_jSL_j, n_iSL_i)(2S+1)(2L_i+1). \quad (4.173)$$

Since this way of splitting an f-value does not require transition energies, the method can be used to approximate fine-structure components when no observed energies are available. Finally, we emphasize that we have considered transitions without change in total spin S, and the algebraic approach is unsuitable for intercombination lines.

[10] A program LSJTOLS [92] for calculating fine-structure components is available at the OSU-NORAD website; for an application to iron and other ions (S. N. Nahar, see [92]).

TABLE 4.6 Recoupling coefficients $C(L_i, L_j; J_i, J_j)$ for relative strengths between components $^{2S+1}L_J$ of multiplets, hence $J_{m[ax]} = 2\min(S, L) + 1$.

	Normal multiplets SP, PD, DF, . . .				
	J_m	J_m-1	J_m-2	J_m-3	J_m-4
J_m-1	x_1	y_1	z_1		
J_m-2		x_2	y_2	z_2	
J_m-3			x_3	y_3	z_3
J_m-4				x_4	y_4

	Symmetrical multiplets PP, DD, FF, . . .			
	J_m	J_m-1	J_m-2	J_m-3
J_m	x_1	y_1		
J_m-1	y_1	x_2	y_2	
J_m-2		y_2	x_3	y_3
J_m-3			y_3	x_4

This chapter (together with Appendix C) was aimed at a general exposition of atomic radiative transitions. While we shall be discussing spectral line formation in greater detail later in Chapters 8, 9, 12 and other chapters, it is worth mentioning the basic types of spectrum observed in astrophysical and laboratory sources. Emission spectra are caused by spontaneous radiative decay from an upper to a lower level; the line intensity is governed by the A-coefficient and upper level population. Populating upper levels may take place through a variety of atomic processes (see Chapter 3), generally via electron impact excitation in the case of low-lying levels, and (e + ion) recombination into high-lying levels, which may cascade downward to affect population distribution among lower levels. If there is an external radiation source, then stimulated emission also contributes via the downward B-coefficient. Emission lines add energy to the continuum, giving the characteristic peaked line profile. Contrariwise, absorption line formation, governed by the upward B-coefficient and the lower level population, removes energy from the ambient radiation field giving a line profile that dips below the background continuum. Chapters 8 and 9 describe the emission and absorption spectra, respectively.

5 Electron–ion collisions

In ionized plasmas spectral formation is due to particle collisions or radiative excitations. In astrophysical situations there is usually a primary energy source, such as nuclear reactions in a stellar core, illumination of a molecular cloud by a hot star or accretion processes around a black hole. The ambient energy is transferred to the kinetic energy of the particles, which may interact in myriad ways, not all of which are related to spectroscopy.

Electron collisions with ions may result in excitation or ionization. The former process is excitation of an electron into *discrete* levels of an ion, while the latter is excitation into the *continuum*, or ionization, as shown in Fig. 3.1 and discussed in Chapter 3. A practically complete description of the (e + ion) excitation process requires collisional information on the ions present from an observed astrophysical source, and for all levels participating in spectral transitions. As the excitation energy from the ground state to the higher levels increases, the ionization energy E_{I} is approached. The negative binding energy of the excited states *increases* roughly as $E \sim -z^2/n^2$, where z is the ion charge. As $n \rightarrow \infty$, $E \rightarrow 0$, i.e., the electron becomes free.

At first sight, therefore, it might seem like a very large number of levels need to be considered for a given atomic system in order to interpret its spectrum completely. While that is true in principle, in practice, the number of levels involved depends on the type(s) of transition(s) and resulting lines depending on physical conditions in the plasma environment, as discussed in detail in later chapters, particularly Chapters 8 and 12. Thus, in actual atomic models rather a limited number of levels are included in representing the collision process. Unlike radiative transitions, collisional transitions are *not* restricted by parity rules. Therefore, electron–ion collisions are the primary mechanism for exciting low-lying transitions within levels of the ground configuration of the same parity, where radiative transitions have extremely small probabilities, such as in ionized gaseous nebulae (Chapter 12).

For non-hydrogenic ions, we need to consider the *LS* terms and fine structure levels *LSJ* of the particular subshell $n\ell$ in the electronic configurations that are energetically accessible to the free electrons in the plasma, so that the kinetic energy of the free electrons can be transferred to the ion in exciting the levels in question. The energies of levels that can be excited in a source are related to the kinetic temperatures of the free electrons, known as the *excitation temperatures*. In the following discussion, we often refer to a generic plasma in an ionized H II region or nebula. Recall (Chapter 1) that these are characterized by low densities and temperatures, $n_e \sim 10^{3-6}$ cm^{-3} and $T_{\mathrm{e}} \sim 10^{3-4}$ K. In nebular plasmas only the forbidden lines, such as the [O II], [O III] and [S II] are collisionally excited. The low-lying participating levels belong to the respective ground configurations; the excitation energies are only a few eV, corresponding to kinetic temperatures $T_{\mathrm{e}} \approx 1000 - 30\,000$ K. Subsequent radiative decays from excited levels give rise to the forbidden emission lines. For higher temperature–density sources, more highly ionized species and more atomic levels come into play, such as the case of the solar corona (including solar flares) with $n_{\mathrm{e}} \sim 10^{8-12}$ cm^{-3} and $T_{\mathrm{e}} \sim 10^{6-7}$ K. The typical coronal ions are Fe X–Fe XXV, with several tens of levels, where collisional excitation can give rise to many forbidden as well as allowed lines.

The complementary process to excitation – collisional ionization – determines the *ionization balance*, or the relative distribution of ionic states of a given element, in collisionally dominated plasmas (without appreciable contribution from a radiative source to ionization). The kinetic temperature of the free electrons also determines which ionization stages of an element would exist in a plasma. However, there is generally a large difference between ionization energies and excitation energies for emission lines. For example, in the nebular case, the excitation energies of emission lines are much lower than the ionization energies of the ions. For example, the ionization

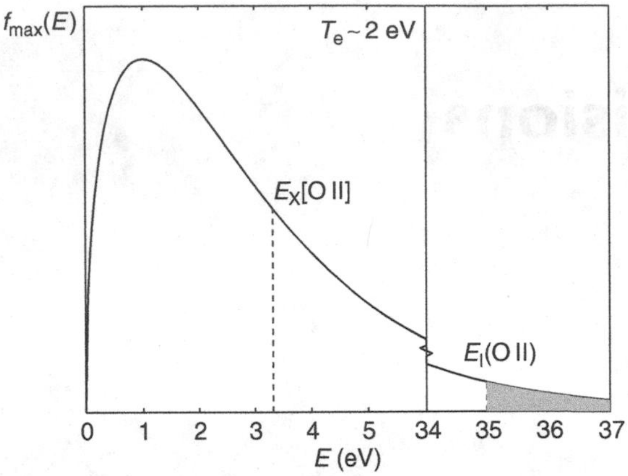

FIGURE 5.1 Maxwellian distribution of electron energies at a kinetic temperature $T_e = 2$ eV or 23 200 K. Excitation and ionization energies of O II are indicated. The excitation energies of the low-lying levels that produce the forbidden [O II] 3726, 3729 lines are much lower than the ionization energy; the ratio of these two fine structure blue lines is a well-known indicator of nebular densities (see Chapters 8, 12).

energy of O II is 35.12 eV, as opposed to the excitation energy of [O II] 3626, 3629 Å lines of about 3 eV. Figure 5.1 exmplifies the situation with respect to excitation and ionization of O II. One might ask how it is that plasmas with $T_e \sim 1$ eV are able to ionize atoms and ions with more than ten times the ionization energy. That is because of external energy sources. Nebular plasmas are ionized by hot stars, with intense UV radiation fields, which produce sufficiently high energies to ionize most elements to one or two stages of ionization.

To determine ionization balance in collisional equilibrium, only total ionization cross sections of the ground state of ions are usually needed (low-lying metastable states may sometimes be important). While both excitation and ionization processes are essential to a complete description of spectral formation, electron impact excitation requires more detailed cross sections for excitation of many specific levels in ions of astrophysical interest. In this chapter, we discuss both processes, with more attention devoted to the excitation process.

5.1 Electron impact excitation (EIE)

Spectral lines due to either (e + ion) excitation or (e + ion) recombination are seen in emission spectra. Emission lines are indicators of temperatures, densities, abundances, velocities and structure of astrophysical objects. Excitation of levels by electrons is followed by radiative decay; these two processes – electron impact excitation and radiative transitions – are responsible for collisionally excited emission lines in optically thin plasmas where photons escape without significant absorption or scattering (discussed in Chapter 9).

If we have the excitation of low-lying levels, it is generally sufficient to consider a relatively small number of levels in an ion. However, the excitation cross sections for these levels, from the ground state as well as among excited states, need to be known accurately, to compute EIE rate coefficients. Most of these cross sections are calculated theoretically using methods decribed in the preceding chapter. But although experimental measurements can be made for only a few transitions in limited energy ranges, they are very important in benchmarking the accuracy of theoretical cross sections. Before going into the calculations in detail, and comparison with experiments, we describe the basic concepts of electron–ion scattering and the relevant quantities to be determined.

5.1.1 Cross section

Collision of an electron with an ion may result in (i) *elastic scattering* – no net exchange of energy – or (ii) *inelastic scattering* – exchange of energy between the electron and the ion, excited or de-excited from an initial level to an energetically different level. For spectral formation, we are interested in process (ii), or the determination of inelastic cross sections for (de-)excitation among the levels of an ion. Figure 5.2 shows a schematic diagram of electron–ion scattering.

An incident beam of free electrons impacts on a positive ion and is scattered at an angle θ between the incident direction and the final (asymptotic) direction into the detector. If the free electrons have sufficient energy to excite ionic levels, there may be transfer of energy ΔE_{ij}, exciting the ion from level i to j. In a real situation, in laboratory experiments or astrophysical environments, most of the electrons are elastically scattered from the ion. Only a relatively small number result in inelastic scattering and excitation of levels. As the name implies, the cross section has units of area. Its simplest definition is

$$Q(E) = \frac{\text{Number of scattered particles}}{\text{Number of incident particles per unit area}} \ \text{cm}^2. \tag{5.1}$$

In laboratory experiments, quantum mechanical effects imply that the number or the flux of scattered electrons depends on the angle where the measurement is made relative to the incident beam, as shown in Fig. 5.2. Electrons are scattered into a solid angle ω about the angle of scattering θ, such that the solid angle $d\omega = sin\theta d\theta d\phi$

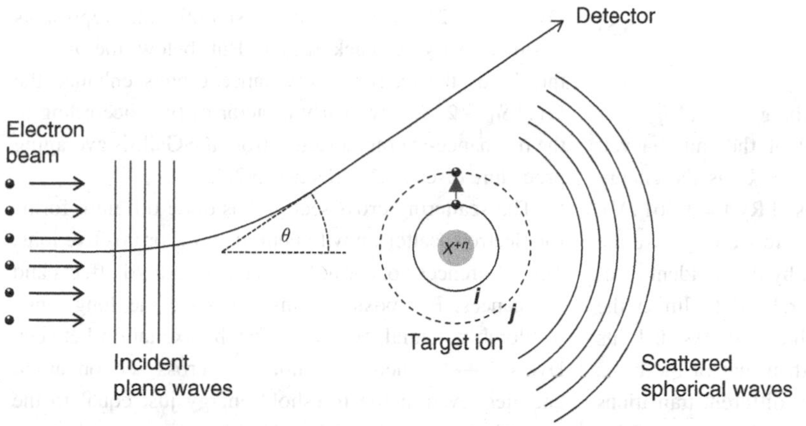

FIGURE 5.2 Electron impact excitation of a positive ion from level $i \rightarrow j$.

(where ϕ is the azimuthal angle). In such a case, the partial *differential cross section* $\mathrm{d}Q$ is written in terms of the *scattering amplitude* $f(\theta)$ as

$$\frac{\mathrm{d}Q(E,\theta)}{\mathrm{d}\omega} = |f(\theta)|^2. \quad (5.2)$$

The scattering amplitude $f(\theta)$ may be related to the amplitude of the outgoing scattered waves, or the scattering matrix **S**, derived from the asymptotic form of the outgoing radial wavefunctions, as discussed in Chapter 3 and in the next section. The total cross section at a given energy E is then the integrated value over all solid angles

$$Q(E) = \int \frac{\mathrm{d}Q}{\mathrm{d}\omega} \mathrm{d}\omega. \quad (5.3)$$

Quantum mechanically, the incident electron wavefunctions may be assumed to be plane waves before scattering, and spherical waves afterwards. The measured cross section of scattering therefore depends on the scattering angle θ. The usual approach to account for the geometry inherent in the scattering process is the decomposition of the incident plane wave into Legendre polynomial functions (Eq. 3.30) $P_\ell(cos\theta)$, or *partial waves* according to the orbital angular momentum ℓ. Equation 4.61 gives the partial wave expansion in terms of the Riccati–Bessel function $j_l(kr)$. For example, $\ell = 0$ refers to radial scattering with the electron directly incident on the target ion. Higher $\ell > 0$ partial waves correspond to a non-radial approach towards the target; also, the higher the ℓ the greater the distance of closest approach to the target.

The total cross section is, therefore, the sum over *partial cross sections* of different angular momenta ℓ

$$Q(E) = \sum_\ell Q_\ell(E). \quad (5.4)$$

As inferred from the description of the *LS* coupling approximation in Chapter 2, and the description of the methods in Chapter 3, this *partial wave expansion* implies that the cross section is generally computed separately for each partial wave. In *LS* coupling, we consider the total symmetry of the (e + ion) system corresponding to each partial wave: $SL\pi$, where S usually refers *not* to the total spin but to the multiplicity $(2S + 1)$.

Experimental measurements may be made at a fixed energy but different angles $0° \leq \theta \leq 180°$ to reveal details of the scattering cross section due to polarized beams or due to angular distribution of partial waves of different ℓ. Such measurements are useful tests of theoretical models, which generally employ partial wave decomposition. Also, since experimental measurements cannot be carried out at all energies, such comparisons can be made at selected energies but a range of angles. Differential cross sections are indicative of details of scattering in different directions ranging from 'forward', $\theta = 0°$, to 'backward' scattering $\theta = 180°$. Intuitively, the direction of the scattered beam tends towards forward scattering as the incident energy E increases; however, the details depend on the quantum mechanical structure of the target and the properties of the incident beam.

5.1.2 Collision strength

Collisional and radiative processes resulting in line formation due to EIE may be viewed in an analogous manner by defining a dimensionless quantity called the *collision strength*,[1] in analogy with the line strength for a radiative transition. The inelastic EIE cross section for excitation of level i to j, in units of the hydrogen atom cross section in the first Bohr radius, is expressed as

[1] This was first defined by D. H. Menzel and named the *collision strength* by M. J. Seaton.

$$Q_{ij}(E) = \frac{\Omega_{ij}}{g_i k_i^2}\left(\pi a_0^2\right), \tag{5.5}$$

where Ω_{ij} is the collision strength, $g_i = (2S_i + 1)(2L_i + 1)$ is the statistical weight of the initial level, $\pi a_0^2 = 8.797 \times 10^{-17}\text{cm}^2$ and $E = k_i^2$ is the energy of the incident electron in Rydbergs (1 Ryd = 13.61 eV). Ω_{ij} is also symmetrical with respect to i and j, like the line strength. Owing to the division by the incident electron energy E (Ry) = k^2 (in the non-relativistic limit), the collision strength varies slowly with energy. As such, its numerical value varies much less than that of the cross section, making it easier to compare different transitions within an ion or in different ions. Throughout most of this and later chapters we will employ the collision strength to discuss the EIE process and applications.

In addition to the *background* collision strength for direct excitation, the presence of autoionizing resonances gives rise to indirect pathways, or channels, that greatly affect its energy dependence. The enhancement due to resonances is seen from Fig. 5.3. The collision strength $\Omega(1^1S_0 - 2^3S_1)$ shown is for the important forbidden X-ray transition in He-like Fe XXV, from the ground state $1s^2\ ^1S_0$ to the metastable level $1s^12s^1\ ^3S_1$ – that gives rise to the z-line discussed in Eq. 4.167. The resonance complexes in Fig. 5.3 lie in between the $n = 2$ and the $n = 3$ states of Fe XXV. Therefore, in this region the autoionizing resonance configurations are generally of the type $1s3\ell n\ell'$. The two groups shown in Fig. 5.3 are $1s3\ell3\ell'$ and $1s3\ell4\ell'$. The dashed line is the resonance-averaged collision strength computed according to the Gailitis procedure described later. Above the $n = 3$ states, $\Omega(1^1S_0 - 2^3S_1)$ is relatively smooth and represents a slowly varying background. But below the $n = 3$ thresholds the extensive resonance groups enhance the $\Omega(1^1S_0 - 2^3S_1)$ by nearly a factor of two, according to the resonance-averaged values from the Gailitis averaging procedure discussed in Section 5.2.3.

The scattering cross section has quite different forms for electron scattering with neutral atoms or positive ions. The differences appear in two ways: threshold effects and resonances. For positive ions, we have the long-range Coulomb potential, therefore, for the excitation between levels $i \rightarrow j$ there is a non-zero cross section at all energies, even at the threshold energy just equal to the excitation energy $E = E_{ij}$. However, for neutral atoms there is no such attractive force and the cross section at the excitation threshold must be zero. Of course the collision strength or the scattering cross section is identically zero for $E < E_{ij}$. Even for positive ions, the incident electron cannot both excite the ion and escape its attractive field unless it has the mininum threshold energy E_{ij}.

For electron scattering with neutral atoms, Rydberg series of resonances such as for (e + ion) scattering do not occur because there is no Coulomb potential that might support an infinite number of (quasi-)bound states or resonances. But in electron–atom scattering we do find *shape resonances* that occur at near threshold energies due to short-range interactions of the electron with the atom. Shape resonances are generally much broader than Rydberg resonances (see Section 6.9.3). Weaker short-range non-Coulomb interactions, such as the dipole and quadrupole polarization potentials, $V_{dip} \sim r^{-2}$ and $V_{quad} \sim r^{-4}$, and the

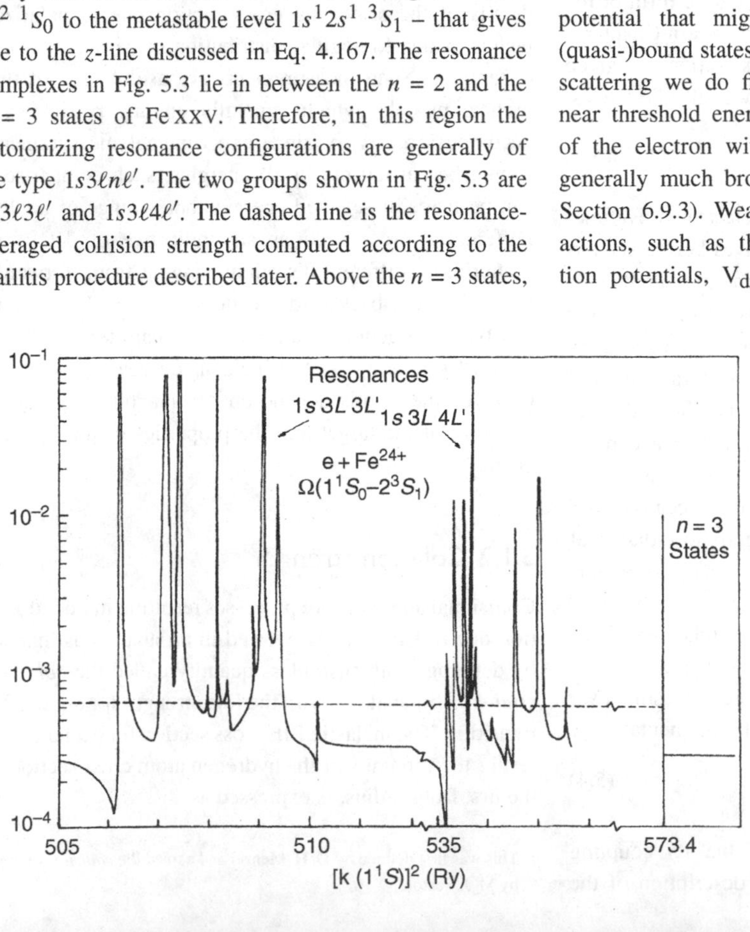

FIGURE 5.3 Resonances in the collision strength for the forbidden transition $1^1S_0 - 2^3S_1$ in Fe XXV (the z-line) Eq. 4.167 [85], showing the Gailitis jump at the $n = 3$ thresholds (Section 5.2.3).

van der Waals potential $V_{vdW}(r) \sim r^{-6}$, assume much greater importance for neutral atoms than for ions where the Coulomb potential $V_{Coul} \sim r^{-1}$ dominates. Therefore, electron–electron correlation effects in electron–atom scattering are much more intricate than in electron–ion scattering. However, in astrophysical sources, atoms are often ionized so excitation of neutral atoms is not as common. Experimental measurements are also difficult for neutral atoms, since there is no Coulomb 'focusing' effect such as that for ions, and at near-threshold energies overall electron–electron repulsion is greater than for ions. For these reasons, there are fewer atoms than ions for which electron scattering cross sections have been calculated or measured.

5.1.3 Variational method

Before we delve into the details of electron–ion collisions, it is worth mentioning the salient features of the unified theoretical framework of atomic processes built up in Chapter 3. Indeed, the coupled channel approximation is based on atomic collision theory, and was developed to describe electron–ion collisions in particular. It was also used to obtain bound states of the (e + ion) system, and hence atomic parameters for bound–bound radiative transitions, bound–free photoionization and free–bound recombination processes. This is because, in principle, the methodology, most effectively exploited by the R-matrix method, solves the fundamental quantum mechanical problem and enables an accurate representation of the total (e + ion) wavefunction in any state of excitation. The derivation of the coupled channel equations (Eq. 3.45) then follows using well-known variational methods.

As is well-known from basic quantum mechanics, a variational principle is needed to find a stationary (minimum) eigenvalue, with respect to variations in a *trial* wavefunction Ψ_t. For bound states we have the *Rayleigh–Ritz variational principle*,

$$E_0 \leq \frac{<\Psi_t|H|\Psi_t>}{<\Psi|\Psi>}. \tag{5.6}$$

E_0 is the upper bound on the minimum (ground state) energy with respect to *first-order* changes $\delta\Psi$ about the exact wavefunction Ψ_{ex} (which is, of course, unknown and to be approximated). In the case of scattering, we have an analogous quantity: the phase shift introduced in a freely propagating wave by an interaction potential (Chapter 3). A corresponding minimum principle in collision theory is the *Kohn variational principle*:

$$\tan\delta = \tan\delta_t - \pi \int_0^\infty F_l^t(r)[H-E]F_l^t(r)\mathrm{d}r. \tag{5.7}$$

The (integro-)differential operator [H − E], including potential interactions relevant to electron–ion collisions as well as the radial form of $F_l(r)$, are as given in Chapter 3. Since $\tan\delta_l$ is defined to be the K-matrix, generalized to Eq. 3.53 for the multi-channel case, the Kohn variational principle yields an upper bound for the minimum phase shift computed from trial wavefunctions Ψ_t with radial function F_l. Using multi-channel close coupling wavefunctions (Eq. 3.42), and for small variations about the exact wavefunctions, Eq. 5.7 may be written as,

$$\delta\left[<\Psi_i|H-E|\Psi_i'> - K_{ii'}\right] = 0, \tag{5.8}$$

where i and i' refer to incoming and outgoing collision channels. The asymptotic form of the trial wavefunctions, Eq. 3.50, yields a trial K-matrix with elements $K_{ii'}^t$. The Kohn variational principle then gives

$$K_{ii'}^{\text{Kohn}} = K_{ii'}^t - <\Psi_i^t|H-E|\Psi_{i'}^t>, \tag{5.9}$$

which differs from the exact form to second order of the error in the trial wavefunctions. The Kohn K-matrix can now be used to compute scattering matrices and cross sections using expressions in Chapter 3, e.g. Eq. 3.56, and as derived in the next section.

5.2 Theoretical approximations

In Chapter 3, we have already described general theoretical methods used to study (e + ion) interactions, in particular the coupled channel approximation implemented in the widely employed R-matrix package of codes.[2] A large number of R-matrix calculations have been carried out, especially under the Opacity Project and the Iron Project collaboration [36, 38]. Prior to the wide use of the coupled-channel approximation in literature, a considerable amount of data for excitation collision strengths had been computed in the distorted wave approximation discussed in Chapter 3, without channel coupling and resonance effects but taking account of the 'distortion'

[2] Another set of close coupling codes was developed at University College, London, by M. J. Seaton, W. Eissner and co-workers called IMPACT [49]). It is no longer employed since IMPACT codes required the solution of the coupled integro-differential equations (Eq. 3.45) *at each incident electron energy*, carried out by a linear algebraic matrix-inversion technique. Therefore, it became computationally more demanding than the R-matrix method, and prohibitively so, in large-scale calculations, such as under the Opacity Project [37] and the Iron Project [38], where delineation of resonances often meant computations at thousands of energies per cross section. Interestingly, however, IMPACT did yield *physical* wavefunctions in the so-called inner region (Fig. 3.8) for both the continuum (open and closed) and the bound (e + ion) channels.

of the incident and outgoing free-electron wavefunctions in the potential due to initial and final states involved in the transition. Going back further in the history of collisional calculations, excitation collision strengths were also computed in simpler approximations such as the Coulomb–Born or, for optically allowed transitions, the Coulomb–Bethe approximation. More recently, the convergent close coupling method, and other variants of the coupled channel approximation, such as the R-matrix with pseudostates, have been employed [93, 94, 95] for a limited number of high-accuracy calculations. In all of the advanced methods, the main quest remains the same: to couple as many of the target levels as possible. In addition to channel coupling, which essentially reflects electron correlation effects, relativistic effects need to be considered as completely as possible. Next, we describe some specialized features often useful in practical computations.

5.2.1 Isoelectronic sequences

Ions with the same number of electrons N but different nuclear charge Z are said to belong to an isoelectronic sequence characterized by (N, Z). Calculations of collision strengths are simplified by considering some elementary properties of the behaviour states in these ions as Z changes. The N-electron Hamiltonian for an ion with nuclear charge Z is (in Ryderg atomic units)

$$H(Z, N) = -\sum_{i=1}^{N}\left(-\nabla_i^2 + \frac{2Z}{r}\right) + \sum_{j=i+1}^{N}\sum_{i=1}^{N-1}\frac{2}{r_{ij}}, \tag{5.10}$$

where the electron–electron repulsion term is written explicitly with sums over pairs of electrons, such that each pair is counted only once. If we transform to variable $\rho = Zr$ then we can write

$$H(Z, N) = Z^2 H_0 + Z H_1, \tag{5.11}$$

where H_0 and H_1 correspond to the one-electron and the two-electron operators in Eq. 5.10. Using first-order perturbation theory, with the expansion parameter 1/Z, the energy is

$$E = Z^2 E_0 + Z E_1. \tag{5.12}$$

E_0 is given by hydrogenic eigenvalues of the principal quantum numbers (with no ℓ dependence)

$$E_0 = -\sum_i \frac{1}{n_i^2}. \tag{5.13}$$

The first-order 'correction' E_1 is the electron repulsion energy obtained by diagonalizing $(\Psi|H_1|\Psi)$, where the eigenfunctions Ψ all belong to the same n. These are the states we referred to in Chapter 2 as an n-complex [9, 12]. Given two energy levels in ions of the same isoelectronic sequence, if the principal quantum numbers are different, then the energy difference,

$$\Delta E_{ij}(Z) = E_i(Z) - E_j(Z) \sim Z^2 \quad (n_i \neq n_j). \tag{5.14}$$

But if the states belong to the same n-complex, then

$$\Delta E_{ij}(Z) \sim Z, \quad (n_i = n_j), \tag{5.15}$$

since the zero-order energy is the same for both i and j. These simple considerations simplify numerical calculations of cross sections along isoelectronic sequence of ions, since (i) the configuration-interaction of electrons in the target ions is similar, if not identical, and (ii) the energy differences between states behave in a predictable manner, as shown above. An important caveat is that, strictly speaking, these arguments are valid if relativistic effects are neglected; fine-structure mixing of levels causes an additional degree of complexity. Nonetheless, in the non-relativistic LS coupling approximation we have

$$Z^2 \Omega_{ij} \rightarrow \text{Constant}, \quad (Z \rightarrow \infty). \tag{5.16}$$

5.2.2 Multi-channel scattering matrix

The most important quantity in scattering theory is the *scattering matrix* **S**, as mentioned in Chapter 3. Elastic and inelastic scattering cross sections are given by the diagonal and the off-diagonal elements of the **S**-matrix, respectively. The dimensionality of the **S**-matrix is the number of scattering channels. Individual channels are labelled by spin and angular quantum numbers of the initial and final levels of the ion $S_i L_i J_i$, $S_j L_j J_j$, and the orbital angular momemtum of the incident and scattered electron ℓ_i, ℓ_j. The symmetry of the **S**-matrix, say in *LS* coupling, is the total spin and angular momenta of the (e + ion) system $SL\pi$. The physical meaning of the **S**-matrix is related to channel coupling achieved by diagonalizing the multi-channel (e + ion) Hamiltonian (Eq. 3.78), say by the program STGH of the R-matrix-coupled channel codes, as explained in Chapter 3. A consequence of this is the *principle of unitarity*, which defines conservation of flux before and after scattering. For a given incident channel i

$$\sum_j |S_{ij}|^2 = 1, \tag{5.17}$$

summed over all outgoing channels j. Therefore, the sum over all *inelastic* channels

$$\sum_{i \neq j} |S_{ij}|^2 = |1 - \mathcal{S}_{ii}^2|, \tag{5.18}$$

subtracting off the flux scattered elastically with no energy exchange between the electron and the ion. The incident flux, or the number of particles crossing unit area in unit time, is normalized to unity. The elastically scattered flux is then given by $1 - \mathbf{S}$, whereas the inelastically scattered flux is given by $\mathbf{S}$. We may define the *transmission matrix* $\mathbf{T} = 1 - \mathbf{S}$. Then, noting that the unit marix is diagonal, both the elastic and inelastic scattering are described by

$$|\mathbf{T}|^2 = |\mathbf{1} - \mathcal{S}|^2. \tag{5.19}$$

In terms of the real reactance matrix $\mathbf{K}$ (Eq. 3.53), *and* the case of weak scattering when the K-matrix elements are small compared with unity, we can write

$$\mathbf{T} = 1 - \mathbf{S} = -2i\mathbf{K}(1 - i\mathbf{K})^{-1} \approx -2i\mathbf{K}. \tag{5.20}$$

Although the above equation violates unitarity, it has been used in the weak-coupling case, such as for highly charged ions where the Coulomb potential is dominant and the K-matrix elements are small.

For inelastic scattering for excitation of level i to j the EIE collision strength corresponding to a given symmetry $SL\pi$ is

$$\begin{aligned} &\Omega_{SL\pi}(S_iL_i\ell, S_jL_j\ell') \\ &= \frac{1}{2}(2S+1)(2L+1)|S_{SL\pi}(S_iL_i\ell, S_jL_j\ell')|^2, \end{aligned} \tag{5.21}$$

where S_iL_i and S_jL_j are the initial and final levels of the ion and ℓ and $\ell\,'$ are the incident and scattered partial waves of the free electron. This is, however, the *partial collision strength*. The total collision strength for the excitation $i \to j$ is the sum over (i) all partial waves of the free electron, and (ii) all contributing total symmetries $SL\pi$ of the (e + ion) system, expressed in shorthand notation simply as

$$\begin{aligned} \Omega_{ij} &= \sum_{SL\pi} \sum_{\ell,\ell'} \Omega_{SL\pi}(S_iL_i\ell, S_jL_j\ell') \\ &= \frac{1}{2}(2S+1)(2L+1) \sum_{SL\pi} \sum_{\ell,\ell'} |S_{SL\pi}(S_iL_i\ell, S_jL_j\ell')|^2 \end{aligned}$$

. (5.22)

Eq. 5.22 is the partial wave decomposition of the collision strength in terms of the S-matrix elements, which characterize the quantum numbers of all channels included in the collision problem.

5.2.3 Multi-channel quantum defect theory

As defined, the $\mathbf{S}$-matrix refers to open channels, at free electron energies capable of exciting an ion from one level to another. But in a multi-level collisional excitation, some channels may be open and some closed, depending on whether the energy of the incident free electron, relative to the level being excited, is higher or lower. As discussed in Chapter 3, (e + ion) resonances are due to coupling between open and closed channels. In a closed channel, the asymptotic wavefunction is an exponentially decaying, (quasi-)bound-state type with an effective quantum number ν_c, such that the resonances occur at energies

$$E = E_c - \frac{z^2}{\nu_c^2}, \tag{5.23}$$

where ν_c is the effective quantum number relative to the threshold E_c of an excited level of the ion. A series of Rydberg resonances thus corresponds to a closed channel; the infinite series of resonances is characterized by ν_c. We may now define the *resonance-averaged* **S***-matrix*

$$<|S_{ij}|^2> = \int_{\nu_c}^{\nu_c+1} |S_{ij}|^2 d\nu. \tag{5.24}$$

The resonance integrated value in Eq. 5.24 is over unit effective quantum number interval (ν_c, ν_{c+1}). We can use analytical expressions from multi-channel quantum defect theory (MCQDT), given in Section 3.6.5, to obtain resonance structures and averaged collision strengths or cross sections *without having to compute the* **S***-matrix in the region of resonances at a large number of energies*. First, we compute the $\mathbf{S}$-matrix (or, usually, the K-matrix which is real, Eq. 3.56) at a few energies above *all* target levels considered in the collision problem. In that case, all channels are necessarily open and the $\mathbf{S}$-matrix is properly defined. In this region, the $\mathbf{S}$-matrix elements may be fitted to a simple analytic function at a few energies. Often a linear or a quadratic fit is sufficient. The fitting function is then extrapolated to energies below threshold(s) in the region of some channels open and some closed. We then write this analytic continuation as the partitioned $\mathcal{S}$-matrix

$$\mathcal{S} = \begin{pmatrix} \mathcal{S}_{oo} & \mathcal{S}_{oc} \\ \mathcal{S}_{co} & \mathcal{S}_{cc} \end{pmatrix}, \tag{5.25}$$

where the sub-matrices refer to open–open, open–closed, closed–open and closed–closed channels. The most useful relation derived from MCQDT now relates the S-matrix elements for the open channels, $\mathrm{S_{oo}}$, in terms of the other sub-matrices. We divide open and closed channels according to energies below threshold $E < E_c$, designated by superscript $\mathcal{S}^<$, and above threshold $E > E_c$, designated as $\mathcal{S}^>$, i.e.

$$\mathbf{S}^<_{oo} = \mathcal{S}^>_{oo} - \mathcal{S}^>_{oc}[\mathcal{S}^>_{cc} - \exp(-2\pi\nu_c)]^{-1}\mathcal{S}^>_{co}. \quad (5.26)$$

The matrix elements $\mathcal{S}^>$ are computed in the region of all channels open above threshold E_c. As mentioned, it is easier to work with the real reactance matrix **K**, for which the analogous MCQDT expression (we omit the superscripts for convenience) is

$$\mathbf{K}_{oo} = \mathcal{K}_{oo} - \mathcal{K}_{oc}[\mathcal{K}_{cc} + \tan\pi\nu_c]^{-1}\mathcal{S}_{co}. \quad (5.27)$$

The closed–closed matrices $\mathcal{S}_{cc}$ or $\mathcal{K}_{cc}$ are diagonalized matrices. The resonance-averaged **S**-matrix elements are, then,

$$<|S_{ii'}|^2> = |S_{ii'}|^2 + \sum_{n,m} \frac{\mathcal{S}_{in}\mathcal{S}_{ni'}\mathcal{S}^*_{im}\mathcal{S}^*_{mi'}}{1 - \mathcal{S}_{nn}\mathcal{S}^*_{mm}}, \quad (5.28)$$

where i, i' refer to open channels and n, m to closed channels. The physical interpretation of this equation may be seen simply for only one closed channel n, in which case

$$<|S_{ii'}|^2> = |S_{ii'}|^2 + \frac{|\mathcal{S}_{in}|^2|\mathcal{S}_{ni'}|^2}{\sum_{i''}|\mathcal{S}_{ni''}|^2}, \quad (5.29)$$

where we have used the fact that the **S**-matrix is unitary, i.e., for any incident channel i

$$\sum_{i''}|S_{ii''}|^2 = 1. \quad (5.30)$$

In Eq. 5.29, $|\mathcal{S}_{in}|^2$ is the probability of capture of the incident electron in open channel i into a resonance state with closed channel n, and $|\mathcal{S}_{ni''}|^2$ is the probability of break-up of that resonance with the electron going away in an open channel i''. The fraction ending up in the final channel i' is $|\mathcal{S}_{ni'}|^2/\sum_i|\mathcal{S}^{''}_n|^2$. The resonance averaging procedure outlined above is known as *Gailitis averaging* [96, 97]. The Gailitis resonance averaging procedure for electron impact excitation was illustrated in Fig. 5.3. It may also be employed for photoionization and recombination (see Chapters 6 and 7).

The MCQDT sketched here is a powerful analytical tool that can, in some instances, obviate the necessity of computing cross sections at a very large number of energies in order to delineate Rydberg series of resonances, resulting in great saving in computing time. However, it must be mentioned that the rigourous derivation of MCQDT is based on the rather drastic approximation that long-range potentials $V_{ii'}(r)$ may be neglected beyond a finite r_0, i.e., $V_{ii'}(r) = 0$ for $r > r_0$. As a consequence, the energy range of application of MCQDT is limited, though approximate procedures may be devised to extend its range of validity. Also, as mentioned earlier in Chapter 3, there are computer codes (e.g., within the R-matrix package, Fig. 3.9, such as JAJOM and STGFJ, which not only implement MCQDT with algebraic transformation from *LS* coupling to pair-coupling to incorporate fine structure, but also to include relativistic effects in the target ion [10, 38, 72].

5.3 Excitation rate coefficients

Given Eq. 5.5, the cross section Q is expressed in terms of the collision strength by Eq. 5.22. However, for practical applications, the quantity of interest is the cross section averaged over the distribution of velocities of all scattering electrons $<v\ Q(v)>$, or $<E\ Q(E)>$ in terms of the distribution of kinetic energy $E = \frac{1}{2}mv^2$. As discussed in Chapter 1, the electron distribution most often encountered in astrophysical environments is a Maxwellian, characterized by a kinetic temperature T_e (or simply T in the discussion below). Since we find it computationally more convenient to use the collision strength, the equivalent quantity of interest is the *Maxwellian averaged collision strength*, sometimes referred to as the *effective collision strength*, given by

$$\Upsilon(T) \text{ or } \gamma(T) = \int_0^\infty \Omega_{ij}(\epsilon_j)\mathrm{e}^{-\epsilon_j/kT}\mathrm{d}(\epsilon_j/kT). \quad (5.31)$$

Now we can define the *excitation rate coefficient*, or the number of excitations per unit volume per second, as

$$q_{ij}(T) = \frac{8.63\times10^{-6}}{g_i T^{1/2}}\mathrm{e}^{-E_{ij}/kT}\Upsilon(T)\ \mathrm{cm^3s^{-1}}, \quad (5.32)$$

with T in K and $E_{ij} = E_j - E_i$ in Rydbergs $(1/kT) = (157885/T)$.

It follows from detailed balance that the *de-excitation* rate coefficient ($E_i < E_j$) is given by

$$q_{ji} = q_{ij}\frac{g_i}{g_j}\mathrm{e}^{E_{ij}/kT}. \quad (5.33)$$

We have seen that, in general, Ω is an energy dependent function, which consists of a slowly varying part, the background or non-resonant Ω, and a rapidly varying part due to myriad series of autoionizing resonances, which can significantly affect the Maxwellian averaged value or the (de-)excitation rate. One of the main points that makes it necessary to compute collision strengths with resonances accurately is that the energy dependence of the collision strength (cross section) introduces a temperature dependence in the rate coefficient, according to Eqs 5.31 and 5.32.

Exercise 5.1 *Use the program from Exercise 1.1, and assuming that Q(E) = C/E (C is a constant), compute and plot the Maxwellian rate coefficient* $q(T_e)$ *at* $\log T = 4$–7 *(K). What is the physical interpretation of the behaviour of* $q(T_e)$*?*

5.4 Atomic effects

The determination of environmental conditions in astrophysical objects, primarily temperature, density and abundances, depends on spectral analysis discussed in later chapters. At low temperature–density conditions, electron impact collision strengths and radiative transition decay rates are of prime importance: electron impact excitation followed by radiative decay in an atom. Since most of the excitation data used in spectral models is theoretically obtained, it is of prime interest to know what factors determine the accuracy of computed parameters. These factors, in turn, depend on the physical effects incorporated in the calculations. Since they also help us understand the underlying excitation process, we discuss some of the most important effects below, with a relatively complex atomic system, Ne-like Fe XVII, as an example.

5.4.1 *Target ion representation*

An accurate representation for the wavefunctions of the target ion is essential to obtain accurate cross sections. Any error in the target wavefunctions propagates through as a first-order error in derived parameters, such as excitation cross sections. The reason is not hard to find. The total (e + ion) wavefunction expansion in the coupled channel (CC) framework described in the preceding chapter (Eqs 3.24, 3.42 and 3.77) assumes that the target ion is *exactly* represented. Let us rewrite wavefunction expansion, Eq. 3.24, as

$$\Psi(\mathrm{e}+\mathrm{ion}) = \mathcal{A}\sum_{i}^{n_f} \Phi_i(\mathrm{ion})\theta_i, \tag{5.34}$$

where the target ion eigenfunctions Φ are assumed to be exactly determined *a priori*, and the electron wavefunction θ is the freely varying parameter in accordance with the Kohn variational principle, Eq. 5.7. That is the reason that the first step in any collision calculation is the determination of an accurate target representation, usually through elaborate configuration interaction (CI) type calculations for the ion (see Chapter 2).

For complex atomic systems, such as the low ionization stages of iron Fe I–Fe V [98], it is necessary to include a large CI basis in order to obtain the proper wavefunctions for states of various target (angular + spin) symmetries $SL\pi$. The accuracy may be judged by comparing the calculated eigenenergies and the oscillator strengths (in the length and the velocity formulations) with experimental or other theoretical data for the states of interest in the collision. The advent of increasingly powerful serial and massively parallel supercomputers has enabled some very large atomic systems to be dealt with. Particular attention has been paid to the important ion Fe II, whose emission lines are prevalent in the spectra of most classes of astronomical object. The largest such calculation is a 262-level representation of the Fe II ion, including relativistic effects in the Breit–Pauli R-matrix approximation [58].

To circumvent the problem of computer memory restrictions, the target state expansion may include *pseudostates* with adjustable parameters in the eigenfunction expansion over the target states for additional CI, i.e., in the first sum on right-hand side of the CC expansion in Eqs. 3.42 and 3.77. Transitions involving target ion pseudostates themselves are ignored. Single-configuration calculations, to be found in older works, are generally less accurate than the ones including CI.

5.4.1.1 Target eigenfunctions for Fe XVII

Table 5.1 is a target ion representation [99] typical of modern calculations on supercomputers.[3] As the first step, the target eigenvalues of Fe XVII are computed prior to the CC calculations using the Breit–Pauli R-matrix (BPRM) codes (Fig. 3.9). The ground configuration $1s^22s^22p^6$ is a closed shell neon-like system with $^1S_0$ ground state. One-electron excitations from the 2s, 2p subshells to the 3s, 3p, 3d orbitals give rise to 37 levels up to the $n = 3$ configurations. The eigenfunctions for these levels included in the CC expansion, and the configurations that *dominate* their composition,[4] are from CI calculations using the code SUPERSTRUCTURE (SS) shown in Fig. 3.9. The Fe XVII configurations employ a one-electron basis set $P_{nl}(r)$ optimized with individual orbital scaling parameters in the Thomas–Fermi–Dirac–Amaldi type central potential

[3] These calculations, like many others reported by the authors and collaborators throughout the text, were carried out at the Ohio Supercomputer Center in Columbus Ohio.

[4] The word 'dominate' here applies to those few configurations *with the largest coefficients in the CI expansion* (Eq. 2.113), which together constitute most of the wavefunction components for a given level. These chosen few configurations are often used to represent each level, as opposed to those that are left out and do not contribute significantly. Note that the CC expansion refers to the target levels included in the wavefunction expansion for the (e + ion) system, whereas the CI expansion refers to the configuration expansion for each level.

TABLE 5.1 Energy levels of Fe XVII included in a coupled channel BPRM calculation [99]. The calculated and observed energies are in Rydbergs. 'OBS' signifies observed values from the NIST website [100], 'SS' and 'MCDF' are results from SUPERSTRUCTURE and GRASP atomic structure codes, respectively. The index i serves as a key for referencing levels.

i	SLJ	jj	OBS	SS	MCDF
1	$2s^22p^6\ ^1S_0$	(0,0)0	0.0	0.0	0.0
2	$2s^22p^53s\ ^3P^o_2$	$(3/2,1/2)^o2$	53.2965	53.3603	53.1684
3	$3s\ ^1P^o_1$	$(3/2,1/2)^o1$	53.43	53.5025	53.3100
4	$3s\ ^3P^o_0$	$(1/2,1/2)^o0$	54.2268	54.2847	54.0957
5	$3s\ ^3P^o_1$	$(1/2,1/2)^o1$	54.3139	54.3772	54.1851
6	$3p\ ^3S_1$	(3/2,1/2)1	55.5217	55.5685	55.3963
7	$3p\ ^3D_2$	(3/2,1/2)2	55.7787	55.8397	55.6606
8	$3p\ ^3D_3$	(3/2,3/2)3	55.8974	55.9463	55.7791
9	$3p\ ^3P_1$	(3/2,3/2)1	55.9804	56.0338	55.8654
10	$3p\ ^3P_2$	(3/2,3/2)2	56.1137	56.1597	56.9950
11	$3p\ ^3P_0$	(3/2,3/2)0	56.5155	56.5820	56.4050
12	$3p\ ^1P_1$	(1/2,1/2)1	56.6672	56.7288	56.5495
13	$3p\ ^3D_1$	(1/2,3/2)1	56.9060	56.9499	56.7855
14	$3p\ ^1D_2$	(1/2,3/2)2	56.9336	56.9817	56.8135
15	$3p\ ^1S_0$	(1/2,1/2)0	57.8894	58.0639	57.9308
16	$3d\ ^3P^o_0$	$(3/2,3/2)^o0$	58.8982	58.9407	58.7738
17	$3d\ ^3P^o_1$	$(3/2,3/2)^o1$	58.981	59.0188	58.8454
18	$3d\ ^3P^o_2$	$(3/2,5/2)^o2$	59.0976	59.1651	58.9826
19	$3d\ ^3F^o_4$	$(3/2,5/2)^o4$	59.1041	59.1821	58.9901
20	$3d\ ^3F^o_3$	$(3/2,3/2)^o3$	59.1611	59.2240	59.0498
21	$3d\ ^3D^o_2$	$(3/2,3/2)^o2$	59.2875	59.3513	59.1797
22	$3d\ ^3D^o_3$	$(3/2,5/2)^o3$	59.3665	59.4471	59.2598
23	$3d\ ^3D^o_1$	$(3/2,5/2)^o1$	59.708	59.7865	59.6082
24	$3d\ ^3F^o_2$	$(1/2,3/2)^o2$	60.0876	60.1439	59.9749
25	$3d\ ^1D^o_2$	$(1/2,5/2)^o2$	60.1617	60.2179	60.0344
26	$3d\ ^1F^o_3$	$(1/2,5/2)^o3$	60.197	60.2627	60.0754
27	$3d\ ^1P^o_1$	$(1/2,3/2)^o1$	60.6903	60.8225	60.6279
28	$2s2p^63s\ ^3S_1$	(1/2,1/2)1		63.3289	63.2125
29	$3s\ ^1S_0$	(1/2,1/2)0		63.7908	63.6986
30	$3p\ ^3P^o_0$	$(1/2,3/2)^o0$		65.7339	65.6346
31	$3p\ ^3P^o_1$	$(1/2,1/2)^o1$	65.601	65.7688	65.6676
32	$3p\ ^3P^o_2$	$(1/2,3/2)^o2$		65.9300	65.8380
33	$3p\ ^1P^o_1$	$(1/2,3/2)^o1$	65.923	66.0724	65.9782
34	$3d\ ^3D_1$	(1/2,3/2)1		69.0163	68.9221
35	$3d\ ^3D_2$	(1/2,3/2)2		69.0352	68.9323
36	$3d\ ^3D_3$	(1/2,5/2)3		69.0674	68.9518
37	$3d\ ^1D_2$	(1/2,5/2)2	69.282	69.4359	69.3247

discussed in Chapter 2. The computed eigenenergies are compared with laboratory data where available [100], as well as theoretical calculations using other codes, such as the fully relativistic code GRASP [11]. Table 5.1 also displays the equivalence between fine-structure designations SLJ in an intermediate coupling scheme, and the fully relativistic jj coupling scheme. The close agreement among the various calculations and the experimentally measured energies for observed levels is an indicator of the accuracy of the target ion representation.

Figure 5.4 is a Grotrian diagram of (a) even parity and odd parity configurations and transitions of astrophysical

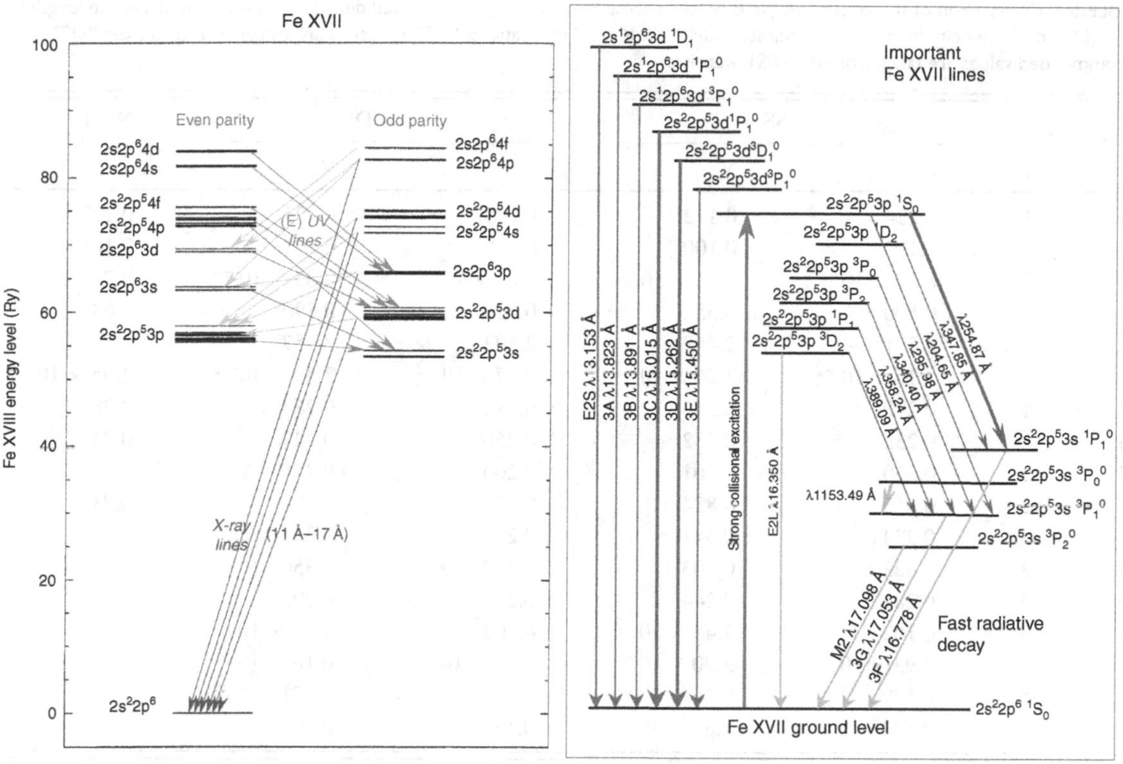

FIGURE 5.4 The Grotrian diagram of Fe XVII energy levels and transitions [99].

interest [99] and (b) important transitions among low-lying levels.[5] Some of the X-ray transitions are labelled as '3C', '3D', etc., whose excitation cross sections have been measured in the laboratory [99]. Line ratios 3C/3D, 3E/3C, and others [101] are used as diagnostics of astrophysical plasmas in the temperature range around a few million K, such as in the solar corona (see Chapter 8). The unusual transition $2s^22p^6\ ^1S_0 - 2s^22p^53p^1\ ^1S_0$ has been of great interest in X-ray laser research. Since both the initial and the final levels have identical symmetry $SLJ \rightarrow {}^1S_0$, radiative decays are strictly forbidden. Therefore, collisional 'pumping' of the upper level can bring about population inversion, and hence laser action via fast dipole transitions to other lower levels as shown in Fig. 5.4.[6]

Apart from a comparison of computed energy levels of the target ion, the accuracy of the wavefunctions may be ascertained by a comparison of the oscillator strengths for transitions among the levels included in the collision problem.[7] In the asymptotic region, the coupling potentials (c.f. Eq. 3.100) are proportional to $\sqrt{f}$, where f is the corresponding oscillator strength. It is therefore particularly important that the wavefunctions used should give accurate results for these oscillator strengths. It is standard practice in most coupled channel calculations to establish first the accuracy of the target ion by a detailed comparison of the energies and oscillator strengths. Table 5.2 compares the SUPERSTRUCTURE (SS) results for f-values with other available data for some dipole allowed transitions among the Fe XVII levels (indexed) in Table 5.1.

Table 5.2 compares results from two different atomic structure codes (Fig. 3.9): SUPERSTRUCTURE [10] in a semi-relativistic intermediate coupling Breit–Pauli scheme, and the fully relativistic multi-configuration Dirac–Fock (MCDF) code GRASP [11]. Comparison is also made with a combination of experimental and

[5] A *Grotrian diagram* is a graphical representation of energies of an atom, sorted according to angular-spin symmetries and transitions in LS or LSJ coupling scheme.

[6] Such a mechanism using high-Z Ne-like ions was proposed as a driver for a space-based X-ray laser system to be deployed for antiballistic missile defence, part of the erstwhile *Strategic Defense Initiative* popularly known as the '*Star Wars*' scenario.

[7] The different types of radiative transitions prevalent in astrophysics, and related quantities such as oscillator strengths, are covered in Ch. 4. Here it suffices to recall that strong transitions between two levels are usually connected by a dipole moment, and the associated radiative rate is large, of the order of $10^9\ s^{-1}$ for neutrals and increases with Z or ion charge z.

TABLE 5.2 Comparison of statistically weighted Fe XVII oscillator strengths *gf* for selected dipole allowed transitions in the length form (L) and the velocity form (V). 'SS' indicates SUPERSTRUCTURE calculations; 'MCDF' results are obtained with GRASP; 'NIST' recommended values are taken from the NIST website.

		SS		MCDF		NIST
i	j	L	V	L	V	
3	1	0.124	0.112	0.125	0.121	0.122
5	1	0.102	0.100	0.106	0.101	0.105
17	1	8.70×10^{-3}	8.11×10^{-3}	1.01×10^{-2}	9.35×10^{-3}	9.7×10^{-3}
23	1	0.590	0.558	0.628	0.590	0.63
27	1	2.571	2.450	2.503	2.357	2.31
31	1	3.15×10^{-2}	3.20×10^{-2}	3.57×10^{-2}	3.59×10^{-2}	2.95×10^{-2}
33	1	0.280	0.296	0.282	0.283	0.282
6	2	0.252	0.232	0.256	0.242	0.248
7	2	0.260	0.261	0.260	0.271	
8	2	0.812	0.824	0.825	0.990	0.819
7	3	0.284	0.341	0.287	0.362	
9	3	0.322	0.303	0.327	0.350	
10	3	0.281	0.244	0.283	0.295	
11	3	0.102	7.43×10^{-2}	0.102	8.70×10^{-2}	
15	3	7.93×10^{-2}	3.90×10^{-2}	7.98×10^{-2}	6.18×10^{-2}	
14	5	0.589	0.562	0.595	0.671	
15	5	0.133	7.82×10^{-2}	0.134	0.104	

theoretical values reported from the NIST compilation [100]. Another instructive comparison is between the *length* (L) and the *veloctiy* (V) forms of the dipole matrix element (Chapter 4), with transition operators **r** and $\partial/\partial r$, respectively. Since exact wavefunctions yield identical results in the length and velocity formulations, it follows that the agreement between the L and V values for the transition strength should indicate the accuracy and self-consistency of the wavefunctions of initial and final levels. However, the discrepancy between the L and V values may not necessarily reflect actual uncertainties. That is because the length operator **r**, and the velocity operator $\partial/\partial r$, sample different parts of the radial wavefunction; the former dominates at large r and the latter at short r (closer to the origin). As such, the f_L-values are to be preferred because they are dominated by the long-range part of the wavefunction where the interaction potentials are less complicated than in the inner region.

5.4.2 Channel coupling

Owing to the fact that coupling among collision channels is often strong, it needs to be included in general. However, its importance varies with collision energy, element (Z), and ionization state. At low incident electron energies near excitation threshold(s), and for low stages of ionization with energetically closely spaced energy levels, electron scattering may couple the probabilities of excitation of several levels. Fine structure is important for all but the light elements, and coupling among those levels is often substantial. When the coupling between the initial and the final levels is comparable to, or weaker than, the coupling with other levels included in the target expansion, then the scattered electron flux is diverted to those other states, and affects the excitation cross section. The S-matrix elements obey Eqs. 5.17 and 5.18 and govern the scattering process. The inelastic cross section corresponding to channels i and j is given by $|S_{ij}|^2$.

The weak-coupling approximations such as the Born or the Coulomb–Born tend to *overestimate* the cross sections, since the unitarity condition, Eq. 5.17, implies that the scattering flux into neglected channels must be accounted for. As the ion charge increases, the nuclear Coulomb potential dominates the electron–electron interaction, and correlation effects, such as coupling of channels, decrease in importance. For neutrals and singly or doubly charged ions, coupling between closely spaced low-lying levels (the energy level separation depends on both Z and the ion charge z), is usually strong, and only a CC calculation may yield accurate results.

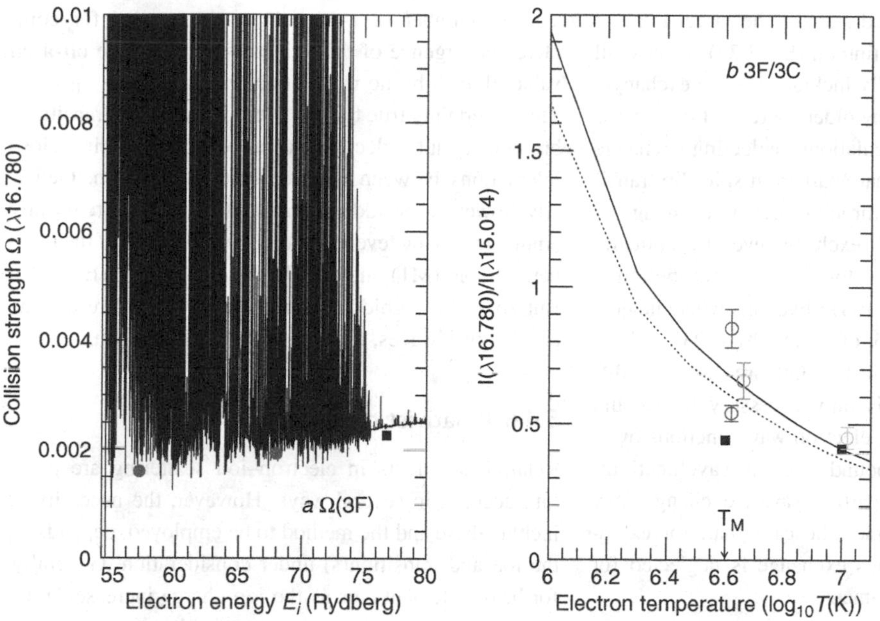

FIGURE 5.5 (a) Resonance structures in the collision strength for a transition in Fe XVII: $\Omega(2s^22p^6\ {}^1S_0 - 2s^22p^53s\ {}^3P^o_1)$ (labelled 3F in Fig. 5.4). The collision strengths are computed in the CC approximation using the BPRM method [101]. The two dots and a square are values in the DW approximation without resonances. (b) The line ratio 3F/3C depends significantly on the resonances, as indicated by the large temperature variation, since the spin-forbidden intercombination transition 3F is enhanced much more by resonances than the dipole allowed transition 3C: $2s^22p^6\ {}^1S_0 - 2s^22p^53d^1\ {}^1P^o_1$. The ratio 3F/3C is a good temperature diagnostic since it is relatively independent of variations in electron density; the solid and dashed curves correspond to $n_e = 10^9$ and $10^{13}\ \mathrm{cm}^{-3}$, respectively. The experimental and observed values are shown with error bars; the filled squares are line ratios obtained using DW collision strengths.

But for multiply charged ions and at relatively high energies, a distorted wave treatment may be sufficient (see Chapter 3). Strong dipole transitions are often not affected by channel coupling since the dipole potential between initial and final levels, and associated channels, dominates the transition.[8]

5.4.3 Resonances

As mentioned in Chapter 3, resonances arise from coupling between open (or continuum) and closed (or bound) channels. Collision strengths from a BPRM calculation for electron impact excitation of a spin-forbidden intercombination transition in Fe XVII are shown in Fig. 5.5(a). In the previous chapter, we discussed various theoretical methods, in particular the coupled channel (CC) and the distorted wave (DW) methods. The extensive resonance structures in the CC collision strengths considerably enhance the effective collision strength $\Upsilon(T)$ (Eq. 5.31) or the rate coefficient $q(T)$ (Eq. 5.32), compared with distorted wave (DW) results without resonances including just the background values. In fact, the distorted wave values are found to lie close to the background, shown at a few energies in the region where the background is discernible in Fig. 5.5(a). The practical importance of including resonances is highlighted by the line ratio in Fig. 5.5(b), which is a sensitive temperature diagnostic. The pronounced variation in resonance-averaged collision strengths seen in Fig. 5.5(a) implies corresponding variation in the rate coefficient with temperature. In Chapter 8, we shall develop emission-line diagnosis based on line ratios, which depend primarily on the electron impact excitation rates computed from collision strengths, such as in Fig. 5.5(a), leading to Fig. 5.5(b).

5.4.4 Exchange

The exchange effect stems from the fact that the total (e + ion) wavefunction should be an antisymmetrized product of the $N+1$ electron wavefunction in the system,

[8] But there are exceptions, such as in Li-like ions with the lowest levels 2s, 2p, 3s, 3p, 3d. The 2p–3s transition is considerably influenced by the stronger coupling between the 2p and the 3d. In such cases, resonance contributions to the cross sections for 2p–3s transition are large, from resonances lying in the energy region between 3s and 3d [102].

with N electrons in a bound state of the target ion and one free electron in the continuum (Eq. 3.24). Almost all present-day calculations fully include electron exchange. However, there are still many older sources of data in the literature that are from calculations neglecting exchange [103]. It has been shown that apart from spin-flip transitions, which proceed only through electron exchange, it may be necessary to include exchange even for optically allowed transitions when the low ℓ-wave contribution is significant, e.g., $1^1S - 2^1P$ in He-like ions where neglect of exchange leads to an error of about 50% [103].

The exchange integrals in Eq. 3.45 are, however, difficult to evaluate. But it is only necessary to account for exchange when the free electron wavefunctions overlap significantly with the bound electron wavefunctions. In other words, for large partial waves, exchange may be neglected. It is sometimes the case that, for example, in R-matrix calculations exchange is neglected for sufficiently high angular momenta ℓ.

5.4.5 Partial wave expansion

Theoretically, the cross section (collision strength) is decomposed in terms of the partial wave expansion according to Eq. 5.22. It follows then that (a) *all* partial waves ℓ of the free electron, and (b) all partial spin and angular symmetries $SL\pi$ of the total (e + ion) system, should be included to obtain the total cross section. In practice, however, this criterion is not easily satisfied. A combination of methods is needed to ensure convergence of the partial wave expansion. This is because different partial waves interact differently with the target ion, and different approximations are then appropriate. For low-ℓ waves the (e + ion) interactions are strong and the CC method is best employed. For high-ℓ waves, however, that is not necessary, particularly since exchange terms are not important and coupled integro-differential equations (Eq. 3.45) need not be solved. Then, simpler weak-coupling methods, such as the distorted wave, Coulomb–Born, or Coulomb–Bethe approximations may be employed, as discussed in Chapter 3. However, the practical problem lies in the fact that different types of transition have different crossover limits, where it is safe to switch over to weak-coupling approximations. For dipole allowed transitions the cross section is dominated by high partial waves, owing to the long-range Coulomb potential. In such cases, the CC method is best for low-ℓ waves that give rise to resonances. Weak-coupling approximations, particularly the distorted wave method, may otherwise yield accurate cross sections for the higher ℓ. On the other hand, for some forbidden transitions only a small number of ℓ-waves may be needed for complete convergence of the total cross section; the oft-used value, though by no means universally valid, is $\ell_0 = 10$. This is certainly true for the forbidden spin-flip transitions brought about by electron exchange, which requires close interactions between the free electron and ion. Generally, however, forbidden transitions of interest are usually among low-lying levels of the same parity, and of the magnetic dipole (M1) and the electric quadrupole (E2) type. But some E2 forbidden transitions may require a large number of ℓ-waves, $\ell > 50$, for full convergence.

5.4.6 Relativistic effects

Relativistic effects in electron–ion scattering are taken into account in several ways. However, the necessity to include these and the method to be employed, depends on the ion and transition(s) under consideration. Generally, for heavy elements, or as the ion charge increases, relativistic effects become prominent and have to be considered explicitly. But there are many cases where relativistic effects per se are not too important, but fine-structure lines are still resolved in practical observations. In those cases, the transition can be described within *LS* coupling, such for low-Z or low-z ions, and the cross sections may be obtained by a pure algebraic transformation from the *LS* to *LSJ*, or in an intermediate-coupling scheme. Within the context of theoretical methods discussed in Chapter 3, computer programs, such as JAJOM or STGFJ offer such solutions as part of the R-matrix package of codes (Fig. 3.9, [38]).

In general, the ratio of the fine structure collision strengths to multiplet collision strengths depends on the recoupling coefficients, but for the particular case of $S_i = 0$ or $L_i = 0$ we have

$$\frac{\Omega(S_iL_iJ_i, S_jL_jJ_j)}{\Omega(S_iL_i, S_jL_j)} = \frac{(2J_j+1)}{(2S_j+1)(2L_j+1)}. \tag{5.35}$$

For example, consider the transitions in p^3 open-shell ions, such as O II or S II with ground configurations $2s^22p^3$ and $3s^23p^3$, respectively, where the terms dominated by the ground configuration are: ${}^4S^o_{3/2}$, ${}^2D^o_{3/2,5/2}$, ${}^2P^o_{1/2,3/2}$. The ratio

$$\Omega\left({}^4S^0_{3/2} - {}^2D^o_{3/2}\right)/\,\Omega({}^4S^o - {}^2D^o) = 4/10,$$

and

$$\Omega\left({}^4S^o_{3/2} - {}^2D^o_{5/2}\right)/\,\Omega({}^4S^o - {}^2D^o) = 6/10.$$

These forbidden transitions are of great importance in emission line diagnostics, as described in Ch. 8.

As the relativistic effects become larger, one may employ three different approaches with increasing levels of sophistication (the approximations refer to the description in Chapter 3). The first approximation to incorporate relativistic effects in the target ion is to generate term coupling coefficients $<S_i L_i J_i|\Delta_i J_i>$, which diagonalize the target Hamiltonian including relativistic terms (Breit–Pauli Hamiltonian); $\Delta_i J_i$ is the target state representation in intermediate coupling. These coefficients are then used together with the algebraic transformation procedure mentioned above to account for relativistic effects (as in JAJOM or STGFJ).

The second approximation entails intermediate coupling, say using the Breit–Pauli R-matrix (BPRM) method [38, 55]. It extends the close-coupling R-matrix method to treat the entire electron–ion scattering process in the Breit–Pauli scheme (and not just the target ion, as in the JAJOM or STGFJ approach). All BPRM scattering calculations thus far use only the one-body operators of the Breit interaction – mass–velocity, Darwin and spin–orbit terms (Section 2.13.2). The two-body terms, usually not too important for (e + ion) scattering, are neglected. However, as mentioned in Chapter 3, recently the BPRM method has been extended to include the two-body Breit terms and is referred to as the 'Full' BPRM approximation, shown in Fig. 3.9. The BPRM method appears particularly well suited for elements up to the iron group ($Z \leq 30$) to account for both the electron correlation and the relativistic effects accurately. Third, for even heavier systems and highly charged ions, high-Z and z, a fully relativistic treatment has been developed based on the Dirac equation, and a package of codes (DARC) has been developed by P. H. Norrington and I. P. Grant [53].

A comparison of the various relativisitic approximations, the Dirac R-matrix, BPRM and term-coupling, for electron scattering with Boron-like Fe XXII shows that the term-coupling approximation may not be very accurate, but that there is good agreement between the DARC and the BPRM calculations, including the fine structure resonances that are important for several transitions [104, 105].

5.5 Scaling of collision strengths

It is often useful to adopt scaling procedures and examine systematic trends in the collision strengths. Transitions may be classified according to the range of the potential interaction ($V_{ii'} \pm W_{ii'}$) in Eq. 3.45. Spin change transitions depend entirely on the exchange term $W_{ii'}$, which is very short range since the colliding electron must penetrate the ion for exchange to occur. Therefore, only the first few partial waves are likely to contribute to the cross section, but these involve quite an elaborate treatment (e.g., close-coupling). For allowed transitions, on the other hand, a fairly large number of partial waves contribute and simpler approximations (e.g., Coulomb–Born) often yield acceptable results. The asymptotic behaviour of the collision strengths for allowed and forbidden transitions is as follows (x is in threshold units of energy $x \equiv E/E_{\rm th}$, for a transition $i \to j$, $E_{\rm th} = \Delta E_{ij}$):

(a) $\Omega(i,j) \underset{x\to\infty}{\sim}$ constant, for forbidden (electric quadrupole) transitions, $\Delta L \neq 1, \Delta S = 0$,

(b) $\Omega(i,j) \underset{x\to\infty}{\sim} x^{-2}$, for spin – change transitions, $\Delta S \neq 0$,

(c) $\Omega(i,j) \underset{x\to\infty}{\sim} a \ln(4x)$, allowed transitions, $\Delta L = 0, \pm 1, \Delta S = 0$.

The slope a in the last equation (c) is proportional to the dipole oscillator strength. The above forms are valid for transitions in LS coupling. For highly charged ions, where one must allow for relativistic effects, through say an intermediate coupling scheme, sharp deviations may occur from these asymptotic forms particularly for transitions labelled as intercombination type, e.g., the transition $1^1S - 2^3P$ in He-like ions (Eq. 4.167). For low-Z ions ($Z < 15$) when LS coupling is usually valid, $\Omega(1^1S - 2^3P)$ behaves as (b). With increasing Z, the fine-structure splitting between $2^3P(J=0,1,2)$ becomes significant and the collision strength $\Omega(1^1S_0 - 2^3P_1)$ gradually assumes form (c).

A useful fact for isosequence interpolation, or extrapolation, is that $Z^2\Omega(i,j)$ tends to a finite limit, as $Z \to \infty$, as a function of k^2/Z^2; i.e., at the Z^2 reduced incident electron energy, $\Omega(i,j)$ is constant or a slowly varying function for large Z. For highly ionized atoms (e.g., H-like, He-like) the Z^2 behaviour is valid even for $Z < 10$, but for many-electron ions (e.g., Ne-like) one needs to go to much higher values of Z.

An elaborate procedure to fit and extrapolate or interpolate effective (Maxwellian averaged) collision strengths $\Upsilon(T)$ has been enabled in a computer program called OMEUPS [86]. The limiting expression of Eq. 5.31 at low energies is

$$\lim_{T\to 0} \Upsilon(T) = \lim_{\epsilon\to 0} \Omega(\epsilon), \tag{5.36}$$

in terms of the threshold collision strength. As we have discussed, in the high-energy limit, simpler approximations may be employed to compute the collision strengths, say in the Coulomb–Born or the Coulomb–Bethe approximations (Chapter 3). Although useful in many cases, the limiting values (i.e., E or $T \rightarrow \infty$) adopted in this approach may not always be accurate, particularly for non-dipole transitions where coupling effects, such as the appearance of high-n resonance complexes, may attenuate the collision strengths even to very high energies, and the Coulomb–Born or Coulomb–Bethe limits may not be applicable. Nonetheless, the scaled collision strength data can be visualized in a systematic manner, for example along isoelectronic sequences [74]. Deviations from the fits may be investigated to explore possible errors, the presence of resonances, convergence of partial wave expansion, etc.

Whereas we have not considered electron scattering with neutral atomic species in this text, excitation of several atoms is quite important, viz. C I, N I, O I and Fe I. However, calculation of neutral excitation cross sections is more difficult, owing to the absence of the Coulomb potential. In that case, long-range potentials, particularly dipole polarization, dominate the scattering process. Theoretically, the threshold collision strength Ω_0 for the excitation of neutral atoms is governed by the *Wigner law*,

$$\Omega_(E \rightarrow 0) = E^{l+1/2}, \tag{5.37}$$

where l is the dominant partial wave. As part of the Iron Project, electron impact excitation of neutral Fe I has been studied [106]. Another form of the Wigner threshold law applies for electron impact ionization, as described in Section 5.8.

5.6 Comparison with experiments

Although experimental measurements of electron–ion scattering cross sections are difficult, a number of advances have been made in recent years. The advent of merged-beam techniques, synchrotron ion storage rings and electron-beam ion traps have made it possible to measure the cross sections to unprecedented accuracy. Experiments are now also capable of the high resolution necessary to resolve resonances up to fairly high-n, approaching Rydberg series limits.

In nearly all cases, the experimental results agree with the most sophisticated coupled channel calculations to within the typical experimental uncertainties of 10–20% – the range of uncertainty often quoted for such theoretical calculations. However, it needs to be emphasized that the details of resonance structures in the experimental data are subject to beam resolution; the theoretical data are convolved over the FWHM beam width in comparing with experiments. A few of the merged-beam cross sections and corresponding theoretical works are presented below.

The cross sections for C II in Fig. 5.6 are for the three lowest transitions from the ground state to different LS states of the next excited configuration: (a) the intercombination transition $1s^22s^22p\ ^2P^o \rightarrow 1s^22s^12p^2\ ^4P$, and the next two dipole allowed transitions to (b) ^2D and (c) ^2S (see Figs 6.10 and 6.6 for energy level diagrams). The latter two final states, ^2D and ^2S, lie relatively close together in energy and are strongly coupled. This is indicated by the fact that their collision strengths may *not* be accurately obtained by simple approximations such as the Coulomb–Bethe method, which essentially treats the collision process as an an induced radiative transition. Hence, in the Coulomb–Bethe approximation the collision strength is related to the oscillator strength (Chapter 3). The proportion of threshold collision strengths for the $^2P^o - {}^2S, {}^2D, {}^2P^o$ transitions in Fig. 5.6 is about 1:3:3, quite different from that of the oscillator strengths which is about 1:1:5 [72]. It follows that simple formulae such as the Van Regemorter formula, or the 'g-bar' approximation – based on the idea that collision strength can be estimated from the oscillator strength – gives incorrect collision strengths.

All three cross sections in Fig. 5.6 show considerable resonance structures. Therefore, approximations neglecting channel couplings, such as the distorted wave, are also likely to yield inaccurate collision strengths in corresponding energy regions.

The measured cross sections [107] in Figs 5.6(b) and (c) for C II are for dipole allowed transitions. Weaker intercombination or LS forbidden transitions are more sensitive to coupling and resonance effects, since the transitions are not dominated by a strong dipole moment but, rather, mediated by weaker non-dipole coupling potentials. Therefore, the background cross section for direct excitation is small, and indirect excitation via resonances assumes greater importance. Recent experimental work has also been carried out for the LS-forbidden transitions in nebular ions such as O II and S II (both ions have the same LS-term structure, see Chapter 8). For example, the measured cross sections for the total $^4S^o - {}^2D^o$ transition in Fig. 5.7 agrees very well with the BPRM calculations in the CC approximation [108].

As the electron impact energy increases, the behaviour of collision strengths differs, depending on the type of transition. For allowed transitions, the collision strengths increase logarithmically, according to the Bethe

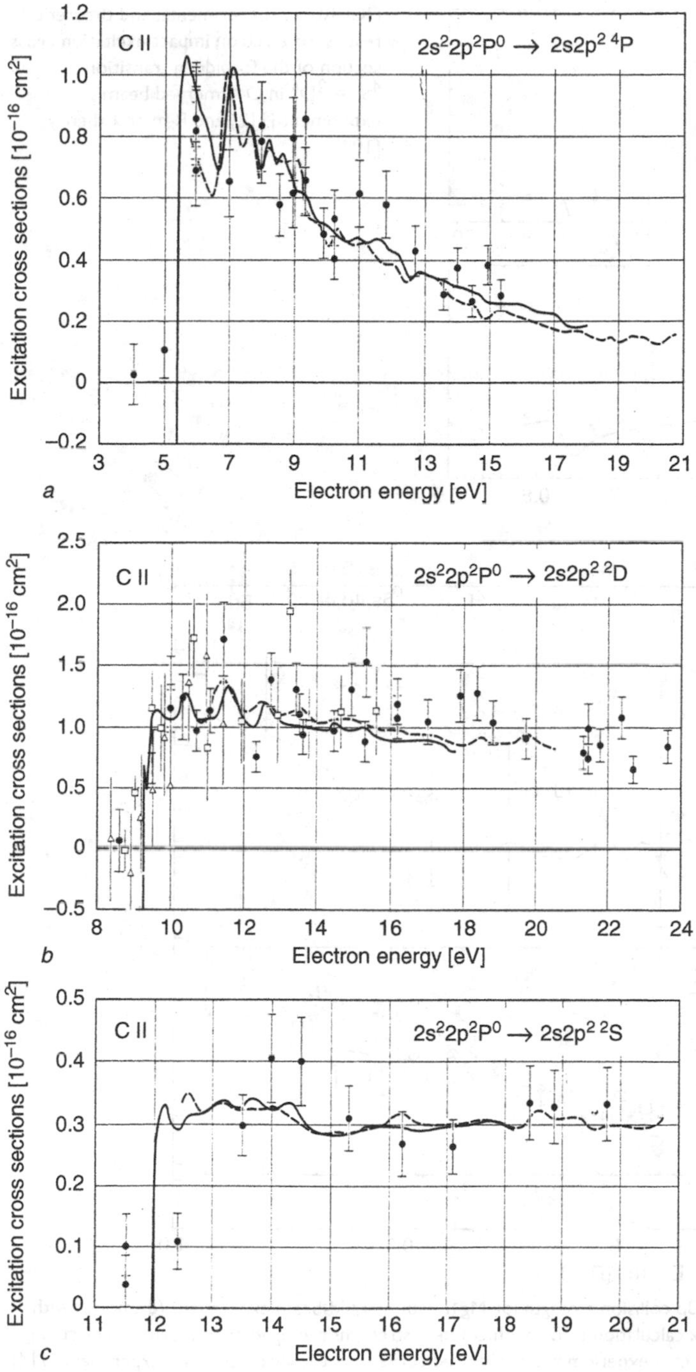

FIGURE 5.6 Experimental (filled circles) and theoretical R-matrix cross sections for C II convolved with a 250 meV FWHM resolution [107]; solid line (experiment), dashed line (theory) [72]; (a) excitation of the intercombination transition ${}^2P^o - {}^4P$, (b) allowed transition ${}^2P^o - {}^2D$ and (c) allowed transition ${}^2P^o - {}^2S$.

asymptotic form $\Omega(E) \sim \ln(E)$. Figure 5.8 presents the measured and theoretical collision strengths for the dipole allowed transtion 3s – 3p in Mg II [109, 110]. The theoretical CC R-matrix results are shown as the solid line, computed using a 10CC approximation including the 10 Mg II states corresponding to the outer orbital configurations 3s, 3p, 3d, 4s, 4p, 4d, 4f, 5s, 5p and 5d. The dashed line is a 5CC calculation including only the first five states [111]. All experimental points except the last one agree with theoretical 10CC R-matrix results [109] within experimental uncertainties [110]. The agreement holds true even at energies that span a number of excited

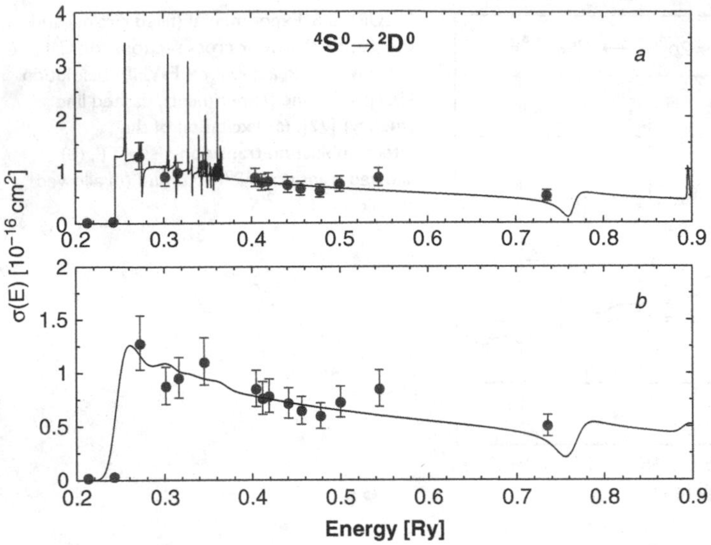

FIGURE 5.7 Experimental and theoretical results for electron impact excitation cross section of the forbidden transition $^4S^o - {}^2D^o$ in O II: merged beam experiment [112] and R-matrix theory [108].

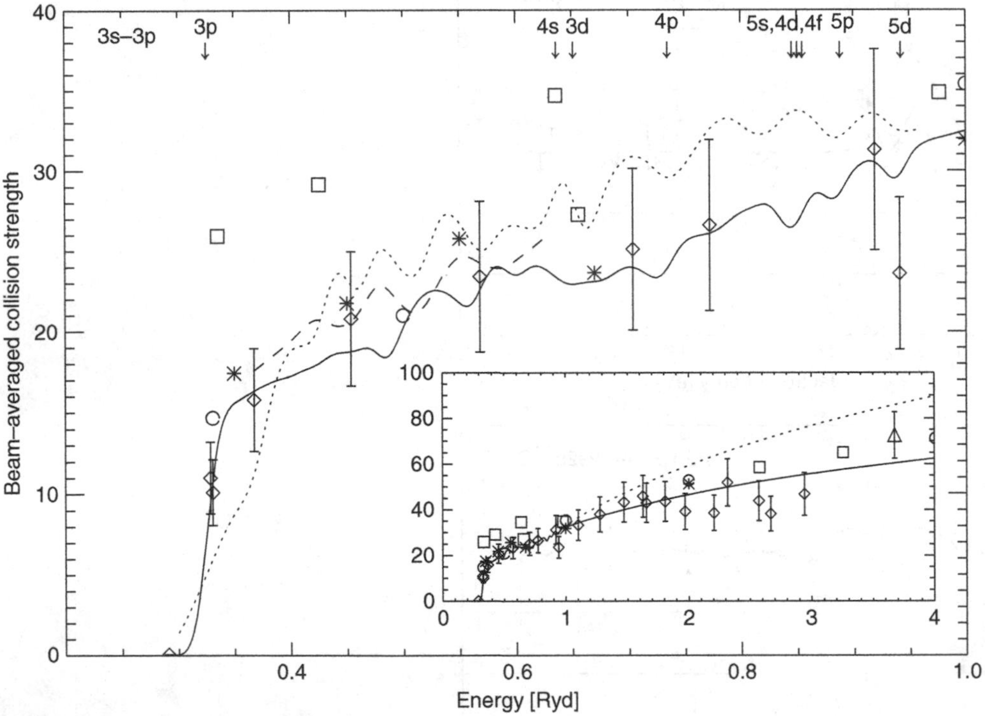

FIGURE 5.8 Comparison of the dipole allowed 3s–3p collision strength for Mg II with merged-beam experiment (diamonds with error bars) [110]. The full solid curve is an R-matrix calculation [109], with collision strengths averaged over a Gaussian electron beam distribution of 0.3 eV FWHM in the merged beam experiement. The dotted curve is an earlier cross-beam experiment [114], and the dashed curve is a previous calculation [115]; both of which are significantly higher. The inset shows a comparison with the Coulomb–Bethe approximation, which overestimates the cross sections.

$n = 4$ and $n = 5$ thresholds out to high energies, as shown in the inset in Fig. 5.8. The R-matrix results fitted to the Bethe form above, and extrapolated to high energies, show a much better agreement with experiment than the Coulomb–Bethe results, which significantly overestimate the cross sections.

Many electron impact excitation cross sections for multicharged ions have been measured using the merged electron-ion beams energy loss (MEIBEL) technique [113]. Measured cross sections include: C^{3+} 2s $^2S - 2p^2P$, O^{5+} $2s^2S - 2p^2P$, Si^{2+} $3s^2$ $^1S - 3s3p^1P$, Si^{2+} $3s^2$ $^1S - 3s3p$ ^3P, Si^{3+} 3s $^2S - 3p$ ^2P, Ar^{6+} $3s^2$ $^1S - 3s3p$ ^1P, Ar^{6+} $3s^2$ $^1S - 3s3p$ ^3P, Ar^{7+}3s ^2S – 3p ^2P, and $Kr^{6+}4s^2$ $^1S - 4s4p$ ^3P. References to the experimental works and comparisons with theory may be obtained from: www.cfadc.phy.ornl.gov/meibel/. There is good agreement, usually within experimental uncertainties, between experimental values and theoretical close coupling calculations using the R-matrix method. It might be noted that the measured transitions are the lowest dipole allowed or intercombination transitions. Some discrepancies between theory and experiment are evident in the figures given, although the overall structures agree.

In summary, the most elaborate theoretical calculations for up to the third row elements generally agree with experiments for the low-lying transitions that have been experimentally studied.

5.7 Electron impact excitation data

Faced with the huge volume of atomic data, calculated with different methods, it is important to analyze, evaluate, compile and disseminate the diverse sets of data in efficient ways. A number of databases and sources are listed in a review of atomic data by A. K. Pradhan and J.-P. Peng [116]. A revised version of their recommended data for a number of ions containing the Maxwellian averaged or effective collision strengths $\Upsilon(T)$, wavelengths and transition probabilities is given in Appendix E.

Two other data sources are also available. First, there are evaluated compilations of theoretical data sources since 1990 by A. K. Pradhan and H. L. Zhang [117], and by A. K. Pradhan and J. W. Gallagher [103] for sources before 1990, and, second, effective collision strengths for all iron ions, for a large number of transitions, taken from the recent Iron Project work or relativistic distorted wave calculations, available from www.astronomy.ohio-state.edu/~pradhan. Other collisional data from data sources, such as the Iron Project database, consisting of high-accuracy R-matrix data, may also be accessed from this website.

5.8 Electron impact ionization

Collisional ionization by electron impact differs fundamentally from collisional excitation. Electron impact ionization (EII) becomes a three-body problem in the final state, as opposed to a two-body (e + ion) problem for excitation. Considering EII of a singly charged ion for simplicity, we have

$$e(E_1) + X^+ \rightarrow e\left(E_1'\right) + e(E_2) + X^{2+}. \tag{5.38}$$

First of all, energy conservation implies

$$E_1 = E_1' + E_2 + E_I, \tag{5.39}$$

where the right-hand side is the sum of the kinetic energies of the two electons, the scattered electron and the ejected electron after EII, and E_I is the ionization energy of the initially bound electron.

The two electrons in the continuum final state interact strongly at low energies, with each other and with the residual ion. These post-collision interactions lead to the behaviour of the cross section at near-threshold energies, which is non-linear and given by the so-called *Wannier form* $\sigma \sim E^a$, where the exponent $a = 1.127$ for EII of neutral atoms and the energy E is the *excess* energy carried away by the two free electrons. The general form of the threshold behaviour of the cross section, with two electrons in the presence of an ion (such as in *double photoionization*), is based on Wigner's theorem and is given by

$$\sigma_0 \propto E^{(2\mu-1)/4}, \tag{5.40}$$

where

$$\mu = \frac{1}{2}\left[\frac{100z - 9}{4z - 1}\right]^{1/2}. \tag{5.41}$$

For $z = 1$, the exponent has the Wannier value 1.127 [118]. However, experimentally, deviations from the Wannier threshold law are observed for many systems and ions.

As in EIE, The direct process of EII in Eq. 5.38 is supplemented by indirect processes involving (e + ion) resonances. We can write both processes as

$$\begin{aligned} e_1 + X^+ &\rightarrow e_1' + e_2 + X^{2+} \\ \rightarrow e_1' + [X^+]^* &\rightarrow e_1' + e_2' + X^{2+}, \end{aligned} \tag{5.42}$$

where $[X^+]^*$ denotes the autoionizing resonance, and e_2 and e_2' distinguish different final electron energies. The indirect steps above are referred to as *excitation–autoionization*, or EA, corresponding to the formation of a resonance in the intermediate step and subsequent breakup in the last step. It may also happen that the final step above may still leave the ion on the right-hand side in another doubly excited resonant state which would again autoionize

$$[X^{2+}]^* \rightarrow e_3 + X^{3+}. \tag{5.43}$$

This double autoionization process is sometimes referred to as *resonant excitation double autoionization* (REDA). Many short-lived intermediate autoionizing states may be involved in the double-ionization REDA process. It is easy to visualize, though difficult to compute or measure, the various pathways that may exist for multiple-ionization via resonant states. These are the Auger processes, which involve many inner-shell transitions leading not only to electron ejections but also to radiative cascades, as discussed at the end of the chapter.

5.8.1 Collisional ionization cross sections

In this section, we describe some theoretical and experimental studies. Let us confine ourselves to single EII leading to two free electrons in the final state. Theoretically, the study of EII is more complicated than EIE, since two-body correlations of the two outgoing electrons must be accounted for in the presence of the residual ion, instead of one continuum electron in the (e + ion) system. That requires, in principle, a double wavefunction expansion for the system

$$\Psi[e_1 + \psi(e_2 + X^{2+})]. \tag{5.44}$$

The $(e_2 + X^{2+})$ term may be regarded as the same as for the EIE problem discussed extensively in earlier chapters. However, now we have the additional continuum electron e_1 interacting with the $(e_2 + X^{2+})$ system. Such a double-continuum wavefunction expansion in the coupled-channel approximation is very involved, and although methods have been developed to implement such expansions, calculations are very difficult. The reason is that (e + ion) channels corresponding to an infinite number of final states are strongly coupled at each incident electron energy. Among the few close coupling approaches, albeit with limited applications, that have been attempted are in [119] and using the *convergent close coupling* method (e.g., [94, 120]). On the other hand, a large number of calculations using the distorted wave approximation without channel coupling, have been made to compute EII cross sections (e.g., [121]).

Fortunately, experimental measurements of total EII cross sections are easier than EIE measurements. This is because excitation experiments are state-selective. For EIE, information is needed on *both* the initial and the final states, requiring the measurement of the electron energy and (often in conicidence), the photon energy from radiative decay of the excited ion. Whereas relatively few measurements have been made for EIE transitions, a considerable body of experimental cross sections for EII have been measured, for nearly all ions of astrophysical importance. Much of the experimental work has also been motivated by applications in fusion plamsa sources [121]. Another factor that makes EII a relatively easier process from a modelling point of view is that only the total collisional ionization cross sections are needed, unlike the EIE cross sections, which are required for all levels of an ion likely to be excited in spectral formation.

Resonant phenomena are fundamentally different, and much more complex, in EII than EIE. Resonances in EIE cross sections occur at definite incident electron energies in the (e + ion) system; the EIE cross section rises rapidly as the resonance energy is approached and then falls equally rapidly once it is crossed. However, in the EII process, resonances are formed with two active(free) electrons and an ion core. One may view the EII resonant process as a function of the incident free electron energy $e_1(E_1)$,[9] corresponding to the energy needed to form a resonance in the (e + ion) system $(e_2(E_2)+X^{2+})$. The key point is that once the incident electron $e(E_1)$ approaches the energy needed to form a resonance at $E_1 = E_{\rm res}$, then that particular resonance can be formed at *all* energies $E_1 > E_{\rm res}$. If the EII cross sections are plotted against the incident energy E_1, then the resonance contribution manifests itself at $E_{\rm res}$, *not* as individual resonance profiles but as *steps* where the ionization cross section increases and remains as a plateau, since that resonance (enhancement) can be affected for all higher energies. In short, resonance enhancements of EII cross sections are generally much larger than in the EIE case. Also, compared with the backgound non-resonant EII cross section, the resonant contributions are often orders of magnitude higher.

As we have mentioned, the DW method does not, in principle account for resonant phenomena inherent in the coupled-channel approximation. However, for EII, the DW method may be extended in a *step-wise* manner to incorporate resonant excitation-autoionization process. Figure 5.9 shows a comparison of the DW results with experiments for electron impact ionization of Fe^{13+} [121]. The excitation–autoionization resonant contributions included in the the total DW cross sections are clearly seen.

5.8.2 Semi-empirical formulae

Historically, there have been a number of analytical formulae to estimate the EII cross sections for near-threshold cross sections (e.g., [24]), and many calculations using

[9] Recall that cross sections are given as functions of the incident particle energy.

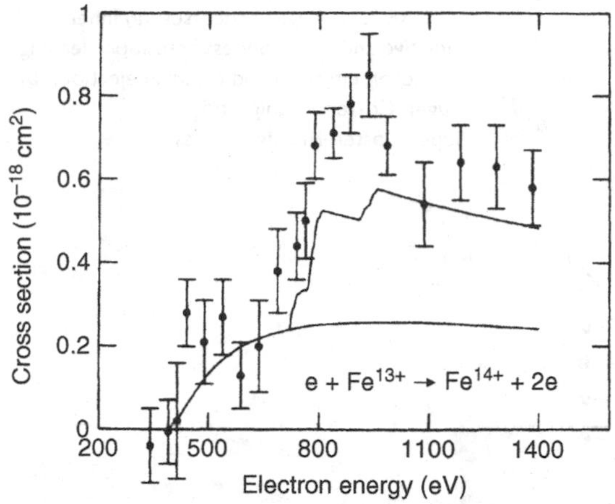

FIGURE 5.9 Electron impact ionization of Fe^{13+} [121]. The experimental data are compared with theoretical DW calculations carried out in a step-wise manner to include direct ionization as well as indirect excitation-autoionization configurations.

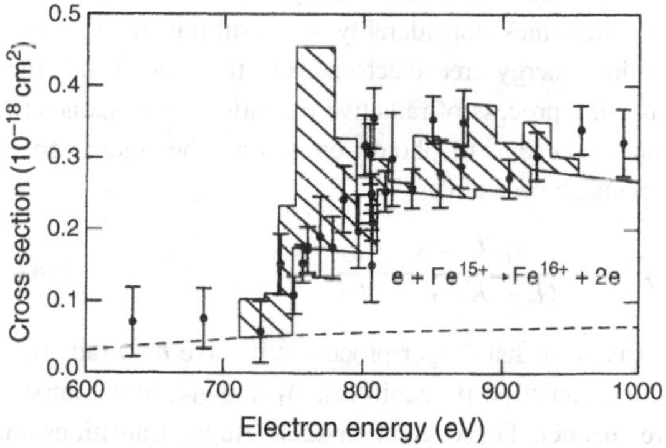

FIGURE 5.10 Electron impact ionization of Fe^{15+} (experimental data from Fig. 5 in [121]). Note the step-wise resonant contributions due to ionization of successive electronic subshells. The difference between the theoretical lower and upper curves (hatched) [125] shows the extent of these REDA contributions. The dashed line is the Lotz fitting formula for direct ionization alone.

relatively simple scaled hydrogenic Coulomb–Born methods and related fitting formulae (e.g. [122]). Owing to the availability of experimental EII cross sections, several other attempts have been made to compile and fit analytical expressions to obtain semi-empirincal formulae. One of the most widely known expression is the *Lotz formula* [123], which includes only direct excitation and does not consider the excitation autoionization or indirect resonant excitations. A. Burgess and M. Chidichimo [124] provided a fitting formula, including fits to resonant excitation–autoionization contributions.

The basic physics of EII is discernible in Fig. 5.10' [121]. The *first* electron impact ionization occurs due to the ionization of the outer valence 3*s* subshell, starting at $E_I = 489$ eV (not shown)

$$e_1 + \mathrm{Fe}^{15+}(1s^22s^22p^63s^1) \rightarrow e_1' + e_2 + \mathrm{Fe}^{14+}(1s^22s^22p^6). \tag{5.45}$$

The EII cross section (experimental points) rises smoothly until the onset of the ionization of the inner 2p subshell at 735 eV, where it rises abruptly, due to the opening of the much larger ionization cross section of the 2p electrons. The step-wise contributions are manifest in the theoretical results. These lie between the lower solid curve in the shaded region in Fig. 5.10, and the upper solid curve, with contributions explicitly added in. The histogram-type contributions are from different subshells that correspond to the ionization of the 2p electrons of sodium-like Fe^{15+}, into not only the ground configuration of neon-like Fe^{16+}, but also the various *excited* levels of Fe^{16+} corresponding to the $2p^5$(3s, 3p, 3d, ...) configurations. In analogy with similar dipole radiative transitions, the 2p–3s and 2p–3d excitations are found to be particularly strong. Taking account of the different angular and spin symmetries associated with these configurations, and fine structure, would obviously give rise to much more structure, but would make relatively little contribution to the total EII cross sections. Finally, note that the Lotz formula, including only direct ionization process, is inaccurate by large factors.

5.8.3 Collisional ionization rate coefficients

It is always desirable to employ accurate theoretical or experimental data. The EII rate coefficients used in astrophysical models are usually obtained from semi-empirical fits that have the advantage of analyticity, if not always of accuracy. More useful and accurate are a number of compilations and fits to experimental (and some theoretical) data. Among the recent ones are those by G. S. Voronov [126], and the extensive database at the National Institute for Fusion Science (https://dbshino.nifs.ac.jp).

5.9 Auger effect

High-energy ionization of inner shells leads to resonant phenomena. We have already seen that in addition to radiative transitions there are non-radiative or radiationless transitions. Electronic transitions involving inner shells are governed by the Auger effect: a vacancy created by inner-shell ionization can be filled via a downward transition from an outer shell, accompanied by the ejection of an electron from the same outer shell, or a higher one.

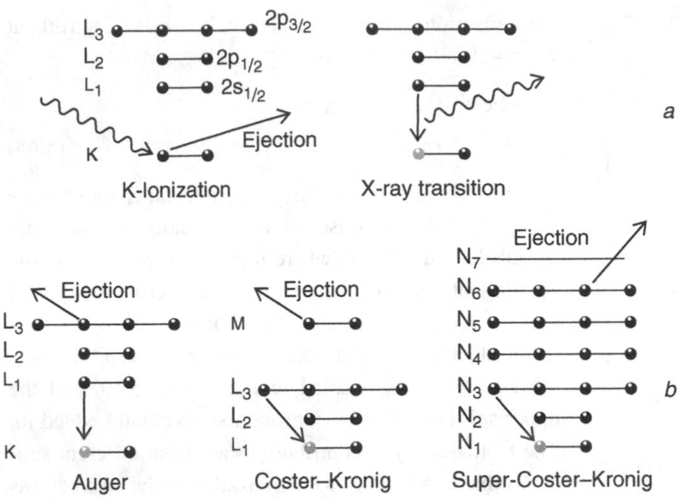

FIGURE 5.11 Auger processes: (a) inner-shell radiative and radiation-less transitions leading to photon emission and electron ejection; (b) Auger, Coster–Kronig; and Super-Coster–Kronig processes.

Such electronic transitions involving closed inner-shells are also referred to as *Auger processes*. Fig. 5.11(a) displays the ionization of an electron from an inner-shell creating a vacancy. The resulting electronic configuration is then in a highly excited atomic state, and undergoes autoionization or radiative decay, or *fluorescence*. The photon emitted by radiative decay may ionize an electron, which then carries away the excess energy as kinetic energy; it is then a radiation-less transition. The Auger processes can take several pathways with which the atom (or ion) relaxes to lower states. A schematic representation of Auger processes is given in Fig. 5.11(b).

We first illustrate in the top panel (Fig. 5.11(a)) the ionization of an electron from the K-shell, creating a vacancy or a hole, which is filled by radiative decay of an L-shell electron, e.g., a 2p → 1s transition. The three types of Auger transition are illustrated in the bottom panel (Fig. 5.11(b)). The first Auger transition corresponds to an *inter-shell* transition, with energy carried away by the an ejected L-electron. The second type of transition, called the *Coster–Kronig* (CK) transition (middle figure in Fig. 5.11(b), corresponds to an *intra-shell* transition, with energy carried away by an electron ejected from a higher (M-)shell. The third type of transition, referred to as the *Super-Coster–Kronig* (SCK) transition, correponds to an intra-shell transition, as well as electron ejection, from the *same* (N-)shell.

In a radiation-less Auger transition $L \rightarrow K$, the energy of the ejected electron

$$\Delta E_e = h\nu(L-K) - E_{\mathrm{IP}}(L), \tag{5.46}$$

where the right-hand side is the difference between the energy of the downward $L \rightarrow K$ transition, minus the energy needed to ionize from the L-shell (ionization potential, IP). This also implies that the kinetic energy of the ejected electrons via an Auger transition is bound to be less than the photon energy in a radiative transition, oftentimes considerably so, resulting in a plasma with low-energy free electrons. On the other hand, the competing process of radiative transitions corresponds to fluorescence due to photon emission. The fluorescence yield may be written as

$$\omega_{LK} = \frac{A_{\mathrm{r}}(L-K)}{A_{\mathrm{r}}(L-K) + A_{\mathrm{a}}(L)}. \tag{5.47}$$

It is clear that Auger processes involve *both* radiative and autoioniztion probabilities, A_{r} and A_{a}, in a competitive manner. For deep inter-shell Auger transitions in high-Z atoms, the radiative probability may exceed the autoionization probability, such as for the aforementioned $L \rightarrow K$ transition. But for the higher shells, M-shell and higher, the autoionization rates are more likely to dominate radiative transmission, i.e., CK and SCK processes become increasingly more important in heavy elements, as one goes to higher Z along the periodic table. These processes can be better understood quantitatively for a complex high-Z atom, for example, gold. The ionization energies E_{I} of the subshells of gold are [127] as given in Table 5.3.

It is clear from the values of E_{I} that a $K\alpha$ photon, with energy due to the transition $\Delta E_{\mathrm{I}}(L_1 - K) = 66\,373$ eV is sufficient to ionize any other subshell (except the K-shell of course), emitting an Auger electron via a radiation-less transition. An example of a CK transition would be the radiative decay, say, $\Delta E_{\mathrm{I}}(L_3 - L_1) = 2433$ eV, which could ionize any of the subshells higher than M_4($E_{\mathrm{I}} = 2295$ eV), as depicted in the middle of Fig. 5.11(b). Now, for a SCK process, both the radiative

TABLE 5.3 Electronic shell ionization energies E_I for gold atoms.

Shell	E (eV)
K	80 729
L_1	14 356
L_2	13 738
L_3	11 923
M_1	3 430
M_2	3 153
M_3	2 748
M_4	2 295
M_5	2 210
N_1	764
N_2	645
N_3	548
N_4	357
N_5	339
N_6	91
N_7	87
O_1	114
O_2	76
O_3	61
O_4	12.5
O_5	11.1
P_1	9.23

decay transition and the ionizing electron, are within the same shell. Thus, an example of a SCK transition would be $\Delta E_I(N_3 - N_1) = 216$ eV, sufficient to ionize subshells N_6 and N_7, with $E_I = 91$ and 87 eV, respectively, as on the right-hand side of Fig. 5.11(b).

Intuitively, one can see that if a deep inner-shell vacancy is created in a heavy atom with several n-shells, then Auger cascades may yield many ejected electrons. Each downward transition from outer- to inner-shells would result in Auger electrons. As seen in Fig. 5.11, the ionization of an inner-shell electron, or the creation of one inner-shell vacancy, may lead to the creation of *two* vacancies in higher or the same electronic shells owing to radiation-less ejection of electrons. Therefore, for each inner-shell vacancy, caused by primary ionization by X-rays or electron impact, several electrons will be emitted at a spectrum of energies. For primary inner-shell ionization energies in the X-ray keV range, most of the Auger electrons will have relatively low energies of the order of ~100 eV or less. Given that autoionization timescales are extremely short, the Auger cascades may take only about $10^{-(14-15)}$ s, or in the femtosecond range for high-Z atoms.

Because the inner-shell transitions at high energies are involved, the Auger processes relate to short-wavelength transitions, often X-ray transitions involving the L-shell and K-shells in low-Z atoms, and the M-and higher shells in high-Z atoms. We have discussed the Auger effect rather heuristcally in terms of overall inter-shell or intra-shell transitions. Of course, a formal computational treatment must take account not only of the electronic configurations, but also of the individual energy levels, partaking in the myriad of transitions that may ensue following inner-shell ionization. Furthermore, the dependence on the atomic number and the ion charge is crucial, since all detailed radiative and autoionization rates depend on both Z and z. Auger fluorescence yields have been computed in a variety of approximations (e.g., [128]).

5.9.1 Z-dependence of X-ray transitions

As mentioned, inner-shell transitions often correspond to X-ray energies. The earliest expression for the wavelength and energies of X-ray transitions due to inner-shell ionization is the semi-empirical *Moseley's law*

$$E(K) = 1.42 \times 10^{-2}(Z-1)^2$$
$$E(L) = 1.494 \times 10^{-3}(Z-1)^2 \qquad (5.48)$$
$$E(M) = 3.446 \times 10^{-4}(Z-1)^2 \quad \text{(keV)}.$$

More precise values than those given by Eq. 5.48 are available from measurements and calculations. Tables of K, L, M,...-shell ionization energies are given in [127], and $K\alpha_1$, $K\alpha_2$ transition energies, for all elements of the periodic table, are available from on-line databases.

6 Photoionization

Most of the observable matter in the Universe is ionized plasma. The two main sources of ionization are collisional ionization due to electron impact as discussed in Chapter 5, and photoionization due to a radiative source. Among the prominent radiation sources we discuss in later chapters are stars and active galactic nuclei. The nature of these sources, and physical conditions in the plasma environments activated by them, vary considerably. The photoionization rate and the degree of ionization achieved depends on (i) the photon distribution of the radiation field and (ii) the cross section as a function of photon energy. In this chapter, we describe the underlying physics of photoionization cross sections, which turns out to be surprisingly full of features revealed through relatively recent experimental and theoretical studies. Theoretically, many of these features arise from channel coupling, which most strongly manifests itself as autoionizing resonances, often not considered in the past in the data used in astronomy. The discussion in this chapter will particularly focus on the nearly ubiquitous presence of resonances in the cross sections, which later would seen to be intimately coupled to (e + ion) recombination (Chapter 7).

The interaction of photons and atoms inducing transitions between bound states has been discussed in Chapter 4. Here we describe the extension to the bound–free transitions. We first revisit a part of the unified picture of atomic processes in Fig. 3.5. When a photon incident on an atom X imparts sufficient energy to an electron for it to be ejected from the atom, leaving it with one additional charge, the bound–free process is called *photoionization*, i.e.,

$$X + h\nu \rightarrow X^+ + e(\epsilon). \tag{6.1}$$

The ejected free electron, often termed as *photoelectron*, is said to be in the continuum with positive energy $\epsilon = 1/2mv^2$, which is equal to the difference between the photon energy $h\nu$ and the ionization potential or the electron binding energy,

$$h\nu - E_{\mathrm{IP}} = \frac{1}{2}mv^2. \tag{6.2}$$

Indirect or resonant photoionization occurs through an autoionizing state as

$$X^{+(n-1)} + h\nu \leftrightarrow (X^{(+n-1)})^{**} \leftrightarrow e + X_c^{+n}. \tag{6.3}$$

The autoionizing resonant state forms when the photon energy is equal to the energy of the compound (e + ion) system given by the Rydberg formula $E_{\mathrm{Res}} = E_{\mathrm{C}} - z^2/\nu^2$, where E_{C} is an excited state of the residual ion X_{C}, also referred to as the 'core' ion, left behind as the doubly excited autoionizing state (indicated by double asterisks) breaks up. With reference to the unified picture in Fig. 3.5, it is clear that detailed balance implies that the photoionization cross section should contain the same resonances as the related or inverse processes of electron impact excitation or (e + ion) recombination.

For a proper treatment of the photoionization process, a number of approaches of increasing sophistication have been adopted over the years. Photoionization cross sections calculated mainly in the hydrogenic or central-field approximations accounted for the background cross section, but not resonances. Many coupled channel calculations included autoionizing resonances in a limited manner. The most thorough and systematic study for photoionization was carried out under the Opacity Project [37], using extensions of the R-matrix codes (Chapter 3). However, relativistic effects and fine structure were not considered in the Opacity Project work. More recent calculations under the Iron Project [38], and other works,[1] employ the Breit–Pauli R-matrix method to compute photoionization cross sections, including relativistic effects with fine structure. Concurrently, sophisticated

[1] The authors' websites refer to many of these up-to-date calculations.

experiments, using accelerator based photon light sources, have enabled high-resolution measurements that reveal resonances in great detail for photoionization of the ground state and low-lying excited levels (e.g., [129, 130, 131, 132]).

Starting with examples of hydrogen and helium, we outline the basic relation between the photoionization cross section and the transition probability. As throughout the text, we shall endeavour to present a unified framework for atomic processes. In that spirit, the emphasis in this chapter is on generalization of the radiative transitions framework developed in Chapter 4, with analogous formulation.

6.1 Hydrogen and helium

We begin with a heuristic depiction of photoionization of hydrogen and helium. The point is to illustrate the differences due to resonances that exist in the photoionization of *all* non-hydrogenic systems. Figure 6.1 compares the photoionization cross sections of H I and He I. Whereas an analytic formula yields the hydrogenic cross section given in Section 6.4.2, the three-body helium problem is not amenable to an exact solution. But it is the physical features that make non-hydrogenic cases far more interesting.

The photoionization cross section for H I, $\sigma_{PI}(\text{HI})$, in Fig. 6.1 is feature-less. Photoionization of the hydrogen atom in the ground state $1s^1\ ^2S$, or in an excited state $n\ell\ ^2L$, leaves a bare ion (proton) with monotonically weakening interaction as a function of the energy of the ejected photoelectron. The threshold value of the ground state $\sigma_0(1s) = 6.4 \times 10^{-18}\ \text{cm}^2$ or 6.4 megabarns (Mb), the usual units for photoionization and recombination cross sections. For excited *ns* levels, $\sigma_0(\text{HI}; ns)$ decreases in value and begins at lower threshold ionization energies with increasing n.

But the photoionization process starts to show rich *resonant* features for systems with more than one electron. This is because autoionization can occur, i.e., the formation of a two-electron doubly excited resonant state via photo-excitation, followed by break-up and ionization. The lower panel of Fig 6.1 shows $\sigma_{PI}(\text{HeI})$, the photoionization cross section of the ground $1s^2(^1S)$ state helium

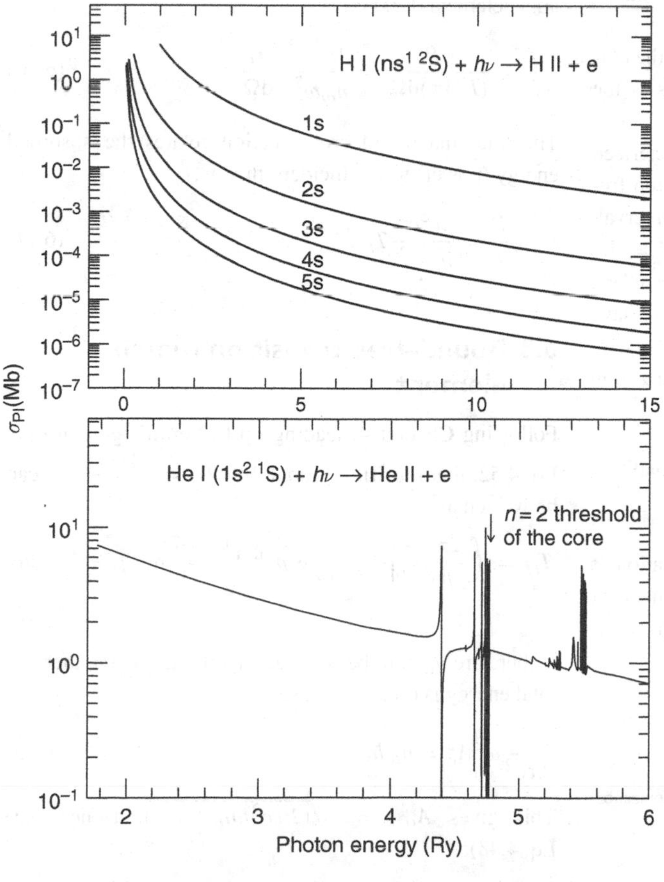

FIGURE 6.1 Photoionization cross section σ_{PI} of the ground state and excited states, $ns^1\,(^2S)$, of H I (top) [133], and the ground state $1s^2(^1S)$ of He I [134] (bottom).

[134]. The two electrons in helium can form a Rydberg series of resonances $E_C \nu l$, where E_C is the energy of the excited core state of the residual H-like ion He II $n\ell(^2L)$. The resonances occur at energies given by the Rydberg formula Eq. 2.66, $E = E_C - 1/\nu_C^2$. The series of autoionizing resonances is $2pn\ell$ shown in Fig. 6.1, corresponding to the excited state $2p$ of the H-like core, *and* the seccond electron also in an excited orbital. To wit: the lowest resonance is $2s2p$, whose eigenvalue is above the ionization energy of He I and which therefore autoionizes. The resonances become narrower and denser, converging on to the excited $n = 2$ core level(s) (marked in Fig. 6.1). The overall cross section follows a monotonic pattern: at the ionization threshold σ_{PI} is high and decays smoothly with higher photon energy. However, the smooth background changes with the appearance of autoionizing resonances that considerably affects the high energy region, particularly below the ionization threshold(s).

6.2 Photoionization cross section

Since the bound–free process involves an initial bound state and a continuum state, the physical quantity that describes photoionization is a cross section σ as a continuous function of incident photon energy $h\nu$, or equivalently, the outgoing photoelectron energy ϵ.[2] Recall that the corresponding quantity for bound–bound transitions is the oscillator strength for discrete bound states.

In nature, a radiation field is generally characterized by a distribution of frequencies. Let a single photon frequency or wave vector be $\mathbf{k}$, and an infinitesimal interval, $\mathbf{k} + \mathbf{dk}$. Consider a cubic enclosure of volume $V = L^3$ with a periodic boundary condition $e^{ik_x(L+x)} = e^{ik_x x}$, indicating that $e^{ik_x L} = e^{ik_y L} = e^{ik_z L} = 1$. Hence, we have $k_x = \frac{2\pi}{L}n_x, k_y = \frac{2\pi}{L}n_y, k_z = \frac{2\pi}{L}n_z$ where n_x, n_y, n_z are integers. The number of modes with a given polarization in the interval $\mathbf{k}$ and $\mathbf{k} + \mathbf{dk}$ is

$$dn = \frac{V}{(2\pi)^3}d^3k = \frac{V}{(2\pi)^3}k^2 dk d\Omega = \frac{V}{(2\pi)^3}\frac{\omega^2}{c^3}d\omega d\Omega, \tag{6.4}$$

where Ω is the solid angle. If the incident radiation is isotropic and naturally polarized, the total number of modes is multiplied by two independent polarizations within the solid angle,

$$dn = \frac{2V}{(2\pi)^3}\frac{\omega^2}{c^3}d\omega d\Omega. \tag{6.5}$$

[2] Having examined (e + ion) scattering in the Chapter 5, photoionization or (e + ion) recombination may also be viewed as a *half-scattering* process [31].

The energy per unit volume for photons in the interval $d\omega$ and $\omega + d\omega$ is,

$$\begin{aligned} u d\omega &= \frac{1}{V}n_\omega \hbar\omega \int dn \\ &= \frac{2n_\omega \hbar\ \omega^3}{(2\pi)^3\ c^3}d\omega \int d\Omega = \frac{n_\omega \hbar\omega^3}{\pi^2 c^3}d\omega. \end{aligned} \tag{6.6}$$

The incident energy crossing unit area per unit time in the frequency interval $d\omega$ and $\omega + d\omega$ is

$$I d\omega = cU d\omega, \text{ i.e., } I = \frac{n_\omega \hbar\omega^3}{\pi^2 c^2}, \tag{6.7}$$

where I is the intensity of radiation,

$$I = \frac{n\hbar\omega^3}{2\pi^2 c^2}. \tag{6.8}$$

If dT_{ij} is the probability per unit time that one quantum of energy in the solid angle element $d\Omega$ is absorbed in transition between states i and j, then the energy (power) absorbed by the atom is

$$dP = \hbar\omega_{ij} dT_{ij}. \tag{6.9}$$

The differential cross section $d\sigma/d\Omega$ is the ratio of energy absorbed dP to the incident flux $(I/4\pi)d\Omega$ in the solid angle element $d\Omega$, i.e.,

$$\frac{d\sigma}{d\Omega} = \frac{dP}{(I/4\pi)d\Omega} = \frac{4\pi^3 c^2}{n_\omega \omega_{ij}^2}\frac{dT_{ij}}{d\Omega}. \tag{6.10}$$

The total integrated cross section relates the absorbed energy (power) to the incident flux as,

$$\sigma = \frac{P}{I} = \frac{\pi^2 c^2}{n_\omega \omega_{ij}^2}T_{ij}. \tag{6.11}$$

6.3 Bound–free transition matrix element

Following Chapter 4, leading up to Fermi's golden rule, Eq. 4.52, the transition probability $T_{ij} = \int \frac{|a_j(t)|^2}{t}dn$ can be written as,

$$T_{ij} = \int \frac{2\pi}{\hbar}A_0^2 \left| <j \left| \frac{e}{mc}\hat{\mathbf{e}}.\mathbf{p}e^{i\mathbf{k}.\mathbf{r}} \right| i > \right|^2 \delta(E_{ij} - \hbar\omega)dn. \tag{6.12}$$

If there are n_ω number of photons of energy $\hbar\omega$, then the total energy is (classically, see Eq. 4.55),

$$\frac{V}{2\pi c^2}\omega^2 A_o^2 = n_\omega \hbar\omega. \tag{6.13}$$

This gives $A_o = \sqrt{(2\pi c^2 \hbar n_\omega/(V\omega)}$. Hence (see Eq. 4.48),

$$T_{ij} = \frac{n_\omega e^2}{2\pi\hbar m^2 c^3}\int |<j|\hat{\mathbf{e}}.\mathbf{p}e^{i\mathbf{k}.\mathbf{r}}|i>|^2\delta(\omega_{ij}-\omega)\omega d\omega d\Omega$$
$$= \frac{n_\omega e^2\omega_{ij}}{2\pi\hbar m^2 c^3}\int |<j|\hat{\mathbf{e}}.\mathbf{p}e^{i\mathbf{k}.\mathbf{r}}|i>|^2 d\Omega. \quad (6.14)$$

The transition probability for absorption of a photon polarized along $\hat{\mathbf{e}}$ in the solid angle $d\Omega$ is then given by

$$dT_{ij} = \frac{n_\omega e^2\omega_{ij}}{2\pi\hbar m^2 c^3}|<j|\hat{\mathbf{e}}.\mathbf{p}e^{i\mathbf{k}.\mathbf{r}}|i>|^2 d\Omega. \quad (6.15)$$

Similarly, it can be shown that for a photon with wave vector $\mathbf{k}$ in the interval between $\mathbf{k}$ and $\mathbf{k}+d\mathbf{k}$, and polarization $\hat{\mathbf{e}}$, energy $\hbar\omega_{ij}$ being emitted in the solid angle $d\Omega$ is

$$dT_{ij} = \frac{(n_\omega+1)e^2\omega_{ij}}{2\pi\hbar m^2 c^3}|<j|\hat{\mathbf{e}}.\mathbf{p}e^{i\mathbf{k}.\mathbf{r}}|i>|^2 d\Omega, \quad (6.16)$$

where n_ω is the average number of photons in $\mathbf{k}$ and $\mathbf{k}+d\mathbf{k}$ and polarization $\hat{\mathbf{e}}$. Equations 6.15 and 6.16 refer to absorption and emission, respectively. Note the factor $(n_\omega + 1)$ in Eq. 6.16. The factor with unity is independent of the intensity of radiation (number of photons n_w) before emission; it gives rise to spontaneous emission (i.e., analogous to the A-coefficient), which is different from zero even if $n_\omega = 0$. However, the term n_ω in the factor gives rise to stimulated or induced emission of radiation (analogous to the B-coefficient).

We employ the dipole approximation as for bound–bound transition, retaining only unity from the expansion of the radiation operator $e^{i\mathbf{k}.\mathbf{r}}$ (Eq. 4.61). For photoionization, it is generally not necessary to consider higher-order non-dipole terms. Using, as in Eq. 4.67, the length and velocity forms of the transition matrix $<j|\frac{e}{mc}|\mathbf{p}|i>$ $= \frac{2i\pi}{c}\nu_{ji} <j|e\mathbf{r}|i>$, the transition probability for absorption is

$$dT_{ij} = \frac{n_\omega e^2\omega_{ij}}{2\pi\hbar m^2 c^3}|<j|\hat{\mathbf{e}}.\mathbf{p}|i>|^2 d\Omega$$
$$= \frac{n_\omega e^2\omega_{ij}^3}{2\pi\hbar c^3}|<j|\hat{\mathbf{e}}.\mathbf{r}|i>|^2 d\Omega, \quad (6.17)$$

where the polarization components are summed over in the integral. The direction of polarization $\hat{\mathbf{e}}$ is perpendicular to $\mathbf{k}$. Let the angle between $\mathbf{k}$ and $\mathbf{r}$ be θ. One component of $\hat{\mathbf{e}}$, say $\hat{\mathbf{e}}_1$, can be in a direction perpendicular to both $\mathbf{k}$ and $\mathbf{r}$. Hence, this component contributes nothing ($\hat{\mathbf{e}}_1.\mathbf{r} = 0$) to the transition matrix. The other component $\hat{\mathbf{e}}_2$ then lies in the plane formed by $\mathbf{k}$ and $\mathbf{r}$, and at an angle $(\pi/2-\theta)$. We then have $|\hat{\mathbf{e}}_2.\mathbf{r}|^2 = |\mathbf{r}|^2\sin^2\theta$, and, therefore for the length form

$$T_{ij} = \frac{n_\omega e^2\omega_{ij}^3}{2\pi\hbar c^3}\int |<j|\mathbf{r}|i>|^2\sin^2\theta d\Omega. \quad (6.18)$$

Using

$$\int d\Omega\sin^2\theta = 2\pi\int_0^\pi \sin^3 d\theta = \frac{8\pi}{3}, \quad (6.19)$$

we get

$$T_{ij} = \frac{4n_\omega e^2\omega_{ij}^3}{3\hbar c^3}|<j|\mathbf{r}|i>|^2 = \frac{4n_\omega\alpha\omega_{ij}^3}{3c^2}|<j|\mathbf{r}|i>|^2, \quad (6.20)$$

where α, as usual, is the fine-structure constant. The photoionization cross section is then

$$\sigma_L = \frac{\pi^2c^2}{n_\omega\omega_{ij}^2}T_{ij} = \frac{4\pi^2\alpha\omega_{ij}}{3}|<j|\mathbf{r}|i>|^2, \quad (6.21)$$

in the length formulation. Similarly, for the velocity form we obtain

$$T_{ij} = \frac{4n_\omega\alpha\omega_{ij}}{3m^2c^2}|<j|\mathbf{p}|i>|^2, \quad (6.22)$$

and the corresponding photoionization cross section,

$$\sigma_V = \frac{\pi^2c^2}{n_\omega\omega_{ij}^2}T_{ij} = \frac{4\pi^2\alpha}{3m^2\omega_{ij}}|<j|\mathbf{p}|i>|^2. \quad (6.23)$$

As shown in Section 4.7 for bound–bound oscillator strengths, both the length and velocity forms should give the same value if the wavefunctions are exact [77], i.e., $\sigma_L = \sigma_V$. However, the wavefunctions for an atomic system beyond hydrogen are all computed approximately with formulations of varying accuracy. Nevertheless, most approximations provide wavefunctions that are accurate in the region beyond the dimensions of the residual core ion. Since the length form of the transition matrix depends on r, it falls off slower than the velocity form (the derivative $\partial/\partial r$ of the wavefunction), giving a larger contribution from the region farther away from the residual ion. It follows that the length form is generally more accurate than the velocity form of the transition matrix for both the bound–bound as well as the bound–free.

In the calculation of the transition probability T_{ij}, and the photoionization cross section σ_{PI}, one significant difference between the bound–bound transition and a bound-free transition is that a continuum wavefunction expansion representing the final free photoelectron needs to be computed in the latter case. Those are, in fact, the continuum wavefunctions we encountered in the (e + ion) collision problem (Chapter 5), and computed by various methods. We first study the relatively simple but illustrative example of the central-field approximation. Although we carry through the formal derivations with both the length and velocity operators, it is generally the length form that is used for photoionization cross sections. Also, higher-order multipole potentials responsible for bound–bound forbidden transitions are not usually important in astrophysical applications of bound–free cross sections, and are not considered.

6.4 Central potential

A suitably constructed radial potential may be employed to obtain the transition probability T_{ij}, Eq. 6.20. The central-field approximation allows us to express an atomic state by independent (i.e., in separable coordinates) radial and angular functions as $R_{nl}(r)Y_{lm}(\hat{r})$. We consider the length form of T_{ij}, and the derivation of the dipole transition matrix in the central-field approxination in Chapter 4. The angular algebra remains the same, given initial bound and final (e + ion) symmetries, and selection rules: $\Delta m = m_j - m_i = 0 \pm 1$, $\Delta l = l_j - l_i = \pm 1$. Also, the radial integrals have a similar form, i.e.,

$$R^{n_j l_j}_{n_i l_i} = \int_0^\infty R_{n_j l_j}(r) R_{n_i l_i}(r) r^3 \mathrm{d}r. \tag{6.24}$$

However, the initial wavefunction is bound but the final wavefunction represents a free electron in the continuum. As before, for a $l_j = l_i + 1$ transition,

$$\sum_{m_j} < \psi_{n_j, l_{i+1}, m_j} |\mathbf{r}| < \psi_{n_i l_i m_i} > |^2 = \frac{l_i + 1}{2l_i + 1} \left| R^{n_j, l_i+1}_{n_i l_i} \right|^2, \tag{6.25}$$

and, for a $l_j = l_i - 1$ transition,

$$\sum_{m_j} < \psi_{n_j, l_{i-1}, m_j} |\mathbf{r}| < \psi_{n_i l_i m_i} > |^2 = \frac{l_i}{2l_i + 1} \left| R^{n_j, l_i-1}_{n_i l_i} \right|^2. \tag{6.26}$$

The total probability for transitions to allowed states with $\Delta l = \pm 1$ and $\Delta m = 0, \pm 1$, without regarding the polarization of radiation field, is

$$\begin{aligned} T_{ij} &= \frac{4 n_\omega \alpha \omega_{ij}^3}{3c^2} | < j|\mathbf{r}|i > |^2 \\ &= \frac{4 n_\omega \alpha \omega_{ij}^3}{3c^2} \frac{1}{2l_i + 1} \left[l_i \left| R^{n_j, l_i+1}_{n_i l_i} \right|^2 \right. \\ &\quad \left. + (l_i + 1) \left| R^{n_j, l_i-1}_{n_i l_i} \right|^2 \right]. \end{aligned} \tag{6.27}$$

This gives the length form of the photoionization cross section,

$$\begin{aligned} \sigma_L &= \frac{4\pi^2 \alpha \omega_{ij}}{3} \frac{1}{2l_i + 1} \left[l_i \left| R^{n_j, l_i+1}_{n_i l_i} \right|^2 \right. \\ &\quad \left. + (l_i + 1) | R^{n_j, l_i-1}_{n_i l_i} | \right]. \end{aligned} \tag{6.28}$$

The velocity form is similar to that for radiative transitions from Chapter 4;

$$\begin{aligned} T_{ij} &= \frac{4 n_\omega \alpha \omega_{ij}}{3m^2 c^2} < j|\mathbf{p}|i > |^2 \\ &= \frac{4 n_\omega \alpha \hbar^2 \omega_{ij}}{3c^2 m^2} \frac{1}{2l_i + 1} \left\{ (l_i + 1) \left| \int_0^\infty P_{n_j l_j} \right. \right. \\ &\quad \times \left. \left[\frac{\mathrm{d} P_{n_j, l_i+1}}{\mathrm{d}r} + (l_i + 1) \frac{P_{n_j, l_i+1}}{r} \right] \mathrm{d}r \right|^2 \\ &\quad + l_i \left. \left| \int_0^\infty P_{n_i l_i} \left[\frac{\mathrm{d} P_{n_j, l_i-1}}{\mathrm{d}r} - l_i \frac{P_{n_j l_i-1}}{r} \right] \mathrm{d}r \right|^2 \right\}, \end{aligned} \tag{6.29}$$

where $P_{nl} = r R_{nl}$. If we define

$$\begin{aligned} (R_+)^{n_j, l_i+1}_{n_i l_i} &= \int_0^\infty P_{n_j l_j} \left[\frac{\mathrm{d} P_{nJ, l_i+1}}{\mathrm{d}r} + (l_i + 1) \frac{P_{n_j, l_i+1}}{r} \right] \mathrm{d}r, \\ (R_-)^{n_j, l_i-1}_{n_i l_i} &= \int_0^\infty P_{n_i l_i} \left[\frac{\mathrm{d} P_{n_j, l_i-1}}{\mathrm{d}r} - l_i \frac{P_{n_j l_i-1}}{r} \right] \mathrm{d}r, \end{aligned} \tag{6.30}$$

then the velocity form of the photoionization cross section is

$$\begin{aligned} \sigma_V &= \frac{4\pi^2 \alpha \hbar^2}{3m^2 \omega_{ij}} \frac{1}{2l_i + 1} \left[(l_i + 1) \left| (R_+)^{n_j, l_i+1}_{n_i l_i} \right|^2 \right. \\ &\quad \left. + l_i \left| (R_-)^{n_j, l_i-1}_{n_i l_i} \right|^2 \right]. \end{aligned} \tag{6.31}$$

An interesting physical feature of photoionization is evident from the bound–free transition matrix element T_{ij}. It involves an exponentially decaying initial bound state wavefunction, and an oscillating continuum wavefunction of the (e + ion) system. That may lead to destructive interference in the integral with initial and final wavefunctions, and thence a *minimum* in the otherwise monotonic photoionization cross section. According to the selection rules, the orbital angular momentum in the dipole matrix element must change by unity between initial and final state, i.e., $\ell \to \ell+1$ or $\ell \to \ell-1$. These minima are prominently seen in photoionization cross section of the ground state of alkali atoms. For example, Na I has a ground configuration $1s^2 2p^6 3s^1$ (to be consistent with the discussion in this section we need not consider the LS term designation of the ground state ^2S). The final (e + ion) system following outer-shell photoionization is Ne-like Na II, and a free electron $e(\epsilon p)$, with a change in the ℓ value of the photoionizing electron from s to p. Owing to a different number of nodes in the radial functions ns and ϵp, integration leads to destructive interference and cancellation in the $3s \to \epsilon p$ matrix element, resulting in a large dip in the cross section referred to as

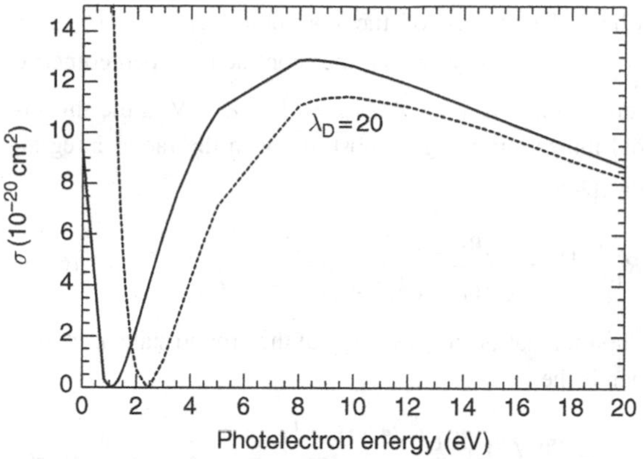

FIGURE 6.2 Alkali photoionization cross section of Na I showing a Cooper minimum (solid line), owing to cancellation between the bound state 3s wavefunction and the ϵp continuum wavefunction. The effect of high plasma density with Debye length $\lambda_D = 20$ is also demonstrated (see Section 9.2.3.5); the 3s orbital becomes more diffuse and the minimum moves outward in energy. (Courtesy, Y. K. Ho, adapted from [136].)

the *Cooper minimum* [135], where it almost approaches zero [136].[3] The solid curve in Fig. 6.2 also shows the effect of Debye screening in a dense plasma [136]. The solid line refers to an isolated sodium atom with Debye length $\lambda_D = \infty$, when the Cooper minimum appears very close to the threshold ionization energy. The dashed line in Fig. 6.2 is with $\lambda_D = 20$ in a heavily screened plasma (i.e., high density), when the 3s orbital becomes quite diffuse, owing to the influence of intervening free electrons within the Debye radius, and the Cooper minimum shifts away from the threshold ([136]; plasma effects and Debye screening are discussed in Chapter 9). More complex examples manifest themselves in the multi-channel treatment of photoioization, as demonstrated later in this chapter.

Historically, several methods have been employed to construct the central potential. They approximate the 'exact' Hartree–Fock non-local electron-electron interaction through a local radial potential $V_{nl}(r)$. Some of these methods are referred to as the *Hartree–Slater* (e.g., [139]), or the *Hartree–Fock–Slater* (e.g., [140]) approximations. The central-field methods have a serious drawback, indicated by the subscript nl in the formulation above. Since only the averaged radial potential is available, they yield the cross section for photoionization of an electronic subshell σ_{nl}, and not for the individual states due to the LS term structure.[4]

6.4.1 Energy dependence

Continuing with the introduction to basic features of photoionization, it is useful to examine the energy dependence. Most of the earlier studies on photoionization of many-electron systems were carried out using simple methods, such as the *quantum defect method*, and, of course, the central-field approximation discussed previously, which yield a relatively featureless background cross section. For many atoms and ions, astrophysical models used the Seaton fitting formula,

$$\sigma_{\rm PI} = \sigma_{\rm T}\left[\beta\left(\frac{\nu}{\nu_{\rm T}}\right)^{-s} + (1-\beta)\left(\frac{\nu}{\nu_{\rm T}}\right)^{-s-1}\right], \quad (\nu > \nu_{\rm T}), \tag{6.32}$$

where σ_T and ν_T are the threshold photoionization cross section and photon frequency, respectively. The fitting parameters β and s are deduced from computed points at several energies. Typically, the cross section first rises in accordance with the first term in Eq. 6.32, and then falls off with a value of $s \sim -3$, which indicates the behaviour of $\sigma_{\rm PI} \sim 1/\nu^3$ at high energies.

This high-energy dependence is given explicitly by the *Kramer's formula*. A Rydberg series of levels, below an ionization threshold where the series converges, becomes denser with increasing n and behaves hydrogenically. The photoionization cross section can then be approximated using hydrogenic wavefunctions that yield the Kramer's formula (e.g., [76]),

$$\sigma_{\rm PI} = \left(\frac{8\pi}{3^{1.5}c}\right)\frac{1}{n^5\omega^3}. \tag{6.33}$$

Equation 6.33 is sometimes used to extrapolate photoionization cross sections in the high-energy region, where other features have diminished contribution with a relatively smooth background. However, it is not accurate. At high photon energies, inner electronic shells and subshells are ionized, and the inner-shell contribution must be included, to obtain total photoionization cross sections. At the ionization energy of the inner (sub)shells there appears a sharp upward jump or edge resulting in sudden enhancement of the photoionization cross section. Figure 6.3

[3] The existence of such a minimum was known in the earliest days of photoionization calculations, viz. D. R. Bates [137] and M. J. Seaton [138].

[4] Although many modelling codes still employ central-field cross sections, they are no longer needed, since more accurate cross sections have been computed under the Opacity Project [37], the Iron Project [38] and improved calculations beyond these two projects (e.g., S. N. Nahar: www.astronomy.ohio-state.edu:~nahar) for nearly all astrophysical atomic species.

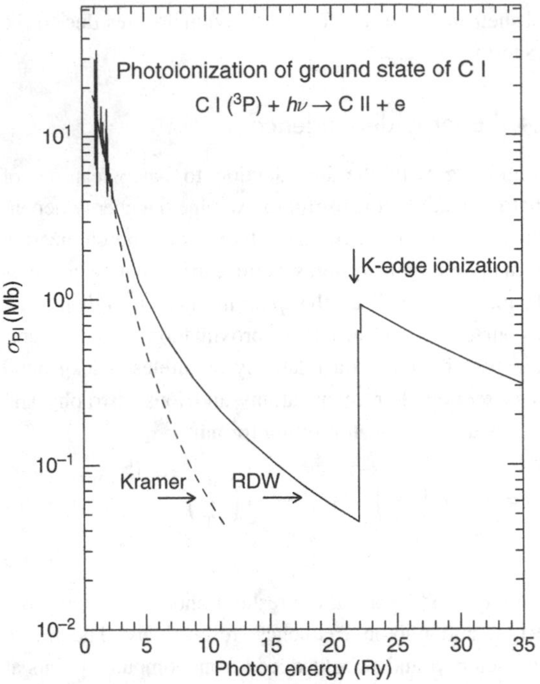

FIGURE 6.3 Photoionization cross section σ_{PI} of the ground state of C I, $1s^22s^22p^2\ ^3P$, computed using the relativistic distorted wave (RDW) approximation [69], compared with the Kramer's hydrogenic formula, Eq. 6.33. The large jump is due to photoionization of the inner 1s-shell or the K-edge. The resonance structures at very low energies are obtained from the coupled channel R-matrix calculations in the Opacity Project. (Courtesy, H. L. Zhang.)

shows results from a relativistic distorted wave (RDW) calculation [69], as compared with the Kramer's formula, Eq. 6.33. The RDW results do not include resonances, and in that sense are akin to the central-field model. Also shown in Fig. 6.3 are the Opacity Project results for resonances in the low-energy near-threshold region $h\nu < 5$ Ry, merged with the RDW results. Resonance phenomena are discussed in detail later.

6.4.2 Hydrogenic ions

The simplest example of photoionization is that of hydrogen or hydrogenic ions. Because of a bound s-electron in the ground state, the selection rule $l_i \rightarrow l_i + 1$ implies that photoionization must be an $s \rightarrow p$ transition. The radial integral involves the bound $1s$ wavefunction and a continuum wavefunction of an electron with $l = 1$ in a Coulomb field $u_l(\nu, r)$ (e.g., [76]),

$$u_1(\epsilon, r) = \frac{2}{3}\left[\frac{\epsilon(1+\epsilon^2)}{1-\exp(-2\pi/\epsilon)}\right]^{1/2} r^2 e^{-i\epsilon r} \times F\left(\frac{i}{\epsilon}+2, 4, 2i\epsilon r\right), \tag{6.34}$$

where the energy of the continuum electron is $\epsilon = \sqrt{\frac{\nu}{\nu_1} - 1}$, $F(\alpha, \gamma, x)$ is the confluent hypergeometric function and $h\nu_1 = Z^2 h\nu_o = 13.6Z^2$ eV is the threshold ionization energy. Substituting in the radial integral, one gets

$$R_{10}^{\epsilon(l=1)} = \frac{8\sqrt{2\pi}}{\sqrt{\epsilon(1+\epsilon^2)^5}} \frac{e^{-(2/\epsilon)\tan^{-1}\epsilon}}{\sqrt{1-e^{-2\pi/\epsilon}}}. \tag{6.35}$$

Photoionization cross section of the ground state of hydrogen is then

$$\sigma_{PI} = \frac{\sigma_0}{Z^2}\left(\frac{\nu_1}{\nu}\right)^4 \frac{e^{4-(4/\epsilon)\tan^{-1}\epsilon}}{1-e^{-2\pi/\epsilon}}, \tag{6.36}$$

where the threshold value is

$$\sigma_0 = \frac{2^8\pi^2\alpha}{3e^4} a_0^2 = 6.4 \times 10^{-18}\text{cm}^2. \tag{6.37}$$

A general expression for the photoionization cross section of level n of a hydrogenic system is [141]

$$\sigma_n = \frac{64\pi\alpha a_0^2}{3\sqrt{3}}\left(\frac{\omega_n}{\omega}\right)^3 \frac{n g(\omega, n, l, Z)}{Z^2}, \tag{6.38}$$

where

$$\omega > \omega_n = \frac{\alpha^2 mc^2}{2\hbar}\frac{Z^2}{n^2}, \tag{6.39}$$

and $g(\omega, n, l, Z)$ is the bound–free Gaunt factor, which is approximately unity at near-threshold energies. A more detailed expression [82] for photoionization of a state nl is

$$\sigma_{nl} = \frac{512\pi^7 m_e^{10}}{3\sqrt{3}ch^6}\frac{g(\omega, n, l, Z)Z^4}{n^5\omega^3}. \tag{6.40}$$

Numerical values of photoionization cross sections for the ground and excited states of hydrogen and hydrogenic ions are available from various sources (e.g., S. N. Nahar [142] and a FORTRAN program by P. J. Storey and D. G. Hummer [143]).

6.5 Generalized bound–free transition probability

To obtain a general form for the cross sections we specify states with their complete quantum designations. Following the formulation in Chapter 4, we define the initial and final states as $l^n(\alpha_1 L_1 S_1)n_i l_i L_i S_i$ and $l^n(\alpha_1 L_1 S_1)n_j l_j L_j S_j$, with degeneracies $g_i = (2S_i + 1)(2L_i + 1)$ and $g_j = (2S_j + 1)(2L_j + 1)$. Following the generalized formulation in Appendix C, the transition probability in the length form is,

$$T_{ij} = \frac{4n_\omega \omega_{ij}^3}{3\hbar c^3} \frac{1}{(2S_i+1)(2L_i+1)} \times \sum_{M_{S_i},M_{S_j}} \sum_{M_{L_i},M_{L_j}} \sum_{l_j} \sum_{L_j,S_j} \times \left| < l^n(\alpha_1 L_1 S_1) n_j l_j L_j S_j M_{L_j} M_{S_j} |\mathbf{D}| \times l^n(\alpha_1 L_1 S_1) n_i l_i L_i S_i M_{L_i} M_{S_i} > \right|^2. \quad (6.41)$$

And in terms of the line strength S_{ij},

$$T_{ij} = \frac{4n_\omega \omega_{ij}^3}{3\hbar c^3} \frac{1}{(2S_i+1)(2L_i+1)} \sum_{l_j} \sum_{L_j,S_j} S_{ij}. \quad (6.42)$$

(It is necessary to guard against the often inevitable confusion between the total spin S_i, which has a single subscript referring to a given atomic level i, and the line strength S_{ij}, which has a double subscript referring to the transition $i \to j$.) For the bound–free photoionization process, or the inverse free-bound (e + ion) recombination process, the line strength may be called the *generalized line strength* with essentially the same form as bound–bound transitions. Since the dipole operator **D** is independent of spin, we have

$$T_{ij} = \frac{4n_\omega \omega_{ij}^3}{3\hbar c^3} \frac{2S_j+1}{(2S_i+1)(2L_i+1)} \sum_{l_j} \sum_{L_j} S_{ij}. \quad (6.43)$$

Since the spin does not change in a dipole transiton, and $S_i = S_j$, the angular algebra remains the same for both the bound–bound and bound–free transitions. So we have the same selection rules for photoionization (Appendix C),

$$\Delta L = L_j - L_i = 0, \pm 1; \; \Delta M = M_{L_j} - M_{L_i} = 0, \pm 1; \; \times \; \Delta l = l_j - l_i = \pm 1. \quad (6.44)$$

With substitution of the line strength, as in the bound–bound case,

$$T_{ij} = \frac{4n_\omega \omega_{ij}^3}{3\hbar c^3} \sum_{l_j = l_i \pm 1} \sum_{L_j} (2L_j+1) \frac{(l_i + l_j + 1)}{2} \times W^2(l_i L_i l_j L_j; L_1 1) | < n_j l_j |D| n_i l_i > |^2, \quad (6.45)$$

where we use $l_{\max} = (l_i + l_j + 1)/2$. The length photoionization cross section is, then,

$$\sigma_L = \frac{\pi^2 c^2}{n_\omega \omega_{ij}^2} T_{ij} = \frac{4\pi^2 \alpha}{3} \omega_{ij} \sum_{l_j = l_i \pm 1} \sum_{L_j} (2L_j+1) \frac{(l_i + l_j + 1)}{2} \times W^2(l_i L_i l_j L_j; L_1 1) | < n_j l_j |r| n_i l_i > |^2. \quad (6.46)$$

With $\alpha = e^2/(\hbar c)$ the cross section is in units of a_0^2 and the constant in the equation is $4\pi^2 a_0^2 \alpha/3 =$ 2.689 megabarns, abbreviated as Mb $= 10^{-18}$ cm^2. The radial integral,

$$< n_j l_j |r| n_i l_i > = \int_0^\infty R_{n_j l_j} r R_{n_i l_i} r^2 \mathrm{d}r, \quad (6.47)$$

involves an exponentially decaying initial bound-state wavefunction and an oscillating continuum free-electron wavefunction as a plane wave. The bound-state wavefunctions are normalized to $|< i|i >|^2 = 1$, and free state wavefunctions are normalized per unit energy in Rybergs as $< E''_j | E'_j > = \delta(E' - E'')$. The velocity form of the cross section σ_V may be obtained in a straightforward manner, noting that it involves the derivative of the final wavefunction, and hence an energy-squared factor in the denominator. But it is worth re-emphasizing that we generally utilize σ_L in practical applications. While the formal equivalence between σ_L and σ_V is a useful check on internal consistency and accuracy of the particular calculation, σ_L is more reliable.

Similar expressions for photoionization cross section can be obtained in intermediate coupling (or jj-coupling) for total angular momentum of atomic states designated as SLJ. The corresponding dipole selection rule is $\Delta J = 0, \pm 1$, with change of parity; other selection rules are as given in Table 4.4, including those for spin-flip intercombination radiative transitions.

6.5.1 R-matrix photoionization calculations

We have formally outlined the theoretical framework in the preceding sections. But practical computations are more involved, owing to the necessity of considering atomic structure precisely, particularly as manifest in channel coupling and resonances. A central potential is inadequate in accounting for these effects, as we illustrate later. The close coupling wavefunction expansion, Eq. 3.42, includes several states of the atomic system. As discussed in previous chapters, the R-matrix method enables channel coupling for both the bound and the continuum (e + ion) states. In analogy with the calculation of bound–bound transition probabilities, the transition matrix for the bound–free process involves an initial bound state wavefunction Ψ_a, which is exponentially decaying, and a continuum free-electron wavefunction $\Psi_b(E)$, which oscillates as a plane wave. Generally in astrophysics we need only consider the dipole approximation since higher-order radiation is insignificant to cause photoionization in astrophysical environments. The transition matrix element with the dipole operator can be reduced to the generalized line strength defined, in either length or velocity form, as

$$S_L = \left| \left\langle \Psi_b \left| \sum_{j=1}^{N+1} r_j \right| \Psi_a \right\rangle \right|^2, \tag{6.48}$$

$$S_V = \omega^{-2} \left| \left\langle \Psi_b \left| \sum_{j=1}^{N+1} \frac{\partial}{\partial r_j} \right| \Psi_a \right\rangle \right|^2, \tag{6.49}$$

where ω is the incident photon energy in Rydberg units, and Ψ_a and Ψ_b are the wavefunctions representing the initial and final states, respectively. The photoionization cross section (σ_{PI}) is proportional to the generalized line strength (S);

$$\sigma_{PI} = \frac{4\pi}{3c}\frac{1}{g_a}\omega S. \tag{6.50}$$

where g_a is the statistical weight factor of the initial state and $\sum_i \mathbf{r_i} \equiv \mathbf{D}$ is the dipole operator.

In the R-matrix method, the reduced dipole matrix is written as the sum of two contributions, as in the case of bound–bound radiative transition (Chapter 4),

$$< a||\mathbf{D}||b >= D^{\mathrm{i}}(a,b) + D^{\mathrm{o}}(a,b), \tag{6.51}$$

where D^{i} represents the contribution from the inner region ($r \leq a$), and D^{o} from the outer region ($r \geq a$). In the outer region, antisymmetrization (i.e., exchange) is neglected. As before, $D = \mathbf{R}+\mathbf{r}$, where $\mathbf{R}$ is the operator for a transition in the target, and $\mathbf{r}$ for a transition by the outer (active) electron. The computation for the bound–free transitions are similar to that of the bound–bound transitions,

$$D^{\mathrm{o}}(a,b) = \sum_{ii'} \alpha_{ii'}(F_{ia}|F_{i'b}) + \beta_{ii'}(F_{ia}|r|F_{i'b}), \tag{6.52}$$

where

$$\alpha_{ii'} =< \Phi||\mathbf{R}||\Phi >, \quad \beta_{ii'} =< \Phi||\mathbf{r}||\Phi >, \tag{6.53}$$

Φ is the target (ion core) wavefunction in the close coupling wavefunction expansion and $F_{ia}, F_{i'b}$ are continuum channel functions, discussed further in Chapter 5 in the context of (e + ion) collisions.

Most of the results presented in this chapter are obtained using the R-matrix method to exemplify a variety of physical effects associated with the photoionization process, and relevant to astrophysical and laboratory plasmas.

6.6 Channel coupling and resonances

Resonance phenomena require a multi-channel description of the atomic system (Chapter 3). In general, it is necessary to include resonances in atomic processes for adequate accuracy. The limitations of the central-field model without consideration of spin and angular momenta (multiplet structure) can be severe, and become apparent when channel coupling, resonances and other electron–electron interactions, are dominant. This is particularly the case for neutral and low ionization stages in the near-threshold energy region, where the ejected electron has low kinetic energy and interacts strongly with the residual ion. Large complexes of resonances appear, due to excitation of core electrons. Coupling between photo-excitation of inner-shell electrons and outer-shell electrons is often strong and gives rise to huge enhancements in the effective cross section.

A striking example of these effects is the photoionizaton of neutral iron, shown in Fig. 6.4. A comparison is made between the coupled channel cross sections computed using the R-matrix method (solid line), and two sets of central-field cross sections (dotted line and squares). It is instructive to study it in some detail. The ground state and electronic configuration of Fe I is $1s^2 2s^2 2p^6 3s^2 3p^6 3d^6 4s^2$ (^{5}D) – note the open 3d inner subshell, in addition to the closed 1s,2s,2p,3s,3p subshells. The 4s subshell ionizes first. The central-field cross section $\sigma(4s)$, predictably, drops rapidly from its value at the ionization threshold (dashed line). However, at the photon energy of ~1.05 Rydbergs, the 3d subshell also ionizes, and the total cross section $\sigma(4s) + \sigma(3d)$ exhibits a sharp jump, due to photoionization of the 3d, whose cross section is larger than that of the 4s.

However, taking explicit account of the energy level (LS term) structure, and channel coupling, changes the picture dramatically. The R-matrix calculations include a large number of LS states coupled together, but in the discussion below we omit the designation of LS terms and refer only to configurations (Fig. 12.5 shows a partial Grotrian diagram of Fe II energy-level structure). The coupled channel cross sections show quite a different behaviour. In addition to resonances, the R-matrix results show no significant decrease in the cross section througout the energy range under consideration, lower or higher than the 3d threshold(s). We consider the total (e + ion) system, photoionization of the ground state of Fe I $3d^6 4s^2(^5D)$, and the residual ion Fe II in several excited configurations, i.e.,

$$h\nu + \text{Fe I}(3d^6 4s^2\ ^5D) \rightarrow \text{Fe II}(3d^6 4s^1, 3d^7, 3d^6 \times (4s, 4p, 4d), 3d^5 4s^2, 3d^5 4s4p) + e. \tag{6.54}$$

The first three configurations of Fe II on the right correspond, in a straightforward way, to the photoionization of

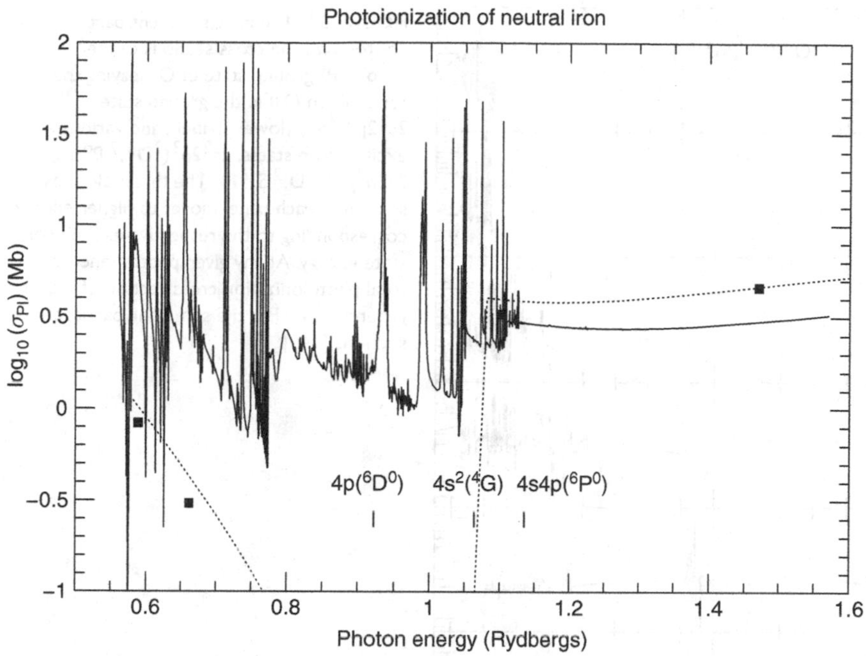

FIGURE 6.4 Photoionization cross sections of Fe I. The coupled channel cross sections with resonances (solid curve) [98, 144], in contrast to those in the central-field approximation without inter-channel coupling (dashed line and solid squares) [140, 139], are up to 1000 times higher (note the log-scale for σ).

the valence subshell 4s, whereas the last two imply photoionization of the 3d. But this straightforward division is complicated by resonances that are superimposed throughout, *below* the energies of the excited 3d configurations $3d^54s^2$, $3d^54s^14p^1$. The difference between the R-matrix and the central-field values below and at the 3d threshold energies is particularly remarkable. Whereas σ (Fe I, ^{5}D) in the central-field model, shows a sudden jump or edge, the R-matrix cross section is slowly varying and interspersed with resonances. Consequently, the total Fe I photoabsorption cross section remains relatively constant, once all coupling amongst levels of the residual ion Fe II are accounted for.

Thus channel coupling includes physical processes not inherent in the central-field model, or that would not be considered in a single-channel treatment. The total photoabsorption *below* the 3d ionization threshold(s) is effectively coupled to the excited 3d configurations. The magnitude of the discrepancy can be very large; the R-matrix cross sections are up to a factor of 1000 higher than the central-field results in the low-energy region, although the two converge reasonably at higher energies. Later, we show that the effect of channel coupling and resonances is found to continue up to high energies, not yet generally considered in astrophysical models.

6.6.1 Partial cross sections

Whereas we omitted the energy level structure in the discussion above for convenience, it is, of course, considered in actual calculations. After photoionization, the residual or core ion may be left not only in the ground state but also in an excited state. As the photon energy approaches an excited core state or threshold, the atomic system can be ionized, leaving the residual ion in that excited state as well as any other lower states. The cross section for leaving the residual ion in a particular state is called the *partial cross section*. The partial cross sections at any given energy must add up to the total cross section. These partial cross sections are useful in astrophysical models, since they may affect the level populations via photoionization.

Figure 6.5 shows an example of partial cross sections for photoionization of O I ground state $1s^22s^22p^4$ (^{3}P) into the ground and excited states of the core ion O II, e.g.,

$$h\nu + (2p^4\,{}^3P) \rightarrow e + \mathrm{O\,II}[(2s^22p^3\,{}^4S^o,\ {}^2D^o,\ {}^2P^o),\\ + (2s^12p^4\,{}^4P,\ {}^2D,\ {}^2S,\ {}^2P). \qquad (6.55)$$

Partial photoionization cross sections of O I, shown in Fig. 6.5, contain resonances that can be seen to be converging onto excited states of the residual ion O II. Note

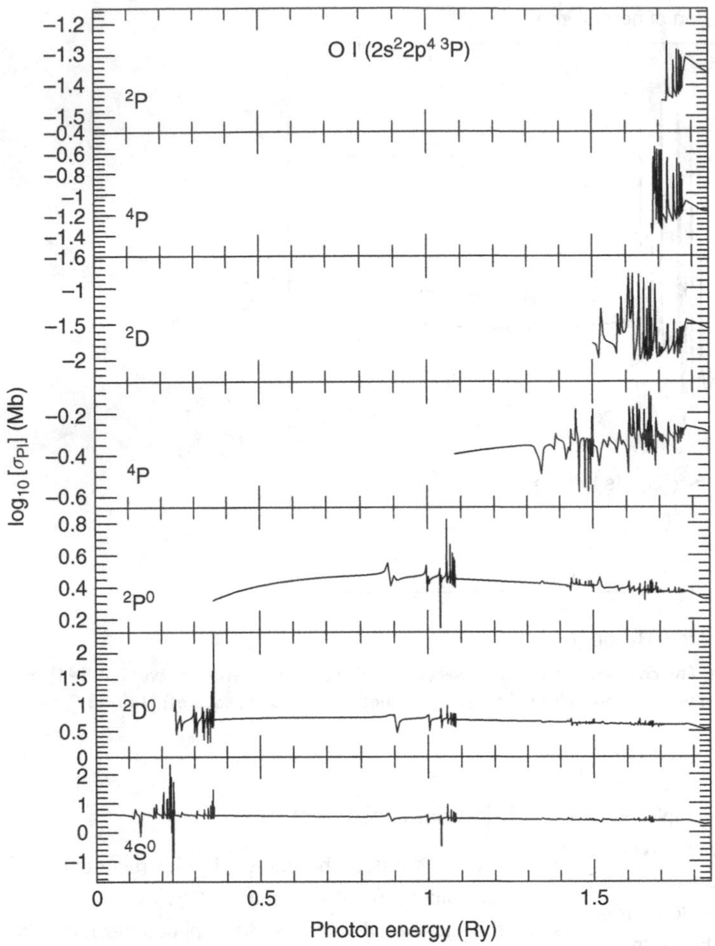

FIGURE 6.5 The panels present partial photoionization cross sections of the $2s^2p^4(^3P)$ ground state of O I leaving the residual ion O II in the ground state $2s^22p^3(^4S^o)$ (lowest panel), and various excited core states, $2s^22p^3(^2D^o, ^2P^o)$, $2s2p^4(^4P, ^2D, ^2S, ^2P)$. The threshold cross section of each panel moves to higher energy, corresponding to the respective excited core state energy. At any given photon energy, the total photoionization cross section of the O I ground state ^3P is the sum of all partial cross sections [145].

that although each panel corresponds to only one ionized state, resonances belonging to other states can still form. Partial photoionization cross sections have another valuable application. Total cross sections are necessary to determine ionization balance, or distribution among ions of an element (Chapter 12), photoionized by a radiation source. But partial cross sections are related, via detailed balance, to (e + ion) recombination cross section *from* a particular state of the residual core ion, as discussed in the next chapter (Chapter 7).

6.6.2 Types of resonance

To study myriad resonant features in photoionization, we need to examine the process in more detail. It is also useful to identify distinct types of resonances and how they manifest themselves. We begin by recapitulating some essential features of the theoretical framework. Resonances are due to coupling of channels, which include core excitations, or the residual ion in excited states. A coupled channel calculation begins with a wavefunction expansion to represent the *residual core* ion in a number of states. The close coupling methodology entails consideration of the bound and continuum states of the (electron + core ion) system, with as many states of the core ion as is computationally feasible. A resonance is formed as a short-lived doubly excited autoionizing state, where (i) the core ion is an excited state, (ii) a loosely bound electron is 'attached' in another excited level and (iii) the total energy of the (e + ion) system is above the ionization energy into the ground state of the residual ion. The symmetry of the resonance is determined by the final continuum state of the (e + ion) system. An initial bound state of a given parity can only be photoionized into an (e + ion) continuum of *opposite* parity.

The first large-scale and systematic study of photoionization, employing the close coupling R-matrix method, was carried out under the Opacity Project [37] for nearly all astrophysically abundant atoms and ions from hydrogen to iron. The Iron Project [38] and subsequent works (e.g. [146]) extended the treatment to include fine structure. The different groups of resonances that have been found to occur are as follows.

6.6.2.1 *Rydberg series of resonances*

Resonances typically occur in a Rydberg series, and become narrower and denser as they approach a series limit of an excited core state E_c. The energy of the resonance is then given by

$$E_{res} = E_c - z^2/\nu_c^2 = E_c - \frac{z^2}{(n-\mu_l)^2}, \tag{6.56}$$

where $\mu_\ell = n - \nu_l$ is the quantum defect with a given ℓ, which is almost constant for all n along the series. Consider the photoionization of the odd-parity ground state $2s^22p(^2P^o)$ of singly ionized carbon C II, with energy levels as in Fig. 6.6.

The Rydberg nature of the resonances is evident from the characteristic pattern illustrated in Fig. 6.7, which shows resonances in photoionization of the C II ground state $2s^22p(^2P^o)$. Two sets of results are presented, in *LS* coupling (top panel), and with fine structure (lower panel).

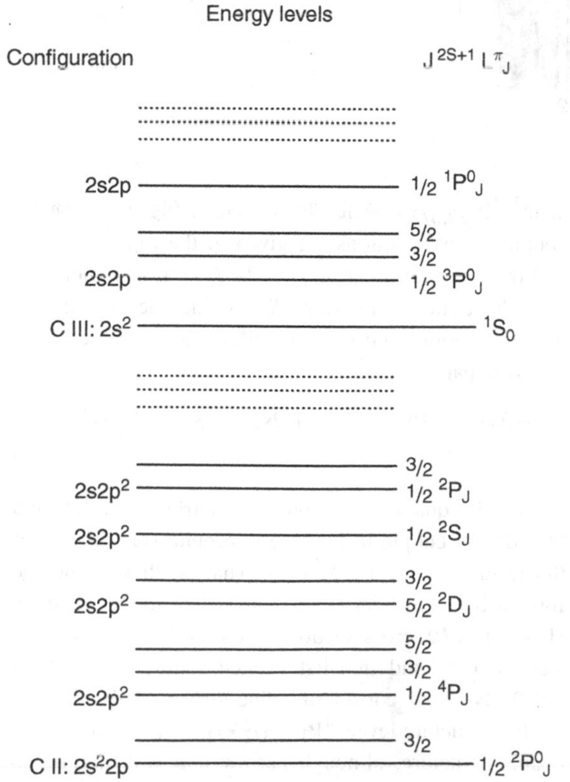

FIGURE 6.6 Low-lying fine structure energy levels $^{2S+1}L_J$ of C II and C III.

Now the ground state of C II has two fine structure components; $^2P^o_{3/2}$ and $^2P^o_{1/2}$. Hence, while there is a single curve in *LS* coupling, there are two corresponding curves with fine structure, essentially a doubling of resonances.

Both panels of Fig. 6.7 show the pattern due to Rydberg series, denoted R1, R2 and R3 in the lower panel. The two series R2 and R3 are common to both panels, and belong to autoionizing states of the type $2s^22p(^3P^o)np(^2D)$ and $2s^22p(^3P^o)np(^2S)$, respectively. The parent core state is $2s^22p(^3P^o)$, and the first resonances corresponds to $n = 4$. A Rydberg series can be identified by the quantum defect $\mu_l = n - \nu$, obtained from the effective quantum number. The resonant energy E_{res} gives $\nu = z(E - E_c)^{-1/2}$, and the first n is typically approximated by the integral value of n that gives a positive μ. For C II the $\mu_l \approx 0.6$ for an s-electron, ≈ 0.4 for a p-electron, ≈ 0.1 for a d-electron and ≈ 0.0 for $\ell > 2$. Resonances of similar shape yield approximately the same quantum defect, while ν_l increases by unity for each successive resonance. Closely spaced core levels may give rise to overlapping Rydberg series of resonances, and may be difficult to identify in practice.

The lower panel of Fig. 6.7 shows one prominent but narrow resonance series R1, corresponding to $2s^22p(^3P^o)np(^2P)$ states, which are missing in the upper panel. This series is not allowed in *LS* coupling, but is allowed to form with fine structure. As mentioned above, in *LS* coupling the even parity autoionizing states $2s^22p(^3P^o)np(^2P)$ cannot decay into the odd parity $[2s^2(^1S) + \epsilon p]$ continuum with the same symmetry as the initial C II ground state $^2P^o$ (Fig. 6.7). However, all three Rydberg series of resonances, $2s^22p(^3P^o)np(^2S, ^2P, ^2D)$, converging on to $2s^22p(^3P^o)$, are allowed to decay to the core ground state if relativistic mixing and fine structure coupling are considered. Through fine structure, the $J = 1/2$ level of ^2P mixes with $^2S_{1/2}$, and the $J = 3/2$ level mixes with $^2D_{3/2}$, which decays via autoionization.

6.6.2.2 *Resonances below threshold in LS coupling*

An important effect of inter-channel coupling is that photoionization may occur through autoionizing channels at energies *below* the threshold for ionization of the optical (active) electron. This has the interesting consequence that there is negligible background, i.e., direct photoionization, which then proceeds indirectly via autoionizing levels only. Figure 6.8 shows the photoionization of the $3s3p^3(^3S^o)$ excited state of S III (or S^{2+}): (a) the total cross sections with positions of various core states of the residual ion S IV are marked by arrows and (b)

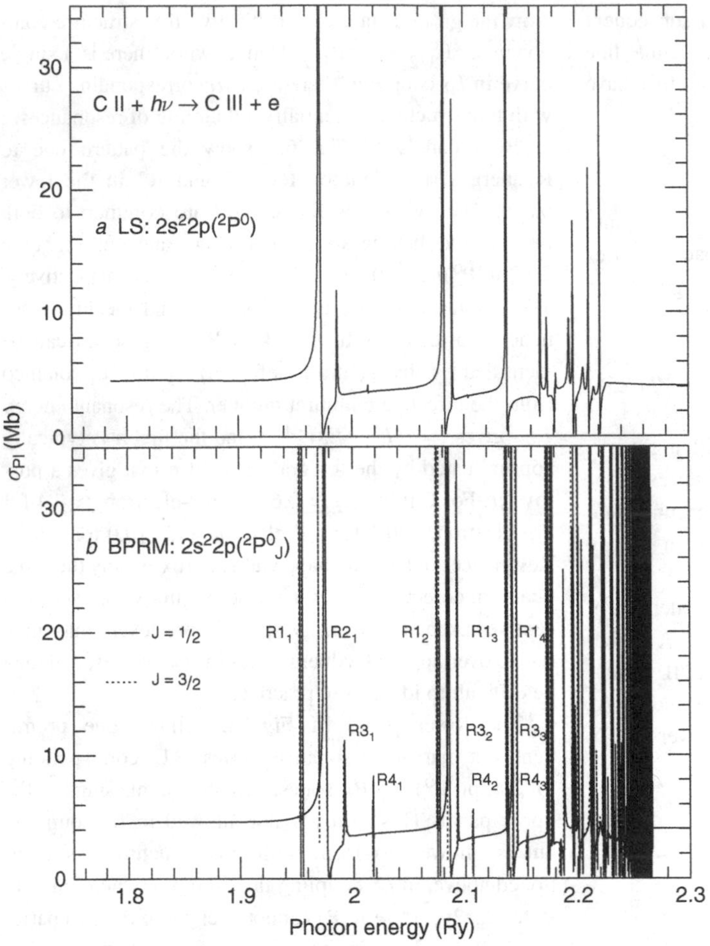

FIGURE 6.7 Photoionization cross section of C II, in (a) *LS* coupling and (b) *SLJ* intermediate coupling. The resonance series are denoted by Rn_i, where n is a series number and i is the resonance number of the series [147].

an expanded version of the region between the ionization threshold $3s^2 3p(^2P^o)$, and the first excited threshold $3s3p^2(^4P)$ of S IV. Since ionization by the 3p optical electron corresponds to leaving the residual ion in the excited ^4P state, without inter-channel coupling the cross section is identical to zero below this threshold. On the other hand, a coupled channel (R-matrix) calculation finds this region to be filled with high-peak and large autoionizing resonances, with practically no background cross section. The resonances of the ^4Pnp series are identified in Fig. 6.8b.

6.6.2.3 *Resonances below threshold due to fine structure*

As shown in Fig. 6.7, fine structure recoupling can introduce narrow resonances below the ionization threshold that are not allowed in pure *LS* coupling. Let us again examine C II, with energy level diagram as in Fig. 6.6, and the photoionization of the excited metastable $2s2p^2(^4P_{1/2,3/2,5/2})$ levels, shown in Fig. 6.9. The *LS* coupling cross sections are given in the topmost panel, and the three fine structure levels $J = 1/2$, 3/2 and 5/2 in the lower three panels. In *LS* coupling, the excited state photoionization process must follow the dipole selection rule with parity change,

$$h\nu + 2s2p^2\,(^4P) \rightarrow e(d,s) + 2s2p^3(^4S^o, ^4P^o, ^4D^o). \tag{6.57}$$

But the quartet (e + ion) symmetries $^4S^o$, $^4P^o$ and $^4D^o$ do not couple to the singlet ground state $2s^2\ ^1S$ of the residual ion C III $2s^2(^1S)$. That is, they cannot be formed by adding an s- or a d-electron to the ^1S state. Hence, the *LS* cross section is zero between the ground state of C III and the first excited state $2s^2 2p(^3P^o)$ at 1.872 Ry. Now considering fine structure, the σ_{PI} for the fine structure levels $^4P_{1/2,3/2,5/2}$ show extensive resonance structure, although narrow and with almost no background. These resonances are formed by relativistic mixing of levels that allow coupling among channels.

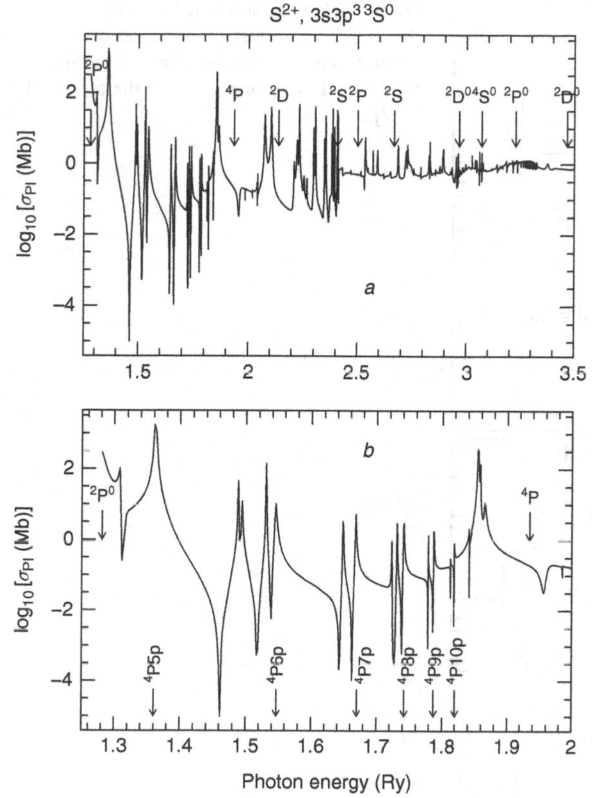

FIGURE 6.8 Photoionization of an excited $3s3p^3(^3S^o)$ state of S III [148]. The top panel (a) shows the total cross sections for leaving the residual ion S IV in various excited states, marked by arrows. The lower panel (b) shows the expanded region between the first ionization threshold, the ground state of S IV $3s^23p\ ^2P^o$, and the excited state ^4P; in this region photoionization proceeds not via direct photoionization, which is disallowed by symmetry considerations, but through autoionizing channels. The cross section therefore consists of resonances with hardly any background. Autoionizing resonances of the ^4Pnp series are demarcated.

For example, fine structure recoupling allows $J\pi$=$(1/2)^o$ and $(3/2)^o$ autoionizing levels of $2s2p(^3P^o)nd(^4D^o)$ to decay into the $2s^2(^1S)\epsilon p(^2P^o{}_{1/2,3/2})$ continua. The ionization thresholds for the $2s2p^2\ (^4P_J)$ levels of C II lie at ~ 1.4 Ry (pointed by the arrow in Fig. 6.9); the ionization energy of the ground state $2s^22p\ (^2P^o)$ state is $\sim$0.48 Ry *below* the ^{4}P.

The dipole selection rule $l_j = l_i \pm 1$ allows the outer 2p electron to be photo-excited only into an even-parity (e + ion) continuum ϵs or ϵd. Hence, the photoionized (e + ion) continua could be: even-parity $2s^2\ [^1S]\ \epsilon s(^2S)$ and $2s^2\ [^1S]\ \epsilon d(^2D)$ (the parent ion *LS* term is often designated within square brackets). However, no even-parity ^{2}P (e + ion) continuum can form in *LS* coupling, since that would require an ϵp electron and would result in an odd-parity continuum when coupled with the $2s^2\ [^1S]$ core state, which is disallowed.

In summary: photoionization alters significantly with the inclusion of fine structure due to relativistic mixing. The selection rule $\Delta J = 0, \pm 1$ not only allows the same *LS* terms, but also associated fine structure levels not possible in *LS* coupling. For photoionization of the C II ground state discussed here, the $^2P^o$ continuum may indeed form through coupling of fine-structure channels. This state then autoionizes, that is, goes through radiationless ionization, during which the total (e + ion) symmetry remains unchanged.

6.6.2.4 *Highly-excited core (HEC) resonances*

Most existing works (e.g., the Opacity Project [37]), include resonances due to relatively low-lying core excitations, often only the ground configuration n-complex of the residual ion. For example, C-like ions have ground configuration $1s^22s^22p^2$, which gives rise to the *LS* term structure shown in Fig. 6.10 (it is the same as the O I ground configuration, $1s^22s^22p^4$, discussed before). Therefore, photoionization of C-like ions may be considered with a similar wavefunction expansion, including the states

$$h\nu + 2s^22p^2(^3P) \rightarrow e + [2s^22p(^2P^o) + 2s2p^2 \times (^4P, ^2D, ^2S, ^2P) + 2p^3(^4S^o, ^2D^o, ^2P^o)]. \quad (6.58)$$

On the right the expansion is restricted to only the $n=2$ ground complex of the residual C II-like ion. The Opacity Project R-matrix calculations mainly used such ground complex expansions.

It might appear that consideration of high-excited core (HEC) resonances, due to core excitations of multiple

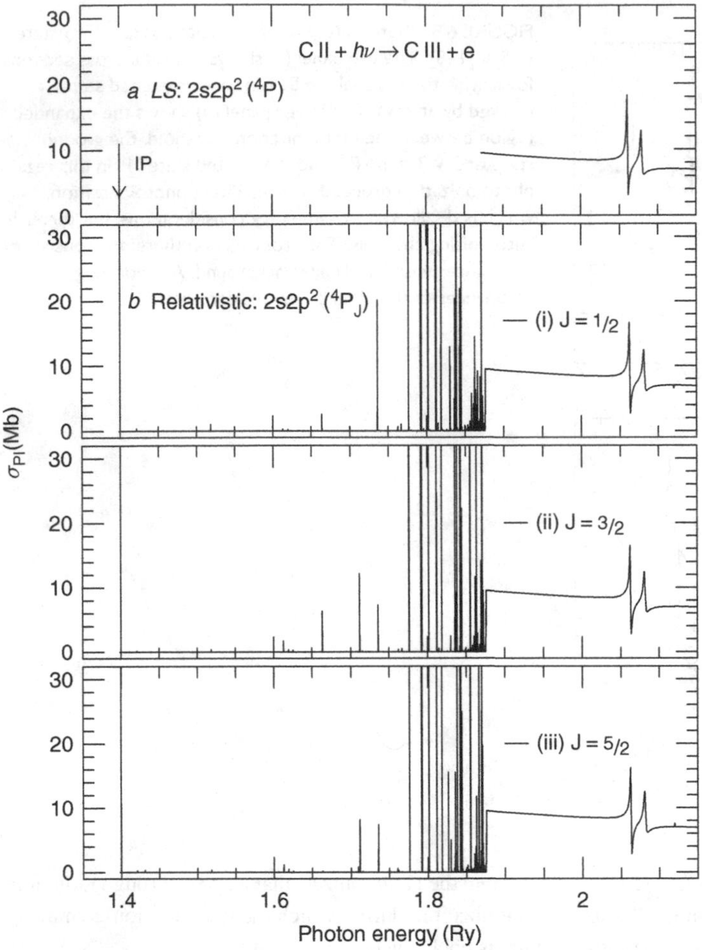

FIGURE 6.9 Photoionization cross section of the $2s2p^2(^4P_J)$ fine-structure levels of C II, illustrating below threshold resonances not allowed in *LS* coupling shown in the top panel [147].

n-complexes, may be ignored since the series of resonances belonging to those might be weak. However, recent studies since the Opacity Project work (see [146]) and references therein), have found that these resonances may be stronger and more prominent than those belonging to the low-lying excited thresholds, especially in highly charged ions. The effect of HEC resonances may also be very important, since highly charged ions exist in high temperature astrophysical plasmas, where excited n-complexes are largely ignored in astrophysical models.

An example highlighting HEC resonances is the photoionization of highly charged C-like iron Fe XXI, shown in Fig. 6.11. The coupled channel R-matrix eigenfunction expansion for the core ion, B-like Fe XXII, includes 29 LS terms (or 29 CC) up to the $n = 3$ complex of configurations: $2s^22p$, $2s2p^2$, $2p^3$, 2s2p3s, 2s2p3p, 2s2p3d, $2s^23s$, $2s^23p$, $2s^23d$. Including only the $n = 2$ complex of configurations gives an eight-term (8CC) expansion, as shown in the energy level diagram in Fig. 6.10. These calculations are very extensive; including photoionization cross sections of 835 LS bound states of Fe XXI [146]. It is particularly noteworthy that the same resonance complexes appear quite differently in photoionization of states with different angular and spin symmetries $SL\pi$. Figure 6.11 shows that resonances belonging to core excitations of the $n = 3$ levels are much stronger than those of $n = 2$ levels.

The HEC resonances particularly affect high-temperature recombination rates, discussed in Chapter 7. While in general the HEC resonances are due to multi-channel couplings to excited-n complexes, there is a particular type of core excitation, which gives rise to huge resonances, described next.

6.6.2.5 *Photo-excitation of core (PEC) resonances*

The most prominent resonance features in photoionization occur via strong dipole allowed photo-excitations for the core ion levels. We refer to these as PEC or Seaton

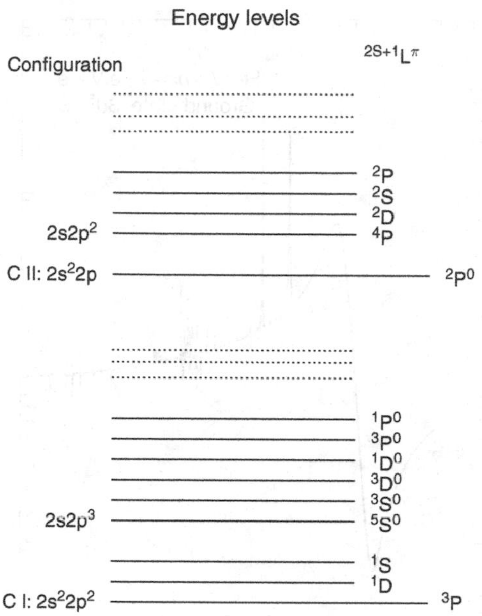

FIGURE 6.10 Low-lying *LS* terms of C I and C II, and respective C-like and B-like ions (see Fig. 6.11).

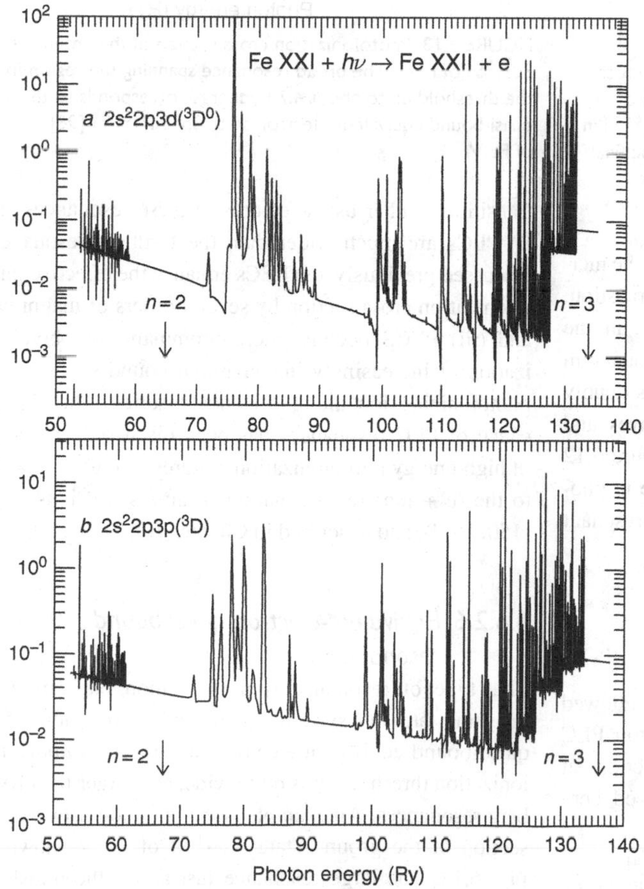

FIGURE 6.11 Photoionization cross sections for two excited states of C-like Fe XXI, with highly excited core (HEC) resonances belonging to the higher *n*=3 core excitations that dominate over those due to the low-lying *n*=2 excitations [146].

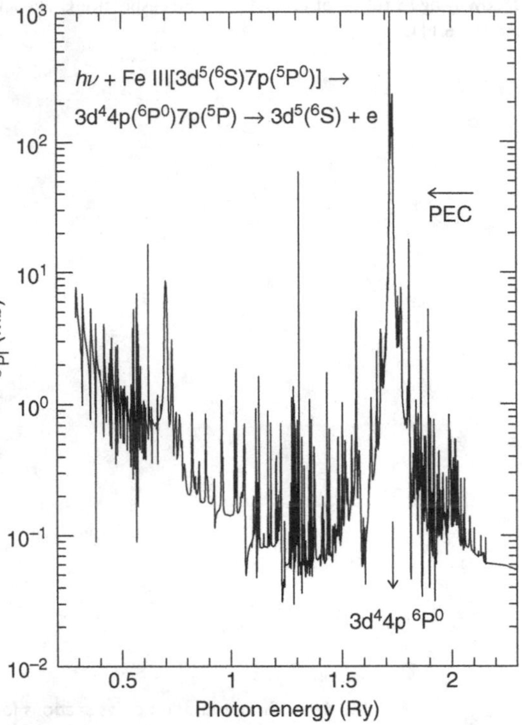

FIGURE 6.12 Seaton resonance due to photo-excitation-of-core (PEC) in the photoionization cross section of the excited state $3d^5[^6S]7p(^5P^o)$ of Fe III. The resonance is due to core excitation $3d^5[^6S] \rightarrow 3d^44p[^6P^o]$ in the residual core ion Fe IV. It is seen centred around the final state energy of the $^6P^o$ state at 1.73 Ry (marked by the arrow) [133].

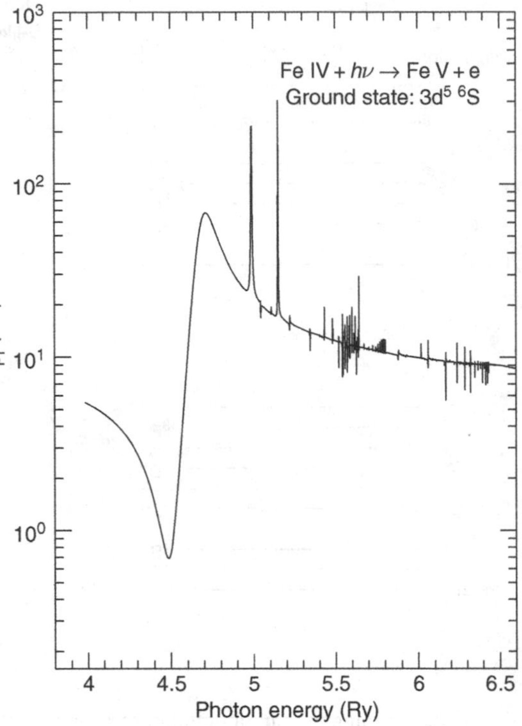

FIGURE 6.13 Photoionization cross section of the ground state $3d^5(^6S)$ of Fe IV. The broad resonance spanning the region from the threshold up to about 4.7 Rydbergs corresponds to the quasi-bound equivalent-electron state $3p^53d^6(^6P^o)$ [98] of Fe IV.

resonances [149].[5] A PEC resonance forms when the incident photon excites the core via a dipole allowed transtion. But the outer electron(s) remains a 'spectator' in the photo-excitation process, i.e., unchanged in its quantum numbers, corresponding to an excited level. This doubly excited state autoionizes, as the outer electron leaves, and the core ion drops back to the ground state. Figure 6.12 shows an example of the Seaton PEC resonance in photoionization of Fe III in an excited state. The resonant excitation process is:

$$h\nu + \text{Fe III } (3d^5[^6S]\,7p[^5P^o] \rightarrow e + \text{Fe IV}(3d^44p[^6P^o]7p), \quad (6.59)$$

corresponding to the strong $[^6S] \rightarrow [^6P^o]$ dipole allowed transition in the S IV ion core. Evidently, Seaton PEC resonances are seen in photoionization cross sections at photon energies corresponding to dipole allowed core transitions. Other usual features of these resonances are: (i) PECs are much wider than the Rydberg resonances discussed previously, (ii) PECs enhance the effective photoionization cross section by several orders of magnitude and (iii) PECS become more prominant for photoionization of increasingly higher initial bound states of the photoionizing ion along a Rydberg series. The significance of PEC resonances has been discussed in studies of high-energy photoionization features, and their relation to the (e + ion) recombination of atoms and ions (e.g., [150, 151]), and described in Chapter 7.

6.6.2.6 *Equivalent-electron quasi-bound resonances*

This type of resonance is not common but can form in some atomic systems with photo-excitation of a quasi-bound equivalent-electron state that lies *above* the ionization threshold. It is quite wide, and larger than Rydberg resonances. An example is the photoionization cross section of the ground state $3d^5(^6S)$ of Fe IV shown in Fig. 6.13. The large resonance just above the threshold is identified as $3s^23p^53d^6(^6P^o)$, an equivalent-electron

[5] M. J. Seaton first explained the general importance of the nearly ubiquitous PEC features in the photoionization work under the Opacity Project [149].

autoionizing state. The extent of the resonance is in stark contrast to the weak and narrow Rydberg series of resonances $3s^23p^53d^5[^7P^o]nd(^6P^o)$ at higher energies. Such a resonance, given its position in the cross section, may be identified through an atomic structure calculation including excited electronic configurations of the photoionizing system itself, in this case Fe IV.

6.7 Experimental measurements

Whereas we have focused largely on a theoretical understanding of photoionization, and the framework used to describe it in detail, new advances have also been made in recent years in measurements of photoionization cross sections with unprecedented resolution by several experimental groups (e.g. [129, 130, 131, 132]). These experiments employ synchrotron-based photon light sources capable of the high resolution necessary to resolve much of the resonance structure. These experiments are essential to benchmark the accuracy of advanced theoretical methods, such as the extensive R-matrix results discussed in the preceding sections.

However, comparison between experiment and theory requires a certain amount of analysis. Most often, the ions in the experimental beam are not simply in the ground state, or pre-specified excited states. Rather, they have a somewhat uncertain population distribution among the ground state and a few low-lying (usually metastable) levels; high-lying levels are ruled out, since they would rapidly decay to lower levels. Therefore, the observed features in photoionization correspond to a combination of those found in the cross sections of ions in several levels. So, theoretical computations require consideration of all such states accessible within the energy range of the experiment, and level-specific cross sections need to be computed at a sufficiently fine mesh to resolve all outstanding features.

Two additional factors are considered for a proper comparison between experiment and theory: (i) the beam width, which determines resolution, and a functional form for convolution of theoretical results, and (ii) estimates of the distribution among excited states in the ion beam, which determine the stength of observed resonances and the background. Figure 6.14 shows a sample comparison between theory and experiment for the photoionization of the ground state $2s^22p^2(^3P)$ of O III, as well as a few low-lying excited states (O III is also C-like, with energy levels as in Fig. 6.6). The experimental results are given in the topmost panel f of Fig. 6.14. The bottom three panels show the individual theoretical cross sections for the three *LS* terms of the ground configuration $2s^22p^2(^3P, ^1D, ^1S)$. It is natural to expect that these three low-lying and long-lived metastable states would be present in the ion beam (as indeed in many astrophysical environments). The ionization thresholds for the excited metastables ^1D and ^1S are *lower* than that of the ground state. Successively higher levels ionize with less energy than those lying lower. Consequently, the beginning of the cross section moves to the left on the energy scale with increasing excitation in (b) and (c), compared with (a), in Fig. 6.14.

However, it turns out that the combined features in an admixture of the three states, $\sigma(^3P + ^1D + ^1S)$, are still not adequate to explain all those observed in the experiment. In particular, the experimental cross section in panel (f) shows features *below* those for photoionization of any of the three metastable states. It was then recognized [132] that another metastable state of the next excited configuration is present in the beam, $2s^1p^3(^5S^o)$. Its signature, the lowest energy resonance, manifests itself when it is included in the theoretical calculations for comparison. Fig. 6.14(d) is a combined sum of the four level-specific photoionziation cross sections $\sigma(^3P + ^1D + ^1S + ^5S^o)$, convolved over the known experimental beam width. Finally, all theoretical cross sections are assigned statistical weights in (e), for a full interpretation of the experimental results in (f). Such studies are of considerable importance in practical applications in laboratory and astrophysical plasmas, where metastable levels, and possibly higher levels, may be significantly populated.

6.8 Resonance-averaged cross section

Rydberg series of resonances converging on to a given level become narrower as their autoionization width decreases as $1/\nu^3$. Resolution of these resonances becomes computationally difficult, as it requires a finer and finer energy mesh. It is therefore desirable to obtain an analytic average in the region just below the threshold of convergence. In many calculations (such as in the Opacity Project) the detailed σ_{PI} are computed to resolve resonances up to $\nu = 10$ in a Rydberg series, and are then averaged over for $10 \leq \nu \leq \infty$. The *Gailitis averaging* procedure [96, 97, 153], discussed in Chapter 5 in connection with multi-channel quantum defect theory applied to (e + ion) scattering resonances (Section 5.2.3), is also implemented for photoionization resonances [154].

Figure 6.15 shows two examples of Gailitis averaging of resonances in the photoionization of the ground state $2s^22p^2(^3P)$ C I, in the small energy regions below

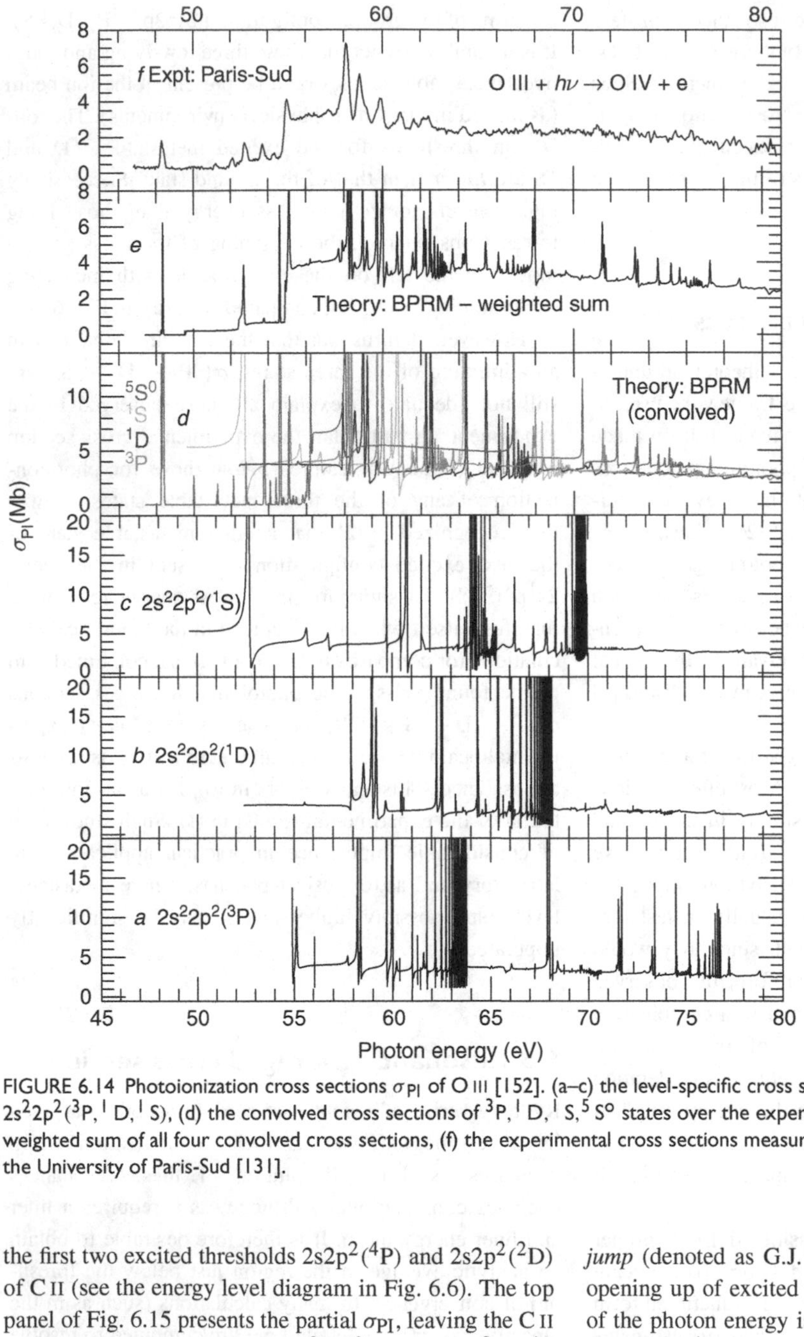

FIGURE 6.14 Photoionization cross sections $\sigma_{\rm PI}$ of O III [152]. (a–c) the level-specific cross sections of the three states $2s^22p^2(^3P, ^1D, ^1S)$, (d) the convolved cross sections of $^3P, ^1D, ^1S, ^5S^o$ states over the experimental beam distribution, (e) the weighted sum of all four convolved cross sections, (f) the experimental cross sections measured using the photon light source at the University of Paris-Sud [131].

the first two excited thresholds $2s2p^2(^4P)$ and $2s2p^2(^2D)$ of C II (see the energy level diagram in Fig. 6.6). The top panel of Fig. 6.15 presents the partial $\sigma_{\rm PI}$, leaving the C II core ion in the ground state $2s^22p(^2P^o)$, in the $^2P^o$–^4P energy range. The resonance series ^4Pnl is fully resolved by direct compuation of cross sections up to $\nu = 10$ (top panel), and then averaged using the Gailitis method. The cross section marked 'averaged resonances', expressed as $<\sigma(^4Pnl)>$, is roughly constant until the Rydberg series limit ^{4}P (top panel). Then the $^4P\epsilon l$ channels open, leaving the ion in the excited ^{4}P state. That loss of flux in the ^{4}P channels manifests itself as a sharp downward *Gailitis jump* (denoted as G.J. in Fig. 6.15). The drop is due to opening up of excited photoionization channels, as most of the photon energy is used up for excitation of the ion core to the ^{4}P. The magnitude of the Gailitis jump in photoionization is precisely equal to the partial cross section for photoionization to the excited state of the residual ion. These jumps are observed only in partial photoionization cross sections, and not in the total cross section, which is a sum of all partial contributions.

The situation in the bottom panel of Fig. 6.15 is rather different. It shows the partial cross section for ionizing into the excited state ^{4}P of C II; the resonances in the

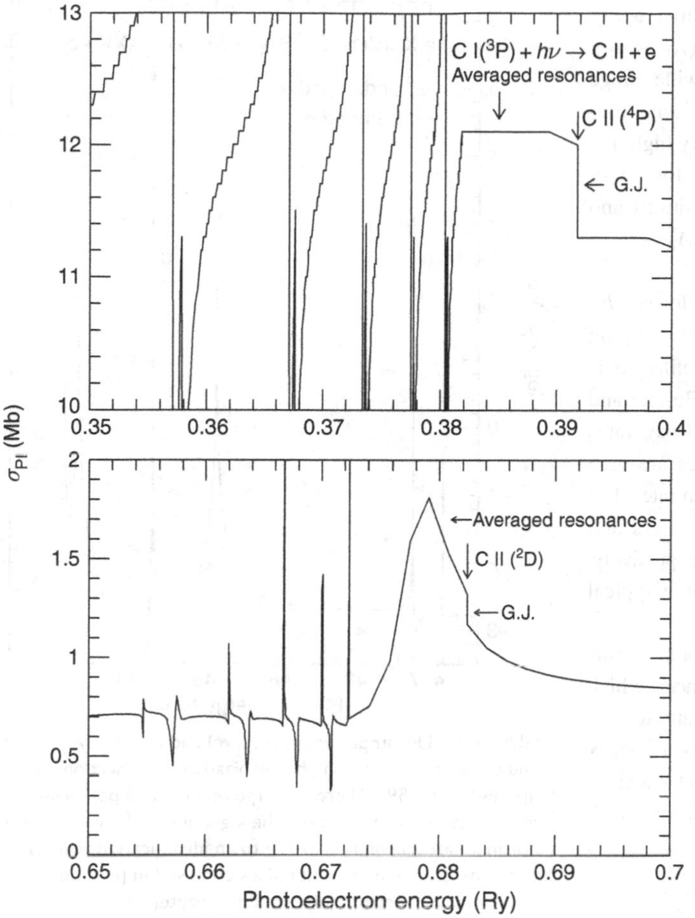

FIGURE 6.15 Partial photoionization cross sections σ_{PI} of the ground state of C I leaving the residual ion C II in the ground $2s^22p(^2P^o)$ state (upper panel) and in the first excited $2s2p^2(^4P)$ state (lower panel). They illustrate Gailitis averaging of resonances in the regions below the first excited threshold $2s2p^2(^4P)$ and the second threshold $2s2p^2(^2D)$ and Gailitis jumps (G.J.) at the thresholds [155].

^4P–^2D region are shown. Following the resolved resonances, Fig. 6.15 shows the averaged resonant cross sections below the ^2D threshold. As can be seen, the resonance-averaged cross section contains a broad peak in the region just below the ^2D. This is because, whereas the ^2Dnl resonances are averaged over, resonances converging on to higher levels still appear, superimposed on the averaged $<\sigma(^2Dnl)>$, i.e., Gailitis averaging has been carried out only for the resonances that converge onto the immediate next excited state. Since resonances correspond to closed channels, this marks the difference between *weakly closed channels*, viz. ^2Dnl, and *strongly closed channels* that belong to excited levels higher than the ^2D, and remain closed even as the $^2D\epsilon l$ channels open up [155].

6.9 Radiation damping of resonances

We know that an autoionizing resonance has combined characteristics of a bound state and a continuum state coupled together (Ch. 3, also discussed later in Section 6.9.3). Therefore, an autoionizing state has a finite probability of breaking up either by decaying into the (e + ion) continuum as a radiation-less transition, with the energy taken up as the kinetic energy of the ejected photoelectron, or emission of a photon via a radiative transition to a bound (e + ion) state. But if the formation of a photo-excited resonance is followed by a radiative transition back to the initial state, then the total autoionization rate is commensurately *reduced*. This results in *radiation damping* of the resonance profile. In practice, including radiation damping results in a decrease in the height of resonances.

However, radiation damping of resonances is of practical importance only for highly charged ions or close to the series limit (high n). This is because the radiative decay rates then compete effectively with autoionization rates. Damping is more prominent in autoionizing states with high effective quantum number when resonances become narrower with longer lifetime. For low-n resonances, the autoionization rates are much larger than radiative decay rates. The total resonance width is given by $(A_r + A_a)$. While typical radiative rates vary over a wide range depending on Z and z, i.e., $A_r = 10^6$–$10^{14}\ s^{-1}$, autoionization rates are roughly the same and generally high, i.e. $A_a = 10^{13}$–$10^{14}\ s^{-1}$. So $A_a(n) \gg A_r(n)$ for low-n resonances. But A_r corresponds to core transitions and remains constant, and therefore must exceed $A_a \sim n^{-3}$ for sufficiently high n.

In addition, for high-z ions, A_r for allowed $E1$ transitions are themselves large, and radiation damping manifests itself even for low-n resonance profiles. It is especially significant for photoionization of He-like and Li-like ions. This is because the radiative decay rates of the respective H-like and He-like ion cores are very high, competing effectively with autoionization rates. For these systems, the A-coefficient for the $2p \to 1s$ and the $1s2p\ ^1P^o \to 1s^2\ ^1S$ dipole transitions, respectively, is of the order of 10^{13}–$10^{14}\ s^{-1}$, approaching typical autoionization rates.

Radiation damping may be taken into account perturbatively [156]. It is applicable to low-n resonances, which can be fully resolved to obtain the *undamped* autoionization rate. The procedure entails fitting the dipole matrix element (as in Eq. 6.20) $<j|\mathbf{r}|i> \equiv D$. Since D is a complex quantity, we represent it is as a functional form with a background and a resonance term, as

$$D(E) = D^0(E) + \frac{C}{E - Z^*}; \quad \left(Z = E_0 - \frac{i}{2}\Gamma_a\right), \tag{6.60}$$

where D_0 represents the background, and the second term is a pole in the complex plane with residue C and resonance energy E_0; Γ_a is the autoionization width in Ry. The radiative width Γ_r is then obtained by

$$\Gamma_r = 4\pi^2 \frac{|C|^2}{\Gamma_a}. \tag{6.61}$$

The *second-order* radiative effects can then be included by considering

$$D(E) \longrightarrow \frac{D(E)}{1 + L(E)}, \tag{6.62}$$

where the operator $L(E)$ is given [157] by

$$L(E) = \pi^2 \mid D^0(E) \mid^2 + 2\pi^2 \frac{C^* D^0(E)}{E - Z} + 2\pi^2 \frac{|C|^2}{(E - Z)(Z - Z^*)}. \tag{6.63}$$

Figure 6.16 shows the effect of radiation damping in the form of reduced peak values of resonance profiles in cross sections computed using the R-matrix method [158].

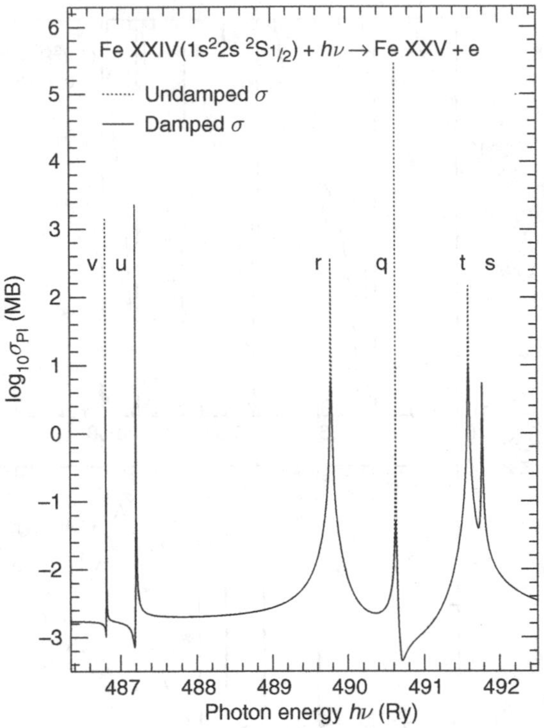

FIGURE 6.16 Undamped (dotted curve) and radiatively damped (solid curve) resonances in photoionization cross sections of Li-like Fe XXIV [158]. There is a large reduction in peak values due to radiation damping (note the log-scale on the y-axis). The resonances are labelled according to an identification scheme for the corresponding emission lines observed in (e + ion) recombination spectrum (discussed in Chapter 7).

While Fig. 6.16 shows photoionization cross sections including radiation damping, the damping contribution to the cross section can also be obtained from the photoexcitation cross section from a level k to a doubly excited level j, absorption radiative rate $A^r_{kj} = \hbar\Gamma_{kj}$,

$$\sigma_{kj} = \frac{4\pi^2 a_0^2}{\alpha^2 E^2} \Gamma_{kj} \delta(E - E_{jk}), \tag{6.64}$$

where E_{jk} is the transition energy, a_0 is the Bohr radius and α is the fine-structure constant.

6.9.1 Differential oscillator strength

In the discussion above we have described the 'absorption line' nature of a resonance in photoionization. In fact, the bound–bound and the bound–free may both be viewed as a continuous process of photoabsorption. We may extend the generalization to treat absorption in lines and the continuum using the same physical quantity, analogous to the oscillator strength. The ionization threshold is the Rydberg series limit for bound–bound transitions from a given initial state to final states with increasing n. The total photoabsorption cross section must then continue smoothly across the ionization threshold into the photoionized continuum. Line absorption below the ionization threshold can be related to the photoionization cross section, by extending the concept of the oscillator strength as follows.

From the usual Rydberg formula, $E = E_c - z^2/\nu^2$, where E_c is the ionization threshold in the core ion and ν is the effective quantum number, $dE = (2z^2/\nu^3)d\nu$. We define the *differential oscillator strength* as

$$\frac{df}{dE} = \frac{df}{d\nu}\frac{d\nu}{dE} = f_{ij}\frac{\nu^3}{2z^2}, \tag{6.65}$$

where $d\nu$ is taken to be unity for discrete line absorption. The photoabsorption cross section was defined (Eq. 6.11) as the ratio of the rate of photon absorption to the flux of incoming photons in a given frequency range. It can also be shown that

$$\sigma_{abs} = 4\pi^2\alpha\frac{df}{dE}a_0^2. \tag{6.66}$$

Substituting for dE, we get

$$\sigma_{abs} = 4\pi^2\alpha\frac{\nu^3}{2z^2}f_{ij}a_0^2. \tag{6.67}$$

Now both the line and the continuous photoabsorption can be expressed in terms of the differential oscillator strength.

$$\frac{df}{d\epsilon} = \begin{cases} \frac{\nu^3}{2z^2}f_{line}, & \epsilon < I, \\ \frac{1}{4\pi^2\alpha a_0^2}\sigma_{PI}, & \epsilon > I, \end{cases} \tag{6.68}$$

where f_{line} is the line absorption oscillator strength, σ_{PI} the photoionization cross section, and I the ionization energy. Figure 6.17 illustrates the continuty of photoabsorption from the bound–bound, and into the bound–free, for three lithium-like ions, C IV, O VI and Fe XXIV. The Rydberg series for discrete line absorption is for transitions $2s \to np$; as n increases towards the series limit, df/dE merges smoothly into the photoionization cross section $\sigma_{PI}(2s \to \epsilon p)$. The oscillator strengths for the $2s \to np$ series, and the photoionization cross sections, are computed using the R-matrix method in *independent* calculations, but with the same coupled channel wavefunction expansion.

6.9.2 Resonance oscillator strength

We now complete the equivalence between absorption lines and resonances by defining and computing oscillator strengths for resonances. As Eq. 6.68 implies, differential oscillator strengths are related to photoionization cross sections. The integrated $df/d\epsilon$ over the entire autoionizing resonance yields the effective photoabsorption in terms of σ_{PI}, i.e., the resonance oscillator strength f_r is defined to be

$$\bar{f}_r(J_i \to J_f) = \int_{\Delta E_r}\left(\frac{df(J_i \to J_f)}{d\epsilon}\right)d\epsilon = \left(\frac{1}{4\pi^2\alpha a_0^2}\right)\int \sigma_{PI}(\epsilon; J_i \to J_f)d\epsilon, \tag{6.69}$$

where J_i, J_f are total angular momenta of the initial bound level and the final bound or continuum (e + ion) system.

Let us consider again the right-hand side of the Fig. 6.17. The df/dE for Fe XXIV shows six resonances identified as $1s2p(^3P^o)2s\ \left[^4P^o_{1/2,3/2}, {}^2P^o_{1/2,3/2}\right]$ and $1s2p(^1P^o)2s\ \left[^2P^o_{1/2,3/2}\right]$ (see Table 7.1 for the labels v, u, r, q, t, s). It is found that the sum over the $\bar{f}_r$ equals 0.782 for all six resonances. That agrees well with the sum of the line absorption oscillator strengths for the E1 transitions in the He-like core ion Fe XXV: $f\left(1s^2\ ^1S_0 - 1s2p\ ^{(3,1)}P^o_1\right)$, which equals 0.772 [158].

The theoretical calculations for the resonance oscillator strengths need highly resolved profiles to enable accurate integration over σ_{PI}. Photoabsorption by resonances manifests itself as absorption 'lines' in the observed spectrum. In Chapter 13, we describe a particular application of resonant and line absorption in the X-ray spectra of active galactic nuclei.

6.9.3 Resonance profiles

We have highlighted the prominent role of resonances in atomic processes, and their astrophysical implications, throughout the text. As explained above, resonances are inherent in photoionization for any atomic system with more than one electron. They can affect the background cross section that cannot generally be expressed as an analytic expression or a fitting formula. It is also useful to look at certain physical features that characterize *shapes of individual resonances*.

The intrinsic line width of a spectral line due to a bound–bound transition is subject to the uncertainty principle; the lifetime is governed by the Einstein decay rates and therefore results in a corresponding energy width. However, a resonance state in the bound–free continuum has an additional broadening mechanism: the autoionization decay rate $A_a \sim 10^{13-14}\ s^{-1}$, which is much larger than most bound–bound A-values. The autoionization

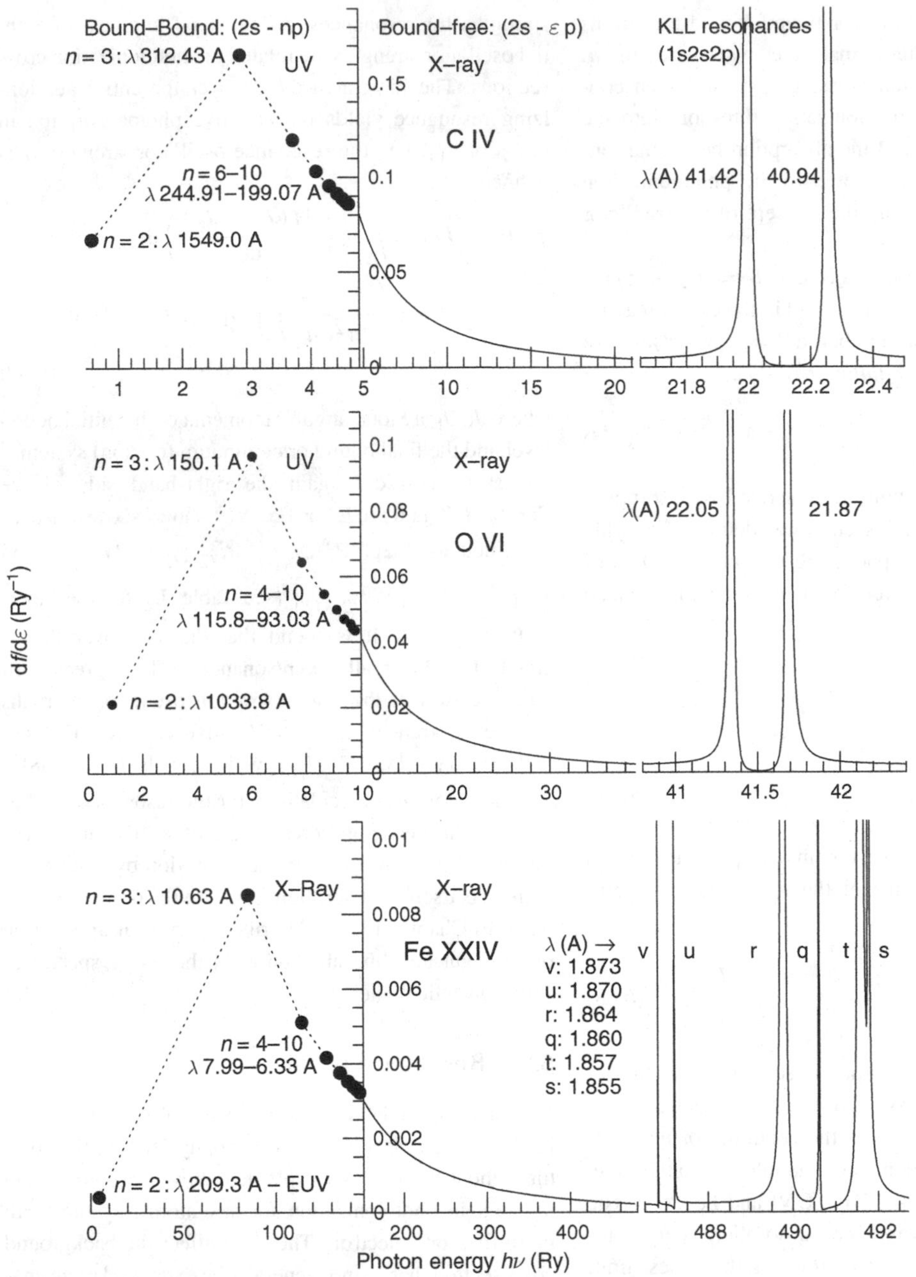

FIGURE 6.17 The differential oscillator strength df/dE for radiative transitions in Li-like ions C IV, O VI and Fe XXIV [158]. The discrete points (filled circles) on the left of the ionization threshold ($\epsilon < I$) are joined by dotted lines, which merge into the photoionization cross section (solid line) on the right ($\epsilon > I$). The matching point between the bound–bound transitions on the left and bound–free transition on the right illustrates the continuity of total photoabsorption.

width is $\Gamma_a = \hbar/\tau_a$, where τ_a is the autoionization lifetime.

Resonance profiles assume various shapes depending on the interplay between the bound state character of the wavefunction, and the continuum wavefunction. Recalling the discussion in Section 3.1, we rewrite Eq. 3.3 as

$$\Psi_{res}(E_r) = \Psi_b(E_b) + \Psi_c(E), \qquad (6.70)$$

where E_r, E_b, E represent the resonance position, the bound state energy and the continuum energies respectively. Operating with the interaction potential in the Hamiltonian,

$$V_E = < E|H|b > \tag{6.71}$$

couples the bound and continuum states to yield a resonance width $\Gamma_a = 2\pi|V_E|^2$. The resonance profile may be parametrized [159] according to the function

$$\frac{(q+\epsilon)^2}{1+\epsilon^2}, \tag{6.72}$$

known as the *Fano* profile (also called the Beutler–Fano profile [159]), where

$$\epsilon \equiv \frac{E - E_r}{A_a/2}, \tag{6.73}$$

and q is a profile parameter, which can assume a range of values describing the shape of the resonance. For large q, Eq. 6.72 approaches the *Breit–Wigner* profile

$$\frac{1}{\epsilon^2 + (\Gamma_a/2)^2} \quad (q \to \infty), \tag{6.74}$$

which has a Lorentzian shape, as discussed in Chapter 9. Using the Fano profile Eq. 6.72, the cross section with a *single* resonance may be expressed as

$$\sigma(\epsilon) = \sigma_{bg} + \sigma_a \frac{(q+\epsilon)^2}{1+\epsilon^2}, \tag{6.75}$$

where σ_{bg} represents the background, non-resonant cross section and σ_a is the interacting continuum. The Fano profile, Eq. 6.72, is especially useful in fitting isolated and well resolved resonances often seen experimentally; the q parameters derived from the fit can be compared with those computed theoretically. However, resonance fitting becomes more complicated when dealing with overlapping Rydberg series of resonances due to interference effects, such as those encountered often in theoretical calculations for complex systems. In addition, modern computational methods and computer codes discussed in Chapter 3 render resonance fitting largely unnecessary, except in specialized cases. Further physical insight may be gained by applying multi-channel quantum defect theory to the analysis of resonances along a Rydberg series.[6]

6.9.4 Shape and Feshbach resonances

In atomic or ionic photoionization one has an (e + ion) system, with an electron in the continuum. We have already seen Rydberg series of resonances for myriad atomic species converging onto excited states of the ion. They arise from the strong *attractive Coulomb potential* between the electron and the ion, which can support an infinite number of (quasi-)bound states. These resonances are relatively narrow. From basic scattering theory, they correspond to a rapid change in phase shift by π. The relation for the reactance matrix $K = \tan\,\delta_l$ (Eq. 3.59), then leads to a sharp change (rise and fall) in the corresponding scattering matrix or the cross section. Furthermore, using Seaton's theorem (Eq. 3.128) $\delta_l = \pi\mu_{nl}$, it follows that there are an infinite series of such resonances converging on to a higher state or threshold of the ion. Resonance energies are given by the Rydberg formula with quantum defect μ_{nl}, and the series corresponds to a closed channel with respect to the threshold of convergence.

But in general, as noted in Chapter 3, resonances occur because of superposition or interaction of bound-state wavefunction(s) and continuum state wavefunction(s). And they can be just as prominent when the continuum state involves a neutral atom (or a molecule), and not an ion. In that case the photoionizing ion is a *negative* ion, and the bound–free process is referred to as *photodetachment*. It is physically distinct from the excitation–ionization processes described hitherto for atomic or ionic photoionization, where the Coulomb potential dominates. An example of great relevance to astrophysics is the photodetachment of the negative ion H^-; a free electron may attach itself to the neutral H I atom with an *electron affinity* or binding potential energy of about 0.055 502 Ry or 0.755 eV [161, 160]. The photodetachment process of H^- is similar to photoionization of the two-electron helium atom, with ground state $1s^2\ ^1S$ and final (e + ion) symmetry $^1P^o$, with a p-electron in the continuum (e.g., [162, 163]), i.e.,

$$H^-(1s^2\ ^1S)) + h\nu \to H(1s\ ^2S) + e(p), \tag{6.76}$$

with the H^- electron affinity as the threshold ionization energy. The photodetachment cross section reveals two distinct resonances shown in Fig. 6.18 close to the $n = 2$ threshold energy of H I. The narrow resonance just below the $n = 2$ threshold is called a *Feshbach resonance* with peak value at 0.748 Ry. It is similar to the ones discussed above for atomic or ionic photoionization, corresponding to a discrete doubly excited state lying in the continuum (such as in a closed channel). In the case of H I the excited state of the core is H I ($n = 2$), i.e., $h\nu + HI \to H(n = 2) + e$. On the other hand, the broad resonance in Fig. 6.18 *above* the $n = 2$ threshold (shaded area) is referred to as a *shape resonance*, on account of the form or shape of the potential in the final state (electron and a neutral atom),

[6] A review article on quatum defect theory by M. J. Seaton deals extensively with resonance phenomena and profiles, including Rydberg series [153]. A resonance fitting algorithm for complex poles in the scattering matrix elements was implemented in a computer code RANAL for (e + ion) scattering (M. J. Seaton and A. K. Pradhan, unpublished), and for photoionization amplitudes in the code PHOTAL (H. E. Saarph and A. K. Pradhan, unpublished).

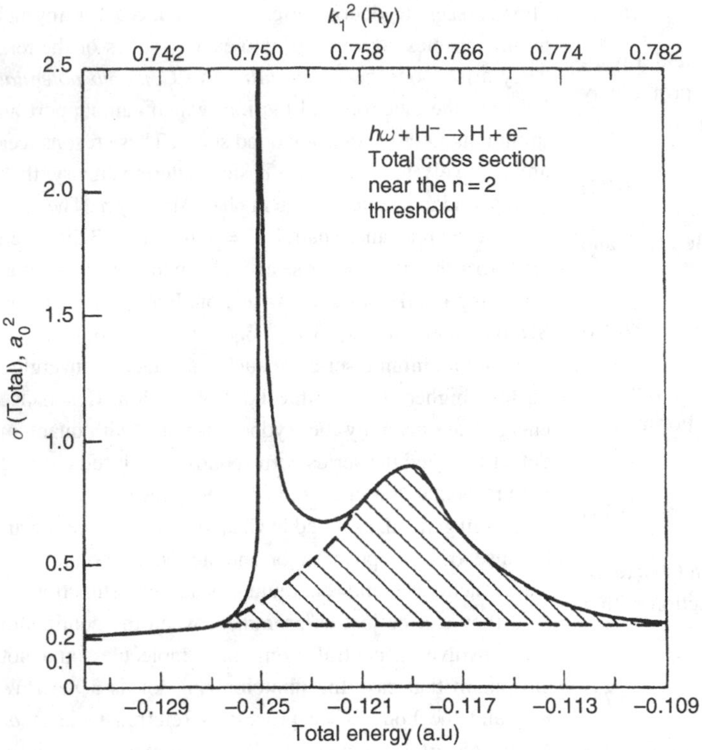

FIGURE 6.18 Bound–free photodetachment cross section of the negative ion H^-, with a Feshbach and a shape resonance near the $n = 2$ threshold [162]. The shaded area approximates the resonant oscillator strength (Eq. 6.69) $\overline{f_r} \sim 0.03$ for the broad shape resonance (see Section 6.9.1).

and is related to the quantum mechanical probability of the continuum *electron tunnelling* through the potential barrier (which is otherwise repulsive). Shape resonances are therefore quite different in character from Feshbach resonances, and are usually found to lie close to threshold photodetachment energies. For H^-, there are also other Feshbach and shape resonances, corresponding to higher thresholds of H I [164], such as at ~11 eV [163].

The H^- resonances are a manifestation of two-electron correlation effects mediated by weak attractive potential(s). In the absence of a dominant Coulomb potential, the two-electron interaction assumes greater importance than for (e + ion) processes hitherto considered. Therefore, very elaborate close coupling calculations, including large basis sets to represent short-range interactions, are necessary (e.g., [160, 162, 165, 166]). In Ch. 10 we will further discuss the importance of the bound–free detachment process of H^- as a source of radiative absorption in rather surprising situations.

6.10 Angular distribution and asymmetry

In addition to the total cross section, a topic of considerable interest in experimental studies of photoionization is the angular dependence manifest in the differential cross section. The outgoing flux is measured at a single energy, but at different angles to reveal details of the potential interactions embedded in otherwise unobserved parameters. Ejected photoelectrons can be measured experimentally at various angles relative to the incident photon beam. The resulting angular distribution provides additional details of the photoionization process. These features are observed in the differential cross section $d\sigma/d\Omega$, since the solid angle $\Omega(\theta, \phi)$ depends on the polar angle θ and azimuthal angle ϕ. We assume a geometrical arrangement with azimuthal symmetry independent of ϕ, and study the angular asymmetry as a function of θ.

Using the expression given earlier for the differential cross section, Eq. 6.10,

$$\frac{d\sigma}{d\Omega} = \left(\frac{4\pi^3 c^2}{n_\omega \omega_{ij}^2}\right) \frac{dT_{ij}}{d\Omega}, \tag{6.77}$$

the differential transition matrix element dT_{ij} is given by $<j|\hat{\mathbf{e}}.\mathbf{r}|i>$, as in Eq. 6.20. Taking the polarization of radiation along the z-axis, and the position vector $\mathbf{r}$ of the photoelectron at an angle θ with the z-axis, $<j|\hat{\mathbf{e}}.\mathbf{r}|i> \equiv <j|r\cos\theta|i>$. Let i be the intitial bound state, and assume j to be the final continuum state in a central potential $V(r)$, satisfying the Schrödinger equation

$$[\nabla^2 + k^2 - U(r)]\psi_j = E\psi_j, \quad (6.78)$$

where $k^2 = \sqrt{2mE}$ and $U(r) = 2mV(r)/\hbar^2$. The solution of this equation is

$$|j\rangle = 4\pi \sum_{l,m} i^l e^{-i\delta_l} G_{kl}(r) Y^*_{lm}(\hat{\mathbf{k}}) Y_{lm}(\hat{\mathbf{r}}), \quad (6.79)$$

where $G_{kl}(r)$ is the radial wavefunction $j_l(kr)/kr$, and δ_l is the phase shift of the lth partial wave. This formulation eventually yields (e.g. [166]) the differential cross section as

$$\frac{d\sigma}{d\Omega} = \sigma_{total}\left[1 + \beta P_2(\cos\theta)\right]. \quad (6.80)$$

The Legendre polynomial term

$$P_2(\cos\theta) = (3\cos^2\theta - 1)/2$$

β is known as the *asymmetry parameter*, describing the angular distribution of the photoelectron with initial angular momentum l as

$$\beta = \frac{l(l-1)R^2_{l-1,l} + (l+1)(l+2)R^2_{l+1,l}}{(2l+1)\left[lR^2_{l-1,l} + (l+1)R^2_{l+1,l}\right]} \times \frac{-6l(l+1)R_{l+1,l}R_{l+1,l}\cos(\delta_{l+1} - \delta_{l-1})}{(2l+1)\left[lR^2_{l-1,l} + (l+1)R^2_{l+1,l}\right]}. \quad (6.81)$$

Since $d\sigma/d\Omega \geq 0$, the limiting values of β are determined from the physical condition

$$1 + \beta P_2(\cos\theta) \geq 0, \quad (6.82)$$

which gives $\beta_{min} = -1$ for $\cos^2\theta = 1$, and $\beta_{max} = 2$ for $\cos^2\theta = 0$. For an s-photoelectron, $\beta = 2$, and the angular distribution is cosine-squared such that it peaks along or about the polarization vector of the radiation. For a photoelectron with $l \neq 0$, the anisotropy of the angular distribution depends on the interference between the two competing $l+1$ and $l-1$ outgoing channels, and the phase difference, $\delta_{l+1} - \delta_{l-1}$. For the limiting case of $\beta = -1$, $d\sigma/d\Omega = \sigma_{total}\left(\frac{3}{2}\sin^2\theta\right)$, with an angular distribution $\sin^2\theta$ that peaks at right angles to the polarization vector. Trivially, for the case of $\beta = 0$, the angular distribution is isotropic.

For an atomic system, in general the derivations are rather involved. Consideration of angular mementa in LS or intermediate LSJ coupling yields general expressions for the differential cross section $d\sigma/d\Omega$, and the asymmetry parameter β [168, 169]. Let us choose an ion X with total angular momentum J_i and parity π_i being ionized by a photon ω of angular momentum $J_\omega = 1$ and parity $\pi_\omega = -1$, i.e.,

$$X(J_i, \pi_i) + \omega(J_\omega, \pi_\omega) \rightarrow X^+\left(J'_c, \pi_c\right) + e(lms, \pi_e). \quad (6.83)$$

To simplify the derivation, U. Fano and D. Dill [168] introduced a momentum transfer vector $\mathbf{J}_t$ for the unobserved reactants as

$$\mathbf{J}_t = \mathbf{J}'_c + \mathbf{s} - \mathbf{J}_i = \mathbf{J}_c - \mathbf{J}_i = \mathbf{J}_\omega - \mathbf{l}, \quad (6.84)$$

in which the unobserved photoelectron spin $\mathbf{s}$ is added to $\mathbf{J}'_c$ to get $\mathbf{J}_c$. The allowed values of $\mathbf{J}_t$ are consistent with

$$\mathbf{J} = \mathbf{J}_i + \mathbf{J}_\omega = \mathbf{J}'_c + \mathbf{s} + \mathbf{l}, \ \pi = \pi_i\pi_\omega = \pi_c\pi_e, \quad \text{i.e.,} \ -\pi_i = (-1)^l\pi_c. \quad (6.85)$$

Hence, the differential cross section is a sum,

$$\frac{d\sigma}{d\Omega} = \sum_{J_t} \frac{d\sigma(J_t, k)}{d\Omega}. \quad (6.86)$$

Further defining the transition matrix in terms of the scattering matrix S,

$$\begin{aligned} &< J_c m_c, lm|S|J_i m_i, J_\omega m_\omega >= (-1)^{J_\omega + m_\omega - J_c + m_c} \\ &\times \sum_{J_t} (J_c - m_c, J_i m_i | J_t, m_i - m_c) \\ &\times (J_t, m_c - m_i | l - m, J_\omega m_\omega) < J_c l|\bar{S}(J_t)|J_i J_\omega > \\ &\times \delta_{m_c + m, m_c + m_\omega}. \end{aligned} \quad (6.87)$$

In the dipole approximation with operator D,

$$\begin{aligned} &< J_c l|\bar{S}(J_t)|J_i J_\omega > \\ &= n(\lambda_o) \sum_J (-1)^{J_i - J - J_\omega} \hat{\mathbf{J}} \begin{Bmatrix} l & J_c & J \\ J_i & J_\omega & J_t \end{Bmatrix} \\ &\times < J_c l, J'|D|\alpha_i J_i J_\omega > \end{aligned} \quad (6.88)$$

where $n(\lambda_o) = (4\pi\alpha\hbar\omega)/\left(3\lambda_o^2\right)$, λ_o is the wavelength of the incident radition divided by 2π, $\hat{\mathbf{J}} = (2J+1)^{1/2}$; the α_i denotes the set of quantum numbers necessary to specify the initial state uniquely. The notation J' is such that the final state is normalized according to the incoming wave boundary condition. Defining the components of the matrix $\bar{S}(J_t)$ as

$$\bar{S}_0(J_t) \text{ for } J_t = l, \ \bar{S}_{\pm 1}(J_t) \text{ for } J_t = l \mp 1, \quad (6.89)$$

one obtains,

$$\frac{d\sigma}{d\Omega} = \frac{1}{4\pi}\left[\sum_{J_t} \sigma(J_t) + \sum_{J_t(uf)} \sigma(J_t)\beta(J_t)P_2(\cos\theta) + \sum_{J_t(f)} \sigma(J_t)\beta(J_t)P_2(\cos\theta)\right], \quad (6.90)$$

where the various quantities are

$$\sigma(J_t) = \pi\lambda_o^2 \frac{2J_t + 1}{2J_c + 1}|\bar{S}_0(J_t)|^2, \quad \beta(J_t) = -1, \quad (6.91)$$

for the parity-unfavoured (pu) case, since $\pi_{\mathrm{i}}\pi_{\mathrm{c}} = -(-1)^{J_{\mathrm{t}}}$, and

$$\sigma(J_{\mathrm{t}}) = \pi\lambda_0^2 \frac{2J_{\mathrm{t}}+1}{2J_{\mathrm{c}}+1}[|\bar{S}_+(J_{\mathrm{t}})|^2 + |\bar{S}_- + (J_{\mathrm{t}})|^2],$$
$$\beta(J_{\mathrm{t}}) = \frac{(J_{\mathrm{t}}+2)|\bar{S}_+|^2 + (J_{\mathrm{t}}-1)|\bar{S}_-|^2 - 3\sqrt{J_{\mathrm{t}}(J_{\mathrm{t}}+1)}[\bar{S}_+\bar{S}_- + \bar{S}_+\bar{S}_-]}{(2J_{\mathrm{t}}+1)(|\bar{S}_+|^2 + |\bar{S}_-|^2)}, \tag{6.92}$$

for the parity-favoured (pf) case, since $(-1)^{l+1} = (-1)^{J_{\mathrm{t}}}$. Expressing

$$\sigma = \sum_{J_{\mathrm{t}}(\mathrm{All})} \sigma(J_{\mathrm{t}}), \quad \beta = \frac{\sum_{J_{\mathrm{t}}(pf)} \sigma(J_{\mathrm{t}})\beta(J_{\mathrm{t}}) - \sum_{J_{\mathrm{t}}(uf)} \sigma(J_{\mathrm{t}})}{\sum_{J_{\mathrm{t}}} \sigma(J_{\mathrm{t}})}, \tag{6.93}$$

the differential cross section can be written as

$$\frac{\mathrm{d}\sigma}{\mathrm{d}\Omega} = \frac{\sigma}{4\pi}\left[1 + \beta P_2(\cos\theta)\right]. \tag{6.94}$$

Here, the net asymmetry parameter β is given by the average of the $\beta(J_{\mathrm{t}})$ values for each value of J_{t}, weighted by the cross sections for each J_{t}. Finally [169], noting that both $\mathbf{J}'_{\mathrm{c}}$ and $\mathbf{s}$ are unobserved, $\mathrm{d}\sigma/\mathrm{d}\Omega$ must be summed over all possible values of $|\mathbf{J}_{\mathrm{c}}| = |\mathbf{J}'_{\mathrm{c}} + \mathbf{s}|$, that is,

$$\sigma = \sum_{J_{\mathrm{c}}, J_{\mathrm{t}}} \sigma(J_{\mathrm{t}}), \quad \beta = \frac{\sum_{J_{\mathrm{c}}}\left[\sum_{J_{\mathrm{t}}(f)} \sigma(J_{\mathrm{t}})\beta(J_{\mathrm{t}}) - \sum_{J_{\mathrm{t}}(uf)} \sigma(J_{\mathrm{t}})\right]}{\sum_{J_{\mathrm{c}}, J_{\mathrm{t}}} \sigma(J_{\mathrm{t}})}. \tag{6.95}$$

7 Electron–ion recombination

The study of (e + ion) recombination has been driven largely by astrophysical applications. In astronomical objects, ionized by a sufficiently intense radiation source, such as a star ionizing a gaseous nebula (Chapter 12), a state of ionization equilibrium is maintained between photoionization on the one hand and (e + ion) recombination on the other hand. It is, therefore, necessary to calculate photoionization and recombination cross sections and rates to obtain the distribution *among* the ionization stages of an element.

From a physical point of view (e + ion) recombination is the inverse of phototionization. A free electron in the continuum can combine with an ion, resulting in the emission of a photon, and also form a bound state of the recombined (e + ion) system with one charge less than that of the recombining ion. This free–bound electron–ion recombination process is also called *photorecombination*. Recalling Eqs 6.1 and 6.3, we may rewrite

$$X^+ + e(\epsilon) \rightarrow X + h\nu. \tag{7.1}$$

As for other atomic processes (Chapter 3), resonances again play a crucial role. In fact, the probability of recombination via an autoionizing state can be very high. As discussed in Chapter 3, a resonant state of the recombined (e + ion) system (Fig. 3.4) can either autoionize, decaying into the continuum where the electron goes free, or form a bound state via radiative stabilisation or dielectronic recombination. The initially free electron becomes bound and the excess energy is emitted via emission of a photon. The two modes are:

$$e + X^{+n} \leftrightarrow (X^{+n-1})^{**} \leftrightarrow \begin{cases} e + X^{+n} & \text{AI} \\ X^{+n-1} + h\nu & \text{DR}. \end{cases} \tag{7.2}$$

Both pathways of recombination are the inverse of photoionization, provided we consider the *same* resonances in a unified and self-consistent formulation. Photoionization and (e + ion) recombination are related by the principle of detailed balance, and both processes can occur directly in a non-resonant manner, or resonantly via autoionizing states. Therefore, the resonant and the non-resonant parts of (e + ion) recombination are *unified and inseparable*, and always occur in nature as such.

However, historically (e + ion) recombination has been considered as two separate processes: (a) *radiative recombination (RR)* due to direct radiative capture and recombination, Eq. 7.1, and (b) *dielectronic recombination (DR)* due to indirect capture and recombination through an autoionizing resonance, Eq. 7.2. This division is a useful approximation for many systems, such as highly charged few-electron ions. Often, RR and DR processes are dominant in different energy-temperature regimes. For example, in low-temperature plasmas free electrons have low energies that may be insufficient to excite any doubly excited autoionizing states (Fig. 3.4). In that case recombination would probably proceed mainly as the non-resonant RR process. On the other hand, in high-temperature sources, resonances may be formed by excitation of the recombining ion to high levels of opposite parity via strong dipole allowed transitions. In this case, DR becomes dominant. However, there is no firm division between the low- and high-energy regimes and both RR and DR must be considered in general. In addition, the energy distribution of resonances is not necessarily confined to relatively high energies; low-energy resonances often occur, leading to a separate low-temperature DR contribution to (e + ion) recombination.

The unification of the RR and DR processes has been explored in previous works (e.g., [170]). In recent years, a comprehensive unified treatment has been developed [151, 171, 172] that combines non-resonant and resonant processes (i.e., RR and DR) in an *ab-initio* manner, while ensuring the self-consistency inherent in the detailed balance between photoionization and (e + ion) recombination. The unified treatment is preferable to methods that consider (e + ion) recombination in parts, employing different approximations in various energy–temperature regimes of varying accuracy. It is also practically useful to have a single unified recombination rate coefficient, rather than separate RR and DR rates from different data sources. Astrophysical applications and plasma modelling also require self-consistency between photoionization and recombination rates that determine the ionic distribution of elements depending on temperature. Another reason we begin with a first-principles discussion of the unified method is because it follows directly from the extensive coupled channel framework we have hitherto developed to describe atomic processes in preceding chapters. In particular, we refer to the schematic diagram of atomic processes (Fig. 3.5) that presents a unified picture. That would be followed by approximations that are often made in computing RR and DR separately.

7.1 Detailed balance

The inverse processes of photoionization and electron–ion recombination (or photorecombination) are related by the principle of detailed balance as (Chapter 3)

$$\sigma_{\rm PI}\frac{g_j}{p_e^2} = \sigma_{\rm RC}\frac{g_s}{p_{h\nu}^2}, \qquad (7.3)$$

where g_j is the degeneracy (statistical weight) of the atom being photoionized, or of a given state of the *recombined* (e + ion) system, and g_s that of the residual ion, or the *recombining ion*, usually in the ground state. The $\sigma_{\rm PI}$ and $\sigma_{\rm RC}$ are cross sections for photoionization and (e + ion) recombination, respectively. The photon momemtum = $h\nu/c = (I+\epsilon)/c$, where I is the ionization energy and ϵ is the ejected photoelectron energy. The photoelectron momentum $p_{\rm e} = mv = \sqrt{2m\epsilon}$. Hence,

$$\sigma_{\rm RC} = \sigma_{\rm PI}\frac{g_j}{g_s}\frac{h^2\nu^2}{m^2c^2v^2} = \sigma_{\rm PI}\frac{g_j}{g_s}\frac{(I+\epsilon)^2}{2mc^2\epsilon}. \qquad (7.4)$$

This expression is known as the Milne relation [173] – a generalization of the Einstein B coefficients for bound–bound absorption to the bound–free regime. Equation 7.4 is the basis for the unified treatment of (e + ion) recombination, since it holds for both non-resonant RR and resonant DR processes, which are, in principle, observationally inseparable. As we shall outline in later sections, the unified treatment may be implemented in practice through the close-coupling approximation using coupled channel wavefunctions, obtained from the R-matrix method [151].

7.2 Total electron–ion recombination rate

Since (e + ion) recombination subsumes both non-resonant and resonant processes, RR and DR, let us denote the total rate coefficients simply as $\alpha_{\rm R}(T)$, as a function of electron temperature (we drop the subscript e and $m_{\rm e}$ for convenience). One may refer to the bound state of the (e + ion) system as the final or recombined state. In terms of the photoionization cross section $\sigma_{\rm PI}$ of any bound state, the Milne relation, Eq. 7.4, yields the recombination cross section $\sigma_{\rm RC}(E)$. Assuming a given distribution of electron velocities, $\sigma_{\rm RC}(E)$ may be averaged over to yield the rate coefficient

$$\alpha_{\rm R}(T) = <\sigma_{\rm RC}v> = \int_0^\infty v\sigma_{\rm RC}f(v,T){\rm d}v, \qquad (7.5)$$

where $v = \sqrt{(2\epsilon/m)}$ is the velocity of the photoelectron and $f(v,T)$ is the velocity distribution function. Typically, in astrophysical plasmas it is the Maxwellian distribution (Eq. 1.10),

$$f(v,T) = \frac{4}{\sqrt{\pi}}\left[\frac{m}{2kT}\right]^{3/2} v^2{\rm e}^{-\frac{mv^2}{2kT}}. \qquad (7.6)$$

Alternatively, in terms of photoelectron energies the substitution of explicit expressions for $f(v,T)$ and using $\epsilon = \frac{1}{2}mv^2$, ${\rm d}\epsilon = mv{\rm d}v$ gives

$$\alpha_{\rm R}(T) = \frac{4}{\sqrt{2\pi m}}\left(\frac{1}{kT}\right)^{3/2}\int_0^\infty \epsilon{\rm e}^{-\frac{\epsilon}{kT}}\sigma_{\rm RC}{\rm d}\epsilon. \qquad (7.7)$$

It is worthwhile here to have an idea of the orders of magnitude involved in these quantities, and also because one frequently finds *both* the atomic units au, or more often Rydbergs (Ry), and cgs units. We express the cross section in Mb (10^{-18} cm^2), and the photoelectron energy in Ry (2.1797×10^{-11} erg). In terms of cgs units, $m = 9.109\times10^{-28}$ g, $k = 1.380\,658\times10^{-16}$ erg K^{-1}. These give $\sqrt{2\pi mk} = 8.889\,316\times10^{-22}$, Ry^{2}Mb $= 4.751\,09\times10^{-40}$ erg^{2}cm^2, and therefore

$$\frac{4}{\sqrt{2\pi m}}\frac{{\rm Ry}^2{\rm Mb}}{k^{3/2}} = 0.015\,484\,\frac{{\rm cm}^3}{\rm s}{\rm K}^{3/2}. \qquad (7.8)$$

Hence, in cgs units,

$$\alpha_{\rm R}(T) = \frac{0.015\,484}{T^{3/2}}\int_0^\infty \epsilon\,{\rm e}^{-\frac{\epsilon}{kT}}\sigma_{\rm RC}{\rm d}\epsilon \;\;{\rm cm}^3{\rm s}^{-1}, \qquad (7.9)$$

where T is in K, ϵ in Ry and $\sigma_{\rm RC}$ in Mb. Note that with ϵ in Ry, $\epsilon/kT = 157\,885\epsilon/T$. This expression is often useful to relate electron energy with kinetic temperature in a plasma; for instance, in relating the recombination cross section measured experimentally in a limited energy range, to the rate coefficient at a given temperature.

Equation 7.9 introduces some uncertainty at and near zero photoelectron energy when σ_{RC} diverges, according to Eq. 7.3. To ensure accuracy, photoionization cross sections σ_{PI} may be used to obtain α_R,

$$\alpha_R(T) = \frac{g_j}{g_s} \frac{2}{kT\sqrt{2\pi m^3 c^2 kT}} \int_0^{\infty} (I+\epsilon)^2 \sigma_{PI}(\epsilon) e^{-\frac{\epsilon}{kT}} d\epsilon$$
$$= 1.8526 \times 10^4 \frac{g_j}{g_s} \frac{1}{T^{3/2}}$$
$$\times \int_0^{\infty} (\epsilon + I)^2 \, e^{-\frac{\epsilon}{kT}} \sigma_{PI} d\epsilon \ \mathrm{cm^3 s^{-1}}, \qquad (7.10)$$

where g_s is the statistical weight of the recombinig ion, and g_j that of the final recombined state. Equation 7.9, and similar expressions for $\alpha_R(T)$ above, correspond not only to the total but to state-specific recombination rate coefficient of individual recombined states j of the (e + ion) system, usually from the ground state g of the recombining ion. The total α_R then comprises of the infinite sum

$$\alpha_R = \sum_j^{\infty} \alpha_R(j). \qquad (7.11)$$

Under typical astrophysical conditions the recombining state is the ground state. Then $\sigma_{PI} \rightarrow \sigma_{PI}(j)$ corresponds to the *partial photoionization cross section* of state j, leaving the residual ion in the ground state in accordance with detailed balance, i.e., the Milne relation, Eq. 7.4, as described earlier in Chapter 6.

It is also useful to express the total recombination rate coefficient at a given energy, or averaged over an energy range, as $\alpha_{RC}(E)$ or $< \alpha_{RC}(E) >$, with photoelectron energy E and velocity v, as

$$\alpha_R(E) = v\sigma_{RC}^{Tot} = v\sum_i^{\infty} \sigma_{RC}(i), \qquad (7.12)$$

where σ_{RC}^{Tot} is the sum of cross sections of an infinite number of recombined states i.

7.3 Independent treatments for RR and DR

It is often computationally convenient to divide the recombination cross sections and rate coefficients into the non-resonant RR and the resonant DR components. Unlike the unified method, RR and DR are treated separately to obtain the RR rate coefficient α_{RR}, and the DR rate coefficient α_{DR}, to get the total

$$\alpha_R(T) = \alpha_{RR}(T) + \alpha_{DR}(T). \qquad (7.13)$$

The separate treatments of RR and DR are a good approximation for atomic systems when there is no significant interference between the two. The basic physics is discussed as outlined in Fig. 7.1, which illustrates a unified rate coefficient, but with the RR-dominated and the resonant DR-dominated portions for electron recombination from He-like to Li-like iron, (e + Fe XXV) → Fe XXIV. The corresponding photoionization cross section for Li-like to He-like ions is similar to the one shown for He I in Fig. 6.1, where the resonances occur at energies far above the threshold ionization energy. The low-energy photoionization cross sections, and the resulting background RR component at low temperatures, are featureless. In contrast, the DR component adds a significant 'bump' in the high-temperature range due to the excitation of resonances at high energies. As is the case for highly charged ions where RR and DR contributions are quite separable, the unified rate coefficient in Fig. 7.1 closely matches the sum of the RR and DR components.

7.3.1 Radiative recombination

Although unphysical, it is possible to use simple approximations to obtain only the background photoionization cross section σ_{PI} without taking into account resonances, which otherwise may dominate the cross section. The Milne relation Eq. 7.4, and Eq. 7.10, then yields the RR component α_{RR} separately in a straightforward manner. Examples of such photoionization calculations are those in the central-field approximation, as described in Chapter 6. It is also possible to use a *one-channel* close-coupling approximation, which would not include resonances, since they arise from channel coupling. However, as explained in Chapter 6, such background cross sections do not include any resonant enhancement that can lead to considerable underestimation of the total rate.

7.3.2 Dielectronic recombination

We recall the discussion in Section 3.1 on electron impact excitation and autoionization, Figs 3.3 and 3.4. Essentially, an autoionizing state is formed by excitation of an ion from level $g \rightarrow p$ by a free electron, where g is the ground state. But if the incident electron has *insufficient* initial kinetic energy $\epsilon < (E_p - E_g)$, then it may not escape the attractive field of the ion, and may become bound to the excited level p in an autoionizing state ($pn\ell$). The captured or recombined electron may either autoionize back into the continuum state ($g\epsilon\ell$), or undergo a

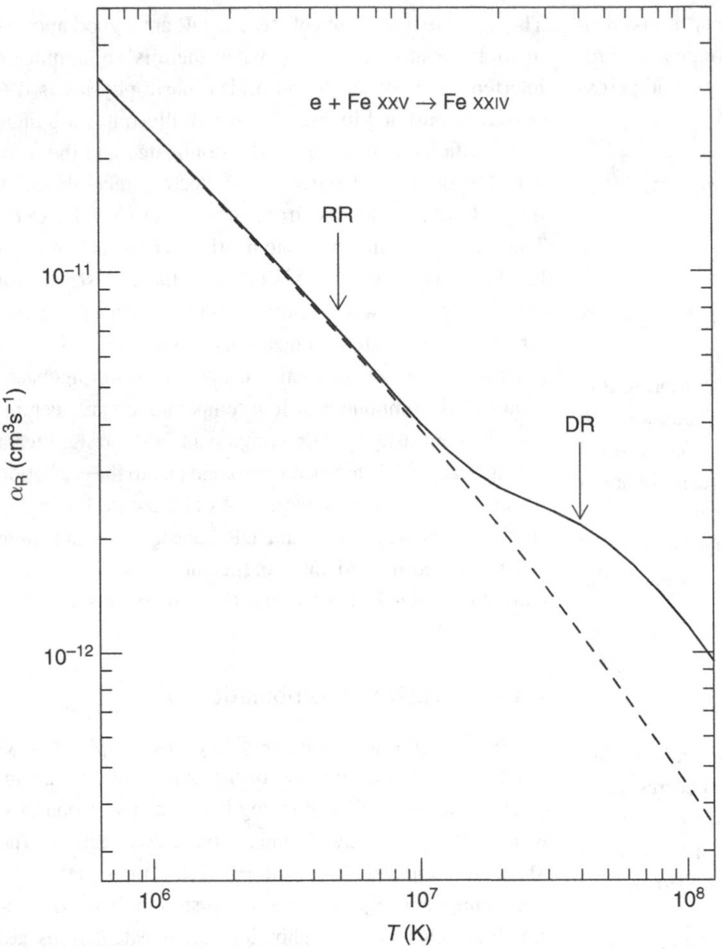

FIGURE 7.1 Radiative and dielectronic components (RR and DR) of the total (e + ion) recombination rate coefficient: solid line – unified rate coefficient [174]; dashed line – RR only [175].

radiative decay to a bound (e + ion) state ($gn\ell$) – the DR process. The captured electron itself remains a 'spectator' in the DR process in the sense that its quantum numbers $n\ell$ remain unchanged during the transition from an autoionizing state to a bound state. In other words, the radiative transition is between the core ion states g and p. As such, the dominant contribution(s) to the DR process stems from strong dipole transitions in the recombining ion, usually from the ground state to the lowest states connected by allowed transitions. For convenience in this discussion, and to avoid carrying the '$n\ell$' throughout, we denote the autoionizing state $pn\ell$ as i, and the bound state as $gn\ell$ as j.

The DR process involves electron recombination via infinite Rydberg series of resonances that converge on to an excited level of the recombining ion. The treatment of the DR process by itself is based on the so-called *isolated resonance approximation* introduced by H. S. W. Massey and D. R. Bates [176], who also named the process as 'dielectronic recombination'. As mentioned above, it is in fact a two-step approximation (Fig. 3.5): (i) an ion X^+ in an initial state g undergoes excitation *and* captures a free electron into a doubly excited autoionizing state i, and (ii) the resonant state i undergoes radiative stabilization to a bound (e + ion) state j with emission of a photon, i.e.,

$$X^+(g) + e \rightarrow X^{**}(i) \rightarrow X(j) + h\nu. \quad (7.14)$$

This expression embodies more complexity than is apparent at first sight. It involves an infinite number of autoionizing levels i, as well as the infinity of final (e + ion) recombined bound levels j. In addition, the initial state g of the recombining ion may not always be the ground state, because the ion may also be present in an excited metastable level in any moderately dense plasma. Hence, a large number of possible electron capture and decay pathways need to be considered.

The isolated resonance apporoximation for DR may be expressed without proof [176]; it depends only on detailed balance and branching between autoionization or radiative decay. Then the DR rate coefficient via autoionizing level i is

$$\alpha_{\rm d}^{i}(T) = \frac{g_i}{2g_s}\frac{h^3}{(2\pi m_e kT)^{\frac{3}{2}}}\exp\left(-\frac{E_{gi}}{kT}\right)A_{\rm r}(i), \quad (7.15)$$

where g_s and g_i are the statistical weights of ground state g, and the autoionizing level i respectively; the $A_{\rm r}(i)$ is the radiative decay rate summed over all final excited recombined state, i.e., all transitions $i \rightarrow j$, and E_{gi} is the energy of doubly excited autoionizing state i relative to initial state g. The total DR rate is then obtained from all contributing doubly excited states considered individually and independently as isolated resonances, i.e., without quantum mechanical interference, which might occur since, after all, they are coupled to the continuum as well as amongst themselves. Typically, both E_{gi} and kT are expressed in Ry or eV, and $\alpha_d^i(T)$ in $\rm cm^3 s^{-1}$ as usual for a rate coefficient. If the density of recombining ions $\rm X^+$ is $n(\rm X^+)$, and the density of free electrons n_e, then $n_e n(\rm X^+)\alpha_d^i$ gives the number of resonant DR occuring via level i per unit volume per unit time.

One may now extend the approximation [177] to include a factor $b(\rm X^{**})$, which is a measure of *departure* from thermodynamic equilibrium. In thermodynamic equilibrium, the detailed balance condition gives

$$n_S({\rm X^{**}})A_{\rm a} = n_e n({\rm X^+})\alpha_{\rm c}, \quad (7.16)$$

where $\alpha_{\rm c}$ is the electron capture coefficient, balanced by the autoionization rate $A_{\rm a}$.[1] In thermal equilibrium, the ionization states are distributed according to the *Saha equation* (discussed in Chapters 9 and 11), which gives the equilibrium number density $n_S(\rm X^{**})$ of the doubly excited ion $\rm X^{**}$ as

$$n_S({\rm X^{**}}) = \frac{g_i}{2g_s}\frac{h^3}{(2\pi m_e kT)^{\frac{3}{2}}}\exp\left(-\frac{E_{gi}}{kT}\right)n_e n({\rm X^+}). \quad (7.17)$$

Replacing the density in the balance condition, Eq 7.16, by the non-thermodynamic equilibrium density as [177]

$$n({\rm X^{**}})(A_{\rm a} + A_{\rm r}) = n_e n({\rm X^+})\alpha_{\rm d}, \quad (7.18)$$

such that

$$n({\rm X^{**}}) = n_S({\rm X^{**}})b({\rm X^{**}}) \quad (7.19)$$

defines the *departure coefficients*

$$b({\rm X^{**}}) = \frac{A_{\rm a}}{(A_{\rm a} + A_{\rm r})}. \quad (7.20)$$

The DR rate is then

$$n_e n({\rm X^+})\alpha_{\rm d} = n_S({\rm X^{**}})b({\rm X^{**}})A_{\rm r}. \quad (7.21)$$

Using Eqs 7.17, 7.18 and the b factor, the DR rate ceofficient is

$$\alpha_{\rm d}^{i}(T) = \frac{g_i}{2g_s}\frac{h^3}{(2\pi m_e kT)^{\frac{3}{2}}}\exp\left(-\frac{E_{gi}}{kT}\right)\frac{A_{\rm r}(i)A_{\rm a}(i)}{A_{\rm a}(i) + A_{\rm r}(i)}. \quad (7.22)$$

The isolated resonance approximation has been further elaborated upon in other works [63, 179, 180]. If the electron capture rate coefficient into level i from an initial state g of the recombining ion is $\alpha_{\rm c}^i$, and B^i is the branching ratio for decays from state i for radiative stabilization, then the DR rate coefficient is

$$\alpha_{\rm d}^{i}(T) = \alpha_{\rm c}^{i}(T)B^i. \quad (7.23)$$

Assuming a Maxwellian velocity distribution of electrons at temperature T, the dielectronic capture rate coefficient (Eq. 7.15) is

$$\alpha_{\rm c}^{i}(T) = \frac{g_i}{2g_s}\frac{h^3}{(2\pi m_e kT)^{\frac{3}{2}}}\exp\left(-\frac{E_{gi}}{kT}\right)A_{\rm a}(i), \quad (7.24)$$

$A_{\rm a}(i)$ is the autoionization rate into the continuum with a free electron, and the residual ion back to the state g. Now the full branching ratio B^i for all possible decays, both radiative and autoionizing, need to include a large number of pathways. But for a particular transition from an autoionizing i state to an excited recombined bound state j the branching ratio is

$$B^i = \frac{A_{\rm r}(i \rightarrow j)}{\left[\sum_{g'} A_{\rm a}(i \rightarrow g') + \sum_{j'} A_{\rm r}(i \rightarrow j')\right]}, \quad (7.25)$$

where the summations on the right are over all possible autoionization decays into (e + ion) continuum states $g'\epsilon\ell$, where g' denotes not only the ground state but also different states of the ion, and ϵ is the energy of the free electron. The denominator also includes all radiative transitions from autoionizing level i to (e + ion) bound states j'. Explicitly, we may write the DR rate coefficient $\alpha_{\rm d}^i$ as $\alpha_{\rm d}^{g\rightarrow i}(T)$ for recombination from an initial ion state g, via an autoionizing state i, to a final (e + ion) state j as

[1] Historically, DR was also referred to as 'inverse autoionization' [178], though that is only a partial description since DR is completed following radiative stabilisation.

$$\alpha_{\rm d}^{g\to i}(j,T) = \frac{g_i}{g_s}\frac{h^3 \exp\left[-\frac{E_{gi}}{kT}\right]}{2(2\pi m_e kT)^{\frac{3}{2}}} \times \frac{A_a(i\to g)A_r(i\to j)}{\sum_{g'} A_a(i\to g') + \sum_{j'} A_r(i\to j')}. \tag{7.26}$$

Usually, the summations on the right in the denominator include the dominant radiative and autoionization transitions. The total DR rate coefficient $\alpha_{\rm d}(DR;T)$ for recombination with an intital state g of the ion is obtained from the summed contributions

$$\alpha_{\rm DR}(T) = \sum_{i,j} \alpha_{\rm DR}^{g\to i}(j,T). \tag{7.27}$$

The isolated resonance approximation, as in Eq. 7.26, implies that the radiative and autoionization rates, $A_{\rm r}$ and $A_{\rm a}$, are computed *independently* neglecting interference between the two. The calculation of bound state and continuum state wavefunctions needed to compute $A_{\rm r}$ and $A_{\rm a}$ may be carried out using a variety of methods using atomic structure or distorted wave methods, such as SUPERSTRUCTURE [10], the Cowan Code [63], relativistic distorted wave code [68] or varients thereof. An important point to note is that in the independent resonance approximation for DR, Eq. 7.26, radiative transitions are considered as lines, whereas they are actually between an autoionizing state i and a bound state j of the (e + ion) system. A consequence of this is that the autoionization widths or resonance profiles, as functions of energy, are not considered.

7.3.2.1 *Quantum probability*

Thus far we have not explicitly invoked the quantum mechanical framework. Much of it remains the same as in the previous chapter on photoionization, owing to the fundamental Milne relation, Eq. 7.4, between photoionization and (e + ion) recombination. Furthermore, we reiterate the connections between $\sigma_{\rm PI}$, the differential oscillator strength, the resonance oscillator strength stated in Eq. 6.68 and the transition probability, Eq. 6.20, that enables a quantum mechanical evaluation of all of these inter-related quantities [181].

In Section 6.9 on radiation damping [178], we considered an isolated resonance as a single pole in the dipole transition matrix element D in the complex energy plane. Rewriting Eq. 6.60 as

$$D = D^{\rm o}(E) + \frac{C}{E - E_{\rm o} - \frac{1}{2}A_{\rm a}}, \tag{7.28}$$

where the pole position is $E_{\rm o} - i/2A_{\rm a}$, and the capture probability is [182],

$$P(E) = \frac{A_{\rm r}A_{\rm a}}{(E-E_{\rm o})^2 + \frac{1}{4}\left(A_{\rm a}^2 + A_{\rm r}\right)^2}, \tag{7.29}$$

where

$$A_{\rm r} = \frac{4\pi^2|C|^2}{A_{\rm a}}. \tag{7.30}$$

The averaged probability for capture over the resonance energy is the integral

$$P = \int_{\rm res} P(E){\rm d}E = \frac{2\pi A_{\rm r}A_{\rm a}}{A_{\rm r}+A_{\rm a}}. \tag{7.31}$$

This expression reflects the branching between autoionization and radiative decay. In the respective limits when one rate dominates the other, viz. $A_{\rm a} \gg A_{\rm r}$, $P \to A_{\rm r}$, and when $A_{\rm r} \gg A_{\rm a}$, $P \to A_{\rm a}$. Equation 7.29 enables the computation of the resonance profile, including *radiation damping* perturbatively, in the independent resonance approximation (discussed later). We can also express the cross section for capture of an electron with velocity v by an ion, with emission of a photon, as

$$\sigma = \frac{\pi}{2g_i}\left(\frac{\hbar}{mv}\right)^2 \sum_J (2J+1)P_J(E), \tag{7.32}$$

where J is the total angular momentum of the electron–ion system and g_i is the statistical weight of the initial ion level; $P(E)$ is the capture probability as above.

The first successful approximation for total DR rate coefficient was given by A. Burgess [178, 179, 183], showing that the dominant contribution to DR is from high-n levels converging on to the excited level(s), via dipole allowed transitions in the recombining ion. Assuming a Maxwellian distribution of velocities, the so-called *Burgess formula* for the total DR rate coefficient $\alpha_{\rm DR}^{\rm B}$ may be expressed as

$$\alpha_{\rm DR}^{\rm B}(T) = 2.5\times 10^{-4}\frac{(z+1)^2}{T^{1.5}} \times \sum_j f_j E_j^{1/2} 10^{-4600\frac{E_j}{T}}\ {\rm cm^3 s^{-1}}, \tag{7.33}$$

where z is charge of the recombing ion, E_j is the excitation energy in eV of the recombined ion and f_j is the absorption oscillator strength of levels j from the ground level; the constant factors are as given in [184]. Whereas Eq. 7.33 includes the contribution of high-n levels to DR, the Burgess formula is not sufficiently accurate for many astrophysical applications (cf. Fig. 7.3), with uncertainties of factors of two or three, or higher. Fortunately, more accurate calculations have been made for most atomic species in the independent resonance approximation, and in recent years, using the unified method developed by S. N. Nahar and A. K. Pradhan [151], which incorporates the precise theory of DR developed by R. H. Bell and M. J. Seaton [157], as outlined in the next section.

7.4 The unified treatment

The unified method [150, 151, 171] is based on (i) the Milne relation, Eq. 7.4, and (ii) the DR formulation in the coupled channel approximation [157]. The total recombination cross section is the sum of cross sections for recombination into the set of infinite number of recombined levels of the (e + ion) system, as illustrated schematically in Fig. 7.2. Consistent with the theoretical framework, we divide the infinite number of recombined (e + ion) levels into two main groups: (A) low-n levels with $n \leq n_0$ and (B) high-n levels with $n_0 \leq n \leq \infty$. The principal quantum number n_0 divides the two groups, and is chosen such that they are independent and complementary in the sense described below.[2]

Level-specific recombination cross sections $\sigma_{\rm RC}(j)$ for levels in group (A) are obtained from the Milne relation using coupled channel photoionization cross sections $\sigma_{\rm PI}(j)$ computed using the R-matrix method, and thereby including resonance contributions. Therefore, Eq. 7.4 ensures that the recombination cross sections thus obtained include both the non-resonant backgruond and the resonant contributions, i.e., RR and DR processes. The physical explanation is that for the low-n levels of group (A), the background cross sections are non-negligible and non-separable from the superimposed resonances. For sufficiently high-n value of n_0, the background contribution becomes negligible. In other words, for group (B) levels RR is negligible and DR is dominant for $n > n_0$.

The DR theory used to compute the contribution from group (B) levels is based on a precise quantum mechanical treatment of electron scattering *and* radiation damping due to recombination [157, 181]. Following the discussion in Chapter 5, we denote the scattering matrix for electron–ion collisions as $\mathcal{S}_{\rm ee}$. Then we incorporate the additional probability of electron capture and radiative decay via emission of a photon by the scattering matrix $\mathcal{S}_{\rm pe}$. Since the total electron–photon flux due to both processes, electron scattering and capture, must be conserved, $\mathcal{S}_{\rm ee}$ and $\mathcal{S}_{\rm pe}$ are in fact sub-matrices of the total scattering matrix $\mathcal{S}$. The unitarity condition for (electron + photon) matrix conserves both the incident electron and the emitted photon flux as

$$\mathcal{S}_{\rm ee}{}^{\dagger}\mathcal{S}_{\rm ee} + \mathcal{S}_{\rm pe}^{\dagger}\mathcal{S}_{\rm pe} = 1. \tag{7.34}$$

We write the *generalized* electron–photon scattering matrix as

$$\mathcal{S} = \begin{pmatrix} \mathcal{S}_{\rm ee} & \mathcal{S}_{\rm ep} \\ \mathcal{S}_{\rm pe} & \mathcal{S}_{\rm pp} \end{pmatrix}, \tag{7.35}$$

where $\mathcal{S}_{\rm ee}$ is the now the sub-matrix for electron scattering *including* radiation damping. The radiatively damped flux, i.e., electron capture leading to photon emisson is given by $\mathcal{S}_{\rm pe}$. It again follows from detailed balance that $\mathcal{S}_{\rm ep}$ refers to the inverse process of photoionization. The last sub-matrix $\mathcal{S}_{\rm pp}$ refers to photon–photon scattering process and is neglected. Of course, in the absence of interaction with the radiation field, $\mathcal{S}_{ee}$ is the usual radiation-free scattering matrix **S** for the electron–ion collision process only, without capture and recombination. Possibility of radiation damping reduces the cross sections for electron–ion collisions [185, 186]. We can write the recombination cross section in terms of the generalized S-matrix just defined as [178],

$$\sigma_{\rm DR} = \frac{\pi}{k^2}\frac{1}{2g_s}\sum_i\sum_j |S_{\rm pe}(i,j)|^2. \tag{7.36}$$

Recall that g_s is statistical weight of the initial state g, i is the autoionizing state into which the free electron is captured, and j is the final recombinated state of the (e + ion) system.

7.4.1 Rate coefficients

As outlined above, the (e + ion) photorecombination cross section $\sigma_{\rm RC}$ may be computed using the Milne relation, i.e., Eq. 7.4 from detailed photoionization cross sections including resonances due to low-n complexes. For the high-n complexes, we obtain the DR cross section, Eq. 7.36. The temperature dependent rate coefficient $\alpha_{\rm R}(T)$ is then calculated using Eqs 7.9 and 7.10. Since resonances lead to the DR contribution, the rate coefficient is also expected to show enhancement due to the *distribution* of resonances in energy. H. Nussbaumer and P. J. Storey first pointed out that near-threshold low-energy resonances give rise to a low-T DR bump [187], in addition to the high-T DR bump shown in Fig. 7.1. Figure 7.3 illustrates the correspondence between the unified rate coefficient $\alpha_{\rm R}(T)$ and separate RR and DR contributions,

[2] Both the effective quantum number ν and the principal quantum number n are used interchangeably for highly excited states with high n *and* ℓ. The two quantities are nearly the same large $\ell > 2$, when the quantum defect $\mu_\ell = n - \nu_\ell$ is very small.

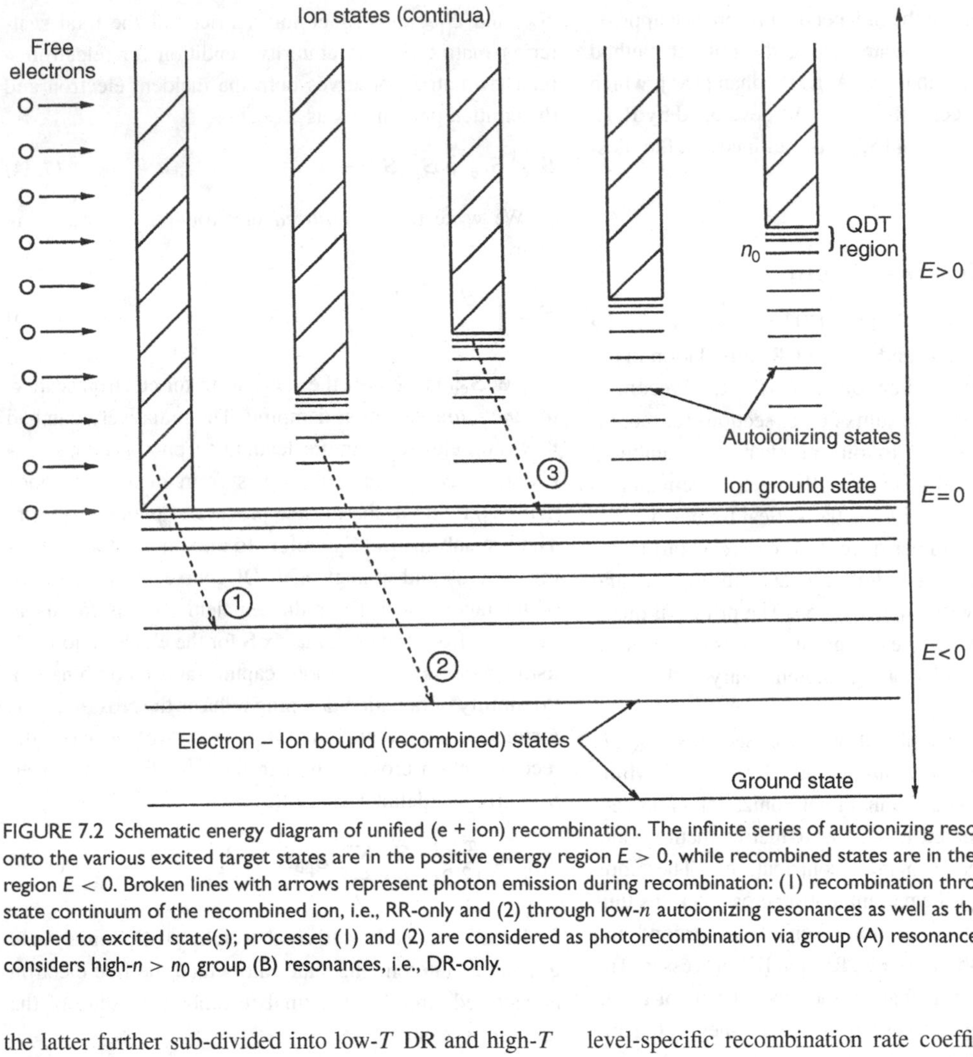

FIGURE 7.2 Schematic energy diagram of unified (e + ion) recombination. The infinite series of autoionizing resonances converging onto the various excited target states are in the positive energy region $E > 0$, while recombined states are in the negative energy region $E < 0$. Broken lines with arrows represent photon emission during recombination: (1) recombination through the ground state continuum of the recombined ion, i.e., RR-only and (2) through low-n autoionizing resonances as well as the continuum coupled to excited state(s); processes (1) and (2) are considered as photorecombination via group (A) resonances. Process (3) considers high-$n > n_0$ group (B) resonances, i.e., DR-only.

the latter further sub-divided into low-T DR and high-T DR. The high-T DR contribution in Fig. 7.3 is computed using the Burgess formula (Eq. 7.33) and overestimates the DR rate, whereas a more recent calculation is in good agreement with the unified DR rate [188].

7.4.2 Level-specific rate coefficients

The $\alpha_R(T)$ are obtained from level-specific recombination rate coefficients of given quantum number n and symmetry $SL\pi$, $\alpha_R(nSL\pi, T)$. These are calculated explicitly from the unified recombination cross sections $\sigma_{RC}(nSL\pi)$ for levels of group (A). However, this is still incomplete, since the high-n DR contribution must be added to obtain the total. Before we outline the DR calculation for group (B) recombined levels, it is useful to illustrate the effect of DR on total level-specific recombination rate coefficients $\alpha_R(nLS\pi)$. Fig. 7.4 presents level-specific recombination rate coefficients of several levels (LS terms) due to electron recombination with B-like Fe XXII into C-like Fe XXI, i.e., (e + Fe XXII) → Fe XXI, discussed in the previous chapter (Fig. 6.11). The ground state $1s^22s^22p^2\ ^3P$, and a few other excited states of Fe XXI, are shown. The $\alpha_R(nLS\pi)$ are computed using level-specific photoionization cross sections of Fe XXI of the type shown in Fig. 6.11, but for partial photoionization into only the ground state of the core ion Fe XXII ($1s^22s^22p\ ^2P^o$) as required by detailed balance. The 'bumps' in $\alpha_R(nLS\pi)$ are from DR, which enhances the rate when the relatively high-lying resonances are excited (Fig. 6.11). It is noteworthy that a separate treatment of RR and DR does not provide total level-specific recombination rate coefficients, as in the unified method.

The contributions to recombination from levels with $n \leq n_0$ of group (A) are obtained from detailed balance using σ_{PI} with autoionizing resonances, as explained in

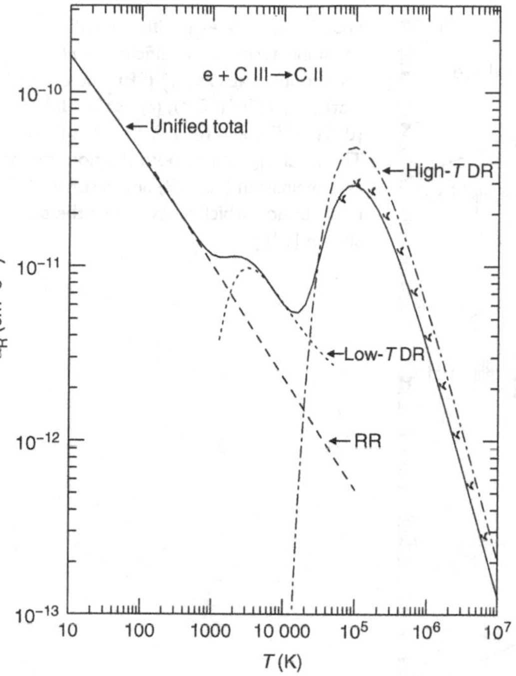

FIGURE 7.3 Unified rate coefficient (solid line) and separate RR and DR contributions for e + CIII → CII (after [188]).

Chapter 6, which gives both the non-resonant and resonant parts of the total recombination rate coefficient (viz. RR and DR) *for each level*. However, there is a relatively minor contribution that needs to be included. Although the integration over $\sigma_{RC}(E)$ to obtain $\alpha_R(T)$ is carried over a large energy range, in principle the integration should be extended up to $E \to \infty$ since the background cross sections may continue to be non-negligible up to high energies. The contribution from some high photoelectron energy ϵ_o up to $\epsilon \to \infty$ can be added through the following rapidly convergent form using the high-energy background photoionization cross sections,

$$\alpha_R^{\epsilon_0 \to \infty} = -\frac{g_i}{g_j} \frac{2}{c^2\sqrt{2\pi m^3 kT}} \int_{x_o}^{0} [1 - kTln(x)]^2 \sigma_{PI} dx, \tag{7.37}$$

where $x = \exp(-\epsilon/kT)$, with limits between 0 and 1, and varies slowly for low temperatures and rapidly at high T. In addition, $x_o = \exp(-\epsilon_o/kT)$ may be set by choosing an energy ϵ_o where resonances are absent, i.e., above all thresholds in the recombining ion considered. So the background σ_{PI} can be approximated simply by Seaton's fiting formula Eq. 6.32, or the Kramer's formula Eq. 6.33.

On a practical note, the unified calculations for photorecombination usually entail all recombined states up to $n \leq 10$. Such a division is necessary, since it is not practical to calculate photoionization cross sections for higher-n levels, but it is sufficient to ensure that DR is a good approximation to the total (e + ion) recombination rate coefficent for $n > 10$ (Section 7.5).

7.4.3 DR cross sections

We now discuss DR into the high-n group (B) levels of the recombined ion (Fig. 7.2). In this region the resonances are dense and narrow, and the background cross section is negligible. Hence the unified method employs the coupled channel theory of DR, and multi-channel quantum defect theory, which enables not only the computation of detailed DR resonant cross sections, but also resonance-averaged probabilities and cross sections [151, 157]. Viewed separately, in the high-n region DR dominates RR. But unlike the independent resonance approximation, where the radiative and autoionization rates A_a and A_r are computed independently of each other, the unified approach includes multi-channel coupling that accounts for overlapping resonance profiles, and hence interference effects.

As a Rydberg series of resonances approaches a threshold of convergence, onto an excited state of the recombining ion as $n \leq \infty$, the autoionization widths decrease as $A_a \sim 1/n^3$ (strictly speaking as z^2/ν^3). These group (B) resonances, with $n_o < n \leq \infty$, lie in a small energy region below each excited target threshold. Their lifetimes are correspondingly longer since their radiative decay rate $A_r \gg A_a$.

The coupled channel DR theory [157] partitions the electron–electron scattering matrix $\mathcal{S}_{ee}$ into sub-matrices of open and closed channels as χ_{oo}, χ_{oc}, χ_{co} and χ_{cc}. Here 'o' denotes open channels and 'c' closed channels, as discussed in Chapter 3. Open channels are those in which incident electrons have sufficient energy to excite the target ion. A closed channel refers to those with electron energies below an inaccessible target level or threshold. The scattering matrix $\boldsymbol{\mathcal{S}_{ee}}$ is then obtained as [157]

$$\boldsymbol{\mathcal{S}_{ee}} = \boldsymbol{\chi_{oo}} - \boldsymbol{\chi_{oc}}[\boldsymbol{\chi_{cc}} - g(\nu)\exp(-2i\pi\nu)]^{-1}\boldsymbol{\chi_{co}}, \tag{7.38}$$

where $g(\nu) = \exp(\pi\nu^3 A_r/z^2)$, ν is the effective quantum number associated with the resonance series, z is the ion charge, A_r is the sum of all possible radiative decay probabilities for the resonances in a series. The A_r remain constant, since they correspond to radiative transitions within the target or the core ion, whereas the autoionization rates decrease as $A_a \sim z^2/\nu^3$.

The factor $\exp(-2i\pi\nu)$ gives rise to rapid variation in **S**, since it introduces functional dependence on the energy

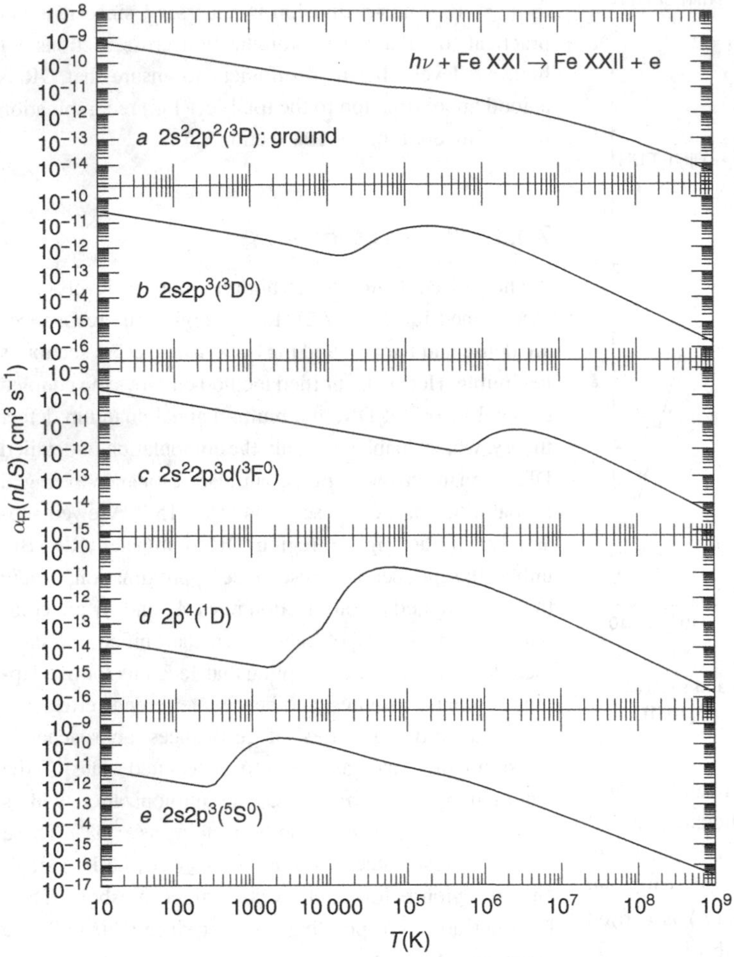

FIGURE 7.4 State-specific unified recombination rate coefficits $\alpha_R(nLS)$ to (a) the ground state $2s^22p^2(^3P)$, and excited states (b) $2s2p^3(^3D^o)$, (c) $2s^22p3d(^3F^o)$, (d) $2p^4(^1D)$, and (e) $2s2p^3(^5S^o)$ of FeXXI. The total α_R include both the non-resonant recombination (i.e., RR) and resonant DR contribution which gives rise to the bumps shown [146].

E via ν, where resonances occur. Owing to coupled closed channels, the scattering matrix has poles occuring at complex energies such that

$$|\chi_{cc} - \exp(-2i\pi\nu)| = 0. \tag{7.39}$$

Resonances corresponding to pole energies repeat themselves serially each time ν increases by unity, with decreasing resonance widths. The electron flux trapped in the closed channel resonances may decay radiatively to bound states of the (e + ion) system. This DR probability, for a given entrance or incident open channel α, is obtained from the unitarity condition as

$$P_\alpha(\mathrm{DR}) = (\mathbf{1} - \mathcal{S}^\dagger_{\mathbf{ee}}\mathcal{S}_{\mathbf{ee}})_{\alpha\alpha}. \tag{7.40}$$

Now, in analogy with the collision strength for electron impact excitation related to the cross section (Eq. 5.5), we define the *collision strength for dielectronic recombination* $\Omega(\mathrm{DR})$ using Eq. 7.40 as

$$\Omega(\mathrm{DR}) = \sum_{SL\pi}\sum_{\alpha}\frac{1}{2}(2S+1)(2L+1)P_\alpha^{SL\pi}. \tag{7.41}$$

The Ω_{DR} are calculated, in a self-consistent manner, using the same coupled channel wavefunction expansion that is used for the calculation of σ_{PI}, and as discussed in Chapter 5. The DR cross section (in Mb) is related to the collision strength Ω_{DR} as

$$\sigma_{\mathrm{DR}}(i \rightarrow j)(\mathrm{Mb}) = \pi\frac{\Omega_{\mathrm{DR}}(i,j)}{g_i k_i^2}\left(a_0^2 \times 10^{18}\right), \tag{7.42}$$

where k_i^2 is the incident electron energy in Ry.

Using multi-channel quantum defect theory the χ matrix can be diagonalized as $\chi_{cc}N = N\chi_{cc}$, where χ_{cc} is a diagonal matrix and N is the diagonalizing matrix with $N^{\mathrm{T}}N = \mathbf{1}$. In terms of N we write $\chi'_{oc} = \chi_{oc}N$

and $\chi'_{co} = N^T \chi_{co}$, where N^T is the transpose of N. The resonance-averaged DR probability [157] is then

$$<P_\alpha(DR)> = G \sum_{\gamma,\gamma'} \frac{\bar{\chi}'_{\alpha\gamma} \bar{\chi}'^{*}_{\gamma'\alpha} (N^T N^*)_{\gamma\gamma'}}{G + 1 - \bar{\chi}_{\gamma\gamma} \bar{\chi}^{*}_{\gamma'\gamma'}}, \quad (7.43)$$

where the factor $G(\nu) = g(\nu)^2 - 1$,

For practical computations of $< P_\alpha(DR) >$, a somewhat more extended formulation yields resonance averaged DR probability [172]

$$<P_\alpha(DR)> = G(\nu) \left[\sum_{\gamma} \frac{|\bar{\chi}_{\alpha\gamma}|^2 \sum_i |\mathbf{N}_{i\gamma}|^2}{G(\nu) + 1 - |\chi_{\gamma\gamma}|^2} + 2 \sum_{\gamma \neq \gamma'} \mathrm{Re} \left(\frac{\bar{\chi}_{\alpha\gamma} \bar{\chi}_{\gamma'\alpha} \sum_i N_{i\gamma} \mathbf{N}_{i\gamma'}}{G(\nu) + 1 - \chi_{\gamma\gamma} \chi_{\gamma'\gamma'}} \right) \right], \quad (7.44)$$

where the summation over i goes through all the closed channels. Equation 7.44 shows the division between direct and interference terms on the right, and gives the resonance-averaged $\Omega(DR)$ [172].

The expressions derived thus far enable us to calculate detailed and averaged DR cross sections or collision strengths using the coupled channel approximation. We illustrate several sets of results obtained using these expressions leading up to Eq. 7.44 in Fig. 7.5(a) for (e + S IV → S III). The averaged DR collision strength <Ω(DR)> is shown in the top panel of Fig. 7.5 (to display the peak magnitudes, the scale in this panel is divided into two parts). The DR contributions are shown in the energy regions ($\nu \leq 10 \leq \infty$) *below* the excited states of the target ion S IV: $3s3p^2(^2D, ^2S, ^2P)$, $3s^23d(^2D)$ and $3s^24s(^2S)$, as noted near the peaks. These excited states decay to the ground state $3s^23p(^2P^o)$ of S IV via dipole allowed transitions. The <Ω(DR)> also shows negligible contribution at the starting energy of $\nu = 10$ (marked by arrows) from the Rydberg series of autoionizing levels below the marked thresholds. But it rises rapidly as it approaches $n \to \infty$, beyond which Ω(DR) drops to zero, since the closed channels open up and resonance contributions to DR vanish. The physical reason is that the electrons trapped in doubly excited autoionizing resonances below threshold are released into the contiuum. Therefore, DR goes over to electron impact excitation (EIE) of the threshold level exactly at the peak energy. The peak values of DR in Fig. 7.5 agree with the EIE collsion strengths, shown as dark circles. The close agreement between independently computed excitation collision strengths, and the peak DR collision strengths, indicates the conservation of photon–electron flux implicit in the unitarity condition of the generalized **S**-matrix, Eq. 7.34.

As discussed already, an important aspect of the unified theory of (e + ion) recombination is the correspondence between the DR process and the electron impact excitation of the recombining target or the core ion. The DR collision strength Ω_{DR} rises exactly up to the EIE collision strength at the threshold energy of the core level. Hence we have the *continuity condition* between DR and EIE [151, 186],

$$\lim_{n\to\infty} \Omega_{DR}(n) = \lim_{k^2\to 0} \Omega_{EIE}(k^2). \quad (7.45)$$

Equation 7.45 is verified by actual computations in Fig. 7.5(a): the peak DR collision strength <Ω(DR)> at each resonance series limit of S IV agrees precisely with Ω(EIE) at that threshold level. Equation 7.45 also provides an accuracy check on both DR and EIE calculations, and the possible importance of long range multipole potentials, partial wave summation, level degeneracies at threshold and other numerical inaccuracies.

Formally, the DR probability from the generalized scattering matrix, Eq. 7.35, used to obtain detailed resonant structures in Fig. 7.5(b), is given by the following expression [133, 172].

$$P_\alpha = G(\nu) \sum_{\gamma} \left\{ \left(\sum_{\gamma'} \chi'_{\alpha\gamma'} \mathbf{N}_{\gamma\gamma'} \right) \times \left[\frac{1}{\chi_{\gamma\gamma} - g(\nu)\exp(-2\pi i\nu)} \right] \times \left[\frac{1}{\chi^*_{\gamma\gamma} - g(\nu)\exp(+2\pi i\nu)} \right] \times \left(\sum_{\gamma'} \chi'^{*}_{\gamma'\alpha} \mathbf{N}^*_{\gamma\gamma'} \right) \right\}. \quad (7.46)$$

The summations go over closed channels $\gamma\gamma'$ contributing to DR. The sum over the diagonal elements of all open channels, linked to the ground state of the target ion, gives the probability of DR through radiative transitions between the excited states and the ground state.

Whereas the high-n levels of group (B) contribute via DR to the (e + ion) recombination rate coefficient at high temperatures, there is also their background RR-type contribution, which arises from photorecombination. The unified method includes this background contribution as 'top-up', obtained from photoionization cross sections computed in the hydrogenic approximation, which is valid for very high-n levels. The background top-up is generally negligible at high temeratures, compared with resonant DR, but significant in the low-temperature

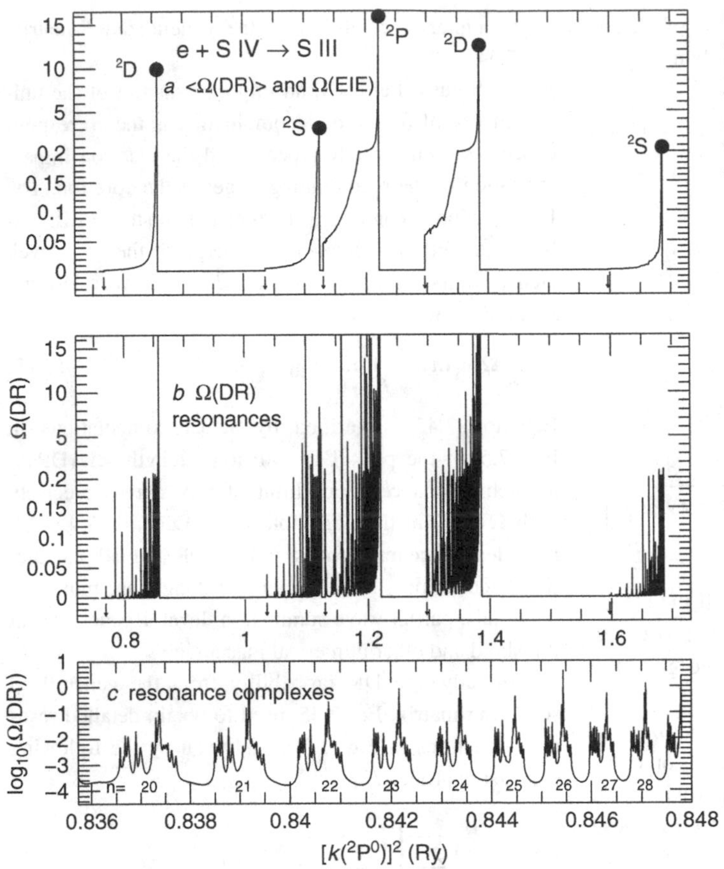

FIGURE 7.5 Dielectronic recombination collision strengths Ω(DR) for (e + SIV) → SIII) [189]. (a) resonance-averaged $<\Omega(DR)>$ in the photoelectron energy regions from $\nu = 10$ (pointed by arrows) where it is negligible, up to the specified excited states (thresholds) where they attain peak values; (b) detailed DR resonances, which become denser as the energy approaches excited state thresholds; (c) an expanded region below the first excited state ^{2}D of the recombining ion SIV to illistrate the extensive complexity of resonant complexes contained in the narrow DR resonances in (b).

range when electrons have very low energies insufficient for core excitation; in that case the electrons recombine into very high-n levels. For an ion with charge z, the z-scaled formula in the hydrogenic approximation is $\alpha_R(z, T) = \alpha_R^H(1, T/z^2)$, in terms of the recombination rate coefficient for neutral hydrogen. The $\alpha_R(z, T)$, for levels with n = 10 to 800 have been computed [133] using H I photoionization cross sections [143], and for levels n = 801 to ∞ using the difference rule [190]

$$\Delta(n) = \alpha_n \left(\frac{n}{n+1}\right)^3 \left(\frac{1+n}{2}\right). \tag{7.47}$$

7.4.4 Multiple resonant features

The total contribution for recombination to the infinite number of recombined levels shown in Fig. 7.2 gives the unified recombination rate coefficients $\alpha_R(T)$, Eq. 7.11. The basic form of $\alpha_R(T)$ is that it starts with a high value of the background RR part of the rate coefficient at very low temperatures, and decreases exponentially until attenuated by a large high temperature, 'DR bump'. At very high temperatures, $\alpha_R(T)$ decreases exponentially and, monotonically (linearly on a log scale). An example of such typical behaviour is seen for recombination of highly charged He-like ions, such as (e + Si XIV → Si III), shown in Fig. 7.6 (a) [191].

However, for more complex systems, resonances can introduce multple bumps, as in recombination of C-like argon (e + Ar XIV → Ar XIII) [192], shown in Fig. 7.6(b). The mutliple DR bumps arise from resonance complexes (as in Fig. 7.7 at high energies), spread across several excited core levels of the recombining ion Ar XIV. The first small group of resonances at 21–25 Ry in Fig. 7.7 gives rise to the first bump in Fig. 7.6(b). Successively higher resonance groups result in further enhancement of $\alpha_R(T)$ at corresponding temperatures. The multiple bumps in the unified $\alpha_R(T)$ are particularly discernible when compared with the DR-only calculations, as shown.

Mutiple Rydberg series of resonances due to several ion thresholds might contribute to $\alpha_R(T)$ over extended energy ranges. That, in turn, enhances the energy-integrated cross sections, and rate coefficients $\alpha_R(T)$, Eqs 7.9 and 7.10, in specific temperature regions. An extension of the concept illustrated in Fig. 7.3, with

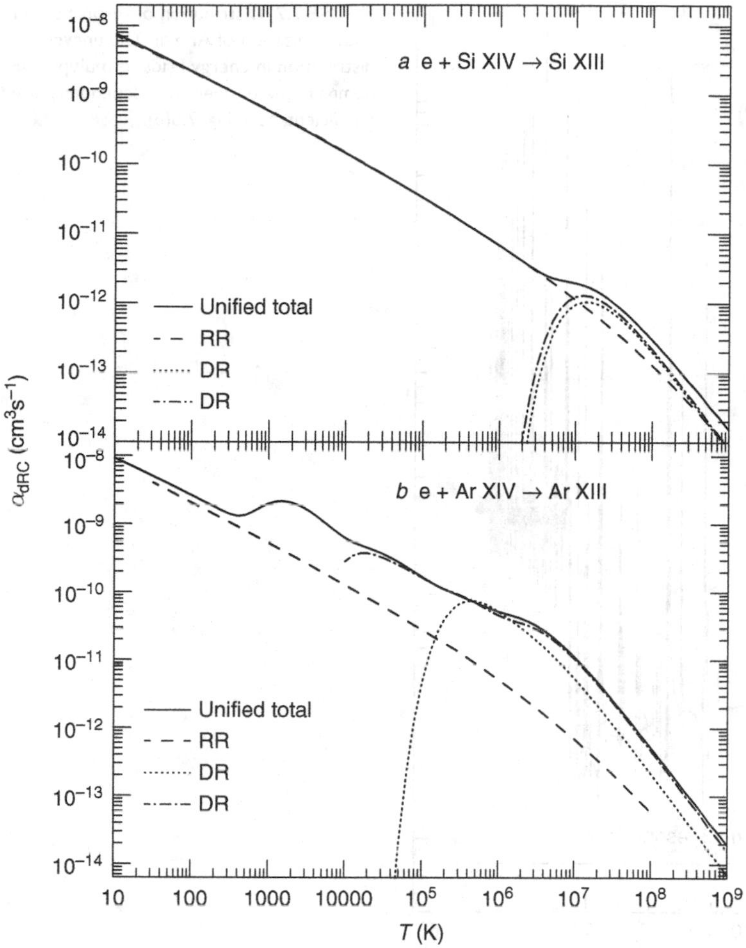

FIGURE 7.6 Unified total recombination rate coefficients $\alpha_R(T)$ (solid curves) for (a) (e + Si XIV) → Si XIII, [146], and (b) (e + Ar XIV → Ar XIII) [192]. While recombination to He-like Si XIII shows a single DR bump, there are multiple bumps for Ar XIII. Other curves represent independently calculated rate coefficients, such as RR (dashed), for Si XIII [175] and Ar XIII [195], DR (dotted), for both ions [196], and (dot-dashed) for Si XIII [197] and Ar XIII [198].

respect to low-T DR enhancement, is seen if several groups of resonances are interspersed throughout, from low to high energies.[3]

7.4.5 Comparison between experiment and theory

As for photoionization, the extensive resonance structures in recombination cross sections can be studied experimentally to verify theoretical results and physical effects. Detailed measurements with high resolution are now possible with sophisticated experimental set-ups using ion storage rings, such as the Test Storage Ring (TSR) in Heidelberg (e.g., [193]), and CRYRING in Stockholm (e.g., [194]). In recent years, a number of *absolute* cross sections have been measured to benchmark theoretical calculations. Experimental results naturally measure the total, unified cross sections, without separation into RR and DR. That is, of course, also the way (e + ion) recombination occurs in astrophysical and laboratory plasmas. The measured cross sections are convolved over the incident electron beam width to obtain an averaged $<v\sigma_{RC}>$, as in Eq. 7.12. Although these can be directly compared with the unified $\alpha_R(T)$, in order to compare them with the many RR and DR calculations in literature the experimental cross sections are processed to separate out the DR contribution. We discuss two examples of comparison between theory and experiment to elucidate several physical features.

In Chapter 5 we had pointed out the astrophysical importance of the Ne-like Fe XVII in many high-temperature sources, such as the solar corona and active galactic nuclei. Figure 7.8 demonstrates recombination to Ne-like Fe XVII from Fe XVIII that was measured experimentally at the TSR facility [199], and theoretically reproduced using the unified method [200].

[3] Though this chapter is aimed at a description of (e + ion) recombination *in toto* using the unified framework, it is common to refer separately to non-resonant and resonant contributions as RR and DR, when one dominates the other.

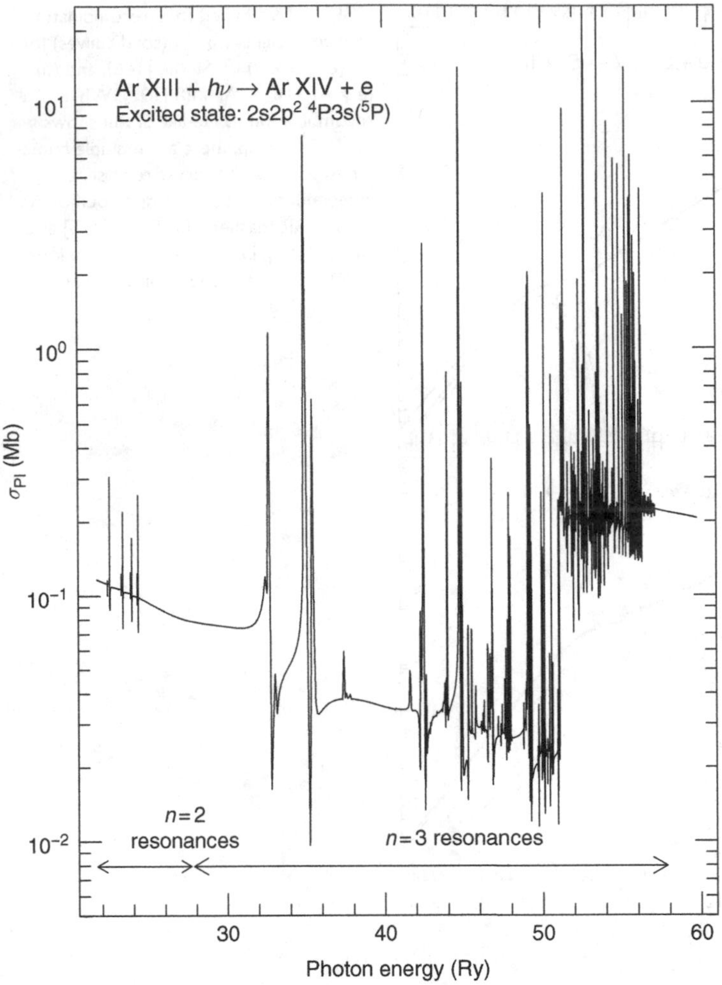

FIGURE 7.7 Distribution of resonances in photoionization of Ar XIII. The uneven distribution in energy leads to multiple DR bumps in the temperature-dependent rate coefficient, as in Fig. 7.6(b).

The ground state of the recombining Fe XVIII core ion is $1s^2 2s^2 2p^5$ $\left({}^2P^o_{3/2}\right)$. The next excited level is the very low-lying upper fine structure level ${}^2P^o_{1/2}$ at 12.8 eV. Also within the $n = 2$ ground complex lies the third level $2s2p^6$ (${}^2S_{1/2}$) at 133.3 eV, which has even parity. Therefore, the Rydberg series of resonances corresponds to dipole transitions from the lower two odd parity levels ${}^2P^o_{3/2,1/2} \rightarrow {}^2S_{1/2}$. The experimental results in Fig. 7.8 (bottom panel) display resonant features up to the excited level ${}^2S_{1/2}$. Since a transition between fine structure levels of the same parity within the ground state are forbidden, the resonances up to the ${}^2P^o_{1/2}$ threshold at 12.8 eV are weak and narrow, and the background RR contribution dominates as the photoelectron (or inversely, recombining electron) energy goes to zero. The remainder of the observed resonance complexes in Fig. 7.8 correspond to Rydberg series of autoionizing resonances converging on to the $2s2p^6({}^2S_{1/2})$ threshold. The resonances are clearly found to be grouped according to n-complexes, which are identified. They become narrower but denser with high n (or ν), as the autoionization width decreases as z^2/ν^3. The recombination cross section, mainly due to DR since the background RR is negligible, peaks at the threshold ${}^2S_{1/2}$.

For comparison, the top panel of Fig. 7.8 shows the detailed unified recombination cross section σ_{RC} from all recombined levels of Fe XVII – photorecombination into the group (A) levels $n \leq n_o$, as well as the DR-only resonances of group (B) with $n_o < n \leq \infty$. For comparison with the experiment (bottom panel) the theoretical cross sections are convolved over the Gaussian beam width of 20 meV to obtain the averaged $< \alpha_R(E) > = < v\sigma_{RC} >$ (middle panel). While it is theoretically feasible to resolve individual resonances (top panel), the monochromatic bandwidth in the experimental detector is much wider than the resonance widths.

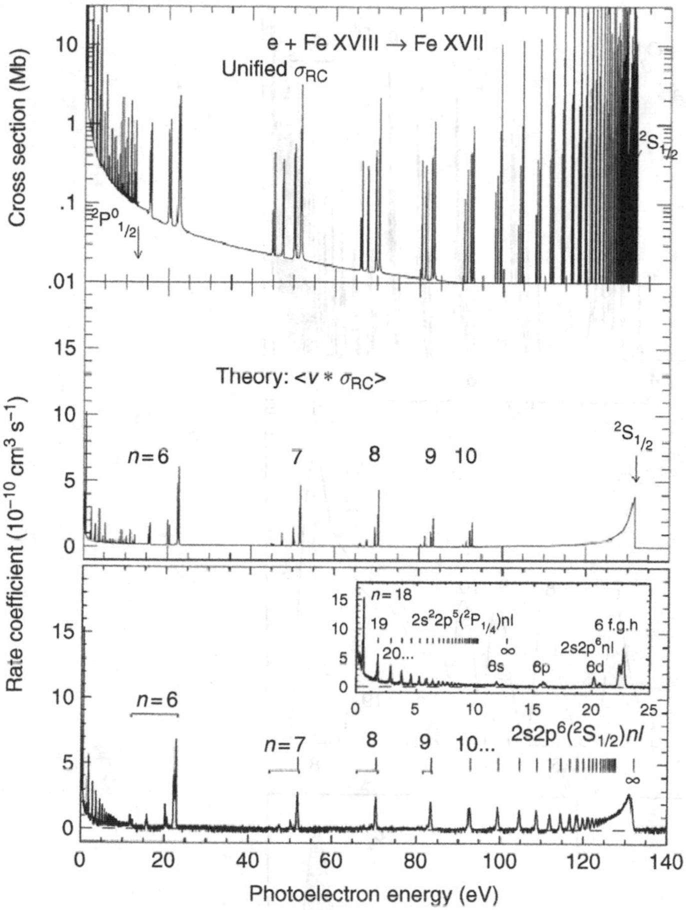

FIGURE 7.8 Unified (e + Fe XVIII) → Fe XVII recombination cross sections (upper panel) with detailed resonance complexes below the $n = 2$ thresholds of Fe XVIII [200]; Gaussian averaged over a 20 meV FWHM (middle panel); experimental data from ion storage ring measurements (bottom panel) [199].

The unified calculations were carried out using the Breit–Pauli R-matrix (BPRM) method described in Chapter 3. The convolved recombination spectrum in Fig. 7.8 shows very good detailed agreement with measurements. The temperature dependent $\alpha_R(T)$ may also be computed, and found to be in good agreement for low temperatures $T_e < 10^6$ K. The temperature of maximum abundance of Fe XVII in collisionally ionized plasmas (discussed later) is around 5 $\times 10^6$ K. The experimental results are in the relatively low-energy range and do not account for the far stronger resonances resulting from the $n = 3$ levels in the core ion Fe XVIII, as has been considered in later works [201].

Another example of reombination, with an apparently simple ion (e + C IV) → C III, again shows that there are complex physical effects related not only to the energy distribution of low-n resonances, but also ionization of high-n levels by external fields in laboratory conditions, and expectedly in astrophysical plasmas. Figure 7.9 displays the measured DR cross section on the ion storage ring TSR [202], compared with the unified BPRM calculations in the relativistic close coupling approximation with fine structure [203]. The bottom panel in Fig. 7.9 shows the experimental results. Of particular interest in this case, recombination with Li-like C IV to Be-like C III, was the presence of the large autoionizing resonance complex $2p4\ell$ of Be-like configuration in the near-threshold region (the inset in the bottom panel). As for Fe XVII discussed earlier, the top panel in Fig. 7.9 is the detailed BPRM cross section, and the middle panel shows those cross sections convolved over the experimental beam width. The effective integrated value of the unified cross sections over the resonance complex $2p4\ell$ agrees with the experimental results [194, 202] (not shown). The unified results lie between the two sets of experiments and agree with each to about 15%, within the uncertainties in measurements.

In addition to the high-resolution needed to study resonances, experimental measurements face another challenge. There is an infinite number of autoionizing resonances that contribute to DR. But as $n \rightarrow \infty$ resonances are further ionized by external fields present in the

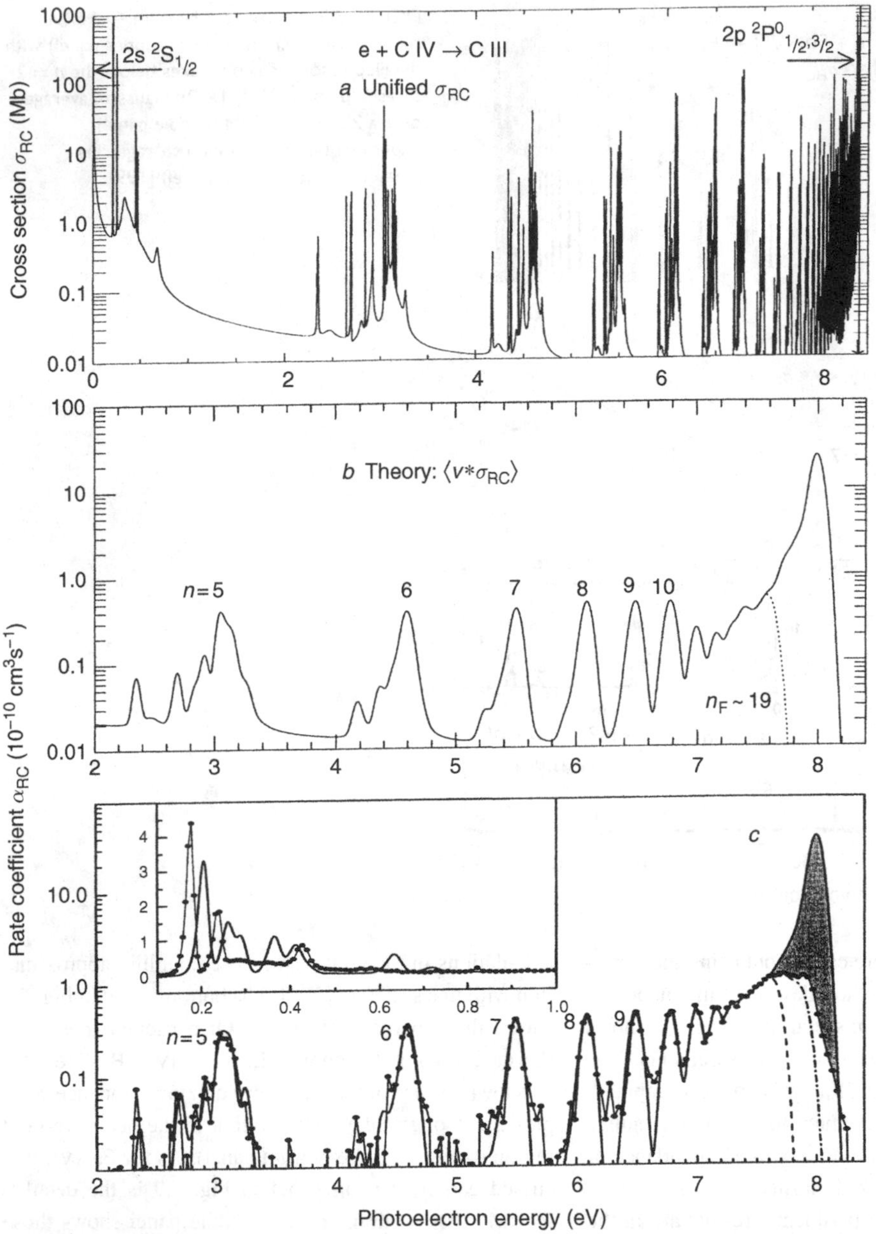

FIGURE 7.9 (a) Unified (e + CIV) → CIII recombination cross section σ_{RC} with detailed resonance structures [203]; (b) theoretical rate coefficient ($v \cdot \sigma_{RC}$) convolved over a Gaussian with experimental FWHM at the Test Storage Ring (TSR) (c) the experimentally measured rate coefficient. The unified σ_{RC} in (a),(b) incorporate the background cross section eliminated from the experimental data in (c). The dashed and dot-dashed lines represent approximate field ionization cut-offs.

apparatus (beam focusing magnetic fields for instance). Therefore, experimental results include contributions only from a *finite* set of resonances up to some maximum value n_F corresponding to field ionization. Theoretically, this implies that a cut-off value of n_F must be introduced to truncate the otherwise infinite sum for comparison with measured cross sections. The high-n resonances in Fig. 7.9, converging onto the $1s^22p\ (^2P^o{}_{1/2,3/2})$ thresholds of the target ion C IV, are found to be in good agreement with the average measured rate coefficient, with $n_F \sim 19$. As shown in the middle panel of Fig. 7.9, the total up to $n = \infty$ also agrees well with the experimental results up to $n_F \approx 19$, but is further augmented by theoretical estimates (shaded portion in the bottom panel).

7.5 Photorecombination and dielectronic recombination

The unified (e + ion) recombination rate coefficients $\alpha_R(T)$ are valid over a wide range of energies and temperatures for all practical purposes. In contrast, separate calculation of RR and DR rate coefficients are carried out in different approximations valid for limited temperature ranges, such as low-temperature DR, high-temperature DR and RR. Moreover, division is sometimes made between $\Delta n = 0$ and $\Delta n \neq 0$ transitions in DR. However, the main problem with separate treatment of RR and DR is more fundamental. Even if the DR treatment is satisfactory, the calculation of RR rate coefficient would require the calculation of unphysical photoionization cross sections *without resonances*, computed in simpler approximations, such as the central-field method that does not include resonances, or a 'one-channel' calculation. On the other hand, a self-consistent and physical treatment of (e + ion) recombination is enabled by the use of coupled channel wavefunctions.

The close-coupling treatment of (electron–ion) recombination is a unified and integrated approach to photorecombination (PR), DR and electron impact excitation (EIE). Figure 7.10 illustrates the inter-relationships required by conservation-of-flux and unitarity conditions for PR, DR and EIE for the (e + C V) → C IV system [204]. The cross sections for all three processes, computed independently but with the same close coupling wavefunction expansion, are continuous functions of energy. The PR cross sections include the background non-resonant contribution as well as the resonances (left of the dashed line in Fig. 7.10), whereas the DR cross sections (right of the dashed line), computed using the coupled channel DR theory, neglect the background contribution. The two cross sections, the PR and DR corresponding to group (A) and group (B) resonances, respectively, match smoothly at $\nu \approx 10.0$, showing that the background contribution is negligible compared to the resonant contribution at high $n > 10$.

Furthermore, the DR cross sections rise exactly up to the EIE cross section at the threshold of excitation, in accordance with the continuity equation, Eq. 7.45, between DR and EIE collision strengths. The DR cross section in Fig. 7.10 at the series limit $2^1P^o{}_1$ agrees precisely with the independently determined value of the electron impact excitation cross section (filled circle) for the dipole transition $1^1S_0 - 2^1P_1$, as required by the unitarity condition for the generalized S-matrix, Eq. 7.35, and conservation of flux, leading up to the continuity condition, Eq. 7.45. The continuous transition between the PR, DR and EIE cross sections serves to validate the accuracy of the unified theory of PR and DR. The DR cross sections are, on the one hand, consistent with an extensively detailed coupled channel treatment

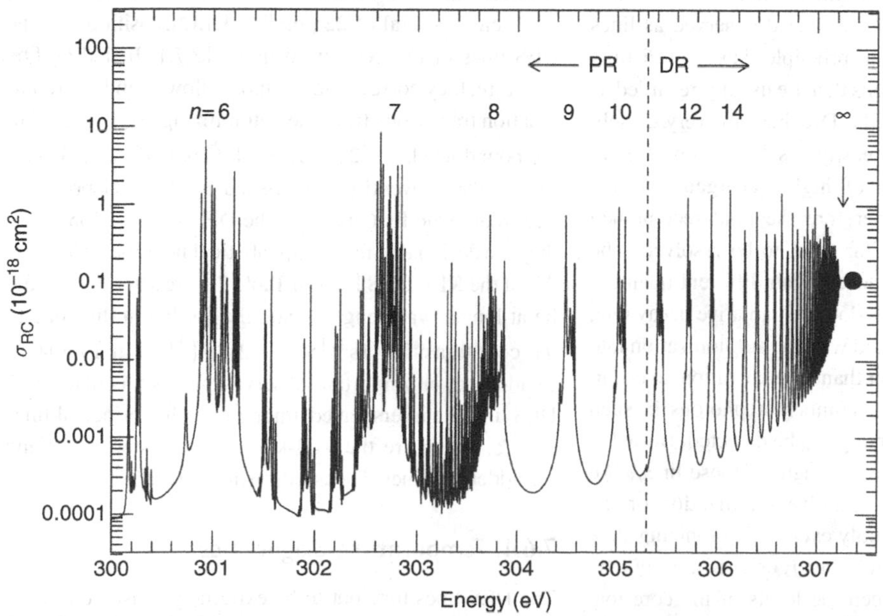

FIGURE 7.10 Photorecombination (PR), DR, and excitation cross sections [204], as derived from photoionization calculations (left of the dashed line), and the dielectronic (DR) cross sections (right of the dashed line) for (e + C V) → C IV; the filled circle represents the near-threshold value of electron impact excitation cross section for the dipole transition $1^1S_0 - 2^1P^o_1$ in C V.

of photorecombination, until an energy region where background recombination is insignificant, and, on the other hand, consistent with the threshold behaviour at the EIE threshold.

The resonances in Fig. 7.10 are radiatively damped using the perturbative technique outlined in Section 6.9. Owing to the interaction with the radiation field, the autoionizing resonances are broadened, smeared and wiped out (in that order) as $n \to \infty$. At sufficiently high-n the resonant contribution (DR) is very large compared with the background, non-resonant photorecombination (PR) cross section. In the unified method of electron–ion recombination, for $n > n_{\max}$, we employ Eqs 7.44 and 7.46 to compute the averaged and the detailed DR cross sections. The agreement and the continuity between the three sets of data in Fig. 7.10 demonstrate the unification of the inter-related processes of PR, DR and EIE. This further underlines the unification of electron–photon–ion processes illustrated in Fig. 3.5.

7.6 Dielectronic satellite lines

The (e + ion) recombination process involves an infinite number of resonances. Recombination through a particular autoionizing state i (Eq. 7.14), gives rise to the emission of a photon corresponding to the energy difference $\Delta E\ (i\text{–}j)$, where j is the final recombined bound state. There is therefore an infinite number of 'lines' due to DR. But while they are indeed observed as lines in the emission spectrum, in principle these are distinct from bound–bound transitions that are usually regarded as line transitions. Generally, the DR lines are very closely spaced in energy and not easily resolved. However, in recombination with low-n of highly charged ions the autoionizing levels, and therefore the final recombined levels, may be sufficiently far apart to be resolved. The energy difference between levels with different quantum numbers increases with z, as does the radiative decay rate, which must be high compared with the autoionization rate for radiative emission, rather than autoionization, to occur.

Emission of a sufficient number of photons in such DR lines may be intense enough to be detected. But what precisely is the energy or wavelength of these lines? We recall from Eq. 7.14 that radiative stabilization of the (e + ion) system from a doubly excited autoionizing state i, or $p(n\ell)$, to a stable bound state j, or $g(n\ell)$, occurs via a radiative transition between the levels of the core ion, i.e., $p \to g$; the electron labelled $n\ell$ remains a spectator with no change in quantum numbers. Primarily, the energy of the DR line is the transition energy $\Delta E\ (p\text{–}g)$, the principal transition in the core ion, which is usually a strong dipole allowed transition with large radiative decay rate. However, it is affected slightly by the presence of the spectator electron, and the energy is marginally less than that of the principal core transition. That is because following DR the (e + ion) system has one more electron than the core ion. For example, the $1s^2 - 1s2p$ dipole transition in He-like Fe XXV is at 6.7 keV, less energy than the $1s - 2p$ transition in H-like Fe XXVI at 6.9 keV. Conversely, the wavelengths of the DR lines are longer than the wavelengths associated with the core transitions. Therefore, they appear as *satellite lines* to the principal line in the recombination spectrum. Such DR lines are referred to as *dielectronic satellite lines* (DES).

The DES lines are commonly observed in the spectra of high-temperature sources, such as solar flares or fusion devices. The most prominent example of DES lines in astrophysics and many laboratory sources is the $K\alpha$ complex of lines formed by DR of the two-electron He-like to three-electron Li-like iron: (e + Fe XXV) $\to$ Fe XXIV. The DR transitions are between the doubly excited autoionizing levels $1s2l2l'$ and bound states of Li-like configuration of Fe XXIV. Because there are two electrons in the n = 2 orbitals, and one in 1s, the autionizing configuration $1s2l2l'$ is designated as KLL, with both L-electrons in excited levels. The recombined state is formed with principal transitions $1s2l \to 1s2s$, $1s2p$ in the He-like core ion Fe XXV, ending up in final bound states of Li-like Fe XXIV with configurations $1s2l2l' \to 1s^22s$, $1s^22p$, $1s2p^2$.

There is a total of 22 possible KLL transitions for the DES lines of Fe XXV, listed in Table 7.1. Including fine structure, they correspond to dipole allowed and intercombination transitions from the autoionizing levels $1s2l2l'$ to the bound levels $1s^22s\ (^2S_{1/2})$, $1s^22p\ \left(^2P^o_{1/2,3/2}\right)$. Following the convention established by A. H. Gabriel and C. Jordan, who first analyzed the DES spectra [205, 206], they are designated by the alphabetical notation $a, b, ..., v$. All of the KLL DES lines in Table 7.1 are satellites of, and lie at longer wavelengths than, the He-like w-line due to the core transition $1s^2(^1S_0) - 2s2p\left(^1P^o_1\right)$ at 1.8504 Å. In addition, an important observational fact is that all 22 DES lines lie interspersed among the other principal lines x, y, z, which are the well-known intersystem (x,y) and forbidden (z) lines discussed earlier in Chapter 4.

7.6.1 Temperature diagnostics

The DES lines turn out to be extremely sensitive to temperature variations. Since KLL DES spectra are formed in high-temperature plasmas, they are often useful as temperature diagnostics of sources, such as solar flares

TABLE 7.1 Dielectronic satelline lines of He-like iron: (e + Fe XXV) → Fe XIV. The 22 KLL autoionizing resonance transitions to Li-like Fe XIV bound levels are labelled *a* to *v* [206]. The last four transitions (*w, x, y, z*) are the principal bound–bound transitions in the core ion Fe XXV. The columns represent the key notation for the DES lines [206], the transition, computed energy E(P) [208], experimental energy E(X) [209], DES line strength S_s: computed from the unified method S(P) [208] and independent resonance approximations S^a [210] and S^b [211].

Key	Transition	E(P)	E(X)	S(P)	S^a	S^b
a	$1s2p^2(^2P_{3/2}) \rightarrow 1s^22p\left(^2P^o_{3/2}\right)$	4685.3	4677.0	6.12	6.40	
b	$1s2p^2(^2P_{3/2}) \rightarrow 1s^22p\left(^2P^o_{1/2}\right)$	4685.2	4677.0	0.21	0.11	0.13
c	$1s2p^2(^2P_{1/2}) \rightarrow 1s^22p\left(^2P^o_{3/2}\right)$	4666.8	4658.6	0.017	0.02	0.02
d	$1s2p^2(^2P_{1/2}) \rightarrow 1s^22p\left(^2P^o_{1/2}\right)$	4666.9	4658.6	0.076	0.07	0.07
e	$1s2p^2(^4P_{5/2}) \rightarrow 1s^22p\left(^2P^o_{3/2}\right)$	4646.6	4639.0	4.85	4.80	4.28
f	$1s2p^2(^4P_{3/2}) \rightarrow 1s^22p\left(^2P^o_{3/2}\right)$	4638.8	4632.9	0.31	0.20	0.26
g	$1s2p^2(^4P_{3/2}) \rightarrow 1s^22p\left(^2P^o_{1/2}\right)$	4638.8	4632.9	0.01	4.5×10^{-4}	4.0×10^{-3}
h	$1s2p^2(^4P_{1/2}) \rightarrow 1s^22p\left(^2P^o_{3/2}\right)$	4629.9	4624.6	6.0×10^{-3}	1.8×10^{-4}	2.1×10^{-4}
i	$1s2p^2(^4P_{1/2}) \rightarrow 1s^22p\left(^2P^o_{1/2}\right)$	4629.9	4624.6	0.08	0.04	0.02
j	$1s2p^2(^2D_{5/2}) \rightarrow 1s^22p\left(^2P^o_{3/2}\right)$	4672.1	4664.1	27.22	29.15	27.22
k	$1s2p^2(^2D_{3/2}) \rightarrow 1s^22p\left(^2P^o_{1/2}\right)$	4664.4	4658.1	18.40	19.60	18.60
l	$1s2p^2(^2D_{3/2}) \rightarrow 1s^22p\left(^2P^o_{3/2}\right)$	4664.4	4658.1	1.44	2.32	1.79
m	$1s2p^2(^2S_{1/2}) \rightarrow 1s^22p\left(^2P^o_{3/2}\right)$	4704.7	4697.7	2.74	2.91	2.56
n	$1s2p^2(^2S_{1/2}) \rightarrow 1s^22p\left(^2P^o_{1/2}\right)$	4704.7	4697.7	0.14	0.13	0.09
o	$1s2s^2(^2S_{1/2}) \rightarrow 1s^22p\left(^2P^o_{3/2}\right)$	4561.5	4553.4	0.89	0.84	0.91
p	$1s2s^2(^2S_{1/2}) \rightarrow 1s^22p\left(^2P^o_{1/2}\right)$	4561.5	4553.4	0.88	0.85	0.92
q	$1s2p^1(P^o)2s\left(^2P^o_{3/2}\right) \rightarrow 1s^22s(^2S_{1/2})$	4624.1	4615.3	0.08	0.11	0.02
r	$1s2p^1(P^o)2s\left(^2P^o_{1/2}\right) \rightarrow 1s^22s(^2S_{1/2})$	4612.6	4604.9	3.80	3.13	3.62
s	$1s2p^3(P^o)2s\left(^2P^o_{3/2}\right) \rightarrow 1s^22s(^2S_{1/2})$	4639.8	4633.2	1.29	0.15	0.90
t	$1s2p^3(P^o)2s\left(^2P^o_{1/2}\right) \rightarrow 1s^22s(^2S_{1/2})$	4637.3	4631.2	5.52	6.35	5.83
u	$1s2p^3(P^o)2s\left(^4P^o_{3/2}\right) \rightarrow 1s^22s(^2S_{1/2})$	4577.5	4570.1	0.16	0.17	0.02
v	$1s2p^3(P^o)2s\left(^4P^o_{1/2}\right) \rightarrow 1s^22s(^2S_{1/2})$	4572.2	4566.3	0.06	0.03	0.02
w	$1s2p\left(^1P^o_1\right) \rightarrow 1s^2(1^1S_0)$					
x	$1s2p\left(^3P^o_2\right) \rightarrow 1s^2(1^1S_0)$					
y	$1s2p\left(^3P^o_1\right) \rightarrow 1s^2(1^1S_0)$					
z	$1s2s(^3S_1) \rightarrow 1s^2(1^1S_0)$					

and laboratory plasmas with $T_e > 10^6$ K. The reason that a given DES line is more sensitive to temperature than, say the principal w-line, is because DES lines are excited *only* by colliding electrons that have precisely the resonance energies corresponding to the DES autoionizing levels, as shown for Fe XXV recombination in Table 7.1. On the other hand, the w-line is due to core excitation of the bound levels, i.e., $1^1S_0 - 1^1P^o_1$, which depends on *all* electron energies in the Maxwellian distribution above the excitation threshold $E(2^1P^o)$. Therefore, the DES line intensity decreases with temperature relative to the w-line. The ratio of a DES line to the w-line is

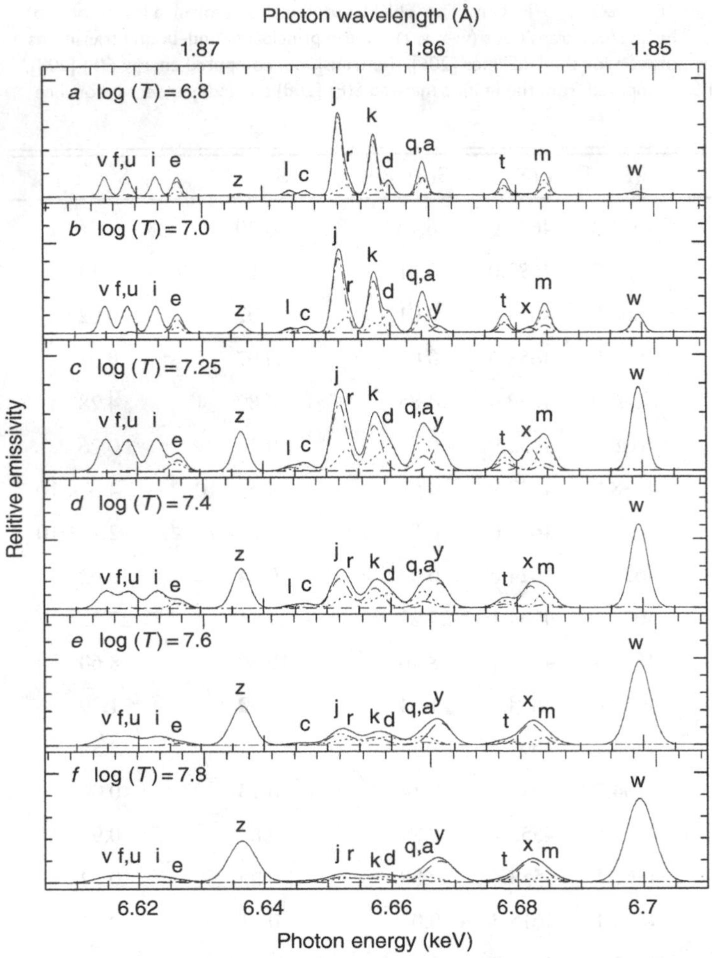

FIGURE 7.11 Theoretically simulated spectrum of the dielectronic satellite lines of He-like FeXXV, and the principal lines w, x, y, z [207].

among the most temperature sensitive diagnostics available. That is particularly so because the w-line is due to the strong dipole transition in He-like ions, which can be excited by electrons out to very high energies (recall that the collision strengths for dipole transitions increase as $\Omega \sim \ln E$, as opposed to those for forbidden or intercombination transitions that decrease with energy, see Chapter 5).

Figure 7.11 shows a theoretically simulated spectrum [207] of the DES and principal lines given in Table 7.1, at different electron kinetic temperatures where Fe XXV is abundant in high-temperature sources. At the lowest temperature shown, log T_e = 6.8 (topmost panel), the DES lines j and k are much stronger than the principal w-line. That is because the Maxwellian distribution of electron energies at $T_e \sim 10^{6.8}$ K is insufficient to excite the w-line substantially. But as T_e increases (from the top panel down in Fig. 7.11), the w-line gains in intensity relative to the DES lines. As the temperature increases, the w-line gets stronger, since there are more high-energy electrons in the tail of the Maxwellian distribution to excite the $2^1P^o_1$ level. We also note in passing that the behaviour of the other principal lines, x, y, z, is very valuable not only in temperature diagnostics, but also that of electron density and ionization equilibrium in high-temperature plasmas (discussed in detail in Chapter 8). The DES spectra will also be discussed for a variety of temperature–density conditions in non-equilibrium time-dependent sources, such as solar flares, as well as photo-excitation by external radiation fields, in Chapter 8.

If we denote the recombination rate coefficient for a DES line by $\alpha_s(T)$, then its intensity (related to emissivity) is

$$I_s(i \rightarrow j, T) = \alpha_s(T) n_i n_e , \tag{7.48}$$

where n_i is the density of the target ion and n_e is the electron density. The emissivity of the w-line[4] is related to the the electron impact excitation rate coefficient $q_w(T_e)$. It is the intensity ratio of a DES line to the w-line that is used for temperature diagnostics. This ratio is then

$$\frac{I_s}{I_w} = \frac{\alpha_s}{q_w}, \tag{7.49}$$

where the calculation of electron impact excitation cross sections and rate coefficients is as in Chapter 5. In the next section we describe the calculations of DES line strengths that yield the rate coefficient $\alpha_s(T_e)$.

7.6.2 Dielectronic satellite line strengths

Since photoionization is the inverse process of (e + ion) recombination, all of the 22 KLL DES resonances in Table 7.1 are found in the photoionization cross sections for $h\nu$ + Fe XXIV → e + Fe XXV [212, 213], as shown in Fig. 6.16. Therefore, we may compute recombination cross sections for the satellite lines in the coupled channel approximation using the unified method. Figure 7.12 shows the σ_{RC} with the KLL resonance profiles delineated as function of the recombining electron, or the photoelectron, energy. On the other hand, the isolated resonance approximation, Eq. 7.26, is commonly employed to calculate the DES line intensites [206, 214]. The autoionization rates $A_a(s)$ and the radiative decay rates $A_r(s)$ are independently computed and substituted in Eq. 7.26 to obtain $\alpha_s(T)$. However, as isolated resonances, the DES lines are treated as a single energy transition, which does not consider the natural shape of the autoionizing resonances, and possible interference due to overlapping resonances, as shown in Fig. 7.12. However, for highly charged ions the isolated resonance approximation is a good approximation because background recombination is usually negligible. In the adaptation of the unified method to compute DES line strengths [213], the KLL resonances appear in recombination cross sections as in Fig. 7.12. They are obtained from photoionization cross sections of the final (e + ion) states of the recombined Li-like levels $1s^2 2s\ (^2S_{1/2})$, $1s^2 2p\ \left(^2P^o_{1/2,3/2}\right)$ (Table 7.1). As evident from Fig. 7.12, the DES lines are obtained with full natural line profiles and display overlapping effects. While the top panel (a) in Fig. 7.12 gives the full observable DES spectrum, the lower three panels (b) to (d) also show the spectroscopic identifications of the individual DES lines corresponding to contributing symmetries $J\pi = (1/2)^e, (1/2)^o, (3/2)^o$. The resonances vary over orders of magnitude in cross section, with often overlapping profiles *within each* $J\pi$*-symmetry*; thus, the interference effects due to channel coupling are manifest in the detailed theoretical spectrum.

In the unified method, the DES recombination rate coefficients $\alpha_s(T)$ are obtained directly from recombination cross sections (Fig. 7.12), as

$$\alpha_s(T) = \frac{4}{\sqrt{2\pi m}} \left(\frac{1}{kT}\right)^{3/2} \int_0^\infty \epsilon\, e^{-\frac{\epsilon}{kT}} \sigma_{RC} d\epsilon. \tag{7.50}$$

For the narrow satellite lines, where the exponential factor varies little over the resonance energy range, the rate coefficient can be written as

$$\alpha_s(T) = \frac{4}{\sqrt{2\pi m}} \frac{e^{-\frac{\epsilon_s}{kT}}}{(kT)^{3/2}} \int_{\epsilon_i}^{\epsilon_f} \epsilon\, \sigma_{RC} d\epsilon, \tag{7.51}$$

where ϵ_s is the centre or the mean energy of the resonance line. The DES line strength may be defined as the temperature-independent integral[5]

$$S_s = \int_{\epsilon_i}^{\epsilon_f} \sigma_{RC} d\epsilon. \tag{7.53}$$

The quantity S_s, expressed in eVcm^2, is also useful for comparisons with experimental measurements of DES spectra, such as experiments on an electron-beam-ion-trap [209, 213]. A convenient expression for the satellite recombination rate coefficient in terms of the line strength is

$$\alpha_s(T) = 0.015484 \frac{\epsilon_s e^{-\frac{\epsilon_s}{kT}}}{T^{3/2}} S_s. \tag{7.54}$$

Equation 7.54 is valid for any satellite line with a narrow energy width. A look at the units is useful: we use ϵ in Ry, σ_{RC} in Mb and T in K. In cgs units, Ry = 2.1797×10^{-11} ergs, Mb = 10^{-18} cm^2 giving Ry^{2}Mb = 4.75109 × 10^{-40} erg^{2}cm^2. Figure 7.13 shows computed ratios of the total intensity of the $K\alpha$ complex of KLL DES lines, relative to the w-line. The ratio I_{KLL}/I_w is obtained from the unified approach [208] and two other calculations in the independent resonance approximation [210, 211].

[4] We carefully avoid the term 'resonance' line to refer to the w-line, which astronomers often use for the first dipole allowed transition in an atom. This is to avoid any confusion with physical resonances, which are an important part of this text. However, readers should be aware that they will undoubtedly encounter references to 'resonance' lines in literature, and should be particularly careful when dealing with the DES lines that actually arise from autoionizing resonances.

[5] At this point we note the analogous expression for *resonance oscillator strengths*, defined in Eq. 6.69 as

$$f_r = \int_r \frac{df}{d\epsilon} = \int_r \sigma_{PI} d\epsilon, \tag{7.52}$$

computed from detailed photoionization cross sections.

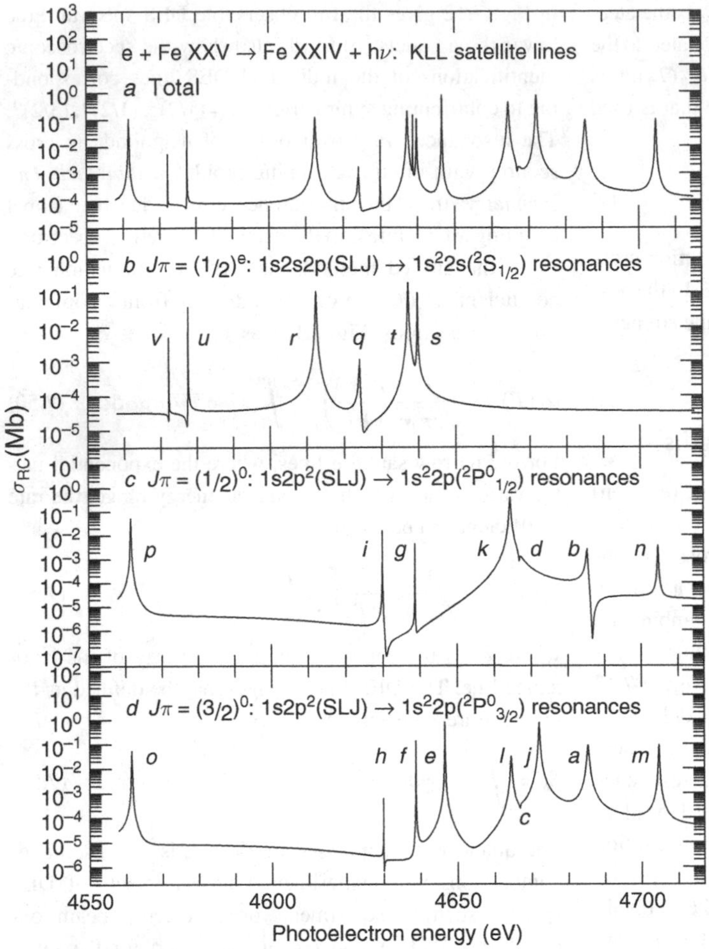

FIGURE 7.12 Satellite lines of Fe XXV in the K_α complex. The top panel (a) shows the total spectrum. The lower three panels show the resolved and identified lines belonging to the final individual recombined level (b) $1s^22s(^2S_{1/2})$, (c) $1s^22p\left(^2P^o_{1/2}\right)$, (d) $1s^22p\left(^2P^o_{3/2}\right)$ [213].

All three curves in Fig. 7.13 appear to converge to good agreement, ~20%, as they approach the temperature of maximum abundance of Fe XXV at $\log(T) \sim 7.4$ in coronal equilibrium (discussed in the next section). However, there are significant differences at lower temperatures.

7.6.2.1 *Correspondence between isolated and unified approximations*

In Eq. 7.53 we made the approximation that the central DES line energy may be given by a single energy ϵ_s, for convenience in comparing with independent resonance approximations where that approximation is inherent. However, the more precise quantity that is computed in the unified method, allowing for energy variation of a resonance, is [213]

$$S_{RC} = \int_{\epsilon_i}^{\epsilon_f} \epsilon\, \sigma_{RC} d\epsilon, \qquad (7.55)$$

where ϵ_i and ϵ_f are the initial and final energies that delimit the extent of the resonance. Such a demarcation is not always possible, or may not be accurate, in the case of weak overlapping resonances, exemplified in Fig. 7.12. Nevertheless, for most DES lines sufficiently strong to be observed in practice, the background contribution is sufficiently small compared with the peak contribution from the resonance that the approximation $S_s = S_{RC}/\epsilon_s$ is valid in most cases.

Using the basic relation Eq. 7.26 for DR, and independent radiative and autoionization rates A_r and A_a, respectively, the satellite recombination rate coefficient correponds to a single autoionizing level into which free electrons may be captured and radiatively decay into a DES line. However, in the independent resonance approximation one needs to consider many pathways for branching between autoionization and radiative decay. Let the capture rate be related to its inverse, the autoionization rate $A_a(i \to m)$ from an autoionizing level i into the continuum m $(\epsilon\ell)$ (Eq. 7.14). Then, the full expression for the

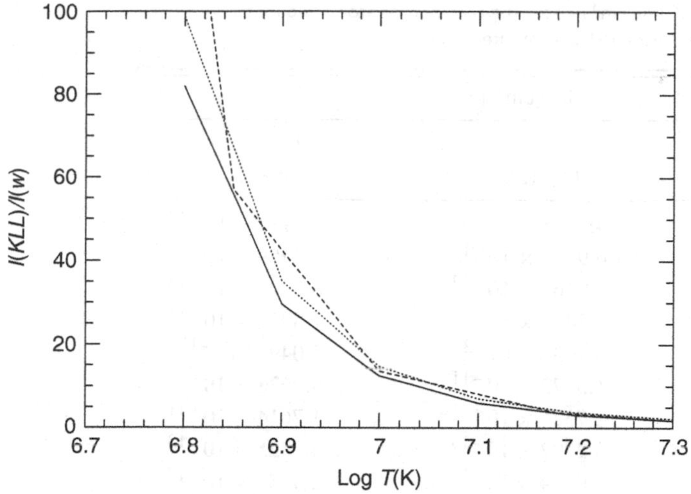

FIGURE 7.13 Comparison of intensity ratios $I(KLL)/I(w)$ of Fe XXV: unified method (solid line [208]) and isolated resonance approximation (dashed [211], dotted [210]).

DES rate coefficient in the independent approximation is,

$$\alpha_s(T) = \frac{g_i}{g_m} \frac{h^3}{2(2\pi m_e kT)^{\frac{3}{2}}} A_a(i \to m) e^{\frac{-\epsilon_s}{kT}} \times \frac{A_r(i \to j)}{\sum_l A_r(i \to l) + \sum_k A_a(i \to k)}. \qquad (7.56)$$

Note that we now specify explicitly the various pathways for autoionization into *excited* levels as final indices; for example, if m denotes the ground level, then k refers to excited levels. Substituting from the unified expression for the DES line strength,

$$4S_{RC} = \frac{g_i}{g_m} \frac{h^3}{4\pi m_e} A_a(i \to n) \times \frac{A_r(i \to j)}{\sum_l A_r(i \to l) + \sum_k A_a(i \to k)}, \qquad (7.57)$$

which gives the autoionization rate $A_a(i \to m)$,

$$A_a(i \to m) = \frac{S_{RC}}{\frac{g_i}{g_m} \frac{h^3}{16\pi m_e} A_r(i \to j) - S_{RC}} \times \left(\sum_l A_r(i \to l) + \sum_{k \neq m} A_a(i \to k) \right). \qquad (7.58)$$

The A_a for other continuum states can be obtained by solving the set of coupled linear equations that arise from Eq. 7.58, provided the unified DES line strengths S_{RC} are known. In the case of KLL lines $\sum_{k \neq m} A_a(i \to k) = 0$, and the expression simplifies to,

$$A_a(i \to m) = \frac{S_{RC}}{\frac{g_i}{g_m} \frac{h^3}{16\pi m_e} A_r(i \to j) - S_{RC}} \times \sum_l A_r(i \to l). \qquad (7.59)$$

7.7 Recombination to H and H-like ions

Before turning our attention to the most important application of photoionization and recombination in astrophysics – ionization balance – we remark on the simplest example of (e + ion) recombination from bare nuclei to H-like ions. Being one-electron systems, only the RR process is relevant. The total recombination rate coefficient of hydrogen can be calculated from photoionization cross sections computed using formulae given in Section 6.4.2. Therefrom, recombination rate coefficients for individual shells up to $n = 800$, and the total, including contributions from the rest of the shells up to infinity, have been computed [133, 215].[6] A sample set of total $\alpha_R(T)$ is given in Table 7.1. For any other hydrogenic ion with charge z, the recombination rate coefficient $\alpha_R(z, T)$ can be calculated using the z-scaled formula $\alpha_R(z, T) = \alpha_R(H, T/z^2)$, where $\alpha_R(H, T/z^2)$ is the recombination rate coefficient of neutral hydrogen at the equivalent temperature T/z^2.

7.8 Ionization equilibrium

Photoionization and (e + ion) recombination determine the ionization balance in low-temperature-density photoionized sources, such as H II regions ionized by an external radiation field (Chapter 12). The assumption is

[6] Hydrogenic recombination rate coefficients are available from the OSU-NORAD Atomic Data website [142].

TABLE 7.2 Recombination rate coefficients $\alpha_R(T)$ (RRC) (in $cm^3 s^{-1}$) for the recombined atoms and ions. Complete sets of $\alpha_R(T)$ are available from the NORAD-Atomic-Data website [142].

Ion	RRC(cm/s)		
$\log_{10} T(K) =$	3.7 5012 K	4 10 000 K	4.3 19 953 K
H	3.1321×10^{-12}	4.1648×10^{-13}	2.5114×10^{-13}
C I	8.6730×10^{-13}	6.9041×10^{-13}	1.2959×10^{-12}
C II	9.0292×10^{-12}	6.0166×10^{-12}	6.0206×10^{-12}
C III	4.1728×10^{-11}	2.3405×10^{-11}	1.8979×10^{-11}
C IV	1.287×10^{-11}	8.103×10^{-12}	5.049×10^{-12}
C V	2.3918×10^{-11}	1.5472×10^{-11}	9.9678×10^{-12}
C VI	4.2088×10^{-11}	2.7326×10^{-11}	1.7614×10^{-11}
N I	8.1021×10^{-13}	5.2797×10^{-13}	4.9325×10^{-13}
N II	4.2084×10^{-12}	3.0754×10^{-12}	2.4883×10^{-12}
N III	1.9095×10^{-11}	1.4609×10^{-11}	1.1918×10^{-11}
N IV	2.8578×10^{-11}	2.2941×10^{-11}	2.0527×10^{-11}
N V	2.267×10^{-11}	1.433×10^{-11}	8.960×10^{-12}
N VI	3.777×10^{-11}	2.432×10^{-11}	1.554×10^{-11}
N VII	5.9457×10^{-11}	3.8764×10^{-11}	2.5028×10^{-11}
O I	4.7949×10^{-13}	3.1404×10^{-13}	2.4927×10^{-13}
O II	3.7524×10^{-12}	2.5156×10^{-12}	1.8260×10^{-12}
O III	1.4113×10^{-11}	1.3332×10^{-11}	1.0768×10^{-11}
O IV	4.7073×10^{-11}	3.9524×10^{-11}	3.0794×10^{-11}
O V	2.5665×10^{-11}	1.9636×10^{-11}	1.8082×10^{-11}
O VI	3.434×10^{-11}	2.185×10^{-11}	1.375×10^{-11}
O VII	5.038×10^{-11}	3.247×10^{-11}	2.078×10^{-11}
O VIII	8.0053×10^{-11}	5.2173×10^{-11}	3.3877×10^{-11}
Fe I	1.4725×10^{-12}	2.2977×10^{-12}	8.7237×10^{-12}
Fe II	4.1778×10^{-12}	4.0159×10^{-12}	3.4231×10^{-12}
Fe III	6.8638×10^{-12}	5.0792×10^{-12}	3.8430×10^{-12}
Fe IV	1.2832×10^{-11}	9.1426×10^{-12}	6.0594×10^{-12}
Fe V	2.5580×10^{-11}	1.6677×10^{-11}	1.0517×10^{-11}
Fe XIII	1.0616×10^{-09}	9.6823×10^{-10}	7.0239×10^{-10}
Fe XVII	4.176×10^{-10}	2.729×10^{-10}	1.778×10^{-10}
Fe XXI	7.4073×10^{-10}	4.6348×10^{-10}	2.9686×10^{-10}
Fe XXIV	7.423×10^{-10}	4.859×10^{-10}	3.160×10^{-10}
Fe XXV	9.105×10^{-10}	6.012×10^{-10}	3.955×10^{-10}
Fe XXVI	1.0703×10^{-09}	7.1077×10^{-10}	4.7048×10^{-10}

not that plasma conditions do not change with time, but that they are sufficiently slowly varying for equilibrium to prevail without affecting spectral or other observable properties. Later, in Chapter 8, we shall also study transient phenomena when non-ionization equilibrium must be considered.

In ionization equilibrium at a given temperature and density, the distribution of an element among various ionization stages can be specified. A considerable simplification occurs in astrophysical models since only a few ionization stages are abundant under a given set of external conditions. The corresponding ionization fractions are the prime output parameters from plasma modelling codes for H II regions: diffuse and planetary nebulae, supernova remnants and broad line regions of active galactic nuclei. The two most common assumptions to describe ionization conditions in such objects are: (i) photoionization equilibrium and (ii) collisional or coronal equilibrium. The dominant ionizing process in the first case is photoionization from the radiation field of an external source, and in the second case, electron impact ionization (Chapter 5). A further assumption is generally made, that the ambient plasma is of sufficiently low density, usually optically thin, so that the ionization balance depends mainly on temperature. That is the case for nebular densities in the range $n_e \sim 10^3 - 10^6$ cm^{-3}, and $n_e \sim 10^8 - 10^{12}$ cm^{-3} in the solar corona and solar flares. We will see later that these densities are well below those needed to achieve thermodynamic equilibrium such as in stellar interiors (cf. Chapters 10 and 11).

While the two ionization processes, photoionization and electron impact ionization, are physically quite different, in low-density plasmas they are both balanced by the same inverse process – (e + ion) recombination – at given electron temperature and usually with a Maxwellian distribution.[7] The total (e + ion) recombination rates may be obtained either as a sum of separately calculated RR and DR rates, or in the unified formulation. In principle, the unified method not only subsumes both the RR and DR processes, but also enables a fundamentally consistent treatment, since both the photoionization and recombination calculations are carried out *using the same set of atomic eigenfunctions*.

7.8.1 Photoionization equilibrium

It is convenient to consider a photoionized nebula (Chapter 12) as an example of photoionization equilibrium. Each point in the nebula is fixed by the balance between photoionization and (e + ion) recombination, as

$$\int_{\nu_0}^{\infty} \frac{4\pi J_\nu}{h\nu} n(X^z)\sigma_{\rm PI}(\nu, X^z)\mathrm{d}\nu = \sum_j n_e n(X^{z+1})\alpha_{\rm R}\left(X_j^z; T\right), \tag{7.60}$$

where J_ν is the intensity of the radiation field; $4\pi J_\nu/h\nu$ is the number of incident photons per unit volume per unit time; $\sigma_{\rm PI}$ is the photoionization cross section of ion X^z at frequency ν; $h\nu_0$ is the threshold ionization energy; $\alpha_{\rm R}\left(X_j^z; T\right)$ is the (e + ion) recombination rate coefficient into level j at electron temperature T; n_e, $n(X^{z+1})$ and $n(X^z)$ are the densities of free electrons and recombining and recombined ions, respectively. In Eq. 7.60, the left side gives the photoionization rate Γ, whereas the right side is the recombination rate. In photoionization equilibrium, the ionic structure of an element X in the ionization stage X^{z+} is given by balance equation (i.e., gain = loss)[8]

$$n(X^{(z-1)+})\Gamma_{z-1} + n_e n(X^{z+1})\alpha_z = n(X^{z+})[\Gamma_z + n_e\alpha_{z-1}]. \tag{7.61}$$

On the left, we have photoionizations from $X^{(z-1)+}$ and recombinations from $X^{(z+1)+}$ *into* X^{z+}, balanced by photoionizations and recombinations *from* X^{z+}. One can map out the ionization structure of a photoionized region through specific ionic stages of key elements. Lines from low ionization stages of iron are commonly observed, and ionization fractions of these ions are of great interest. For instance, we consider a typical planetary nebula (Chapter 12) with a black body ionizing source at an effective temperature of $T_{\rm eff} = 100\,000$ K, an inner radius of 10^{10} cm, and particle density of 3600 cm^{-3}. Using the atomic data for photoionization cross sections from the Opacity Project [37], and recombination rate coefficients from the unified method, the ionization structure of Fe ions obtained from the photoionization modelling code CLOUDY [218], is shown in Fig. 7.14 (solid curves, [219]). It is seen that, under the assumed conditions, Fe IV is the dominant ionization state in most of the nebula. Significant differences are found using photoionization cross sections computed in the central-field model [140], and previously computed RR and DR rate coefficients [220] (dotted and dashed curves). In particular, the relative fraction of Fe V is reduced by nearly a factor of two in the region near the illuminated face of the cloud, compared with less accurate atomic data. This is directly related to some of the features responsible for this difference, such as the presence of a large near-threshold resonance in $\sigma_{\rm PI}$(Fe IV), shown in Fig. 6.13. Proper inclusion of such features also showed that the fraction of Fe VI is increased by almost 40% over previous results.

Table 7.3 gives numerical values of computed averaged ionic fractions, and presents the Fe I–Fe VI fractions averaged over the whole volume of the nebula. Using

[7] However, electron density effects pertaining to level-specific recombination rates, due to collisional redistribution among high-lying levels, are important for recombination line spectra, such as H-like and He-like ions [216, 217].

[8] Readers need to be aware of confusion that may arise due to terminology used for recombination of, or recombination into, an ion. Here, we employ the convention consistent with the unified treatment of photoionization and (e + ion) recombination: the rate coefficient for forming the recombined ion z is α_z from ions X^{z+1}. But often in literature the convention refers to recombining ion and rate coefficient α_{z+1} instead. To wit: in the unified formulation we speak of photoionization and recombination of the single ion C I. Separately however, one may refer to photoionization of C I, but recombination of C II.

TABLE 7.3 Averaged ionic fractions of iron in a model planetary nebula (cf. Chapter 12) at $T_{eff} = 10^5$ K. Unlike the new data, the earlier data employ photoionization cross sections without resonances, and the sum of RR and DR rate coefficients.

	Fe I	Fe II	Fe III	Fe IV	Fe V	Fe VI
Earlier σ_{PI}, α_R	3.3×10^{-4}	0.166	0.109	0.495	0.145	0.080
New σ_{PI}, α_R	1.5×10^{-4}	0.145	0.127	0.470	0.112	0.140

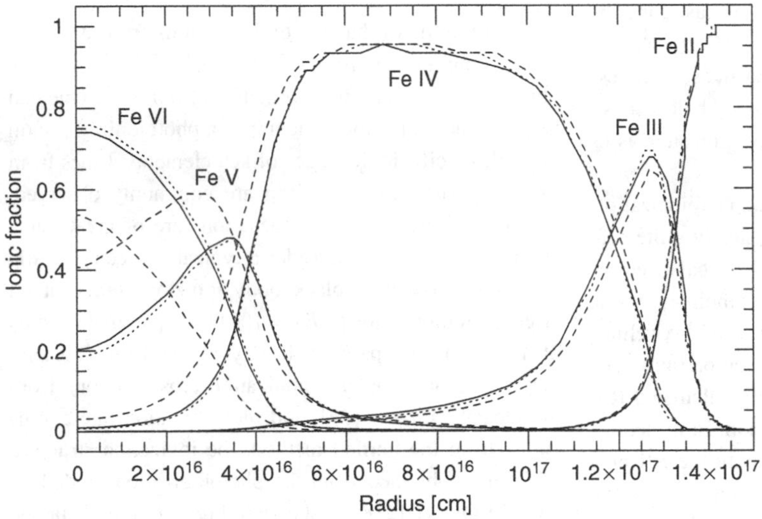

FIGURE 7.14 Computed ionization structure of iron in a typical planetary nebula (see text). Solid curves are ion fractions using the photoionization and recombination data from the Opacity Project and unified method, respectively [98]. The dotted curves are with the same photoionization data, but independently computed RR and DR rate coefficients [220]. The dashed curves are employ the central-field type photoionization data [140], and the independent RR and DR rates.

photoionization cross sections with resonances, and unified recombination rates, the averaged fraction of Fe V decreases by approximately 30%, while the fraction of Fe VI is almost doubled. Furthermore, previous data predicted that the fraction of Fe VI should be about half the fraction of Fe V, but the new data indicate that Fe VI should actually be ~1.3 times more abundant than Fe V. Whereas the details may vary, the main point here is that coupled channel R-matrix photoionization cross sections, and (e + ion) recombination rate coefficients obtained self-consistently with those calculations, yield quite different ionization balance than data from simpler approximations.

7.8.2 Collisional equilibrium

In the absence of an external photoionizing source and radiation field, collisional ionization by electrons is the primary mechanism for ionization. Such a situation occurs in the prototypical case of the solar corona, with moderately high electron densities, $n_e \sim 10^9$ cm^{-3}, that establish what is known as collisional equilibrium, commonly known as *coronal equilibrium*. At first sight, one might think that the radiation field of the Sun might also play a role in (photo)ionizing the corona. But the effective temperature of the Sun is only about 5700 K (recall the discussion of the Sun as a black body in Chapter 1), whereas the kinetic temperatures in the corona exceed 10^6 K, nearly a thousand times higher. Therefore, photoionization by the relatively 'cold' solar radiation field has no appreciable effect on the ionization balance of highly charged ions that are present in the solar corona.

In coronal equilibrium, the relative concentration of the ions of a given element X are determined by the condition

$$C_I^z(T, X^z)n_e n(X^z) = \sum_j n_e n(X^{z+1})\alpha\left(X_j^z; T\right), \quad (7.62)$$

where C_I^z is the rate coefficient for electron impact ionization of an ion of charge $z = Z - N$ into an ion with charge $z+1$, with Z and N as the atomic number and the number of electrons, respectively. Let $\alpha_z = \sum_j \alpha\left(X_j^z\right)$ be the total recombination rate from the ground state of X^{z+1}. The normalization of ionic densities requires that the total density be equal to the sum of those in all ionization stages,

$$n_T(X) = \sum_{z=0}^{z_{max}} n\left(X_g^z\right). \quad (7.63)$$

The ionization rates on the left-hand side of Eq. 7.62 refer to the ground state of the ion. Though recombinations occur to an infinite number of levels on the right side of Eq. 7.62, cascading radiative transitions return the ion core to its ground state. This condition is predicated on the assumption that the radiative and collisional processes proceed on faster timescales than photoionization and recombination. Substantial departures from these equilibrium conditions may result in high densities where some excited states are significantly populated, or in LTE.

In practice, to calculate the ionization fraction we must consider *all* ionization stages. Equation 7.62 implies a set of coupled simultaneous equations. For example, for carbon ions with $Z = 6$, there are seven ionization stages; from $z = 0$ for neutral, to $z = +6$ for fully ionized carbon. For fractions of different ionization stages relative to the total, we have seven simultaneous equations,

$$\begin{aligned}
&\frac{n_0}{n_T} C_0 - \frac{n_1}{n_T}\alpha_1 = 0 \\
&\frac{n_1}{n_T} C_1 - \frac{n_2}{n_T}\alpha_2 = 0 \\
&\ldots \\
&\ldots \\
&\ldots \\
&\frac{n_5}{n_T} C_5 - \frac{n_6}{n_T}\alpha_6 = 0 \\
&\frac{(n_0+n_1+n_2+n_3+n_4+n_5+n_6)}{n_T} = 1
\end{aligned} \qquad (7.64)$$

where the notation n_z represents $n(X^z)$. Now these equations can be solved for any ionization fraction n_i/n_T. If D is the determinant,

$$D = \begin{vmatrix}
C_0 & -\alpha_1 & 0 & 0 & 0 & 0 & 0 \\
0 & C_1 & -\alpha_2 & 0 & 0 & 0 & 0 \\
0 & 0 & C_2 & -\alpha_3 & 0 & 0 & 0 \\
0 & 0 & 0 & C_3 & -\alpha_4 & 0 & 0 \\
0 & 0 & 0 & 0 & C_4 & -\alpha_5 & 0 \\
0 & 0 & 0 & 0 & 0 & C_5 & -\alpha_6 \\
1 & 1 & 1 & 1 & 1 & 1 & 1
\end{vmatrix}, \qquad (7.65)$$

then, n_1/n_T, for example, is

$$\frac{n_1}{n_T} = \frac{1}{D}\begin{vmatrix}
C_0 & 0 & 0 & 0 & 0 & 0 & 0 \\
0 & 0 & -\alpha_2 & 0 & 0 & 0 & 0 \\
0 & 0 & C_2 & -\alpha_3 & 0 & 0 & 0 \\
0 & 0 & 0 & C_3 & -\alpha_4 & 0 & 0 \\
0 & 0 & 0 & 0 & C_4 & -\alpha_5 & 0 \\
0 & 0 & 0 & 0 & 0 & C_5 & -\alpha_6 \\
1 & 1 & 1 & 1 & 1 & 1 & 1
\end{vmatrix}. \qquad (7.66)$$

Similarly, all other ionization fractions of carbon may be obtained. Figure 7.15 presents ionization fractions of oxygen ions in coronal equilibrium using unified recombination rate coefficients (solid curves) [221], compared with those from the individual treatment of RR and DR [222].

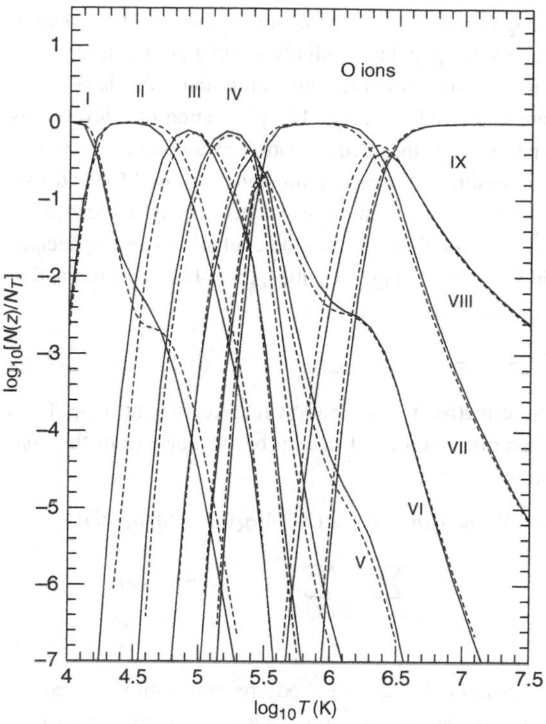

FIGURE 7.15 Ionization fractions of oxygen ions in coronal or collisional equilibrium (solid curves [221]) using unified (e + ion) recombination rate coefficients, compared with earlier results (dashed curves [222]).

The basic features in the two sets of results in Fig 7.15 are similar. However, there are significant differences in the numerical values, particularly at the rapidly varying transition boundaries between adjacent ionization stages. Differences can be seen in the depths of the 'dip', and the high temperature behaviour of ion fractions. For example, the unified recombination rates imply a faster decrease in the abundance of O II with temperature, while that of O III and O VIII rises faster with temperature. These differences could affect the computation of spectral line intensities in astrophysical models.

Finally, when three-body recombination and the density dependences of dielectronic recombination can be neglected, the relative abundances of various stages of ionization are independent of density and functions only of the electron temperature.

7.9 Effective recombination rate coefficient

The effective recombination rate coefficient $\alpha_{\mathrm{eff}}(nl)$ for a recombined level $n\ell$ of an ion is often needed in astrophysical plasmas. Recombination is an important process for populating highly excited atomic levels, and hence the strength of resulting emission lines via radiative transitions and cascades. The population of a level $n\ell$ is dependent on the recombination rate coefficients $\alpha(nl)$, the densities of recombining ions, say $n(X^+)$, and free electrons n_e, in addition to radiative decay A-coefficients and collisional (de)-excitation coefficients among recombined levels. At high densities, three-body recombination can also be important,

$$X^+ + e + e \rightarrow X + e' \quad (7.67)$$

The effective recombination rate coefficient, α_{eff}, for a recombined ion level nl can be obtained from the condition,

$$n_e n(X^+)\alpha_{eff}(nl) = n_e n(X^+)[\alpha_R(nl) + n_e \alpha_t(nl)] + \sum_{n'=n+1}^{\infty} \sum_{l'=l\pm1} [A_{n'l',nl} + n_e C_{n'l',nl}] n_{n'l'}. \quad (7.68)$$

where $\alpha_t(s)$ is the three-body recombination rate coefficient for the level nl, $A_{n'l',nl}$ is the radiative decay rate, $C_{n'l',nl}$ is the (de-)excitation rate coefficient due to electron collision from level $n'l'$ to level nl, and $n_{n'l'}$ is the population density of level $n'l'$. The left-hand side of Eq 7.68 gives the total number of recombined ions formed in state nl per unit time per unit volume.

Extensive work has been carried out [143, 187, 223] in the hydrogenic approximation. At high densities, DR from high-n levels increases, and contributions from highly excited autoionizing levels are often included through extrapolation using $1/\nu^3$ scaling. Most of the available $\alpha_{eff}(s)$ are obtained in LS coupling. They can be used for fine structure compoments through algebraic transformation as [224],

$$\alpha_{eff}(SL_iJ_i \rightarrow SL_fJ_f) = \alpha_{eff}(SL_i \rightarrow SL_f)b(J_i, J_f), \quad (7.69)$$

where

$$b(J_i, J_f) = \frac{(2J_i+1)(2J_f+1)}{2S+1} \begin{Bmatrix} J_i & J_f & 1 \\ L_f & L_i & S \end{Bmatrix}. \quad (7.70)$$

From the $\alpha_{eff}(nl)$, the rate coefficient for a recombination line from transition $i \rightarrow j$ at wavelength λ_{ij} can be obtained as

$$\alpha_{eff}(\lambda_{ij}) = B_{ij}\alpha_{eff}(i), \quad (7.71)$$

where for convenience we replace i for level nl. The B_{ij} is the branching ratio

$$B_{ij} = \frac{A_r^{ij}}{\sum_{k<i} A_r^{ik}}. \quad (7.72)$$

$n_e n(X^+)\alpha_{eff}(\lambda_{ij})$ gives the number of photons emitted in the line cm^{-3} s^{-1}. The *recombination line emissivity* is then

$$\epsilon(\lambda_{ij}) = N_e N(X^+)\alpha_{eff}(\lambda_{ij})h\nu_{ij} \ [\text{erg cm}^{-3}\text{s}^{-1}]. \quad (7.73)$$

7.10 Plasma effects

The effect of external fields and high densities is rather complicated to include in an *ab-intio* treatment. But they are important in practical situations, such as for DR from high-n levels near the Rydberg series limits of resonances. We have already seen an example in Fig. 7.9, where the field-ionization cut-off in the experiment is estimated at $n_F \approx 19$ from theoretical results. However, a general treatment of plasma fields and densities for the calculation of total (e + ion) recombination cross sections is yet to be developed.

8 Multi-wavelength emission spectra

The origin of spectral lines depends on the matter and radiation fields that characterize the physical conditions in the source. However, the lines actually observed also depend on the intervening medium towards the observer. The wide variety of astrophysical sources span all possible conditions, and their study requires both appropriate modelling and necessary atomic parameters. The models must describe the extremes of temperature, density and radiation encountered in various sources, from very low densities and temperatures in interstellar and intergalactic media, to the opposite extremes in stellar interiors and other environments. As such, no single approximation can deal with the necessary physics under all conditions. Different methods have therefore been developed to describe spectral formation according to the particular object, and the range of physical conditions under consideration.

This is the first chapter devoted mainly to astrophysical applications. The theoretical formulation of atomic spectroscopy described hitherto is now applied to the analysis of emission-line observations in three widely disparate regions of the electromagnetic spectrum: the visible, X-ray and far-IR. Examples include some of the most well-known and widely used lines and line ratios. Emission line analysis depends on accurate calculations of emissivities, which, in turn, are derived from fundamental parameters such as collision strengths for (e + ion) excitation and recombination, and radiative transition probabilities. However, spectral models in complicated situations, such as line formation in transient plasmas and in the presence of external radiation fields, assume a level of complexity that requires consideration of a variety of processes and parameters. We will discuss additional examples of emission-line physics in specific cases, such as nebulae and H II regions (Chapter 12), stars (Chapter 10), active galactic nuclei (Chapter 13) and cosmological sources (Chapter 14).

But we first discuss emission lines, in the so-called *optically thin* approximation, where freely propagating radiation is not signficantly attenuated by the material environment. It corresponds to media with sufficiently low densities to enable radiation to pass through without much interaction. In this chapter, we assume that such a situation prevails in the source under observation, which may be exemplified by nebulae and H II regions in general, and the interstellar medium (Chapter 12). Absorption line formation and radiative processes in *optically thick* media, such as stellar atmospheres, are treated in the next two chapters, in Chapters 9 and 10. There we consider several topics, such as radiative transfer and line broadening, that also affect emission lines.

Having described the fundamental atomic processes responsible for spectral formation in previous chapters, we are now in a position to describe the elementary methodology for spectral diagnostics with the help of well-known lines in a wide variety of astronomical sources that can be treated in the optically thin approximation. The two kinds of line observed from an astrophysical source are due to emission from excited atomic levels, or absorption from lower (usually ground state) to higher levels. Observationally, both emission and absorption line fluxes are usually measured relative to the *continuum*, which defines the background radiation field without the emission or absorption line feature(s). Therefore, an emission line may be defined as the *addition* of energy to the continuum flux due to a specific downward atomic transition, and an absorption line as the *subtraction* of energy from the continuum due to an upward atomic transition, as in Fig. 8.1. In this chapter, we describe spectroscopic analysis of some of the most commonly observed emission lines in optically thin sources. The spectral ranges are in the optical, X-ray, and far-infrared (FIR).

The examples discussed are the well-known forbidden optical lines [O II], [S II] and [O III], formed in nebulae and H II regions (Chapter 10). They are characteristic of plasmas with low ionization states of elements at low to moderate temperatures and densities: $T_e \sim 10\,000$–$20\,000$ K

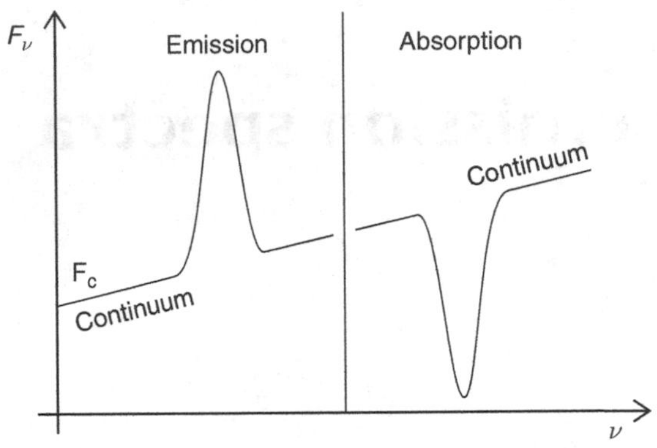

FIGURE 8.1 Emission and absorption line flux relative to the continuum.

$\approx$ 1–2 eV, and $n_e \sim 10^{2-6}$ cm^{-3}. On the other hand, in high temperature (optically thin) plasmas, the He-like ions of many elements give rise to the most prominent lines in X-ray spectroscopy, as described herein. Towards the opposite extreme of low temperatures, forbidden FIR lines from boron-like ions [C II], [N III], etc., are observed in the cold and tenuous interstellar medium and H II regions.

The atomic processes involved in emission and absorption are quite different. Discussion of several specialized, but nevertheless frequent and important, phenomena will be described in other chapters in the context where they occur. Both emission and absorption lines may also be formed from autoionizing levels excited by electron or photon impact. As we saw in the previous chapter, atomic resonances can decay radiatively, giving rise to dielectronic satellite lines. Likewise, resonances in photoionization appear in absorption spectra of astrophysical sources, such as active galactic nuclei (Chapter 13).

8.1 Emission line analysis

Emission lines primarily depend on collisional processes and radiative decay. In low-temperature nebular plasmas with $T_e \approx 1$ eV (=11 600 K) only the low-lying levels are excited. Some of the most prominent lines in nebular spectra are from excited metastable levels within the ground configuration of singly ionized ions of oxygen, sulphur, etc. Therefore such transitions are between levels of the same parity, which implies forbidden lines in the spectra with very small A-coefficients, typically 10^{-4}–10^{-1} s^{-1}, corresponding to magnetic dipole (M1) and electric quadrupole (E2) transitions. The line ratio analysis requires the knowledge of the relevant radiative and collisional rates, discussed next.

8.1.1 Atomic rates and lifetimes

Let us begin with the simple definition of the *rate of a collisional process* with two reactants, say electrons and ions. The rate is defined as the *number of events (reactions) per unit time per unit volume*. In the case of electron–ion scattering (or electron impact excitation)

$$\text{rate}(\text{s}^{-1}\text{cm}^{-3}) = n_e(\text{cm}^{-3}) \times n_i(\text{cm}^{-3}) \times q(\text{cm}^3\text{s}^{-1}). \tag{8.1}$$

The three quantities that determine the rate are the densities of the electrons and ions, n_e and n_i respectively, and the *rate coefficient* q. The units of q are related to the rate defined in Chapter 5 in terms of the cross section or the collision strength averaged over an electron distribution, generally assumed to be a Maxwellian, as

$$q_{ij}(\text{cm}^3\text{s}^{-1}) = <\text{v}_e(\text{cm/s})\ Q_{ij}(cm^2)>. \tag{8.2}$$

The rate per second is inversely related to the lifetime for a given atom. Here, we mean the lifetime *against* a transition to another state. The lifetime against collisional excitation $i \to j$ is

$$\tau_{\text{coll}} = \frac{1}{q_{ij}n_e}, \tag{8.3}$$

where $q_{ij}n_e$ is the collisional rate for a single atom, obtained by dividing Eq. 8.2 by the ionic density n_i. Similarly, we may define the recombination lifetime

$$\tau_{\text{recomb}}(\text{X}) = \frac{1}{\alpha_r(X)n_e}, \tag{8.4}$$

where $\alpha_r(\text{X})$ is the recombination rate coefficient for an ionic species X (Chapter 7). Similar considerations apply to, say, the photoionization rate and lifetime

$$\tau_{\text{phot}} = \frac{1}{\Gamma_{\text{phot}}(\text{X})}, \tag{8.5}$$

where, given a radiation field intensity J_ν and photoionization cross section $\sigma_\nu(\text{X})$ (discussed further in Chapter 10),

$$\Gamma(X) = \int \frac{4\pi}{h\nu} J_\nu \sigma_\nu(\text{X}) d\nu. \tag{8.6}$$

A comparison of rates and lifetimes is useful in the analayis of astrophysical plasmas, as it indicates the relative efficiencies of atomic processes. For example, we may compare the relative timescales for an atom to be collisionally excited from i to j, or for it to be photoionized in a radiation field; a comparison of lifetimes as above should give the answer (cf. Chapter 10).

8.1.2 Collisional and radiative rates

We have already considered the Einstein A and B coefficients that relate the inverse processes of radiative absorption and emission. We shall also need to relate collisional excitation and de-excitation cross sections and rate coefficients between any two levels. From the principle of detailed balance, we have

$$g_i v_i^2 Q_{ij}(v_i) = g_j v_j^2 Q_{ji}(v_j), \tag{8.7}$$

where the kinetic energy of the incident electron is relative to levels i and j, such that

$$\frac{1}{2} m v_i^2 = \frac{1}{2} m v_j^2 + E_{ij}. \tag{8.8}$$

The collisional excitation and de-excitation rates per unit volume per unit time are balanced as

$$n_e n_i q_{ij} = n_e N_j q_{ji}, \tag{8.9}$$

and therefore,

$$q_{ij}(T_e) = \frac{g_j}{g_i} q_{ji} e^{-Eij/kT_e} = \frac{8.63 \times 10^{-6}}{g_i T^{1/2}} \Upsilon(T_e), \tag{8.10}$$

in cm^3 s^{-1}, as obtained in Chapter 5 in terms of $\Upsilon(T_e)$, the Maxwellian averaged collision strength Ω_{ij} (related to the cross section Q_{ij}).

8.1.3 Emissivity and line ratio

Emissivity is the energy emitted in a given line $j \to i$ per unit volume per unit time

$$\epsilon_{ji} = \frac{1}{4\pi} N_j A_{ji} h\nu_{ij}, \tag{8.11}$$

where N_j is the level population per unit volume, A_{ji} is the rate of radiative decay per unit time and $h\nu_{ij}$ is the photon energy. The division by the total solid angle 4π assumes that the emission is isotropic. The nature of temperature- and density-sensitive lines is illustrated by simply considering a three-level system i,j,k with $E_i < E_j < E_k$. If levels j and k are excited from i, usually the ground state where most of the atoms are found, then, according to the Boltzmann equation, the level populations in thermal equilibrium N_j and N_k are in the proportion

$$\frac{N_j}{N_k} = \frac{g_j}{g_k} e^{(E_j - E_k)/kT}. \tag{8.12}$$

If the levels j and k are well separated in energy then a variation in temperature T will manifest itself in the variation in the emissivities of lines $j \to i$ and $k \to i$, which depend on N_j and N_k, respectively. Therefore, the line emissivity ratio should be an indicator of the temperature of the plasma. On the other hand if the levels j and k are closely spaced together, i.e., $(E_j - E_k) \approx 0$, then the level populations are essentially independent of temperature (since the exponential factor is close to unity). Now if the j and k are also metastable, connected via forbidden transitions to the lower level i with very low spontaneous decay A-values, then their populations would depend on the electron density. This is because the collisional excitation rate for the transition $j \to k$ can compete with spontaneous radiative decay. Thus, depending on the difference in energy levels ΔE_{ij}, we can utilize line emissivity ratios to determine temperature or density.

For low-density optically thin plasma sources, the ratio of emissivities for a pair of lines may be compared directly with observed intensity ratios. We note the advantage of a line ratio, as opposed to an individual line intensity: *the line ratio does not generally depend on external factors other than the temperature and density.* Two lines from the same ion would have the same abundance and ionization fraction of a given element. Furthermore, it is usually a good approximation that the particular ions exist in a region with the same temperature, density and other physical conditions (such as velocity fields, spatial extent, etc.).

Whereas a line ratio may depend on *both* temperature and density, we can particularize it further, such that it is predominantly a function of either temperature *or* density. Indeed, that is the primary role of line ratio analysis, and we look for ions with a pair of lines whose ratio varies significantly with T_e or n_e, but not both. There are a few well-known ions and lines that are thus useful. The main characteristics may be understood by examining the three-level atom again, but now shown in two different ways in Fig. 8.2: (i) when the two excited levels, 2 and 3, are

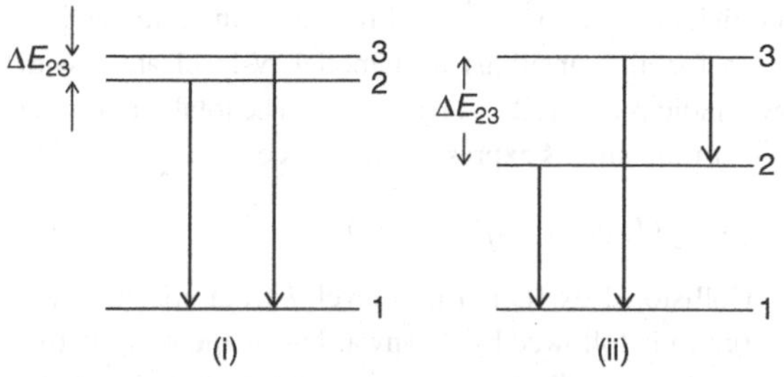

FIGURE 8.2 Energy level schematics of (i) density and (ii) temperature diagnostics. Closely spaced levels yield lines relatively independent of temperature, such as the [OII], [SII] lines, and levels spaced farther apart yield lines sensitive to electron temperature, owing to the Boltzmann factor in Eq. 8.12, such as in [OIII] (see text).

closely spaced, and (ii) when they are significantly apart in energy. From the Boltzmann equation the line ratio will depend on the temperature and the energy difference E_{23}, and the density via the excitation rates q_{12}, q_{13} and q_{23} that determine the distribution of level populations among the three levels. But when the energy difference E_{23} is small, $E_{23} \approx 0$, then the T_e dependence via the exponential factor will also be small and the line ratio should depend mainly on the electron density. To determine the line ratio at all densities, the level populations (N_1, N_2, N_3) must be explicitly computed by solving the three-level set of equations and calculating the emissivity ratio[1]

$$\frac{\epsilon_{21}}{\epsilon_{31}} = \frac{N_2 A_{21} E_{21}}{N_3 A_{31} E_{31}}. \tag{8.13}$$

8.2 Collisional-radiative model

To determine the emissivity of a line we need to calculate the level populations of ground and excited levels. Although in principle all atomic levels are involved, it is neither practical nor necessary to include more than a limited number. In simple cases, usually only few levels are explicitly considered. They are coupled together via excitation and radiative transitions among all levels included in the (truncated) model atom. A coupled set of equations, therefore, can be written down, involving the *intrinsic atomic parameters* that determine the intensities of observed lines.

In the simplest case, we consider only electron impact excitation and spontaneous radiative decay (further specializations may include radiative or fluorescent excitation, discussed later). The primary dependence on *extrinsic variables* is through the electron temperature T_e and density n_e. For given (T_e, n_e), the level populations are obtained by solving a set of simultaneous equations. The two atomic parameters needed are the excitation rate coefficient q_{ij} (cm^3 s^{-1}) and the Einstein A values A_{ji} (s^{-1}) for all transitions in a model N-level atom with level indices $i = 1, 2, \ldots, j, \ldots, N$. The total rate for the $i \rightarrow j$ transition is expressed in terms of

$$P_{ij} = q_{ij}(T_e) n_e + A_{ij} \quad (j \neq i). \tag{8.14}$$

Collisional excitation into level j from all other levels (gain) is followed by downward spontaneous radiative decay (loss) to all levels $j < i$. In steady state, the level populations N_i are calculated by balancing the number of excitations into level i (per unit volume per unit time), with radiative decays out of it ($A_{ji} \equiv 0$ if $j < i$), i.e.,

$$N_i \sum_{j \neq i} P_{ij} = \sum_{j \neq i} N_j P_{ji}. \tag{8.15}$$

A more general form of Eq. 8.15 involves time-dependence, and *all* collisional and radiative processes that affect each level population N_i. But if we have statistical equilibrium then the rate equations are time-independent, i.e.,

$$\frac{dN_i}{dt} = \sum_{j \neq i} N_j P_{ji} - N_i \sum_{j \neq i} P_{ij} = 0. \tag{8.16}$$

These *collisional-radiative (CR) equations* may be cast in matrix notation as $\mathbf{P} = \mathbf{C} + \mathbf{R}$, where the matrix elements C_{ij} involve collisional (de)-excitation rate coefficients q_{ij}, and the R_{ji} involve the Einstein radiative decay rates. In the presence of a radiation field of intensity J_ν, we can write the radiative sub-matrix as

$$R_{ji} = A_{ji} + B_{ij} J_\nu. \tag{8.17}$$

The CR equations can be solved by any simultaneous-equation solver, such as matrix-inversion or Gauss–Jordan elimination and back substitution. Eq. 8.15 constitutes a CR model with a finite (usually small) number of levels (for the time being we are neglecting photo-excitation by an external radiation source). We illustrate the essential physics by considering a simple three-level atom (Fig. 8.2), and write down the populations of levels 2 and 3;[2]

$$\begin{aligned} N_2 A_{21} + n_e N_2 (q_{21} + q_{23}) \\ = n_e N_1 q_{12} + n_e N_3 q_{32} + N_3 A_{32}, \end{aligned} \tag{8.18}$$

and similarly for level 3,

$$\begin{aligned} N_3 (A_{31} + A_{32}) + n_e N_3 (q_{31} + q_{32}) \\ = n_e N_1 q_{13} + N_e N_2 q_{23}. \end{aligned} \tag{8.19}$$

One can also do this for level 1. That would give three simultaneous equations which, in principle, can be solved for the three-level population. However, the normalization would still be arbitrary. Therefore, we generally normalize the level populations to unity, i.e., $(N_1 + N_2 + N_3) = 1$. But even before we solve the set of equations exactly, it is instructive to simplify the above equations by physical arguments for a given ion in a particular source, such

[1] Here one should distinguish between 'line ratio', which may refer to observed quantities, and 'emissivity ratio', which refers to theoretically computed parameters.

[2] We follow the convention that the densities are denoted as lower case n (cm^{-3}), such as n_e or n_i, and level populations as upper case N (cm^{-3}), such as N_2, N_3, etc. But in some cases it will be necessary to use different notation, viz. Chapter 9.

as low-density H II regions, where the excited state populations are negligible and almost all the ions are in the ground level. In that case, the excited levels are populated only by excitation from the ground level, and radiative decay from above. Then we can drop all collisional terms from excited levels, such as $n_e N_2(q_{21}+q_{23})$ and $n_e N_3 q_{32}$ from Eq. 8.18, and $n_e N_3(q_{31}+q_{32})$ and $n_e N_2 q_{23}$ from Eq. 8.19 to obtain

$$\begin{aligned} N_2 A_{21} &= n_e N_1 q_{12} + N_3 A_{32} \\ N_3(A_{31}+A_{32}) &= n_e N_1 q_{13}. \end{aligned} \tag{8.20}$$

If we are only interested in the ratios of the three lines from excited levels, (i) 3→1, (ii) 3→2 and (iii) 2→1, then we can divide out N_1 and obtain *relative* level populations,

$$\frac{N_3}{N_1} = \frac{n_e q_{13}}{(A_{31}+A_{32})}. \tag{8.21}$$

Substituting in the first Eq. 8.20 also gives N_2/N_1. We can now obtain the emissivities of the three lines. For example,

$$\epsilon_{31} = \frac{n_e q_{13}}{(A_{31}+A_{32})} * (A_{31} * h\nu_{31}). \tag{8.22}$$

The other two line emissitives may be obtained similarly, and the emissivity ratios such as $\epsilon_{31}/\epsilon_{21}$ can be obtained. Yet another simplification occurs in cases where we have two lines arising from the same upper level, say from level 3. In that case the ratio is *independent* of the level populations, since it is the same for both lines and cancels out. For example,

$$\frac{\epsilon(3\rightarrow 1)}{\epsilon(3\rightarrow 2)} = \frac{A_{31} h\nu_{31}}{A_{32} h\nu_{32}}, \tag{8.23}$$

which is the ratio of the respective A-values and energy differences. Furthermore, if the lines are close in wavelengths with $h\nu_{31} \approx h\nu_{32}$, then the line ratio is simply the ratio of A-values alone. Observations of line intensities I_{31} and I_{32} originating from the same upper level thereby provides an empirical check on the A-values, which are usually calculated theoretically. To ensure accuracy, however, it is best to solve the coupled statistical equations by setting up the CR model (Eq. 8.15) with a reasonably complete subset of levels likely to contribute to line formation in an ion, and solve them exactly. To summarize the discussion thus far: we are now left with the task of determining level populations in excited states of an ion, which depend essentially on the rate of excitation of levels and radiative decays. Since the excitation rates are functions of temperature and density alone, and determine level populations, it follows that line ratios are indicators of T_e and n_e. Recall that we neglected radiative excitations. However, more complicated situations may also involve external radiation fields and time-dependence, considered later.

8.3 Spectral diagnostics: visible lines

In this and following sections we discuss some of the most common and basic diagnostic lines and emission line ratios in three widely different regions of the electromagnetic spectrum: optical, UV/X-ray, and far-infrared. These diagnostics will later be employed in the analysis of optically thin plasmas in astrophysical objects described in subsequent chapters on stellar coronae (Chapter 10), nebulae and H II regions (Chapter 12) and active galactic nuclei (Chapter 13), etc. More complicated species, such as Fe II, and more specialized effects, such as resonance fluorescence and photo-excitation, will also be dealt with in those chapters pertaining to the actual astrophysical situations where they are likely to occur.

8.3.1 Optical lines: [O II], [S II], [O III]

Among the most prominent lines in the optical range are those due to forbidden transitions in O II, O III and S II. The [O II] and [S II] lines form a pair of similar fine structure 'doublets' at λλ 3726, 3729 and λλ 6716, 6731, respectively.[3] These lines are the best density diagnostics in H II regions, and they conveniently occur at the blue and the red ends of the optical spectrum. On the other hand, the strongest [O III] lines act as temperature diagnostics and occur in the middle of the optical range, with three main transitions at λλ 4363, 4959, 5007 Å. Figure 8.3 is a spectrum of the Crab nebula [225], and encompasses the optical range displaying not only these lines, but also a number of quintessential nebular lines.

The [O II] lines are also referred to as 'auroral' lines, since they are formed in the Earth's ionosphere and seen in Aurorae near the geomagnetic polar regions; neutral oxygen is photoionized by the Sun's UV radiation to form O II. The [S II] lines are traditionally referred to as 'nebular' lines, and observed in most H II regions. For these ions, the set of CR equations may be as small as five

[3] In a classic paper in 1957, M. J. Seaton and D. E. Osterbrock first demonstrated the utility of the forbidden [O II] and [S II] line ratios as nebular diagnostics [226]. The crucial dependence of the line ratios on the *accuracy* of the fundamental atomic parameters was, and continues to be, one of the main drivers behind the development of state-of-the-art atomic physics codes based on the coupled channel approximation described in Chapter 3 and elsewhere.

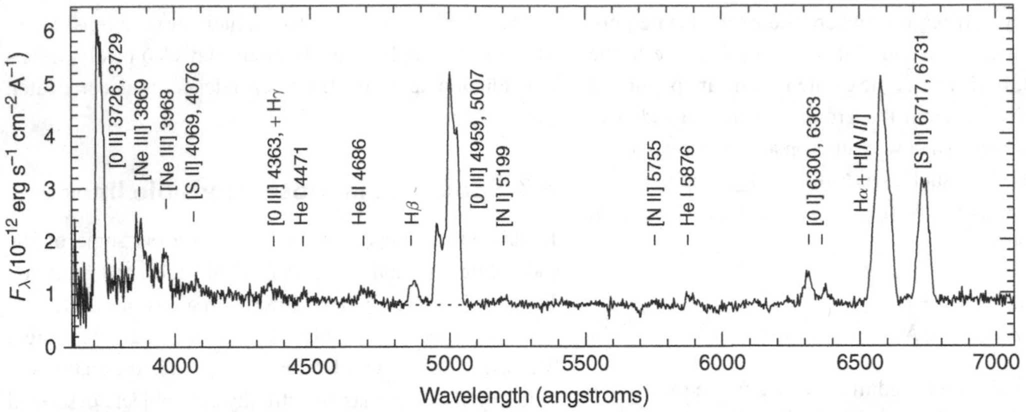

FIGURE 8.3 Typical optical emission lines from the Crab nebula. The spectrum covers the entire optical range: forbidden [OII] lines in the blue, [SII] lines in the red, and [OIII] lines in the middle (see also Chapter 12). The fine structure components are discernible in the observed line profiles. Emission lines from other common nebular ions, including the allowed lines of H (Hα and Hβ) and He, are also shown. ([225], Courtesy: N. Smith).

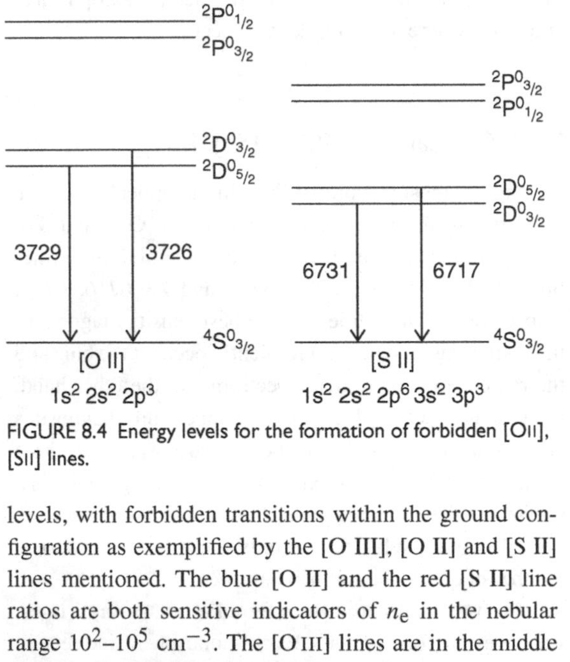

FIGURE 8.4 Energy levels for the formation of forbidden [OII], [SII] lines.

levels, with forbidden transitions within the ground configuration as exemplified by the [O III], [O II] and [S II] lines mentioned. The blue [O II] and the red [S II] line ratios are both sensitive indicators of n_e in the nebular range 10^2–10^5 cm^{-3}. The [O III] lines are in the middle of the optical region and provide the 'standard' nebular temperature diagnostics.

8.3.1.1 *Density dependence of [O II] and [S II] lines*

As shown in Fig. 8.4, these lines originate from transitions within the same type of ground configuration: $2p^3$ and $3p^3$, respectively. The energy level structure consists of five levels $\left({}^4S^o_{3/2}, {}^2D^o_{3/2,5/2}, {}^2P^o_{1/2,3/2}\right)$; their energies are given in Table 8.1. The transitions of interest for density diagnostics are from the ${}^2D^o{}_{3/2,/5/2}$ levels to the ground level, at λλ 3729, 3726 Å for [O II], and λλ 6717, 6731 Å for [S II]. Note that the ${}^2D^o_{3/2,/5/2}$ and the ${}^2P^o_{1/2,3/2}$ levels are switched around in O II and S II. Hence the line ratio corresponding to the *same* transitions is 3729/3726 in [O II], higher-to-lower wavelength, but the reverse in the [S II] 6717/6731 ratio. Since the two fine-structure levels are closely spaced, the temperature dependence for collisional excitation is small, and the line ratios depend nearly entirely on electron density. The main transitions of interest are from the ground level $S^o_{3/2}$ to the two fine structure levels ${}^2D^o_{5/2,3/2}$, that give rise to the density-sensitive lines via collisional mixing. The detailed collision strengths for [O II] as a function of electron energy are shown in Fig. 8.5 [108]. The top two panels show the collision strengths for electron impact excitation ${}^4S^o_{3/2} \rightarrow {}^2D^o_{5/2,3/2}$. The collision strengths divide in the ratio 6:4 according to statistical weights, even at resonance energies (Chapter 5). But the actual sensitivity to electron density depends primarily upon the collision strength $\Omega\left({}^2D^o_{5/2} - {}^2D^o_{3/2}\right)$ for mixing between the upper levels ${}^2D^o_{5/2}$ and ${}^2D^o_{3/2}$, which contains considerable resonance structure, as shown in the bottom panel of Fig. 8.5. Although these two levels are reversed in [O II] 3729/3726 and [S II] 6717/6731 ratios (Table 8.1), it makes no qualitative difference to the density dependence, since $\Omega\left({}^2D^o_{5/2} - {}^2D^o_{3/2}\right)$ is symmetrical with respect to energy order (Chapter 5).

Although in practice the line ratios need to be calculated from a CR model (Eq. 8.15) at all densities (and temperatures) of interest, it is instructive to consider the low- and high-density limits of the [O II] and [S II] line ratios from physical considerations alone. We consider two limiting cases.

TABLE 8.1 Energies of O II and S II ground configuration levels (Rydbergs).

Level	O II ($2p^3$)	S II ($3p^3$)
$^4S^o_{3/2}$	0.0	0.0
$^2D^o_{5/2}$	0.244 315 7	0.135 639 6
$^2D^o_{3/2}$	0.244 498 1	0.135 349 9
$^2P^o_{3/2}$	0.368 771 6	0.223 912 3
$^2P^o_{1/2}$	0.368 789 8	0.223 486 7

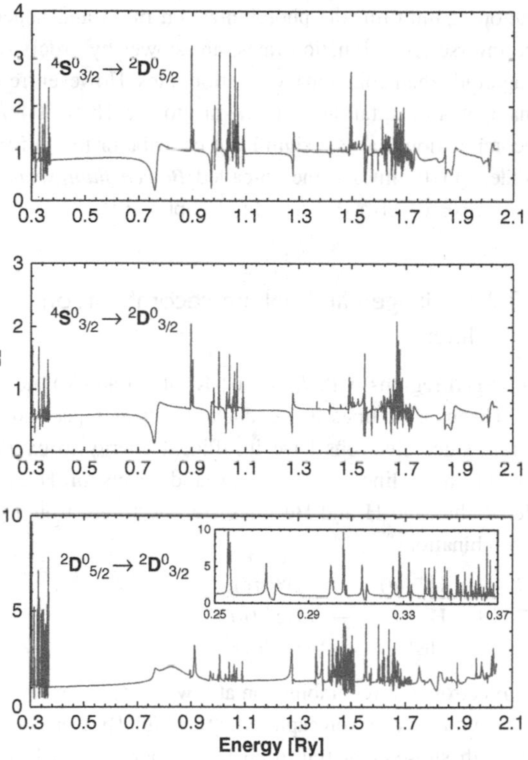

FIGURE 8.5 The collision strengths for [O II] transitions sensitive to electron density [108]. The near-threshold resonances in $\Omega\left({}^2D^o_{5/2} - {}^2D^o_{3/2}\right)$ (inset) enhance the effective Maxwellian averaged rate coefficients that determine the collisional mixing between the two fine-structure levels.

The low density limit $n_e \to 0$

At very low densities, most ions are in the ground level. The other excited levels in Table 8.1 are metastable, and downward transitions are dipole (E1) forbidden. But they decay via magnetic dipole (M1) or electric quadrupole (E2) transitions with A-values ($\sim 10^{-1} - 10^{-4}\ \mathrm{s}^{-1}$), that are about eight to ten orders of magnitude smaller than is typical for dipole allowed transitions. However, after excitation from the ground $^4S^o_{3/2}$ level to the upper metastable levels, it is highly improbable that the ion will be collisionally de-excited to the ground level by electron impact, since the number of electron collisions would be very small. Thus, in a low-density plasma, the excited ion is likely to remain so, until radiative decay (the probability that it will is nearly unity). Thus the ratio of lines corresponding to the relevant transitions $^2D^o_{5/2,3/2} \to {}^4S^o_{3/2}$ is governed only by the ratio of the excitation rate coefficients $q\left({}^4S^o_{3/2} - {}^2D^o_{5/2}\right)$ and $q\left({}^4S^o_{3/2} - {}^2D^o_{3/2}\right)$. Now, if we further neglect the temperature dependence, since the two excited $^2D^o_{5/2,3/2}$ levels have negligible energy differences, the ratio of the rate coefficients is simply the ratio of the (effective) collision strengths. Therefore,

$$\lim_{n_e \to 0} \frac{N\left({}^2D^o_{5/2}\right)}{N\left({}^2D^o_{3/2}\right)} = \frac{\Omega\left({}^4S^o_{3/2} - {}^2D^o_{5/2}\right)}{\Omega\left({}^4S^o_{3/2} - {}^2D^o_{3/2}\right)} = \frac{6}{4}. \tag{8.24}$$

The collision strength ratio divides according to the statistical weights of upper levels, because in LS coupling if the total L or $S = 0$ for the initial term, then the fine-structure collision strengths can be obtained algebraically from the LS collision strength (Chapter 5). Given that the intial LS term is $^4S^o$ with $L = 0$, the low-density limit for both the [O II] and [S II] line ratios I(3729)/I(3726) and I(6717)/I(6731) is 3/2.

The high density limit $n_e \to \infty$

At sufficiently high densities, the level populations assume the Boltzmann distribution, and the [O II] and [S II] line ratios are given by

$$\lim_{n_e \to \infty} \frac{\epsilon\left({}^2D^o_{5/2} - {}^4S^o_{3/2}\right)}{\epsilon\left({}^4S^o_{3/2} - {}^2D^o_{3/2}\right)} = \frac{g\left({}^2D^o_{5/2}\right) A\left({}^2D^o_{5/2} - {}^4S^o_{3/2}\right)}{g\left({}^2D^o_{3/2}\right) A\left({}^2D^o_{3/2} - {}^4S^o_{3/2}\right)}, \tag{8.25}$$

which depends only on the ratio of the intrinsic A values (given in Appendix E). The high density limits are: [O II] I(3729)/I(3726) = 0.35, and [S II] I(6716)/I(6731) = 0.43.

Figure 8.6 shows the line ratios obtained from a full solution of the five-level atom in Table 8.1. It is seen that the theoretical line ratios approach their respective limiting values. There are slight deviations, owing to coupling with other levels and associated rates, from the limiting values derived from physical consideration alone. But such simple limits are not easily applicable to other ions, since they are subject to certain caveats, assuming (i) relativistic effects are small so that LS coupling is valid,

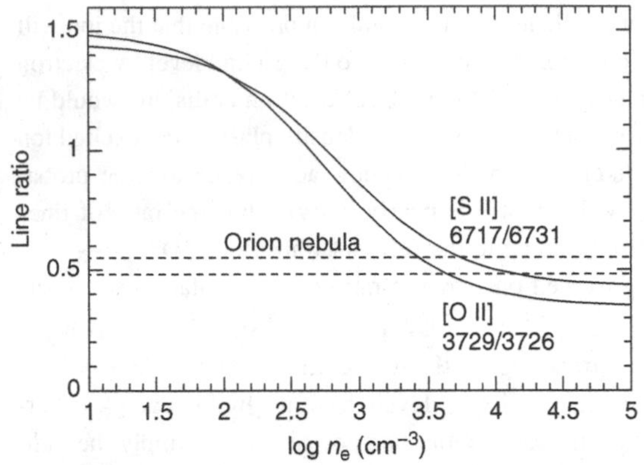

FIGURE 8.6 Forbidden [OII] and [SII] line ratios as function of electron density. The observed intensity ratios from the Orion nebula are also shown for both ions [227]. The intersection of the observed values with the line ratio curves demarcates the density.

(ii) the lower term has total L or $S = 0$, (iii) resonance effects do not preferentially affect one excited fine structure level over the other. If (i) holds but (ii) does not, then algebraic rules may still yield fine structure collision strengths from *LS* coupling values. However, in complicated cases such as Fe II, discussed in Chapter 12, none of these caveats apply, and no simple limits can be obtained. The [O II] and [S II] lines are immensely valuable diagnostics in H II regions since the low- and high-density limits discussed above always hold. The theoretical limits derived above are well-established, and verified by many observational studies of nebulae with no observed deviations from these 'canonical' values. In Fig. 8.6, the observed intensity ratios from the Orion nebula, obtained from Echelle spectrophotometry using the Very Large Telescope [227], are also shown; they constrain the densities fairly precisely to log $n_e \approx 3.7$ ($n_e \approx 5 \times 10^3 \mathrm{cm}^{-3}$) from both the [O II] and [S II] line ratios.

8.3.1.2 *Temperature dependence of* [O III] *lines*

The energetics of [O III] emission lines are quite different from [O II] and [S II] (Fig. 8.2), and highly suitable for temperature diagnostics. The five levels of the ground configuration are $1s^2 2s^2 2p^2 (^3P_{0,1,2}, ^1D_2, ^1S_0)$. The three *LS* terms are separated by a few eV, so that the level populations due to electron impact excitation are dependent on temperatures typical of nebular H II regions. Often, the strongest lines in the optical nebular spectra are the three lines λλ 4959, 5007, 4363 Å due to transitions $^1D_2 -^3 P_1, ^1D_2 -^3 P_2, ^1S_0 -^1 D_2$, respectively. Combining the fine structure transitions, the line ratio [I(4959) + I(5007)/I(4363)] is a very useful diagnostic of temperatures ~1–2 eV ≈ 10 000 20 000 K. Using known excitation rate coefficients and *A*-values for [O III] lines (see Appendix E) emissivities can be parametrized in terms of T_e and n_e as in Eq. 8.22. For the [O III] line ratio ([228]),

$$\frac{I(\lambda 4959 + \lambda 5007)}{I(\lambda 4363)} = \frac{8.32\, e^{3.29\times 10^4/T}}{1 + 4.5 \times 10^{-4} n_e/\sqrt{T}}. \tag{8.26}$$

Similar expressions may be constructed for other line ratios in ions with a few-level CR model [228]. The O III lines are discussed in Chapter 12 in the context of photoionized H II regions. It is shown that in ionization equilibrium the photoionization rates, and hence the inverse recombination rates, are slower by orders of magnitude than collisional excitation rates. However, resonant photo-excitation of O III due to the He II 304 Å recombination line is a significant contributor to the forbidden [O III] lines – the so-called *Bowen fluorescence mechanism* described in detail in Chapter 10.

8.3.2 Hydrogen and helium recombination lines

In all H II regions with $T_e \sim 10^4$ K (about an eV) the H and He emission lines are *not* due to electron impact excitation, since electrons have insufficent energy to excite *any* of those lines from the ground levels of H and He. Rather, the H and He lines are due to electron–ion recombination:

$$\begin{aligned} e^- &+ H^+(p) &\rightarrow\ & H^o(n\ell), \\ e^- &+ He^{++} &\rightarrow\ & He^+(n\ell), \\ e^- &+ He^+ &\rightarrow\ & He^o(n\ell). \end{aligned}$$

Electron–ion recombination at low temperatures takes place with the electron captured into a high Rydberg level $n\ell$, with subsequent transition to another (lower) level, or to the ground state. For example, Hα (6563 Å) is emitted following capture (or cascades) into the $n = 3$ levels followed by radiative transition to the $n = 2$ levels. Such radiative cascades usually proceed via the fast dipole allowed (E1) transitions with high *A*-values. Also, note that there are a number of levels in any excited *n*-complex in H or He; therefore, there are several transitions associated with recombination lines of H and He. For example, the Hα line is due to transitions

$$3s(^2S_{1/2}), 3p\left(^2P^o_{1/2,3/2}\right), 3d(^2D_{3/2,5/2})$$
$$\rightarrow 2s(^2S_{1/2}), 2p\left(^2P^o_{1/2,3/2}\right) : \ H\alpha. \tag{8.27}$$

Similarly, visible helium lines are produced in nebulae by radiative transitions among many levels, following electron recombination of H-like He II into He ions. For example, the 5876 Å line from He I is due to the transition among the following levels,

$$1s3d(^3D_{1,2,3}) \rightarrow 1s2p\left(^3P^o_{0,1,2}\right).$$

Hydrogen recombination lines are discussed further in Chapter 12 under nebular conditions.

8.4 X-ray lines: the helium isoelectronic sequence

The atomic physics of X-ray emission line diagnostics is considerably more involved than the optical forbidden emission lines discussed above. The most useful atomic system is helium-like ions of nearly all astrophysically abundant elements, from carbon to nickel. Although it has only one more electron, the atomic physics of helium is fundamentally different from hydrogen. The two-electron interaction, absent in H, dominates atomic structure and spectral formation in not only neutral He but also in He-like ions that are of great importance in X-ray spectroscopy of high-temprature astrophysical and laboratory plasmas.

As discussed earlier, in Chapters 4 and 5, He-like ions have a closed-shell ground configuration $1s^2$. Excitations out of this K-shell into the excited levels of configurations 1s2l – the He $K\alpha$ transitions – require high energies, typically in the X-ray range. The downward radiative decays entail several types of transitions as shown schematically in the energy-level diagram in Fig. 8.7, which is a simplified version of the more elaborate diagram in Fig. 4.3 with fine structure, as discussed in Section 4.14.

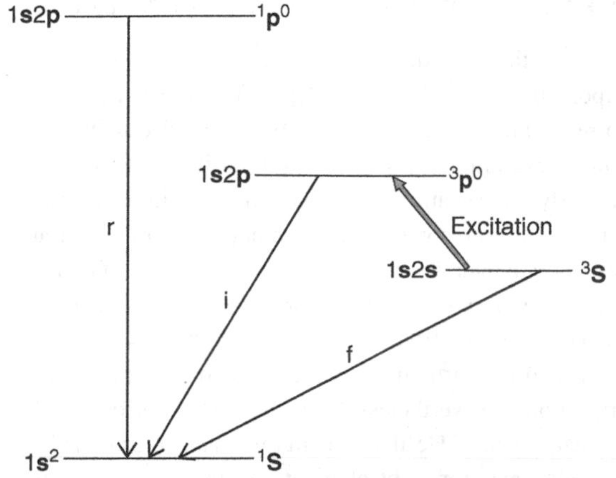

FIGURE 8.7 Basic energy level diagram for the formation of He $K\alpha$ X-ray lines.

Referring to *LS* coupling term designations to begin with, appropriate for low-*Z* elements, the primary X-ray transitions are

$$\begin{array}{llllll} 1s^2(^1S) & \rightarrow & 1s2p(^1P) & \text{dipole allowed} & (r) & \\ 1s^2(^1S) & \rightarrow & 1s2p(^3P) & \text{intersystem} & (i) & \\ 1s^2(^1S) & \rightarrow & 1s2s(^1S) & \text{two-photon continuum} & 2h\nu & (8.28) \\ 1s^2(^1S) & \rightarrow & 1s2s(^3S) & \text{forbidden} & (f) & \end{array}$$

The parenthetical (r) on the right refers to the common astronomical notation for the first dipole transition in an atom, and stands for 'resonance' transition.[4] All of these transitions, and their line ratios, are the primary diagnostics for density, temperature, and ionization balance in high temperature plasmas [205, 206, 231, 232, 233]. Figure 8.8 shows the quintessential (though simplest) specrum of the X-ray He $K\alpha$ lines from O VII, as observed from the star Procyon.[5] The primary diagnostics from He-like ions depend on two line ratios, discussed below.

$$R = \frac{f}{i}, \quad \text{and} \quad G = \frac{i+f}{r}. \qquad (8.29)$$

8.4.1 Density diagnostic ratio $R = f/i$

As shown in Fig. 8.7, the forbidden lines arises from the metastable state 1s2s (^{3}S) (or 2^3S), which has a very low *A*-value for decay to the ground state $1s^2(^1S)$. On the other hand the 2^3S state has a high excitation rate to the nearby higher state $2(^3P^o)$: the two states are connected via strong dipole transition. As the electron density increases, electron impact excitation from 2^3S transfers the level population to 2^3P^o via the transition $2^3S \rightarrow 2^3P^o$, thereby decreasing the f-line intensity and increasing the i-line intensity. This transition is similarly affected by photo-excitation, if there is an external radiation source, such as a hot star or active galactic nucleus. The downward radiative decay from the 2^3P^o back to the 2^3S partially makes up for the upward transfer of population, but (a) the 2^3S is metastable and likely to get pumped up again and (b) part of the 2^3P^o level population decays

4 This term, though common, is inaccurate and is unrelated to *resonances* as described elsewhere in this text.

5 Stellar classes and luminosity types are discussed in Chapter 10. Procyon is a main Sequence F5 IV–V star, not too different from the Sun (G2 V) with similar coronal spectra; although Procyon is a binary system with a white dwarf companion. The similarity reflects common microscopic origin under coronal plasma conditions, unaffected by the overall macroscopic peculiarities of the source, per se.

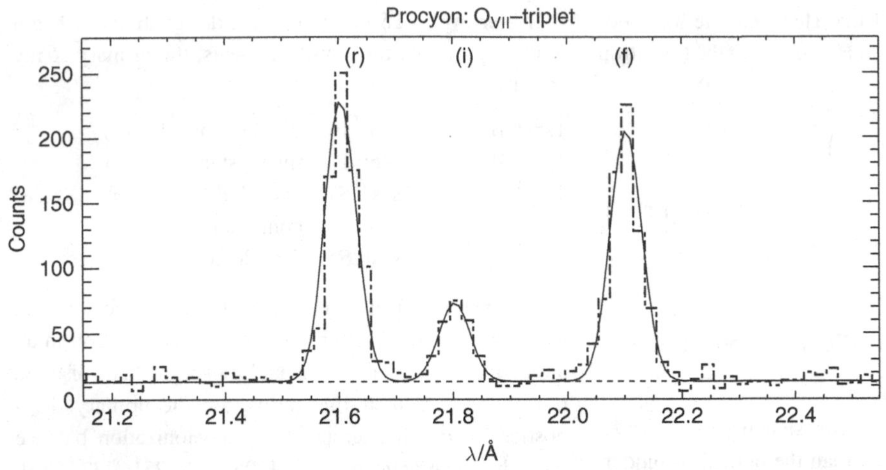

FIGURE 8.8 Resonance (r), intercombination (i) and forbidden (f) X-ray lines of He-like O VII from the corona of Procyon ([229], reproduced by permission). The 'triplet' here refers simply to three lines, and not to the spin-multiplicity (2S + 1), which is the standard spectroscpic usage (viz Chapter 2), and throughout this text.

to the ground state. Hence, the ratio R = f/i is a sensitive density indicator beyond some critical density n_c, determined by the competition between the radiative and excitation rates for the relevant transitions. Specifically,

$$n_c \sim \frac{A(2^3S - 1^1S)}{q(2^3S - 2^3P^o)}.$$

For $n_e > n_c$ the f-line intensity decreases in favour of the i-line (as shown later in Fig. 8.10).

8.4.2 Temperature and ionization equilibrium diagnostics ratio G = (i + f)/r

If we add together the lines from the excited triplet $n = 2$ levels $2(^3S, ^3P^o)$, then the density dependence vanishes since collisional excitations would re-distribute the level population among the two levels, but the sum would remain constant. Since the singlet 2^1P^o level lies higher in energy than the triplets, the ratio G = (i + f)/r is expected to depend on the temperature. However, there is another dependence of G that is highly useful – on ionization balance in the ambient plasma. This occurs because of recombinations from H-like to He-like ions, and on ionization from Li-like into He-like ions. The recombinations preferentially populate the high-lying triplet (spin multiplicity 2S + 1 = 3) levels, generally according to statistical weights, but then cascade down to populate the $n = 2$ levels (Fig. 8.7). However, the triplets are far more likely to cascade down to the 2^3S via $\Delta S = 0$ dipole allowed transitions, than to the ground state 1^1S, which would be via the less-probable spin-change transitions. The metastable 2^3S state thus acts as a 'pseudo-ground state' for recombination cascades from excited triplets. The excited singlets n^1L on the other hand are most likely to radiate quickly to the ground state 1^1S, following the most direct singlet cascade pathways. Therefore, the $2(^3S, ^3P^o)$ states are populated far more than the $2(^1P^o)$ by recombinations from a H-like ionization stage. It follows that the ratio G would be highly sensitive to the ionization fraction H/He of an element. The ratio G is also sensitive to inner-shell ionization from the Li-like to He-like stages: $1s^22s \rightarrow 1s2s(^3S) + e$. Therefore, ionizations also enhance the population of the metastable $2(^3S)$ level (again acting as a 'pseudo-ground state'). Recall that inner-shell excitations into the singlet level $1s2s(^1S)$ do not result in line formation, since it decays into a two-photon continuum (Fig. 4.3, Eq. 4.167).

8.4.3 Electron impact excitation of X-ray lines

Why is the forbidden f-line nearly as strong as the dipole allowed w-line (Fig. 8.8)? The answer lies partly in recombination cascade transitions, but also in electron impact excitation rates that are similar in magnitude. Particularly so, because the forbidden transitions are more susceptible to resonance enhancement than allowed transitions (Chapter 5). While the discussion thus far illustrates the basics of He-like spectra, in non-relativistic *LS* coupling notation, it is not precise because we have neglected fine structure and the detailed nature of atomic transitions. Nevertheless, it is useful to introduce the primary lines of He-like ions in this manner, as done historically, and for light elements up to neon, where the fine structure is unresolved. But it is necessary to consider the full complexity of He-like spectra to determine

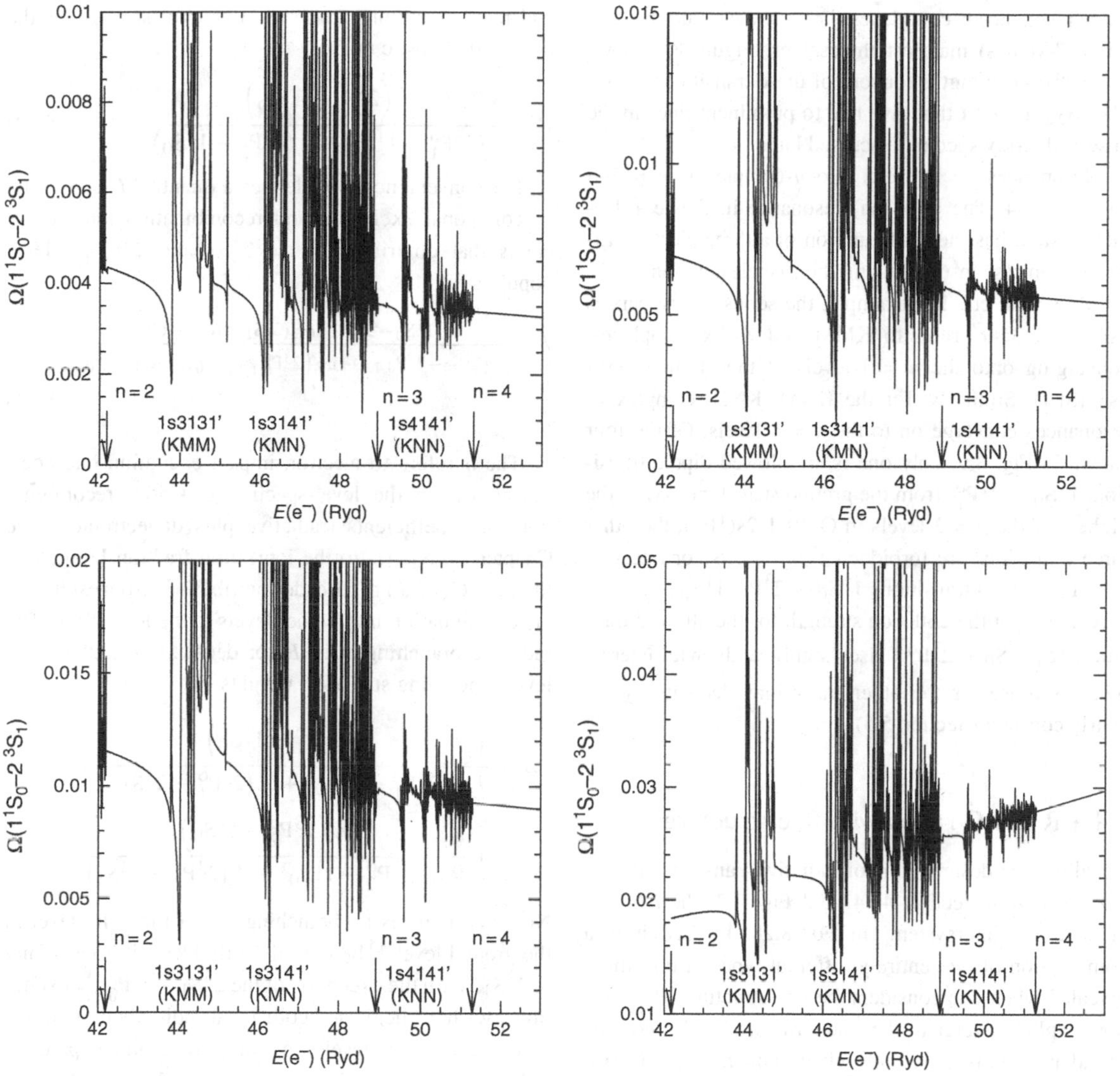

FIGURE 8.9 Collision strengths for X-ray transitions in He-like O VII [234].

the atomic transitions and rates needed to be solved for level populations and line intensities.

To introduce fine structure fully, we recall the discussion in Chapter 4 of the different types of radiative transitions that occur in He-like ions. In addition, we also need to understand electron scattering with helium-like ions, which is the process primarily responsible for the formation of X-ray lines.[6] We consider the excitation from the ground level $1s^2$ ($^1S_0$) to the n = 2,3,4 levels 1sns $\left(^3S_1, ^1S_0\right)$ and 1snp $\left(^3P^o_{0,1,2}, ^1P^o_1\right)$. One of the main features of the excitation cross sections is the extensive appearance of series of resonances that often dominate the energy range of interest.[7] As we have seen, for positive ions the electron may be captured into Rydberg series of infinite resonant states belonging to a higher threshold with energy $E_k > E_{ij}$, temporarily bound to the ion in autoionizing states $E_k(n\ell)$. As $n \to \infty$, these resonances form a Rydberg series converging on to the higher threshold E_k. We illustrate the practical situation where

[6] The atomic astrophysics of He-like systems is discussed throughout the text as a prime example of astrophysical spectroscopy. In addition to the formation of these lines covered in this chapter, particular aspects are described in radiative transitions (Chapter 4), and X-ray spectra of active galactic nuclei (Chapter 13).

[7] This section also highlights the reason for employing the coupled channel (CC) approximation to compute collision strengths, since that fully accounts for resonance effects.

many resonance n-complexes (i.e., same n-values but different ℓ-values) manifest themselves. Figure 8.9 shows the collision strengths for four of these transitions in He-like oxygen O VII that give rise to prominent lines in the observed X-ray spectra, discussed later.

Resonances lie between two n-complexes: n = 2–3 and n = 3–4. The (e + ion) resonance in the (e + He-like) system has the configuration of a three-electron Li-like system $1sn\ell n'\ell'$, or using the n-shell notation, KLL, KLM, KMM, etc. For example, the series of resonances $1s3\ell3\ell'$, $1s3\ell4\ell'$ refer to KMM and KMN complexes, converging onto the n = 3 levels of the He-like O VII $1s3\ell$(SLJ). Similarly, for the KNM, KNN, complex of resonances converge on to the n = 4 levels. Of the four shown in Fig. 8.9, only one is an allowed dipole transition, $1^1S_0 - 2^1P^o_1$, from the ground state $1s^2(^1S_0)$ to the highest of the n = 2 levels in O VII $1s2s(^1P^o_1)$; the other three transitions are forbidden $1^1S_0 - 2^3S_1$ or intersystem (Eq. 4.167) transitions $1^1S_0 - 2^3P^o_1$, $1^1S_0 - 2^3P^o_2$.[8] Now note that the collision strength for the allowed transition $\Omega\left(1^1S_0 - 2^1P^o_1\right)$ rises significantly with energy, whereas those for the other transitions decrease or are nearly constant (Section 5.5).

8.4.4 R and G ratios with fine structure

Recalling the designations of radiative transitions in He-like ions from Section 4.14, and Fig. 4.3, the i-line is labelled as 'intersystem' in Eq. 8.28. In fact, it is a combination of two entirely different types of transition, revealed when one considers the fine structure $^3P^o_{1,2}$. We now explicitly refer to the fine structure levels and type of radiative transitions, and switch to the more distinctive notation already introduced in Fig. 4.3; and Eq. 4.167

$$\begin{aligned} f \rightarrow\ z\ &:\ \ 2^3S_1 - 1^1S_0(M1)\\ i \rightarrow x+y&:\ \ \left(2^3P^o_2, 2^3P^o_1\right) - 1^1S_0(M2, E1)\\ r \rightarrow\ w\ &:\ \ 2^1P^o_1 - 1^1S_0(E1) \end{aligned} \tag{8.30}$$

and the line ratios

$$R = z/(x+y) \qquad G = (x+y+z)/w. \tag{8.31}$$

While the emissivities are calculated by a detailed solution of the full collisional–radiative model (Eq. 8.15) for He-like ions, comprising all seven levels shown in Fig. 4.3, it is again instructive to write down the diagnostic line ratios in physical terms with contributing transitions and rates of atomic processes. Let us first consider the density diagnostic ratio

$$R = \frac{I\left(2^3S_1 - 1^1S_o\right)}{I\left(2^3P^o_1 - 1^1S_0\right) + I\left(2^3P^o_2 - 1^1S_0\right)}. \tag{8.32}$$

For convenience, one defines a quantity F consisting of collisional excitation and recombination rate coefficients that contribute to the 2^3S_1 and $2^3P^o_{0,1,2}$ level populations;

$$F = \frac{q(1^1S_0{-}2^3S_1) + \alpha_R(2^3S_1)X_{H/He} + C(2^3S_1)}{q(1^1S_0{-}2^3P_{0,1,2}) + \alpha_R\left(2^3P^o_{0,1,2}\right)X_{H/He} + C(2^3P^o{}_{0,1,2})}. \tag{8.33}$$

The q refers to electron impact excitation rate coefficient, α_R to the level-specific (e + ion) recombination rate coefficients (radiative plus dielectronic – see Chapter 7), $X_{H/He}$ to the ionization fraction H-like/He-like, and $C(SLJ)$ to cascade contributions from excitation or recombination into higher levels to the level SLJ. The radiative branching ratio B for decays out of the $2^3P^o_J$ levels, including statistical weights, is

$$B = \frac{3}{9}\frac{A\left(2^3P^o_1 - 1^1S_0\right)}{A\left(2^3P^o_1 - 1^1S_0\right) + A\left(2^3P^o_1 - 2^3S_1\right)} + \frac{5}{9}\frac{A\left(2^3P^o_2 - 1^1S_0\right)}{A\left(2^3P^o_2 - 1^1S_0\right) + A\left(2^3P^o_2 - 2^3S_1\right)}. \tag{8.34}$$

Note that there is no branching out of the $2^3P^o_0$ level to the ground level 1^1S_0 (strictly forbidden); it decays only to 2^3S_1. With the exception of the $2^3S_1 - 2^3P^o_{0,1,2}$ excitations, we have neglected collisional redistribution among other excited $n = 2$ levels since they are small compared to radiative decay rates. The density dependence of the ratio R can now be expressed as

$$R(n_e) = \frac{R_o}{1 + n_e/n_c}, \tag{8.35}$$

where n_c is the critical density given by

$$\frac{A(2^3S_1 - 1^1S_0)}{(1+F)q(2^3S_1 - 2^3P_{0,1,2})}, \tag{8.36}$$

and

$$R_o = \frac{1+F}{B} - 1. \tag{8.37}$$

As the electron density n_e exceeds or is comparable to n_c, the ratio R decreases from its low-density limit R_o. Figure 8.10 shows a plot of R vs. n_e for O VII.

Now we examine the G ratio between the lines from the triplet levels $(x+y+z)$ to the singlet line w, which

[8] We often abbreviate the configuration $[n,\ell]$ by writing only the n-prefix when writing the transition from one $(n\ell)SLJ$ to another.

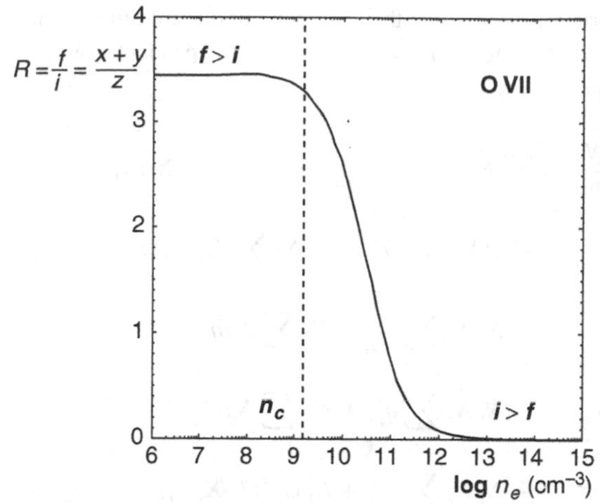

FIGURE 8.10 X-ray density diagnostic line ratio $R = f/i \equiv (x+y)/z$ for O VII (modelled after [229]). The forbidden-to-intercombination line ratio varies from $R_o = R > 1$ at low densities, to $R < 1$ at high densities (Fig. 8.8).

is primarily a function of temperature and ionization balance. The density dependence due to redistribution transitions within triplets is taken out by adding all triplet lines:

$$G(T_e, X_{H/He}, X_{Li/He}) = \frac{q(1^1S_0-2^3S_1)+q\left(1^1S_0-2^3P^o_{0,1,2}\right)+\left[\alpha_R(2^3S_1)+\alpha_R\left(2^3P^o_{0,1,2}\right)\right]X_{H/He}+C(2^3S_1)+C\left(2^3P^o_{0,1,2}\right)+C_I(2^3S_1)X_{Li/He}}{q\left(1^1S_0-2^1P^o_1\right)+\alpha_R(2^1P^o{}_1)+C(2^1P^o_1)}. \quad (8.38)$$

Here, we explicitly express G as function of T_e, recombination from H- to He-like ionization stage, and inner-shell excitation/ionization from Li- to He-like ionization stage, $1s^22s + e \rightarrow 1s2s(2^3S_1) + 2e$ (C_I denotes the ionization rate), which can be a significant contributor to the 2^3S_1 population and hence the forbidden z-line intensity. Detailed calculations yield a range of values for the generic behaviour of the line ratio $G(T)$ under different plasma conditions:

$$\begin{aligned} 0.7 < G < 1.5 &\quad \text{coronal equilibrium} \quad (T \sim T_m) \\ G > 1.5 &\quad \text{recombining plasma} \quad (T < T_m) \\ G_o < G < 0.7 &\quad \text{ionizing plasma} \quad (T > T_m) \end{aligned} \quad (8.39)$$

The range of G ratios given above is approximate, and several points need to be noted. While $G \lesssim 1.0$ is generally valid for sources in collisional equilibrium around the temperature of maximum abundance T_m of the He-like stage, the G values may vary widely under non-equilibrium and transient conditions. $G(T)$ is particularly sensitive in the low-temperature range, and is a useful indicator of the nature of the plasma source. Photoionized plasmas with a radiation source as the dominant ionization mechanism have lower electron temperatures than collisionally ionized plasmas. Examples of the former include H II regions, such as planetary nebulae ionized by hot stars, and active galactic nuclei ionized by a central radiation source powered by supermassive black hole activity.

The ratio $G(T)$ begins with a characteristically high value G_o in the low temperature limit $T \ll T_m$ for each ion, and decreases with temperature, as shown in Fig. 8.11 for several He-like ions. This is because, at low temperatures, collisional excitation from the ground state to the energetically high $n = 2$ levels is small compared with (e + ion) recombination, which preferentially populates the triplet levels. As the temperature increases, the high excitation rate coefficient of the w-line begins to outweigh the triplet $(x + y + z)$ excitation rate, and $G(T)$ decreases. Another reason for the high value of G_o is that at lower temperatures the Li-like ionization state is more abundant (albeit much more transient than the He-like stage), and therefore inner-shell excitation–ionization into the 2^3S_1 level contributes to the enhancment of the z-line. For the reasons discussed above the observed $G(T)$ vs. the theoretical G_o $(T \ll T_m)$ is an excellent discriminant between photoionization vs. collisionally dominated sources. Another trend that is discernible from Fig. 8.11 is that at $T \gg T_m$, the temperature of maximum ionic abundance in collisional (coronal) equilibrium, the G ratio tends to rise, owing to ensuing recombinations from the increasing fraction of the H-like ionization state.

8.4.5 Transient X-ray sources

We have described the essential astrophysics inherent in the R and G ratios, and their diagnostic value for density and temperature. However, the limits on temperature refer to extreme conditions of ionization, and do not provide a complete temporal evolution of a plasma source with temperature changing with time. Examples of these sources may be found in X-ray flares and in laboratory sources for magnetic and inertial confinement nuclear fusion [21]. Once again, we continue the reference to the useful He-like ions. In the discussion below, we select Fe XXV, which covers the high-temperature range and a wide range of plasma conditions.

First, we write down the complete set of coupled equations for (i) ionization balance and (ii) collisional–radiative

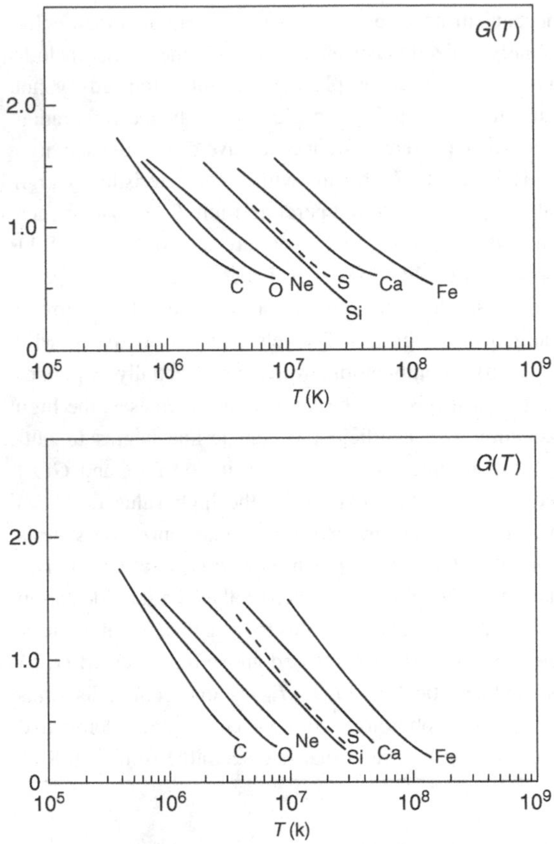

FIGURE 8.11 The line ratio $G = (f + i)/r = (x + y + z)/w$: temperature and ionization diagnostics with X-ray lines of He-like ions [233]. Upper panel – collisional equilibrium; lower panel – neglecting recombination (ionizing plasma).

models of He-like ions. In addition, we can also account for non-ionization equilibrium and the presence of a radiation source characterized by an *ionization parameter* U, defined as the ratio of local photon to electron densities[9]

$$U = \frac{1}{N_e} \int_{I_H}^{\infty} \frac{\Phi(\nu)}{j\nu} d\nu, \qquad (8.40)$$

where $\Phi(\nu)$ is the radiation field intensity. For a black body $\Phi(\nu) \equiv B_\nu$, given by the Planck function and a reasonable approximation for a stellar source. However, for active galactic nuclei with a non-thermal radiation source, the $\Phi(\nu)$ may be taken to be a power law $\Phi(\nu) = \nu^{-\alpha}$, where α is the so-called photon index typically in the range of 1.0–2.0 (discussed in Chapter 11). In transient sources, such as X-ray flares, the temperature variations imply that the level populations are time dependent. The general form of the time-dependent coupled equations is

$$\begin{aligned} \frac{dN_i}{dt} = & \sum_{j>i} N_j A_{ji} - N_i \sum_{j<i} A_{ij} + \alpha_i^{H}(T) X_{H/He} \\ & - \alpha^i X_{Li/He}(T) N_i + N_e \sum_{j \neq i} N_j q_{ji}^{e}(T) \\ & - N_e N_i \sum_{j \neq i} q_{ij}^{e} + N_e \sum_{j \neq i} N_j q_{ji}^{p}(T) \\ & - N_e N_i \sum_{j \neq i} q_{ij}^{p} + N_e \sum_{j \neq i} N_j q_{ji}^{\alpha}(T) \\ & - N_e N_i \sum_{j \neq i} q_{ij}^{\alpha} + N_e C_i^{Li}(T) X_{Li/He} \\ & - N_e C_H^{i}(T) N_i + \sum_{j \neq i} N_j \Phi(h\nu_{ji}) B_{ji} \\ & - N_i \sum_{j \neq i} \Phi(h\nu_{ij}) B_{ij}. \end{aligned} \qquad (8.41)$$

In addition to the processes discussed thus far, we have introduced 'particle' impact rate coefficients, denoted as q^e (electron collisions), q^p (proton impact), and q^α (α-particle impact). Although at first sight collisions between positively charged protons or helium nuclei, on the one hand, and highly (positively) charged He-like ions, on the other hand, may seem very improbable, they are in fact quite significant in *redistribution* of population among excited levels [23, 25, 235]. Proton impact excitation is particularly effective in mixing level populations among closely spaced levels, such as the fine-structure levels of positive ions when the ratio of transition energy to kinetic temperature $\Delta E/kT \ll 1$. This condition is satisfied in high-temperature X-ray plasmas with $T_e > 10^6$ K. At high energies, protons have much larger angular momenta, owing to their mass ($\ell\hbar = mvr$), which is 1836 times heavier than electrons. Therefore, the partial wave summation over ℓ (Chapter 5) yields a large total cross section. The proton impact rate coefficient for fine structure transitions $2^3S_1 \rightarrow 2^3P_J$ can exceed that due to electron impact [235]. Morever, the proton density $n_p \sim 0.9\ n_e$ in fully ionized plasmas, so the total rate is quite comparable to that due to electron impact.

For plasma sources in equilibrium, the left-hand side of Eq. 8.41 is zero, and it corresponds to quiescent plasmas in, for example, active but non-flaring regions of stellar coronae. For transient sources, we may parametrize the electron temperature as a function of time: T_e (t). Of course, the electron density may also vary, but we confine

[9] There are several analogous definitions of ionization parameters in the literature with reference to differing particle densities, such as the number density of H-atoms.

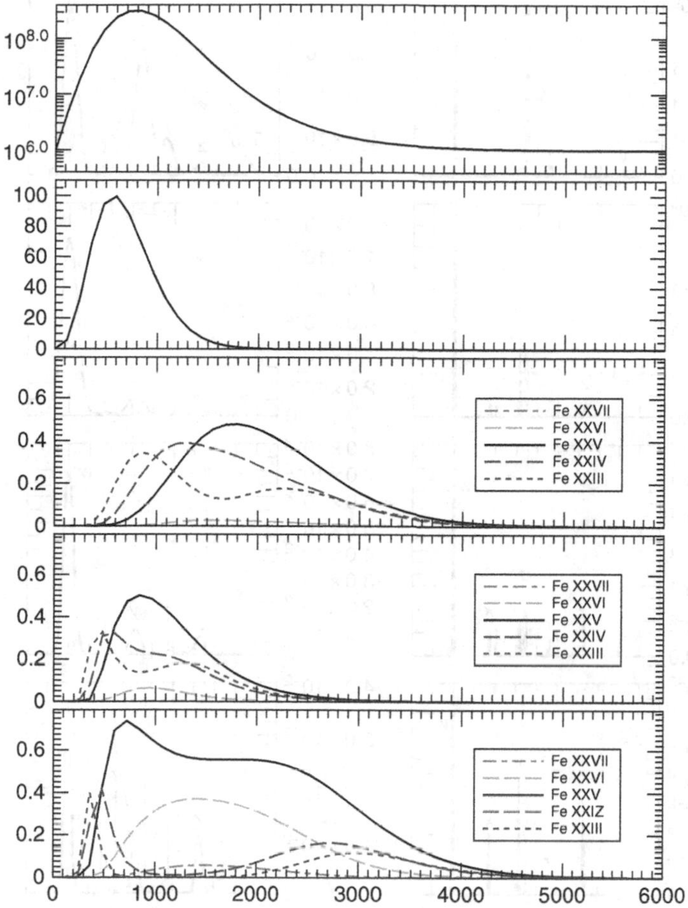

FIGURE 8.12 Time-dependent parameters and ionization fractions for He-like FeXXV and other ionization stages, with variations up to approximately one hour (3600 s). The topmost panel is the electron temperature profile $T_e(t)$. The second panel from the top is the ionization parameter $U(t)$. The middle panel shows corresponding ionization fractions from neighbouring ionization stages of Fe XXV, from Fe XXIII to fully ionized Fe XXVII, for the collisional ionization case with $T_e(t)$; the second panel from the bottom is for photoionization by an external source (using $U(t)$, as shown), and the bottom-most is the hybrid case with both $T_e(t)$ and $U(t)$. Note that the ionization states in a purely photoionized plamsa exist at much lower temperatures (times) than the pure collisional case.

this illustrative discussion to temporal temperature variations only.[10] With respect to the He-like lines and the $K\alpha$ complex, three general categories of plasma conditions may be considered: (i) collisional ionization, without a photoionizing external radiation field, (ii) photoionization, characterized by an ionization parameter (essentially the ratio of photon/electron densities), and (iii) a 'hybrid' situation with collisional ionization *and* photoionization. Note that conditions (ii) and (iii) may both be 'controlled' numerically by adjusting the ionization parameter. Figure 8.12 is the temperature profile T_e(t), ionization parameter $U(t)$, and ionization fractions, for a numerical simulation of time-dependent plasmas for all three cases [236]. The resulting spectra may correspond to an X-ray flare of an hour duration (say, a solar flare).

Figure 8.13 displays the intensity variations of the Fe XXV $K\alpha$ complex, including the dielectronic satellite lines (Chapter 7), which range over 6.6–6.7 keV. The rather complex Fe XXV spectrum in Fig. 8.13, in contrast with the simple O VII spectrum in Fig. 8.8, is largely the result of the presence of a number of KLL dielectronic satellite (DES) lines (Chapter 7), interspersed among the primary lines w, x, y and z. At low temperatures, the DES lines dominate the spectrum, and the total $K\alpha$ intensity is shifted towards lower energies by up to 100 eV below the principal Fe XXV w-line at 6.7 keV. At high temperatures, the DES diminish, relative to the intensity of the w-line (see Chapter 7). This has important consequences in ascertaining the nature of the plasma in the source. Iron ions, Fe I–Fe XVI, in ionization stages less than Ne-like with a filled 2p-shell, give rise to the well-known fluorescent emission $K\alpha$ line at 6.4 keV, which is formed in relatively 'colder' plasmas, such as in accretion discs around black holes $T < 10^6$ K ([239], Chapter 13). On the other hand, in higher temperature central regions of active galactic nuclei, the 6.7 keV $K\alpha$ complex of He-like Fe XXV is observed. For example, from the proximity of the star Sagittarius A* at the galactic centre of the

[10] The He-like line ratios in transient astrophysical and laboratory plasmas have been discused in considerable detail in [207, 236, 237, 238].

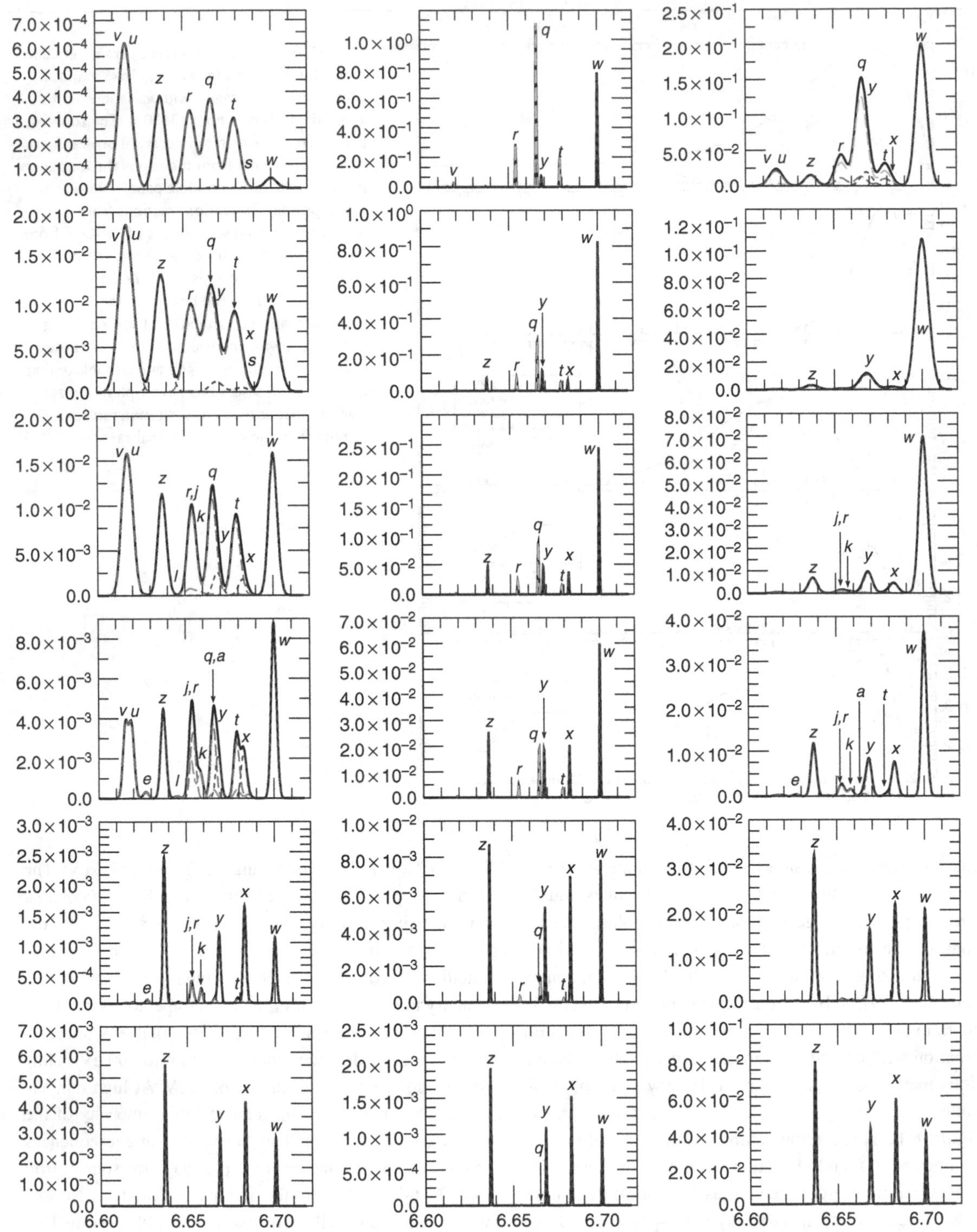

FIGURE 8.13 Spectral simulation of a transient plasma in the 6.6–6.7 keV $k\alpha$ complex of Fe XXV, with time-dependent parameters as in the previous figure. The left panels correspond to the pure collisional case applicable to flaring activity in stellar coronae, the middle panels to a photoionized plasma, and the right-hand side panels to a hybrid photo-excitation–photoionization *and* collisional case. The forbidden (f or z), intersystem (i or $x+y$), and the resonance (r or w) lines are shown, together with the dielectronic satellite lines (a–v). The six panels from top to bottom correspond to $t = 480$, 1080, 1320, 1560, 1920 and 2400 seconds. Note the enhanced dielectronic satellite spectra in the collisional case (left-hand side top two panels, $t = 480$ s, 1080 s) corresponding to low T_e during 'rise' times, and the 'spectral inversion' between the forbidden z-line and the resonance w-line at late times (left-hand side bottom two panels, $t = 1920$ s, 2400 s) also at low T_e [236].

Milky Way [240], the observed X-ray flux lies in the range 6.6–6.7 keV, which could be due to the DES of Fe XXV. Thus the precise observed energy of the $K\alpha$ complex, dependent on the DES, is potentially a discriminant of the temperature, dynamics, composition and other extrinsic macroscopic variables in a variety of sources and regions therein.

In low-resolution spectra, the KLL DES lines are blended together with the x, y, z lines, and the higher-n KLn satellites are unresolved from the w-line [211, 241, 242]. It is then useful to redefine the G ratio as $GD(T)$

$$GD \equiv \frac{(x+y+z+KLL)}{(w+\sum_{n>2} KLn)}. \tag{8.42}$$

The ratio $GD(T)$ for Fe XXV is compared for a variety of astrophysical situations in Fig. 8.14 [207]. The reference temperature T_m is that of maximum abundance of Fe XXV in collisional equilibrium. The top panel in Fig. 8.14 shows that while $G(T)$ remains constant, $GD(T)$ shows a considerable enhancement and sensitivity at low T, owing to the DES contributions. The other two panels represent simulations for three different plasma environments, with different ionization fractions $X_{H/He}$ but no population in Li-like ionization state (middle panel, $X_{Li/He} = 0$), and with different $X_{Li/He}$ and no H-like population ($X_{H/He} = 0$). In spite of large differences in different environments, the one thing that stands out is that at low temperatures $GD \gg 1$, more so and over a much wider temperature range in photoioized plasmas than in coronal ones.

8.5 Far-infrared lines: the boron isoelectronic sequence

We have discussed the utility of using ratios of two lines as diagnostics of physical conditions because most of the systematic effects, such as detector sensitivity and interstellar reddening, affect both lines in the same way. But often in astrophysical sources, we may be able to observe only a single line in a given wavelength band. In this section, we look at one such well-known transition.

Fine structure transitions between closely spaced levels of the ground state in several ions are of immense diagnostic value. That is especially the case when the atomic system can be treated as a simple two-level problem, decoupled from higher levels. Lines from B-like

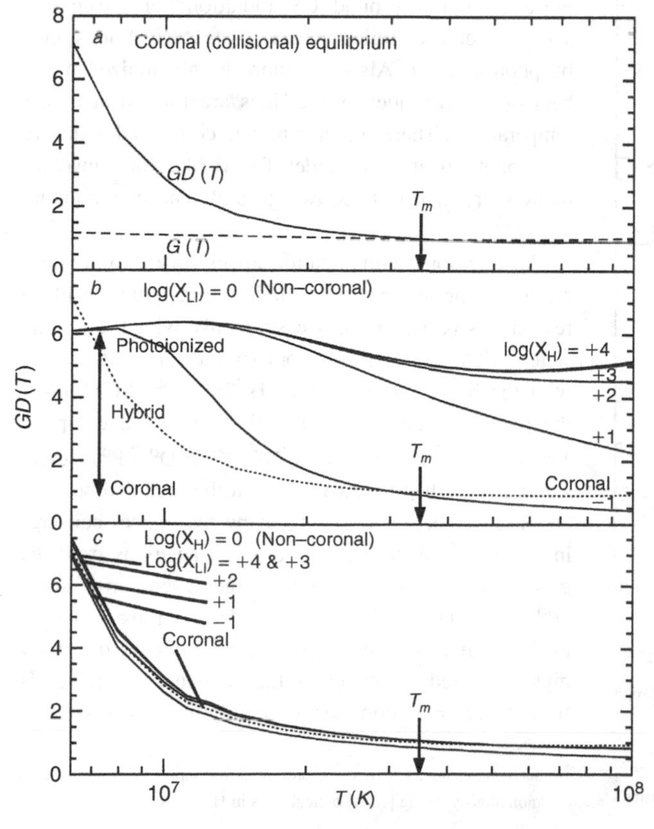

FIGURE 8.14 X-ray line ratios $G(T)$ and with dielectronic satellites $GD(T)$ for FeXXV in coronal ionization equilbrium and non-coronal plasmas.

ions are particularly useful, ranging from the FIR lines [C II] at 157 μm, [N III] at 58 μm, [O IV] at 26 μm, etc., up to short wavelengths in the EUV for [Fe XXII] at 846 Å. They arise from the same ground-state fine-structure transition $1s^2 2s^2 2p\left({}^2P^o_{3/2} - {}^2P^o{}_{1/2}\right)$ in B-like ions (see energy level diagram in Fig. 6.6). Since the electron kinetic energy in low-temperature sources, such as the interstellar medium is much smaller than the *LS* term differences, we may ignore higher terms (Fig. 6.6) in writing down the line emissivity as simply a function of upper level population within the ground state ${}^2P^o_{1/2,3/2}$, i.e.,

$$\epsilon\left({}^2P^o_{3/2} - {}^2P^o_{1/2}\right) = N\left({}^2P^o{}_{3/2}\right) A\left({}^2P^o_{3/2} - {}^2P^o_{1/2}\right) \times h\nu\left({}^2P^o{}_{3/2} - {}^2P^o_{1/2}\right)/4\pi. \tag{8.43}$$

Figure 8.15 is the collision strength for the [C II] 157 Å transition. The $\Omega\left({}^2P^o_{1/2} - {}^2P^o_{3/2}\right)$ is dominated by a large

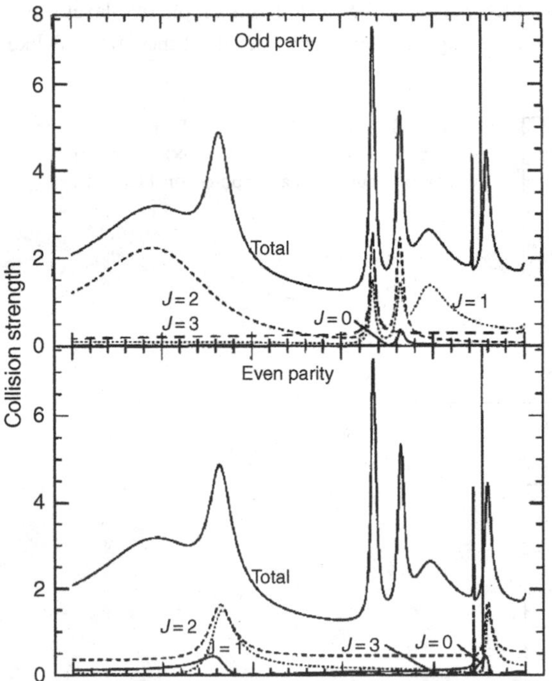

FIGURE 8.15 The collision strength $\Omega\left({}^2P^o_{1/2} - {}^2P^o_{3/2}\right)$ for the FIR [CII] 157 Å line (Fig. 1 from [243]). The two panels show a comparison of partial wave contributions of odd and even symmetries $J\pi$ to the total collision strength, dominated by low-J angular momenta. The broad near-threshold resonance enhances the rate coefficient considerably at low temperatures.

near-thresold resonance, which enhances the effective rate coefficient by several factors [243]. If we assume that no other levels are involved in the collisional–radiative model, then the probability of downward decay is unity: every excitation upwards must be followed by the emission of photon following downward decay. Then the flux emitted in the line, or the intensity, is

$$I\left({}^2P^o_{3/2} - {}^2P^o_{1/2}\right) = n_e n_i q\left({}^2P^o_{3/2} - {}^2P^o_{1/2}\right) h\nu/4\pi, \tag{8.44}$$

where q is the rate coefficient in $cm^3\ s^{-1}$, which determines the upper level population of ${}^2P^o_{3/2}$, and $h\nu$ is the transition energy (strictly speaking the Einstein rate $A\left(2P^o_{3/2} - {}^2P^o_{1/2}\right)$ should be included on the right, but since all decays are to the ground level only the intensity depends on the number of ions excited to the upper level alone, see Eq. 8.45 below). At low temperatures where the FIR lines of [C II] and other B-like ions are formed, Eq. 8.44 is a good approximation since (i) other higher levels are not collisionally excited and (ii) ionization balance, and hence recombinations to higher levels from higher ionization stages, may be neglected compared with the fast collisional excitation rate for ${}^2P^o_{1/2} \rightarrow {}^2P^o_{3/2}$ transition within the ground state. However, in the presence of background UV radiation fields from hot stars in nebulae higher levels of boron-like ions may be photo-excited. Also, for more highly ionized members of the B-sequence, the lines are formed at higher temperatures. Therefore, a complete collisional–radiative plus photo-excitation model for B-like ions involves many more than just the two ground state fine structure levels.[11]

An additional complication arises owing to configuration mixing and channel coupling effects that result in resonances (Chapter 5). We start only with the ground state $1s^2 2s^2 2p\ {}^2P^o$, but configuration mixing occurs with the higher configuration $1s^2 2p^3\ {}^2P^o$. In addition, the other two configurations of the $n = 2$ complex, $1s^2 2s2p^2({}^4P, {}^2D, {}^2S, {}^2P)$, $1s^2 2p^3({}^4S^o, {}^2D^o, {}^2P^o)$ – eight *LS* terms in all (Fig. 6.6) – need to be included because coupled channels give rise to many resonances converging on to higher terms. The dipole transitions from the ground state to the excited doublet even-parity terms ${}^2P^o \rightarrow ({}^2D, {}^2S, {}^2P)$ are particularly strong and coupled together. The situation is even more involved for highly ionized members of the sequence, since levels from the $n = 3$ complex of configurations also enter

11 Such an extended CR model is presented in Chapter 12 to explain the anomalously strong [Ni II] optical lines in H II regions.

into the picture. Again, there are strong dipole transitions among the ground state and several of the $n = 3$ configurations (cf. Fig. 6.11). All of these effects manifest themselves most prominently as resonances in collision strengths (and photoionization and recombination cross sections) lying beween the $n = 2$ and the $n = 3$ levels. Collisional calculations of rate coefficients q need to include these resonance structures for accuracy at high temperatures.

Once the basic atomic physics has been incorporated by including as many of the higher levels as necessary to determine the rate coefficients, we may use the two-level model, Eq. 8.44, to obtain single line emissivities. However, since we do not have a line ratio, the emissivities are necessarily dependent on ionic abundances n_i, which must be estimated from other means or derived from observations. For the nearly ubiquitous [C II] 157 μm line in the interstellar medium, the intensity is

$$I([\text{C II}];\ 157\text{Å}) = \frac{h\nu A\left({}^2\text{P}^{\text{o}}{}_{3/2} - {}^2\text{P}^{\text{o}}_{1/2}\right)}{4\pi} \times \frac{N\left({}^2\text{P}^{\text{o}}_{3/2}\right)}{\sum_i N_i(\text{C II})} \times \frac{n(\text{C II})}{n(\text{C})} \times \frac{n(\text{C})}{n(\text{H})} \times n(\text{H})\ \text{erg cm}^{-3}\text{s}^{-1}. \qquad (8.45)$$

Note that the above expression is the same as Eq. 8.44, except that we have included the A-value to ensure a dimensionally correct expression and respective fractions of level population $N\left({}^2\text{P}^{\text{o}}_{3/2}\right)$, as opposed to the sum of all other C II levels (i.e., n(C II)). The ionization fraction C II/C, abundance ratio C/H, and hydrogen density n(H), introduce uncertainties in the absolute determination of line intensity. Nevertheless, the intensity of just one line in an ion may be related to the total atomic density in the medium, as well as its physical state as reflected in excitation and ionization fractions.

9 Absorption lines and radiative transfer

Among the most extensive applications of atomic physics in astronomy is the precise computation of transfer of radiation from a source through matter. The physical problem depends in part on the bulk temperature and density of the medium through which radiation is propagating. Whether the medium is relatively transparent or opaque ('thin' or 'thick') depends not only on the temperature and the density, but also on the atomic constituents of matter interacting with the incident radiation via absorption, emission and scattering of radiation by particular atomic species in the media. Since optical lines in the visible range of the spectrum are most commonly observed, the degree of transparency or opaqueness of matter is referred to as *optically thin* or *optically thick*. However, it must be borne in mind that in general we need to ascertain radiative transfer in all wavelength ranges, not just the optical. Macroscopically, we refer to optical thickness of a whole medium, such as a stellar atmosphere. But often one may observe a particular line and attempt to ascertain whether it is optically thick or thin in traversing the entire medium.

Radiative transfer and atomic physics underpin quantitative spectroscopy. But together they assume different levels of complexity when applied to practical astrophysical situations. Significantly different treatments are adopted in models for various astrophysical media. At low densities, prevalent in the interstellar medium (ISM) or nebulae, $n_e < 10^6$ cm^{-3}, the plasma is generally optically thin (except for some strong lines, such as the Lyα, that do saturate), and consideration of detailed radiative transfer effects is not necessary. Even at much higher densities $n_e \approx 10^{9-11}$ cm^{-3}, such as in stellar coronae (Chapter 10) or narrow-line regions of active galactic nuclei (AGN, Chapter 13), the plasma remains optically thin except, of course, in some strong allowed lines which readily absorb radiation at characterstic wavelengths. Forbidden and intercombination lines are not affected, since their transition probabilities are several orders of magnitude lower than the allowed lines.

At higher densities in the $n_e \approx 10^{12-15}$ cm^{-3} range, such as in stellar atmospheres or even broad-line regions of AGN, radiative transfer effects become crucial in determining spectral properties. At densities higher than stellar atmospheres, moving into the stellar interior, collisional–radiative physics simplifies, owing to eventual dominance of *local thermodynamic equilibrium (LTE)*. In LTE, collisional processes dominate and establish a statistical distribution of atomic level populations at a *local temperature*. These distributions are the well-known *Saha–Boltzmann equations*, employed to understand radiative diffusion in stellar interiors where LTE generally holds (Chapter 11).

In this chapter we begin with the basic concepts and terminology of radiative transfer in astrophysics. We sketch the basic ideas underlying the advanced computational mechanisms from the point of view of applied atomic physics. This area entails an enormously detailed framework, which forms the field of radiative transfer dealt at length in many other works, especially on stellar atmospheres (e.g. [244, 245, 246]).

9.1 Optical depth and column density

When radiation propagates through matter, it is not the *geometrical distance* that is crucial, but rather the *amount* of matter encountered. The attenuation of radiation also depends on the *state* of matter in the traversed medium between the radiation source and the observer. Therefore, we need a measure that includes the macroscopic state (temperature, density) of the intervening material, and microscopic atomic properties thereof (absorption, emission, scattering). We define a dimensionless quantity called the *optical depth* τ, such that the probability of transmission of radiation (photons) decreases exponentially as $e^{-\tau}$. Starting from a given point $\tau = 0$, as τ increases the probability $P_{esc}(\tau)$ that a photon escapes the medium (and possibly be observed) behaves as

$$\tau \to 0, \quad P_{esc}(\tau) \to 1. \tag{9.1}$$

$$\tau \to \infty, \quad P_{esc}(\tau) \to 0. \tag{9.2}$$

The canonical value of optical depth that divides plasmas into two broad groups based on particle density is taken to be $\tau = 1$: a plasma source is characterized as *optically thin* if $\tau < 1$ and *optically thick* if $\tau > 1$. Physical considerations leading to the definition of the *macroscopic* optical depth require that τ depend on the geometrical distance s, and the *opacity* of the matter κ,

$$\tau = s \times \kappa, \tag{9.3}$$

or

$$\tau = \int \kappa \mathrm{d}s. \tag{9.4}$$

Since τ is dimensionless (note that it appears in an exponential function that determines the escape probability), the dimensions of the opacity coefficient κ are in units of inverse path length, say cm^{-1}. The optical depth and opacity in a given line with frequency ν is

$$\tau_\nu = s \times \kappa_\nu. \tag{9.5}$$

Further reconsideration of the optical depth is useful in particular environments, such as the ISM where densities are very low, typically of the order of a few particles per cm^3. However, the distances are large and the total amount of material, i.e., the number of particles along a given line of sight, may be quite large. The local density or temperature at any given point along the line of sight may vary significantly. But we are primarily concerned with the total number of particles taking part in the attenuation of radiation. Further simplifying the picture, we consider only radiative absorption at a given frequency in one line due to a transition in one ion. Instead of the number density per unit volume, it is useful to define another quantity called the *column density* N_i of just that one ionic species i. The column density N_i is the number of ions in a column of area of 1 cm^2 over a given distance, from the observer to the source. The units of N_i are cm^{-2}, unlike that of volume number density in cm^{-3}, and

$$N_i(\mathrm{cm}^{-2}) = \int n_i(\mathrm{cm}^{-3})\, \mathrm{d}s(\mathrm{cm}). \tag{9.6}$$

The next thing to consider is the *cross section* σ_ν for absorption of the photon in a line at frequency ν by the ion. We can then write down another expression for the optical depth as

$$\tau_\nu = N_i \times \sigma_\nu, \tag{9.7}$$

where σ (cm^2) is the cross section; note that given the units of $N_i(\mathrm{cm}^{-2})$, τ is again dimensionless. Equation 9.7 may be considered as a *microscopic* definition of the optical depth in terms of the cross section of the atoms of the material along the path. Restricting ourselves to a single-line transition in the ion, we can express σ in terms of the absorption oscillator strength f and a line profile factor $\phi(\nu)$ over a range of frequencies

$$\sigma(\nu) = \left(\frac{\pi e^2}{m_e c}\right) f\,\phi(\nu), \tag{9.8}$$

where $(\pi e^2/m_e c)$ has units of cm^2-sec. Although in principle the line profile factor is normalized as

$$\int_{-\infty}^{+\infty} \phi(\nu)\, \mathrm{d}\nu = 1, \tag{9.9}$$

in practice the range of integration is small, and depends on several line-broadening mechanisms, discussed later in this chapter.

As the oscillator strength and the Einstein A and B coefficients are all related to one another (Chapter 4), we can readily write down a number of expressions for the optical depth in terms of any of these quantities, as well as the dominant line-broadening mechanism(s) in the plasma under consideration. For example, for a transition at wavelength λ, the *line centre optical depth* is

$$\tau_0(\lambda) = \int \kappa \, \mathrm{d}\ell = \frac{\sqrt{\pi}\, e^2\, \lambda\, f_\lambda}{m_e c} \left(\frac{M_i}{2kT}\right)^{1/2} n_i\, L, \tag{9.10}$$

where L is the total path length, f_λ is the oscillator strength, M_i is the mass of the ion, and n_i is the average ion density. Note that $n_i(\mathrm{cm}^{-3})\, L(\mathrm{cm}) = N_i(\mathrm{cm}^{-2})$, the column density defined above.

However, the simple concepts sketched above need to be generalized and refined with much more precision, as in the following sections. First, an important set of processes concerns line broadening, due to temperature (thermal broadening), and density of electrons and ions (collisional or pressure broadening). We begin with the basics of line broadening theory that deal with the mechanisms responsible for the oft-appearing line profile factor $\phi(\nu)$.

9.2 Line broadening

All spectral lines have a finite width and a particular profile. It is generally defined as a function of energy removed from, or added to, the observed background or the continuum (Fig. 8.1). Line broadening is an important area of atomic astrophysics since the width and shape of a line is directly dependent on the atomic transitions in question, on the one hand, and the plasma environment on the other hand. The complexity of the subject derives from the intricate interplay between the atomic physics and the plasma physics that must be considered, as manifest in several physical mechanisms responsible for line widths and shapes. To begin with, in the quantum mechanical treatment, there is the fundamental or natural breadth of a line due to the uncertainty principle. The line width must reflect this uncertainty by way of broadening, as well as other radiative and collisional effects: (i) finite lifetimes of energy levels, resulting in natural line width or damping given by radiative decay rates, (ii) the temperature of the ambient plasma in the line formation region, leading to Doppler broadening due to thermal velocity distribution of ions, and (iii) the particle density, involving collisions among electrons, ions and neutrals.

The natural line width is related to the uncertainty principle $\Delta E \Delta t \geq \hbar$. An upper level has a characteristic decay lifetime Δt, which is inversely proportional to the broadening, ΔE (assuming, that the lower level has a much longer lifetime). The calculation of radiation damping, therefore, requires the calculation of all possible radiative decay rates for levels involved in a line transition. The characteristic line profile for radiation damping is a Lorentzian function defined by a specific radiation damping constant.

Thermal Doppler broadening of a line reveals the velocity fields present in the line formation region. In many astrophysical sources it provides a direct measure of the temperature. It can be shown that for a Maxwellian distribution of particles at a given temperature the line shape due to Doppler broadening is a Gaussian function, characterized by a constant parameter depending on the temperature and the mass of the particles. Therefore, a Doppler broadening profile is more constrained to the line centre than a Lorentzian profile.

Collisional broadening (also called pressure broadening) is a more complex phenomenon. Different physical effects account for scattering by electrons, ions (mostly protons) and neutral atoms. Qualitatively, elastic electron–ion scattering tends to broaden energy levels, whereas ion–ion scattering induces Stark broadening due to splitting into sublevels. Whereas elastic electron–ion collisions are mainly responsible for broadening of lines from non-hydrogenic ions, Stark broadening is particularly effective for hydrogenic ions where the ℓ-degeneracy is lifted in the presence of the electric field of another ion. This is because the Stark effect is *linear* for hydrogenic systems, with the energy levels splitting into a number of sublevels. It follows that for highly excited states of any ion, with many $n\ell$ levels, both the electron collisions and Stark broadening would be effective. In relatively cold plasmas broadening by collisions of neutral atoms may also be significant, mediated by the long-range van der Waals interaction. The collisional broadening line shape is also a Lorentzian, with characteristic damping constants for each of the processes mentioned.

Finally, in cases where several types of broadening mechanisms manifest themselves, given by Gaussian and Lorentzian functions, the total line profile is obtained by a convolution over both functions resulting in a Voigt profile. In this chapter we discuss these basic concepts.

9.2.1 Natural radiation damping

A spectrum consists of electromagnetic energy at a range of frequencies. For a spectral line, the distribution of energy spans a certain range as a function of time. First, we describe the fluctuations (oscillations) of energy with time in general, before particularizing the discussion to individual modes of line broadening.[1]

A wavetrain of photons passing a given point in space (or impinging on a detector) may be specified by the conjugate variables, angular frequency ω and time t. It is convenient to think of a radiating oscillator emitting the energy spectrum as a function of ω and time dependent amplitude $a(t)$. The value at a definite ω involves integration over all time, and the value at time t involves integration over all ω. Mathematically, such functions are transforms and inverse transforms of each other. The most appropriate transform in our context is the Fourier transform (FT), which entails decomposition of a sinusoidal wave in terms of a basis set of sines and cosines, or generally by $\exp(i\omega t) = \cos \omega t + i \sin \omega t$, involving both functions. At frequency ω the FT is defined as

$$F(\omega) \equiv \int_{-\infty}^{+\infty} a(t)\, \mathrm{e}^{-i\omega t}\, \mathrm{d}t, \tag{9.11}$$

and the reciprocal inverse transform is

$$a(t) = \frac{1}{2\pi} \int_{-\infty}^{+\infty} F(\omega)\, \mathrm{e}^{i\omega t}\, \mathrm{d}\omega. \tag{9.12}$$

Note that since we are dealing with both variables t and ω, the FT and its inverse yield the spectrum as a function of time and the frequency; the latter is the *energy spectrum*, defined as

[1] This section essentially follows the treatment in *Stellar Atmospheres* by D. Mihalas [244].

$$E(\omega) \equiv \frac{1}{2\pi} F^*(\omega) F(\omega)$$
$$= \frac{1}{2\pi} |F(\omega)|^2 = \frac{1}{2\pi} \left| \int_{-\infty}^{+\infty} a(t) e^{-i\omega t} dt \right|^2 . \quad (9.13)$$

It is easily seen that

$$\int_{-\infty}^{+\infty} E(\omega) d\omega = \int_{-\infty}^{+\infty} a^*(t) a(t) dt$$
$$= \frac{1}{2\pi} \int_{-\infty}^{+\infty} F^*(\omega) F(\omega) d\omega . \quad (9.14)$$

Similarly, the *power spectrum* is defined in terms of the product $a^*(t)a(t)$, which is the power at a given instant t. The power spectrum or the radiation intensity is defined in terms of energy per unit time, and obtained by integrating Eq. 9.14 over a full cycle of oscillation and divided by the time period T of long duration,

$$P(\omega) \equiv \lim_{T \to \infty} \frac{1}{2\pi T} \left| \int_{-T/2}^{+T/2} a(t)\, e^{-i\omega t}\, dt \right|^2 . \quad (9.15)$$

The integral averages to zero for a single radiating oscillator emitting a light pulse of short duration, but averaged over an infinite time interval. However, in reality there is usually an ensemble of many oscillators radiating incoherently, so that the net emitted power is non-zero.

9.2.1.1 *Damping constant and classical oscillator*

The interaction of the electron in a radiation field of frequency ω may be treated as a damped classical oscillator with the equation of motion,

$$m_e \ddot{x} + \gamma \dot{x} + \omega_0^2 x = 0. \quad (9.16)$$

Here the second term represents the 'friction', in terms of a damping constant, γ, and the third term the 'restoring force'.

Exercise 9.1 *Show that the solution of the above equation is* $x = x_0 e^{i\omega t} e^{-\gamma t/2}$, *neglecting terms quadratic terms in the classical damping constant* $\gamma \equiv (2/3)(e^2\omega_0^2)(m_e c^3)$; ω_0 *is the central frequency. Hence, derive*

$$F(\omega) = \frac{x_0}{i(\omega - \omega_0) + \gamma/2}, \quad (9.17)$$

and the energy spectrum of a single oscillator

$$E(\omega) = \frac{x_0^2}{2\pi} \frac{1}{(\omega - \omega_0)^2 + (\gamma/2)^2}. \quad (9.18)$$

The characteristic line profile due to radiation damping of a given oscillator is manifest in Eq. 9.18, and is the well-known Lorentzian function. The total power spectrum $P(\omega)$, or the radiated intensity $I(\omega)$, of a whole ensemble of many oscillators is proportional to that of a single one, and should have the same basic form, i.e.,

$$I(\omega) = \frac{C}{(\omega - \omega_0)^2 + (\gamma/2)^2}$$
$$= \frac{\gamma/2\pi}{(\omega - \omega_0)^2 + (\gamma/2)^2}, \quad (9.19)$$

where the constant C is obtained on imposing the normalization

$$\int_{-\infty}^{+\infty} I(\omega)\, d\omega. \quad (9.20)$$

For a *Lorentzian profile* from natural radiation damping, the power or the intensity $I(\omega)$ drops to half its peak value at $\Delta\omega = \omega - \omega_0 = \pm\gamma/2$, which defines γ as the *full width at half maximum (FWHM)* of the line. Now we note that the expression above is simply the spectral distribution or the oscillator strength per unit frequency

$$I(\omega) = \frac{df}{d\omega}, \quad (9.21)$$

which is also directly related to the photoabsorption (or photoionization) cross section (see Chapter 6).

The uncertainty principle can be used to relate the classical treatment outlined thus far to the quantum mechanical picture. The radiating oscillator corresponds to a transition from an excited state k to a lower state i in an atom. The time-dependent probability of radiative decay of state k is

$$|\psi_k|^2 \exp(-t/\tau_k) = |\psi_k|^2 \exp(-\Gamma t), \quad (9.22)$$

where τ_k characterizes the decay time of the excited state, defined as the *lifetime* (τ_k is not to be confused with the same usual notation for the optical depth τ). The reciprocal of the lifetime is inversely proportional to the decay rate, the analogue of the classical damping constant γ, given by $\Gamma_k = 1/\tau_k = A_{ki}$, the spontaneous decay rate for transition $k \to i$. As already mentioned, the uncertainty principle leads to natural broadening of a spectral line since the upper and the lower levels do not have exactly the energy E_k and E_i respectively (except when i is the ground state), but non-zero energy widths associated with each level. Following through the FT analysis of the decaying probability amplitude, we obtain an expression for the intensity spread in the quantum mechanical case in frequency space,

$$I(\omega) = \frac{\Gamma/2\pi}{(\omega - \omega_0)^2 + (\Gamma/2)^2}. \quad (9.23)$$

Γ is the total width of the line, depending on the individual widths of the lower and the upper levels. For multiple levels,

$$\Gamma_k = \sum_{j<k} A_{kj}, \quad \Gamma_i = \sum_{\ell<i} A_{i\ell}. \tag{9.24}$$

It follows that $\Gamma = \Gamma_i + \Gamma_k$, and $\Gamma/2$ is the natural half-intensity width or FWHM of the Lorentzian spectral profile.

Another way to relate the damping of energy from a classical oscillator to quantum mechanical transition probabilities is to consider the decay rate

$$\frac{dE}{dt} = -\Gamma E, \tag{9.25}$$

where $\Gamma = (-2/3)(e^2\omega^2/m_e c^3)$, and $E = E_0 e^{-\Gamma t}$. If Γ is evaluated classically from the constants given, then we obtain the FWHM $\Gamma/2 = \Delta\lambda = 0.6 \times 10^{-4}$ Å for all lines. But this classical width is far smaller than observed, which depends on several quantum mechanical effects. Since the radiated energy depends on the upper level population, i.e., $E = N_k\, h\nu_{ik}$, setting $dE/dt = d(N_k h\nu)/dt$, we again obtain $\Gamma = A_{ki}$, the quantum mechanical probability of spontaneous decay.

Three other important points may be emphasized about line shapes. First, the line profile function is normalized to unit intensity according to $\int \phi(\nu) d\nu = 1$ (Eq. 9.20). Second, if we assume detailed balance, then both the emission and the absorption profiles have the same form. Since, for random atomic orientations in a plasma, a photon may be emitted in any direction, the observed emission in a given direction is divided by a factor of 4π (whereas absorption of energy along a line of sight is obviously one-dimensional or uni-directional). Third, since natural radiative broadening depends on the spontaneous decay rates A_{ki}, the stronger dipole transitions would have the broadest profiles so long as other broadening mechanisms are small; contrariwise, forbidden lines with small A-values are generally narrow. The first condition also means that the normalized line profile factor can be used to multiply the oscillator strength to obtain the spread in frequency (energy) of the absorption coefficient for a given line, i.e.,

$$a_\nu = \left(\frac{\pi e^2}{m_e c}\right) f\,\phi(\nu)$$
$$= \left(\frac{\pi e^2}{m_e c}\right) f \left[\frac{\Gamma/4\pi^2}{(\nu-\nu_0)^2 + (\Gamma/4\pi)^2}\right]. \tag{9.26}$$

9.2.2 Doppler broadening

The Doppler effect implies that the frequency of radiation emitted by an atom is higher if the relative movement of the source and the observer is towards each other, and the frequency is shorter if they are receding from each other; the effect on the wavelength is the inverse of that on the frequency. The frequency or the Doppler shift depends on the velocity of that atom, which in turn depends on the ambient temperature T. In most astrophysical plasmas the velocity distribution of particles is characterized by a Maxwellian function $f_{\rm Max}$ at that temperature. The probability of atoms with a line-of-sight velocity between v and $(v + dv)$ is given by

$$f_{\rm Max}(v)\, dv = \left(\frac{1}{v_0 \pi^{1/2}}\right) e^{-(v^2/v_0^2)}\, dv. \tag{9.27}$$

Kinetic theory relates the root mean square velocity to temperature as $m\langle v^2\rangle/2 = (3/2)kT$, and therefore we have

$$\langle v^2\rangle = \int_{-\infty}^{+\infty} v^2 f_{\rm Max}(v)\, dv = \frac{v_0^2}{2}, \tag{9.28}$$

with

$$v_0 = \sqrt{\frac{2kT}{M}} = 1285 \sqrt{\frac{T(\mathrm{K})}{10^4 A_M}}\ \mathrm{m/s}. \tag{9.29}$$

where M and A_M are the mass and the atomic weight of the atom, respectively. We need to describe the total absorption in a line at a given frequency ν, absorbed by an atom in its rest frame at the Doppler-shifted frequency $\nu(1 - v/c)$. The total absorption coefficient is obtained by integrating the absorption coefficient of an atom $a_\nu[\nu(1 - v/c)]$ over all velocities,

$$a_\nu = \int_{\infty}^{+\infty} a_\nu[\nu(1 - v/c)]\, f_{\rm Max}(v)\, dv. \tag{9.30}$$

As expected, the absorption coefficient of a line is related to the oscillator strength f of the corresponding transition as

$$a_\nu = \left(\frac{\pi e^2}{mc}\right) f. \tag{9.31}$$

The Doppler shift in frequency of a photon radiated by an atom at line-of-sight velocity v is $\Delta\nu = \nu v/c$. Therefore, the symmetric *Doppler width* about the central frequency ν_0 of a line is defined as

$$\Delta\nu_D \equiv \left(\frac{v}{c}\right) \nu_0, \tag{9.32}$$

and the normalized Gaussian line profile is

$$a_\nu = \frac{1}{\sqrt{\pi}\,\Delta\nu_D} \exp\left[-\left(\frac{\nu-\nu_0}{\Delta\nu_D}\right)^2\right]. \tag{9.33}$$

The functional form is a Gaussian, describing thermal motions of particles given by a Maxwellian velocity distribution characterized by the parameter v_0 at temperature T. However, in general we also have the natural radiation damping profile factor, which is a Lorentzian (Eqs 9.19 and 9.23). Therefore, the combined (radiation + Doppler) profile is a convolution of a Gaussian and a Lorentzian function (Fig. 9.1). This leads to the full representation of a line profile subject to both forms of line broadening mechanisms (but no others) as

$$a_\nu = \left(\frac{\pi^{1/2}e^2}{m_e c}\right)\left(\frac{f}{\pi v_0}\right) \times \int_{-\infty}^{+\infty}\left[\frac{(\Gamma/4\pi)\,e^{-v^2/v_0^2}}{(\nu-\nu_0-\nu_0 v/c)^2+(\Gamma/4\pi)^2}\right] \times e^{-v^2/v_0^2}\,dv. \tag{9.34}$$

The convolution of a Gaussian and a Lorentzian function yields a *Voigt function* (heuristically: Voigt → Gaussian ⊗ Lorentzian). The above expression may be rewritten using the variables $x \equiv (\nu-\nu_0)/\Delta\nu_D$, $y \equiv \Delta\nu/\Delta\nu_D = v/v_0$ and $b = \Gamma/(4\pi\Delta\nu_D)$,

$$a_\nu = \left(\frac{\pi^{1/2}e^2}{m_e c}\right) f \left(\frac{H(b,x)}{\Delta\nu_D}\right), \tag{9.35}$$

where the Voigt function $H(b,x)$ is defined as

$$H(b,x) \equiv \frac{b}{\pi}\int_{-\infty}^{+\infty}\frac{e^{-y^2}dy}{(x-y)^2+b^2}. \tag{9.36}$$

The total line width is the sum of the partial widths due to all line broadening mechanism $\Gamma = \Gamma_{rad} + \Gamma_{coll}$. Algorithms for computing Voigt functions have been developed for a variety of applications (e.g., [247]).

Exercise 9.2 *Show that the normalization of $H(b,x)$ is such that the integral over all x is $\pi^{1/2}$. Also, show that for $(x^2+b^2) \gg 1$, $H(b,x) \approx \pi^{1/2}/(x^2+b^2)$.*

The physical nature of the two effects embodied in the Voigt function manifests itself clearly in the limiting cases of the two variables: $a \ll 1$, when the Doppler–Gaussian width is large compared with the natural Lorentzian width, and $x \gg 1$ when the Lorentzian component of the Voigt profile dominates; thus we may represent,

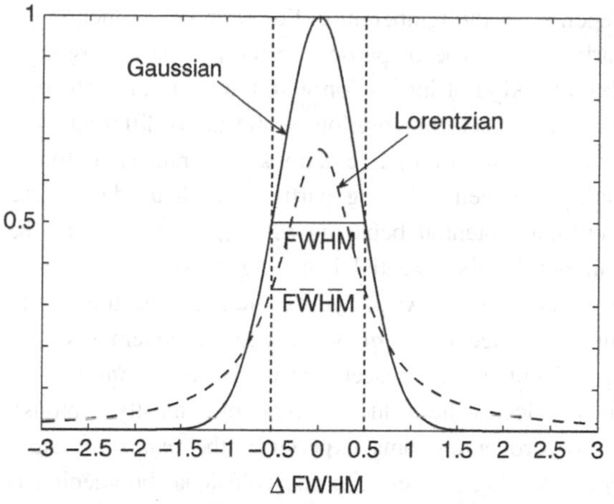

FIGURE 9.1 The Gaussian and Lorentzian line shapes as function of the full width at half maximum (FWHM) separation from line centre. The Lorentzian falls off much more slowly than the Gaussian. Whereas the Gaussian dominates the line core (or is confined to it), the Lorentzian dominates in the line wings out to several times the FWHM.

$$H(b,x) \to e^{-x^2} + \frac{b}{\sqrt{\pi}x^2}. \tag{9.37}$$

Since x measures the separation in frequency from the line centre, the first term (Gaussian) dominates in the line core, decaying exponentially with x, and the second term (Lorentzian) dominates in the line wings, decaying much more slowly as x^{-2}. In other words, thermal broadening dominates the line core, whereas natural radiative damping manifests itself in the line wings. Figure 9.1 shows schematically the forms of line broadening discussed thus far. The relative intensity (in arbitrary units) is plotted as a function of FWHM units from line centre. As we shall see, collisional or pressure broadening, which dominates at high densities, is also Lorentzian in nature, and may considerably alter the picture developed thus far at high densities.

9.2.3 Collisional broadening

The interaction of an atom with other particles in the plasma leads to broadening of spectral lines, since such interactions perturb upper and lower levels to broaden the energy range of the transition. The magnitude of broadening depends on the particle density in the source as well as the temperature. Since the pressure in a gas depends on the temperature and density, collisional broadening is also referred to as *pressure broadening*. The excitation energy of atomic levels comes into play, since higher levels are less strongly bound to the nucleus than lower ones and are more perturbed; their excitation energy itself depends on the temperature. Furthermore, as mentioned earlier, each type of particle in the plasma undergoes a different kind of interaction with the atom. Free electron impact or electron–atom(ion) scattering is different from atom(ion)–atom(ion) interactions. The ranges of forces acting between colliding particles, such as due to the Coulomb potential behaving as $V_{Coul} \sim r^{-1}$, up to the van der Waals potential behaving as $V_{vw} \sim r^{-6}$, need to be considered. Also, hydrogenic ions, and those with highly excited levels nl, subject to ℓ-degeneracy within an n-complex, are susceptible to the *linear Stark effect* in the electric field due to other ions (mostly protons); non-hydrogenic atoms experience the *quadratic Stark effect*. Owing to these factors, collisional broadening is particularly difficult to treat precisely. Nevertheless, we begin with the simple classical impact approximation to illustrate some basic features.

9.2.3.1 *The classical impact approximation*

If the atom is treated as a radiating oscillator, then a sufficiently close interaction with another particle leads to an abrupt phase change in the wave train, otherwise freely propagating and described with a monochromatic frequency ω_0 and the function

$$a(t) = e^{i\omega_0 t}. \tag{9.38}$$

Let t be the time interval between two successive collisions. Then the Fourier decomposition in terms of the frequency variable ω is

$$F(\omega, t) = \int_0^t e^{i(\omega_0-\omega)t'} dt' = \frac{e^{i(\omega-\omega_0)t} - 1}{i(\omega - \omega_0)}. \tag{9.39}$$

However, the time interval t may vary considerably about a *mean collision time* t_0 for collision between any two particles in the ensemble. The probability that a collision takes place within a time interval dt is dt/t_0, or

$$P(t)\, dt = \frac{e^{-t/t_0}}{t_0} dt. \tag{9.40}$$

The energy spectrum averaged over t_0 is then

$$E(\omega) = \langle E(\omega, t)\rangle_{t_0} = \frac{1}{2\pi}\int_0^\infty F^*(\omega, t)\, F(\omega, t)\, P(t)\, dt. \tag{9.41}$$

Normalizing,

$$\int E(\omega)\, d\omega = 1, \tag{9.42}$$

and integrating we obtain the frequency dependent spectrum as

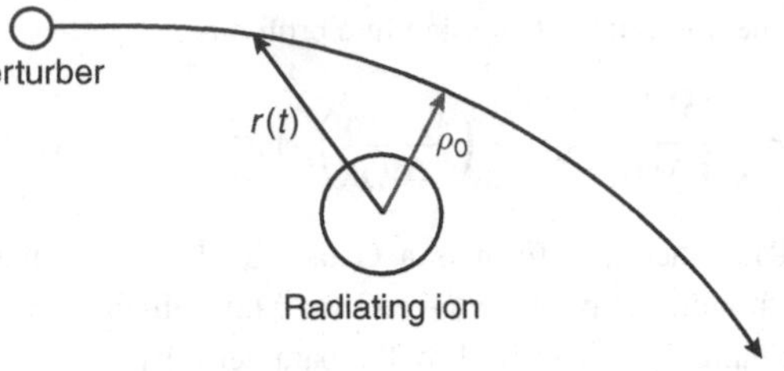

FIGURE 9.2 Distance of closest approach ρ_0 of a colliding perturber to the radiating ion. The classical collision cross section is $\pi\rho_0^2$.

$$E(\omega) = \frac{1/(\pi t_0)}{(\omega-\omega_0)^2 + (1/t_0)^2} = \frac{\Gamma_{col}/2\pi}{(\omega-\omega_0)^2 + (\Gamma_{col}/2)^2}, \tag{9.43}$$

with $\Gamma_{col} = 2/t_0$. Thus we see that the simple impact approximation – instantaneous change of phase on impact – yields the same Lorentzian profile as natural radiation damping, *irrespective of the type of particle interaction*. Ergo: all collisional broadening leads to Lorentzian profiles. Furthermore, both the natural damping and the collisional broadening Lorentzian profiles may be convolved into a single Lorentzian, giving a total $\Gamma = \Gamma_{rad} + \Gamma_{col}$.

Although collisional interactions have the same functional form, especially manifesting themselves in the line wings (Fig. 9.1), each interaction is different and must be considered quantum mechanically. For the present, however, we extend the impact approximation to define the *impact parameter* ρ_0, which is the distance of closest approach of a perturbing particle to the atom, as shown in Fig. 9.2.

If the particle density is n and the mean velocity v_0, then $v_0 t_0$ is the mean distance travelled before a collision. We can think of ρ_0 in terms of the cross section $\pi\rho_0^2$ for particle scattering. Then the total number of collisions in time t_0

$$\text{number of collisions} = v_0 \left(\pi\rho_0^2\right) n\, t_0, \tag{9.44}$$

and hence the mean time t_0 for one collision is defined as[2]

$$n \left(\pi\rho_0^2\right) v_0 = \frac{1}{t_0} \tag{9.45}$$

[2] Compare the total number of collisions with the *rate*, which is the number of collisions per unit volume per unit time in terms of the rate coefficient $\langle vQ\rangle$ and particle densities (where Q is the cross section in cm^2).

and the width as

$$n\left(\pi\rho_0^2\right)\, v_0 = \Gamma/2. \tag{9.46}$$

For a Maxwellian distribution the *relative mean velocity* v_0 between colliding particles in a plasma depends on the temperature T and their masses, say M_1 and M_2,

$$v_0 = \sqrt{\langle v^2\rangle} = \left[\frac{8kT}{\pi}\left(\frac{1}{M_1}+\frac{1}{M_2}\right)\right]^{1/2}. \tag{9.47}$$

9.2.3.2 *Quantum mechanical treatment of the impact approximation*

The impact parameter ρ_0 depends on the range of interaction and the types of collisions; electron–ion, proton–ion, atom–atom, etc. In collisional broadening, the atomic levels are perturbed due to interactions with free particles in the plasma (transitions to other bound levels are treated as radiation damping). We shall also discuss these later in the calculation of plasma opacities. In general, the change in energy or frequency, and hence the phase shift, depends on the interaction potential, which affects a change in the total energy of the (free particle–atom) system from one energy to another. The behaviour of the potential as function of range r is then the crucial quantity, and is related to the collision time, and hence the change in frequency or energy as

$$\Delta E = h\Delta\nu = C_p \times r^{-p}. \tag{9.48}$$

As already mentioned, the three most common types of perturbation for collisional broadening of lines are: charged particles, mostly protons and electrons, impacting on hydrogenic ions via the linear Stark effect with $p=2$, on non-hydrogenic ions (most lines) via the quadratic Stark effect with $p = 4$, and neutral particles (mostly neutral H) via the van der Waals interaction V_{vw} with $p = 6$. To examine this point further, we recall from classical electrodynamics the expression for the electrical field between a charged particle and a dipole, which is $F_d = e/r^2$. As an electric charge approaches the target atom (or the ion) it induces an electric dipole, since the electron cloud and the nucleus tend to be displaced in opposite directions. The linear Stark effect is most important for the prominent lines of hydrogen due to the perturbing electric field(s) of the free protons in the plasma. This is because the ℓ-degeneracy in the hydrogenic energy levels n is lifted in the presence of an external electric field, leading to mixing (transitions) among the $n\ell$ sublevels. We have $p = 2$ for the Stark potential, and therefore the Stark line shift of the sublevels is $\Delta\lambda_s \sim e/r^2$.

Since the interaction potential is described by the same form as Eq. 9.48, we may employ the *Wentzel–Kramers–Brillouin* (WKB) approximation to obtain the induced phase shift δ as

$$\delta(t) = C_p \int_{-\infty}^{t} \frac{dt'}{r(t')^p} = C_p \int_{\infty}^{t} \frac{dt'}{(\rho_0^2 + v^2 t'^2)^{p/2}}, \tag{9.49}$$

where we relate the instantaneous variable $r(t)$ to the impact parameter ρ_0 as in Fig. 9.2. We also note the usual convention, whereby the time before impact is denoted asymptotically from $t \to -\infty$, up to $t \to +\infty$ after impact and far away from the interaction region and potential. The integral above may be evaluated to yield the phase shift

$$\delta(t \to +\infty) = \frac{C_p I_p}{v \rho_0^{p-1}}, \tag{9.50}$$

where $I_p = \pi, \pi/2, 3\pi/8$ for $p = 2, 4$ and 6, respectively. Assuming an arbitrary value of the critical phase shift, beyond which line broadening would be significant as $\delta_0 = 1$, results in the *Weisskopf approximation*, which gives the collisional broadening width

$$\Gamma \approx 2\pi n v \left(\frac{C_p I_p}{\delta_0 v}\right)^{2(p-1)}, \tag{9.51}$$

corresponding to the impact parameter

$$\rho_0 = \left(\frac{v}{C_p I_p}\right)^{p-1}. \tag{9.52}$$

However, the calculation of phase shift accurately requires a more detailed quantum mechanical treatment than the Weisskopf theory, which has an arbitrary assumption for the critical phase shift and does not address small phase shifts or predict line shifts in addition to broadening.

9.2.3.3 *The nearest-neighbour approximation*

Since collisional line broadening depends on interactions with many particles, a general treatment therefore needs to be based on quantum statistics in the sense that a suitably averaged effect of all particles can be determined. The basic idea is that the effect of each kind of perturbing particle, electrons, ions or atoms, on the radiating atom (ion) can be characterized by an *averaged* distribution. The averaged field effect can be approximated by a probability function, which depends, apart from density, on the type and range of interaction potential between the perturber and the radiator. The methods employed begin with the simple approach that the frequency shift leading to collisional broadening is assumed to be due to the closest (nearest) perturber, to the exclusion of all others. We will first consider broadening related to Stark splitting of energy levels of the target ion by the perturbing ions, usually protons, which move much slower than the electrons by a velocity ratio $\sqrt{m_p/m_e} = 42.85$.

Since the frequency shift is related to the interaction potential as $\Delta\omega = C^p/r^p$, the intensity profile is a function of frequency difference from the line centre $\Delta\omega$. It is related to the probability $P(r)$ of finding a perturber at a distance r from the radiating atom

$$I(\Delta\omega)\,\mathrm{d}(\Delta\omega) \sim P(r)\,\mathrm{d}r. \tag{9.53}$$

The problem then is to obtain as precisely as possible the probability function that describes the averaged net effect of all perturbations that affect atomic transitions. Given particle density n per unit volume, the interparticle distance is defined as the radius r_0 such that there is one particle in volume $4\pi r_0^3/3$; therefore

$$r_0 = \left(\frac{4\pi n}{3}\right)^{-1/3}. \tag{9.54}$$

Now we consider the situation that there is no perturbing ion up to a distance r. Then the probability P(r) that this *nearest neighbour* lies between r and $(r + \mathrm{d}r)$ is

$$P(r) = \left[1 - \int_0^r P(r')\,\mathrm{d}r'\right] (4\pi r^2)\,n\,\mathrm{d}r, \tag{9.55}$$

where the square bracket is the probability that there is no other particle at $r' < r$, and $4\pi r^2 n dr$ is the probability that the particle lies in the shell $(r, r + \mathrm{d}r)$. Note that as r increases the probability that a nearest neighbour exists within the shell decreases. Therefore, the incremental probability behaves as $P(r + \mathrm{d}r) = P(r)$ $[1 - 4\pi r^2 \mathrm{d}r]$, and from the Taylor expansion to first order, $P(r + \mathrm{d}r) \approx P(r) + (\mathrm{d}P/\mathrm{d}r)\mathrm{d}r$. From the arguments above, the first derivative is negative, i.e., $\mathrm{d}P/\mathrm{d}r = -4\pi r^2 n P(r)$. We also note that $P(r = 0) = 1$.

Exercise 9.3 *Show that*

$$P(r) = \exp\left(-\frac{4}{3}\pi r^3 n\right) 4\pi r^2 n. \tag{9.56}$$

Calculate the expectation value $\langle r\rangle \equiv \int r P(r)\,\mathrm{d}r$, *which is the mean perturber distance, and compare with the mean interparticle distance* r_0 *at a given density* n *to show that the difference is small.*

Given that the difference between r_0 and $\langle r\rangle$ is small, we may obtain a characteristic frequency shift using the former, as $\Delta\omega_0 = C_p/r_0^p$, and hence $\Delta\omega/\Delta\omega_0 = (r_0/r)^p$. Then

$$P(r)\,\mathrm{d}r = \mathrm{e}^{-(\Delta\omega_0/\Delta\omega)^{3/p}}\,\mathrm{d}\left((\Delta\omega_0/\Delta\omega)^{3/p}\right). \tag{9.57}$$

This relation expresses the probability of the nearest perturber to collisional broadening in terms of the range index p of the perturbation. For example, for the linear Stark effect, the perturbing field strength is $F^s = e/r^2$, and therefore the characteristic strength particular to the given interaction and density is

$$F_0^s = (e/r_0^2) = e\left(\frac{4}{3}\pi n\right)^{2/3} = 2.5985\,e\,n^{2/3}. \tag{9.58}$$

Since, in general $(r/r_0)^p = F/F_0$, we measure the field strength in terms of variable $\beta \equiv F/F_0$. For Stark broadening we write $\beta = F^s/F_0^s$ and the probability function as

$$P_s(\beta)\,\mathrm{d}\beta = (3/2)\,\beta^{-5/2}\,\exp(\beta^{-3/2})\,\mathrm{d}\beta. \tag{9.59}$$

As $\beta \to \infty$, $P_s(\beta) \sim \beta^{-5/2}$. This implies that the Stark profile falls off as $\Delta\omega^{-5/2}$ in the line wings as opposed to $\Delta\omega^{-2}$ in the simple impact formula.

In simple terms, the impact approximation may be valid for fast-moving electrons, while the nearest-neighbour approximation is suitable for the nearly static (by comparison) ions. However, the nearest neighbour theory is inadequate, since it does not consider the overall effect of perturbations from an ensemble of particles in the plasma, but only from the closest perturber.

9.2.3.4 *The Holtsmark distribution*

The effect of quantum statistical fluctuations of the net electric field created by an ensemble of particles was developed by Holtsmark. The theory yields a *microfield* distribution, for an interaction potential C_p/r^p, given by the probability function

$$P(\beta) = \left(\frac{2\beta}{\pi}\right) \int_0^\infty \mathrm{e}^{x^{3/p}}\,x\,\sin\beta x\,\mathrm{d}x. \tag{9.60}$$

As before, $\beta = F/F_0$ with characteristic field strength

$$F_0 = G_p\,C_p\,n^{p/3}, \tag{9.61}$$

with

$$G_p = \frac{(2\pi^2 p)}{[3(p+3)\,\Gamma(3/p)\,\sin(3\pi/2p)]^{p/3}}. \tag{9.62}$$

(Note that on the right-hand side we have a Gamma function, not the line width. The Gamma function is also obtained on integrating the error function in Exercise 9.3.) In the Holtsmark theory for the Stark effect, with $p = 2$, we have $F_0^s = 2.6031\,e\,n^{2/3}$, nearly the same as obtained from the nearest-neighbour approximation. Examining the limiting cases we have, for small β,

$$P(\beta) = \left(\frac{4}{3\pi}\right) \sum_{\ell=0}^{\infty} (-1)^{\ell}\, \Gamma\left(\frac{4\ell+6}{3}\right) \frac{\beta^{2\ell+2}}{(2\ell+1)!}, \tag{9.63}$$

which simplifies to $P(\beta) \sim \beta^2$ ($\beta \ll 1$). The asymptotic expansion for $\beta \gg 1$ is

$$P(\beta) = \left(\frac{1.496}{\beta^{5/2}}\right) \times \left[1 + 5.107\beta^{-3/2} + 14.43\beta^{-3} + \cdots\right], \tag{9.64}$$

where the first term is the same as in the nearest-neighbour approximation. Tables of $P(\beta)$ have been computed (see [244] for references).

9.2.3.5 *Debye screening potential*

Before we compare the Holtsmark distribution with the nearest-neighbour approximation, it is useful to discuss the theory of Debye screening of the electron–nuclear potential ($-Z/r$) in an ion by perturbing particles in a plasma. The effect of a positively charged ion is to polarize the surrounding plasma by attracting free electrons and repelling free protons. As the particle density n increases, the bound electrons in the ion are increasingly 'screened' from the nucleus by free electrons. The screening effect on the Coulomb potential (Z/r) is studied by introducing a characteristic *Debye length* d, which is the radius of the *Debye sphere* centred at the ion and defined in such a way that for r > d the potential decreases exponentially. The *Debye potential* due to the nucleus of the ion then behaves as function of r

$$V_{\rm d}(r) = \frac{Ze^{-r/d}}{r}. \tag{9.65}$$

The Debye potential of the ion $V_{\rm d}$ $(r \to 0) = Z/r$, and decreases with increasing r, vanishing rapidly for $r > d$. An explicit expression for d may be derived by considering a charge distribution described by the Poisson equation. As with all plasma properties, the Debye length depends on the temperature and the density. For an electron–proton plasma, we have

$$d = 6.9 \left(\frac{T_{\rm e}}{n_{\rm e}}\right)^{1/2} \text{ cm.} \tag{9.66}$$

Exercise 9.4 *Calculate and compare in tabular form the Debye lengths at a wide range of temperatures and densities that characterize various astronomical objects, as in Fig. 1.3.*

One can now relate the nearest-neighbour approximation and the Holtsmark theory using the Debye length. The number of perturbers inside the volume of the Debye sphere is 4/3 $\pi d^3 n \equiv N_{\rm d}$. The effective plasma microfield depends on $N_{\rm d}$. We can therefore calculate not one but a number of Holtsmark distributions $P(\beta, N_{\rm d})$ corresponding to different $N_{\rm d}$ as shown in Fig. 9.3. At high densities $N_{\rm d} \to \infty$, $P(\beta, N_{\rm d})$ approaches the Holtsmark distribution, and for low densities it approaches the nearest-neighbour approximation.

Also, using the Debye potential, we can estimate the weakening of the electron–nuclear potential in the ion due

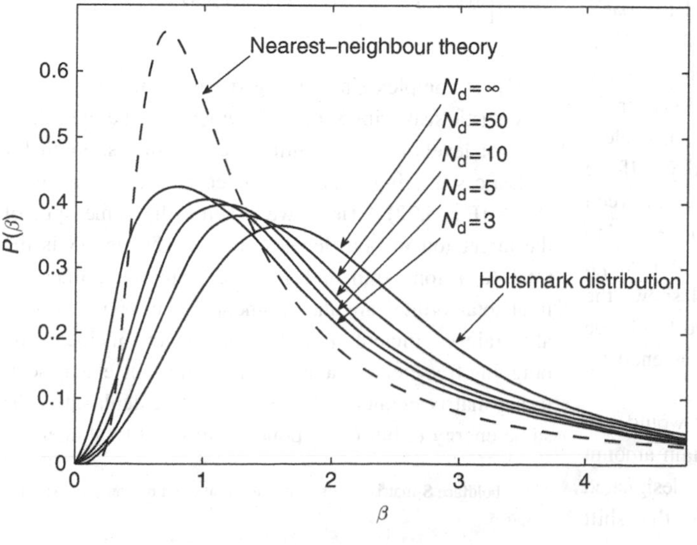

FIGURE 9.3 Field strength probabilities in different approximations, as function of field strength parameter $\beta = F/F_0$ and number of perturbing particles $N_{\rm d}$ within the Debye sphere (as in [244]).

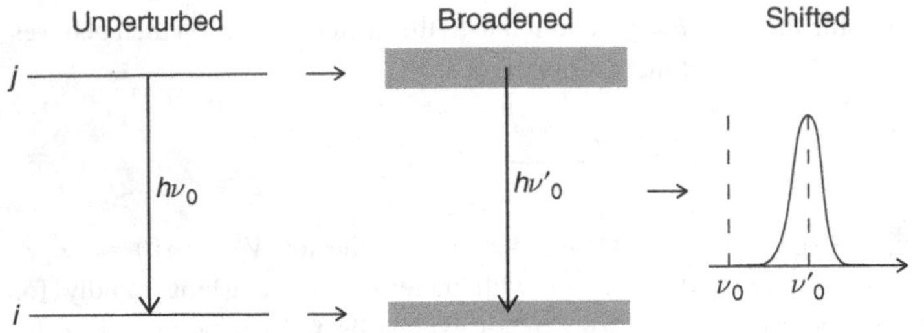

FIGURE 9.4 Line broadening and line shift

to plasma screening. The binding energy of the bound electrons is reduced by

$$\Delta E = \frac{e^2 Z}{r}\left(e^{-r/d} - 1\right). \tag{9.67}$$

In particular, the *ionization potential* of a bound electron in the ion will be reduced by an amount equal to ΔE_{IP}, which we can approximate as

$$\Delta E_{\mathrm{IP}} = -\frac{e^2 Z}{d} = 2.2 \times 10^{-9}\, Z \left(\frac{T}{n}\right)^{1/2} \mathrm{Ry}. \tag{9.68}$$

That Debye screening can significantly affect atomic parameters was pointed out in Fig. 6.2, showing the photoionization cross section of Na I at high densities. The presence of a screened atomic potential by the free electons leads to a more diffuse 3s valence orbitals for neutral sodium. That results in a significant shift in the signature feature in photoionization of alkali atoms and ions: the Cooper minimum moves to higher energies than those for the unpertubed atom.

9.2.3.6 *Electron impact broadening*

As shown above, the probability function of microfield distribution in the plasma is directly related to the line widths and shifts. So far, we have discussed the basic concepts underlying the theory of collisional broadening, with particular reference to Stark broadening by the electric field distributions of free protons. However, in addition to Stark broadening $\Delta\lambda_s^{-5/2}$, there is also electron impact broadening which behaves as $\Delta\lambda_e^{-2}$. If we consider broadening of a line due to a transition between levels i and j of an ion X and energy $h\nu_{ij}$, then we need to calculate the cross sections for electron impact $(e + X_{i,j})$, *separately for each level i and j*. Figure 9.4 shows the physical nature of impact broadening: the two levels are broadened, but they are not likely to be broadened by exactly the same widths.

It follows that the resulting line shape would not only be broadened, but also *shifted* by a certain amount related to $(h\nu_0 - h\nu_0')$. We have already designated the line width by the parameter γ; now the shift is denoted as x. The effect of many perturbers is considered in the so-called *impact approximation* [248]. Whereas the details are rather complicated, the expression for the line shape ϕ_e due to electron impact broadening is

$$\phi_e(\nu) = \frac{(\gamma/2\pi)}{(\nu - \nu_0 + x/2)^2 + (\gamma/2)^2}. \tag{9.69}$$

The width γ and the shift x are related to the thermally averaged collisional damping rate coefficient q_D, similar to the one we have encountered in electron–ion collisions (Chapter 5). However, it is now a *complex number* defined as

$$\gamma + \mathrm{i}x = n_e\, q_D. \tag{9.70}$$

That this must be so becomes clear when we examine the form of the damping collision strength $\Omega_D(\epsilon)$ in terms of the scattering matrix elements for *elastic scattering* in the initial and final levels i and j [249],

$$\begin{aligned}\Omega_D(i, j; \epsilon) &= \frac{1}{2S_I + 1} \sum_S \sum_{L\pi} \sum_{L'\pi'} \sum_{\ell\ell'} (2S+1)(2L+1)(2L'+1) \\ &\times W(L_i L_j L L'; 1\ell)\, W(L_i L_j L L'; 1\ell') \\ &\times \left[\delta(\ell, \ell') - \mathbf{S}_i(SL\pi; \ell, \ell')\, \mathbf{S}_j^*(SL'\pi'; \ell\ell')\right].\end{aligned} \tag{9.71}$$

Ω_D is complex since it is given in terms of the complex elastic scattering **S**-matrix elements on the right-hand side for levels i and j, unlike the collision strength for inelastic scattering which is related to the real quantity $|S_{ij}|^2$ (Eq. 5.22).[3] Here, we assume the same spin of the target ion in both levels (no spin change), S is the total (e + ion) spin and $L\pi$, $L\pi'$ are the initial and final total orbital angular momenta. The W is a Racah algebraic coefficient; ℓ and ℓ' denote the incident and outgoing free electron angular momenta. The elastic scattering matrix elements $\mathbf{S}_i$, $\mathbf{S}_j$ are both calculated at the same energy ϵ, but correspond to different total energies

[3] The boldface **S**-matrix, or its elements, should not be confused with the spin S.

for each level $E = E_i + \epsilon$ and $E' = E_j + \epsilon$. Now, the *effective damping collision strength* is the Maxwellian average at temperature T,

$$\Upsilon_D(T) = \int_0^\infty \Omega_D(\epsilon)\, e^{\epsilon/kT}\, d(\epsilon/kT), \tag{9.72}$$

which is dimensionless, as before. The collisional damping rate coefficient is (in atomic units $\hbar = m_e = 1$)

$$q_D(i, j; T) = \left(\frac{2\pi}{kT}\right)^{1/2} \Upsilon_D(T). \tag{9.73}$$

At high densities, in stellar interiors and similar plasma environments, collisional or pressure broadening is the most important component of line profiles for most atomic systems. Since fully quantum mechanical calculations, as outlined above, are extremely difficult, very approximate formulae have been used, basically exploiting the fact that the atomic scattering cross section of an electronic n-shell is proportional to $\langle r_n \rangle^2 \sim n^4$, where $\langle r_n \rangle$ is the expectation value, or the mean radius. One such formula, for line widths in frequency units, is [250]

$$\Delta\nu_e(ij) = 3.2 \times 10^{16} \frac{\left(n_i^4 + n_j^4\right) \rho}{(z+1)^2\, T_4^{1/2}} \text{ s}^{-1}, \tag{9.74}$$

where n_i, n_j are the principal quantum numbers of the initial and final levels, ρ is the mass density (g cm^{-3}), z is the ion charge, and $T_4 = T(K) \times 10^{-4}$. However, this expression is not quite in accord with elaborate quantum mechanical calculations; for example, the approximate line width from the above expression is multiplied by a factor of four in [251]. A number of approximations for astrophysical applications are discussed in [251, 252, 253]. Rather more detailed treatments are required to obtain accurate expressions for the computation of these quantities. We shall further address this topic in our description of the plasma equation-of-state and opacities in Chapter 11. For the time being, we re-emphasize that Stark broadening due to ions (protons), and that due to electron impact, lead to two distinct components of total collisional broadening, behaving as

$$\Delta\lambda \sim C_s \Delta\lambda_s^{-(5/2)} + C_e \Delta\lambda^{-2}, \tag{9.75}$$

both of which lead to a slower fall-off of line wings than the line centre, which is affected predominantly by thermal Doppler broadening (Eq. 9.33) that falls off exponentially.

Finally, to encapsulate the general functional form of line broadening profile: it may be expressed as a convolution over functions representing the different forms discussed in this section, i.e.,

$$\phi(\Delta\lambda) = [\phi_s(\Delta\lambda^{-5/2}) \otimes \phi_e(\Delta\lambda^{-2})]_{coll} \times \otimes\, \phi_r(\Delta\lambda^{-2}) \otimes \phi_D\, e^{-(\Delta\lambda)^{-2}}, \tag{9.76}$$

where ϕ_s, ϕ_e, ϕ_r refer to Stark, electron impact and radiation damping profiles respectively, all of which are Lorentzian (with collisional terms in the square bracket), and ϕ_D to Doppler profile, which is Gaussian.

9.2.3.7 Escape probability

It is often useful to introduce the concept of *escape probability* in physical conditions, such as under particular temperatures and densities in nebulae or active galactic nuclei, that do not necessitate a full radiative transfer treatment, as required in stellar atmospheres. The escape probability is expressed in terms of the optical depth, with the assumption that the frequency distribution profile for absorption by an atom matches the emission profile, reflecting *complete frequency redistribution*. This assumption applies to varying extent for strong 'resonance' lines.

There are myriad expressions used in practice for escape probability in a line (e.g., [254]). For a thermalized line with a Gaussian distribution, the escape probability has been given by Zanstra [255] as

$$P_{esc}(\tau_{\nu 0}) \approx \frac{1}{\tau_0(\pi \ln \tau_0)^{1/2}}. \tag{9.77}$$

A rough approximation for radiative transfer effect of self-absorption (by the same ionic species within the plasma), at frequency ν, is to modify the Einstein A coefficient for a transition $j \to i$ as

$$A'_{ji} = A_{ji} \times P_{esc}(\tau). \tag{9.78}$$

The idea is that the spontaneous decay probability, and hence the number of photons or the intensity of the line, is reduced ($P_{esc} < 1$) by an amount dependent on the optical depth of the medium. The *escape probability* method is employed in several photoionization modelling codes for nebular and active galactic nuclei models (e.g., [256]).

Now that we are able to obtain the line profile factor introduced in Eq. (9.8), we can move on to absorption line diagnostics in the next section.

9.3 Absorption lines

Absorption lines, like emission lines, are also used as density and abundance indicators in a variety of objects, e.g., the ISM, active galactic nuclei and stellar photospheres. But whereas emission lines probe *local* physical conditions where they are predominantly formed, absorption lines entail the extent of the absorbing medium in its entirety. The atomic processes involved in emission and absorption are also quite different. Owing to the low temperatures generally prevalent in the ISM, emission spectra due to collisional excitation (or recombination) are less common than in other sources with higher temperatures (although many exceptions abound, even including highly ionized species, such as Li-like O^{5+} in the ISM and AGN; see Chapter 13). Absorption spectra are therefore useful in ISM studies because low temperatures imply that all ions are likely to be in the ground level, and only a few levels might be excited. Absorption lines are extremely useful in ascertaining abundances in extended objects – especially the interstellar and intergalactic media (IGM). Their utility is manifest over large distances and low densities, especially cosmological distances in the IGM out to high redshift (Chapter 14). In some ways, absorption line analysis is easier than the analysis of emission line ratios. One usually considers absorption in only one line at a time. The only atomic parameter required is the absorption oscillator strength, and the main dependence is on the number of ion absorbers along the line of sight. However, there are additional dependencies on the thermal and turbulent velocity distribution of ions and saturation effects. The monochromatic flux from a source absorbed by the intervening medium up to the observer is directly proportional to the oscillator strength and the total number of absorbers, i.e.,

$$F_\nu(\text{abs}) \propto N_i \times f_{ij}(\nu) \tag{9.79}$$

where N_i is the number of ions along the line of sight, i.e., the column density.

A problem commonly encountered in observations is that if the number of absorbers is large enough then *all* of the flux at line centre from a given source is absorbed, leading to saturation in the absorption line profile. Therefore, in practice *weak* lines are particularly valuable since they are less amenable to saturation effects than strong lines. For example, intercombination lines have f-values that are orders of magnitude smaller than dipole allowed lines. A correspondingly larger number of ions would absorb the same amount of flux in an intercombination $E1$ transition than a dipole allowed $E1$ transition. Of course, one does need sufficient absorption in a line in order to be observed. Hence, forbidden $E2$ or $M1$ lines are not usually good candidates, since their f-values are far too small.

The following subsections lay out the methodology employed in absorption line analysis.

9.3.1 Equivalent width

A spectral line always has a finite width. Observations may not often resolve the line profile in complete detail, and the frequency dependence of emission or absorption via the line transition may not be ascertained exactly. A quantitative measure of the amount of energy in a spectral line relative to the continuum (Fig. 9.5) is given by the area contained in the line profile under the continuum line.

Although we now know that line broadening is determined by several processes, for the purpose of *abundance determination* it is deemed sufficient to know that the line is formed by a given ionic species via a well-identified atomic transition. The area between the line profile and the continuum must then be related to (i) the number of ions or the abundance and (ii) the strength of the transition. It is useful to define a quantity corresponding to this area, the *equivalent width*, which we write formally as

$$W_\nu = \int \frac{F_c - F_\nu}{F_c}\, d\nu. \tag{9.80}$$

The F_c and the F_ν are the measured continuum and line fluxes or intensities respectively, and W denotes the equivalent width. In this chapter, unless otherwise specified, we shall describe equivalent widths as derived from observed absorption spectra. Thus in Eq. 9.80, F_ν denotes the energy transmitted from the continuum radiation of the source through absorption by an ion in a given transition at frequency ν. If we consider F_c to be the continuum source intensity then,

$$\frac{F_\nu}{F_c} = e^{-\tau_\nu}. \tag{9.81}$$

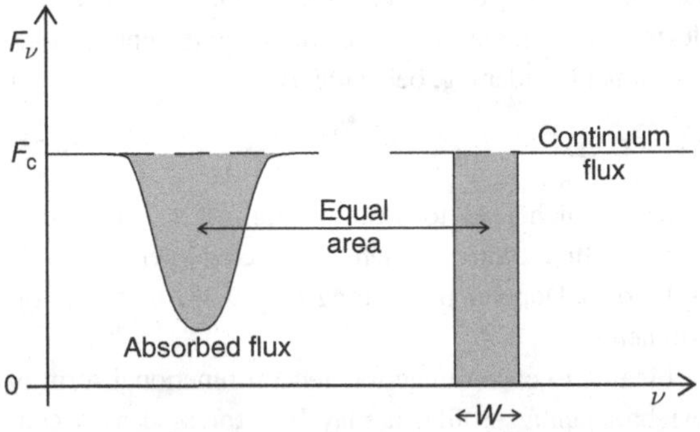

FIGURE 9.5 Spectral line energy relative to the continuum; W is the equivalent width.

Since spectral observations are usually in wavelengths, and given $\nu = c/\lambda$, we can express W in wavelength notation as

$$W_\lambda = \frac{\lambda^2}{c} \int \left(1 - e^{-\tau_\nu}\right) d\nu. \tag{9.82}$$

In the ISM, optical depths are generally small, $\tau \ll 1$ (with outstanding exceptions such as for the Lyα), and therefore we can approximate $e^{-\tau} \approx 1 - \tau_\nu$, so that,

$$W_\lambda = \frac{\lambda^2}{c} \int \tau_\nu \, d\nu. \tag{9.83}$$

In terms of the column density the optical depth is

$$\tau_\nu = \left(\frac{\pi e^2}{m_e c}\right) N_i \, f_\nu \, \phi(\nu), \tag{9.84}$$

where $\phi(\nu)$ is the line profile. So long as $\tau \ll 1$, the equivalent width is commonly parametrized in terms of wavelength as [257]

$$W_\lambda = \left(\frac{\pi e^2}{m_e c^2}\right) N_i \, f_\lambda \, \lambda^2 = 8.85 \times 10^{-13} \, N_i \, f_\lambda \, \lambda^2. \tag{9.85}$$

Note that the denominator contains the factor c^2, and both sides have dimensions of length. A measurement of the equivalent width in a line (for $\tau \ll 1$) directly yields the column density N_i, which for a given element E may be simply related to the hydrogen column density $N_{\rm H}$ as

$$N_i = \frac{N_i}{N_E} \times \frac{N_E}{N_{\rm H}} \times N_{\rm H}, \tag{9.86}$$

multiplying with its ionization fraction and the element abundance ratio on the right-hand side. The average H-density is related to the column density as $N_{\rm H} = n_{\rm H} \times \ell$, where ℓ is the distance between the observer (e.g., the Earth) and the source. It is more useful to rewrite Eq. 9.85 in a slightly different way as

$$W_\lambda/\lambda = 8.85 \times 10^{-13} \, N_i \, f_\lambda \, \lambda. \tag{9.87}$$

The derivation of the equivalent width above is not quite complete. While the integrated value of the line profile (in general, as in Eq. 9.76) is unity (Eq. 9.9), its detailed form indicates the nature of the plasma along the line of sight, i.e., the length-averaged Doppler shifts, temperatures and densities of the medium, determine the overall line shape. Therefore, the equivalent width W depends on the line shape in a more involved manner when τ becomes significantly large (it was assumed to be small in the preceding discussion). Although the ISM densities are generally very low, the column densities can be very large, depending on the path length to the source along the line of sight. The temperature dependence for a plasma with a Maxwellian velocity distribution can be characterized by a Gaussian function. The mean kinetic energy at temperature T given by the rms velocity is called the *b-parameter* [257]

$$b = \sqrt{\frac{2kT}{M}} = 1.29 \times 10^4 \left(\frac{T(K)}{M_A(\text{amu})}\right)^{1/2} \text{cm s}^{-1}, \tag{9.88}$$

where M is the atomic weight of the element. Considering the Doppler profile factor only[4]

$$\phi(\nu) = \begin{cases} \dfrac{\lambda}{\pi^{1/2} \, b} \, e^{-(v/b)^2} \\[2ex] \dfrac{\lambda}{\pi^{1/2} \, b} \, e^{-c\Delta\nu/b\nu}, \end{cases} \tag{9.89}$$

since $v/c = \Delta\nu/\nu$. The velocity–temperature dependence of the optical depth is

$$\tau(v, T) = N_i \, \sigma_\lambda \, \phi(v) = \left(\frac{N_i \, \sigma_\lambda \, \lambda}{\pi^{1/2} \, b}\right) e^{-(v/b)^2}. \tag{9.90}$$

or more simply,

$$\tau = \tau_0 \, e^{-(v/b)^2}, \tag{9.91}$$

where $\tau_0 = (N_i \, \sigma_\lambda \, \lambda/\pi^{1/2} \, b)$ is the maximum line centre optical depth at the central wavelength λ. There are several useful expressions for τ_0 that may be particularized to different astrophysical situations, e.g.,

$$\tau_0 = 1.497 \times 10^{-15} \, \frac{f_\lambda \, N_i \, \lambda}{b}, \tag{9.92}$$

where λ is in Å and b in km/s. Such expressions are generally used for extended plasma sources, such as the ISM or the IGM, with low velocity fields.

Using Eq. 9.82 we obtain the more exact approximation for the equivalent width

$$W = \frac{2\lambda b}{c} \int_0^\infty \left[1 - \exp\left(\tau_0 e^{-x^2}\right)\right] dx$$
$$= \frac{2\lambda b}{c} \, F(\tau_0). \tag{9.93}$$

Exercise 9.5 *Derive the above equation in terms of the functional $F(\tau_0)$ using the velocity dependent expression for the optical depth, and the Doppler broadening parameter $x = (c/b\nu)d\nu$, without making the approximation in Eqs 9.82 and 9.83. Note that Doppler broadening means $\Delta\nu/\nu_0 = b/c$. Write a computer program to evaluate W at different temperatures.*

[4] The densities in the ISM are generally too low to cause significant pressure broadening. Also, there may be clouds with slightly different Doppler shifts along the line of sight.

9.3.2 Curve of growth

Now that we have the exact expression for the equivalent width (Eq. 9.93) we may evaluate it as a function of temperature and density. These functions are very useful in ISM studies and are called *curves of growth.* For each chosen absorption line at a given wavelength λ, W_λ may be plotted with respect to abundance (column density) of the ionic species. A family of such curves may be obtained, one for each temperature related to thermal velocity characterized by the b-parameter. The physics of the temperature and density dependence of line broadening plays a crucial role in determining the observed profile and the overall behaviour of W_λ.

First, we consider the connection between the observed line profiles and the dependence of W_λ vs. N_i, the number density of ion absorbers. Figure 9.6 shows a schematic diagram of this curve of growth. The line profiles in Fig. 9.6 corresponds to deepening and broadening in direct proportion to the energy removed from the continuum by an increasing number of absorbers. There are three distinct segments of the curve of growth, related to different line profiles. (i) *The linear part:* according to Eqs 9.85 and 9.87 the equivalent width should increase directly (linearly) as the number of ions N_i of element X in optically thin regions $\tau \ll 1$, as

$$W \sim \tau_\nu \sim N_i \sim \frac{N_i}{N_{\rm X}} \frac{N_{\rm X}}{N_{\rm H}} N_{\rm H} \sim A_x, \qquad (9.94)$$

where A_x is the abundance ratio $N_{\rm X}/N_{\rm H}$. The line profile in the linear part in Fig. 9.6, $W \sim N_i$, corresponds to deepening and broadening in direct proportion to the energy removed from the continuum by an increasing number of absorbers. (ii) *The saturated part:* corresponds to the saturation of the line profile when the density of ions is sufficient to absorb nearly *all* of the continuum photons at the line centre wavelength; any further increase in density results in a slow increase in $W \sim \sqrt{\ln N_i}$, related mainly to Doppler broadening. (iii) *The damped high-density part:* when the line profile at the central Doppler core is saturated, ions absorb photons in the line wings, which are then seen to be 'damped' beyond the line centre. The line wings on either side of the line centre increase (are enhanced) with column density, and $W \sim \sqrt{N_i} \sim A_x^{1/2}$ (see Exercise 9.6). With an increasing number of absorbers, the line profile growth is slower than linear, and continues to expand sideways with damped line wings, eventually assuming a 'square' shape as absorption continues to manifest itself in the line wings (such as in absorption spectra of 'damped' Lyα clouds; see Chapter 14).

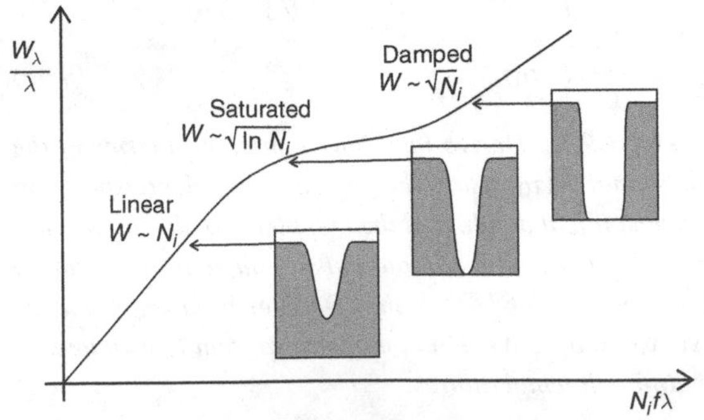

FIGURE 9.6 The curve of growth relating the equivalent width of an absorption line to the number density along the line of sight towards the source: W_λ/λ vs. $\log(Nf\lambda)$.

As described above, the line profile is a convolution of natural and pressure broadening represented by a Lorentzian function, and thermal Doppler broadening given by a Gaussian. The line absorption cross section may be represented[5] as

$$\sigma_\nu = \left(\frac{\pi e^2}{m_e c}\right) f_\nu \left[\frac{\Gamma/4\pi^2}{\Delta\nu^2 + (\Gamma/4\pi^2)}\right] \times \left[\frac{1}{\sqrt{\pi}\,\Delta\nu_{\rm D}} e^{-(\Delta\nu/\Delta\nu_{\rm D})^2}\right], \qquad (9.95)$$

where $\Delta\nu$ is the line width, Γ is the total damping constant due to natural, radiative and pressure broadening and $\Delta\nu_{\rm D} = b\nu/c$ is the Doppler width. Note that we distinguish between the total width $\Delta\nu$, which includes all processes, and $\Delta\nu_{\rm D}$ due to thermal broadening only. It is clear that part (iii) of the curve of growth in Fig. 9.6 is the most complicated. The equivalent width then depends on the temperature, the density and the abundance of the element. All of these external factors cannot be ascertained from one measurement of W alone. But for the time being we only wish to determine the behaviour of W in this regime with respect to density. We make two approximations for part (iii). One; the saturated line core dominated by thermal Doppler broadening is narrow, and the Gaussian component is then like a Delta function. Two; for sufficiently large spread in the line wings, $\Delta\nu^2$ in the denominator is large compared with Γ, which depends only on atomic parameters and is comparatively smaller. Thus Eq. 9.95 simplifies to

$$\sigma_\nu \approx \left(\frac{\pi e^2}{m_e c}\right) f_\nu \left(\frac{\Gamma/4\pi^2}{\Delta\nu^2}\right), \qquad (9.96)$$

[5] Again, with the caveat that we do not consider turbulence or clouds with different radial velocities in this formal expression.

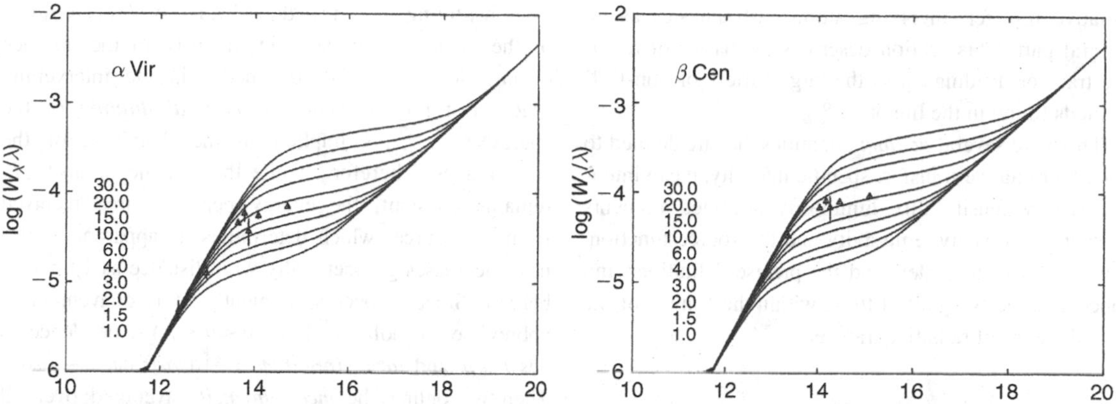

FIGURE 9.7 The Si II abundances in the ISM determined from curves of growth with absorption in Si II lines towards two different stars [258]: log $N(\text{Si II})$ = 14.07 (α Vir) and 14.46 (β Cen). A χ^2 procedure is employed to ascertain the best values of the column density and the b-parameter.

and the optical depth

$$\tau_\nu = \left(\frac{\pi e^2}{m_e c}\right) \frac{f_\nu}{4\pi^2 \, \Delta\nu^2} \int_0^\ell N_i \, \Gamma \, ds. \tag{9.97}$$

The ion column density N_i is related to the abundance of the element $A_x = N_X/N_H$ and the ionization fraction N_i/N_X, as shown above. Letting the integral represent an average over the whole path length ℓ as $\langle N_i \Gamma \rangle$, we approximate

$$\tau_\nu = \left(\frac{\pi e^2}{m_e c}\right) f_\nu \frac{\langle N_i \Gamma \rangle}{4\pi^2 \, \Delta\nu^2}. \tag{9.98}$$

Evaluating W with this expression of τ yields its dependence on the ion density for lines on the damped part of the curve of growth

$$W_\nu \sim \sqrt{\langle N_i \Gamma \rangle \, f_\nu} \sim \sqrt{N_i}. \tag{9.99}$$

Exercise 9.6 *Derive Eq. 9.99. Hint: use the change of variable* $x^2 = \Delta\nu^2 / \left[\langle \Gamma N_i \rangle (\pi e^2/m_e c) f_\nu\right]$.

In part (ii) of Fig. 9.6, when the density dependence is still small, the main variation in W is due to the temperature, as reflected in the thermal velocity distribution of ions, i.e., through the b-parameter. However, the velocity profile is unknown a priori, since the local temperature varies or is otherwise unknown in an extended object. Theoretical curves of growth are therefore computed at a range of b-parameters to compare with observations and deduce column densities. Figure 9.7 displays the curves of growth using Si II lines, with lines of sight through the ISM towards two relatively bright stars α Virginis and β Centauri. Note that while the linear and damped parts of the curves of growth merge together for different b-parameters, the main variation in the saturated part is due to thermal velocity distribution. The observed values are also shown in Fig. 9.7, and yield a fairly well constrained range of column densities $N(\text{Si II}) \sim 10^{14}$ cm^{-2} [258].

Another example where the curve of growth technique may be used for abundance determinations is with the well-known pair of strong Na D lines in stellar atmospheres (see solar spectra, Figs 1.1 and 9.13). The two fine-structure components are D1 and D2: $3s^2S_{1/2} - 3p^2P^o_{1/2,3/2}$ at $\lambda\lambda$ 5895.924 and 5889.950, with f-values of 0.320 and 0.641, respectively, in the proportion 1:2. Since the Na D lines are readily observed and their two components easily resolved, they enable accurate determination of photospheric densities under various conditions and models [259].

Absorption spectroscopy can also be extended to include absorption into resonant levels, or excitation from a bound state into an autoionizing level lying in the continuum. The corresponding 'resonance line strengths' (defined in Eq. 6.69) may be analyzed in much the same way as above. However, the computation of resonance oscillator strengths, required to compute equivalent widths, is more complicated, since these resonances appear in photoionization cross sections. This topic is discussed in the context of resonant X-ray photoabsorption in AGN in Chapter 13 (in particular, see Exercise 13.1 on calculating X-ray column densities).

9.4 Radiative transfer

The previous discussion has introduced some of the terminology of radiative transfer in spectral analysis. But a far more extensive and rigorous methodology has been developed for large-scale numerical computations for radiative transfer, involving atomic astrophysics as an integral part. This section describes the basics of radiative transfer, leading up to the large variety of non-LTE methods found in the literature.[6]

There are several *distinct* quantities that are defined to describe radiative transfer: specific intensity, mean intensity, energy density, flux, luminosity, averaged moments of intensity, opacity, emissivity and the source function. It is necessary to understand the precise definitions and concepts underlying all of these within the context of the formal theory of radiative transfer.

9.4.1 Intensity and flux

A source of radiation is characterized by a *specific intensity*, I, defined to be *constant* along a ray of light propagating in free space from the source to the receiver. The source may be a differential area dA_1 on the surface of a star, and the receiver may be another differential area dA_2 on an object like an ionized gas cloud or the detector at a telescope. The monochromatic intensity I_ν of radiation emitted by the source is defined in ergs per unit solid angle ω, per unit area, per unit time, per unit frequency

$$I_\nu = I\,(\mathrm{erg/(cm^2\ s^1\ str^1\ Hz^1)}). \tag{9.100}$$

Let us consider a given amount of energy dE passing through two surface elements (one at the source and the other at the receiver) dA_1 and dA_2, with geometrical disposition as shown in Fig. 9.8 and separation r. The solid angles subtended by each, at a point on the other, are: $d\omega_1/4\pi = dA_2 \cos\theta_2/4\pi r^2$ and $d\omega_2/4\pi = dA_1 \cos\theta_1/4\pi r^2$. Then the energy per unit time is

$$\begin{aligned} dE_\nu &= I_\nu(1)\, dA_1\ \cos\theta_1\ d\omega_1\ d\nu\ dt \\ &= I_\nu(2)\, dA_2\ \cos\theta_2\ d\omega_2\ d\nu\ dt\,. \end{aligned} \tag{9.101}$$

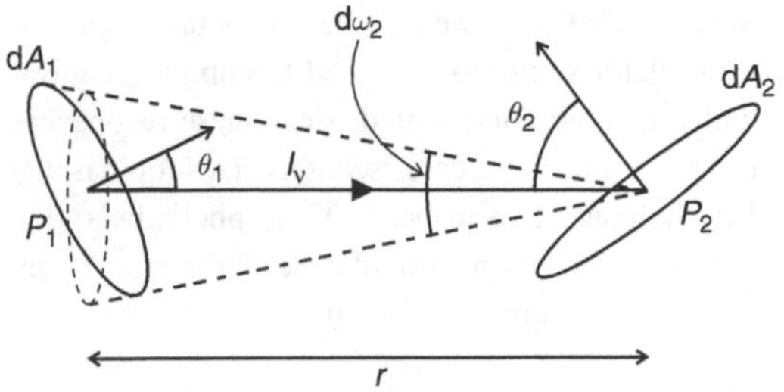

FIGURE 9.8 Constancy of specific intensity I_ν along a ray from the emitting area dA_1 at the source to the receiving area dA_2 at the observer (detector), irrespective of the arbitrary geometrical configuration shown.

[6] The subject is discussed in many excellent textbooks and monographs, such as [244, 245, 246].

Substituting for $d\omega_1, d\omega_2$ leads to $I_\nu(1) = I_\nu(2)$, or the *invariance* of specific intensity in the absence of any source or sink of energy in the intervening space. But note that the *geometrical dilution* of the energy as r^{-2} is implicit in the definition of the solid angles. Therefore, while the specific intensity I_ν remains constant, the energy per unit area received from the source, which determines its apparent brightness, decreases geometrically with distance as $1/r^2$ (see below). Since it occurs frequently, it is convenient to abbreviate the polar angle θ $(0 \le \theta \le \pi)$ dependence as $\cos\theta \equiv \mu$, and $d\omega = sin\theta d\theta d\phi = -(dcos\theta)d\phi = -d\mu d\phi$. Then we define the *mean intensity* averaged over all directions

$$\begin{aligned} J_\nu &= \frac{1}{4\pi}\int I_\nu d\omega = \frac{1}{4\pi}\int_0^{2\pi} d\phi \int_{\theta=0}^{\theta=\pi} I_\nu sin\theta d\theta \\ &= \frac{1}{2}\int_{-1}^{+1} I_\nu d\mu. \end{aligned} \tag{9.102}$$

The *monochromatic energy density* is related to the mean intensity as

$$u_\nu = (4\pi/c)\ J_\nu. \tag{9.103}$$

Note the analogy from fluid mechanics with the flow of a fluid of density ρ and velocity v such that ρv is the amount at or through a given point. The *monochromatic flux* is

$$F_\nu = \int I_\nu\ \cos\theta\ d\omega = 2\pi \int_{-1}^{+1} I_\nu\ \mu\ d\mu. \tag{9.104}$$

The total luminosity L of an object is defined to be the net energy flowing outward through a sphere of radius R, per unit time, per unit frequency (erg/(s^{1}Hz^1)). Therefore,

$$L = (4\pi R^2) \int I_\nu\ \mu\ d\omega = \pi F_\nu (4\pi R^2). \tag{9.105}$$

This equation involves integration over all directions, i.e., outward flow of radiation as well as inward through a surface. If, however, the flow is only outward, and the source is such that I_ν is isotropic, then the integration implies that at the surface of the source

$$F_\nu = I_\nu \quad \text{(outward flow)}. \tag{9.106}$$

The total energy or luminosity L of an isotropic source with outward flux F, as it flows through successive spheres of radii r_1, r_2, etc., can be written as

$$L = \pi\ F(r_1)\ 4\pi\ r_1^2 = \pi\ F(r_2)\ 4\pi\ r_2^2, \tag{9.107}$$

and hence the flux decreases geometrically as

$$\pi F = \frac{\text{const}}{r^2}. \tag{9.108}$$

Exercise 9.7 *Assuming a spherical source to be uniformly bright, show that $F = I\,(R/r)^2$, where R is the radius of the source and r is the distance to the observer. [Hint: I is independent of $\mu = \cos\theta$ for a uniform source.]*

The surface flux is π times the μ-weighted average of I_ν. If the mean intensity J_ν is thought of as the zeroth moment of I_ν, then by analogy the first moment is

$$H_\nu = \frac{1}{2}\int_{-1}^{+1} I_\nu\,\mu\,d\mu = \frac{1}{4\pi}\,(\pi F_\nu), \tag{9.109}$$

which is called the *H-moment*. This expression for the flux is referred to as the *Eddington flux*. Similarly, the second moment of specific intensity is the *K-moment*,

$$K_\nu = \frac{1}{2}\int_{-1}^{+1} I_\nu\,\mu^2\,d\mu. \tag{9.110}$$

Let us re-state the zeroth, first and second *moments of specific intensity* $I_{\mu\nu}$, respectively, in terms of their physical meaning.

$$J_\nu = \frac{1}{2}\int_{-1}^{+1} I_{\mu\nu}\,d\mu \quad \text{mean intensity,} \tag{9.111}$$

$$H_\nu = \frac{1}{2}\int_{-1}^{+1} I_{\mu\nu}\,\mu\,d\mu \quad \text{Eddington flux,} \tag{9.112}$$

$$K_\nu = \frac{1}{2}\int_{-1}^{+1} I_{\mu\nu}\,\mu^2\,d\mu \quad \text{radiation pressure.} \tag{9.113}$$

All of these moments are related to general physical characteristics of astrophysical objects. The Eddington flux is related to the observed astrophysical flux $\mathcal{F}_\nu/4\pi = F_\nu/4 = H_\nu$. It has a simple physical interpretation: the radiated flux from the surface depends on the specific intensity I_ν and the angle between the direction of propagation of radiation at the surface and the line of sight to the observer (Fig. 9.9); the factor 4π results from integration over all solid angles ω, i.e.,

$$\mathcal{F}_\nu = \int I_\nu\,\cos\theta\,d\omega = 4\pi\,H_\nu. \tag{9.114}$$

The (monochromatic) radiation pressure is

$$P_\nu = \frac{1}{c}\int I_\nu\,\cos^2\theta\,d\omega = \frac{4\pi}{c}\,K_\nu. \tag{9.115}$$

If the source is unresolved, and assuming the radiation to be isotropic, integration over $\cos^2\theta$ yields a factor of $\frac{1}{3}$. Radiation pressure, in units of dyne/(cm^2 Hz), is related to the energy density as $P_\nu = u_\nu/3$. Isotropic radiation means integration over all angles, and therefore $I_\nu = J_\nu$, and $u_\nu = (4\pi/c)J_\nu$. The total energy density u and the

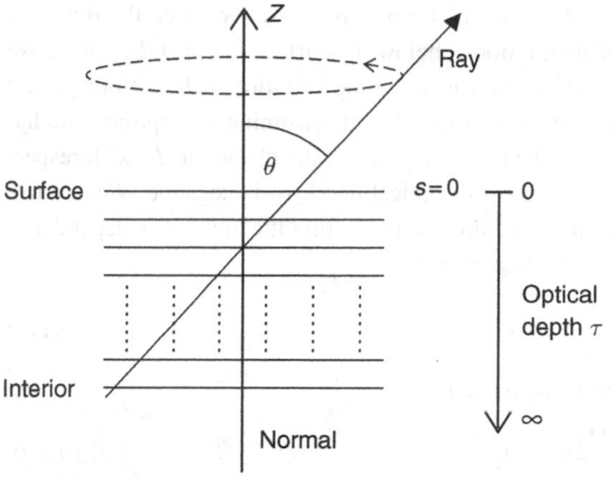

FIGURE 9.9 Schematic variation of the optical depth τ from the surface where radiation escapes and $\tau = 0$, into the deep interior of a source (e.g., star) where $\tau \to \infty$.

total photon number density $N_{h\nu}$ are obtained by integrating over all frequencies. For a black body, for example, the temperature T determines the density

$$u = \int u_\nu\,d\nu = \frac{1}{c}\int B_\nu\,d\omega\,d\nu = \frac{4}{c}\,\sigma T^4, \tag{9.116}$$

and the number of photons,

$$N_{h\nu} = \int_0^\infty \frac{u_\nu}{h\nu}\,d\nu \approx 20\;T^3\;\text{cm}^{-3}, \tag{9.117}$$

where the number density $u_\nu/h\nu$ at each frequency is the energy density divided by the energy of the photon $h\nu$; Eq. 9.117 also gives an approximate value for the total number of photons in a black body at temperature T.

9.4.2 Transfer equation and the source function

For the purpose of the definitions above, we have thus far assumed that there is no source of absorption or emission of energy (i.e., matter) along the path of flow of radiation. But of course the whole point of radiative transfer theory is to deal with situations where we do have material of various kinds acting as absorbers or emitters of energy (viz. sink or source). As we have seen, it is not the geometrical distance but the optical depth τ_ν that is the meaningful quantity for the propagation of radiation; if $\tau_\nu = 0$ then I_ν remains constant. But τ_ν depends on the absorption coefficient κ_ν, related to the opacity and the distance s according to

$$\tau_\nu = -\int_0^s \kappa_\nu(s')\,ds'. \tag{9.118}$$

Note that τ is a positive quantity, since the direction of integration is below the surface at $s = 0$ (Fig. 9.9). We therefore seek to study the variation in I_ν with respect to τ_ν, i.e., $(dI_\nu/d\tau_\nu)$. Again, assuming absorption of radiation to be the only process, the change in I_ν with respect to distance s or optical depth τ_ν is negative. We can formally write this as (neglecting the angular μ-dependence for the time being),

$$\frac{dI_\nu}{ds} = -\kappa_\nu I_\nu, \tag{9.119}$$

or, using $d\tau = \kappa ds$,

$$\frac{dI_\nu}{d\tau_\nu} = -I_\nu. \tag{9.120}$$

To recover the exponential dependence as before,

$$I_\nu = I_0(\nu)\, e^{-\tau_\nu}, \tag{9.121}$$

where $I_0(\nu)$ is some constant initial value of intensity. In general the intervening material not only absorbs radiation but also (re-)emits energy. Therefore, an *emission coefficient* η_ν must be considered in the transfer equation, i.e.,

$$\frac{dI_\nu}{ds} = -\kappa_\nu I_\nu + \eta_\nu. \tag{9.122}$$

Note that the signs of the opacity and the emissivity coefficients are opposite. We may consider absorption as the loss and emission the gain of energy, subtracting or adding to the intensity of radiation as it propagates through a material medium. The units of the emission coefficient are η_ν (erg/(cm^3 s^1 str^1 Hz^1)). Recall that earlier we had defined the units of the opacity (absorption) coefficient κ in inverse units of length, cm^{-1}. It may also be defined as[7]

$$\kappa_\nu\ (\mathrm{cm}^{-1}) = \rho\ (\mathrm{g/cm}^3) \times k_\nu\ (\mathrm{cm}^2/\mathrm{g}) \tag{9.123}$$

in terms of the *mass absorption coefficient* in k_ν (cm^2 g^{-1}). Since $d\tau_\nu = -\kappa_\nu ds$, we can write the formal transfer equation as

$$\frac{dI_\nu}{ds} = -\kappa_\nu (I_\nu - S_\nu), \tag{9.124}$$

or

$$\frac{dI_\nu}{d\tau_\nu} = I_\nu - S_\nu, \tag{9.125}$$

where we have introduced the most important quantity in radiative transfer theory, the *source function* defined as $S_\nu \equiv \eta_\nu/\kappa_\nu$, which is basically the ratio of emissivity to opacity. The source function carries all the information about the material medium, and how it affects the radiation at each frequency.

[7] One needs to guard against the confusion that arises because the total or the mean opacity κ of matter usually incorporates the density ρ, and is also measured in cm^2 g^{-1} (i.e., re-define κ as k; see the discussion on stellar opacities in Chapter 11).

9.4.3 Spectral lines

We are particularly interested in atomic transitions for spectral lines and their source function. The reason we have carried through the subscript ν in the definitions of physical quantities of radiative transfer theory is to note the explicit frequency dependence of radiation. That, in turn, depends on radiative transitions in atoms and molecules: interaction of radiation and matter at discrete frequencies associated with bound quantized states. Therefore, we need to examine how the physical processes and quantities enter into transfer calculations.

Spectra generally consist of a background continuum 'c', and superimposed lines 'l_ν' at specific frequencies. Since the underlying physical processes are sufficiently different for continuum as opposed to line radiation, it is convenient to divide the source function into S_ν^c and S_ν^ℓ. The line source function is then defined as the ratio of the *monochromatic emissivity* to the opacity. At this point we may particularize the meaning of opacity κ_ν in terms of the absorption coefficient α_ν at a given frequency for a bound–bound transition. From above, the line source function is

$$S_\nu \equiv \frac{\eta_\nu}{\alpha_\nu}, \tag{9.126}$$

and the total (continuum + lines) source function is

$$S_\nu^t = \frac{\eta_\nu^c + \eta_\nu^\ell}{\alpha_\nu^c + \alpha_\nu^\ell}. \tag{9.127}$$

We next examine the line source function for the simplest case.

9.4.4 The two-level atom

Disregarding the continuum contribution, it is instructive to consider the source function for a two-level atom simply in terms of the Einstein A and B coefficients. With reference to Fig. 4.1, the monochromatic emissivity[8] related to the A-coefficient for spontaneous emission, in a line at frequency ν corresponding to a transition between two levels $j \to i$ $(i < j)$ is

$$\eta_\nu = \frac{h\nu_{ij}}{4\pi} N_j A_{ji} \phi_\nu, \tag{9.128}$$

[8] Note that in Chapter 8, on emission lines, we had used the notation ϵ_ν for emissivity of a line and therefore $\eta_\nu \equiv \epsilon_\nu$.

where the line profile function ϕ_ν depends on surrounding plasma conditions, as discussed earlier in this chapter. The monochromatic opacity is related to the Einstein B coefficients for absorption, and the inverse process of stimulated emission, as

$$\kappa_\nu = \frac{h\nu_{ij}}{4\pi}(N_i B_{ij} - N_j B_{ji})\,\phi_\nu. \tag{9.129}$$

Note that the stimulated emission term on the right-hand side 'corrects' the absorption term, which is effectively reduced by that same amount, to give the net absorption.[9] Since these expressions generally apply to any two-level transition, the line source function

$$S_\nu(ij) \equiv \frac{\eta_\nu}{\kappa_\nu} = \frac{N_j A_{ji}}{N_i B_{ij} - N_j B_{ji}}. \tag{9.130}$$

In LTE (see Section 9.5), the level populations are given by the Boltzmann equation,

$$\frac{N_i}{N_j} = \frac{g_i}{g_j}\,\mathrm{e}^{-h\nu/kT}, \tag{9.131}$$

and recall that (Chapter 4) the ratio

$$\frac{A_{ji}}{B_{ji}} = \frac{2h\nu^3}{c^2}, \tag{9.132}$$

and $g_i B_{ij} = g_j B_{ji}$. Putting these together,

$$\frac{\eta_\nu}{\kappa_\nu} = \frac{2h\nu^3}{c^2}\left(\mathrm{e}^{h\nu/kT} - 1\right)^{-1} \equiv B_\nu(T). \tag{9.133}$$

Thus we arrive at the simple result that the source function in thermal equilibrium is the Planck function. This is known as *Kirchhoff's law*: $\eta_\nu/\kappa_\nu \equiv B_\nu$. Alternatively, considering the level populations and the mean intensity of the radiation field J_ν, the number of transitions into the state i is equal to the number going out, and therefore

$$N_i\, B_{ij}\, J_\nu(ij) = N_j\, A_{ji} + N_j\, B_{ji}\, J_\nu(ij). \tag{9.134}$$

Solving for $J_\nu(ij)$,

$$J_\nu(ij) = \frac{A_{ji}/B_{ji}}{(g_i B_{ij}/g_j B_{ji})\,\mathrm{e}^{h\nu/kT} - 1}, \tag{9.135}$$

which is equal to the Planck function B_ν, given Eq. 9.133 and the Einstein relations: $g_i B_{ij} = g_j B_{ji}$ and $A_{ji} = (2h\nu^3/c^2) B_{ji}$.

The monochromatic opacity κ_ν in general encompasses not only the bound–bound transitions via the absorption coefficient α_ν, but all other processes related to absorption, such as bound–free and free–free processes. We will incorporate those explicitly in the calculation of opacities, as discussed in Chapter 11.

[9] What happens if the stimulated 'correction' exceeds absorption? That is how a laser or maser is formed, with population inversion so that $N_j > N_i$. Usually the population ratio exceeds the statistical weight ratio, $N_j/N_i > g_j/g_i$.

9.4.5 Scattering

The number of times a photon scatters before absorption is a measure of the optical depth τ. The number of scatterings N is related to the mean free path of the photon. Applying the *random walk* principle, the actual distance travelled is the vectorial sum of the N individual paths the photon undergoes before escape,

$$\mathbf{r} = \sum_i^N \mathbf{r}_i. \tag{9.136}$$

The magnitude of this distance can be approximated by assuming all path lengths on the right-hand side to be equal to the *photon mean free path* $\langle r\rangle$. Then $|\mathbf{r}|^2 = N\langle r\rangle^2$, which gives the root mean square distance travelled by the photon in terms of the mean free path and the number of scatterings

$$|\mathbf{r}| = \sqrt{N}\,\langle r\rangle. \tag{9.137}$$

If D is the total geometrical thickness of the medium the photon needs to travel before escaping, then $D \equiv |\mathbf{r}| \sim \sqrt{N}\langle r\rangle$. The optical depth may be assumed to be $\tau \approx D/\langle r\rangle$, and therefore the number of scatterings before absorption for an optically thick medium $N \sim \tau^2$. On the other hand, for an optically thin medium the mean free path would be of the order of its geometrical length, or less, and the number of scatterings $N \sim \tau$.

How does scattering enter into the source function, which thus far has included absorption and emission? In thermal equilibrium the amount of energy absorbed is equal to the amount of energy emitted. This is how Kirchoff's law applies, and the transfer equation and the source function have the simple forms discussed before. However, if the photons have a certainly probability of being 'destroyed' upon absorption, instead of all being (re-)emitted, then we need to separate the source function into two parts: absorption and *coherent or elastic scattering* of photons by electron scattering. Typical scattering processes are Thomson, Rayleigh and Compton scattering (Chapter 11). In terms of the rate coefficients for the two processes respectively, α_ν and σ_ν^{sc}, the radiative transfer equation is

$$\frac{\mathrm{d}I_\nu}{\mathrm{d}s} = -\left(\alpha_\nu + \sigma_\nu^{\mathrm{sc}}\right)(I_\nu - S_\nu). \tag{9.138}$$

If we denote the photon destruction probability (absorbed, and *not* coherently re-emitted) by ϵ_ν then

$$\epsilon_\nu = \frac{\alpha_\nu}{\alpha_\nu + \sigma_\nu^{sc}}. \tag{9.139}$$

The emitted intensity J_ν in the source function is reduced by the amount $-\epsilon_\nu J_\nu$. On the other hand the amount absorbed increases by $+\epsilon_\nu B_\nu$, where $B_\nu(T)$ is the Planck function distribution at a temperature T. Putting the two together,

$$S_\nu = J_\nu - \epsilon_\nu J_\nu + \epsilon_\nu B_\nu = (1 - \epsilon_\nu) J_\nu + \epsilon_\nu B_\nu. \tag{9.140}$$

Another way of expressing the photon destruction probability ϵ is in terms of the number of scatterings N before destruction or escape, i.e., $N = 1/\epsilon$. Therefore, the net distance travelled $D = \epsilon^{-1/2}\langle r\rangle$. We regard the *effective mean free path* $\langle r\rangle$ due to random walks as the *thermalization* or the *diffusion length* for a photon before being absorbed (destroyed). Then

$$\tau_\nu = \frac{D}{\langle r\rangle} = \sqrt{N} = \epsilon_\nu^{-1/2}. \tag{9.141}$$

It is in this sense – random walk and number of scatterings before escape or absorption – that the optical depth is generally understood. For thermalization length compared with the size of the medium D, $\tau > 1 \rightarrow$ optically thick medium, and $\tau < 1 \rightarrow$ optically thin medium. Also, for the optically thin case, the escape probability is high ($\epsilon_\nu \rightarrow 1$), and $S_\nu \approx \epsilon B_\nu$, from Eq. 9.140.

For the two-level atom, i and j, the collisional destruction probability may be expressed as the de-excitation $j \rightarrow i$,

$$\epsilon_\nu(ij) = \frac{n_e\, q_{ji}}{n_e\, q_{ji} + A_{ji}}, \tag{9.142}$$

where $n_e\, q_{ji}$ is the electron-collision rate, or the number of electron impact de-excitations per second at electron density n_e, as discussed in Chapter 5. If level j is collisionally de-excited before radiative decay with probability A_{ji}, then a net absorption of the photon takes place. Note also that if $\epsilon \sim 1 \rightarrow n_e q_{ji} \gg A_{ji}$, collisions are likely to de-excite the upper level j and photons are effectively absorbed (destroyed) in *each* interaction, since $N \sim 1$.

9.4.6 Plane-parallel approximation

In principle, the specific intensity I_ν is a function of the angles (θ, ϕ), and is a three-dimensional quantity with a specified direction, like a vector (Fig. 9.9). This specification is needed because astrophysical objects are extended, and have several intrinsic physical processes responsible for the creation or destruction of radiation in different parts of the object. A simple example is sunspots, which differ from (are cooler than) other parts of the Sun's surface, owing to enhanced magnetic activity. The specific intensity of emitted radiation varies according to the surface element. As such, the problem of determining the $I_\nu(r, \theta, \phi)$ or $I(x, y, z)$ appears to be very difficult. Furthermore, it can only be ascertained in sources that can be resolved, such as the Sun or bright diffuse nebulae, where we are able to study different regions. Most astrophysical sources are unresolved. Therefore, we must necessarily integrate $I_\nu(r, \theta, \phi)$ over the entire surface.

A great simplification in radiative transfer theory is to assume the geometry to consist of one-dimensional parallel planes of infinite extent, as shown in Fig. 9.9. With the optical depth in the normal direction, μ is constant with height along the direction of I_ν. Here we consider the optical depth in the normal direction, increasing inwards in a stellar atmosphere. This is often a physically reasonable approximation. Consider a typical main-sequence star, where the temperature and density vary only with height denoted by z. The horizontal variation in the x-y plane is negligible, with non-local variations only with z. Then the specific intensity in either the spherical or the rectangular coordinates can be considered to be simply a function of z,

$$I_\nu(r, \theta, \phi) = I(x, y, z) = I_\nu(z). \tag{9.143}$$

Now we revert to include the θ- or μ-dependence, so that the intensity $I^+_{\mu\nu}$ directed outward corresponds to $0 < \mu < +1$ $(-\pi/2 < \theta < \pi/2)$, and $I^-_{\mu\nu}$ $(-\pi < \theta < \pi)$ directed inward corresponds to $-1 < \mu < 0$. The formal solution of the transfer equation, Eq. 9.124, including the μ-dependence, a first-order linear differential equation, is straightforward. Using the integrating factor $e^{-\tau/\mu}$, we can write

$$\frac{d}{d\tau}(I_{\mu\nu} e^{-\tau/\mu}) = \frac{-S_\nu\, e^{-\tau/\mu}}{\mu}. \tag{9.144}$$

We obtain the outgoing intensity at the surface ($\tau \rightarrow 0$), from an object with practically infinite optical depth in the deep interior ($\tau \rightarrow \infty$), as

$$I_{\mu\nu}(\tau_\nu = 0) = \int_0^\infty S_\nu\, e^{-t_\nu/\mu}\, \frac{dt_\nu}{\mu}. \tag{9.145}$$

Now consider a plane at optical depth τ. This divides the whole body of star into two regions in τ-space: $0 - \tau$ from the surface to the layer in question, and $\tau - \infty$ from the plane to regions of practically infinite opacity. Solving Eq. 9.145, the intensity of radiation I^+ flowing outward $(0 \leq \mu \leq 1)$ from the τ_ν-layer is given in terms of

$$I^+(\tau_\nu) = \int_\tau^\infty S_\nu \, e^{-(t-\tau_\nu)/\mu} \, \frac{dt}{\mu}. \tag{9.146}$$

Similarly, the radiation intensity I^- flowing inward $(-1 \le \mu \le 0)$ from the τ-layer is

$$I^-(\tau_\nu) = \int_\tau^0 S_\nu \, e^{-(t-\tau_\nu)/\mu} \, \frac{dt}{\mu} = -\int_0^\tau S_\nu \, e^{-(t-\tau_\nu)/\mu} \, \frac{dt}{\mu}. \tag{9.147}$$

Exercise 9.8 *Work out the detailed steps in deriving $I^\pm_{\mu\nu}$.*

A physically illustrative result (Fig. 9.10) is obtained by approximating the source function to vary as a linear function of the optical depth: $S_\nu(t) \approx a + bt$ (say, for example, if we approximate $S(\tau)$ in Fig. 9.10 by a straight line). Then the emergent intensity is simply

$$I_{\mu\nu}(\tau = 0) = S_\nu(\tau_\nu = \mu). \tag{9.148}$$

This is known as the *Eddington–Barbier relation*: $S_\nu(\tau_\nu = \mu)$, which approximates the emergent intensity for a ray at an angle θ (note that $\tau_\nu/\mu = 1 \rightarrow \tau_\nu = \mu$). If the direction of the emergent ray is normal to the surface ($\mu = 1$), the specific intensity represents the source function at unit optical depth.

9.4.7 Schwarzschild–Milne equations

The equations for the moments of $I_{\mu\nu}$ given in the previous section are in terms of both $I^\pm$. In particular, the mean intensity is

$$J_\nu = \frac{1}{2}\int_0^1 I^+_{\mu\nu} \, d\mu + \frac{1}{2}\int_{-1}^0 I^-_{\mu\nu} \, d\mu. \tag{9.149}$$

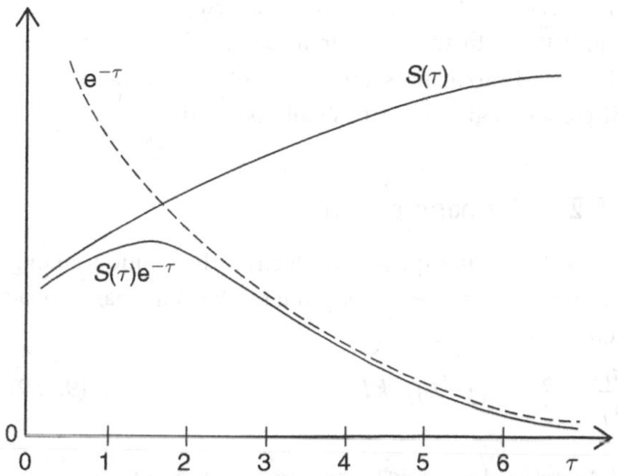

FIGURE 9.10 Convolution of the source function and the optical depth

The natural boundary condition for a star is that the incoming radiation on the stellar atmosphere is negligible, and radiation flows only outward; hence the second term on right-hand side is zero. The moment equations of $I^\pm_{\mu\nu}$ can be rewritten in terms of exponential integrals, and are known as the *Schwarzschild–Milne relations*. For example,

$$J_\nu(\tau_\nu) = \frac{1}{2}\left[\int_{\tau_\nu}^\infty S_\nu(t_\nu) \, E_1(t_\nu - \tau_\nu) \, dt_\nu + \int_0^{\tau_\nu} S_\nu(t_\nu) \, E_1(\tau_\nu - t_\nu) \, dt_\nu\right]. \tag{9.150}$$

9.4.8 The Λ-operator

We now have expressions for the physical quantities given by the Schwarzschild–Milne relations, Eq. 9.150. Their mathematical form suggests an important and powerful method for calculating those quantities. Since the mean intensity of the source is the most useful such quantity, we rewrite Eq. 9.150 as

$$J_\nu(\tau_\nu) = \frac{1}{2}\int_0^\infty S_\nu(t_\nu) \, E_1(|t_\nu - \tau_\nu|) \, dt_\nu. \tag{9.151}$$

It is now useful to convert this equation to an 'operator' form – the Laplace transform – by defining the so-called Λ-operator as $J = \Lambda\,[S]$, or

$$J_\nu(\tau) = \Lambda_\tau[S_\nu(t)] \equiv \frac{1}{2}\int_0^\infty S_\nu(t) \, E_1(|t - \tau|) \, dt. \tag{9.152}$$

The Λ-operator yields the mean intensity J_ν, given an approximate source function S_ν. The exponential integral satisfies the property $\int_0^\infty E_1(x)dx = 1$. Therefore at large τ, say in the semi-infinite case with no incident radiation [246], $J = S$; and at the surface $J = S/2$. It follows that for sufficiently large τ where LTE prevails, $\Lambda[S(\tau)] \rightarrow B(\tau)$, the Planck function.

We know that we cannot ascertain the source function $S_\nu \equiv \eta_\nu/\kappa_\nu$ exactly, or even sufficiently precisely in many cases. That would entail (i) taking account of *all* atomic processes and transitions that determine the emissivity η_ν and the opacity κ_ν at all frequencies, and (ii) measuring all atomic parameters that correspond to (i) with sufficient accuracy. A priori, the source function therefore has a certain degree of 'error'. The relevant question then is: can one 'improve' on an initial estimate of the source function? Before answering the question, it is useful to put it in perspective. One often encounters such problems in physics. Among the notable ones is the calculation of atomic wavefunctions themselves. As we have seen, much of the theoretical atomic physics machinery rests on the

Hartree–Fock approximation, which is a self-consistent *iterative* procedure involving the wavefunctions and the atomic potential that generates them. This is how the Λ-operator form becomes useful – in deriving a numerically self-consistent iterative procedure to solve for J_ν, given a ('trial') source function S_ν, i.e., $J_\nu = \Lambda_\nu[S_\nu]$.

Elaborate numerical procedures have been devised to solve the operator equation $\mathbf{J} = \Lambda\,\mathbf{S}$, and implement the Λ-operator formalism. Advanced versions of the method are aimed at 'accelerating' its convergence and are referred to as the *accelerated lambda iterative* (ALI) schemes (or for 'multi-level' atoms, as MALI) (e.g. [260]). Specialized expositions are, for example, calculations of NLTE line-blanketed stellar atmospheres [261]. These methods are highly specialized and form the basis of state-of-the-art research on NLTE spectral models. This section is intended only as a brief introduction leading into the NLTE methodologies, described in detail in these and other works [244, 245, 246]. Although they are outside the scope of this text, the atomic physics input into the vast computational framework of radiative transfer and spectroscopy are very much the focus of our effort.

Next, we point out the connection between LTE and NLTE approximations, which relates to quantum statistics and quantum mechanics.

9.5 LTE and non-LTE

The physical concepts underlying LTE, and departures therefrom, are of fundamental significance in astrophysics. Non-LTE situations are the *raison d'etre* for a detailed radiative transfer treatment when the assumptions underlying LTE are not valid. Heretofore, we have referred to LTE without going into the statistical foundation upon which it is based. This is now addressed in the next section.

9.5.1 Maxwell–Boltzmann statistics

We begin with quantum statistics formulae that describe the distribution of quantized (bound) and free particles in thermodynamic equilibrium. The probabilities are subject to three rules: (i) all quantum states of equal energy have equal probability of being populated, (ii) the probability of populating a state with energy ϵ at kinetic temperature T is $\exp -\epsilon/kT$ and (iii) no more than one electron may occupy a quantum state (spin is considered explicitly).

For free electrons with velocity v and energy $\epsilon = mv^2/2$, the wavenumber is $k = mv/\hbar$. Then the number of electron states per unit energy per unit volume – the statistical weight of free electrons – is

$$g_e(\epsilon) = 2 \times \frac{m^2 v}{2\pi^2 \hbar^3}, \qquad (9.153)$$

where the factor of two refers to spin degeneracy. With n_e as the number of electrons per unit volume, we denote the number of electrons in the energy range $(\epsilon, \epsilon + d\epsilon)$ as $n_e\, F(\epsilon)\, d\epsilon$, where the probability distribution function $F(\epsilon)$ is normalized as

$$\int_0^\infty F(\epsilon)\, d\epsilon = 1. \qquad (9.154)$$

According to the three rules mentioned just now we can express

$$F(\epsilon) = \frac{g_e(\epsilon)\, e^{-\epsilon/kT}}{U_e}, \qquad (9.155)$$

with the fractional population of free electrons defined by the electron *partition function*

$$U_e = 2\left(\frac{m_e kT}{2\pi\hbar^2}\right)^{3/2}. \qquad (9.156)$$

For a Maxwellian distribution, we have

$$F(\epsilon) = 4\pi m_e^2\, (2\pi m_e kT)^{-3/2}\, v^2\, e^{-\epsilon/kT}. \qquad (9.157)$$

Since there cannot be more than one electron per quantum state, the fraction of occupied states is

$$\frac{n_e\, F(\epsilon)}{g_e(\epsilon)} = \frac{n_e}{U_e}\, e^{-\epsilon/kT}. \qquad (9.158)$$

Maxwell–Boltzmann statistics expressed by this equation refers to *non-degenerate* electrons subject to the three rules,[10] and is valid only if the number of available quantum states is much greater than the number of electrons, i.e., $n_e\, F(\epsilon)/g_e(\epsilon) \ll 1$, or $U_e \gg n_e$. This is true for most of the interior of the star (at T_e and n_e given in Table 11.1). But it is not entirely valid for some important cases, such as the central cores of stars where the densities reach $\sim 100\ \mathrm{g\,cm^{-3}}$, even at temperatures greater than 10^7 K. In the degenerate cores of stars therefore, Maxwell–Boltzmann statistics is not strictly valid and a different equation-of-state should be used.

9.5.2 Boltzmann equation

As we know, the quantized electron distribution among the atomic energy levels is given by the Boltzmann equation,

$$\frac{n_i}{n_j} = \frac{g_i}{g_j}\, \exp(-E_{ij}/kT). \qquad (9.159)$$

10 As mentioned in Chapter 1, for degenerate fermions we need Fermi–Dirac statistics, and for bosons (naturally degenerate) we employ Bose–Einstein statistics.

Let the total number of ions in all occupied levels be

$$N = \sum_i n_i. \tag{9.160}$$

We then define the *internal atomic partition function*,

$$U = \sum_i g_i \, e^{-E_i/kT}, \tag{9.161}$$

so that the level populations are

$$n_i = g_i \, e^{-E_i/kT} \frac{N}{U}. \tag{9.162}$$

The partition function describes the level-by-level population of the occupied levels in an atom in a plasma in thermal equilibrium characterized by local temperature T. The atomic partition function U is a difficult quantity to compute, since it embodies not only the atomic structure of all levels in the atom, but is also a *divergent* sum over the infinite number of levels with statistical weights increasing as $g_i \sim 2n_i^2$. However, in real situations the truncation of the partition function naturally occurs, owing to the interaction of the atom with surrounding particles. The more highly excited an atomic level, the higher the magnitude of plasma perturbations, since the radii of atomic orbitals also increase as $\sim n^2$. Likewise, the higher the plasma density the stronger the perturbations, and consequent truncation of the atomic partition function at a limited number of levels. We return to the proper solution of this problem, by considering the probability of occupation of an atomic state as part of defining an equation-of-state for opacities calculations, in the next chapter.

9.5.3 Saha equation

Whereas the Boltzmann equation relates the populations of levels within an atom (ion) of any one ionization state, the *Saha equation* relates the distribution *among* different ionization states of an element, or the *ionization fractions*, in thermodynamic equilibrium. Using the terminology of the Boltzmann equation developed here, the Saha equation also entails energies above the ionization energy E_{im} of a level i of ionization stage m. The population distribution therefore needs to specify both the ionization stage and the atomic level. Given two successive ionization stages m and $m+1$

$$\frac{n_{i,m+1} \, n_e \, F(\epsilon)}{n_{j,m}} = \frac{g_{i,m+1} \, e^{-E_{i,m+1}/kT} \, g_e(\epsilon) \, e^{-\epsilon/kT}}{g_{j,m} \, e^{-E_{j,m}/kT}}, \tag{9.163}$$

where $E_{i,m+1}$ and $E_{j,m}$ are the energies of levels i and j in each ionization stage, and $g_{i,m+1}$ $g_{j,m}$ are their statistical weights. The Saha equation can now be written as

$$\frac{n_{m+1} \, n_e}{n_m} = \frac{U_{m+1} \, U_e}{U_m}. \tag{9.164}$$

The ionization fraction is specified as n_m/N, with the total number distributed among all ionization stages of an element, $N = \sum_m n_m$ (Eq. 9.164). A simplified form of the Saha equation that is often used is the assumption that adjacent ionic populations (n_m, n_{m+1}) are in their ground states only. This allows us to replace the partition functions with the statistical weights of the ground states of respective ionization stages, i.e., $U_{m+1} \to g_{1,m+1}$ and $U_m \to g_{1,m}$. Referring only to level index '1' then,

$$\frac{n_{m+1} \, n_e}{n_m} \approx \frac{g_{1,m+1}}{g_{1,m}} \, U_e \, e^{-I_m/kT}, \tag{9.165}$$

where $I_m \equiv (E_{1,m+1} - E_{1,m})$ or the ionization energy from the ground state of ion m into the ground state of ion $(m+1)$. Since we know the free-electron partition function U_e (Eq. 9.156) we can rewrite this equation more explicitly, as

$$\frac{n_{m+1} n_e}{n_m} = 2 \left[\frac{m_e kT}{2\pi\hbar^2} \right]^{3/2} \left(\frac{g_{1,m+1}}{g_{1,m}} \right) e^{-I_m/kT}. \tag{9.166}$$

Of particular interest is the ionization stage with the *maximum abundance* of an element X (also denoted as m for convenience), i.e., $n(X^{m+})/n(X)$. We note that (i) usually the ratio of statistical weights $g_{1,m}/g_{1,m+1} \approx 1$ (or of the order of unity) and (ii) close to maximal abundance, the ionization state distribution would tend to be $n_{m+1}/n_m = g_{1,m}/g_{1,m+1} \approx 1$. Then from Eq. 9.165,

$$n_e \approx U_e \, \exp(-I_m/kT). \tag{9.167}$$

Now we recall one of the basic characteristics of Maxwell–Boltzmann statistics: $U_e \ll n_e$, and thence,

$$I_m \gg kT. \tag{9.168}$$

This implies that the ionization state with the maximum abundance of an element occurs at a kinetic temperature much lower than that corresponding to the ionization energy of that state, typically $I_m \approx 10kT$. The LTE temperature distribution is such that most photon energies are insufficient to ionize the ground states of ions, but are capable of inducing transitions by absorption from excited levels. This, in turn, has far-reaching implications in calculations of fundamental quantities, such as the plasma opacities discussed in Chapter 10. Excited states contribute significantly to photon opacities even though the abundances of ions in excited states may be low.

Put another way, the Planck function is such that most photons have $h\nu \approx kT$, and therefore the (black-body) radiation is absorbed by excited states that have low ionization or excitation energies – which is preferentially the case for highly excited levels, as opposed to low-lying ones.

The combination of the Saha–Boltzmann equations specify the quantum statistical nature of the plasma, or the equation-of-state (EOS), in LTE. They yield both (i) the ionization fractions of a given element and (ii) their level populations. However, we still need to consider plasma effects. The electrons, being lighter than ions, are the dominant colliding species in a plasma, and are assumed to follow a Maxwellian distribution of velocities. Local thermal equilibrium is usually a safe assumption at sufficiently high electron densities when collisions would establish a Boltzmann distribution. Referring to the two-level atom, if the densities are not sufficiently high and the upper levels are not collisionally de-excited before radiative decay, then the photons could escape and the level populations would deviate from their Boltzmann statistical values. In the next chapter we shall discuss the Saha–Boltzmann equations further, modified to describe stellar interiors in LTE.

9.5.4 Non-LTE rate equation and equation-of-state

As mentioned, departures from LTE – and hence from the *analytical* Saha and Boltzmann formulae – lead to a huge increase in the level of complexity of the atomic physics, coupled with the radiation field and plasma effects. Thus a proper non-LTE (NLTE) formulation requires elaborate *numerical* methods needed to solve the coupled atom–radiation problem. Since the Saha–Boltzmann equations are no longer valid in NLTE, the level populations must be ascertained explicitly. That means taking account of all radiative and collisional processes that determine level populations. Often though, one saving grace is that we may generally assume time invariance, i.e., the population of a given level does not change with time in most astrophysical situations. As we have seen, the rate equations in statistical equilibrium can be quite involved if a number of physical processes are to be considered (see Eq. 8.15). The NLTE problem is further complicated by the fact that the radiation field J_ν is itself to be derived self-consistently from an iterative process. It is instructive first to study each situation physically to identify the dominant processes, and then to write down the rate equations correspondingly. If we begin simply as before, with the time-independent level population for a given ionization state (as in Chapter 8),

$$\frac{dN_i}{dt} = 0 = \sum_{j \neq i} N_j P_{ji} - N_i \sum_{j \neq i} P_{ij}, \tag{9.169}$$

where $P_{ij} = R_{ij} + n_e q_{ij}$ is the sum of individual radiative and collisional rates, respectively. The radiative term R_{ij} on the right-hand side involves the radiation field for bound–bound transitions at frequency ν_{ij}, i.e.,

$$R_{ij} = A_{ij} + J_{\nu_{ij}} B_{ij}. \tag{9.170}$$

The coupling of the atomic level populations to the radiation field is clear: each depends on the other. The atomic collisional and radiative rates (the Einstein coefficients) must be computed explicitly for all levels likely to be of interest in the model.

An important point to note is that in general the monochromatic radiation field intensity may not be given by the Planck function that characterizes LTE, i.e., $J_\nu \neq B_\nu$ in NLTE. But we may attempt to treat this inequality by modifying the source function S_ν to include departure from LTE. Writing the line source function as

$$S_\nu(ij) \approx \frac{b_i}{b_j} B_\nu(ij), \tag{9.171}$$

or more explicitly,

$$S_\nu(ij) = \left(\frac{2h\nu^3}{c^2}\right)\left(\frac{b_i/b_j}{e^{h\nu/kT} - 1}\right), \tag{9.172}$$

where b_i, b_j are the *departure coefficients* for the lower and upper levels takes account of the deviation of their populations from Boltzmann statistical values in LTE.

Now for the NLTE equation-of-state, we also need to consider explicitly the ionization and recombination processes that determine the ionization fractions of a given element. First, let us consider ionization – photoionization by a radiation field J_ν, and electron impact ionization – and the deviation from LTE given by B_ν. The population ratio affecting ionization of a level i (neglecting all other levels) is

$$\frac{N_i}{N_i^{\mathrm{LTE}}} = \frac{n_e C_I(i) + \int_{\nu_i}^{\infty} \sigma_\nu(i)\,[4\pi B_\nu/h\nu)]\,d\nu}{n_e C_I(i) + \int_{\nu_i}^{\infty} \sigma_\nu(i)\,[4\pi J_\nu/h\nu]\,d\nu}. \tag{9.173}$$

Here the electron impact ionization rate $n_e C_I(i)$ remains the same in both cases; $\sigma_\nu(i)$ is the photoionization cross section of level i with ionization threshold energy $h\nu_u$.

Written as above, it is trivial to conclude that if $J_\nu \rightarrow B_\nu$ then $N_i \rightarrow N_i^{\rm LTE}$. But this result is quite significant physically. It also implies that when the radiation field is relatively insignificant compared to the collisional rate, the LTE limit would be approached. That would always be the case at high electron densities $n_{\rm e} \rightarrow \infty$, even if $J_\nu \neq B_\nu$, such as when the optical depth τ may not be too large and radiation escapes. In other words, the fractional ionization populations would be in local equilibrium at a temperature T, and may be given by the Saha equation (or variant thereof, see Section 11.4.1).

In Chapters 6 and 7, on photoionization and recombination, we had written down expressions for two particular cases of plasma equilibrium: coronal or collisional equilibrium, and nebular or photoionization equilibrium. They represent the opposite limits of the equation above (Eq. 9.173). When the radiation field is negligible, then successive ionization states are determined by a detailed balance between collisional ionization and (e + ion) recombination in the coronal approximation (characteristic of conditions in the solar corona). Then, for any two successive ionization stages,

$$\frac{N({\rm X}^{m+})}{N({\rm X}^{m+1})} = \frac{\alpha_{\rm R}(T_{\rm e};{\rm X}^{m+1})}{C_{\rm I}(T_{\rm e};{\rm X}^{m+1})}. \qquad (9.174)$$

Here $C_{\rm I}(T_{\rm e})$ and $\alpha_{\rm R}(T_{\rm e})$ are total ionization and recombination rate coefficients for the ionization stages m + and (m +1) respectively. We note that $\alpha_{\rm R}(T_{\rm e})$ here is assumed to include both the radiative and the dielectronic recombination processes, as in the unified formulation of (e + ion) recombination given in Chapter 7, or by adding the two rate coefficients separately. In collisional or coronal equilibrium the ionization fractions depend only on the temperature, and not the density.

In the nebular case, when electron densities are low and photoionization occurs predominantly by the radiation field of a hot star, we may have significant deviations of ionization fractions from that in LTE, depending on the difference between J_ν impacting on the nebular gas at a distance from the star, and the ideal black-body radiation B_ν. Since atomic levels in a given ion may have widely varying level-specific photoionization or recombination rate coefficients, level populations (or ionization fractions) may be overpopulated or underpopulated. Again, an approach in terms of departure coefficients may be adopted for level populations to account for deviations from the Saha–Boltzmann equations, and a combined excitation–ionization form of the two equations above can be written down explicitly. For example, in photoionization equilibrium (neglecting collisional ionization) the level population for level j is

$$\sum_i (A_{ji} + n_{\rm e} q_{ji})\, N_j + \int_{\nu_j}^{\infty} \frac{4\pi J_\nu}{h\nu}\, a_\nu(j)\, {\rm d}\nu$$
$$= n_{\rm e}\, \alpha_{\rm R}(j) + \sum_k n_{\rm e}\, \alpha_{\rm R}(k)\, C_{kj} + (A_{kj} + n_{\rm e} q_{kj})\, N_k, \qquad (9.175)$$

where C_{kj} are the cascade coefficients for all indirect radiative transitions from $k \rightarrow j$, A_{kj} is the direct radiative decay rate, and q_{ji}, q_{kj} are electron impact (de-)excitation coefficients (Chapter 5).

But generally, how do we ascertain the radiation field intensity J_ν, which, while acting locally, must depend on non-local input from other atoms in different regions? That is the crux of the non-LTE problem: to derive the source function that yields the radiation intensity, as outlined in this brief treatment of radiative transfer. But we are faced with the formidable problem of considering *all* collisional and radiative processes that determine ionic distributions and level populations. The radiative transfer equations must then be solved self-consistently for the mutually interacting radiation field J_ν and the level populations. The problem obviously becomes more complicated as the number of atomic levels that need to be considered in the NLTE model increases. Thus, the needs of atomic data for radiative transfer models are vast. While experimental measurements may be made for selected transitions and processes, most of the atomic parameters need to be computed theoretically (much of this textbook is aimed at a detailed exposition of those processes). The primary processes that determine the line source function in NLTE are: (i) bound–bound radiative transitions (Chapter 4), (ii) bound–free transitions, viz. level-specific and total photoionization and (e + ion) recombination (Chapters 6 and 7), (iii) collisional excitation and ionization (Chapter 5) and (iv) free–free transitions and plasma line broadening (this chapter and Chapters 10 and 11).

10 Stellar properties and spectra

Stars exist in great variety. They are among the most stable, as well as occasionally the most unstable, objects in the Universe. While extremely massive stars have short but very active lifetimes of only millions of years after birth, the oldest stars have estimated ages of up to 14 billion years at the present epoch, not much shorter (though it must be) than the estimated age of the Universe obtained by other means, such as the cosmological Hubble expansion. In fact, the estimates of the age of the Universe are thereby constrained, since the Universe cannot be *younger* than the derived age of the oldest stars – an obvious impossibility.[1] Stellar ages are estimated using well-understood stellar astrophysics. On the other hand, variations in the rate of Hubble expansion may depend on the observed matter density in the Universe, the gravitational 'deceleration parameter', the 'cosmological constant', 'dark' (unobserved) matter and energy, and other exotic and poorly understood entities. Needless to say, this is an interesting and rather controversial area of research, and is further discussed in Chapter 14.

But stars are the most basic astronomical objects, and astronomers are confident that stellar physics is well-understood. This confidence is grounded in over a century of detailed study of stars, with the Sun as the obvious prototype. Most of this knowledge is derived from spectroscopy which, in turn, yields a wealth of information on nearly every aspect of stellar astrophysics; stellar luminosities, colours, temperatures, sizes, ages, composition, etc. Most of these depend on one single physical quantity of overwhelming importance: the mass of a star. Other stellar parameters, such as the luminosity, size, and surface temperature, largely depend on the mass. To be sure, other parameters such as the chemical composition are also critical. Radiation transport through stellar matter – characterized by the fundamental quantity, the opacity – determines or affects observed properties such as stellar spectra, evolutionary paths, stellar ages, etc.

In this chapter, we first describe the categorization of stars based on their overall luminosity and spectra. This is followed by a brief and general discussion of overall stellar structure: the stellar core, the envelope and the atmosphere. This is underpinned by a detailed exposition on the radiative opacity, which depends on the atomic and plasma physics in the stellar interior (described in the next chapter). We also discuss in some detail spectral formation in the upper layers of stellar atmospheres, exemplified by the Sun and its immediate environment, the solar corona and associated phenomena, such as winds, flares and coronal mass ejections. We begin with the basics of stellar classification scheme. That also embodies a discussion of stellar evolution, and the precursors of sources such as planetary nebulae (PNe) and supernovae (SNe) in later stages.

10.1 Luminosity

The energy output of a star and its spectrum are related. Its luminosity is a measure of the total emitted energy and depends on the effective temperature by the black-body Stefan–Boltzmann law, as well as on its spherical surface area with radius R,

$$L = 4\pi R^2 \sigma T_{\rm eff}^4. \tag{10.1}$$

For example, a star with twice the radius and twice the effective temperature of the Sun is 64 times more luminous. The total energy output, referred to as the *bolometric luminosity*, may be measured by a *bolometer* – a detector of emitted energy at all wavelengths. Figure 10.1 shows a black-body function at $T_{\rm eff} = 5770$ K, which provides the best fit to the radiation field of the Sun [81]. The differences with a pure Planck function arise due to *line blanketing* caused by attenuation of

[1] After all, one can't be older than one's parents!

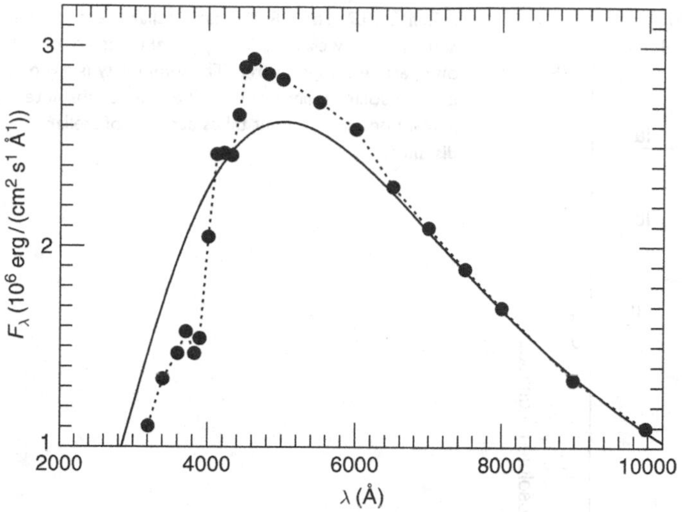

FIGURE 10.1 Black-body Planck function at 5770 K (solid curve) compared with the solar radiation distribution (dashed curve with data points [81].

radiation by the multitude of radiative transitions in stellar atmospheres. Spectral formation by the ionic, atomic and molecular species dominant at the particular T_{eff} of a star, on the one hand, and its total luminosity L, on the other hand, are found to be connected in a phenomenological manner, which provides the foundation for stellar classification.

10.2 Spectral classification – HR diagram

The relationship between L and T_{eff} for *all* of the great variety of stars is revealed remarkably by the so-called Hertzsprung–Russell (HR) diagram in Fig. 10.2.[2] Most stars lie in a fairly narrow (curved) strip in the middle of the diagram diagram called the *main sequence*. The L–T relation manifests itself in the spectra of stars and provides the basis for the spectral classification scheme. The energy emitted by the star as a black body is absorbed by the atomic species in the cooler regions of the stellar atmosphere through which radiation escapes. The observed stellar spectra therefore resemble a Planckian radiation field distribution modulated by absorption and emission at discrete wavelengths. Recall that the peak wavelength of emission from a black body at effective temperature T_{eff} is given by Wien's law (Chapter 1) $\lambda_p = 0.0029/T_{eff}$. Wien's law can be used to approximate the wavelength at which the star at surface temperature T_{eff} emits maximum light.

It is natural to begin with the observed strength of absorption lines of hydrogen, as in Table 10.1. Initially, stars were classified in decreasing order of the strength of Balmer lines, beginning as A, B, etc. But it was realized that the temperature does not quite correlate with the strength of H-lines. Although B stars have weaker H-lines than A stars, they have higher temperature, and so need to be placed ahead of A stars. This is where other atomic species enter the picture. Neutral helium lines are strong in B stars, owing to the higher temperature, which correlates with the higher ionization potential of He I, 24.6 eV, as opposed to that of H I, 13.6 eV. Another type of star is even hotter than the B stars. These are labelled as O stars and exhibit ionized helium lines; the ionization energy of He II is 4 Ry or 54.4 eV. In general, the ionization fraction of an element increases with hotter spectral types. Several such departures from the originally intended sequence based on H-lines resulted in the final classification scheme as we have it today (Table 10.1): O, B, A, F, G, K, M, R, N, S.[3] It ranges from the hot O stars to the cool M (and R, N, S) stars, and is based on bright optical lines from characteristic atomic species. About 90% of all stars lie on the main sequence. This is because most stars spend most of their lives in the hydrogen-to-helium fusion phase in their cores, which defines the physical characteristic of main-sequence stars. As hydrogen fusion wanes, stars evolve away, and upward, from

[2] We again note the seminal role played by Henry Norris Russell in the development of atomic physics itself, as pointed out in Chapter 1. It is astonishing, and not well-known, that even as the new science of quantum mechanics was being developed in 1925, Russell teamed up with physicist Frederick Albert Saunders to develop the Russell–Saunders or *LS* coupling scheme. It was devised to explain atomic structure and spectra of stars, discussed in this chapter.

[3] The well-known mnemonic for spectral classes of stars is: Oh Be A Fine Girl(Guy) Kiss Me Right Now (Smack). In addition to these, some cooler spectral classes have recently been added.

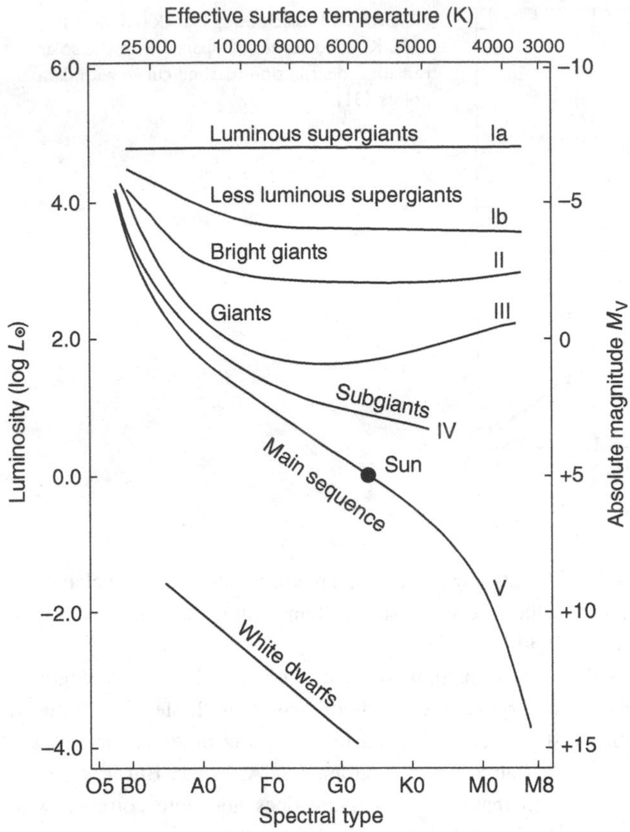

FIGURE 10.2 The HR diagram: stellar spectral types and luminosity classes. The spectral types are based on effective temperatures. The luminosity is given in units of solar luminosity L⊙. The scale of absolute magnitudes on the right takes account of stellar distances.

the main sequence as a result of further nucleosynthesis (Section 9.2.1). Subsequent evolution is usually on a rapid timescale compared with their main sequence lifetime. It is customary to trace stellar *evolutionary tracks* on the HR diagram to depict the *post-main-sequence* phase of stellar lifetimes (Fig. 10.6, discussed later).

The spectral types are further subdivided according to the strengths of spectral lines, indicated by numerals following the letter, on a scale of zero to nine. Peak absorption from atomic species can be associated with each spectral type. A more detailed scheme is the division into spectral subtypes by W. W. Morgan and P. C. Keenan, known as the MK system, as in Table 10.1. The hottest stars, with $T_{\rm eff}$ = 30 000–40 000 K, and spectral types O 3–9, show strong lines of He II (λλ 4026, 4200, 4541 Å); doubly or triply ionized elements are also seen, along with a strong UV continuum. Next, B 0–9 stars, in the 15 000–28 000 K range, have no He II, but strong He I (λ 4471 Å), and stronger H I lines than in O stars; strong lines from some ions are also present, with Ca II making its appearance weakly. Type A stars are distinguished by strong H I lines, strongest of all stellar types; the strength is maximum for A0 and weakens towards higher A subtypes approaching A9; Mg II and S III are strong, and Ca II is present but still weak. In F stars, H I begins to weaken, ions of lighter elements are generally weaker, but Ca II strengthens. Solar-type G stars (5500–6000 K) have the strongest Ca II lines, the so-called H and K lines due to valence electron excitations: $4s\ ^2S_{1/2} \rightarrow 4p\ ^2P^o_{1/2,3/2}$ transitions; neutral atomic and molecular lines appear. K stars (3500–4000 K) have the weakest H I, but strong neutral metals and stronger molecular bands. M stars (2500–3500 K) have the strongest molecular bands, especially TiO. Carbon stars C (R, N) are distinguished by carbon compounds, such as CO. The last three classes, S, L and T, of low-temperature stars, have molecular bands of compounds (metal oxides and sulphides) with heavy elements up to the Lanthanides (nuclear charge Z = 57–71).

In additon to the dominant ionic, atomic or molecular species for each spectral type, the relative strengths of lines are also indicated qualitatively in Table 10.1. The usual convention is that the lines of a given species appear weakly in the preceding and succeeding types. As mentioned, H lines are strongest in spectral type A0, and weakening towards A9. To some extent there is bound

TABLE 10.1 Morgan–Keenan (MK) system of spectral types and features in optical spectra.

Spectral type	Characteristic spectral features
O	He II, He I (weaker), C III, N III, O III, Si IV, H I (weak)
B	No He II, H I (stronger), He I (strong) C II, O II, Si III
A	H I (strongest), Mg II and S III (strong), Ca II (weak)
F	H I (weaker), Ca II (stronger), neutrals appear
G	H I (weaker), Ca II (strongest), neutral atoms (stronger), ions (weaker), molecules (CH)
K	Ca I (strong), H I (weakest), neutrals dominate, molecular bands (CN, CH)
M	Molecules (strong), TiO, neutrals (Ca I strong)
C (R, N)	Carbon stars, no TiO, carbon compounds C_2, CN
S	Heavy-element stars, molecular bands of LaO, ZrO, etc., neutrals
L, T	Molecules CO, H_2O, CaH, FeH, CH_4, NH_3; substellar masses

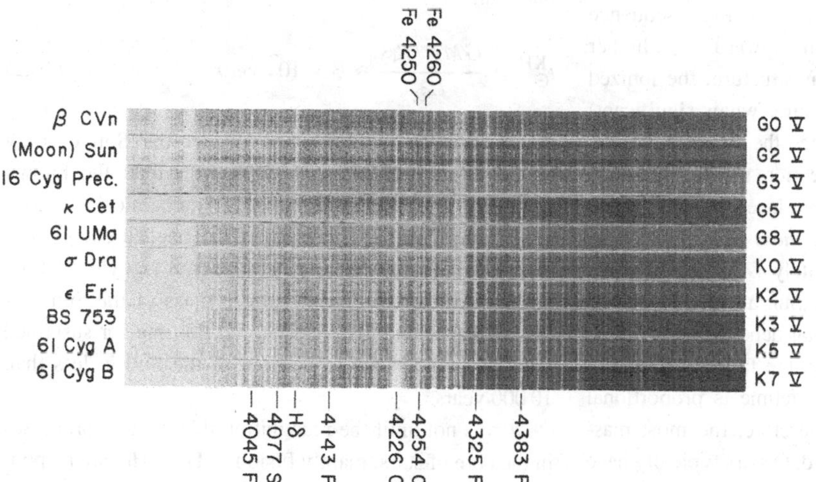

FIGURE 10.3 Progression of stellar spectra from G to K spectral types according to the MK classification scheme – luminosity Class V stars in the lower main sequence in Fig. 10.2. The legends on the left refer to the (abbreviated) name of the constellation and star. The similarities in the spectra imply near-solar composition. In the early G stars of normal solar composition, the temperature types are determined by the ratio of H to metal lines, for example, through the Hδ/Fe 4143 ratio.

to be some ambiguity and overlap in attempting to fit a huge number of stars into this rather limited scheme. In addition, there are categories of stars that have unexpected spectral features, although they may satisfy the general criteria for a spectral type. For example, there are 'peculiar' metal-rich stars denoted by a suffix 'p', such as Ap stars.

There is another division, according to *luminosity class*, that extends to post-main sequence stages of evolution on the HR diagram: Ia – bright supergiants, Ib – supergiants, II – bright giants, III – giants, IV – subgiants, and V – main-sequence stars (also referred to as 'dwarfs' on the part lower than the Sun). These luminosity classes are denoted above the lines or curves in Fig. 10.2, running almost horizontally across the HR diagram, with the exception of the main sequence. This scheme is imprecise, since, for example, stars of all spectral types are said to have the same luminosity class (V); many of those that lie lower than, and including, the Sun are called 'dwarfs'. But their physical characteristics are distinct from white dwarfs that are more than a hundred times smaller in radius and less in brightness than the Sun, and in fact do not lie on the main sequence at all but form a separate sequence lying below the main sequence. The spectral classification of the Sun is G2 V. On the other hand, a G2 III star would be a larger and more luminous star of similar temperature. A K4 III star is a red giant, whereas an O5 Ia star would be a luminous supergiant and an O5 Ib less so.

As an example of the extensive stellar spectroscopy necessary to ascertain the spectral and luminosity types precisely, Fig. 10.3 is a collection of spectra of a number of stars around the solar type G2 V. Figure 10.3 shows the *progression* in spectral features between the G and K stars along the lower main sequence. The similarity in their spectra implies near-solar chemical composition (note the spectrum of the Sun as inferred from light reflected from the Moon). The strength of various lines and elements may be followed from each sub-class to adjoining ones; for instance, the H/Fe ratio may be tracked by the Balmer Hδ to Fe 4143 ratio. The legends on the ordinate of Fig. 10.3 are the names of the stars, as they progressively fit into the MK classification scheme.[4]

10.3 Stellar population – mass and age

The most important physical quantity of a star is its mass. In the HR diagram, the masses of main sequence stars are higher towards the left and upwards, i.e., higher T or L. To enable a stable stellar structure, the ionized gas pressure (and radiation pressure, when significant) in the interior must counterbalance the force of gravity $M(r)/r^2$ at any given point r. The more massive the main sequence star, the more energy it produces in the core, and the more luminous it is. But the *mass–luminosity relationship* is non-linear, approximately $L \sim M^{3.5}$. Stars ten times more massive than the Sun are nearly 10 000 times brighter. The lifetime of a star depends on its mass, since the more luminous the star, the faster it burns up the fuel in the core. The stellar lifetime is proportional to the ratio $M/L \sim M^{-2.5}$. Therefore, the most massive stars are also the shortest lived. O stars typically have lifetimes of tens of millions of years, compared with the almost 10 billion years for the Sun. Stellar populations are also roughly divided into young or old stars. Young stars with relatively high metal abundances, found in the disc or the spiral arm of a galaxy, are called *Population I* or Pop I stars. On the other hand, stars in the halo of a galaxy and globular clusters are older, and called *Population II* or Pop II stars, which are metal-poor. Therefore, Pop I or II broadly refers to stellar age and metallicity. At 4.5 billion years in age, and with its location in a spiral arm of our Milky Way, the Sun is an old Pop I star. Pop II stars have metal abundance of the order of one percent that of the Sun.[5]

[4] We are grateful to R. F. Wing for making available the original plates of stellar spectra, used to establish the MK scheme, from the collection of P. C. Keenan and R. C. McMeil (Ohio State University Press).

[5] There is another class of stars, Pop III, which refers to the first-generation stars formed early after the big bang and containing only primordial elements. The isotopes produced were $^1H_1, ^2H_1, ^3He_2, ^4He_2, ^7Li_3$. Nucleosynthesis of all heavier elements occured in subsequent generations of stars.

Exercise 10.1 *From Eq. 10.1 and the effective temperature, estimate the luminosity ratio of an O5 star to an M dwarf star, say M8, and compare with that in Fig. 10.2.*

The rate at which the Sun radiates energy is approximately 4×10^{26} watts (joule/second). Attempts to explain such a huge generation of energy by means other than nuclear fusion – gravitational contraction or chemical reactions – were unsuccessful. Kelvin and Helmholtz had proposed that gravitational potential energy was the source of the Sun's energy. But the total gravitational potential energy of the Sun would yield no more than about 6×10^{41} J. Radiating at the given rate (present luminosity), the Sun would radiate its entire energy in the *Kelvin–Helmholtz timescale*

$$t_{\odot}^{KH} = \frac{GM_{\odot}^2/R_{\odot}}{L_{\odot}} \approx 3 \times 10^7 \ years. \quad (10.2)$$

The Kelvin–Helmholtz timescale for the Sun of only about 30 million years is far shorter than the known geological age of the Earth. Similarly, chemical reactions are far too inefficient to produce the needed energy. For example, burning ordinary carbon-based fuel yields about 10^{-19} J per atom. Given that the mass of the Sun corresponds to about 10^{57} atoms, the lifetime for sustained radiation would lead to burnout of the Sun in less than 10 000 years.

It was not until the recognition that nuclear processes in the core of stars, mainly fusion of H $\rightarrow$ He, are responsible for the continuous generation of energy, that the problem was solved. Thus a star may be thought of as a continuously exploding hydrogen bomb through most of its lifetime. The primary nuclear reactions are discussed later in this chapter. Here we note only that we can use them to calculate the 'nuclear lifetime' of the Sun, balancing energy generation with hydrogen fuel, against expenditure of energy via radiation at its current luminosity. The fractional amount of energy released by fusing a given mass of hydrogen M_H into helium is $0.007 M_H c^2$, so that all of the Sun's hydrogen will be exhausted in

$$t_{\odot}^{fusion} = \frac{0.007 M_{\odot} c^2}{L_{\odot}} \approx 10^{11} \text{ years}. \quad (10.3)$$

This is actually not too far from the projected lifespan of the Sun of about 10 billion years, calculated from more elaborate models, especially if one allows for the increase in the Sun's luminosity when it expands into the red giant phase, about five billion years from now.

10.4 Distances and magnitudes

The observed brightness of a star obviously depends on its distance, and does not indicate its true luminosity. Therefore, a distinction between apparent vs. intrinsic luminosity needs to be made, and quantified in terms of magnitude *and* distance. The relative apparent brightness B of two stars is measured according to their *apparent magnitude m* on a logarithmic scale. Each magnitude (often abbreviated as 'mag') change corresponds to a 2.5-factor change in brightness, i.e.,

$$m_2 - m_1 = 2.5 \log \left(\frac{B_1}{B_2} \right). \tag{10.4}$$

This scale is based on the ability of the human eye to judge brightness, which happens to be logarithmic. In ancient times, the Greek astronomer Hipparcus divided the brightness of stars, as they appear to the naked eye, into six categories, which are now known to span a factor of 100. The relation above follows that convention, since $2.5^5 \approx 100$ or, more precisely, $2.512^5 = 100$. In recognition of this fact, Eq. 10.4 was adopted as the *exact* relationship, such that a change of 5 mag $\rightarrow \times 100$. In other words, an increase in brightness by a factor of 100 implies a *decrease* in magnitude m by a factor of five.

But we also need to establish an *absolute scale* that relates the apparent brightness to its intrisic brightness. Clearly, this must involve the distance. The *absolute magnitude M* is defined as the apparent magnitude the star would have if it were located at a distance of 10 parsecs (pc) from the Earth (1 pc = 3.26 light years = 3.09 $\times 10^{13}$ km). Thus the absolute and apparent magnitudes are related as

$$m = M + 5 \log \left(\frac{d(\mathrm{pc})}{10} \right). \tag{10.5}$$

Note that (i) the factor $5 = 2 \times 2.5$ arises from the geometrical dilution of apparent brightness with distance, according to the inverse square law $\sim d^{-2}$. Also note that when $d = 10$ pc, $m = M$. Thus we can rewrite,

$$m - M = 5 \log d - 5, \tag{10.6}$$

where $(m - M)$ is referred to as the *distance modulus*. The more negative the magnitude, the brighter the object. The Sun is the brightest astronomical source, owing to its proximity, with m(Sun) $= -26.9$; however, its absolute magnitude is M(Sun) = +4.85, a star of rather ordinary brightness. A star a thousand times more luminous than the Sun has $M = 4.85 - 7.5 = -2.65$. Some of the brightest stars in the sky, such as Betelgeuse, Deneb and Rigel, are over 10 000 times brighter than the Sun, with $M < -5$.

10.5 Colour, extinction and reddening

Colour and temperature are qualitatively related in the visible range as blue–white–red hot, with decreasing temperature. In stellar astronomy, the HR diagram represents the most useful *phenomenological* relationship used to analyze the *colour-luminosity* of stellar groups. A black body at an effective temperature apparently emits energy according to the Planck distribution (Chapter 1). Stars emit radiation at all wavelengths, with a distribution that peaks at a particular wavelength that characterizes the dominant 'colour' of the star. Figure 10.1 shows that the Sun, with T_{eff} = 5770 K has peak emission at around 5500 Å, corresponding to the colour yellow.[6] But while the overall colour of a star reflects its temperature, its spectral energy distribution spans other wavelength bands as well. The measured value of apparent or absolute magnitudes is with reference to a particular photometric band of energy or colour in which the star is observed. For example, M_V implies an absolute magnitude in the visual band. Therefore, it is useful to observe and measure the energy emitted in several bands, and quantified by observed or apparent magnitudes in each band. Most common are the UBV magnitudes: ultraviolet (m_U or U), visible (m_V or V) and blue (m_B or B). The observations are carried out using filters in each band, with central wavelengths as: $\lambda_U \approx 3650$ Å, $\lambda_B \approx 4400$ Å, $\lambda_V \approx 5500$ Å. In addition to the visible and near-UV, other bands often employed towards the red end of the spectrum include R ($\sim 0.7\,\mu$m) and I ($\sim 1\,\mu$m). Three other near IR bands are particularly useful in obervations: J ($\lambda_J \approx 1.2\,\mu$m), H ($\lambda_H \approx 1.6\,\mu$m) and K ($\lambda_K \approx 2.2\,\mu$m); their utility stems from the fact that they correspond to spectral 'windows' in the water vapour absorption in the Earth's atmosphere (shown in Fig. 1.5).

However, the spectral energy observed in each photometric band may be affected by the presence of intervening interstellar material, mainly dust grains. The observed magnitude is thus a lower bound on the actual value which, assuming no interstellar matter, may be represented by Eq. 10.6. The effect of interstellar matter on magnitude and colour is referred to as *extinction*. The absorption A (in magnitudes) enters the equation as a positive quantity, since it implies a reduction in brightness and an increase in m, i.e.,

$$m - M = 5 \log d - 5 + A, \tag{10.7}$$

[6] It is not a conincidence that human beings evolved so as to have the human eye most sensitive to yellow colour, right in the middle of the visible band, flanked by the red and the blue extremities.

where A refers to the same wavelength band as m and M. Roughly, $A_\lambda \sim 1/\lambda$, because shorter wavelength, or higher energy, blue or UV light suffers more extinction than red light, owing to preferential scattering by dust; wavelengths longer than the size of dust particles tend not to scatter or undergo significant extinction.

Having associated colours with wavelength bands in (logarithmic) magnitudes, it follows that, quantitatively, the 'colour' of an object is actually a magnitude difference, i.e., a slope measured between two wavelength bands. The overall extinction of radiation from a star may also be attenuated differentially according to colour. We therefore define a quantity called the *colour excess E*. For example, the difference in the B and V bands of the light from a star, relative to what it would be without colour attenuation by interstellar matter, is

$$E(B-V) = (B-V) - (B-V)_0. \quad (10.8)$$

For the same reason as higher extinction in the UV, this colour excess results in shifting the observed colour towards the red, and is called *reddening*. Dust or grains of matter are the predominant source of reddening. But spectroscopically, whereas absorption line strengths in stellar spectra are unaffected, the observed intensities of emission lines from sources such as nebulae do depend on extinction in their wavelength region. Measured nebular spectral line intensities therefore need to be *de-reddened* to obtain the true intensities from the source. A recent discussion of uncertainties in extinction curves and de-reddening of optical spectra is given in [262]. For sources within the Galaxy[7] we may express both the visual extinction and reddening or colour excess together, by the approximate relation

$$A_V = 3.1\ E(B-V), \quad (10.9)$$

where A_V is the extinction in the visual band.

10.6 Stellar structure and evolution

We see a star only through the light that escapes. This implies that only the uppermost part of the star, the atmosphere, is visible. The energy generated in the core of the star by nuclear processes takes a long time to make its way to the surface, as it is repeatedly reprocessed by stellar material in the main body of the star. Photons scatter coherently and incoherently a large number of times in the interior, in a random-walk behaviour, before escaping to the outer (visible) layers of the atmosphere. In the Sun, for example, it takes of the order of a million years (!) for the energy of photons produced in the core to cover the distance out to the surface. The radiation–matter interactions that underlie this process determine the *opacity* of the matter inside the star. Based on elementary considerations, the stellar interior is divided into three regions: (I) nuclear core, (II) radiation zone and (III) convection zone, shown in Fig. 10.4. Together, regions (II) and (III) comprise what is referred to as the *stellar envelope*, discussed in the next chapter. In addition, the outermost layers of the star constitute the stellar atmosphere through which radiation escapes, and which are therefore most relevant to spectroscopy. As nuclear energy produced in the core diffuses outward, there is a huge variation in temperatures and densities within the star. Figure 10.5 displays the temperature and density profiles in the Sun as a function of the radius. For example, central temperatures are in excess of 10 million kelvin in the Sun, whereas the atmospheric temperatures are a few thousand kelvin. Solar core densities range to over 100 g cm^{-3}, but densities outside the core and throughout the envelope and atmosphere are orders of magnitude less, down to 10^{-9} g cm^{-13} (see Table 11.1). The mean density of the Sun still works out to about 1.4 g cm^{-13}. Exactly how the nuclear energy from the core is processed through the vast middle envelope, the radiative and convective zones (Fig. 10.4), and released through the atmosphere, depends mainly on the opacity and composition (abundances) of the stellar material (Chapter 11). Qualitatively, the three regions of the star may be described as follows.

The extremely high temperatures and densities in stellar cores enable substantial energy to be produced via nuclear reactions. Fusion in stellar cores is extremely sensitive to the *central temperature* T_c of the star, that has no direct relation to the effective 'surface' temperature, which corresponds to the effective black-body temperature of the surface. Since the temperatures T_c required

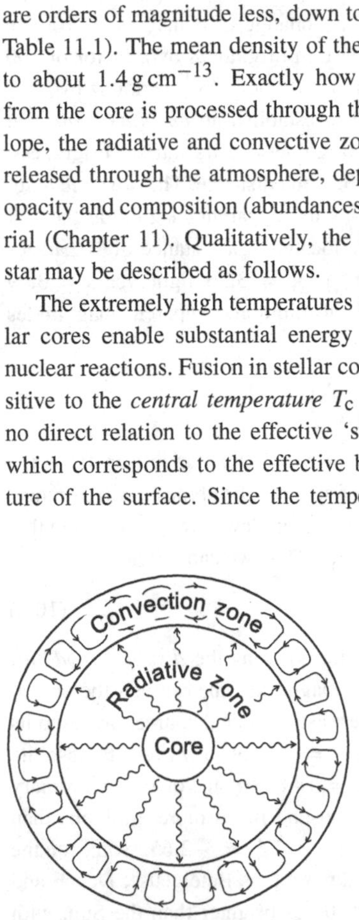

FIGURE 10.4 The main regions of the stellar interior.

[7] It is customary to refer to our own galaxy, the Milky Way, as the Galaxy.

for fusion are greater than 10^7 K, only the central core is sufficiently hot for the plasma to undergo nuclear burning of even the lightest element hydrogen into helium through *pp reactions* or *CNO reactions*. Atoms of all elements in the core are highly or fully ionized nuclei. Most belong to different isotopes, such as ^{1}H (ordinary hydrogen), ^{2}H (deuterium or heavy hydrogen), ^{3}He and ^{4}He. Heavier nuclei than the proton would, of course, require higher temperatures for fusion, owing to greater Coulomb repulsion. Main sequence stars produce energy via the proton–proton (*pp*) reactions at core temperatures $T_c <$ 16 million K, and via the CNO cycle at higher temperatures. Table 10.2 gives the primary nuclear rections, as well as related cyclic processes. The CNO cycle converts H to He without change in the ^{12}C abundance, since carbon acts as a catalyst and remains after each cycle. But it requires higher temperatures to overcome the greater Coulomb repulsion of nuclei than the *pp* reaction. Since T_c(Sun) is about 15 million K, the *pp* chain accounts for 85% of the solar energy and the carbon cycle the rest. Main sequence stars more massive than the Sun have higher central temperatures and produce energy mainly via the CNO cycle.

Nuclear reactions given in Table 10.2 are read as follows: $[A+a \to b+B]$, with reactants on the left side and products on the right, formatted as $A(a, b)B$. In addition, the decay of an unstable nuclear isotope with very short lifetime is designated as: $A \to B + a + b$ (here ν stands for a neutrino, γ for a photon, β^+ for a positron and β^- for an electron). Note that the two protons produced in Step 3 of the *pp* reaction continue the cycle with deuterium production as in Step 1. Likewise, the ^{12}C isotope in Step 6 of the CNO cycle feeds back to Step 1. The complementary *triple-α* process – the fusion of three helium nuclei or α-particles into carbon – while not significant during the main sequence phase, is the dominant source of energy in the 'helium burning' phase of a red giant, whose core is exhausted of hydrogen. Thermonculear reactions with fusion of carbon, oxygen, etc., continue in massive stars

TABLE 10.2 Basic stellar nuclear reactions.

Step	Reaction	Energy (MeV)
	pp reaction	
1	$2[^1H(p, \beta^+\nu)^2H]$	1.2
2	$2[^2H(p, \gamma)^3He]$	5.5
3	$^3He(^3He, 2p)^4He$	12.9
	CNO cycle	
1	$^{12}C(p, \gamma)^{13}N$	1.9
2	$^{13}N \to ^{13}C + \beta^+ + \nu$	1.5
3	$^{13}C(p, \gamma)^{14}N$	7.5
4	$^{14}N(p, \gamma)^{15}O$	7.3
5	$^{15}O \to ^{14}N + \beta^+\nu$	1.8
6	$^{15}N(p, \alpha)^{12}C$	5.0
	Triple-α process	
1	$2(^4He) + \gamma \to ^8Be$	
2	$^4He +^8 Be \to ^{12}C$	7.7 MeV

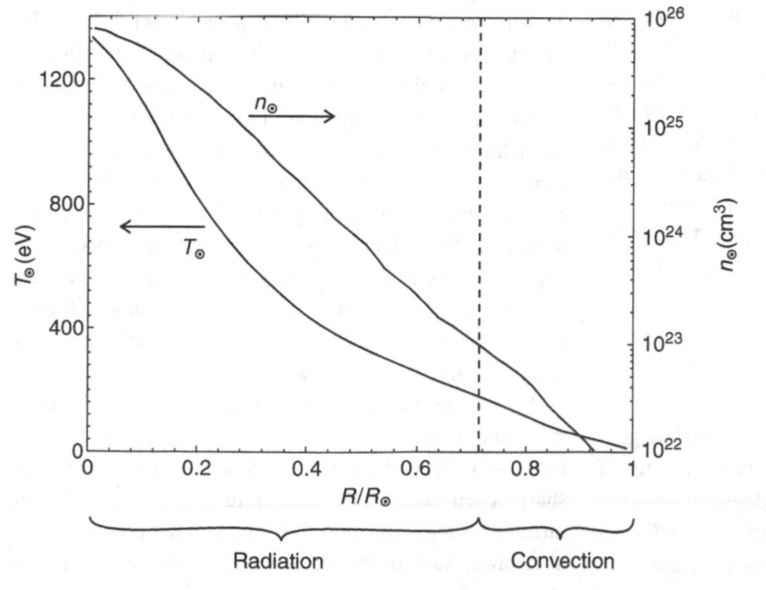

FIGURE 10.5 Temperature and density profiles in the Sun. The solar temperature $T_\odot$ and density $n_\odot$ decrease outward with radius, as shown. The temperature in the solar atmospheric layers is less than 1 eV. But, though not shown in the figure, the kinetic temperature rises rapidly above the atmosphere, through the transition region, into the solar corona to over 100 eV at densitites of about 10^9 cm^{-3}.

at the very high temperatures in their cores up to the production of iron, which does not fuse further to provide nuclear energy by exothermic reactions.

10.6.1 Nucleosynthesis and evolutionary stages

All natural elements except hydrogen are made in stars during their myriad phases of evolution. Thus far we have described only the fusion processes that underlie most of the stellar energy generation mechanism. But nuclear reactions are also responsible for nucleosynthesis of other natural elements in the periodic table (Appendix A). Nucleosynthesis and stellar evolution are inextricably linked, and affect stellar structure in gradual (adiabatic) to explosive scenarios. Nucleosynthesis first begins with the *primordial elements*, H, D (^{2}H) and He, that were made during the big bang. These would have been the constituent elements of the very first stars or objects in the history of the Universe. Later generations of stars, such as the Sun, contain all of the natural elements a priori (although not all have been detected), having been produced in previous stellar cycles. The initial mass and the *chemical composition* – a given mixture of elements – are the main determinants of stellar evolution. We sketch this in Fig. 10.6, which depicts the evolutionary sequence of events known as *evolutionary tracks* on the HR diagram through different phases of the life of stars.

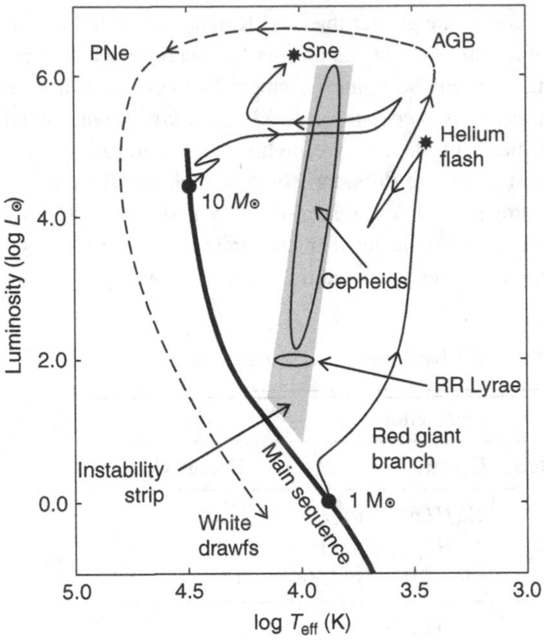

FIGURE 10.6 Stellar evolutionary tracks across the HR diagram. The shaded area contains pulsating stars, the Cepheids and the RR Lyrae stars. The horizontal branch stars, following the helium flash, are in the He-burning phase with the triple-α process; the RR Lyrae are a subset thereof. The 'SNe' refer only to massive stellar core-collapse supernovae, and not those with white dwarf progenitors (discussed further in Chapter 14). To elucidate a number of stellar phases, the diagram above is not drawn to any scale.

10.6.2 Red giants

The HR diagram reflects the fact that stars spend most of their lifetimes on the main sequence, characterized by H $\rightarrow$ He burning in the core. As the H supply is exhausted, stars evolve away from (*not* along) the main sequence. If $M_* < 8\,M_\odot$ the evolution proceeds along evolutionary tracks similar to that shown in Fig. 10.6 for a solar mass ($1\,M_\odot$) star. As the He in the core builds up, and the H burning shell moves outward, the star expands and becomes more luminous but cooler (redder). This is the *red giant* phase, which continues until the star ignites He $\rightarrow$ C fusion.

10.6.3 White dwarfs

The He core is extremely dense and consists of 'degenerate' matter (all available quantum mechanical states of fermions are filled). Degenerate matter does not obey the ideal gas law $PV = nkT$, i.e., a rise in temperature T does not result in expansion of the core, nor does a decrease in T lead to further contraction of the core. At some point along this part of the evolutionary track, temperatures in the core reach $T \approx 10^8$ K, when the He core ignites into a more rapid and energetic He $\rightarrow$ C fusion reaction via the triple-α process, which is responsible for 85% of the energy production at this stage. This point on the HR diagram is referred to as the *helium flash*, which terminates the ascent along the red giant branch. Initially, although the helium flash raises T_c, it does not manifest itself in increased surface luminosity of the star, since the energy produced by helium burning is confined to the otherwise inert core. The electron pressure of the non-relativistic degenerate gas is $P \approx n_e^{5/3}$, independent of T. However, the electrons may attain relativistic speeds, in which case the gas law becomes $P \approx n_e^{4/3}$; such a situation prevails at the extreme densities in white dwarfs.

During the helium flash, the temperature in the core rises rapidly, further enhancing the energy generation rate of He $\rightarrow$ C through the triple-α process, which has a very sharp dependence on the temperature: $E \approx T^{18}$. This, in turn, raises the temperature further and the core gets into a 'runaway fusion' mode. Nearly $\frac{3}{4}$ of the helium in the degenerate core can be ignited within minutes during the helium flash. However, paradoxically, the luminosity of the star *decreases* somewhat in the immediate aftermath of the helium flash (Fig. 10.6). This is because the outer layers are still cooling and the ignited core does not yet provide additional energy to maintain the luminosity. So the star rapidly moves down its evolutionary track in the HR diagram. Subsequently, steady He burning in the core raises the internal temperature, though not the luminosity, and stars bunch up horizontally on a leftward track, called *the horizontal branch* (not shown in Fig. 10.6, but near the arrow below the 'helium flash' mark). Eventually, the rise in temperature lifts the degeneracy of the core, as electrons gain energy and the highest-energy electrons begin to behave more like an ideal gas, leading to expansion and a rise in the star's luminosity. Another way of saying it is that the electrons evaporate from the Fermi sea (Fig. 1.7), towards forming an ideal gas.

10.6.4 Asymptotic giant branch and planetary nebulae

Following the helium flash, the star has a double-shell structure, with a helium-burning core and a hydrogen-burning surrounding shell. As the helium in the inner core becomes depleted, and fusion moves to outer regions, the star rises in luminosity and again ascends the HR diagram along the so-called *asymptotic giant branch* (AGB). The AGB is a short-lived phase, since the star ejects its outer layers of ionized material, which form a surrounding quasi-spherical shell of ionized gas called the *planetary nebula* (PNe).[8] Eventually, the inert carbon core (with some oxygen) cools and goes on to form a *white dwarf*. Gravity in low-mass stars ($M < 8\,M_\odot$) is not able to compress the carbon core further to ignite fusion to heavier elements. Most white dwarfs are therefore made up mostly of carbon, with significant amounts of oxygen.

10.6.5 Massive stars

Higher-mass stars, with $M > 8\,M_\odot$, can, however, continue fusion to produce elements heavier than carbon. Such stars may traverse the HR diagram along horizontal tracks, fusing successively heavier elements in their cores. The evolutionary track for a $10\,M_\odot$ star is shown (albeit approximately) in the upper part of Fig. 10.6. Nuclear reaction rates with α-particles, at temperatures in excess of 100 million K, are much higher than the *pp rates* at lower temperatures. Therefore, the so-called *α-elements*, with even numbered nuclei starting with oxygen, are synthesized with a higher abundance than elements with odd atomic number. In the heaviest stars (supergiants) the fusion process continues up to iron. Those massive stars thus accumulate a number of layers of α-elements – C, O, Ne, Mg, Si, S, Ar, Ca and Ti (not all of which are produced evenly). However, the reaction rates are low and the energy released during the production of α-elements is relatively small. Consequently, the overall luminosity of a high-mass star does not rise appreciably, even as elements heavier than carbon are synthesized in their cores. It criss-crosses the HR diagram horizontally, expanding and cooling in a relatively unstable state as different types of nuclear reactions take place in the core and heavy elements are produced. Such stars are often observed in a region of HR diagram called the *instability strip* (the shaded region in Fig. 10.6).

[8] No relation to planets; the name historically originated with low-resolution images that showed blobs of matter surrounding the hot central star, which were misinterpreted as planet-like structures. The PNe are discussed in Chapter 11.

10.6.6 Pulsating stars

Massive stars within the instability strip have an additional and extremely useful property. They 'pulsate' with remarkable regularity. Their luminosity varies by factors of up to two or three periodically, usually within a matter of days (see Fig. 14.12). Such stars are known as *Cepheid variables*. There is an empirical relation, known as the *period–luminosity* relation between the pulsation period and the absolute luminosity of Cepheids (discussed in Chapter 14, in connection with the cosmic distance scale). A measurement of their observed periods thereby gives their intrinsic or *absolute luminosities*, as opposed to their apparent magnitudes, which depend upon their distance via the distance–modulus relation, Eq. 10.6. As they are very bright stars, the Cepheids are prominently observable to large distances and, as reliable distance indicators, they serve to establish the *distance scale* in astronomy, acting as 'standard candles' of known luminosity. The Cepheids are massive and high-metallicity stars; their pulsation periods depend on their elemental composition or the heavy metal-content that is related to the opacity (discussed in Chapter 11). Metal-poor low-mass stars may also pulsate, and are called *RR Lyrae* stars. These are confined to a small region on the horizontal branch in the HR diagram (Fig. 10.6). The RR Lyrae stars generally have periods of about a day, and nearly the same absolute magnitude. While not as bright as the Cepheids, the period–luminosity relation of the RR Lyrae stars is particularly useful in determining distances to older Pop II stars, such as in globular clusters.

10.6.7 Supernovae

A high-mass star may repeatedly reach up to the AGB, in the red supergiant phase, until available nuclear fuel is exhausted. Since fusion reactions to nuclei heavier than iron are mostly endothermic, evolutionary stellar nucleosynthesis beyond iron no longer yields energy to sustain stellar luminosity or structure. With an inert iron core, the star cannot generate energy and resulant pressure to support the 'weight' of the outer massive layers. Sufficiently massive stars may begin their ascent up the AGB as red supergiants, but then at some point the iron core undergoes gravitational collapse, ending up as what is known as a Type II supernova (SN). Although Fig. 10.6 shows massive stars ($M > 10\ M_\odot$) ending up as core-collapse Type II SNe, there is a caveat. Massive stars in late stages of evolution may show extreme volatility and become extremely luminous, apparently in an attempt to ward off impending collapse. A famous example is that of one of the most massive stars known: *Eta Carinae* or ηCar. It is estimated to be $\sim 100\ M_\odot$, with massive outflows forming the dumb-bell-shaped Homunculus nebula, shown on the cover, and discussed later (also in [263] and references therein).

Here, it is useful to mention an important and fairly precise astrophysical concept that applies to white dwarfs as well as to high-mass stars. It was shown by S. Chandrasekhar that there is a natural gravitational limit associated with a mass of $1.4\ M_\odot$, known as the *Chandrasekhar limit*. This mass corresponds to internal pressure due to quantum mechanical degeneracy of electrons, following from the Pauli exclusion principle. Owing to high densities in the cores, the degeneracy pressure builds up to a maximum value, which can support *at most* the weight (downward pressure) of $1.4\ M_\odot$. Low-mass stars, $M < 8\ M_\odot$, eventually end up as while dwarfs with core masses below the Chandrasekhar limit. When a degenerate mass exceeds the Chandrasekhar limit, electron degeneracy pressure is not sufficient to counteract gravitational pressure, and electrons fall back onto nuclei forming neutrons. The iron core thus collapses into what becomes a *neutron star*. Immediately following core collapse there is (as must be) a 'core bounce' as the infalling matter from the outer layers hits the extremely dense neutron core, and bounces back with tremendous force to eject nearly the entire stellar envelope – observed as a Type II supernova (see the stellar evolutionary track of a $10\ M_\odot$ star in Fig. 10.6). But since neutrons are also fermions, they are also subject to degeneracy pressure like electrons, and have a particular degeneracy limit. If the mass of the neutron core exceeds $3\ M_\odot$, then the core further collapses into a *black hole* since neutron degeneracy pressure cannot support $M > 3\ M_\odot$. Gravitational core collapse, and the explosive end of massive supergiants as progenitors, gives rise to the Type II SNe, leaving behind either a neutron star or a black hole. Spinning neutron stars are known as *pulsars*, and emit radiation from radio waves to gamma rays. For example, the Crab pulsar at the centre of the Crab nebula has a spin rate of about 30 revolutions per second (the optical spectrum is shown in Fig. 8.3).

Novae and Type Ia supernovae

Whereas massive blue and red supergiants are likely to end up as core-collapse Type II supernovae, less massive stars that form white dwarfs may also become supernovae – if they are binary stars or undergo a cataclysmic merger with another white dwarf. There is another type of supernovae, the *Type Ia*, also related to the Chandrasekhar limit. Although the mass of white dwarfs $M \sim M_\odot$, their compact formation ensures high densities and surface gravity $g \sim 10^6 g_\odot$. If a white dwarf is in a close binary formation with another star then stellar matter from the other star can be gravitationally drawn to and accrete onto the surface of the white dwarf. Such a situation often occurs when the other star is a giant with an extended envelope.

One evolutionary scenario is that the increasing pressure on the surface of the already dense white dwarf leads to the onset of thermonuclear fusion. Since this occurs on the surface, the stellar system dramatically increases in luminosity and becomes a *nova*. Novae are often recurring phenomena since the surface fusion is automatically shut off when the accreting gas fuses and is no longer available as nuclear fuel. Also, as the degeneracy of the outer layers of the white dwarf is lifted, following fusion on the surface, they expand and cool.

The other scenario is more extreme. Owing to accretion, if the mass of the white dwarf increases beyond the Chandrasekhar limit of $\sim 1.4\ M_\odot$, then the ensuing gravitational collapse ignites thermonuclear fusion. The entire star is rapidly engulfed by fast-paced fusion reactions and is blown up in a gigantic explosion, referred to as a Type Ia supernova. Since the masses of all white dwarfs that reach the Chandrasekhar limit are similar $\approx 1\ M_\odot$, the energy released in Type Ia SNe is similar, i.e., their *absolute luminosity* is roughly the same. This is of great importance since Type Ia SNe events are so powerful and luminous that they can be observed out to cosmological distances at high redshift. With known absolute luminosities, Type Ia SNe also act as standard candles at much greater distances than the Cepheid stars, and are the preferred means of studying deviations from the Hubble law and related quantities, such as the deceleration parameter and matter–energy density, that underlie the cosmological model. The situation becomes rather more complicated if the Type Ia supernovae occur as a result of merger of white dwarfs. In that case, there is a range of progenitor masses close to the Chandrasekhar limit, and the absolute luminosity is more uncertain. Chapter 14 provides a more extended discussion of SNe spectra. There are also supernovae classified as Type Ib and Ic, which are similar to Type II and are also discussed in Chapter 14.

10.7 High-Z elements

If evolutionary stellar nucleosynthesis terminates at iron, then how are heavier *high-Z* elements of the periodic table produced? They must be produced during events whereby energy may be available for nuclear fusion beyond iron to occur, which, as noted above, are endothermic and require external energy input. We note two such scenarios. During the AGB phase, with a twin-shell H and He fusion in progress, the structure of the star is in considerable turmoil. Deep convective motions in AGB stars can *dredge up* the synthesized elements from the interiors. Nuclear reactions involving heavy nuclei may now occur via 'slow neutron capture', called the *s-process*, producing elements along a chain of elements in the periodic table terminating with (and including) bismuth, $^{209}Bi_{83}$. Successive capture of neutrons by nuclei is slow and takes about a year each; unstable nuclei may decay during this time. The s-process does not proceed beyond Bi, since the nuclei undergo radioactive decay back to Bi as fast as they form. However, the AGB phase is sufficiently long to synthesize high-Z elements, such as Cu, Pb, Ag and Au. Another nuclear process, called the 'rapid nuclear capture', or the *r-process*, occurs in core-collapse Type II SNe, producing heavier elements beyond iron along a chain terminating with the end of the periodic table, up to thorium and uranium, $^{232}Th_{90}$ and $^{238}U_{92}$ [264]. The r-process takes place within a few minutes of the onset of the supernova explosion, when a copious supply of fast neutrons is available. Nucleosynthesis of high-Z elements via the s-process or the r-process occurs under different physical conditions and sources, such as evolutionary epochs of AGB stars or supernova activity early in the history of galaxy formation [265, 266, 267]. The abundances of the s-process and r-process elements along distinct branches of the periodic table are ascertained by studying photospheric lines due to the corresponding elements; this requires rather elaborate three-dimensional non-LTE models with a large number of atomic parameters [268, 269].

Stellar evolution results not only in the formation of elements other than H and He, but also in their dispersal into the interstellar medium following either the AGB phase or supernovae explosions. The heavy elements thereby 'seed' cold gas in molecular clouds which, when they undergo stellar formation, give birth to newer generations of stars. It is thus that stars like the Sun contain trace elements of all stable naturally produced elements. While nuclear physics and plasma physics determines the origin of elements, the measurement of abundances in a star depends on atomic physics, radiative transfer and spectroscopy. Assuming evolutionary nucleosynthesis in stars as described above, it is a non-trivial effort to ascertain quantitatively the amount of each element formed in a given object. Observational spectroscopy provides the vital clues to the strengths of spectral lines from which relative abundances may be deduced by modelling based on the physical conditions in the source. But before stellar energy from the nuclear core escapes the surface, it traverses the rest of the star through widely varying physical conditions, eventually manifest in the characteristics of stellar atmospheres.

10.8 Atmospheres

Above the convection zone in lower-mass stars lies a relatively thin layer that constitutes the stellar atmosphere. It is physically distinct in that convective pressure is no longer sufficient to generate bulk motions. Thus, once photons make their way through the convection zone, their mean free path is much longer as they escape from the atmosphere. Stellar spectroscopy is confined to the outer layers comprising the stellar atmosphere, and the material surrounding the main body of the star, primarily the corona. While radiation escapes through the visible layer of stars – the *photosphere* – it is not quite optically thin. But it is also far from the assumption of LTE that is valid in most of the interior of the star. The Boltzmann–Saha equations, which are the operational forms of LTE, are no longer sufficient to describe the population distribution of an element among atomic levels and the different ionization fractions (Chapter 11). However, the atmospheres are sufficiently dense that radiation transport through them, and spectral analysis thereof, requires a thorough understanding of radiative transfer and related atomic physics (Chapter 9).

Owing to non-LTE effects, quantitative modelling of stellar atmospheres requires elaborate radiative transfer codes. Sophisticated computational mechanisms to implement departures from LTE – essentially microscopic coupling of radiation field with atomic excitation or ionization – result in intricate numerical problems related to convergence of the relevant operators that yield the source function and the radiation field.[9] Suffice it to say, that radiative transfer is the link that connects atomic spectroscopy to the astrophysics of a vast number of astrophysical situations, from stellar atmospheres to black hole environments, where thermodynamic equilibrium defined at a local temperature no longer holds. For example, the analysis of stellar spectra including non-LTE radiative transfer is crucial to the accurate determination of the abundances of elements.

Stellar spectroscopy not only underpins the analysis of stellar atmospheres, but also yields indirect information that can be derived from spectra about activity in the interior. Since the Sun naturally forms the 'standard' for stellar astrophysics, and provides the most detailed and comprehensive test of stellar models, it is with reference to the Sun that we describe spectral formation.

10.9 Solar spectroscopy

The solar atmosphere shows a multi-layered structure. In addition, the atmosphere generates or reflects solar activity that manifests itself not only in the environs surrounding the disc of the Sun, but also out to the farthest reaches of the solar system. In particular, the solar magnetic field underlies phenomena that are inextricably linked to, if not the cause of, much of the Sun's activity, such as solar flares and mass ejection of copious amounts of ionized matter. In this section, we describe the various layers of the solar atmosphere, and the spectroscopic analysis that is employed to study related features.

10.9.1 Photosphere

The visible layer of the atmosphere is the photosphere, which characterizes the colour-temperature of the star. The black-body temperature of the star determines the colour of the photosphere, and hence that of the star. The photosphere is the effective 'surface' of the star. The surface temperature of the Sun is $T_\odot = 5850$ K,[10] corresponding to the colour of a predominantly yellow object. Stars cooler than this are red-hot or redder, and those at higher temperature are blue-hot or bluer. The solar photosphere is a thin layer of only about 600 km as opposed to the solar radius $R_\odot = 700\,000$ km, further supporting the analogy between a surface and the photosphere.

But why is it that the Sun appears to have such a sharp surface boundary? The answer lies partly in the peculiar source of opacity in the solar atmosphere. One might expect ionized H in the interior to recombine to neutral H in the cooler atmosphere, and thus H I to be the main source of opacity. But photoabsorption and photoionization by H I is in the UV range; visible photons are not effectively absorbed by H I except in the Balmer lines. This is because the lowest atomic photoabsorption transition in H is the Lyα (1215 Å) at about 10 eV, and the photoionization energy of the ground state is 13.6 eV (912 Å); both energies (wavelengths) are in the UV range. Clearly, photo-excitation or photoionization of H I is not going to be an important opacity process, since the solar flux peaks in the visible, as shown in Fig. 10.1.

In regions where neutral H is the dominant species, the black-body radiation field, at a few thousand kelvin, ionizes only a small fraction of H. But free electrons are still produced from ionization of metals with lower ionization energy than H I. These free electrons attach to H forming (e + H)$\rightarrow$H$^-$, via long-range attractive potentials (Eq. 6.76). Now, low energy photons with energies of a few electron volts are transparent to H, since they are insufficient to excite or ionize H. But the photons in the visible region of the spectrum, red (7000 Å)–blue (3500 Å), are in the range 1.8–3.5 eV. These visible and lower-energy photons are instead absorbed by H$^-$, starting in the near-infrared region at $\lambda < 1.63\,\mu$m, corresponding to a threshold ionization energy of H$^-$ of only 0.75 eV. Therefore, H$^-$ is the major source of stellar opacity in the visible to near-infrared range. In the Sun, the H$^-$ opacity determines the phenomenon that causes the Sun to appear as a disc, with a fairly well-defined boundary, rather than a more diffuse object.[11] Thus, instead of H I, the dominant source of opacity in the solar atmosphere is the bound–free *photodetachment*

$$h\nu + \mathrm{H}^- \rightarrow \mathrm{e} + \mathrm{H}, \qquad (10.10)$$

[9] The underlying physics and the computational infrastructure is described by D. Mihalas [244] in *Stellar Atmospheres* (latest edition in press)

[10] One often finds values quoted in literature that differ by about 100 K regarding the effective temperature of the Sun (c.f. the black-body temperatures in Fig. 10.1 and Fig. 1.5).

[11] The Sun is a perfect disc to a remarkable 0.002% [270].

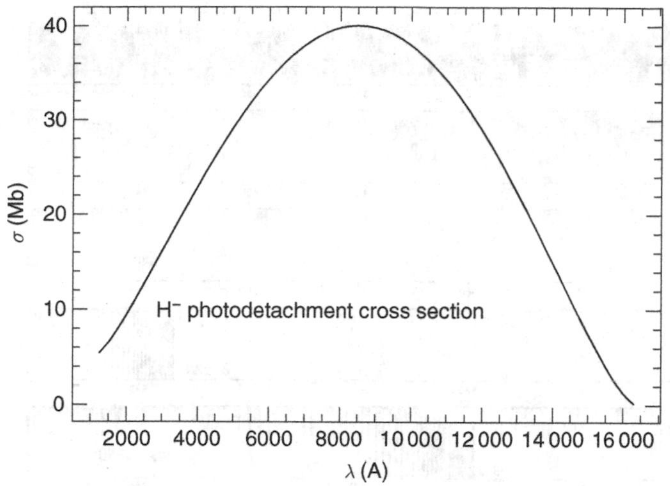

FIGURE 10.7 The bound–free photodetachment cross section of H^- [161]. The H^- opacity peaks in the visible to near-infrared range (Fig. 10.1).

which occurs from photon eneriges as low as 0.75 eV, the *electron affinity* of H^- (Chapter 6). The photodetachment cross section of the He-like ground state ^1S of H^- is given in Fig. 10.7 [161] around 1 eV, just above the threshold photodetachment energy of 0.75 eV. It is worth noting that the Shape and Feshbach resonances, disussed in Chapter 6 (Fig. 6.18), lie at much higher energies around 10 eV, or in the UV range, which does not directly correlate with the peak output of solar flux in the visible (Fig. 10.1).[12] At those lower energies, H^- accounts for the absorption of visible solar radiation effectively. The H^- cross section is appreciable throughout the near-infrared, visible and near-ultraviolet range, and peaks at 8500 Å, where it is 40 Mb. By contrast, the neutral H I photoionization cross section starts out well into the UV, at 912 Å at 6.4 Mb (Chapter 6). In additon to H^- opacity, the free–free process

$$e + h\nu + H \rightarrow e' + H, \qquad (10.11)$$

is also a contributor to the solar atomospheric opacity dominating in the IR beyond 1.6 μm. So the total H^- opacity is related to the sum of the bound–free cross section in Fig. 10.7 and the free–free cross section (e.g., Fig. 4.2 in [244]).

Temperature decreases with height in the solar photosphere, from about 7000 to 4400 K. Since the lower photosphere is hotter, with cooler material in front, as seen from the Earth, the spectrum of the photosphere contains prominent absorption lines. As noted in Chapter 1, the absorption lines from the solar photosphere were among the earliest spectrocopic observations in astrophysics – the Fraunhofer lines corresponding to a number of elements. It is worth emphasizing that while none of the elements, including hydrogen, absorb solar radiation sufficiently to affect its overall shape compared to a black body (Fig. 10.1), the solar spectrum shows signficant attenuation of the underlying photon flux via absorption in many atomic transitions in the range 3000–10 000 Å by essentially all elements up to iron. Determination of photospheric abundances of elements is made from these photospheric absorption lines seen at high resolution, superimposed on the broad black-body shape (note the blue side of the solar flux in Fig. 10.1).

High resolution spectroscopic observations, and numerical simulations based on the underlying atomic physics, yield a much more detailed picture. Figure 10.8 is a synthetic spectrum computed by R. L. Kurucz and collaborators [271], simulating the monochromatic solar flux received at one particular point on the Earth's surface, the Kitt Peak National Observatory in Arizona. Some of the prominent features and atomic transitions are described here (however, note that in Fig. 10.8 the x-scale is in nanometres, and that 10 Å = 1 nm).

The Hα feature at 6563 Å (656.3 nm), due to absorption $n = 2 \rightarrow 3$, lies in the middle of panel 4 in Fig. 10.8. In addition to the first few members of the Balmer series of H, Hα–Hδ in the optical, we see the well-known lines of sodium (Na D lines in Fig. 10.8) at λλ 5890, 5896 due to the fine structure doublet transitions $3s^2S_{1/2} \rightarrow 3p^2P^o_{3/2,1/2}$. Similarly, isoelectronic

[12] Physicists and astronomers have rather different interests regarding H^-. The 10 eV resonance features shown in Fig. 6.18 are important to both experimentalists and theorists in physics to study the *simplest* two-electron system. On the other hand, and quite fortuitously, H^- happens to be an important opacity source in astronomy, but at an energy about an order of magnitude lower, ∼ 1 eV.

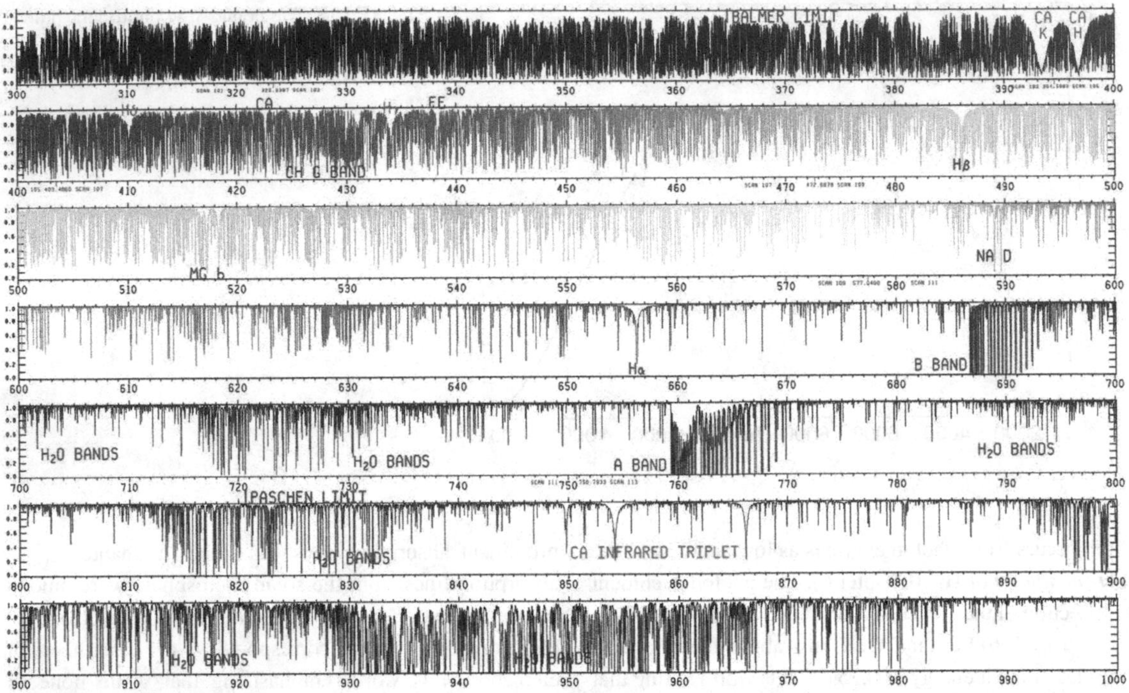

FIGURE 10.8 Simulated solar spectrum at the surface of the Earth transmitted through the atmosphere, computed in correspondence with observations at the Kitt Peak National Observatory [271] (Courtesy: R. Kurucz). Note that the left side of the top panel has maximum line blanketing with $\lambda < 400$ nm (4000 Å), and the right side of the third panel from top, with $600 > \lambda > 550$ nm, has about the least. The latter is around the yellow band (before the orange Na D lines), which determines the characterstic colour of the Sun.

with Na I, Mg II has the same transitions at $\lambda\lambda$ 2795.528, 2802.704 Å. These Mg II lines are called the *h* and *k* lines, in analogy with the Ca II H and K lines, due to transitions $3p^64s\ ^2S_{1/2} \rightarrow 3p^64p\ ^2P^o_{3/2,1/2}$ at $\lambda\lambda$ 3933.663, 3968.469 Å (Fig. 10.8). However, the low-lying energy levels of Ca II are different from those of Mg II. While the ground configuration of Ca II is $3p^64s$ (analogous to the Mg II $2p^63s$), the next higher levels are due to $3p^63d$, which lies below the $3p^64p$. A schematic diagram of the Ca II H and K lines and three near-IR lines is shown in Fig. 10.9. In the near-infrared stellar spectra of spectral type A–M stars lie these so-called 'calcium triplet' lines in Ca II at $\lambda\lambda$ 8498.02, 8542.09 and 8662.04 Å, due to transitions within the $3p^6(3d\ ^2D - 4p\ ^2P^o)$ multiplet: $^2D_{3/2} -^2 P^o_{3/2}, ^2D_{5/2} -^2 P^o_{3/2}, ^2D_{3/2} -^2 P^o_{1/2}$, respectively.[13] Note that there is no allowed transition between the $J = 5/2$ and 1/2 levels, since $\Delta J > 1$. This set of three Ca II lines is sometimes labelled 'CaT' (e.g., [272]). From an atomic physics point of view, the interesting point about the CaT lines is that they involve an excited initial state, $3p^64d\ ^2D$, which may be significantly populated, since it is metastable, owing to same parity as the ground state $3p^64s\ ^2S$. Therefore, the CaT transition array is quite prominent in stellar atmospheres, in addition to the 'resonance doublet', H and K. Furthermore, it follows that the $3p^6(4s - 3d - 4p)$ system of energy levels shown in Fig. 10.9 would be highly coupled in terms of collisional and radiative calculations [273].

The near-ultraviolet part of the solar spectrum in Fig. 10.8 is dominated by iron lines, as is the spectrum of other stars and astronomical objects. The predominant ionization state of iron in many astrophysical objects is singly ionized Fe II, which has a multitude of transitions ranging from the near-infrared one-micron (1 μm) lines (Chapter 13), to far-ultraviolet below the Lyman limit at 912 Å (the Fe II lines are discussed in detail in Chapters 8 and 13). But in the Sun, and cooler stars, Fe I dominates and has hundreds of lines. These many lines in stellar models are referred to as *line blanketing* of the underlying black-body continuum. This explains the significant

13 We again note that the use of the terms 'doublet' and 'triplet' in astronomy refers only to the set of two or three lines respectively, not to spin multiplicity (2s + 1), which is the common spectroscopic usage.

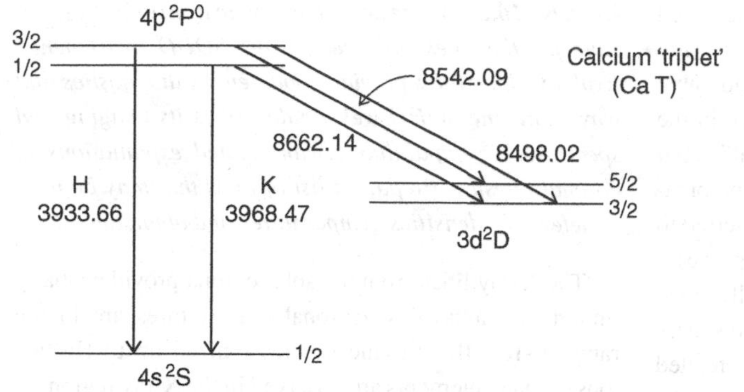

FIGURE 10.9 The Ca II H and K lines in the visible, and the 'calcium triplet' lines in the near-infrared.

deviation from the Planck function on the blue side, shown in Fig. 10.1.

10.9.2 Chromosphere

Above the photosphere, there are other regions related to the stellar surface and interior activity. These are the *chromosphere* and the *corona*, connected by a *transition region*. The chromosphere is actually cooler than the photosphere. But contrary to the situation in the photosphere, the temperature rises with height, from 7000 K at the lower interface with the photosphere, to about 20 000 K at the upper interface with the transition region which, in turn, leads into the corona at very high temperatures of the order of 10^6 K. The chromosphere may be thought of as the 'bubbling up' of the solar material, seen as spikes, called 'spicules', above the 'surface' (photosphere). The chromosphere, and the hot plasma above in the corona, are intimately related to the solar magnetic field. The Sun is a rotating ball of hot electrically charged particles. As such, it acts like a dynamo, generating a strong magnetic field. The chromosphere and the corona are visible during total solar eclipses, when the solar disc is occulted by the moon. Satellite observations of the outer regions can be similarly made by creating an artificial 'eclipse' by blocking out the solar disc. Such observations provide valuable information on the magnetically driven solar activity.

As the temperature rises with height above the photosphere, the chromosphere is the source of emission lines, as opposed to mainly absorption lines from the photosphere. Since both layers have much the same composition, the emission and absorption lines are from the same elements. Chromospheric lines include atomic species such as H I, Ca II, Ca I, Mg II, He I, and also a few doubly ionized ions, C III, Si III and Fe III. Hα is seen in emission, produced by electron–proton recombination into high-lying levels and cascades downward from the $n = 3 \rightarrow 2$ (Hα is also seen in absorption at the limb against the disc). The Ca II H and K lines discussed above are also seen in emission. In addition, several other lines from excited levels with radiative transitions in the ultraviolet are observed, such as the C II 'resonance' multiplet at ~1335 Å due to dipole allowed transition in the array $2s2p^2(^2D_{5/2,3/2,1/2}) \rightarrow 2s^22p\left(^2P^o_{3/1,1/2}\right)$.

10.9.3 Transition region

As mentioned already, the transition region between the chromosphere and the corona is a relatively sharp boundary with large temperature gradients. At the lower end towards the chromosphere the temperatures are of the order of 10^3 K, but at the higher end into the corona they jump to more than 10^6 K.

10.9.4 Solar corona

The most intriguing part of the solar atmosphere is the corona. It consists of a tenuous and optically thin hot plasma surrounding the Sun, rising millions of kilometres into space. Characteristic of hot gas at millions of kelvins, the coronal spectrum is dominated by emission lines of highly ionized atomic species that radiate predominantly in the EUV and X-ray regions. It is important here to again distinguish the sense in which we define a black-body temperature. It refers to the total amount of energy radiated by an object in thermal equilibrium given by the Planck distribution. On the other hand, the kinetic temperature of a gas is determined by particle velocities, generally as a Maxwellian distribution. In the corona the temperature refers to the electron temperature T_e, assuming a Maxwllian distribution, whereas in the photosphere temperature refers to the surface temperature approximated by a Planckian function with $T_{eff} \approx 5800$

(Fig. 10.1). In addition, ionization balance is also achieved by different physical processes. In the interior of the Sun (and most other stars) we have LTE, but considerable departures from LTE in the atmosphere. However, in the solar corona the ionization balance is between collisional ionization, on the one hand, and electron–ion recombination on the other. In earlier discussions we have referred to it by the generic term 'coronal equilibrium' (Chapter 6).

The crucial question, to a significant extent still a mystery, is: what is the exact nature of the energy source and kinematics of the corona? A partial answer, related to magnetic activity, is that gigantic flares rise from the base of the solar atmosphere into the corona, and provide the energy and the material. Additionally, there is thought to be a continuous supply of energy to the corona from *microflares* all over the Sun. The answer is also related to such important phenomena as *coronal mass ejections* (CMEs); streams that carry away subtantial mass and energy in the form of high-energy charged particles and radiation. Owing to the tremendous activity and temperature in the corona, there is a continuous *solar wind* blowing out of the Sun into the *heliosphere*, and way beyond to the farthest planets. Flares and other coronal activity can be readily observed, and affect the atmosphere of the Earth in ways that are not entirely understood.

Solar activity is also governed by an 11-year cycle, wherein it undergoes a maximum and a minimum. The next solar maximum is due in 2012, and it is obvious that solar astronomers would particularly like to observe the Sun during the solar maximum. Among the most extensive recent observations of the Sun are those by the satellite *Solar and Heliospheric Observatory* (SOHO), which has several spectroscopic instruments (e.g., [274]): the Coronal Diagnostic Spectrometer (CDS), the Solar Ultraviolet Measurement of Emitted Radiation (SUMER) and the Ultraviolet Coronograph Spectrograph (UVCS). The high-resolution ultraviolet wavelength coverage of SOHO instruments is about 300–1500 Å. For example, the CDS observes a number of pairs of density-sensitive lines, such as the Mg VIII multiplet with lines around 430 Å and 335 Å; the UVCS obtains data for a variety of lines that can be used to determine O, Mg, Si and Fe abundances, such as O VI 1032, 1037 Å, Mg X 610, 625 Å, Si XII 495, 521 Å and Fe XII 1242 Å. The electron densities derived from these and other measurements are about 10^9 cm^{-3} for the active but non-flaring regions of the corona, and up to 10^{12} cm^{-3} in the flares.[14]

Exercise 10.2 Write down the atomic transitions responsible for the lines observed by the SOHO instruments. Explain which lines provide good density diagnostics and why. Hint: the SOHO website describes its imaging and spectroscopic capabilities, with related explanations of coronal physics. Prepare a list of lines that may be used to determine densities, temperatures and abundances.

The X-ray lines from the solar corona provide equally important diagnostics. Coronal temperatures are in the range 1–10 MK. Atomic species ionized up to He-like ions of many elements are observed in the X-ray region. In Chapter 8, we have discussed spectral formation of O VII and Fe XXV line due to excitation of $K\alpha$ complexes. The density sensitivity of the O VII line ratios lies in the critical range for coronal densities $n_e \sim 10^9$ cm^{-3} (Figs 8.8 and 8.10). Similarly, the $K\alpha$ complexes of He-like Ca Fe, and Ni lie at 3.8, 6.7 and 7.1 keV, respectively. The complex spectral analysis relevant to the flaring and rapidly transient coronal activity, including dielectronic satellite lines, has been described in Chapter 8, using the $K\alpha$ complex of Fe XXV. Another particularly useful example of a strong X-ray feature is that due to L-shell excitations in Ne-like Fe XVII, shown in Fig. 5.5. A more complex spectral analysis relevant to the flaring and rapidly transient coronal activity involves dielectronic satellite lines of He-like ions, described in Chapter 8.

10.10 Cool and hot stars

Given the elaborate stellar classification and complex mechanisms for spectral formation (some described herein), it might seem outlandish to refer simply to groups of 'cool' stars or 'hot' stars. Yet, there are some general characteristics that do warrant such apparently simple terminology, albeit with many caveats. First, cool stars form the largest such group, including all spectral classes later than A in age, and cooler than about 7000 K in surface temperature [276]. The Sun is an ordinary G-type cool star, and in many ways prototypical of this group, with much of the structure and characteristics discussed in the previous section on solar spectroscopy. One prominent example is the magnetic phenomena that link the activity on the relatively cool stellar surface to the hot corona, leading up to X-ray emission.

Hot stars have much higher temperatures than cool stars, and range up to 40 000 K for main sequence O stars. Massive hot stars are very luminous, up to 10^5–10^6 $L_\odot$.

[14] A collisional–radiative model, as described in Chapter 8, may be used for emission-line diagnostics in collisional equilibrium or the coronal approximation (Chapter 6). Several codes and databases are devoted to such efforts, e.g., *CHIANTI* (http://chianti.nrl.navy.mil/chianti.html. *APEC/APED* (http://hea-www.harvard.edu/APEC) and *MEKAL* [275].

Relating wavelength and colour to temperature simply according to Wien's law (Eq. 1.7), hot stars are blue and cool stars are red. One property that distinguishes cool stars from the hot O and B stars is *stellar wind and mass loss*. Hot stars generate high-speed and dense stellar winds that carry sufficient matter to cause significant mass loss, up to 10^{-8}–10^{-4} $M_\odot$/year. In contrast, stellar winds from cool stars are much weaker, and carry relatively little material; the bulk of stellar matter is magnetically confined to the surface, with the exception of occasional coronal mass ejections and flaring activity that maintains the corona. A particular class of ultraluminous hot stars, the *Wolf–Rayet stars* (WR), may have winds up to about 2000 km s^{-1}, and temperatures up to 50 000 K.

The radiation-driven winds of hot stars largely obscure the underlying atmospheres. Their spectral analysis requires consideration of dynamical phenomena, particularly as related to line formation in rapidly expanding media – quite different from the relatively quiescent atmospheres of cool stars. Non-LTE line blanketing models in spherically expanding outflows (e.g., [277]) may be used to analyze prominent CNO lines. A significant effect is the two to five-fold increase in the strengths of optical lines, such as the C III 5696 and C IV 5805 in WR stars, owing to extensive line blanketing by Fe lines.

The most common signature of outflows is the characteristic line shape known as the *P-Cygni profile*, first analyzed from the hot B star P Cygni. As one might expect in a symmetric outflow from a source, the Doppler effect in the fast winds, moving both away from and towards the observer, leads to large broadening widths. The resulting line profile is asymmetrical, and exhibits both absorption and emission features, as shown in Fig. 10.10. The trough at wavelengths less than the line centre $\lambda < \lambda_0$, is caused by absorption by atoms in the material moving towards the observer. Photons emitted by those atoms are blueshifted, and are likely to find other atoms ahead moving in the same direction; hence, the blueshifted absorption. On the other hand, redshifted photons are emitted from atoms moving away from the observer and are not absorbed by the intervening material, which is largely moving towards the observer. Therefore, the red wing of the line is unaffected, and dominates the emission line profile.

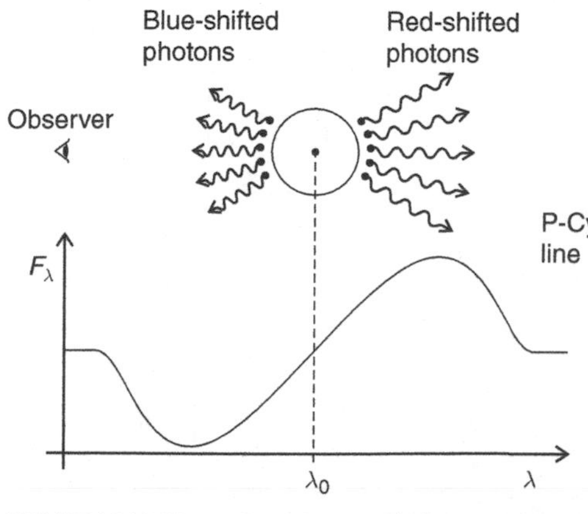

FIGURE 10.10 The P-Cygni line profile from a spherically symmetrical outflow.

The P-Cygni profile is a general feature of outflows, seen not only from stars (and not only in the optical) but from all sources with winds carrying significant material. An interesting example is the outflow from a Galactic X-ray binary stellar system called Circinus X-1, comprising a neutron star and a main-sequence hydrogen burning star.[15] The outflow is driven gravitationally by matter accreting from the main-sequence star on and around the surface of the compact neutron star. As the stars orbit each other, X-ray line emission is highly variable and exhibits P-Cygni profiles. A time-dependent animation of the H-like Si XIV $K\alpha$ line at 6.18 Å observed by the Chadra X-ray Observatory may be viewed at *http://www.astro.psu.edu/users/niel/cirx1/cirx1.html.*

10.11 Luminous blue variables

Because luminosity increases rapidly in a highly non-linear way with mass, approximately $L \propto M^{3.5}$, very massive stars are naturally extremely luminous. Such stars occupy the upper left-hand corner of the HR diagram (Fig. 10.2). The sub-class of stars with $M \gg 10 M_\odot$ are referred to as *luminous blue variables* or LBVs. We have already noted that such massive stars are expected to undergo core collapse and end up as Type II SNe. However, LBVs are thought to avoid this fate (albeit temporarily, according to astronomical timescales) by ejecting large amounts of their masses. The mass loss may be continuous in the form of high-velocity stellar winds (as in Wolf–Rayet stars), about 10^{-4}–$10^{-1} M_\odot$, or periodic outbursts of massive ejections of up to 1 $M_\odot$ or more for extreme LBVs.

Currently the most prominent LBV is *Eta Carinae* (ηCar), shown on the cover page [278]. It is one of most massive stars known, with an estimated mass $>100 M_\odot$, and luminosity $>10^6 L_\odot$. It is thus highly unstable, with gigantic outflows from equatorial regions that constitute

[15] Conventionally, the upper case 'Galactic' refers to our own galaxy, the Milky Way. Otherwise it is lower case 'galactic' for galaxies in general.

the so-called *Homunculus nebula*, with its characteristic dumb-bell shape (as shown in the HST image on the cover). There is considerable evidence based on spectroscopic and morphological analysis [184] that it is a close interacting binary star in a symbiotic formation – a common envelope with two stars in different stages of evolution; a compact object, such as a white dwarf, and a voluminous massive main-sequence O star. The periodic variation in luminosity and kinematics expected in such a situation has been observed in several wavelength ranges, including in the X-ray region, from the Chandra X-ray Observatory. However, spectroscopic measurements are rendered uncertain by significant variations and anomalous line intensities. For example, a number of forbidden and allowed lines from Fe II and Fe III are seen from the infrared to the ultraviolet. But the Fe II UV line ratios are difficult to reconcile with a straightforward collisional–radiative analysis of the Fe II atomic model (Fig. 12.5). Hubble Space Telescope observations of $\lambda\lambda$ 2507, 2509 lines, arising from transitions between the *excited LS* multiplets, require a model based on Lyα pumping and laser-like transitions, coupled with interesting geometry, to explain their anomalous strengths [279]. In any event, ηCar is close to the limit of stability, and it is likely that ηCar would end up as a Type II supernova. Meanwhile Eta Carinae continues to be the most extensively studied LBV, and a remarkable laboratory of atomic physics and astrophysical processes.

11 Opacity and radiative forces

An elaborate radiative transfer treatment (Chapter 9) is necessary for stellar atmospheres through which radiation escapes the star. But that, in a manner of speaking, is only the visible 'skin' of the star, with the remainder of the body opaque to the observer. Radiation transport throughout most of the star is therefore fundamentally different from that through the stellar atmosphere. Since radiation is essentially trapped locally, quite different methods need to be employed to determine the opacity in the interior of the star. However, since there is net outward propagation of radiation from the interior to the surface, it must depend on the variation of temperature and pressure with radius, as in Fig. 10.5.

Perhaps nowhere else is the application of large-scale quantum mechanics to astronomy more valuable than in the computation of astrophysical opacities.[1] Whereas the primary problem to be solved is radiation transport in stellar models, the opacities and atomic parameters needed to calculate them are applicable to a wide variety of problems. One interesting example is that of abundances of elements in stars, including the Sun. Observationally, the composition of the star is inferred from spectral measurements of the atmospheres of stars, i.e. *surface abundances*, because most of the interior of the star is not amenable to direct observation. However, radiative forces acting on certain elements may affect surface abundances that may be considered abnormal in some stars. In previous chapters, we laid the groundwork for the treatment of a specialized, but highly important, topic of opacities in the stellar context. We begin with the definitions of physical quantities and elucidation of basic concepts, and end with a description of the state-of-the-art electronic facilities for on-line computation of stellar opacities, and radiative forces or accelerations.

11.1 Radiative and convective envelope

Most of the stellar interior consists of the envelope region through which energy generated in the nuclear core propagates upwards to the surface. Depending on the stellar type and mass, the envelope is further subdivided into two regions, depending on the relative dominance of the two competing physical processes of *energy transport*: (i) radiative diffusion and (ii) convection.[2] Generally, stellar envelopes can be defined as '*those regions of stellar interiors where atoms exist and are not markedly perturbed by the plasma environment*' [36]. The envelope densities are much less than that of water, $\rho(\text{envelope}) < 1\ \text{g cm}^{-3}$, and decreasing towards the outer-most atmospheric regions. Typical, though approximate, temperatures and densities in stellar envelopes are given in Table 11.1. The free-electron partition function U_e, and the ionization potential I_m of the ionization state close to the maximum abundance of an element, are discussed in Section 11.4.1. The phrase 'markedly perturbed' implies that we may retain the isolated atom(ion) description of quantized levels up to a certain quantum number where the plasma effects cause broadening, dissolution, and ionization of high-lying levels. The internal dynamics and structure of stellar

[1] Atomic physics assumes a central role in opacities of *all* high-energy density (HED) plasma sources, in particular laboratory fusion devices. Of course, the dynamics is quite different in the laboratory. Non-equilibrium, time-dependent and three-dimensional hydrodynamics all play important roles in radiation transport on extremely small spatial–temporal scales, but otherwise at temperatures and densities required to achieve nuclear fusion, as in stellar cores.

[2] For energy transport through the low densities outside stellar cores, the third process of energy transport, conduction, is not viable for most stars, since it requires metallic densities that are only found in stars with degenerate cores, such as white dwarfs, where conduction is in fact the dominant mechanism. However, even at low densities, where the electron mean free path is large, conduction of electrical charges may play an important role, such as energy transfer from magnetic activity in the chromosphere into the corona, and in stellar flares and coronal mass ejections.

TABLE 11.1 Typical stellar temperatures and densities (solar composition).

Log T (K)	ρ (g cm^{-3})	n_e (cm^{-3})	n_e/U_e	I_m/kT
4.5	3.2×10^{-9}	2.1×10^{15}	7.9×10^{-8}	16
5.0	1.0×10^{-7}	6.8×10^{15}	4.0×10^{-7}	15
5.5	3.2×10^{-6}	2.2×10^{18}	3.2×10^{-6}	13
6.0	4.0×10^{-4}	2.2×10^{20}	5.0×10^{-5}	10
6.2	2.0×10^{-2}	1.0×10^{22}	6.3×10^{-4}	8
6.5	2.0×10^{-1}	1.0×10^{23}	7.9×10^{-4}	7
7.0	2.0×10^{1}	1.0×10^{25}	–	–
7.2	2.0×10^{2}	1.0×10^{26}	–	–

envelopes is governed by a set of equations discussed in the next section.

11.2 Equations of stellar structure

A theoretical model of a star depends on several physical quantities inter-related by four equations of stellar structure, supplemented by an equation-of-state of the plasma in the interior. We consider stars to be the idealized geometrical shape of a sphere (neglecting stellar rotation and magnetic fields, not because they are not important but because they require specialized treatment not covered in this text).

The first equation is simply *mass conservation.* At any radius r, measured from the centre at $r = 0$, the mass inside a spherical shell with radius r is

$$M(r) = \int_0^r \rho(r')4\pi r'^2 \mathrm{d}r', \tag{11.1}$$

where $\rho(r')$ is the local mass density in shell r'. The second relation governs *energy generation*, equal to the energy flowing outward from a similar spherical shell,

$$L(r) = \int_0^r \epsilon(r')\rho(r')4\pi r'^2 \mathrm{d}r', \tag{11.2}$$

where ϵ is the rate of energy generation (e.g., W g^{-1}). The basic dynamics of the star is that each layer in the star is balanced by the gravitational pressure inward and the gas pressure outward. The equation of this *hydrostatic equilibrium* is then

$$\frac{\mathrm{d}P(r)}{\mathrm{d}r} = -\frac{\rho(r)GM(r)}{r^2}. \tag{11.3}$$

Exercise 11.1 *Using order-of-magnitude estimates for typical stars, show that $P(r)$ has generally negligible contribution from radiation pressure due to the radiated photon flux. Hint: black-body radiation pressure is $(4\sigma/3c)T^4$.*

The fourth equation is related to the mechanism of radiation transport. Since conduction is not a viable mechanism, it is the *competition* between radiative diffusion and convective motions that governs energy transport at any given point r in the interior of the star. The *equation of energy transport* depends on the local temperature gradient, which for the radiative mode can be written as

$$\frac{\mathrm{d}T(r)}{\mathrm{d}r} = -\frac{L(r)}{K4\pi r^2} \quad \text{(diffusion)}. \tag{11.4}$$

Here we introduce a *diffusion constant* K dependent on material properties, mainly the opacity. Alternatively, the other mode for energy transport is governed by the gradient

$$\frac{\mathrm{d}T(r)}{\mathrm{d}r} = \left(1 - \frac{1}{\gamma}\right)\frac{T(r)}{P(r)}\frac{\mathrm{d}P(r)}{\mathrm{d}r} \quad \text{(convection)}. \tag{11.5}$$

The convective mode depends on the quantity γ in the adiabatic equation-of-state $P\rho^{-\gamma}$ = constant, and is equal to the ratio of specific heats. For a perfect gas $P = k\rho T/\mu$, where μ is the mean molecular weight.

Exercise 11.2 *Considering the Sun to be made of ionized hydrogen only, and using the equations of stellar structure in Section 11.2 and the perfect gas law, show that the order of magnitude of the central temperature is about 10 million kelvin (sufficient to ignite thermonuclear fusion of H to He).*

If the radiative diffusion temperature gradient Eq. 11.4 is smaller than the one for adiabatic convection, Eq. 11.5, i.e., $(\mathrm{d}T/\mathrm{d}r)_{\mathrm{diff}}(r) < (\mathrm{d}T/\mathrm{d}r)_{\mathrm{conv}}(r)$, then energy transport by radiative diffusion is *more* efficient and convective bulk motions of stellar material would not start. If the temperature gradient at a given r is large, say due to large local opacity, then pressure from inside builds up and the bulk material rises; energy transport by convective motions becomes more efficient. Thus deep inside the star radiative diffusion transport is important, but gradually gives way to convection, which becomes more efficient at increasing radius (except in high-mass main-sequence stars where the situation may sometimes be the opposite, owing to fusion of successively heavier elements, He, C, etc., in the core). This balance determines a well-defined boundary known as the *depth of the convection zone* (CZ). In the Sun, the CZ boundary lies at about 0.73 $R_\odot$, i.e., more than a quarter of the Sun is convective rather than radiative in terms of energy transport outward. It follows that, by volume, half the Sun is convective, but by mass only a small fraction. In more massive sars, the radiative zone is much bigger, and extends throughout the star. This is the reason that in classical Cepheid stars, with masses several times $M_\odot$, the opacity in the radiative zone is the crucial determinant of pulsation properties that drive the periodic variation in luminosity but in very massive stars with ongoing CNO cycles, convective interiors dominate.

We have introduced the equation-of-state in the form of the perfect gas law – a relationship among *macroscopic* quantities pressure, temperature and density defined in accordance with the equations of stellar structure. But the quantity K, the radiative diffusion constant, is still undefined. It depends on the local opacity, which, in turn, is due to *microscopic* absorption and scattering of photons by atoms and ions of elements in the star. Therefore, the diffusion constant or the opacities need to be computed taking into account all the microscopic physical processes throughout the star.

11.3 Radiative flux and diffusion

Nuclear fusion energy produced in the core of the star is transported through stellar matter via *radiative diffusion* and *convection*. As we have seen, the structure of the star is divided into stellar interior (core), envelope and atmosphere. The conditions of the matter in these three regions are sufficiently distinct that different approximations need to be employed. From an atomic physics point of view we adopt the following characterizations.

Stellar core:

(a) highly ionized ions (e.g., bare nuclei, H- and He-like),

(b) energy levels of ions perturbed by plasma interactions to an extent that the isolated atom approximation may not be valid,

(c) densities and temperatures: $T > 10^{6.5}$ K and $\rho > 11\,\mathrm{g\,cm^{-3}}$).

Stellar envelope:

(a) atoms and ions exist and may be considered free of bulk plasma effects,

(b) radiation field may be treated by the black-body Planck function (also, of course, in the stellar core), and first-order deviations therefrom, but essentially in LTE,

(c) (T, ρ) as in the range given in Table 11.1.

Stellar atmospheres:

(a) LTE is not valid and non-LTE approximations must be employed,

(b) atomic structure and *radiative transfer* are coupled and need to be treated in detail.

In the previous chapter, we introduced the opacity κ to define the optical depth τ, but without recourse to the underlying physics. The physical nature of the opacity depends on several parameters – temperature, density and composition. We can introduce the atomic density dependence n_{A} explicitly by redefining κ as

$$\rho(\mathrm{g\,cm^{-3}})\kappa(\mathrm{cm^{-2}\,g}) = n_{\mathrm{A}}(\mathrm{cm^{-3}})\sigma_\nu. \quad (11.6)$$

The left-hand side provides the *macroscopic* definition of κ, whereas the right-hand side includes the *microscopic* quantity, the cross section. The units of κ on the left-hand side imply an opacity cross section per unit gram of material. The mass density $\rho = m_{\mathrm{A}} n_{\mathrm{A}}$, where $m_{\mathrm{A}}(g)$ is the atomic mass. The monochromatic opacity is, similarly,

$$\kappa_\nu = \frac{\sigma_\nu}{m_{\mathrm{A}}} \quad (\mathrm{cm^2\,g^{-1}}), \quad (11.7)$$

where σ_ν is the photoabsorption cross section. The optical depth is then expressed as

$$\tau_\nu = \int_r^\infty \rho(r')\kappa_\nu(r')\mathrm{d}r', \quad (11.8)$$

for a photon to escape from a distance r inside the star to the surface (or $r \to \infty$). The main topic of this chapter is to develop an understanding of the astrophysical opacity, and its application to stellar astrophysics in regions where the optical depth $\tau_\nu \gg 1$.

11.3.1 Diffusion approximation

Radiative diffusion is the primary mechanism for radiation transport throughout most of the body of main-sequence stars. The radiation field in the interior is described as a black body and its intensity I_ν given by the Planck function, so long as the photon mean free path is shorter than the distance over which the temperature varies significantly. The deeper one goes into the star the photons are trapped with smaller mean free paths and hence escape probability. As we discussed in Chapter 9, in a perfect black body there is no escape of radiation, i.e., $dI/d\tau = 0$ and $I_\nu = S_\nu$ is the Planck function B_ν. The emissivity and the opacity in the radiative transfer equation yield the simple form given by Kirchhoff's law: $j_\nu/\kappa_\nu = B_\nu(T)$. The radiative transfer problem would be particularly simple if the source function were indeed B_ν. However, this limit is not strictly realized through much of the star, since, after all, there is a net outward flux. But a working assumption can be made: the source function approaches the Planck function, with relatively small deviations therefrom. To obtain the actual source function S_ν at frequency ν, taking account of deviations from the ideal black-body form B_ν, we employ the usual method of expansion at a point x in a Taylor series within some depth τ, i.e.,

$$S_\nu(x_\nu) = \sum_k^\infty \frac{(x_\nu - \tau_\nu)^k}{k!} \left(\frac{d^k B_\nu}{d\tau^k} \right). \tag{11.9}$$

The specific intensity (Chapter 9) is, then,

$$I_\nu(\tau, \mu) = \sum_k^\infty \mu^k \frac{d^k B_\nu}{d\tau_\nu^k}$$
$$= B_\nu(\tau_n u) + \mu \frac{dB_\nu}{d\tau_\nu} + \mu^2 \frac{d^2 B_\nu}{d\tau_\nu^2} + \cdots. \tag{11.10}$$

Assuming the deviations from a black body source function in the interior to be small, we retain only the first derivative. Then the moments of the radiation field are

$$\begin{aligned} I_\nu &\approx B_\nu + \mu \left(\frac{dB_\nu}{d\tau\nu} \right) \\ J_\nu &\approx B_\nu \\ H_\nu &\approx \frac{1}{3} \frac{dB_\nu}{d\tau_\nu} \\ K_\nu &\approx \frac{1}{3} B_\nu(\tau_\nu). \end{aligned} \tag{11.11}$$

It follows that in near-LTE situations, with small deviations from a black-body source function are, at large optical depths, $\lim_{\tau \to \infty} J_\nu \to 3K_\nu$. This is known as the *Eddington approximation*. Assuming the Eddington approximation, $J = 3K$, to be valid in the atmosphere, explains the so-called *limb darkening* effect quite well. It can be seen that the emergent radiation from the centre of a stellar disc appears more intense to the observer than the radiation from the circular edge of the stellar disc. The edge of the solar disc or the limb, corresponding to a viewing angle $\mu = \cos 90° = 0$, is significantly darker than the brightest part, the centre of the disc corresponding to $\mu = \cos 0° = 1$. As the temperature decreases with height in the photosphere, the observer's line of sight towards the edge of the disc passes through higher, cooler and therefore less bright material, than when it passes through the central regions with hotter matter radiating more brightly. It can be shown that limb darkening depends on the derivative of the source function with optical depth and, using the Eddington approximation, the limb/centre intensity ratio is 0.4. In addition, the Sun appears disc-like with a discernible boundary due to the absorption by the negative ion H^- found above the solar atmosphere, discussed in the previous chapter.

Now we recall (Chapter 9) that the Eddington flux H_ν determines the outward radiation flow. If we express the optical depth in terms of the density ρ_ν and the opacity κ_ν explicitly, i.e., $d\tau_\nu = \kappa_\nu \rho_\nu dr$, the transfer equation becomes

$$H_\nu = -\frac{1}{3} \frac{1}{\rho_\nu \kappa_\nu} \frac{dB_\nu}{dT} \frac{dT}{dr}. \tag{11.12}$$

In the equations of stellar structure we had introduced the *diffusion constant* K in the radiative temperature gradient (Eq. 11.4). The standard form of a diffusion equation is: flux (flow) = (rate of change) × (diffusion coefficient). In the present case the radiative diffusion coefficient K depends on microphysical atomic processes. Diffusion of radiation through ionized matter is governed by scattering and absorption of photons by electrons and ions, and determines the opacity. Since the flux $F = (4\pi/3)\ H$, when radiative diffusion is the dominant form of radiation transport, the diffusion approximation discussed above yields

$$F_\nu = -\frac{4\pi}{3} \frac{1}{\rho \kappa_\nu} \frac{dB_\nu}{dT} \frac{dT}{dr}. \tag{11.13}$$

This equation pertains to photons at a single frequency ν, and to the monochromatic opacity κ_ν of the plasma encountered by those photons. Note that the κ_ν appears in the denominator, as one might expect physically. Wherever the opacity is lower, the radiation flow is greater, hence the inverse relation between the radiative flux and the opacity. But in describing the total flow of radiation through a star we must integrate over all frequencies, i.e.,

$$F = \int F_\nu d\nu = -\frac{4\pi}{3} \frac{1}{\rho \kappa_R} \frac{dB}{dT}. \tag{11.14}$$

The interesting quantity on the right-hand side, $1/\kappa_R$, is defined as

$$\frac{1}{\kappa_R} \frac{dB}{dT} = \int \frac{1}{\kappa_\nu} \frac{dB}{dT} d\nu. \tag{11.15}$$

The symbol κ_R stands for the *Rosseland mean opacity* (RMO). The physical analogy of radiation flow through a plasma is that it is like the flow of current in an electrical circuit with many resistors in parallel at different resistances. In that case, the total resistance of the circuit is given by the harmonic mean $1/R = \sum_i (1/R_i)$. Moreover, the flow of current is obviously the greatest through the resistor with the least resistance. Similarly, the escape of radiation from a star would occur most efficiently through 'windows' at frequencies with the lowest material opacity κ_ν, so long as there is sufficient flux at ν. But note that the mathematical form of RMO includes the factor $1/\kappa_\nu$. Therefore, if $\kappa_\nu \rightarrow 0$ the integral Eq. 11.15 for RMO diverges. In practical terms, this implies that care must be exercised for bound–bound or bound–free transitions, where the cross sections might have deep depressions or minima, corresponding to very low opacities. Of course, in reality the radiative cross sections of many atomic species and levels overlap, and such 'zeros' in the opacity do not occur.

Another mean opacity, the *Planck mean opacity* (PMO), is the usual mean, defined as

$$\kappa_P B(T) = \int \kappa_\nu B_\nu \mathrm{d}\nu \,. \tag{11.16}$$

The PMO is the integrated opacity and related to the *total* radiation absorbed; as such, it is complementary to the RMO, which determines the amount of radiation that gets through. The PMO is required to calculate the radiation pressure exerted on matter, and thereby radiative forces or accelerations induced, as discussed later. Finally, recall that the integrated Planck function $B(T) = \int B_\nu \mathrm{d}\nu$ = $(2\pi^4/15c^2h^3)(kT^4)$.

We have formulated simplified radiation transport through the radiative envelope, using small departures from the black-body radiation field and the diffusion approximation. But we are now faced with a major problem with no easy solution. How do we know the opacity?

11.4 Opacity

The flow of radiation through matter depends on its opacity. The opacity of a medium is related to plasma properties on the one hand, and intrinsic atomic physics on the other hand. The plasma conditions determine the atomic species encountered by the photons, as affected by particle interactions and defined by the local temperature, density and composition. The atomic properties come into play to determine the microscopic quantities due to absorption via atomic transitions and scattering. The atomic and plasma effects are both responsible for the ionization states of an element, at a given temperature and density, and the number of levels in an ion that are effective in absorption. Since an atom(ion) has an infinite number of levels, we need to ascertain how many and how much they are populated, so as to impose a physically and computationally realistic limit. Such a prescription is the *equation-of-state* (EOS) of the plasma. Whereas macroscopically, the ideal gas law is approximately the equation-of-state, we have seen that the microscopic atomic and plasma properties in LTE are described by the *Saha–Boltzmann equations*. But it needs to be modified to account for plasma effects explicitly.

The atomic physics is the biggest problem. Since a star, and most astrophysical sources, span a huge range of temperatures and densities, all atomic quantities related to absorption and scattering of radiation at all frequencies must be computed, for all astrophysically abundant elements in all ionization states. Furthermore, and consistent with the EOS, a large number of levels required in each atom (ion) must be explicitly taken into account.

The most recent calculations on astrophysical opacities have been carried out by two groups: the *OPAL* group at the Lawrence Livermore National Laboratory ([280] and references therein) and the *Opacity Project (OP)* [36]. The final results from the two projects are in very good broad agreement for the mean opacities, averaged over all frequencies using the Planck function $B_\nu(T)$. However, there are some significant differences in monochromatic opacities ([281], see also Section 9.5.2). In our discussion we will describe the OP methods since they place their main emphasis on using state-of-the-art atomic physics methods, including the R-matrix method described in Chapter 4. Large amounts of atomic data have been computed using the atomic codes developed under the OP, and a follow-up project called the Iron Project [38]. Before we compare the results and physical interpretation of the differences, we outline the steps and nature of the opacity calculations.

11.4.1 Equation-of-state

The basic EOS in stellar and many non-stellar sources in LTE is the Saha–Boltzmann equation, which combines, at a local temperature T, (i) the population among different atomic levels according to the Boltzmann equation, and (ii) the distribution among different ionization states according to the Saha equation. There are, however, several physical considerations in adapting the combined Saha–Boltzmann equation for the calculation of opacities.

But first we describe the physical components of the opacity.

11.4.1.1 *Partition function and occupation probability*

As noted in Chapter 9 in the discussion of the partition function, the problem with using the Saha–Boltzmann equation to compute numerical values of ionization fractions and level populations is immediately apparent: the partition function diverges since E_{ij} approaches a constant value equal to the ionization potential, but the sum is over an infinite number of atomic levels with the statistical weight increasing as $2n^2$ as $n \to \infty$ at the ionization limit, although $\mathrm{d}f/\mathrm{d}E$ does not diverge and the total absorption oscillator strength f is finite. In reality, the mean radii of atomic levels also increase as n^2, and interparticle effects perturb these levels, depending on the density and hence the mean interparticle distance. Therefore, we expect that plasma effects would lead to a 'cut-off' at some critical value of n_c, where levels with $n > n_\mathrm{c}$ are not populated but are replaced by free electrons. Intuitively, we also expect that, rather than an abrupt cut-off, the levels might be 'dissolved' as $n \to n_\mathrm{c}$ due to plasma fields and effects, such as line broadening (Section 9.2). Thus the cut-off is not altogether sudden but gradual, as the increasingly excited levels broaden, dissolve or ionize (in that order). The effective level population in a plasma environment at a given temperature and density needs to be computed with a realistic EOS that also determines the ionization fractions of an element.

There are several ways to modify the Saha–Boltzmann equations to incorporate the effect of plasma interactions into the EOS. The latest work on astrophysical opacities is based on two approaches. One is the so-called *physical picture*, adopted by the OPAL group [280], that formulates the EOS in terms of fundamental particles, electrons and nuclei, interacting through the Coulomb potential. An elaborate quantum-statistical treatment then enables the properties of composite particles (atoms, ions, molecules) to be computed. Negative energy solutions of the Hamiltonian correspond to bound states of these systems. The physical picture is appealing from a plasma physics point of view, without distinction between free or loosely bound states, and therefore with no modification of the internal partition function.

The alternative approach, adopted by the OP group is the Mihalas–Hummer–Däppen (MHD[3]) EOS [282, 283, 284, 285, 286, 287]. It pertains to the so-called *chemical picture*, which begins with *isolated* atoms and ions and discrete (spectroscopic) energy levels as the basic entities [282]. The atomic energy levels are then modified according to prevailing plasma interactions as a function of temperature and density. The chemical picture is more amenable to advanced atomic physics methods, which compute atomic properties for isolated atoms. Below, we describe the EOS in the chemical picture, but re-emphasize that the final mean opacities differ little between the two EOS formulations.

The modified Saha–Boltzmann equation is based on the concept of *occupation probability* w of an atomic level being populated, taking into account perturbations of energy levels by the plasma environment. We rewrite

$$N_{ij} = \frac{N_j g_{ij} w_{ij} \mathrm{e}^{(-E_{ij}/kT)}}{U_j} \tag{11.17}$$

The w_{ij} are the occupation probabilities of levels i in ionization state j. The occupation probabilities do not have a sharp cut-off, but approach zero for high-n as they are 'dissolved' due to plasma interactions. The partition function is redefined as

$$U_j = \sum_i g_{ij} w_{ij} \mathrm{e}^{(-E_{ij}/kT)}. \tag{11.18}$$

The EOS adopted by the OP, based on the chemical picture of isolated atoms perturbed by the plasma environment, entails a procedure for calculating the w_{ij}. Here, we recall the discussion for collisional line broadening as it perturbs atomic levels (Section 9.2). In particular, the Stark effect in the presence of the plasma microfield of nearby ions broadens the bound states, which would eventually ionize either as a function of n or sufficiently intense microfields. One combines both of these criteria by requiring that the occupation probability w_n is such that a bound state exists only if the field strength F is smaller than a critical value F_n^c for level n. Then

$$w_n = \int_0^{F_n^\mathrm{c}} P(F)\mathrm{d}F, \tag{11.19}$$

where $P(F)$ is the microfield distribution.

In the presence of an external Coulomb field, owing to other slowly moving and sufficiently nearby ions, the break-up of levels into sublevels constitutes *Stark manifolds* for each n; these are labelled as n, m. However, oscillations in the plasma microfield lead to overlapping Stark manifolds, as shown in Fig. 11.1. In particular, Fig. 11.1 shows the effect of interaction among the extreme members of consecutive manifolds, n and $(n+1)$. If the field strength exceeds some critical value, i.e., $F > F_n^\mathrm{c}$, then the highest m-sublevels of the n manifold could

[3] We remind the readers to keep in mind the confusion with the widespread terminology for magnetic hydrodynamics!

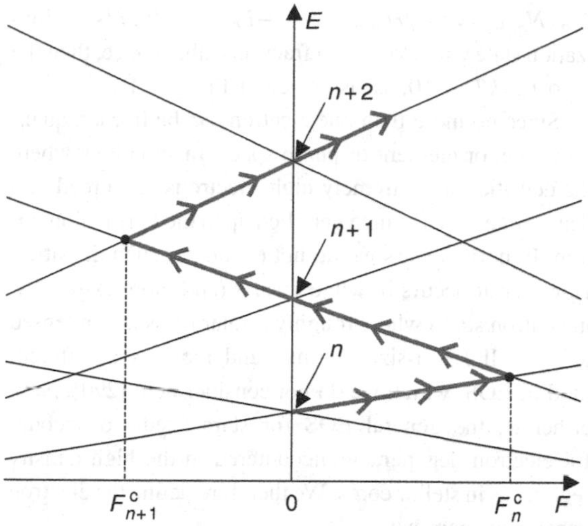

FIGURE 11.1 Stark manifolds of successive principal quantum numbers n, $(n+1)$, etc., showing level crossings among extreme states (adapted from [288]). The field strengths F^c_n, F^c_{n+1} are such that the manifolds n and $n+1$ intersect and the electrons move along arrows as shown.

cross over to the lowest ones in the $(n+1)$ manifold. The process could go on in a similar manner, driven by microfield oscillations, and the electron could move on to the $(n+2)$ manifold, and so on, until it is ionized. In this way, the m-sublevels are mixed or ionized together by oscillations in the plasma microfield. The critical field value decreases rapidly as n increases; higher states ionize in weaker fields, since $F^c_n \sim n^{-5}$. With these assumptions made, dissolution of states occurs when the highest Stark component of level n has an energy equal to the lowest component of level $(n+1)$. It may be shown that Stark splitting is an efficient mechanism for bound state dissolution for $n > 3$. For a hydrogenic ion with nuclear charge Z, and applying first-order perturbation theory, we have

$$F^c_n = \frac{Z^3(2n+1)}{6n^4(n+1)^2}. \tag{11.20}$$

One may employ several approximations of varying complexity to compute F^c_n, and hence w_n, such as the nearest-neighbour or the Holtsmark theory discussed in in Chapter 9. We use the Holtsmark microfield distribution P_H discussed in relation to collisional broadening. Indeed, dissolution of bound states may be viewed as one extreme of line broadening; eventually, dissolution leads to ionization. Another way of looking at ionization due to plasma microfields and electron impact is that effectively we have *lowering of the continuum*. The continuum refers to the energy needed to ionize an electron leaving a residual ion; plasma effects reduce this energy relative to that in an isolated atom. In the OP EOS formalism, the occupation probability is

$$w_n = \int_0^{\beta^c_n} P_H(\beta)\mathrm{d}\beta, \tag{11.21}$$

where

$$P_H(\beta) = \left(\frac{2\beta}{\pi}\right)\int_0^\infty \mathrm{e}^{-y^{3/2}} \sin\beta y \mathrm{d}y \tag{11.22}$$

is the Holtsmark distribution function and $\beta \equiv F^c_n/F_0$, where F_0 is the field strength due to a perturber ion of charge Z_p at the mean inter-ionic distance r_p. Then

$$F_0 = \frac{Z_p}{r_p^2} = Z_p\left(\frac{4\pi N_p}{3}\right)^2 a_0^2. \tag{11.23}$$

For the sake of developing approximate formulae, in astrophysical situations we may assume a hydrogen plasma with protons as the dominant perturbing ionic species, i.e., $Z_p = 1$, and that the proton density equals the free-electron density $N_p = N_e$. A modified treatment taking account of all ionic species z yields, with F^c_n as above,

$$\beta = K_n\left(\frac{I_n^2}{4Z_i}\right)\left(\frac{4\pi a_0^3}{3}\right)^{-2/3} N_e^{-1}N_{ion}^{1/3}, \tag{11.24}$$

where $N_{ion} = \sum_z N_z$, and

$$K_n = \begin{cases} 1 & \text{for} \quad n \le 3 \\ 16n/[3(n+1)^2] & \text{for} \quad n > 3. \end{cases} \tag{11.25}$$

The different values are adopted to reflect that Stark manifolds are assumed not to overlap for $n \le 3$. For non-hydrogenic ions we replace the nuclear charge Z with the effective charge z_j in ionization state j, and n with the effective quantum number ν_{ij} for level i. Then the result for the occupation probability is,[4]

$$w_{ij} = \exp\left(-\frac{64\pi}{3}\left[\frac{(z_j+1)^{1/2}e^2}{K_{ij}^{1/2}I_{ij}}\right]^3 \sum_j z_j^{3/2}N_{jk}\right) \tag{11.26}$$

This treatment involves several approximations, and it is therefore important to verify the results. Experimental data are very sparse for non-hydrogenic systems. However, the lowest hydrogen Balmer line profiles have been measured [291], and are in good agreement with those predicted by the OP EOS [286, 292]. Figure 11.2, from [288], shows the Balmer lines lying between λ = 3600 and 5200 Å, at a temperature 10^4 K and electron density

[4] A more explicit formulation is provided by [289]; see also [290].

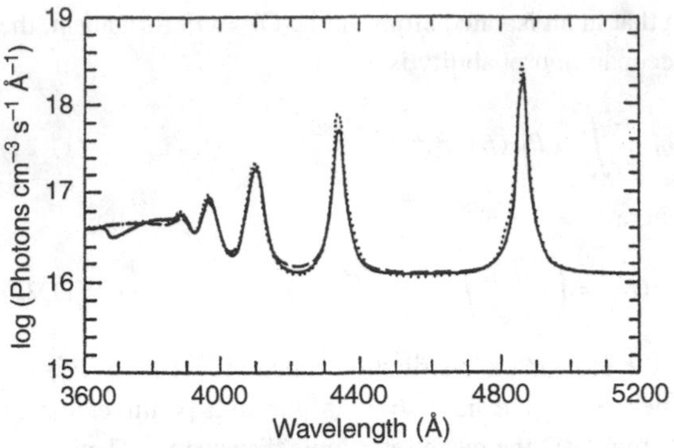

FIGURE 11.2 Density–temperature dependence of line widths and emissivities (Fig. 4 from [288]) – experiment vs. theory (EOS): hydrogen Balmer line emissivities at $N_e = 1.8 \times 10^{16} \text{cm}^{-3}$ and $T = 10^4$ K. Dotted line – experiment (absolute values) [291], Solid line – unit probabiliites and sharp cut-off: $w_n = 1$ for $n \leq 30$ and $w_n = 0$ for $n > 30$, long dashed line – chemical picture EOS [286].

1.8×10^{16} cm^{-3}, typical of stellar photospheres. Figure 11.2 also compares the emissivities derived using occupation probabilities from (i) the OP EOS formalism in the chemical picture described above, and (ii) an ad-hoc approximation with a sharp cut-off in occupation probabilities as mentioned. The various EOS approximations are in good agreement with the experimental results in Fig. 11.2 for the Balmer line broadening.[5]

11.4.1.2 *Electron degeneracy*

In addition to divergent level populations, another problem with the basic Saha–Boltzmann equation is that the plasma density effects, and interparticle potential interactions at high densities, are not considered in the distribution of ionization states of a given element. To formulate a realistic EOS valid over a large range of densities and temperatures, these effects must be taken into account. Throughout most of the star, except inside the core, at any temperature T, the electron density N_e is much less than the total number of quantum states available to the non-degenerate electrons in phase space, i.e., $N_e \ll U_e$, or $N_e/U_e \ll 1$, as given in Table 11.1 for typical stellar temperatures and densities [293]. We also recall from Chapter 9 that assuming most ions to be in the ground state, the Saha–Boltzmann equation gives the ratio to two successive ionization states, $N_m/N_{m+1} \approx (g_m/g_{m+1}) \exp(-I_m/kT)$. If m is the ionization state with maximum fractional abundance, then the ratio $I_m/kT \approx 10$, as also given in Table 11.1.

Since no more than one electron can be in each quantum state, or element of phase space, in situations where the densities are extremely high, electrons are forced into degenerate states, which get filled up to the Fermi momentum. In normal stars we do not encounter such densities. However, it occurs in white dwarfs (and more extremely, in neutron stars) where roughly a solar mass is condensed to a small Earth-size volume, and requires a different kind of EOS, which we do not consider here [290]. Nevertheless, the general EOS for stars needs to account for electron degeneracy encountered in the high density regime, as in stellar cores. We therefore define the electron degeneracy parameter

$$\eta = \ln\left(\frac{N_e}{U_e}\right), \tag{11.27}$$

introduced into the EOS formulation. In addition to electron degeneracy, the Saha–Boltzmann equation also needs to take account of Coulomb interactions of free charged particles in the plasma when calculating ionization balance. This efffect is parametrized with another parameter ϕ_j (not to be confused with line profile). In the chemical picture of the EOS, both the η and the ϕ are obtained by minimizing the total free energy [282, 286]. We then obtain the equation

$$ln\left(\frac{N_{j+1}/U_{j+1}}{N_j/U_j}\right) + \frac{I_j}{kT} + \eta + \phi_j = 0. \tag{11.28}$$

Neglecting the electron degeneracy and Coulomb interaction parameters, η and ϕ_j, yields the orginial form of the Saha ionization equation. With the general outline of EOS formulation as above, we now turn to the atomic processes relevant to opacity calculations.

11.4.2 Radiative atomic processes

The main atomic processes related to photon absorption are: bound–bound (bb), bound–free (bf) and inverse bremsstrahlung free–free (ff). In addition, photon–electron scattering (sc) is a contributor to overall opacity, mainly Thomson scattering by free electrons and Rayleigh scattering by bound atomic and molecular electrons. The total *monochromatic opacity* consists of these four components, i.e.,

$$\kappa_\nu = \kappa_\nu(\text{bb}) + \kappa_\nu(\text{bf}) + \kappa_\nu(\text{ff}) + \kappa_\nu(\text{sc}). \tag{11.29}$$

The quantities of interest related to the first three components are the photoionization cross section σ_ν

[5] A full discussion is given in [286], and the EOS in the physical picture used by the OPAL group is discussed in [282, 283, 284, 285, 286, 288].

for bound–free transitions, the oscillator strength f_ν for bound–bound transitions and collisional damping constants for free–free transitions. These are discussed next individually.

11.4.2.1 *Bound–bound opacity: spectral lines*

For a transition between two atomic levels, 1 and 2

$$X_1 + h\nu_{12} \rightarrow X_2, \tag{11.30}$$

by an atomic species X in lower state 1, the contribution to monochromatic opacity is expressed in terms of the absorption oscillator strength f_{12} as

$$\kappa_\nu^{12}(\text{bb}) = \left(\frac{\pi e^2}{m_e c}\right) N_1 f_{12} \phi_\nu, \tag{11.31}$$

where ϕ_ν is the line profile factor that distributes the line oscillator strength over a certain frequency range according to the plasma environment via line broadening mechanisms (Section 9.2). The f-value is related to the line strength **S** in terms of the wavefunctions as (Chapter 3),

$$f_{12} = \left(\frac{4\pi^2 m_e}{e^2 h}\right)\left(\frac{\nu_{12}}{3g_1}\right)\mathbf{S_{12}}, \tag{11.32}$$

where the dipole line strength[6]

$$\mathbf{S_{12}} = e|<\Psi_2||\mathbf{D}||\Psi_1>|^2, \tag{11.33}$$

with the usual dipole operator $\mathbf{D} = \sum_i \mathbf{r_i}$. Since the line oscillator strength can be related to an absorption cross section over a frequency range, given its line profile, it is useful to express the line opacity in terms of a cross section as well,

$$\kappa_\nu(\text{bb}; 1\rightarrow 2) = N_1 \sigma_\nu(1\rightarrow 2), \tag{11.34}$$

where

$$\sigma_\nu^{\text{bb}}(1\rightarrow 2) = \left(\frac{\pi e^2}{m_e c}\right) f_{12}\, \phi_\nu. \tag{11.35}$$

This expression puts the bound-bound line opacity $\kappa_\nu(\text{bb})$ in the same units (area) as the bound–free opacity $\kappa_\nu(\text{bf})$ described next.

6 We ignore higher multipole moments in the calculation of opacities since their magnitudes relative to the dipole moment are orders of magnitude smaller, as discussed in Chapter 4.

11.4.2.2 *Bound–free opacity: photoionization*

The bound–free transition occurs when the photon is sufficiently energetic to ionize an electron from atomic species X in initial bound state b:

$$X_b + h\nu_{b\epsilon} \rightarrow X^+ + e(\epsilon) \tag{11.36}$$

where $h\nu_{b\epsilon}$ is the photon energy in excess of the ionization energy E_I, and ϵ is the energy carried away by the ejected photoelectron, i.e.,

$$h\nu_{b\epsilon} = E_I + \epsilon. \tag{11.37}$$

The bound–free opacity is therefore expressed in terms of the photoionization cross section

$$\sigma_\nu(\text{b}\rightarrow\epsilon) = \left(\frac{2\pi\nu}{3g_b c}\right)\mathbf{S}(\text{b}\rightarrow\epsilon), \tag{11.38}$$

where **S** is now generalized to refer to the dipole matrix element between the initial bound state wavefunction Ψ_b and a free-electron wavefunction Ψ_ϵ at an energy ϵ in the continuum

$$\mathbf{S}(\text{b}\rightarrow\epsilon) = |<\Psi_\epsilon||\mathbf{D}||\Psi_b>|^2. \tag{11.39}$$

Now we can express the bound–free component of the opacity as

$$\kappa_\nu(\text{bf}; \text{b}\rightarrow\epsilon) = \frac{2\pi\nu N_b}{3g_b c}\mathbf{S}(\text{b}\rightarrow\epsilon), \tag{11.40}$$

in terms of the generalized dipole line strength **S**.

11.4.2.3 *Free–free transitions: inverse bremsstrahlung*

Radiation is emitted when an electric charge accelerates in an electromagnetic field: the *bremsstrahlung* process. A free electron scattering from a positive ion X^+ can, in general, result in the emission of a photon. The principle of detailed balance requires the existence of the *inverse bremsstrahlung process*: absorption of radiation by a free electron and ion (e + ion) system, or

$$h\nu + \left[X_1^+ + \text{e}(\epsilon)\right] \rightarrow X_2^+ + \text{e}(\epsilon'), \tag{11.41}$$

with the ion X^+ in the initial and final states 1 and 2, respectively. The total (e + ion) initial and final energies, E and E', are

$$E = E\left(X_1^+\right) + \epsilon, \tag{11.42}$$

$$E' = E\left(X_2^+\right) + \epsilon', \tag{11.43}$$

and the photon energy hν = $E - E'$. Both the initial and final states of the (e + ion) system have a free electron, and therefore this process is said to entail a free–free transition. Since the free electron(s) can be at a wide continuum of energies, we write the *differential* contribution to the free–free opacity from a range $\mathrm{d}E$ as

$$\mathrm{d}\kappa_\nu(\mathrm{ff}) = \left(\frac{2\pi}{3c}\right) \frac{N_1}{g_1} \frac{N_e F_e}{U_{e(\epsilon)}} \mathbf{S}(E'2, E1)\mathrm{d}E, \quad (11.44)$$

where the factor $N_e F_e / U_e(\epsilon)$ is the fraction of initial electron continuum states occupied, as computed using the partition functions in the Boltzmann–Saha equations. The total free–free opacity is

$$\kappa_\nu(1,2;\mathrm{ff}) = \frac{N(\mathrm{X}^+)}{U(\mathrm{X}^+)} \frac{N_e}{U_e} \frac{2\pi\nu}{3c} \sum_{1,2} \int \exp(-E/kT) \times \left[\sum_{1,2} \mathbf{S}(E+h\nu, 2; E, 1) \right] \mathrm{d}E, \quad (11.45)$$

where the line strength is summed over all initial and final bound states 1 and 2, and the integral ranges over all continuum energies. It is also computed in the dipole approximation as

$$\mathbf{S}(E'2; E1) = \sum_{1,2} | < \Psi(E'2)||\mathbf{D}||\Psi(E1) > |^2. \quad (11.46)$$

Note that we have expressed all the atomic transition probabilities for bb, bf and ff processes in terms of the same quantity, the line strength $\mathbf{S}$ (Chapter 4). Accurate coupled channel or configuration interaction type wavefunctions Ψ may be used to compute $\mathbf{S}$. For example, the ff transition strength may be computed using coupled-channel wavefunctions for the bound state and a continuum electron, as in the electron–ion scattering problem Chapter 5. Explicit calculations may be made using the *elastic scattering matrix* elements for electron impact excitation of ions ([292], Chapter 9). For comparison, an approximate expression for free–free transitions is given in terms of a Gaunt factor,

$$\kappa_\nu^{\mathrm{ff}}(1,2) = 3.7 \times 10^8 N_e N_i g_{\mathrm{ff}} \frac{Z^2}{T^{1/2}\nu^3}. \quad (11.47)$$

Note the pseudo-hydrogenic form $\kappa \sim 1/\nu^3$, similar to the Kramer's bound–free hydrogenic cross section for photoionization.

11.4.2.4 *Photon–electron scattering*

Most of the scattering contribution to opacity is accounted for by Thomson scattering, using

$$\kappa(\mathrm{sc}) = N_e \sigma_{\mathrm{Th}}, \quad (11.48)$$

where the Thomson cross section is

$$\sigma_{\mathrm{Th}} = \frac{8\pi e^4}{3m_e^2 c^4} = 6.65 \times 10^{-25}\ \mathrm{cm}^2. \quad (11.49)$$

The Rayleigh scattering of photons by bound electrons may be approximated by the expression

$$\sigma_\nu^{\mathrm{R}} \approx f_t\, \sigma_{\mathrm{Th}} \left(\frac{\nu}{\nu_I}\right)^4, \quad (11.50)$$

where $h\nu_I$ is the binding energy and f_t is the *total oscillator strength* associated with the bound electron, i.e., the sum of all possible transitions, such as the Lyman series of transitions 1s→np in hydrogen (Chapter 2). The Rayleigh opacity for H I is

$$\kappa_\nu^{\mathrm{R}} = n_{\mathrm{H}} \sigma_\nu^{\mathrm{R}}(\mathrm{H}). \quad (11.51)$$

Eqs. 11.50 and Eq. 11.51 imply that the cross section or the opacity increases inversely and rapidly with wavelength.[7] We note that Eq. 11.51 expresses the opacity in the oft-used units of inverse length (cm^{-1}).

Once the radiative atomic processes discussed above have been taken into consideration, the calculation of detailed opacities proceeds as follows.

11.4.3 Monochromatic opacities

The MHD EOS prescription described previously gives the ionization fractions and level populations of each ion of an element in levels with non-negligible occupation probability. The opacity of the plasma results from the interaction of photons with ions via absorption, as well as scattering by free particles. We are now in a position to express the monochromatic opacity κ_ν in terms of basic atomic quantities: bound–bound oscillator strengths, bound–free photoionization cross sections and free–free (inverse bremsstrahlung) cross sections for each ion. To relate the EOS and opacities calculations, we define, with element k in ionization state j and level i: level population fraction $F_{ijk} = N_{ijk}/N_k$, ion fractions $F_{jk} = N_{jk}/N_k$, electrons per atom $\epsilon_k = \sum_j z_j F_{jk}$, and chemical abundance fractions $A_k = N_k/N$, where N is the total density. Electrons in the plasma exist either bound to ions or free; the latter are ionized from atoms of all elements present. Then the free electron density is

$$N_e = \sum_k \epsilon_k N_k = N \sum_k \epsilon_k A_k, \quad (11.52)$$

and the *mass density* is

$$\rho = \sum_k M_k N_k = N \sum_k M_k A_k, \quad (11.53)$$

[7] The ν^4 dependence implies that blue light is scattered more than red light; the phenomenon that makes the sky appear blue.

where M_k is the atomic mass of element k. From the two equations above, we can eliminate N to obtain

$$\rho = N_e \left(\frac{\sum_k M_k A_k}{\sum_k \epsilon_k A_k} \right), \tag{11.54}$$

which relates the mass density ρ and the electron density N_e (the T_e dependence is implicit through the number of free ionized electrons per atom). Table 11.1 lists the ρ and N_e so derived. In opacity calculations, it is convenient to introduce a dimensionless parameter $u \equiv h\nu/kT$. Monochromatic opacities $\kappa(u)$ are then computed at a range of log u, given by

$$\kappa(u) = \left(1 - e^{(-u)}\right) \left[\sum_{ijk} N_{ijk}\sigma_{ijk}(u) + \sum_{jk} N_{jk} N_e \sigma_{ff}(u) \right] + N_e \sigma_e(u), \tag{11.55}$$

where $(1 - e^{-u})$ is the correction factor for stimulated emission, and σ_{ff} and σ_e are the cross sections for free–free transitions and electron scattering, respectively. The σ_{ijk} is the total absorption cross section from level i, due to both lines (bound–bound) and photoionization (bound–free). Given the monochromatic opacities $\kappa(u)$ the Rosseland and the Planck mean opacities, κ_R and κ_P in units of $cm^2\,g^{-1}$, are calculated from expressions given earlier (Eqs 11.15, 11.16), at a mesh of temperature–density $[T(r), N_e(r)]$, all along the stellar radius r. When radiation pressure is dominant, the ratio of matter pressure to radiation pressure is essentially constant. Then the ideal gas law and the Stefan–Boltzmann law ensure that the quantity ρ/T^3 is also approximately constant. Therefore, a convenient variable for tabulating opacities that describes both the density and the temperature is

$$R(\rho, T) = \frac{\rho}{T_6^3}, \tag{11.56}$$

where the density is in $g\,cm^{-3}$ and T_6 is in units of 10^6 K, i.e. $T_6 = T * 10^{-6}$. The parameter R is a combination of density and temperature, since both physical quantities vary similarly in stellar interiors (Fig. 10.5). Then, $\log_{10} R$ is a small number lying between -1 and -6 for the conditions in the Sun (Fig. 11.4). For example: for $\log R = -3$, at $T = 10^6$ K, the corresponding density is $\rho = 0.001\,g\,cm^{-3}$. In the core of the Sun, with nuclear fusion energy production environment via pp reactions (Table 10.2), $T \sim 1.5 \times 10^7$ K and $\rho \sim 80\,g\,cm^{-3}$, so that $\log R = -1.625$. We next describe the manner in which atomic data for the processes discussed above are utilized in opacity calculations.

11.4.4 Abundances, mixtures and atomic data

It is clear that an immense amount of atomic data are needed for opacity calculations. The main problem is not only the generality – all transitions in various ionization stages of the elements in different types of stellar compositions – but also the fact that in specific regions of a star different atomic species and processes could be vitally important. Since hydrogen is the most abundant element, approximately 90% by number and 70% as mass fraction in the Sun, it is usual practice to measure the abundances of all other elements relative to hydrogen. The next most abundant element is helium, approximately almost 10% by number and 28% by mass fraction in the Sun.[8] The remainder is all other elements, generically called 'metals' but their abundance by number is less than 1%, and about 2% by mass fraction.

The stellar element mixtures are often specified by X, Y and Z. For instance, the solar H abundance is denoted as $X = 0.7$, the He abundance as Y = 0.28, and the overall metal abundance, in its totality, as Z = 0.02. The abundances (A_k) of elements vary over several orders of magnitude, and are usually expressed on a $\log_{10}$ scale. It is traditional to take $\log(A_H) = 12$. Then the abundances of other elements, on a log scale up to 12, are given relative to H. For example, a representative mixture of the 'standard' abundances for the Sun is given in Table 11.2 [36]. Stellar abundances are relative to solar abundances, and are sometimes taken to be representative of cosmic values as well. A few points deserve mention. The C and O abundances are the highest among metals, as expected from nucleosynthesis processes; O is the most abundant element of all metals, with O/C $\sim$ 2.0. That is followed by the α-elements, Ne, Mg, Si and S, whose abundances are lower by a factor of 10 to 20. Also note that A(Fe) is comparable to these elements.

However, the accuracy of stellar opacities is being examined in connection with rather large discrepancies found recently in solar photospheric abundances determined spectroscopically [269, 294]. Serious differences arise with helioseismological data in stellar interior models when the standard or the new solar abundances (Table 11.2) are employed. Helioseismology is capable of

[8] Measured cosmological abundances are somewhat different, hydrogen about 73% and helium about 25%.

TABLE 11.2 Solar photospheric abundance mixture [36]. Columns 2 and 3 compare the standard solar abundances with the new abundances recently proposed [294]. The uncertainties in each set are generally within a few percent. Calculated opacities presented in this text use standard solar abundances (Figs 11.3 and 11.4).

Element (k)	Log A_k (standard)	Log A_k (new)	$A_k/A_{\rm H}$ (standard)
H	12.0	12.0	1.0
He	11.0	10.93	1.00×10^{-1}
C	8.55	8.43	3.55×10^{-4}
N	7.97	7.83	9.33×10^{-5}
O	8.87	8.69	7.41×10^{-4}
Ne	8.07	7.93	1.18×10^{-4}
Na	6.33	6.24	2.14×10^{-6}
Mg	7.58	7.60	3.80×10^{-5}
Al	6.47	6.45	2.95×10^{-6}
Si	7.55	7.51	3.55×10^{-5}
S	7.21	7.12	1.62×10^{-5}
Ar	6.52	6.40	3.31×10^{-6}
Ca	6.36	6.34	2.29×10^{-6}
Cr	5.67	5.50	4.68×10^{-7}
Mn	5.39	5.43	2.46×10^{-7}
Fe	7.51	7.50	3.24×10^{-5}
Ni	6.25	6.22	1.78×10^{-6}

measuring solar oscillations to high accuracy and is potentially an accurate probe of internal solar material and structure. Stellar models thereby constrain solar abundances and crucial stellar parameters, such as the sound speed, depth of the convection zone, and surface abundance ratios of elements. The solar abundance problem has been discussed by several researchers [295, 296]. The recent work [269, 294], based on revised spectroscopic analysis and new three-dimensional time-dependent hydrodynamical NLTE models, yields significantly lower abundances for the light volatile elements, especially C, N, O and Ne. To wit: the oxygen abundance is over 40% lower. The third column in Table 11.2 also lists the new solar abundances [294]. There is considerable controversy over these new solar abundances, and it is a very active area of contemporary research in stellar astrophysics, with a crucial role concerning the precision of currently available opacities [297].

The data needed for line opacity consist of oscillator strengths for all bound–bound transitions in elements in the stellar mixture used to model a particular type of star. Generally, the H and He abundances in normal main-sequence stars are primordial in nature, and therefore largely invariant. It is often the small but significant abundances of other elements (metals) that characterize the properties of stars and other astrophysical objects. Moreover, the role of metals and heavy elements can be crucial in the interior of stars in driving stellar phenomena, such as pulsation in metal-rich bright stars, e.g., Cepheid variables. The connection between opacity and pulsation becomes evident on considering the *sound speed* $c_{\rm s}$ in ionized material;

$$c_s = 9.79 \times 10^3 \left[\frac{\gamma Z T_e}{\mu}\right]^{1/2} \quad {\rm m\,s^{-1}}. \tag{11.57}$$

As evident from the kinetic relation $\frac{1}{2}mv^2 = \frac{3}{2}kT$, the sound speed increases with temperature as $c_{\rm s} \sim T_{\rm e}^{1/2}$. That is quite physical, since particle velocities increase with temperature and 'sound', or any material disturbance, such as pulsation, is transmitted more readily through the medium. But, as mentioned, the temperature distribution is governed by the local opacity (see Fig. 11.3, discussed next). Since the Cepheid pulsation period is proportional to the diameter or the stellar radius, we have the relation [219]

$$P \propto \frac{R_*}{c_s} \propto \frac{R_*}{T^{1/2}} \propto \frac{R_*}{\kappa^{n/2}}, \tag{11.58}$$

where R_* is the stellar radius. Strictly speaking, the $\kappa \sim T^{-n}$ proportionality between the opacity and temperature inherent in this equation is not quite accurate; the

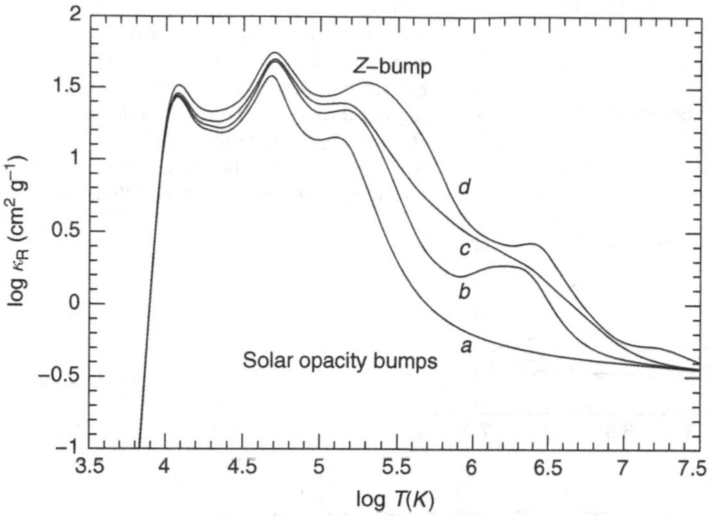

FIGURE 11.3 Rosseland mean opacities at $R = -3$ for element groups (a) H and He, (b) H to Ne, (c) H to Ca and (d) H to Ni. The composition is as in Table 11.2, normalized with $X = 0.7$ and $Z = 0.2$, with addition of elements going from (a) to (d). The different 'bumps' reflect corresponding opacity enhancements. The temperature may be related to the density and radius according to Fig. 10.5. The figure is drawn using on-line opacities from the publicly accessible electronic database OPserver at the Ohio Supercomputer Center in Columbus Ohio ([300], http://opacities.osc.edu).

general period–opacity relation is more complicated and depends on the elemental composition (see Fig. 11.3). The Cepheid pulsation periods are discussed further in Chapter 14, in connection with the universal distance scale and expansion. Other examples where detailed opacities are important are radiatively driven winds, which depend critically on the radiation absorbing metal content of outflowing material. Surface abundances of stars depend on the interplay of hydrodynamic (convective) and radiative forces on constituent elements in stellar atmospheres. For these reasons and more it is necessary to carry out opacity calculations for a variety of mixtures and at a sufficiently fine grid of temperature–density points to enable accurate interpolation in (T, N_e).

Figure 11.3 [36] shows the Rosseland mean opacity κ_R as a function of temperature, which is related to various depths r in a star. The several bumps are associated with excitation or ionization of different atomic species at those temperatures. The first (lowest κ_R) curve (a) has three bumps corresponding to the ionization of neutral H, He and He II, at log $T \approx 4$, 4.6 and 5.2, respectively. The temperatures associated with these bumps mark the ionization zones at corresponding depths in the star. Including elements up to Ne raises the opacity towards higher temperatures, as shown in curve (b). In addition, another bump appears at log $T \approx 6-6.5$. This because the second row elements from Li to Ne have two electronic shells that ionize succesively, the L-shell (n = 2) and the K-shell (n = 1); the latter typically ionizes at a million degrees or more depending on the atomic number. Addition of further elements up to Ca raises the overall opacity significantly as shown in curve (c); in particular, we no longer have a dip in opacity seen in curve (b) just below log $T = 6$. It is, however, the topmost curve (d), due to the further inclusion of the iron group elements up to Ni, that gives rise to a considerable increase in opacity for all $T > 10^4$ K. The most outstanding enhancement in opacity due to iron occurs around log $T = 5.2-5.6$, referred to as the Z-bump. It is mainly due to excitation and ionization of Fe ions with a partially filled M-shell (n = 3): Fe IX–Fe XVI. We also notice that the high-temperature K-shell bump due to inner-shell processes is also signficantly increased and moved to higher temperatures, log $T \approx 6.4$–6.5, compared with (b) [298].

Figure 11.4 shows a more detailed behaviour of the OP Rosseland mean opacity for the Sun. The standard mixture of elements given in Table 11.2 is used, and computations carried out for several values of log R at temperatures that range throughout the Sun, from photospheric temperatures of a few thousand degrees to core temperatures of tens of millions of degrees.[9]

Whereas the frequency integrated Rosseland mean opacities show only a few bumps in the T–ρ plane, the monochromatic opacities can be extremely complex even for single ions. Figure 11.5 shows the monochromatic opacity of Fe II at log $T = 4.1$ and log $N_e = 16.0$, plotted as a function of the wavelength in the range $\sim 1000 < \lambda < 100\,000$ Å, in atomic units a_0^2. At that temperature–density Fe II is the dominant ionization species with ionization fraction Fe II/Fe = 0.91. The calculations include 1242 bound levels of Fe II, with over

[9] The OP values generally agree with those from OPAL to within a few percent [298].

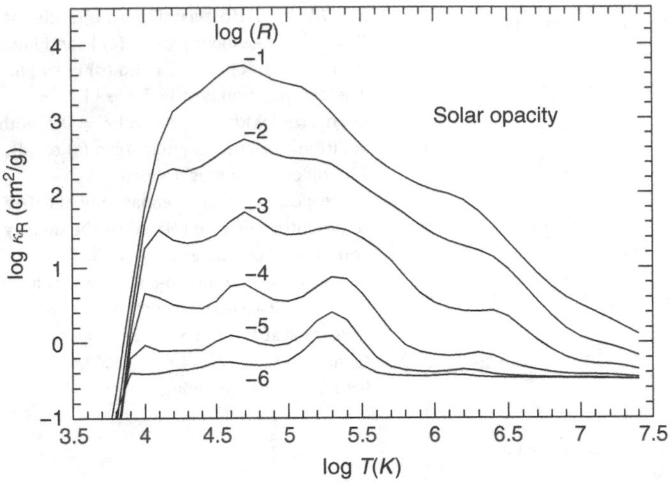

FIGURE 11.4 Solar opacity from the Opacity Project [36, 37] using the 'standard' mixture of elements in Table 11.2. The Rosseland mean opacity κ_R in several temperature-density regimes throughout the solar interior, characterized by the parameter $R = \rho(\mathrm{g\,cm^{-3}})/T_6^3$. The figure is drawn using on-line opacities from OPserver [300].

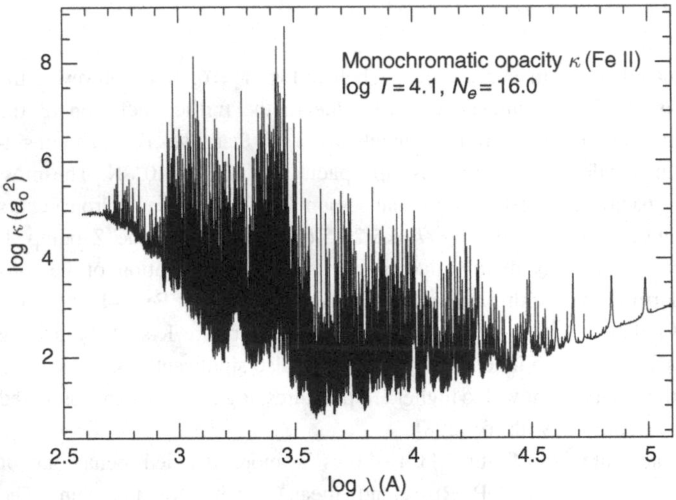

FIGURE 11.5 Monochromatic opacity spectrum of Fe II resolved at 10^5 frequencies (cf. [299]).

34 007 lines (bb-transitions) [299]. Photoionization or bf cross sections for all levels are also included, together with scattering and ff contributions. The RMO (Eq. 11.15) is $\kappa_R(\mathrm{Fe\,II}) = 63.7\ \mathrm{cm^2\ g^{-1}}$. Note that radiative absorption or the opacity cross section for Fe II ranges over seven orders of magnitude in a_0^2, the hydrogen atom cross section. Moreover, the Fe II absorption occurs mostly in the wavelength range from near-ultraviolet $\lambda < 1000$ Å to near-infrared $\lambda \sim 30\,000$ Å.

Of crucial importance are the wiggles and bumps in the RMOs and PMOs at certain temperatures corresponding to specific radii in the star (see Fig. 10.5). These, in turn, depend on the enormously complex structure of the total monochromatic opacity from all ions prevalent in those T–ρ regimes. If there is sufficient enhancement of opacities in these regions then radiative forces can cause acceleration of matter in the stellar interior, as discussed next.

11.5 Radiative forces and levitation

The opacity of matter at any point inside the star is related to the radiation force or pressure. Given the mass and composition, the internal dynamics and structure of a star is determined by the balance between gravitational and radiative forces according to the equations of stellar structure (Section 11.2). The bulk radiative force on matter (plasma) is due to photon–atom interactions, discussed above, which taken together constitute the opacity. The competition between gravity and radiation can manifest itself in interesting ways.

We consider photon–atom interactions of individual elements. If the gravitational force downward (towards the centre of the star), on the atoms of a given element, dominate over kinetic gas pressure and radiation pressure outward, then *gravitational settling* occurs. Stratification of elements tends to take place: heavier elements move down towards the centre, and separate from the ligher elements, which should move up correspondingly. But we now need to consider radiation pressure, in addition to normal gas pressure that ensures hydrostatic equilibrium in the stellar interior. When the differential perturbation due to an additional radiative force or acceleration, combined with the gas pressure, exceeds the gravitational force then the element would be *levitated*, and rise or move up towards the surface – the opposite of gravitational settling. At first, one might assume that heavier elements would be more subject to gravitational settling, rather than *levitation*, in proportion to their atomic masses. However, this is where atomic structure and detailed atomic physics come into play, with interesting consequences. Whereas the gravitational force is certainly proportional only to the atomic mass of the element, the radiative force depends on the total absorption of radiation, or its opacity. The monochromatic opacity of an element varies greatly with the ionization stage, which determines the number of electrons interacting with the radiation field. The ionization stage, in turn, depends on the local temperature and density. Therefore, the balance of radiative and gravitational accelerations of an element fluctuates according to local physical conditions at different points in the star. So a heavier atomic system, with more active electrons than a lighter one, can absorb sufficiently more radiation to be relatively levitated.

It is simple to see how gravity and radiation would compete differently in the interior of the star. For example, the gravitational acceleration downward on iron, with atomic mass $A(\text{Fe}) = 55.85$ amu, is little affected by its ionization stage since the masses of electrons are irrelevant. But radiative absorption depends almost entirely on the number of electrons in the ion, and the atomic transitions that the electrons can undergo. Radiative forces on the highest ionization stages, H-like or He-like, are vastly smaller than in lower ionization stages. However, that does imply that the neutral atom would experience the largest radiative force. The frequency distribution of the radiation field depends on the particular type of star. Therefore, certain ionization stages of an element absorb stellar radiation more effectively than others. In most stars, the ionization stages of Fe with L- and M-shells open are the most efficient absorbers. It follows that in the interior regions of the star, where those ionization stages exist, the opacity, and consequently radiative levitation, would be the greatest. Indeed, the Z-bump in the Rosseland mean opacity at $T_{\text{eff}} \approx 2 \times 10^5$ K corresponds to the region of maximum radiative accelerations (Figs 11.3 and 11.4), via Fe ions Fe IX–Fe XIX with open M-shell electrons.

That radiative levitation affects the internal dynamics and structure of the star can now be inferred: *convective motions driven by radiative forces can start deep within a star, and thereby affect even the surface abundances of elements*. There are classes of stars, e.g., *mercury-manganese (HgMn) stars*, where high-Z elements beyond the Fe-peak elements are observed. Large abundance anomalies are found in heavy elements up to Pt, Au, Hg, Tl and Bi [301, 302]. Heavy-particle transport in such stars occurs due to radiative accelerations, which manifest themselves in anomalous photospheric abundances relative to solar values. Perhaps the best-studied star showing the high-Z element spikes is χ Lupi [303], observed in the ultraviolet with the Goddard High Resolution Spectrograph (GHRS) aboard the *Hubble Space Telescope*. High-resolution spectral observations of high-Z elements are often in the ultraviolet, since the lowest allowed transition ('resonance' line) corresponds to relatively high energies, in contrast to low-Z elements, where the lowest transitions may be in the optical or infrared. Another example of radiative levitation is the observed overabundance of iron in the *atmospheres* of hot, young white dwarfs despite their high surface gravity. The spectra of young white dwarfs born out of the hot central stars of PNe are predominantly in the EUV, and were observed extensively by the Explorer class satellite launched by NASA, the *Extreme Ultraviolet Explorer (EUVE)* ([304], see also [290] and references therein).

11.5.1 Atomic processes and momemtum transfer

The four processes contributing to the opacity (bb, bf, ff, sc) have somewhat different forms for transferring radiative momentum to the free electrons and ions in the plasma. We express the total cross section as for opacities,

$$\sigma_\nu = \sigma_\nu(\text{bb}) + \sigma_\nu(\text{bf}) + \sigma_\nu(\text{ff}) + \sigma_\nu(\text{sc}). \tag{11.59}$$

These components are computed taking into accout the following physical factors.

(i) In bound–bound (bb) line transitions

$$h\nu + \text{X}_i \rightarrow \text{X}_j, \tag{11.60}$$

the entire momentum of the incident photon is absorbed within the atom in transition $i \rightarrow j$. Therefore, the transfer cross section is related to the full absorption oscillator strength distributed along a line profile; we compute $\sigma_\nu(\text{bb})$ to obtain $\kappa_\nu(\text{bb})$.

(ii) The photoionization bf process is, however, different. After photoionization,

$$h\nu + \mathrm{X} \rightarrow \mathrm{X}^+ + \mathrm{e}, \quad (11.61)$$

the photon momentum is carried away by the ion, and the ejected electron. Therefore, the cross section for net radiative momentum transfer σ_ν^{mt} that accelerates the atom(ion) of an element is related to the total bf cross section, minus the cross section for the momentum transferred to the free electron $\sigma_\nu(\mathrm{e})$,

$$\sigma_\nu^{\mathrm{mt}} = \sigma_\nu(\mathrm{bf}) - \sigma_\nu(\mathrm{e}). \quad (11.62)$$

The $\sigma_\nu(\mathrm{e})$ can be calculated in a variety of approximations. A simple expression is obtained by introducing a factor K, which represents the fraction of the total bf cross section, and

$$\sigma_\nu(\mathrm{e}) = \sigma_\nu(\mathrm{bf}) K \times \left(1 - \frac{\nu_0}{\nu}\right), \quad (11.63)$$

with $K = 1.6$ and ν_0 the ionization threshold. The treatment in the OP work is more sophisticated, but with similar values that include the dependence on the principal quantum number and angular distribution of the photoionized electrons.

(iii) As we have seen in the calculation of $\kappa(\mathrm{ff})$, the free–free process is more complicated, since it involves the initial and final states of the (e + ion) system, as well as the electron density

$$\left[\mathrm{X}^+(i) + \mathrm{e}\right] + h\nu \rightarrow \mathrm{X}^+(j) + \mathrm{e}'. \quad (11.64)$$

However, for momentum transfer the ff process is not significant, and a hydrogenic approximation is sufficient, assuming that nearly all of the photon momentum is absorbed by the ion.

(iv) The photon–electron scattering component is related to the electron density and the Thomson scattering cross section

$$\sigma_\nu(\mathrm{sc}) = \left(\frac{n_\mathrm{e}}{N}\right) \sigma_{\mathrm{Th}}, \quad (11.65)$$

where $\sigma_{\mathrm{Th}} = (8\pi e^4/3m_\mathrm{e}c^4) = 6.65 \times 10^{-25}\ \mathrm{cm}^2$, and N is the total number of atoms per unit volume.

11.5.2 Radiative acceleration

The theory of radiative accelerations and numerical calculations has been described, for example, in [305, 306, 307]. The OP data have been used in some of these earlier references. More recently, M. J. Seaton ([308] and references therein) recomputed radiative accelerations using more extensive and consistent sets of OP data. We first obtain an expression for radiative acceleration. Energy flow is governed by the monochromatic radiative flux

$$\mathcal{F}_\nu = -\left(\frac{4\pi}{3}\right) \frac{1}{\rho\kappa_\nu} \frac{\mathrm{d}B_\nu}{\mathrm{d}T} \frac{\mathrm{d}T}{\mathrm{d}r}, \quad (11.66)$$

and the total radiative flux

$$\mathcal{F} = \int \mathcal{F}_\nu \mathrm{d}\nu = -\left(\frac{4\pi}{3}\right) \frac{1}{\rho\kappa_\mathrm{R}} \frac{\mathrm{d}B}{\mathrm{d}T} \frac{\mathrm{d}T}{\mathrm{d}r}, \quad (11.67)$$

per unit area per unit time. The momentum associated with radiation in frequency range $\mathrm{d}\nu$ is

$$\frac{1}{c}\mathcal{F}_\nu \mathrm{d}\nu. \quad (11.68)$$

The momentum actually transferred to the atom depends on its absorption cross section. We denote the radiation momentum transfer cross section (in units of area) as $\sigma_\nu^{\mathrm{mt}}(k)$, for an atom of element k. Then the momentum transfer per atom per unit time by all photons is

$$G(k) = \frac{1}{c} \int_\nu \sigma_\nu^{\mathrm{mt}}(k) \mathcal{F}_\nu \mathrm{d}\nu, \quad (11.69)$$

which is the *radiative force* per atom. We can relate this to the *gravitational acceleration* g_{rad} and atomic mass M_k as

$$g_{\mathrm{rad}} = \frac{G(k)}{M_k}. \quad (11.70)$$

This equation refers to levitation due to radiation pressure, as a perturbation to hydrostatic equlibrium between kinetic gas pressure and gravitational pressure (Section 11.2). Keeping in mind the units of the opacity and mass, κ ($\mathrm{cm}^2\,\mathrm{g}^{-1}$) and M (g), we define the opacity cross section $\sigma_\nu = \kappa M_k$, and similarly the Rosseland cross section as $\sigma_\mathrm{R} = \kappa_\mathrm{R} M_k$. From Eqs 11.64 and 11.65,

$$\mathcal{F}_\nu = \left(\frac{\mathrm{d}B_\nu/\mathrm{d}T}{\mathrm{d}B/\mathrm{d}T}\right) \frac{\mathcal{F}\kappa_\mathrm{R}}{\kappa_\nu}. \quad (11.71)$$

The total radiative flux at depth r in a star with temperature T_{eff} and radius R_* is

$$\mathcal{F}(r) = \pi B(T_{\mathrm{eff}}) \left(\frac{R_*}{r}\right)^2. \quad (11.72)$$

Combining Eqs 11.68–11.70, we obtain the radiative acceleration

$$g_{\mathrm{rad}} = \left(\frac{1}{c}\right) \frac{M}{M(k)} \kappa_\mathrm{R} \gamma(k) \mathcal{F}, \quad (11.73)$$

where M is the mean atomic mass $M = \sum_k A_k M_k$, with normalized fractional abundances A_k, such that

$\sum_k A_k = 1$. An important new dimensionless quantity γ, for each element k in the above equation, is defined as

$$\gamma_k = \int \frac{\sigma_\nu^{\mathrm{mt}}(k)}{\sigma_\nu} \left(\frac{\mathrm{d}B_\nu/\mathrm{d}T}{\mathrm{d}B/\mathrm{d}T} \right) \mathrm{d}\nu. \quad (11.74)$$

The Rosseland cross section, analogous to the Rosseland opacity, Eq. 11.15, is

$$\frac{1}{\sigma_\mathrm{R}} = \int \frac{1}{\sigma_\nu} \left(\frac{\mathrm{d}B_\nu/\mathrm{d}T}{\mathrm{d}B/\mathrm{d}T} \right) \mathrm{d}\nu. \quad (11.75)$$

Using the variable $u = h\nu/kt$ used to define the frequency–temperature mesh in opacity computations, we obtain

$$\frac{\mathrm{d}B_\nu/\mathrm{d}T}{\mathrm{d}B/\mathrm{d}T} = \frac{15hu^4}{4\pi^4 kT} \mathrm{e}^{-u}[1 - \mathrm{e}^{-u}]^{-2}. \quad (11.76)$$

It is the quantity γ_k that contains all the information about the atomic physics of radiative acceleration in terms of detailed atomic cross sections for momentum transfer from a radiation flow.[10] With M as the mean atomic mass, κ_ν ($\mathrm{cm}^2\,\mathrm{g}^{-1}$) = $\sigma_\nu(\mathrm{cm}^2)/M$ (g). Then the mass density $\rho = M \times N$, where N is the total number of atoms per unit volume, and the *opacity per unit length* is $\rho\kappa_\nu = n\sigma_\nu$ (cm^{-1}).

The calculation of the gravitational acceleration g_rad, for a given element k, may be made approximately (e.g., [301, 307]). For instance, we may include only the bound–bound lines and set $\sigma_k^{\mathrm{mt}} = \sigma_\nu^{\mathrm{bb}}$, and the Rosseland mean for the background, $\sigma_\nu = \sigma_\mathrm{R}$. These approximations yield

$$\gamma_k \approx \int \frac{\sigma_\nu^{\mathrm{bb}}(k)}{\sigma_\mathrm{R} + \chi A_k \sigma_\nu^{\mathrm{bb}}(k)} \left(\frac{\mathrm{d}B_\nu/\mathrm{d}T}{\mathrm{d}B/\mathrm{d}T} \right) \mathrm{d}\nu, \quad (11.77)$$

where we also introduce a factor χ, of importance in practical calculations, which multiplies the abundance A_k of element k, keeping fixed the relative abundances of all other elements. That enables the study of radiative forces that accelerate (levitate) a given element inside the star. For comparison, an approximate expression for g_rad in stellar interiors at radius r is [301],

$$g_\mathrm{rad}(k) = \left[\frac{\mathcal{F}(r)}{4\pi r^2 c} \right] \frac{\kappa_\mathrm{R}}{A_k} \frac{15}{4\pi^4} \times \int_0^\infty \frac{\kappa_u(k)}{\kappa_u(\mathrm{total})} \frac{u^4 e^u}{(e^u - 1)^2} \mathrm{d}u, \quad (11.78)$$

in terms of the monochromatic and Rosseland mean opacities. The detailed OP results for radiative accelerations are electronically archived, as described in the next section.

11.6 Opacities and accelerations database

As we have seen, the opacity is the fundamental quantity in stellar models. It determines radiation transport, in addition to convection in the outer regions, and thereby stellar structure and evolution. The opacities are also interconnected with surface elemental abundances and internal physical processes, such as the sound speed and the depth of the convection zone (Fig. 10.4). Furthermore, our understanding of chemical evolution and stellar ages also depends on the underlying opacities. Thus, accurate calculation of opacities throughout the stellar interiors is a vital necessity in astrophysics.[11] Stellar opacities need to be computed at all temperatures and densities, which may be transformed as a function of stellar radius, as shown in Fig. 10.5. In addition, since different stars may have quite different elemental compositions, opacities need to be correspond to a variety of abundance mixtures, that may deviate considerably from the solar abundances listed in Table 11.2.

The OP team has established an interactive on-line database called *OPserver* to compute such 'customized opacities' [300].[12] Rosseland and Planck means may be computed for an arbitrary mixture of elements, and a fine mesh in temperature and density. Monochromatic opacities are tabulated as a function of $u = h\nu/kT$, at photon frequencies relevant to the Planck function at a temperature T, for each element. These opacity spectra or cross sectons can be immensely complicated,

[10] One may consider the quantity γ_k as the *radiation strength*, in analogy with other dimensionless quantities, viz. the collision strength and the oscillator strength.

[11] With the advent of nuclear fusion devices, such as the Z-pinch machines and laser-induced inertial confinement facilities, stellar interior conditions may be recreated in the laboratory (e.g., [309, 310]). Monochromatic opacities can now be measured in LTE, at temperatures and densities close to, or deeper than, the solar convection zone. Benchmarking laboratory and astrophysical opacities to high precision is of great interest in the emerging field of *high-energy density (HED)* physics, as well as for the solution of outstanding problems in astronomy, such as the anomalous solar abundances [294].

[12] The OPserver website is: http://opacities.osc.edu.

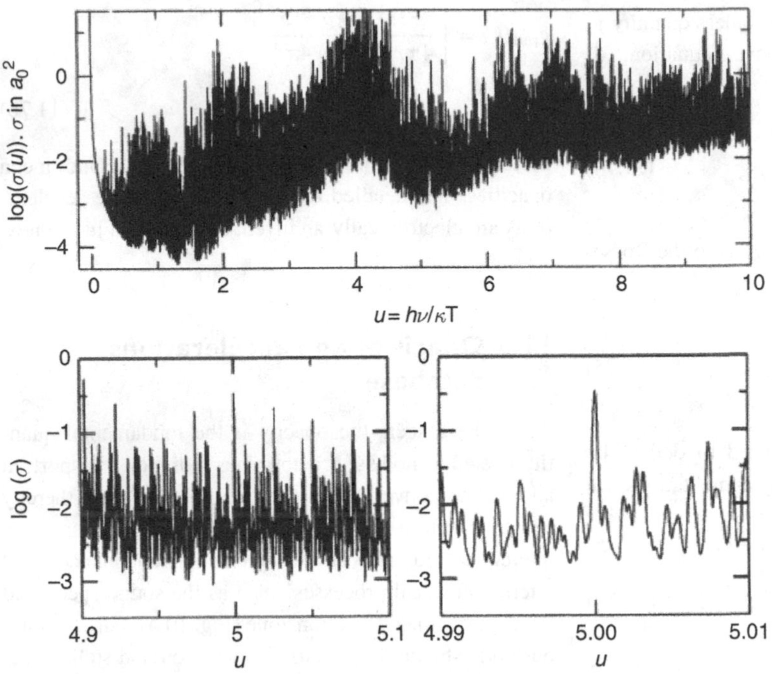

FIGURE 11.6 Iron monochromatic opacity and its complex structure at high resolution (Fig. 10 from [36], at the Z-bump temperature and density (Fig. 11.3). The two lower panels show progressively higher resolution.

since they incorporate the contributions to absorption by all bound–bound and bound–free transitions of all ionization states of an element. Figure 11.6 shows the opacity cross section $\sigma\left(a_0^2\right)$ for iron, the most complex of all species with high abundance. The temperature is $\log(T_e) = 5.3$ ($T = 2 \times 10^5$ K), and the electron density is $\log(N_e) = 18.0$; these correspond to $\log(R) = -3.604$ [36], close to values at the Z-bump shown in Figs 11.3 and 11.4. The top panel shows the full opacity spectrum, whereas the lower two panels are enlargements with successively finer resolution, from 10 000 to a million points on the u-mesh.

The interactive computations in OPserver depend on the parameters

$$\left[\kappa_R, \frac{\partial \kappa_R}{\partial \chi_j}, \gamma, \frac{\partial \gamma}{\partial \chi_j}\right](T, N_e, \chi_j) \tag{11.79}$$

where χ_j is known as the *abundance multiplier* for element j. The output consists of

$$[\log \kappa_R, \log \gamma, \log g_{rad}]\,(\chi_j), \tag{11.80}$$

at each point r in the user-specified stellar depth profile $(T, \rho, r/R_*)$. The variable χ_j enables stellar models to experiment with element abundances, which may depend on stellar opacity due to radiative accelarations.

12 Gaseous nebulae and H II regions

Ionized hydrogen or H II regions and gaseous nebulae are generally low-density objects that appear as extended and diffuse clouds. Typical electron temperatures are of the order of 10^4 K, or $\sim$1 eV, and densities are between 10^2 and 10^6 cm^{-3}. But ionizing sources of H II regions in general are quite diverse. Among the most common variety are those found in giant molecular clouds photoionized by newly formed hot stars with sufficient UV flux to ionize hydrogen and several other elements to low ionization states. Similar H II regions are commonplace in astronomy, as part of otherwise unrelated objects, such as active galactic nuclei (Chapter 13) and supernova remnants. Such regions are also easily observable, since they are largely optically thin. Furthermore, a number of nebular ions are commonly observed from a variety of gaseous objects. In fact, in Chapter 8 we had developed the spectral diagnostics of optical emission lines, as observed from the Crab nebula in Fig. 8.3. That nebula is the remnant of a supernova explosion, in the constellation of Taurus, witnessed in AD 1054 by Arab and Chinese astronomers. The central object is a fast spinning neutron star – pulsar – energizing the surrounding nebula. Nebular spectroscopy therefore forms the basis of most spectral analysis in astrophysics.

We describe the essentials of nebular astrophysics with emphasis on spectroscopic analysis, and address the pervasive problem of atomic data sources of varying accuracy. For more advanced studies, a knowledge of specialized photoionization and radiative transfer models is necessary (e.g., [256, 311]). Moreover, observational aspects of line measurements and abundance determination in nebulae require special attention, particularly with respect to their interpretation (e.g., [312, 313].

12.1 Diffuse and planetary nebulae

Two kinds of nebula are ionized by stars. The first kind are the *diffuse* nebulae created in star forming regions with young O and B stars. The most prominent example is the great nebula (NGC 1976 or M42)[1] in the constellation Orion, shown in Fig. 12.1. It is the brightest H II region in the sky, at a relatively close distance of 389 pc, or 1279 light years. Its central region consists of four very young stars in a trapezium formation about 300 000 years old. The dominant ionizing flux in Orion is from a single O star in the Trapezium, $^1\theta$ Orionis, with surface temperature somewhat less than 40 000 K, and which provides over 90% of the UV ionizing flux.

The other kind of stellar nebulae are called the *planetary nebulae* (abbreviated as PNe, and no relation to planets).[2] The PNe are the ejected shells of circumstellar material from old low-mass AGB stars (discussed in Chapter 10) that are in late stages of evolution. Thus the PNe are the transitional phase from such AGB stars to white dwarfs (see the HR diagram, Fig. 10.2), undergoing extended radiative cooling. The central stars of PNe are hot and bright stellar cores with surface temperatures much higher than main sequence O stars, from 50 000 to over 10^5 K. Their cores usually contain mostly carbon, produced in the helium burning phase. The ejected envelope of ionized material is driven by radiation pressure across the cavity between the radiatively cooling central star and the expanding nebular gas.

12.2 Physical model and atomic species

The most well-known, and one of the most well-studied astrophysical objects, is the Orion nebula – the prototypical H II region and diffuse nebula shown in Fig. 12.1. The Orion nebula in its entirety is a rather complex

[1] The identifications refer to the two most common catalogues of astronomical objects, the *New General Catolog (NGC)* and *Messier (M) Catalog*.

[2] The name 'planetary' associated with the PNe historically arose from this apparent disc-like configuration, viewed at low resolution, surrounding the central star.

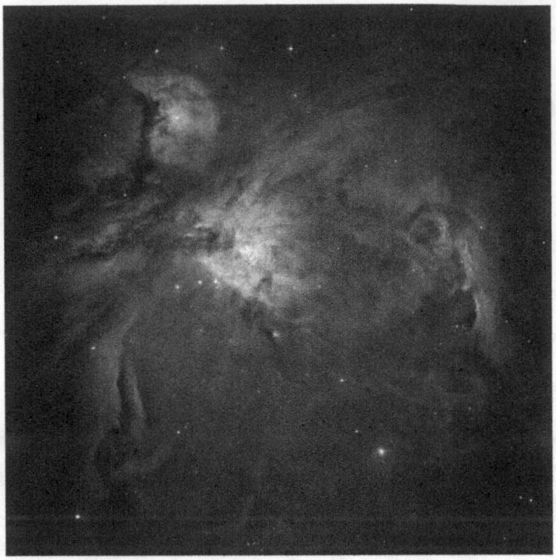

FIGURE 12.1 HST image of the Orion nebula, a diffuse ionized H II region created by photoionization of a giant molecular cloud by hot, young stars.

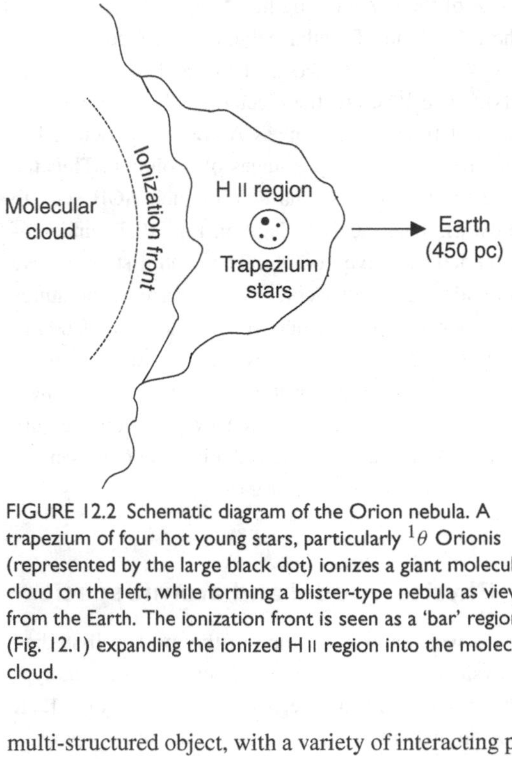

FIGURE 12.2 Schematic diagram of the Orion nebula. A trapezium of four hot young stars, particularly $^1\theta$ Orionis (represented by the large black dot) ionizes a giant molecular cloud on the left, while forming a blister-type nebula as viewed from the Earth. The ionization front is seen as a 'bar' region (Fig. 12.1) expanding the ionized H II region into the molecular cloud.

multi-structured object, with a variety of interacting physical processes. We confine our discussion to atomic and plasma physics with reference to the simplified sketch in Fig. 12.2. The dynamical aspect of the nebula constitutes an ionization front driven by radiation from the Trapezium stars (mainly $^1\theta$ Orionis), into the giant Orion molecular cloud. The Orion nebula itself appears as a *blister* (burnt by the hot stars) on the surface of the molecular cloud, which is far bigger in size than the nebula. The ionization front is observed as a bright bar-shaped region at the interface between the nebula and the molecular cloud. As the densities build up close to the ionization front, towards the cold side of the nebula, the optical depth rapidly increases. There is a *photodissociation region* (PDR), between molecular H and ionized H, where molecular H_2 first dissociates into atomic H and then ionizes, forming the H II region. The ionized region is further sub-divided into a partially ionized zone (PIZ) and a fully ionized zone (FIZ). Atomic species are found in either zone, depending on their ionization potentials relative to that of H I (see Table 12.1, also discussed later).[3]

TABLE 12.1 Ionization potentials of nebular atomic species (eV).

Element	I	II	III	IV
H	13.6	–	–	–
He	24.59	54.42	–	–
C	11.26	24.38	47.89	64.49
N	14.53	29.60	47.45	77.47
O	13.62	35.12	54.93	77.41
Si	8.15	16.35	33.49	45.14
S	10.36	23.33	34.83	47.30
Cr	6.77	16.50	30.96	49.10
Fe	7.90	16.16	30.65	54.80
Zn	9.39	17.96	39.72	59.40

Initially, and for convenience, one may invoke an idealized model known as the *Strömgren sphere*. Its radius depends on the spectral type of the ionizing star, and hence the intensity and frequency distribution of the stellar radiation field. The radius of the idealized Strömgren sphere may be derived from simple analytic arguments regarding photoionization of H I [228, 311]. The Strömgren model divides the nebula into a fully ionized H II region and a neutral H I zone, separated by a very thin boundary consisting of the ionization front. In reality, however, diffuse nebulae are not spherically symmetric, and contain elements other than H, most notably He. Nevertheless, the Strömgren radius vs. the spectral type of the ionizing star provide a useful conceptual picture. This is particularly

[3] The ionization energy of the ground state is referred to as the first ionization potential, sometimes abbreviated as FIP. The so-called 'FIP effect' refers to relative ionization energies of different atomic species. Lighter atoms generally have a higher FIP than heavier ones. The FIP differential between the second- and third-row elements has important consequences for the distributions of ionization stages and elemental abundances in different parts of ionized regions.

relevant to atomic species, such as Fe II, that are spectrally quite prominent but are largely confined to a relatively thin region near the FIZ / PIZ boundary, as discussed later.

Generally, in addition to stellar nebulae, H II regions are also produced in other environments dominated by strong ionizing sources. For example, supernova remnants expand and evolve into gaseous nebulae as they cool down to nebular temperatures and densities. Nebular diagnostics developed herein are basically applicable to those environments, albeit the kinematics and elemental abundances may be quite different. At typical temperatures and densities, H II regions are virtually transparent to emergent optical radiation, i.e., optically thin. Cool and tenous as they are, they are still copious sources of well-known spectral lines, forbidden and allowed, from singly and doubly ionized species of many astrophysically abundant elements. But nebular plasmas exist in highly energetic environments as well, such as the luminous blue variable Eta Carinae that we discussed in Chapter 10.

Nebulae are excellent laboratories for the observation and development of spectral diagnostics. Stellar radiation fields from hot stars, basically described by a black-body Planck function (viz. Chapter 10) contain sufficient UV flux up to about 54 eV, the ionization potential of He II. Some of the prominent nebular ionic species are He II, C II, C III, O I, O II, O III, S II, S III, Fe II, Fe III and others, with strong emission lines in the infrared, Optical and ultraviolet. In Chapter 8 we had discussed emission line diagnostics of the forbidden lines of [O II], [S II] and [O III], shown in Fig. 8.3. The dominant physical processes in H II regions are photoionization and (to a much lesser degree) collisional ionization, electron–ion excitation and recombination, and radiative excitations and decays. We first discuss ionization and recombination processes that determine fractional populations in different ionization states of a given element, and then collisional and radiative processes that determine the emissivities or intensities of emission lines, depending on level populations of a given ion. Such a decoupling of the ionization state from the excitation of an ion, is generally valid since the ionization and recombination rates are much slower than excitation and decay rates. Therefore, nebular modelling codes usually compute ionization fractions and emissivities independently, without a full radiative transfer solution in NLTE (see [254]). The next section outlines the ionization structure of abundant elements. The ionization models often refer to the well-known prototype, the Orion nebula. Having already described spectral formation from light atomic species prominent in nebulae, such as O II, O II, S II, etc. (Ch. 8), we focus on the rather complex example of Fe ions in the following discussion. Iron is prevalent in low ionization stages in nebulae. A description of emission mechanisms of iron ions requires much more extensive atomic models than those of lighter elements. The concluding section also refers to a comprehensive compilation of atomic parameters needed to construct collisional–radiative models for the nebular emission lines given in Appendix E.

12.3 Ionization structure

Gaseous nebulae are usually photoionized by stellar radiation fields. Although the underlying stellar radiation is basically black body, the radiation field J_ν generally has considerable energy variation superimposed on the Planck function B_ν, owing to *line blanketing*. For example, the emergent flux of the Sun in Fig. 10.1 shows a *deficit* in the ultraviolet, owing to absorption in a multitude of lines of abundant elements. Such a 'blanket' of UV absorption lines in the solar spectrum is evident in the simulated spectrum in Fig. 10.8 (top panel). The ionizing stellar continuum is therefore significantly attenuated by ionization and excitation edges, and the features of dominant elements in low ionization stages. Iron ions are prime contributors to the UV opacity that causes a reduction in the stellar flux. Photoionization modelling codes attempt to employ a realistic J_ν (in general $J_\nu \neq B_\nu$) particular to the type of star(s) ionizing the nebula, in order to construct resultant ionization structures. The two quantities that characterize the incident photon flux are the frequency distribution and the intensity of the source. These depend on the temperature of the ionizing star, the geometrical $1/r^2$ dilution, and attenuation by atomic species in the intervening nebular material. The ionizing field of a hot main-sequence star at 30 000 K in a diffuse nebula produces up to three-times ionized stages (I–IV) of abundant elements (see Table 12.1). On the other hand, the central stars in planetary nebulae are at higher temperatures around 100 000 K, and are capable of producing up to IV–VI ionization stages of elements.

Given a radiation source, the next step is to construct the ionization structure, as a function of the distance r from the source, in terms of the run of temperature and density (pressure). This involves modelling the nebula using pressure equilibrium conditions that would yield these parameters at each r, and hence the ionization fractions of an element. Since H I is the dominant atomic species, we write first the ionization balance between photoionization of H and (e + ion) recombination as described in Chapter 7,

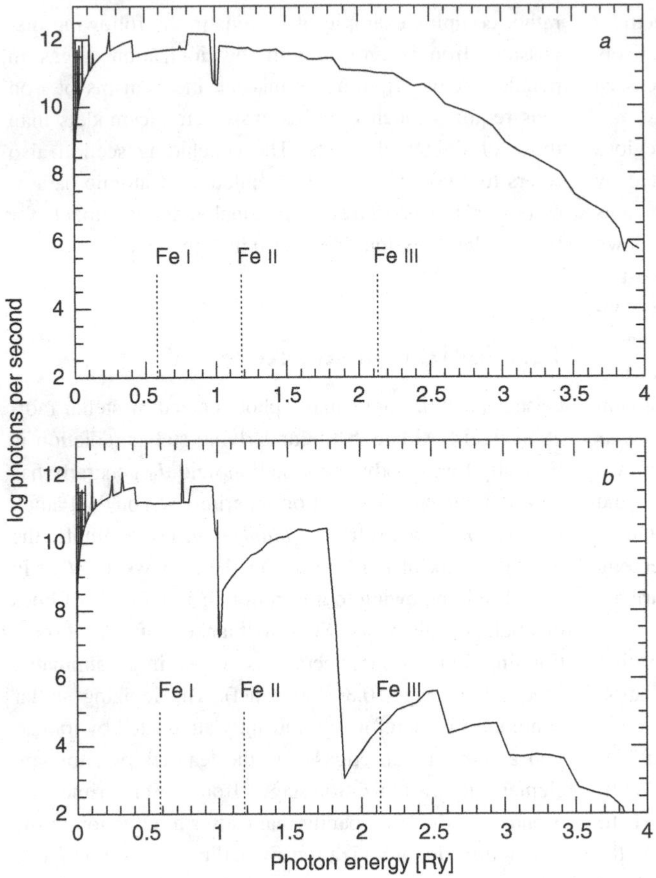

FIGURE 12.3 The ionizing flux vs. photon energy (in Rydbergs): (a) close to the illuminated face and (b) after attenuation by material roughly halfway into the cloud. Ionization energies of Fe I–Fe III are indicated, relative to the sharp drops in flux corresponding to the H I, He I ionization edges at 1 Ry and 1.9 Ry, respectively. One can therefore ascertain the spatial zones where different ionic species are prevalent in the nebula.

$$\int_0^\infty \frac{4\pi J_\nu}{h\nu} N(\mathrm{H}^0)\sigma_\nu(\mathrm{H}^0) = n_\mathrm{e} n_\mathrm{p} \alpha(\mathrm{H}^0, T_\mathrm{e}), \qquad (12.1)$$

where J_ν is the monchromatic flux from the radiation field appropriate to the source. For another species X we replace H^0 with X and atomic parameters correspondingly. Figure 12.3 shows a sample of flux from a hot star as a function of energy ionizing a cloud; thresholds for ionization of Fe I–Fe III are also marked. Figure 12.3(a) is the flux near the illuminated face of the cloud, and Fig. 12.3(b) is the flux deeper into the cloud, with the edges corresponding to ionization of Fe ions. Figure 12.3(a) is representative of the black-body spectrum from the star before absorption or ionization of the cloud. The large drops in flux at 1 Ry and 1.9 Ry in Fig. 12.3(b) correspond to the ionization thresholds of H and He, respectively.

Before proceeding further, let us examine the ionization potentials of some common nebular elements given in Table 12.1. The ionization energies and the ambient temperature in the medium determine which ionization states are likely to exist. Therefrom one can infer the spatial coincidence of different ionic species. For example, Fe II has an ionization energy of 16.16 eV, only somewhat higher than H I. Therefore, we expect Fe II–H I to be spatially co-existent in the nebula, or in the PIZ together with significant amounts of neutral hydrogen. On the other hand, and at energetically higher ionization potentials (Table 12.1), the Fe III–He I zones coincide in the FIZ. Fe III ionizes further into Fe IV at 30.65 eV. But the high-energy stellar flux diminishes rapidly to avoid significant ionization of Fe beyond Fe IV, whose ionization energy is over 54 eV, close to that of He II (Table 12.1). In fact, using Wien's law (Chapter 1), the wavelength corresponding to He I ionization edge is 504 Å, which would correspond to the peak wavelength of a black-body distribution of 57 500 K – much hotter than even the O stars (see the HR diagram, Fig. 10.2).

Photoionization rates are calculated by integrating the cross sections over the ionizing flux at each point in the nebula. With reference to the atomic physics of photoionization and recombination (Chapters 6 and 7), it is important to include autoionizing resonances that can significantly alter (generally enhance) the cross sections in certain frequency (energy) ranges. Electron–ion

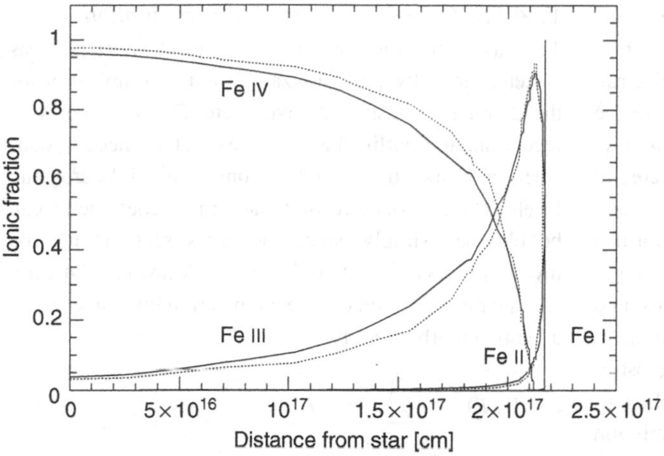

FIGURE 12.4 Ionization structure of Fe ions for conditions in the Orion nebula [98]. The solid lines are results with new photoionization cross sections and unified electron–ion recombination rate coefficients computed using the R-matrix method (Chapter 3); the dotted lines use earlier data.

recombination can also be treated self-consistently using the same (e + ion) wavefunctions as photoionization, to yield total unified (e + ion) recombination rate coefficients, including non-resonant radiative recombination, and resonant dielectronic recombination (Chapter 7). Figure 12.4 shows the ionization fractions of Fe ions obtained using the R-matrix photoionization and unified (e + ion) recombination cross sections (solid lines), and compared with earlier models. The models shown assume constant gas-pressure clouds as function of distance (cm) from the illuminated face up to the PIZ. The distance shown in Fig. 12.4 is less than 0.1 pc (1 pc = 3×10^{18} cm). For comparison, the idealized Strömgren sphere for an O7 star is about 70 pc [228, 311].

Several features of ionic distribution in the Orion nebula are apparent from Fig. 12.4. The dominant ionization stage in the fully ionized zone is Fe IV. As mentioned, this is related to the fact that the ionization potential of Fe IV is 54.8 eV, slightly higher than that of He II at 54.4 eV; helium absorbs much of the high energy flux. There is relatively little ionizing flux beyond these energies even from very hot stars, such as θ^1 Ori C at ~40 000 K, to further ionize Fe IV to higher ionization stages.

Photoionization models of H II regions require photoionization cross sections and (e + ion) recombination rate coefficients as in (Eq. 7.60),

$$\sum_i \int_{\nu_0}^{\infty} \frac{4\pi J_\nu}{h\nu} n(X^z)\sigma_{\rm PI}(X_i)(\nu, X^z)d\nu = \sum_j n_{\rm e} n(X^{z+1})\alpha_{\rm R}\left(X_j^z; T\right). \tag{12.2}$$

As we noted in Chapters 6 and 7 on photoionization and (e + ion) recombination, the summation *inconsistency* in the photoionization balance equation Eq. 12.2, can largely be redressed by including photoionization on the left-hand side not only from the ground state but also excited levels with significant populations, such as the low-lying metastable levels. The right-hand side involves recombination into *all* levels of the recombining ion. Although for very high-n levels hydrogenic approximation may be used to achieve convergence to $n \to \infty$, it is nonetheless necessary to obtain level-specific recombination rate coefficients for many excited levels that are quite non-hydrogenic for complex ions (Chapter 7). The modelling of ionization structure depends on several other assumptions, apart from considerations of atomic physics. These relate to radial density dependence (constant, exponential or power law), temperature profile, or thermal or radiative pressure (constant or varying). Thus, the ionic ratios, say Fe I/Fe II/Fe III, can differ by up to several factors (c.f. [314, 315]).

12.4 Spectral diagnostics

We refer back to the discussion in Chapter 8 on optical emission lines, as shown in Fig. 8.3 from a typical nebular source, the Crab nebula. Emission line diagnostics of nebular ions yield not only temperatures and densities, but also kinematical information and elemental abundances. Recapitulating the discussion in Chapter 8, the basic physical mechanisms for emission lines in optically thin environments may be divided into two main categories: (i) (e + ion) recombination and cascades, and (ii) collisional excitation and radiative decay. For instance, the first category is responsible for the formation of H I, He I and He II recombination lines seen in emission spectra. Collisional excitation is not significant, since the excited energy levels lie too high to be excited by ~1 eV electrons at ambient temperatures of ~10^4 K. The $n = 2$ levels lie at about 10 eV for H I, and about 24 eV for He I. On the other hand, collisional excitation is primarily responsible for the forbidden spectral lines of low-ionization species, such as [O II], [O III] and [S II] (see Fig. 8.3). Since the electron temperature is low, only the low-lying levels *within* the ground configuration are collisionally excited. These levels decay via forbidden transitions. In addition, there are recombination lines of ions, such as (e + O III) → O II and (e + Ne III) → Ne II. In these instances, they, are formed by recombination cascades from doubly ionized to singly ionized species, i.e., recombinations into highly excited levels and then cascading via strong dipole allowed radiative decays into lower levels. Nebular plasma diagnostics involve either allowed recombination lines, or collisionally excited forbidden lines, both observed in emission spectra.[4]

12.4.1 Hydrogen recombination lines

The most common examples of recombination lines seen in emission spectra are the spectral series of hydrogen: Lyman, Balmer, etc. Electron–proton recombination in ionized plasmas entails

$$e + p \rightarrow h\nu + H^*(n\ell), \tag{12.3}$$

followed by radiative decays or cascades (Fig. 2.1),

$$n = 2, 3, 4, ... \rightarrow 1 \quad Ly\alpha\ , Ly\beta\ , Ly\gamma, ... \quad : Lyman\ series, \tag{12.4}$$

$$n = 3, 4, 5, ... \rightarrow 2 \quad H\alpha\ , H\beta\ , H\gamma, ... \quad : Balmer\ series. \tag{12.5}$$

Each series of recombination lines converges onto a *recombination edge*, corresponding to the onset of a continuum at $n \rightarrow \infty$. The Lyman series goes from Lyα at 1215 Å up to the Lyman recombination (or ionization) edge at 912 Å, and the beginning of the *Lyman continuum* $\lambda < 912$ Å. The *optical recombination lines* are in the Balmer series and span the wavelength range from Hα at 6563 Å to the Balmer recombination edge at 3646 Å, leading into the *Balmer continuum* $\lambda < 3646$ Å. The electron temperature in plasmas is too low for *any* of the H-lines to be excited by electron impact, since H would be largely ionized before excitation, and therefore recombination dominates line formation.

12.4.1.1 Case A and Case B recombination

The basic ionization structure of astrophysical plasmas is determined by photoionization and recombination of the dominant constituent, hydrogen. The schematics of recombination outlined above show that we need to consider level-specific recombinations into all hydrogenic levels. The hydrogen recombination rate coefficients can be obtained simply, since the cross sections for the inverse process of photoionization are known analytically (Chapter 6). The sum over recombination into all levels of a hydrogen orbital $n\ell$ is

$$\alpha_A(H, T) = \sum_{n\ell;n=1}^{\infty} \alpha_{n\ell}(T). \tag{12.6}$$

All H-levels are doublets with multiplicity $2S+1 = 2$. We implicitly group together levels $n\ell(SLJ)$, i.e.,

$$\alpha_{n\ell;SLJ} = \alpha(n, \ell;^2 L_J), \tag{12.7}$$

and $\ell \equiv L$ and $J = L \pm ½$. Recombinations into the excited levels rapidly cascade down to the ground state $1s(^2S_{1/2})$ via dipole allowed transitions, viz. $nf \rightarrow n'd \rightarrow n'p \rightarrow n's$.[5] The total recombination rate coefficent into all levels Eq. 12.6 is designated as α_A, and refers to what is known as *Case A recombination.* The sum in Eq. 12.6 includes the ground state of H I. But recombination directly into the ground level produces a photon with sufficient energy, $E \geq 1$ Ry or $\lambda \leq 912$ Å, to ionize another H atom. Now if the environment is basically optically thin, and the likelihood of encountering another H atom small, then α_A would correspond to the total recombination rate. But on the other hand, assuming the ionized region to be surrounded by a sufficiently dense neutral medium, recombination into the H I ground state would *not* affect the net ionization balance, since the emitted photon will be immediately absorbed elsewhere in the surroundings. In that case, the *net* recombination rate coefficient is obtained simply by omitting recombination into the ground level, or

$$\alpha_B(H, T) = \sum_{n\ell;\ n=2}^{\infty} \alpha_{n\ell}(T), \tag{12.8}$$

referred to as the *Case B recombination* rate coefficient. Hydrogen recombination lines in Case A encounter low

4 There are several widely used codes for nebular diagnostics in photoionization equilibrium (Chapter 7). Among these are: *MAPPING* [256], (*CLOUDY* (http://www.nublado.org), *XSTAR* (http://heasarc.gsfc.nasa.gov/docs/software/xstar/xstar.html), ION [345] and TITAN [254].

5 Note that in the reverse case, the oscillator strengths for *upward* radiative transitions $nl \rightarrow n'(l+1)$ are much higher than for $nl \rightarrow n'(l-1)$, though both are allowed by dipole selection rules for angular mometum l (readers may wish to verify it by examining tables of f, A-values, and line strengths S). Some recombination cascades end up in the $2s^2S_{1/2}$ level, but those are a small fraction of the ones decaying to the ground state (compare the A-values).

optical depths, at low densities when the photon emitted during recombination escapes the plasma. Case B reflects high optical depths in H-lines at high densities when those photons are trapped in the source. The number of H-recombination line photons emitted in Case A is equal to the total recombination rate $n_e n_p \alpha_A(T_e)$, and in Case B it is $n_e n_p \alpha_B(T_e)$, both in $cm^3\ s^{-1}$.

The calculation of individual recombination line intensities due to particular transitions between two levels requires a much more detailed consideration of all pathways in terms of *level-specific recombination rate coefficients*, and the A-transition rates that determine downward cascade coefficients for (to and from) a given level. Recombination line intensities also depend on the intrinsic atomic physics via the ℓ-distribution within each n-complex. The actual $n\ell$ level population may deviate significantly from Boltzmann distribution at densities sufficiently high to 'mix' different ℓ-levels. That would occur at a critical density, particular to each level, where the collisional rate begins to match that due to spontaneous radiative decay; the critical density is obviously different for each level. Whereas the collisional mixing effect is small at nebular densities, $n_e \leq 10^6\ cm^{-3}$, it is discernible in computations of recombination line intensities including collisional rates (e.g., [311]). For simplicity, we omit explicit consideration of collisional ℓ-redistribution in describing the nature of cascade matrices (Chapter 8).

In the following discussion, we again refer to all levels of a given orbital $n\ell(SLJ)$, and implicitly assume that the transitions $n\ell((SLJ) \to n'\ell'(S'L'J')$ follow the appropriate selection rules. Let a given H-level be designated by $n\ \ell$, and a line transition as $n\ell \to n'\ell'(n' < n)$. The probability or the branching ratio for photon emission in the line is then $A(n\ell - n'\ell')/\sum_{n''\ell''} A(n\ell - n''\ell'')$ for $n'' < n$. The ℓ-values range from 0 to n–1, and the transitions are dominated by fast dipole transitions with selection rules $\Delta S = 0$, $\Delta\ell = 0, \pm 1$ (Section 4.13). The level population $N(n\ \ell)$ is only partially determined by direct recombination into the level $n\ \ell$, i.e., by $\alpha_R(n\ell)$ (the subscript R refers to recombination, under Case A or Case B). In fact the population depends considerably on recombinations into all upper levels, and cascades therefrom, into $(n\ell(SLJ)$. Hence, we need to compute *cascade coefficients* involving a large number of transitions in terms of A-values and successive branching ratios.[6] Line emissivities and other parameters for H-recombination lines have been extensively tabulated under both Case A and Case B conditions [217, 256] (He I recombination rate coefficients are discussed in [316]).

There are also physical conditions intermediate between the optically thin and optically thick approximations, viz. Case A and Case B, which respectively correspond to no absorption in Lyman lines or full absorption. The intermediate cases are referred to as Case C, which pertains to sources with a background continuum that also has significant intensity in Lyman lines. Then complete absorption in Lyman lines may not occur, and they may not be entirely optically thick as in Case B. In fact, Lyman lines as part of the continuum may also be instrumental in exciting other ions, such as Fe II via fluorescent excitation, discussed in Chapter 13.

12.4.1.2 *Cascade coefficients and emissivities*

For any given level i in atom X, the population N_i due to recombinations from ion X^+ is

$$N_i \sum_{m<i} A_{im} = n_e n(X^+)\left[\alpha_R(i) + \sum_{j>i} \alpha_R(j) C_{ji}\right], \tag{12.9}$$

where the last term on the right represents cascades from all higher levels j, and α_R is the level-specific recombination rate coefficient. The *cascade* coefficients C_{ji} are

$$C_{ji} = \frac{A_{ji}}{\sum_{i'<j} A_{ji'}} + \sum_{k>i} C_{jk} \frac{A_{ki}}{\sum_{i'<k} A_{ki'}}. \tag{12.10}$$

The first term on the right is the branching ratio for direct decay from $j \to i$, and the second one represents indirect decays via cascade routes $j \to k \to i$. A *cascade matrix* $\mathbf{C}_{ij}$ can thus be constructed with the cascade coefficients, given all the A-values (see Chapter 4.) Computations of level populations N_j also require level-specific rate coefficients $\alpha_R(j)$, whose computations are described in Chapter 7. Isotropic emissivity in a line i→f is written as usual[7]

$$j_{if} = N_i A_{if} \frac{h\nu_{if}}{4\pi}. \tag{12.11}$$

Provided an appropriately complete set of A-values and level-specific $\alpha_R(i, T_e)$ is known, the calculation of cascade matrices is straightforward. One property of cascade coefficients is helpful in this regard. It is known that the $C\left(n'S_i'L_i'J_i' \to nS_iL_iJ_i\right)$ for cascades from upper levels along a given Rydberg series characterized by $n'S_i'L_i'J_i'$, to a particular lower level $nS_iL_iJ_i$ $(n' > n)$,

[6] A tabulation of H-recombination rate coefficients is given in the database NORAD: www.astronomy.ohio-state.edu/~nahar.

[7] The 'standard' notation for recombination line emissivity is often j_{if}, whereas for collisionally excited emission lines it is ϵ_{if}, as in Chapter 8.

quickly converge to a slowly varying behaviour with n'. For example, in recombination cascades from H-like ions to the $n = 2$ levels of He-like ions (Chapter 8) the cascade coefficients converge to a relatively constant value for $n' \geq 5$ (a more graphical discussion is given in [233]).

One can define an *effective recombination coefficient* for a line $\alpha_{\mathrm{if}}^{\mathrm{eff}}$ in terms of the emissivity j_{if}, which must equal

$$j_{\mathrm{if}} = (n_{\mathrm{p}} n_{\mathrm{e}}) \alpha_{\mathrm{if}}^{\mathrm{eff}} \frac{h\nu_{\mathrm{if}}}{4\pi}, \tag{12.12}$$

and therefore

$$\alpha^{\mathrm{eff}}(\mathrm{if}) = \frac{4\pi j_{\mathrm{if}}}{h\nu_{\mathrm{if}}(n_{\mathrm{p}} n_{\mathrm{e}})}. \tag{12.13}$$

We can now relate the luminosity emitted in a line by the nebula to the total luminosity of the ionizing star. In other words, the intensity in a given line – the number of photons times the photon energy emitted per second per unit volume – is due to emissions from within the ionized zone in photoionization equilibrium, between ionizations by the stellar radiation field and recombinations in the nebula. Since the nebula has a finite volume with a thin boundary (recall the Strömgren sphere), and owing to continuous ionization of a cold molecular cloud, it is optically thick to the ionizing radiation. Therefore, Case B recombination is appropriate; all Lyman series and Lyman continuum ($h\nu > 1Ry \equiv \nu_{\mathrm{H}}$) photons are absorbed. Then

$$\int (n_{\mathrm{p}} n_{\mathrm{e}}) \alpha_{\mathrm{B}}(H) \mathrm{d}V = \int_{\nu_{\mathrm{H}}}^{\infty} (L_\nu / h\nu)\, \mathrm{d}\nu. \tag{12.14}$$

The stellar luminosity per frequency is denoted as L_ν, and the integration is over the whole nebular volume. Further denoting the luminosity in a particular line as $L(\nu_{\mathrm{if}})$, it is straightforward to approximately relate this ratio to the ratio of recombination rate coefficients

$$\frac{L(\nu_{\mathrm{if}})/h\nu_{\mathrm{if}}}{\int_{\nu_{\mathrm{H}}}^{\infty} L_\nu / h\nu \mathrm{d}\nu} = \frac{\int n_{\mathrm{p}} n_{\mathrm{e}} \alpha^{\mathrm{eff}}(\nu_{\mathrm{if}}) \mathrm{d}V}{\int n_{\mathrm{p}} n_{\mathrm{e}} \alpha_{\mathrm{B}}(\mathrm{H}) \mathrm{d}V} \approx \frac{\alpha^{\mathrm{eff}}(\nu_{\mathrm{if}})}{\alpha_{\mathrm{B}}(\mathrm{H})}. \tag{12.15}$$

This approximation implies constant density profiles of electrons and protons in the nebula. Extending the discussion from an individual recombination line flux emitted by the nebula to the emitted flux due to recombinations in a given wavelength range, another useful relation may be derived to yield the effective temperature T_* of the ionizing star. Assuming a stellar radiation field to be described by the Planck function, we can write the ratio in a specified or observed energy range (ν_1, ν_2) as

$$\frac{L(\nu_1 - \nu_2)}{B(\nu_1 - \nu_2; T_*)} = \frac{\int_{\nu_{\mathrm{H}}}^{\infty} (L_\nu / h\nu)\, \mathrm{d}\nu}{\int_{\nu_{\mathrm{H}}}^{\infty} (B_\nu(T_*)/h\nu)\, \mathrm{d}\nu}. \tag{12.16}$$

The observed luminosities in, say, the UV band, compared with the total stellar luminsity, then yield the effective temperature of the star, called the *Zanstra temperature*, following the method proposed by Zanstra [255]. For example, the central temperature of PNe may be estimated by comparing the Hα line intensity with the background continuum underlying Hα. This is because the Hα strength is related to the UV ionizing flux of the central star, since photoionization of H I atoms results in recombination cascades, leading to photon emission in the Hα line. Thus the Zanstra method is a measure of the UV flux to the 'red' continuum, corresponding to the black-body temperature of the central star.

12.4.2 Departures from LTE

Level populations in LTE can be obtained analytically by combining the Saha equation for ionization fractions and the Boltzmann equation for level populations. For any two ionization stages of an element, say neutral X and ion X^+, the relative densities[8]

$$\frac{n_{\mathrm{e}} n(\mathrm{X}^+)}{n(\mathrm{X})} = \left(\frac{2\pi mkT}{h^2}\right)^{3/2} \mathrm{e}^{-E_I/kT_{\mathrm{e}}}, \tag{12.17}$$

where E_{I} is the ionization energy of neutral X into the ground state of the ion X_1^+. Now consider the level population of an excited state $N(\mathrm{X}_i)$, given by the Boltzmann equation

$$\frac{N(\mathrm{X}_i)}{N(\mathrm{X}_1)} = \frac{g_i}{g_1} \mathrm{e}^{-E_{1i}/kT}. \tag{12.18}$$

From the two equations above we have the excited level populations of X_i in LTE as

$$N(\mathrm{X}_i) = \frac{g_i}{g_{\mathrm{X}_1^+}} n(X^+) n_{\mathrm{e}} \left(\frac{h^2}{2\pi mkT}\right)^{3/2} \mathrm{e}^{E_i/kT}, \tag{12.19}$$

where E_i is the ionization energy of level i. For hydrogen, the LTE level populations are

$$N_{n\ell} = \frac{g(n\ell;{}^2L_J)}{2} (n_{\mathrm{p}} n_{\mathrm{e}}) \left(\frac{h^2}{2\pi mkT}\right)^{3/2} \mathrm{e}^{E_n/kT}, \tag{12.20}$$

where $E_n = 1/n^2$ Rydbergs, and the statistical weight of the ground state $g({}^2S_{1/2}) = 2$. However, at low densities

[8] Here we remind ourselves of the earlier discussion in Chapter 11 on the Boltzmann and Saha equations. For simplicity, we approximate the atomic partition function U over all levels with the ground state statistical weight. Note also the convention we have generally followed through most of the text of denoting level populations of an ion as N, and ionic densities as n, with appropriate subscripts (with the exception of Chapter 11 on opacity).

in nebulae it is unlikely that populations would remain in LTE for highly excited levels with large decay rates, i.e., given by the Saha–Boltzmann equation (Eqs 12.19 and 12.20). Collisional or recombination rates at nebular densities (or even coronal, $n_e \sim 10^{9-10}$ cm^{-3}) are insufficient to maintain statistically populated excited levels. For instance, the level-specific rate coefficient for an $n = 3$ level of helium $\alpha(3^3P^o) \approx 2 \times 10^{-14}$ at 10^4 K; with $n_e < 10^6$ cm^{-3} and $n_p \approx n_e$ the recombination rate is about $\sim 10^{-2}$ cm^{-3} s^{-1}. On the other hand, the A-values for dipole transitions are $>10^8$ s^{-1}, and depopulation by the large radiative decay rate is orders of magnitude greater than population via recombination. In contrast, stellar interiors are generally in LTE (though not stellar atmospheres [244]), owing to high densities $n_e > 10^{15}$ cm^{-3}.

Progressive deviation from LTE is taken into account by introducing *departure coefficients* b_i for each level, multiplying the LTE level population given by the Saha–Boltzmann equation (by definition $b_i = 1$ in LTE). For hydrogen, we have

$$N_{n\ell} = b(n\ell;{}^2L_J)\frac{g(n\ell;{}^2L_J)}{2}(n_p n_e)\left(\frac{h^2}{2\pi mkT}\right)^{3/2} e^{E_n/kT}. \qquad (12.21)$$

Assuming (e + p) recombination and radiative cascades to be the only processes forming H-recombination lines, the statistical equilibrium equations may be written as (we omit the actual fine structure level designations SLJ as before, and refer only to n and ℓ, although the equations are strictly valid for each level),

$$n_e n_p \alpha_R(n\ell) + \sum_{n'\ell'} N_{n'\ell';n'>n} A(n'\ell' \to n\ell) = N_{n\ell} \sum_{n''\ell''} A(n\ell \to n''\ell''). \qquad (12.22)$$

Exercise 12.1 *Derive the expression for hydrogenic departure coefficients from the two equations above, also including collisional excitation in addition to recombination.*

12.4.3 Collisional excitation and photoionization rates

Before proceeding to a detailed consideration of nebular emission lines for complex atomic species, it is instructive to compare the magnitudes of the collisional rate with the photoionization rate. The first point is to ascertain the relative efficacy of the electron impact excitation process that is responsible for emission lines of a given ion, and photoionization of that ion in the nebula ionized by a stellar radiation field. Let us consider the excitation rate of the well-known green line of [O III], 5007 Å, due to the transition $2p^2(^3P_2 - ^1D_2)$. The rate coefficient q is calculated using the Maxwellian averaged collision strength (also called the effective collision strength) Υ as follows:

$$q(\lambda, T) = \frac{8.63 \times 10^6}{g_i \times \sqrt{T}} e^{(-\Delta E/kT)}\ \Upsilon(T), \qquad (12.23)$$

where ΔE (Ry) $\approx 912/\lambda$ (Å), and

$$\Delta E[2p^2(^3P_2 - ^1D_2)] = 912/5007 = 0.18 \text{ Ry}$$

$$\Upsilon(10^4 K) = 2.29$$

$$\exp(-\Delta E/kT) = \exp(-0.18 \times 157885/10000) = 0.058$$

$$q(5007; 10^4 K) = \frac{8.32 \times 10^{-6}}{5 \times 10^2} \times 0.058 \times 2.29 = 2.29 \times 10^{-9}\ \text{cm}^3\text{s}^{-1}. \qquad (12.24)$$

Note that it is entirely fortuitous that $\Upsilon = 2.29$ and $q = 2.29 \times 10^{-9}$. They are quite different; Υ is a dimensionless quantity, and q is the rate coefficient related to collisional excitation rate (cm^3 s^{-1}) $= q$(cm^3 s^{-1}) $\times n_e$(cm^3) $\times n_{ion}$(cm^{-3}). For the [O III] λ 5007 line, the collisional rate at $T_e = 10^4$ K and $n_e = 10^4$ cm^{-3} is $2.29 \times 10^{-9} \times 10^4 \times n(O\ \text{III}) = 2.29 \times 10^{-5} n(O\ \text{III})$ cm^{-3} s^{-1}.

On the other hand, the photoionization rate can be approximated by estimating the radiation field of the ionizing star in numbers of photons per second, which for an O8 star with effective temperature of ~40 000 K is (as in [228])

$$n_{h\nu} = \int_{\nu_0}^{\infty} \frac{L_\nu}{h\nu} d\nu \sim 10^{50}\ \text{photon s}^{-1}. \qquad (12.25)$$

The radiation field J_ν dilutes geometrically as

$$4\pi J_\nu = \frac{L_\nu}{4\pi r^2} (\text{erg (cm}^{-2}\ \text{s}^{-1}\ \text{Hz}^{-1})). \qquad (12.26)$$

At a distance of 5 pc (1 pc = 3 $\times 10^{18}$ cm) from the ionizing star, the photoionization rate is

$$n_{O\ III} \int_{\nu_o}^{\infty} \frac{4\pi J_\nu}{h\nu} \sigma(\text{O III}) \approx n_{O\ III} 10^{-8}\ \text{s}^{-1}. \qquad (12.27)$$

Dividing the collisional and the photoionization rates cancels out the O III density n(O III), and with the photoionization cross section value at the ionization threshold $\sigma_o(\text{O III}) \approx 5 \times 10^{-18}$ cm^2 [152], we have

$$\frac{\text{Collisional rate [O III(5007, } 10^4\text{ K)]}}{\text{Photoionization rate (O III)}} \approx \frac{10^{-5}}{10^{-8}} = 10^3, \qquad (12.28)$$

i.e., collisional excitation of [O III] dominates photoionization of O III a thousand-fold. We obtain essentially the same result even after refining some of the rates, which alter their values at most be a factor of a few. Relative to H density,

$$n_{\mathrm{O\,III}} = \frac{n_{\mathrm{O\,III}}}{n_{\mathrm{O}}} \times \frac{n_{\mathrm{O}}}{n_{\mathrm{H}}} n_{\mathrm{H}}. \tag{12.29}$$

If n_{H} is ten atoms per cm^3, and the O/H abundance ratio is given by $n_{\mathrm{O}} \approx 10^{-4} n_{\mathrm{H}} = 10^{-3}\ \mathrm{cm}^{-3}$, then we obtain the O III/O II ionization fraction as $n_{\mathrm{O\,III}} \approx 10^{-1} n_{\mathrm{O}} = 10^{-4}\ \mathrm{cm}^{-3}$. In Eq. 12.28 we had taken a fixed value of the threshold photoionization cross section σ_0(O III) = 5 Mb. But, of course, a proper calculation would entail the detailed frequency-dependent cross section over the entire energy range of practical interest, including autoionizing resonances such as in the R-matrix close coupling calculations in [152].

To summarize the comparison: *collisional excitation of forbidden lines is faster than photoionization by orders of magnitude.* In equilibrium, therefore, this justifies decoupling photoionization or recombination with much slower rates than the collisional–radiative line formation, which is much faster. In optically thin nebular or coronal plasmas we may *independently* solve the ionization balance problem for the ionization structure, and the collisional–radiative problem for line intensities.

12.4.4 Iron emission spectra

In Chapter 8 on spectral formation we described the basics of emission line diagnostics for light ions O II, S II, O III, etc., with relatively simple atomic physics. Important as these ions are, real observed spectra of nebulae contain lines from a number of other ions that are much more complex and require large collisional–radiative (hereafter CR) models for optically thin plasmas. In contrast, we need non-LTE radiative transfer models for higher-density systems, such as stellar atmospheres, broad line regions of active galactic nuclei, expanding (radiatively driven) ejecta of supernove, etc. Although the optically thin approximation is usually sufficient for gaseous nebulae in general, it is known that specialized radiative transfer effects, particularly continuum and line fluorescence, can be of great importance in excitation of some prominent lines. The next few sections will be devoted to discussing CR diagnostics, and line ratio analysis, of Fe ions.

12.4.4.1 [FeII] *lines*

Singly ionized iron is one of the most prevalent atomic species observed from a wide variety of astrophysical sources: the interstellar medium, stars, active galactic nuclei and quasars, supernova remnants, etc. A number of lines arise from both allowed and forbidden transitions ranging from the far-infrared to the far-ultraviolet. The enormous complexity of Fe II makes it necessary to understand the underlying atomic physics in detail. The rich spectral formation in astrophysical plasmas from Fe II is due to the many levels that give rise to several complexes of lines, which are so numerous as to often form a pseudo-continuum (see Chapter 13). In gaseous nebulae, Fe II exists in the PIZ, as opposed to ions of lighter elements, such as O II, O III and S II, which exist in the FIZ. This is because the respective ionization potentials are (Table 12.1): E_{IP} (Fe II) = 16.16 eV, E_{IP} (O II) = 35.117 eV, E_{IP} (O III) = 54.934 eV and E_{IP} (S II) = 23.33 eV. As the ionization energy of Fe II is somewhat above H I (13.6 eV), it is partially 'shielded' by neutral hydrogen in the PDR. As such, the Fe II spectrum is affected by the changing temperatures and densities to a more significant extent than other ions.

In this section, we describe only those lines due to forbidden [Fe II] transitions among relatively low-lying levels in the infrared and optical regions that may be used for nebular temperature, density and abundance diagnostics. However, the allowed lines and the overall Fe II emission is also of great interest in active galactic nuclei (Chapter 13), particularly in a sub-class of quasi-stellar objects known as 'strong Fe II emitters'. In those sources, Fe II emission is abnormally intense, with many more highly excited levels than in nebulae. Consequently, more powerful radiative transfer non-LTE methods need to be employed (Chapters 9 and 13).

A partial Grotrian diagram of Fe II levels and lines is shown in Fig. 12.5. The ground configuration LS term and fine structure levels of Fe II are: $3d^64s(a^6D_{9/2,7/2,5/2,3/2,1/2})$. The next higher term and levels belong to the first excited configuration: $3d^7(a^4F_{9/2,7/2,5/2,3/2})$, followed by two others: $3d^64s(a^4D_{7/2,5/2,3/2,1/2})$ and $3d^7(a^4P_{5/2,3/2,1/2})$. As given, these 16 low-lying level energies are in ascending order, with the ground level $a^6D_{9/2}$ at zero energy (recall that the prefix 'a' denotes the lowest term of even parity). This group of 16 levels is separated by approximately 0.9 eV from the next higher term of even parity and quartet multiplicity $3d^64s(b^4P)$. In this energy gap there are four $3d^7$ *doublet* multiplicity terms (a^2G, a^2P, a^2H, a^2D), and one quartet term, a^4H. However, these doublet terms and the relatively high angular momentum term ^4H (L = 5) are weakly coupled to the quartet/sextet system considered above. This provides some justification for neglecting the doublets initially to restrict the CR model, and enables us to proceed with a small 16-level system that gives rise to several strong NIR and FIR lines of Fe II,

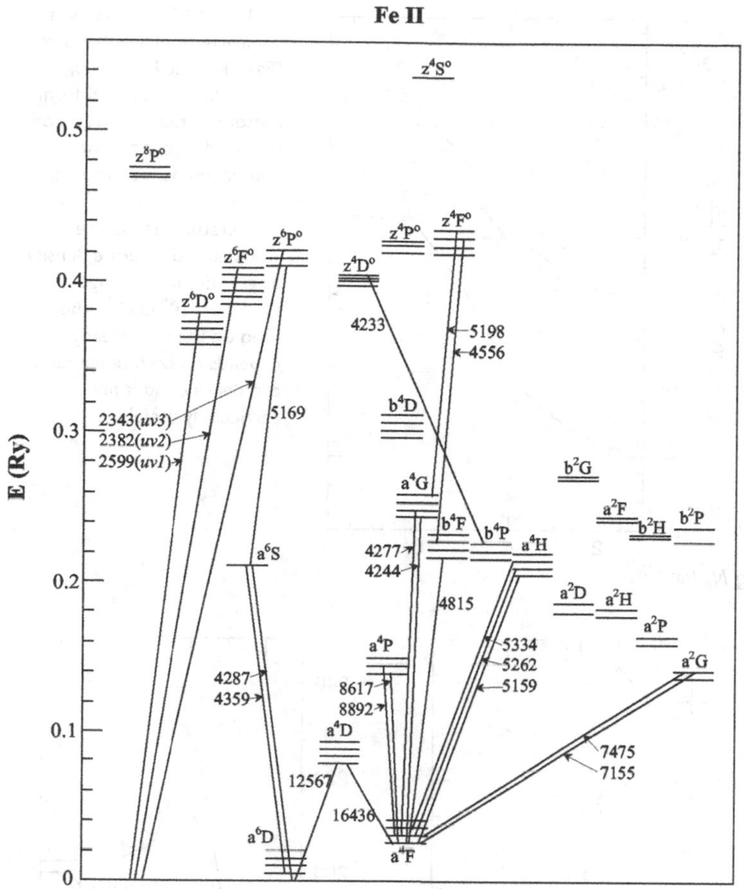

FIGURE 12.5 Fe II lines in the infrared, optical and ultraviolet. The ground term 3d^{6}4s(^{6}D) has five fine structure components, $J = 9/2, 7/2, 5/2, 3/2, 1/2$, with ground level energy $E(^6D_{9/2}) = 0$.

such as the 1.533 μm and 1.644 μm NIR lines, and the 17.93 μm and 25.98 μm FIR lines (cf. [317].)

For nebular conditions, we note a simple fact: the intensity ratio of two lines *orignating with the same upper level* depends only on the ratio of the A-values, since the population of the upper level in the expression for emissivity (cf. Eq. 8.22) cancels out. Hence, a full CR calculation is not necessary. The ratios of A-values can be compared directly with observations *regardless of the physical conditions in the source*, which affect the upper level equally for both lines. Then the observed line ratio for a three-level system should be

$$\frac{\epsilon(3\to 1)}{\epsilon(3\to 2)} \approx \frac{A(3\to 1)}{A(3\to 2)} \quad : \text{ same upper level.} \qquad (12.30)$$

Owing to the small energy separations among the first 16 levels, and because the forbidden A-values are extremely small, the infrared and optical forbidden lines are good density indicators, with weak temperature dependences. Figure 12.6 shows a comparison between two different [Fe II] emissivity ratios of near-infrared lines calculated as a function of n_e at representative T_e from 3000 K to 12 000 K. The line ratio $I(1.533\,\mu\text{m})/I(1.644\,\mu\text{m})$ should be a very good density diagnostic in the range 10^{3-5} cm^{-3}, since the T_e dependence is negligible over a wide range of nebular temperatures. On the other hand, the $I(8617\,\text{Å})/I(1.257\,\mu\text{m})$ ratio is *not* amenable to accurate spectral analysis since it varies significantly with both n_e and T_e. In the nebular temperature–density regime, several Fe II line ratios of observed forbidden lines provide useful and sensitive diagnostics of density and temperature, determined only by collisional excitation and radiative decay.

The lowest [Fe II] transitions among the fine structure levels of the ground state ^6D_J lie in the FIR. These are currently of much interest in space astronomy since these lines lie in the spectral range 5–38 μm covered by the high-resolution spectrograph aboard the Spitzer space observatory (www.spitzer.caltech.edu), or possibly, even the more recent Herschel observatory (http://herschel.esac.esa.int) with coverage over 240–650 μm for high-redshift objects.

Figure 12.7 shows three such line ratios. It is apparent that the line ratio at the lowest T_e (3000 K) differs significantly from those at higher temperatures. This is because the Maxwellian averaged $\Upsilon(T)$ varies more sharply as

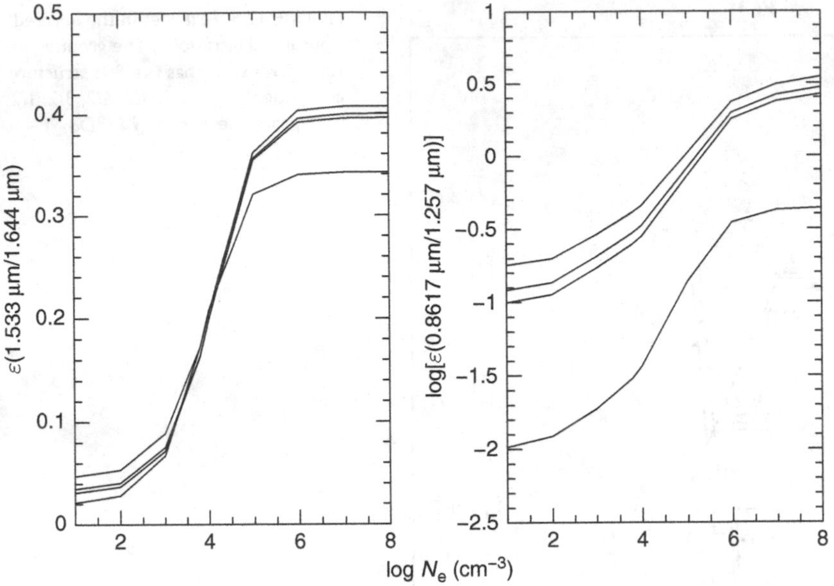

FIGURE 12.6 Line ratios: a comparision of line ratios of [Fe II] lines at T_e = 3000, 6000, 10 000, 12 000 K from bottom to top). The ratio on the left depends almost entirely on density alone and not much on the temperature, and hence would afford accurate density diagnostics in the range $n_e = 10^3$–10^5 cm^{-3}. The ratio on the right clearly depends on both temperature and density, and is not particularly useful.

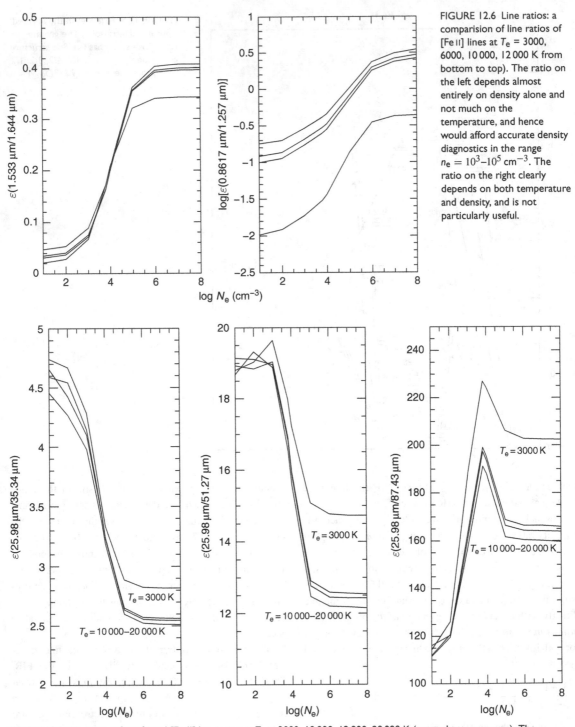

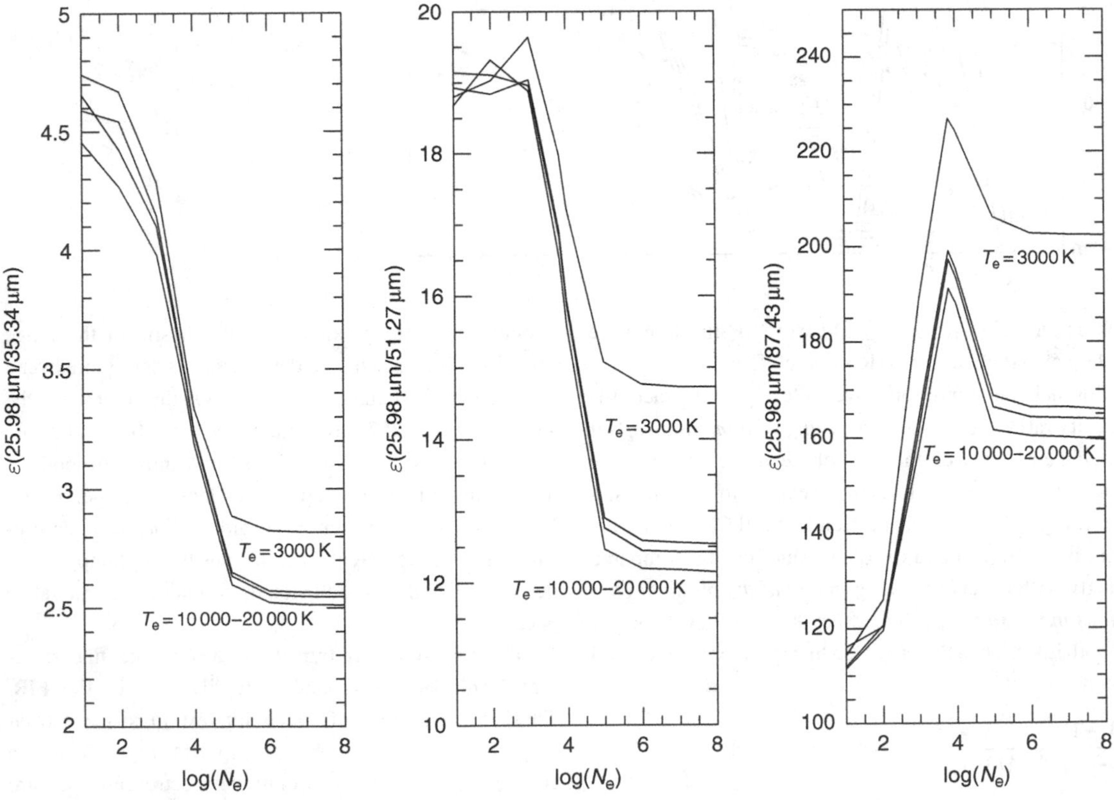

FIGURE 12.7 Line ratios: far-infrared [Fe II] line ratios at T_e = 3000, 10 000, 12 000, 20 000 K (top to bottom curves). The temperature dependence at low temperatures $T_e \leq 3000$ K is highly sensitive to atomic rate coefficients (the sharp edges are due to inadequate density resolution).

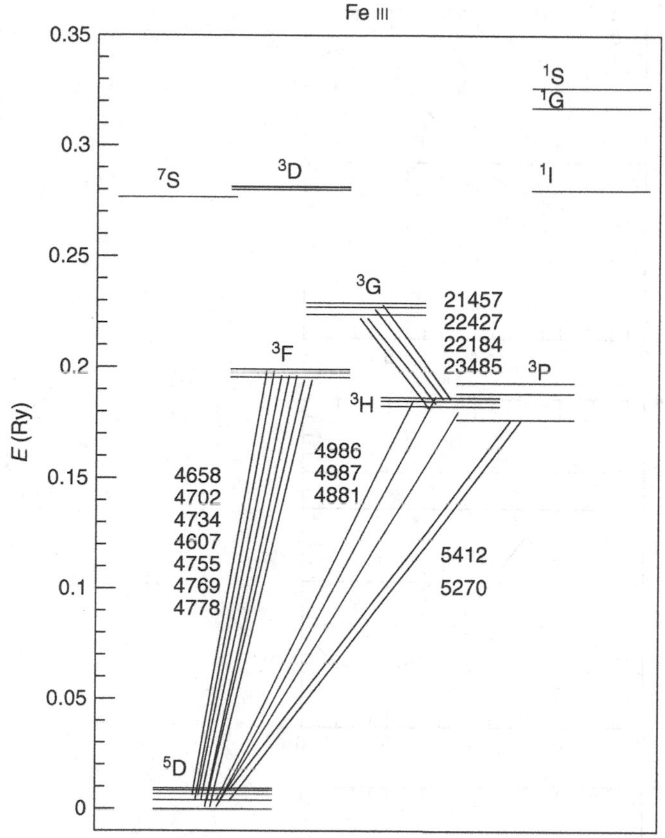

FIGURE 12.8 [Fe III] levels and transitions for infrared and optical lines.

the temperature decreases, since the Maxwellian functions sample a smaller energy range above the excitation thresholds of collision strengths; at $T_e \sim 3000$ K it is a fraction of an eV. The collision strengths for FIR transitions in Fig. 12.7 were computed in *LS* coupling, and transformed algebraically to fine-structure levels, without explicitly including relativistic effects. Also, near-threshold resonances in fine-structure collision strengths are most likely to affect $\Upsilon(T)$ at low temperatures. Therefore, there is considerable uncertainly in the FIR line ratios at $T_e \leq$ 3000K.[9]

12.4.4.2 [Fe III] *lines*

Many forbidden optical and infrared [Fe III] lines are due to transitions among levels of the ground configuration terms $3d^6(^5D, ^3P, ^3H, ^3F, ^3G)$ shown in Fig. 12.8. Since it is twice ionized, the optical [Fe III] lines within low-lying multiplets are of shorter wavelength than the [Fe II] lines. The 2 μm near-infrared [Fe III] lines originate from higher excitation levels ^3G_J, compared with optical lines from $3d^5(^3F_J, ^3H_J, ^3P_J)$. Therefore, the near-infrared nebular lines from Fe III are weaker than the optical lines. The maximum difference among the fine structure levels of the three terms responsible for optical lines is only about 0.02 Rydberg units, or ~ 3000 K, which makes the relative line intensities largely insensitive to temperature. In Fig. 12.9 we plot several optical line ratios as a function of electron density, and use observed line ratios to infer electron densities in Orion. Most observed line ratios yield $n_e \sim 10^{3\text{–}4}\ \mathrm{cm}^{-3}$, consistent with densities in the spatial Fe III zone in the FIZ, and those obtained from the [O II] and [S II] line ratios (Fig. 8.6). These densities are typical of the main body of the ionized nebula, and lower than those expected in the PIZ, which is closer to the ionization front moving into the partially dissociated plasma in the PDR.

12.4.4.3 [Fe IV] *lines*

The Grotrian diagram in Fig. 12.10 shows that the lowest transitions in [Fe IV] are among the ground state $3d^5(^6S_{5/2})$ and fine structure levels of the excited terms

[9] The 16-level CR model given here was computed using the *A*-values and Maxwellian averaged collision strengths for all 120 transitions ($N = 16 : N \times (N_1)/2$) from [317]. A more extensive and up-to-date tabulation of Fe II collision strengths from the Breit–Pauli R-matrix calculations is given in [58]; the new data should be used for future work.

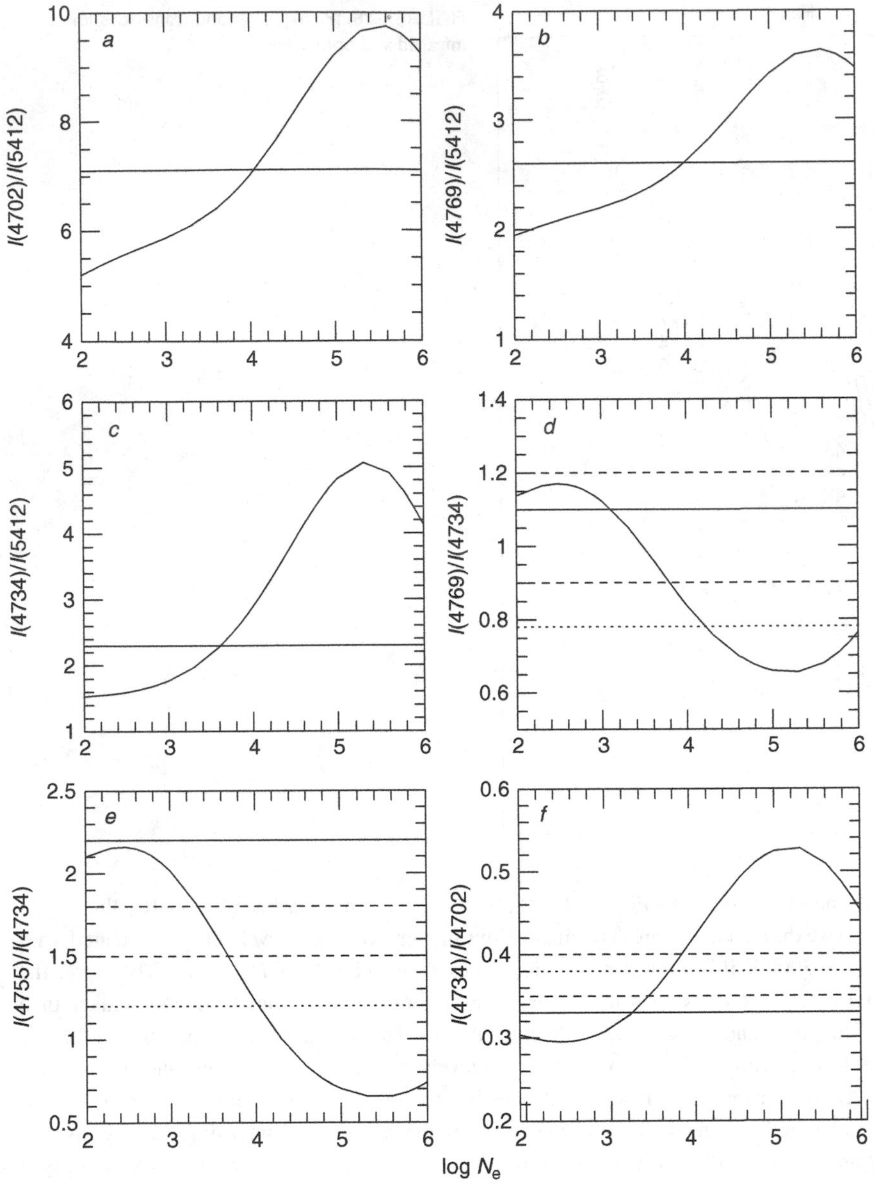

FIGURE 12.9 Fe III line ratios vs. $\log_{10} N_e(\mathrm{cm}^{-3})$ at $T = 9000$ K. The horizontal lines are various observed values from the Orion nebula [98].

$^4(P, D, G)_{3/2,5/2,7/2}$ (we have grouped together the *LS* and fine-structure levels). The corresponding lines lie in the UV around λλ 2500–3100. The forbidden optical lines are from transitions among excited multiplets. In Fig. 12.11 we compare [Fe IV] and [O III] optical line ratios [98] to derive electron densities using the observed value from a high-excitation bipolar young planetary nebula M2-9 (also known as the 'Butterfly Nebula'), with a symbiotic' star core [318] and a fast and dense stellar wind.[10] The inferred densities from the [Fe IV] optical line ratio 5035/3900 are about 5–6 $\times 10^6$ cm^{-3}, and the temperatures are $\sim$ 9000–10 000 K. This indicates that a high-density plasma environment is possibly in dense knots, expected in the common stellar envelope but at

[10] As noted in the case of LBV Eta Carinae in Chapter 10, a *symbiotic stellar system* consists of two stars in different stages of evolution with a common nebular envelope.

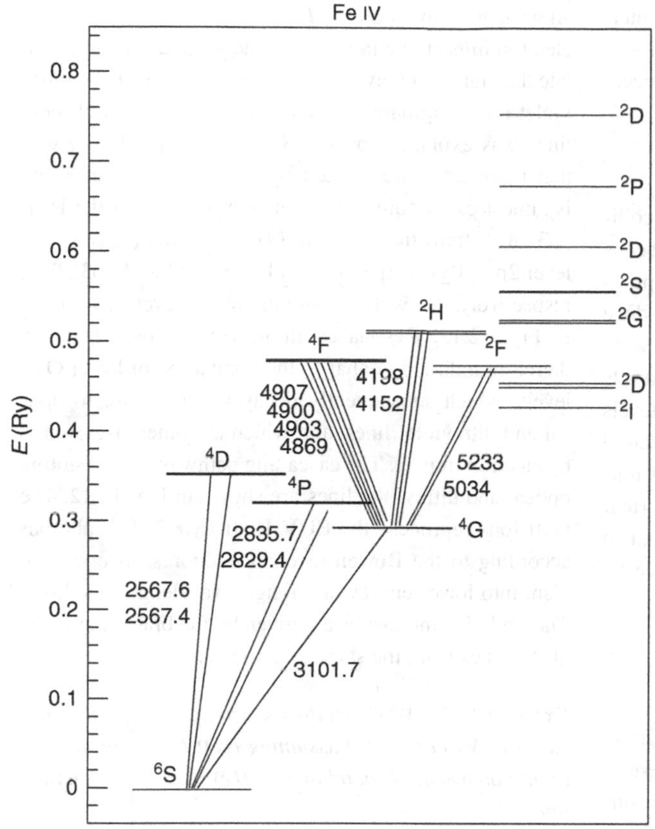

FIGURE 12.10 Energy levels of Fe *IV* with transitions in the ultraviolet and optical regions.

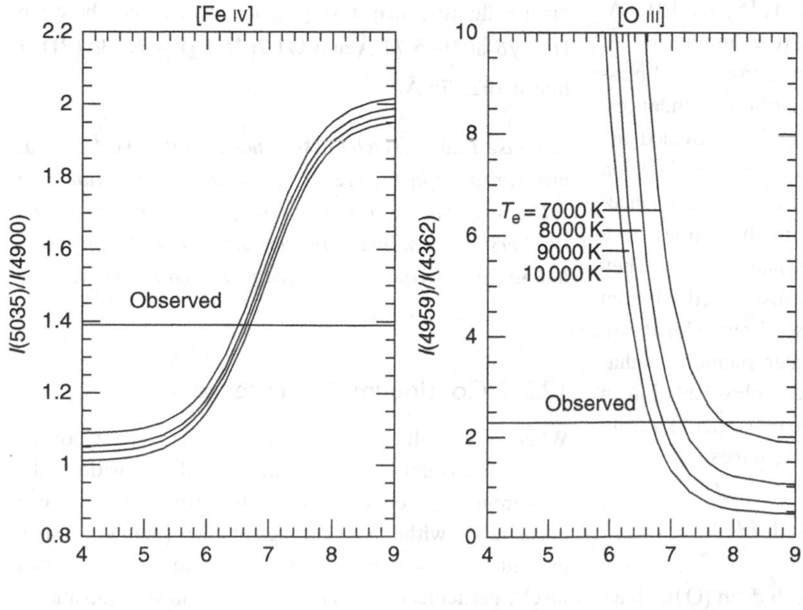

FIGURE 12.11 Density diagnostics using optical [Fe IV] and [O III] line ratios vs. $\log_{10} n_e$, for the 'Butterfly Nebula' M2-9.

typical nebular temperatures. On the other hand, the interpretation of [O III] line ratios and density determination is ambiguous, owing to the sharp temperature dependence.

12.5 Fluorescent photo-excitation

In the discussion so far we have considered only collisional excitation and radiative decay. However, gaseous nebulae are ionized by a radiative source and the proximity of the emitting region to that source could induce photo-excitation of specific level populations as well. Such radiative excitation may populate high-lying levels via UV transitions, which can then decay to lower levels thereby contributing to the forbidden optical and infrared line intensities. Such *fluorescent excitation* (FLE) of ions results from (i) line fluorescence from strong H I and He II lines, owing to the more abundant H and He present in the emitting region, and (ii) continuum fluorescence by background flux from a radiation source.

12.5.1 Line fluorescence

There are several interesting coincidences between transitions in two different ions where the transition energies are nearly equal. In such cases, line emission from one (more abundant) ion can be absorbed by the other. Since H and He ions are the most abundant, the main fluorescence mechanisms depend on the usually strong Lyα, Lyβ recombination lines from H, at 1215 and 1016 Å, respectively, and the hydrogenic He II α line at 304 Å. Since line radiation from these ions in astrophysical plasmas is relatively intense, owing to their high abundances, it can 'pump' a transition in another ion provided the energy difference between the levels in the excited ion matches closely, i.e., the transitions are in 'resonance' according to wavelengths. Photons in these lines have significant to large optical depths in many objects. They may be scattered many times and redistributed, or even trapped entirely (viz. Case B or Case C recombination). We shall discuss Lyα fluorescence later, including radiative transfer and excitation of the complex Fe II ion in AGN (Chapter 13). Here, we describe two other relatively simpler cases in optically thin nebular sources.

12.5.1.1 *Bowen fluorescence: O III–He II excitation*

As we have seen in Chapter 8, the forbidden [O III] lines are among the strongest lines seen in nebulae. They are excited by electron impact from the ground level. However, O III lines are also seen in the optical and ultraviolet. At nebular temperatures ($kT_e \sim 1$ eV) it is not possible for electron impact excitation from the ground state to populate the high-lying levels from where the optical and ultraviolet lines originate. The anomalous intensities of these lines was explained by I. S. Bowen [319], who showed that these lines are excited by line fluorescence. There is almost exact coincidence in energy between the He II 303.78 Å transition and the O III transition(s) from the level $2p^2(^3P_2) - 2p3d\left(^3P^o_{1,2}\right)$ at 303.62 and 303.80 Å, respectively, as well as among other levels as shown in Fig. 12.12. Resonance fluorescence from He II can therefore indirectly enhance the population of lower O III levels, which subsequently decay and give rise to optical and ultraviolet lines, and which are otherwise excited by electron impact. The cascading pathways and resulting optical and ultraviolet lines are shown in Fig. 12.12. The O III ions reprocess the EUV He II Lyα 304 Å photons according to the Bowen resonance–fluorescence mechanism into lower energy and longer wavelength UV lines. Their relative intensities are given by the branching ratios of A-values from the same upper level.

Exercise 12.2 *Work out the branching ratios of the optical and UV lines of O III assuming He II FLE to be the only excitation mechanism, relative to He II Lyα recombination line.*

We note another useful line conicidence, analogous to the O III–He II combination. There is also the O I–Lyβ resonance fluorescence, owing to the coincidence between H I Lyβ at 1025.72 Å and O I $2p^4(^3P_2) - 2p^33d\left(^3D^o_3\right)$ line at 1025.76 Å.

Exercise 12.3 *Sketch the schematics of the O I–Lyβ FLE mechanism as in Fig. 12.12, and calculate the branching ratios as in the previous exercise. [Note: A more exhaustive version of the two problems above would be to use a line ratios program to compute line intensities and ratios].*

12.5.2 Continuum fluorescence

Whereas line fluorescence owing to accidental coincidences in wavelengths can excite particular transitions, the background UV continuum radiation from a hot star can excite levels within a wide range of energies in ions. Such excitations may occur from the ground state to higher levels, particularly via strong dipole allowed transitions. Also, in that case, the number of exciting photons, i.e., the background photon density or the radiation flux from a source, can compete with the ambient electron density

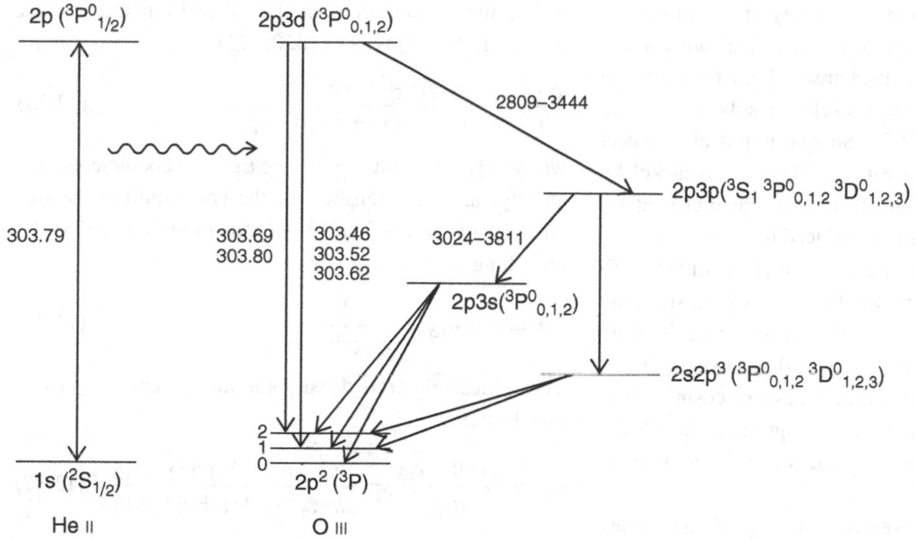

FIGURE 12.12 The Bowen fluorescence mechanism: 'resonant' excitation of O III lines by He II. Note that several transitions are grouped together because they are observationally unresolved, or for clarity.

in exciting a particular level. But unlike the local electron density, the photon density profile has a distribution defined by geometrical dilution as $1/r^2$, where r is the distance from the source. Additionally, the continuum photon flux depends on the luminosity or the temperature of the source. Therefore, the line intensities are a function of four variables (T_e, n_e, T_* r), instead of just the first two in the case of collisional excitation alone without FLE. Many nebulae are excited by hot stars with a strong UV continuum. A number of high-lying levels of an ion can thereby be excited by continuum fluorescence, in contrast to just one particular level by line or resonance fluorescence.

We can extend the CR model and rate equations to include continuum FLE in optically thin situations. Recall that for each bound level i, with population N_i and energy E_i, the equations of statistical equilibrium are (Chapter 8).

$$N_i \sum_j R_{ij} = \sum_j N_j R_{ji}, \tag{12.31}$$

where the sums are over all other bound levels j. As before, the quantity R_{ij} denotes $R_{ij} = q_{ij} + A_{ij}$ for $E_i > E_j$, or $R_{ij} = q_{ij}$ for $E_i < E_j$ (here q_{ij} is the electron impact excitation or de-excitation rate coefficient). This CR model does not consider fluorescence due to the background continuum radiation of an external source. The line emission photons created by the ions are also assumed to escape without absorption. The effect of diffuse background radiation is introduced in the CR model by assuming a thermal continuum radiation pool, in which all ions can be excited by photon pumping. The thermal radiation field may be assumed to be a black body at temperature T. Then the rate coefficient R_{ij} becomes

$$R_{ij} = q_{ij} + A_{ij} + U_\nu B_{ij} \quad (E_i > E_j), \tag{12.32}$$

or

$$R_{ij} = q_{ij} + U_\nu B_{ij} \quad (E_i < E_j),$$

with $h\nu = |E_i - E_j|$. Here, B_{ij} is the Einstein coefficient and U_ν is the radiation density of photons of frequency ν. If we assume a black-body radiation field with temperature T_* we have

$$\frac{c^3 U_\nu}{8\pi h\nu^3} = \frac{w}{e^{(h\nu/kT_*)} - 1}, \tag{12.33}$$

where w is the geometrical dilution factor at a distance r from a star with radius R,

$$w = \frac{1}{4}\left(\frac{R}{r}\right)^2. \tag{12.34}$$

The resulting rate equations in the CR + FLE model (Eq. 12.32) can be solved in the usual manner, outlined in Chapter 8. We apply the CR+FLE model to a particularly interesting example below.

12.5.2.1 *Fluorescence of Ni II*

The ionization potentials of Fe and Ni are similar. The ground-state ionization energies are 7.9 and 16.2 eV for Fe I and Fe II, respectively, and 7.6 and 18.2 eV for Ni I and Ni II, respectively. Thus one might expect Fe and Ni ions to co-exist in the PIZs. One indication of the co-existence of both ions comes from the correlation between

[Fe II] and [Ni II] emission in a variety of gaseous nebulae. However, in a number of astrophysical objects, the observed nickel line intensities are far higher than would be commensurate with the nickel/iron solar abundance ratio of ~0.05 (Table 11.2). So one might also expect the cosmic iron abundance to be higher than nickel by about a factor of 20. But apparently, the nickel abundance is anomalously high, as deduced from [Ni II] optical lines. For example, in the circumstellar ejecta of the luminous Be star P Cygni (with the famous P Cygni signature line profiles (Chapter 10), the [Ni II] line intensities are enhanced by up to a factor of 1000 relative to [Fe II] lines, over what is expected on the basis of cosmic abundances. Many other H II regions also appear to show Ni/Fe enhancements that range over orders of magnitude (e.g., [320, 321]).

Since such a large overabundance of nickel cannot be realistic, another physical explanation needs to be explored. The mechanism invoked to explain such an enhancement is photo-excitation by the strong UV radiation background [320]. The observed intensity of the λ 7379 [Ni II] line is employed to deduce the Ni abundance. But UV FLE enhances the intensity of this line indirectly as follows. The relevant [Ni II] lines are due to transitions within the three-term system, all with doublet spin-multiplicity (2S + 1), shown in Fig. 12.13. Whereas several transitions are possible, we focus on transitions between the three levels shown by arrows. Continuum UV radiation pumping via the strong dipole transition 1→3, from the ground state $a\,^2D_{5/2}$ (level 1) to the opposite odd parity level $z^2D^o_{5/2}$ (level 3), occurs at 1742 Å. This is followed by spontaneous decay 3→2 to $a\,^2F_{7/2}$ (level 2) via another UV transition at 2279 Å, which thereby enhances the population of level 2. The final step in the FLE mechanism is the radiative decay 2→1, which gives rise to the λ7379 line in the optical. The otherwise forbidden 7379 line is thereby strengthened by the FLE mechanism via allowed transitions within the doublet system. In this

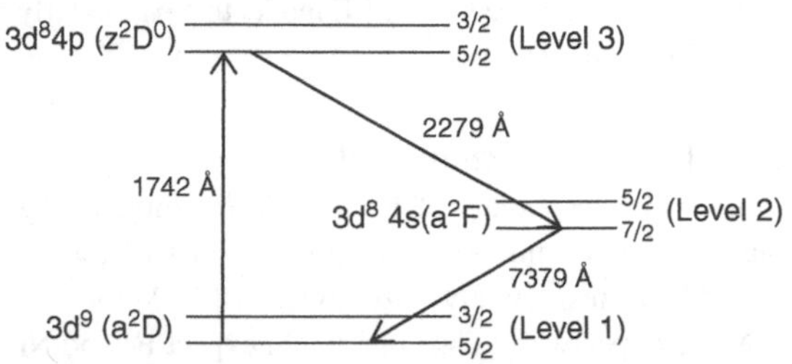

FIGURE 12.13 Continuum UV fluorescence of optical lines in Ni II.

model, the population of level 2, N_2, with respect to that of level 1, N_1, is given by [320, 321]

$$\frac{N_2}{N_1} = \frac{n_e q_{12} + b_{32} B_{13} J_{13}}{n_e q_{21} + A_{21} + b_{31} B_{23} J_{23}}, \tag{12.35}$$

where A_{21}, B_{13} and B_{23} are the Einstein coefficients, J_{13} and J_{23} are the intensities of the continuum at the frequencies of the 1→3 and 2→3 transitions, and b_{32} is the branching ratio, defined by

$$b_{32} = 1 - b_{13} = \frac{A_{32}}{A_{32} + A_{31}}. \tag{12.36}$$

The critical electron density for fluorescence (n_{cf}) can now be defined,

$$n_{cf} = \frac{b_{32} B_{13} J_{13}}{q_{12}} = \frac{c^2}{2h\nu_{13}^3} \frac{\omega_3}{\omega_1} \frac{A_{31} A_{32}}{A_{31} + A_{32}} \frac{J_{13}}{q_{12}}. \tag{12.37}$$

If the electron density $n_e < n_{cf}$ the emission is dominated by fluorescence. Conversely, if $n_e > n_{cf}$, the line is predominantly excited by electron collisions. In this equation, ν_{13} is the frequency of the 1→3 transition. The critical density n_{cf} decreases as ν_{13}^{-3}, and only the lowest odd parity terms of the ion coupled to the ground state are likely to contribute significantly to the FLE mechanism. In the case of a black-body radiation field, the ν_{13}^{-3} dependence of N_{cf} cancels out, and it drops exponentially with ν_{13}.

Two other lines in the $^2D_J - ^2F_{J'}$ multiplet (called the 2F multiplet; cf. [100]) are also pumped similarly: $^2D_{3/2} - ^2F_{5/2}$ at 7413.33 Å and $^2D_{5/2} - ^2F_{5/2}$ at 6668.16 Å. Both of these transitions, and the 7379 Å transition illustrated above, are forbidden $E2$ and $M1$ transitions, but the $E2$ dominates by about four orders of magnitude over $M1$. The $A(E2)$ values for the λλ 7379, 7413 and 6668 Å lines are 0.23, 0.18 and 0.098, respectively. Another transition $^2D_{5/2} - ^2F_{7/2}$ is much weaker; it occurs only as an $E2$ transition with $A(E2) = 0.013$. Generalizing the model to calculate level populations and line emissivities in this multi-level system, including photo-excitation, in Fig. 12.14 we plot the ratio [Ni II] λ7413/λ7379 vs. N_e, with fluorescence (dashed line) and without (solid line, collisional excitation only). Note that the critical density with FLE $n_{cf} \sim 10^7\,\mathrm{cm^{-3}}$, indicative of the denser PIZ compared with the FIZ. The line ratio is enhanced by more than a factor of three for $n_e < n_{cf}$, but as $n_e \to n_{cf}$ the two line ratios approach each other and merge for $n_e > n_{cf}$.

The enhancement due to FLE over collisional excitation alone in Fig. 12.14 implies reduced dependence on electron density, since photo-excitation plays a major, if not dominant, role. The value for this ratio from a different calculation at $n_e = 600\ \mathrm{cm^{-3}}$ [320] shows a ~15%

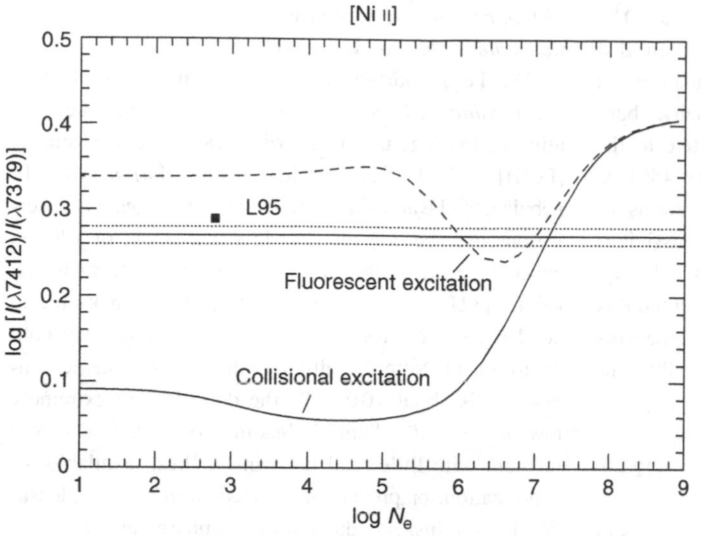

FIGURE 12.14 Enhancement of line intensities due to continuum UV fluorescence. The figure shows an optical [Ni II] line ratio vs. n_e with and without continuum UV fluorescence (see text). The solid line is without FLE and the dashed line includes both collisional excitation and FLE. The solid square is an earlier calculation [320].

difference, which is not significant and most likely due to different atomic data. Also shown are the flux ratios of these lines as measured in P Cygni [322], as horizontal lines, including the range of uncertainty as dotted lines. To simulate the radiation field responsible for FLE due to P Cygni, the basic parameters have been taken to be $T_* = 20\,000$K, $R_* = 89.2\ R_\odot$, and the illuminated ejecta to be at a distance of 0.08 pc from the star, which yields a dilution factor $w = 1.6 \times 10^{-10}$, (Eq. 12.34). The inferred densities again lie in the range 10^{6-7} cm^{-3}.

We have seen that FLE of [Ni II] lines is important in regions with strong UV background, owing to the strong transitions among doublet spin-symmetry levels. On the other hand, for Fe II the fluorescence excitation of either the IR or the optical lines is much less important because nearly all of the lines observed in these spectral ranges correspond to transitions across different multiplicities. The transitions occur among quartet and doublet spin-multiplicy levels, which, in turn, cannot be pumped and connected by dipole allowed transitions from the sextet $3d^64s(^6D_J)$ ground state levels (Fig. 12.5). Also, intercombination transitions from the ground state to odd parity quartet levels are relatively inefficient, as their transition probabilities are at least one or two orders of magnitude smaller than for the dipole transitions. Perhaps, the greatest fluorescence effect may be seen for the level $a\ ^6S_{1/2}$ that gives rise to the λ4287 line in Fig. 12.5. Photo-excitation of this level could occur via pumping of the $z\ ^6P^o_J$ levels from the ground term; however, inspection of the energy of the $z\ ^6P^o$ multiplet relative to the ground state, and of the A-values for the transition involved in the process, indicates that the critical fluorescence densities in Fe II are an order of magnitude lower than those for Ni II, and hence collisions would be that much more dominant than fluorescence. Likewise, it is clear from Figs. 12.8 and 12.10 that FLE is not likely to be of importance for [Fe III] and [Fe IV] lines.

12.6 Abundance analysis

Abundances of elements may be derived from emission lines in nebulae. In principle, one needs to know individually the abundance of each ionization state of an element X with nuclear charge Z, i.e.,

$$n(\mathrm{X}) = n(\mathrm{X}^0) + n(\mathrm{X}^+) + n(\mathrm{X}^{2+}) + \cdots + n(\mathrm{X}^{Z+}), \qquad (12.38)$$

from neutral to fully ionized.[11] However, the lines observed are often from only one or two ionization stages. Therefore, a knowledge of ion densities is needed in each observed ionization stage, or *ionization fractions* relative to the total abundance of the element. The way one obtains that information is (i) to assume that the observed ionization states are the dominant states, given the physical conditions in the source (primarily the temperature), and (ii) by comparison with another element whose abundance can be ascertained with better accuracy. For example, we may approximate the total abundance of oxygen as

$$n(\mathrm{O}) \approx n(\mathrm{O}^+) + n(\mathrm{O}^{2+}), \qquad (12.39)$$

[11] It is customary to use the notation indicating ion charge as superscript, rather than roman numerals, when referring to ion densities or abundances.

assuming that no further ionization states, such as O^{3+} and higher, are present, or are negligible. Then that requires the measurement of two ionic abundances on the right-hand side, O^+ and O^{2+}. Observationally, abundances are inferred from line intensities relative to the optical H I recombination lines, such as the Hβ 4861 Å. We can write the ionic abundance ratio in terms of the measured flux ratio and the intrinsic emissivities (Chapter 10). Considering the example of the well-known nebular ion O III and its forbidden lines, the observed flux ratio is related to the abundance ratio in terms of the (theoretically computed) emissivities of the [O III] 5007 line and Hβ as

$$\frac{I(5007)}{I(4861)} \times \frac{\epsilon[\mathrm{OIII}, {}^1D_2 - {}^3P_2]}{\epsilon[\mathrm{HI}, n = 4 \to 2]} = \frac{n(\mathrm{O}^{2+})}{n(\mathrm{H}^0)}. \tag{12.40}$$

Similarly, we may determine the relative intensities of the [OII] doublet 3617, 3631 lines to estimate the O II density $n(\mathrm{O}^+)$. Thus the total oxygen abundance may be approximately obtained from the [O II] and [O III] lines. But that is not to say that other ionization stages, such as O IV, are not actually present to some extent. Therefore, ionization balance calculations, and observational analysis, are necessary to ascertain the ionization fractions of *all* O ions with the specific physical conditions in the nebula.

A somewhat more complicated analysis is needed for heavier elements, where observations of a given ionization species are not readily obtained. For example, although Fe IV is the dominant ionization state in the fully ionized zone of nebulae (Figs. 7.14 and 12.4), it has few observable diagnostic lines. The lowest Fe IV transitions lie in the ultraviolet (Fig. 12.10), and they are difficult to observe and analyze, owing to extinction and other effects (see Section 12.4). Similar considerations apply to other Fe peak elements. Note that the ionization potential of O III is very close to that for Fe IV (Table 12.1). This implies that O III is likely to co-exist spatially in the regions with Fe IV, and as discussed above, in the FIZ. Since [O III] lines are generally more intense and better measured, the O III/O ratio may be used to derive a *correction* for the Fe IV/Fe ratio as follows

$$\frac{n(\mathrm{Fe})}{n(\mathrm{H})} \times \frac{n(\mathrm{O}^{2+})}{n(\mathrm{O})} \approx \frac{n(\mathrm{Fe}^{3+})}{n(\mathrm{H}^+)}. \tag{12.41}$$

Here we adopt the *ionization correction factor* (ICF) = O^{2+}/O, the fractional abundance O III relative to O I. It is amply evident from Table 12.1 that the second-row elements have much higher ionization potentials than the Fe-peak elements. Therefore, *lower ionization species of lighter elements spatially co-exist with higher ionization states of heavier elements.*

The Fe abundance in stars is usually referred to as the *metallicity*. It is the ratio of the iron abundance relative to hydrogen, defined with respect to the Sun as $[\mathrm{Fe/H}] = \log[n(\mathrm{Fe})/n(\mathrm{H})]_{\mathrm{star}} - \log[n(\mathrm{Fe})/n(\mathrm{H})]_{\odot}$. In nebulae, the Fe abundance derived from *ionized gas* observations may be considerably less than solar, or what is expected *if* all of the iron were in ionized form in the nebula [321, 323]. This results from the formation and condensation of iron onto grains in the relatively cold environment. Note that although the electron kinetic temperature is about 10 000 K, the densities are extremely low, $n_e \sim 10^{3-6}\mathrm{cm}^{-3}$, leading to a relatively cold medium with little total mechanical 'heat' available for vaporization, or preventing ionized 'iron gas' condensation into grains. A reduction of gas-phase abundances of an element by condensation on to grains is referred to as *depletion* of that element. Thus, the total Fe abundance is that observed from ionized regions of nebulae in the gas phase, plus that condensed in dust grains. The depletion factors are difficult to measure, since dust grains are not readily observable. They may also involve the formation of complicated molecular species, such as iron oxides like Fe_2O_3, whose concentrations are not easy to ascertain. Nevertheless, another Fe-peak element, Zn, may be used as a surrogate for Fe, since it is not depleted onto dust grains significantly. Also, the ionization energies are similar (Table 12.1), so observations of Zn I–IV could be used to infer Fe depletion factors by measuring Zn/Fe abundance ratios approximately, as described above.

Direct spectroscopic observations of the gas phase iron abundance in the Orion nebula have been carried out using the well-known forbidden optical and infrared lines of [Fe II] and [Fe III] shown in Figs 12.5 and 12.8, respectively, as well as the faintly observed ultraviolet lines of [Fe IV] $\lambda\lambda$ 2568.4, 2568.2, due to transitions $3d^6(^4D_{5/2,3/2} \rightarrow {}^6S_{3/2})$ (Fig. 12.10). However, there are discrepancies by large factors, ranging from a few times less than solar, to up to 200 times less than solar [321, 323]. The problem lies not only in the difficulty of making accurate measurements from ions that may exist in spatially distinct regions with different physical conditions, but also in the interpretation of observations based on atomic data of inadequate precision. The determination of nebular abundances is a challenging problem that needs not only improved and varied measurements, but also high accuracy atomic parameters, such as collision strengths, radiative transition probabilities and (photo)ionization and recombination rate coefficients.

12.7 Atomic parameters for nebular emission lines

Given the physical processes in H II regions, the primary atomic parameters of interest are: photoionization cross sections, (e + ion) recombination and electron impact excitation rate coefficients, and the spontaneous decay rates for all relevant transitions. Nearly all of these data need to be computed quantum mechanically, not only for the levels corresponding to the observed lines but for all levels that might significantly contribute to those levels. However, in nebular sources we may often restrict the CR models of atomic spectra to collisional excitation and radiative decay of low-lying levels. A vast body of literature exists on collision strengths and transition probabilities, as reviewed in [324, 325]. Appendix E is an extensive and up-to-date tabulation of recommended excitation rate coefficients and A-values for most nebular ions.

The inaccuracy or the incompleteness of available atomic data from theoretical calculations (or experiments, albeit limited in scope) is of serious concern. Even for relatively simple atomic systems, such as O II [326], there have been problems that severely plagued the interpretation of astronomical observations [108]. Fe II lines remain difficult to analyze in many sources. One of the most important reasons for disentangling the accuray of basic atomic physics from astrophysical phenomena themselves is to determine elemental abundances. While H II regions appear to be well-understood in terms of physical conditions, the abundances of elements derived from measurements in different wavelength regions vary widely by up to several factors (and not only for vastly anomalous cases, such as the Ni/Fe abundance ratio discussed). For example, $[Fe/S]_{UV}$ (from UV lines) $= -0.35 \pm 0.12$, but $[Fe/S]_{vis}$ (from visible lines) $= -1.15 \pm 0.33$, a discrepancy of a factor of six [327]

More generally, at the present time there is a perplexing discrepancy between abundances derived from collisionally excited emission lines (CEL) on the one hand, and recombination emission lines (REL) on the other hand *for the same element* [312, 313]. Of course, the physical processes are different: CEL arise from electron impact excitation of low-lying levels of ions such as [O II], [O III], [Fe II], etc., while the REL lines are due to electron recombinations into high-lying levels cascading into observed lines. Therefore, it is possible that the atomic rate coefficients for the two processes are discrepant. Whereas the collisional excitation data are relatively well-determined in terms of accuracy, the level-specific recombination rate coefficients have been calculated with adequate precision only for a few ions, and not yet widely employed in astrophysical models. It is essential to resolve this issue in order to disentangle the atomic physics, which, in principle, can be addressed to sufficient accuracy, from astrophysical phenomena in nebulae, such as temperature fluctuations and abundance variations within nebulae.

13 Active galactic nuclei and quasars

The number of stars in a galaxy varies over a wide range. Whereas dwarf galaxies may contain as 'few' as $\sim 10^7$ stars, giant galaxies at the other end of the numerical stellar count are five orders of magnitude higher, into trillions of stars $\sim 10^{12}$. Our own Milky Way is a collection of about 100 billion $\sim 10^{11}$ stars, which we may take to be the number in a 'normal' galaxy. Given the luminosity of the Sun as the benchmark $L_\odot = 3.8 \times 10^{33}$ erg s^{-1}, the luminosity of a normal galaxy $L_G \sim 10^{44}$ erg s^{-1}. Large as that number is, it turns out that a good fraction of galaxies, at least 10%, are much brighter. To be a bit more precise, the *central regions* of such galaxies are extremely bright, with central luminosity at least equal to that of the rest of the galaxy. At first sight, this might not seem illogical, given the expectation of a greater concentration of stellar systems towards the centre of a galaxy. However, these ultrabright central regions exhibit a number of outstanding observational facts.

(i) The emergent luminosity can be extremely intense with a range of 10^{12-15} $L_\odot$, or more than 10 000 times the luminosity of an entire galaxy L_G.

(ii) The central source is highly concentrated in an extremely small volume of a small fraction of a parsec (pc), on the spatial scale of no more than our solar system, whereas the galaxy itself may be tens of kpc.

(iii) The distribution of emitted spectral energy of the galactic nucleus is non-thermal, quite different from that of stars.

(iv) The observed energy distribution across all wavelength ranges does *not* decrease exponentially with increasing energy, as expected for a Planck function for stellar energy; rather, it can be essentially constant or decreasing slowly as a *power law* in energy, $L(E) \sim E^{-\Gamma}$ ($\Gamma \sim 1$), even out to high energies with strong X-ray emission.

(v) About one tenth of these bright nuclei are also intense radio sources, implying a non-thermal origin of radiation.

(vi) Significant variability of the observed flux, also implying that the power source must be compact; normal galaxies do not exhibit such variability over measurably short timescales.

Something drives enormous activity at the centre of many galaxies, with tremendous output of non-stellar form of energy. A particularly interesting and useful fact is that this activity is reflected in an enormous variety of emission and absorption spectra from most astrophysically abundant atomic species, from neutral hydrogen to highly ionized iron and nickel. This, of course, enables the means to study the physical conditions and to probe the central regions of active galaxies.

As mentioned, if galaxies were simply a collection of stars (albeit billions of them) one might expect their spectra to be not fundamentally different from stars. And since the total luminosity should be dominated by the brightest and most massive stars, the expected spectra of *normal* galaxies should have similar spectral components weighted rather towards the near-ultraviolet and the optical. This is indeed the case for most ordinary galaxies, *not* otherwise active in the sense described in the criteria above. For any black body (or collection thereof), the Planck function ensures some output of electromagnetic radiation in all wavelength ranges. However, deviations from this basic fact begin to manifest themselves, as one examines the central regions of active galaxies and finds that luminosity is quite non-Planckian in nature. The term generally employed for centres of galaxies with energy output fuelled by some tremendous activity is *active galactic nuclei* (AGN). However, there are a number of sub-classes of objects that fall under the AGN characterization. At one end of extreme luminosities are the *quasi-stellar objects*, or QSOs, which are identified by their non-zero redshifts at great distances out to $z > 6$.

Historically, a number of QSOs were observed by their copious radio emission, and called *quasars* (quasi-stellar radio sources).The QSO spectra show significant redshifts of lines compared to rest or laboratory wavelengths. That means that QSOs are at large distances from us and, owing to the cosmological expansion, must have originated at earlier times in the history of the Universe. Quasi-stellar objects may be among the first large-scale objects formed at the earliest epochs. Many catalogued quasars are at the most luminous end of AGN luminosity, partly due to selection effects in observing far away objects, which would tend to be the brightest. Again, something or some process is needed to generate the stupendous amount of energy put out by QSOs. While we don't quite know the precise nature of this *central engine* of AGN, QSOs, and maybe even normal galaxies, the working paradigm that has emerged is a gravitationally accreting supermassive black hole (SMBH), which apparently unifies most of the observed phenomena mentioned above.

This chapter is mainly devoted to spectral characteristics of the general types and structures of AGN. As we shall see, spectral features of different parts of AGN reveal (or conceal!) intriguing phenomena, many of which are not presently understood.

13.1 Morphology, energetics and spectra

Active galactic nuclei phenomena manifest themselves in a huge variety of ways which, in turn, result in myriad classifications and sub-classifications of AGN and QSOs. The main problem in AGN research is to understand otherwise unrelated characteristics in some sort of a *unified* scheme, which reflects the essential underlying physics. The over-arching phenomenology rests on AGN activity fuelled by supermassive black holes (SMBH). These central engines are thought to drive the observed energetics and structures. The morphology of AGN comprises a number of apparently disparate regions, as shown schematically in Fig. 13.1. The central SMBH is surrounded by an *accretion disc* formed by infalling matter. Conservation of angular momemtum requires the formation of an accretion disc (not unlike water swirling around a drain). The accretion disc converts gravitational energy of infalling matter into mechanical energy – heat and light. The emergent radiation shows enhanced flux of UV radiation in the continuum together with ionized gas outflows, as ascertained by broad absorption in strong lines of well-known atomic species. The thermal emission from the the accretion disc can be modelled as a black body, albeit modifed according to geometry and opacity effects (e.g. [328]).

The unifying structure of AGN is also thought to be governed by the angular momemtum of accreting matter. The angular momemtum of infalling material is conserved, as it is removed by centrifugally driven winds from accretion discs, which, in turn, account for the observed bipolar outflows and jets [329, 330]. The accretion disc is not directly visible, but a 'jet' of material moving out at relativistic velocities, arising from the interaction of the SMBH and infalling matter, is often visible at radio frequencies out to tens of kpc. Matter outflow at relativistic velocities results in the jet around the polar axis of an intensely magnetic spinning black hole and accretion disc. As a result of jet emission and material interactions, the electromagnetic spectrum of AGN ranges from radiowaves to TeV energies. While radio emission would be expected from all AGN, the intensity and extent of observed radio emission determines whether the AGN is classified as *radio-loud* or *radio-quiet*.

At the high-energy end, the X-ray background continuum is generated in three processes: (i) *bremsstrahlung radiation* due to electrons accelerating in the proximity of ions, (ii) *Compton scattering* of photons by electrons, and consequent shift in wavelength, and (iii) *synchrotron radiation* due to electrons accelerating in a magnetic field. All three processes play a role in the formation of AGN X-ray continuum driven by the central black hole.

Beginning with our current understanding of AGN from the inside out, the most direct evidence of a relativistic accretion disc around a SMBH is an iron line due to $K\alpha$ transition(s) at about 6.4 keV ([239], discussed later). The line is seen to have an extremely wide but asymmetrical broadening towards the red, down to about 5.7 keV. This indicates gravitational redshift, lowering of the energy of photons emitted, predicted by the theory of general relativity. Later, we discuss the 6.4 keV line, together with the whole group of other *Fe $K\alpha$ X-ray lines*. The material interactions in the vicinity of the disc also result in hot highly ionized matter prominently visible in X-ray spectra. The gas outflow, which shows overall X-ray absorption, is referred to as the *warm absorber* (WA). Both the disc-corona and the warm absorber are prominent X-ray emitters and absorbers in atomic species ranging from O VII, Fe XVII in the soft X-ray ($\sim$0.5–2 keV) to Fe XXV, Fe XXVI in the hard X-ray (6.6–7.0 keV).

Farther away from the central source, at distances of tens of pc, but influenced by the gravity of the SMBH, are the *broad-line-region* (BLR) clouds of relatively less ionized gas. The BLR are characterized by bulk Doppler

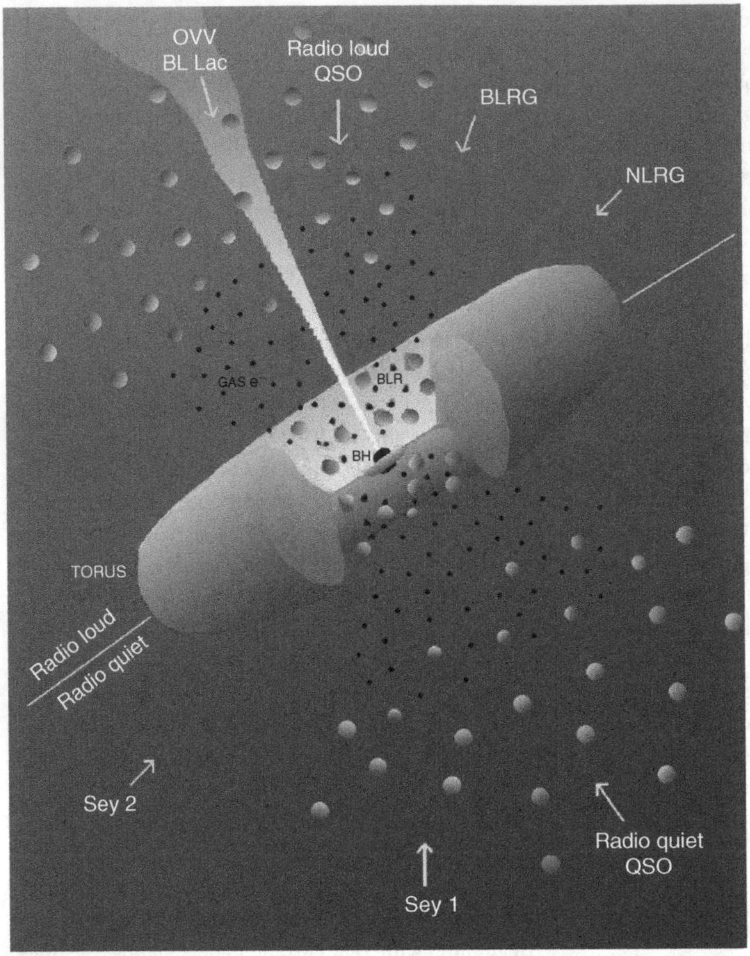

FIGURE 13.1 Schematic model of an AGN showing a central black hole (BH), accretion disc, radio-jet, clouds of ionized gas and a molecular torus surrounding the accretion disc (modelled after Urry and Padovani [331]). The presence of a relativistic jet implies that the AGN would be radio-loud. The diagram also illustrates the unification scheme. The obscuring torus, and its orientation with respect to the observer, determines the view along the line of sight to the central region. A more-or-less direct view manifests itself in spectral characterstics of a Sey 1 galaxy (also radio-quiet QSOs), whereas the obscured view corresponds to a Sey 2. Quasars and Seyferts differ in luminosity and distance (redshift); classifications Type 1 and 2 refer to orientation. Very few Type 2 quasars (edge-on) are observed, presumbly owing to the difficulty of observing obscured objects at large distances. The dark dots represent ionized gas. The broad line regions are clouds of gas (BLRG) influenced by the central source. The narrow line regions of ionized gas (NLRG) are relatively farther away and visible when the central source is obscured from the line of sight. The outflowing gas acts as a warm absorber, which heavily attenuates the X-ray continuum radiation from the hot corona surrounding the central source.

motions with velocities up to thousands of km s^{-1}, inferred from the width of emission lines such as the Hβ (4860 Å) in the optical or C IV (1541 Å) in the ultraviolet. The BLR gas shows spectral features of typical nebular stages of ionization temperature, but much higher electron densities of 10^{9-12} cm^{-3} and much broader emission lines from species such as O III and Fe II. Still farther out are the *narrow line region (NLR)* clouds, quite similar to ionized gas in H II regions. The NLR spectra have typically nebular forbidden lines of low-ionization atomic species, such as [O II], [O III], [S II] and [Fe II] (cf. Fig. 8.3), but at lower densities than the BLR. The central region of the AGN is shrouded in a large *molecular torus* rotating about the SMBH, as inferred by Doppler blue- and redshifts of *H_2O masers*: the maser action in the water molecule is pumped by AGN activity. For example, a sub-parsec maser disc is observed from NGC 4258 (Messier M106) in the strong 22 GHz radio line [332].

The immense variety of radiative and material processes occurring within the AGN makes it obvious that their spectral features should encompass practially the entire electromagnetic spectrum, from hard X-ray to radio. The global spectral energy distribution is such that it remains large and flat across all wavelength ranges, as measured by the quantity νF_ν (or equivalently λF_λ) in any specified range (as shown later in Fig. 13.5). But actual observations reveal large differences in properties of AGN. Nevertheless, the current paradigm unifying the AGN phenomenology is determined by the essential geometry of a SMBH and an accretion disc. The orientation of the accretion disc relative to the observer (on the Earth), and associated activities, determines the morphological, spectral, and temporal (time variability) properties. The unified model rests on the same underlying physical processes for all AGN.

13.1.1 Seyfert classification

The black hole and an accretion disc at the centre determine the morphology of an AGN. The orientation towards the observer, or the viewing angle, would therefore be the crucial factor in establishing the line of sight towards different parts of the AGN (Fig. 13.1). At the two extremes one has (i) a full face-on view that enables direct observations of the nucleus, and (ii) an edge-on view that obscures most of the nuclear activity. Seyfert [333] first identified six galaxies with extremely bright nuclei, and discovered the essential discriminant between the two types: the spectral widths of emission lines. Seyfert found the first kind to be associated with broad lines in their spectra, and narrow lines in the second kind. These are respectively referred to as Seyfert 1 and Seyfert 2 AGN.

The reason for the differences in the width of the lines is relatively straightforward. In Seyfert 1 AGN, one observes the central source more-or-less directly, thereby sampling the high velocity BLR clouds, moving under the gravitational influence of the SMBH with Doppler broadened emission lines. Observed velocity dispersions of up to 10 000 km s^{-1} or more are seen. By contrast, in Seyfert 2 AGN the central source is obscured and only the outer, relatively colder regions farther away than the BLR clouds are seen by the observer. There appears to be another distinction between 'classical' Seyfert 2s where the BLR is obscured but its presence can be inferred from polarized light, and those that do not appear to have a BLR. The unification scheme divides Seyferts 1 and 2 according to the *face-on* or *edge-on geometry* of the BH-disc combination, respectively. Seyfert 1s are generally characterized by the presence of broad emission lines, whereas Seyfert 2s contain narrow lines. The atomic physics of these lines is a clear discriminant between the underlying spectral environments: the narrow lines in Seyfert 2s are associated with forbidden transitions, and the broad lines with strong dipole allowed or intercombination transtions, as given in Tables 13.1 and 13.2. However, there is a virtual continuum of sub-classes between Seyferts 1 and 2, ascertained from detailed spectroscopy [228, 311]. It is common to find Seyfert galaxies classified in the literature according to the width of commonly observed lines, as Seyfert 1.1, 1.2, etc., up to Seyfert 2.

In addition to the Seyfert progression from 1 to 2, there are other subclasses, such as low-ionization narrow-line emission radio (LINER) galaxies [311]. Likewise, another sub-class is especially interesting. These are the so-called narrow-line Seyfert 1 galaxies (NLSy1) that have been the subject of extensive study in recent years [334]. As the name implies, they are distinguished by significantly narrower lines than the BLR lines from classic Seyfert 1 galaxies. Some of their prime characterstics are extreme by normal AGN criteria, steep hard and soft X-ray spectra and variability, and strong Fe II emission [303, 335, 336, 337, 338, 339, 340]. The properties of NLSy1s can be explained as young AGN with high accretion rate, implying a small but growing black hole mass (e.g., [341, 342]).

13.1.2 Supermassive black holes: the central engines

Why a black hole, instead of some other physical process, such as nuclear fusion, to create the energy emanating from the central regions? After all, stars create immense amount of energy for long epochs via thermonuclear fusion. It is therefore logical to ask if extremely massive stars can produce the observed luminosities, especially since L rises as a steep power of mass, i.e., $L \sim M^{3.5}$ (Chapter 10). However, elementary considerations show that no other process except gravitational infall can result in the observed energy output observed from AGN. Eddington first provided the key considerations that relate the maximum mass of a radiating star, held together by inward force of gravity balanced by outward radiation pressure. The minimum interaction between radiation and matter, i.e., the *minimum* opacity, is due to scattering of photons and electrons in a fully ionized gas given by the Thomson cross section

$$\sigma_T = \left(\frac{8\pi e^4}{3m_e^2 c^4}\right) = 6.65 \times 10^{-25}\ \mathrm{cm}^2 \qquad (13.1)$$

Since both the radiative force, related to luminosity L, and the opposing gravitational force, related to the mass

TABLE 13.1 Forbidden lines and H, He lines in narrow-line regions.

Ion	λ (Å)	Transitions
[Ne V]	3345.8, 3425.9	$2p^2(^3P_{1,2} - ^1D_2)$
[O II]	3726.0, 3728.8	$2p^3\left(^4S^o_{3/2} - ^2D^o_{3/2,5/2}\right)$
[Ne III]	3867.5, 3968.8	$2p^4(^3P_{1,2} - ^1D_2)$
[S II]	4076.35	$2p^3\left(^4S^o_{3/2} - ^2P^o_{1/2}\right)$
$H\delta$	4101	2–6
$H\gamma$	4340	2–5
[O III]	4363.2	$2p^2(^1D_2 - ^1S_0)$
He I	4471	$1s2p(^3P_{1,2}) - 1s4d(^3D_{1,2})$
He II	4686	3–4
Hβ	4861	2–4 (3s,3p,3d – 4s,4p,4d,4f)
[O III]	4958.9, 5006.6	$2p^2(^3P_{1,2} - ^1D_2)$
[N I]	5179.9, 5200.4	$2p^3\left(^4S^o_{3/2} - ^2D^o_{3/2,5/2}\right)$
[Fe XIV]	5302.86	$3s^23p\left(^2P^o_{1/2} - ^2P^o_{3/2}\right.$
[Fe VII]	5721.11	$3p^63d^2(^3F_2 - ^1D_2)$
[N II]	5754.6	$2p^2(^1D_2 - ^1S_0)$
He I	5875.6–5875.97	$1s2p\left(^3P^o_{0,1,2}\right) - 1s3d(^3D_{1,2,3})$
[Fe VII]	6086.92	$3p^63d^2(^3F_3 - ^1D_2)$
[O I]	6300.30, 6363.78	$2p^4(^3P_{2,1} - ^1D_2)$
[Fe X]	6374.53	$3s^23p^5\left(^2P^o_{3/2} - ^2P^o_{1/2}\right)$
[N II]	6548.1, 6583.4	$2p^2(^3P_{1,2} - ^1D_2)$
Hα	6563	2 – 3 (2s,2p – 3s,3p,3d)
[S II]	6716.5, 6730.8	$2p^3\left(^4S^o_{3/2} - ^2D^o_{5/2,3/2}\right)$
[Ar III]	7135.8	$3p^4(^3P_2 - ^1D_2)$
[O II]	7319.9–7329.6	$2p^3\left(^4D^o_{5/2,3/2} - ^2P^o_{3/2,1/2}\right)$
[Ar III]	7751.1	$3p^4(^3P_1 - ^1D_2)$

of the central object M, fall off in the same way as $1/r^2$, their ratio

$$\frac{F_{\rm rad}}{F_{\rm grav}} = \frac{(\sigma_T L/4\pi c r^2)}{(GMm_e/r^2)} \tag{13.2}$$

is independent of the distance r from the central object. When the gravitational and radiative forces are equal,[1] this simple relation provides a specific limit on the luminosity of the object, called the *Eddington luminosity*

$$L_{\rm Edd} = \frac{GM4\pi c m_e}{\sigma_T}. \tag{13.3}$$

For protons, assuming the same Thomson opacity as electrons, $L_{\rm Edd} = 1.26 \times 10^{38} (M/M_\odot)$ erg s^{-1}. This is also the lower limit on the Eddington luminosity for the H atom, with the proviso that the opacity would be higher

[1] A discussion of radiative accelerations is given in Chapter 11.

TABLE 13.2 Allowed and intercombination lines in broad-line regions.

Ion	λ(Å)	Transitions
Lyγ	972.5366, 972.5370	$1s(^2S_{1/2}) - 4p\left(^2P^o_{1/2,3/2}\right)$
Lyβ	1025.7218, 1025.7229	$1s(^2S_{1/2}) - 3p\left(^2P^o_{1/2,3/2}\right)$
Lyα	1215.6682, 1215.6736	$1s(^2S_{1/2}) - 2p\left(^2P^o_{1/2,3/2}\right)$
C II	1334.53,1335.66,1335.71	$2s^22p\left(^2P^o_{1/2,3/2}\right) - 2s2p^2(^2D_{3/2,5/2})$
C IV	1548.20, 1550.77	$2s(^2S_{1/2}) - 2p\left(^2P^o_{3/2,1/2}\right)$
He II	1640	2–3 (2s,2p – 3s,3p,3d)
C III	1908.73	$2s^2(^1S_0) - 2s2p\left(^3P^o_{1,2}\right)$
C II	2324.21, 2328.84	$2s^22p\left(^2P^o_{1/2,3/2}\right) - 2s2p^2(^4P_{1/2,3/2,5/2})$
He II	4686	3–4 (3s,3p,3d – 4s,4p,4d,4f)
Hβ	4861	2–4 (3s,3p,3d–4s,4p,4d,4f)
He I	5876	$1s2p\left(^3P^o_1\right) - 1s3d(^3D_{1,2})$
O I	6300.30, 6363.78	$2s^22p^4(^3P_2 -^1 D_2,^3 P_1 -^1 D_2)$
Hα	6563	2–3 (2s,2p–3s,3p,3d)
O I	8446.5	$2s^22p^3(^4S^o)\left[3s\left(^3S^o_1\right) - 3p(^3P_{1,2})\right]$
N V	1238.82, 1242.80	$2s(^2S_{1/2}) - 2p\left(^2P^o_{3/2,1/2}\right)$
O IV	1397.23, 1407.38	$2s^22p\left(^2P^o_{1/2,3/2}\right) - 2s2p^2(^4P_{1/2,3/2,5/2})$
S IV	1393.76, 1402.77	$3s(^2S_{1/2}) - 3p\left(^2P^o_{3/2,1/2}\right)$
Mg II	2795.53, 2802.71	$3s(^2S_{1/2}) - 3p\left(^2P^o_{3/2,1/2}\right)$

than the Thomson value. Therefore, matter in a radiative source must have $L \leq L_{\rm Edd}$ to ensure gravitational stability against radiation pressure, for the object to exist. Despite this simple theoretical argument, there are many instances of AGN with $L > L_{\rm Edd}$, particularly NLSy1s with high accretion rates. In solar units of luminosity and mass, $L_\odot$ and $M_\odot$, respectively, the maximum luminosity is given by

$$\frac{L}{L_\odot} \leq \frac{L_{\rm Edd}}{L_\odot} = \frac{4\pi G m_{\rm p} M}{\sigma_{\rm T} L_\odot} = 3.2 \times 10^4 (M/M_\odot). \tag{13.4}$$

Since $L(\text{AGN}) \sim 10^{12} L_\odot$, massive stars that may produce this much energy from thermonuclear reactions must be 10^8 times more massive than the Sun. Such stars do not exist! Even stars less than 100 times as massive than the Sun, such as the Wolf–Rayet stars, have strong and fast radiatively driven winds that carry away a significant fraction of stellar mass, and are therefore on the verge of radiative instability (recall the discussion on the LBV η Car discussed at the end of Chapter 10). So, single massive stars cannot be the source of the amount of energy generated within AGN, both from a dynamical and a spectral point of view.

But what about clusters of massive stars at the centre of galaxies? Again, quite logical to ask, but ruled out by considerations of the size of the emitting region and stellar spectral properties. As shown below, the size of the nucleus is estimated from spectral-time variablity studies to be of the order of 0.01 pc, or 0.03 light years. Thus apart from the fact that the observed spectra of AGN are quite different from thermal stellar spectra, the small physical size makes it impossible for stars, or clusters thereof, to constitute the nucleus. Therefore, the source of energy needs to be from conversion of much more mass into energy than can be achieved by nuclear fusion. Hence the SMBH paradigm, which rests simply on the assumption that the observed luminosity

$$L = \epsilon \dot{M} c^2, \tag{13.5}$$

is due to conversion of mass $\dot{M}$ into energy via some *unknown* efficiency fraction ϵ. Exactly how this happens is at the heart of the AGN conundrum. For instance, the efficiency of graviational-to-mechanical energy conversion depends on the rotation the black hole; ϵ is ~5% for a non-rotating Schwarzschild black hole, but can be up to ~50% for a a rotating Kerr black hole, since the last stable orbit for accreting matter is much farther in the latter case. With the working SMBH paradigm then, the task is to understand black-hole physics and the matter–light interactions occurring under the influence of extreme gravity.

13.1.3 Black hole masses and kinematics

The spectra of AGN in several lines and wavelength bands show variations attributed to black hole activity. This variability or 'reverberation' may be exploited to ascertain black hole masses and other properties of the emitting region – such studies are called *reverberation mapping* [343, 344, 345, 346, 347]. A straightforward manifestation is exemplified by the simple picture of 'orbital' motion of BLR clouds around a SMBH. In that case, the observed Doppler velocity dispersions may be inferred by the widths of bright hydrogen lines Hα or Hβ typical of ionized H II regions (Fig. 8.3). If the centrifugal force of the BLR with mass $m_{\rm BLR}$ is balanced by the gravitational force due to the SMBH of mass $M_{\rm BH}$ then

$$\frac{(m_{\rm BLR})(v^2_{\rm FWHM})}{R_{\rm BLR}} = \frac{(GM_{\rm BH})(m_{\rm BLR})}{R^2_{\rm BLR}}, \qquad (13.6)$$

where $v_{\rm FWHM}$, dispersion at full width half maximum, is inferred from the width of the line formed in a BLR cloud at a distance of $R_{\rm BLR}$ from the central black hole, whose mass would then be

$$M_{\rm BH} = \frac{< v^2 > R_{\rm BLR}}{G}. \qquad (13.7)$$

Measured velocity dispersions of BLR clouds of up to $0.1c$, at a distance of ~0.1 pc from the black hole, gives an order of magnitude estimate of its mass to be $\sim 10^8 M_\odot$, radiating at the Eddington limit of $\sim 10^{12} L_\odot$. Observed velocities and AGN luminosities imply the SMBH masses to be in the $10^{6-9} M_\odot$ range.

A further refinement is due to the fact that the geometry of an AGN is determined by the orientation of the BH-disc-BLR viewing angle i. Therefore, the BLR velocity is not likely to be isotropic. We may assume that it comprises two components, $v_{\rm p}$ parallel to the disc, and $v_{\rm r}$ accounting for random (radial) motions. Then it may be shown that the black hole mass is given by

$$M_{\rm BH} = \frac{1}{4(v_{\rm r}/v_{\rm p})^2 + \sin i}\left(\frac{v^2_{\rm FWHM} R_{\rm BLR}}{G}\right). \qquad (13.8)$$

The parameters on the right, including the geometical anisotropy factor, are determined observationally and by modelling. There are several methods for determining black hole masses, relying on the *virial theorem*, which embodies the basic kinematics outlined above in the relation

$$U = \frac{1}{2}\Omega, \qquad (13.9)$$

where U is the kinetic energy and Ω is the gravitational potential energy.

13.1.4 The $M_\bullet - \sigma_*$ relation

A remarkable correlation has now been established between the masses of central black holes and the dispersion in velocity of stars in the bulge of galaxies (e.g., [348, 349, 350]), as shown in Fig. 13.2. The tight correlation is clear evidence that every galaxy with a bulge has a SMBH. It suggests that the growth of the central black hole is related via some *feedback mechanism* to the mass of the galaxy, and consequently its growth and evolution. This correlated growth in both the black hole and the galaxy now provides a framework and a connection between AGN activity, as determined by the SMBH and the host galaxy (albeit those with a bulge). Observational techniques for determining black hole masses [351] relying on spectral properties and the virial theorem are: reverberation mapping [343, 344, 345, 352], the bulge/BH mass ratio [353, 354], disc luminosity [353, 356], Hα line analysis, the IR calcium 'triplet' CaT (Chapter 10) measurements [357], and X-ray variability or power-spectrum break ([351, 358, 359] (also as discussed later with reference to Fig. 13.6).

The correlation in Fig. 13.2 has been quantitatively determined emprirically as

$$M_\bullet({\rm BH}) \propto \sigma_*^\alpha. \qquad (13.10)$$

A more refined analyis assumes a log-linear $M_\bullet - \sigma_*$ relation: $\log (M_\bullet/M_\odot) = \alpha + \beta \log \sigma_*/\sigma_0$, where $\alpha = 7.96 \pm 0.03$, $\beta = 4.02$ and $\sigma_0 = 299$ km s^{-1} [351]. The scatter does *not* increase with SMBH mass, although black holes may grow via several processes, such as galaxy mergers and gas accretion. This implies the existence of a feedback mechanism. Such a scenario had been proposed on theoretical grounds for SMBH formation *preceding* first stellar formation [349]. Gravitational collapse of primoridal giant clouds of cold dark matter could form

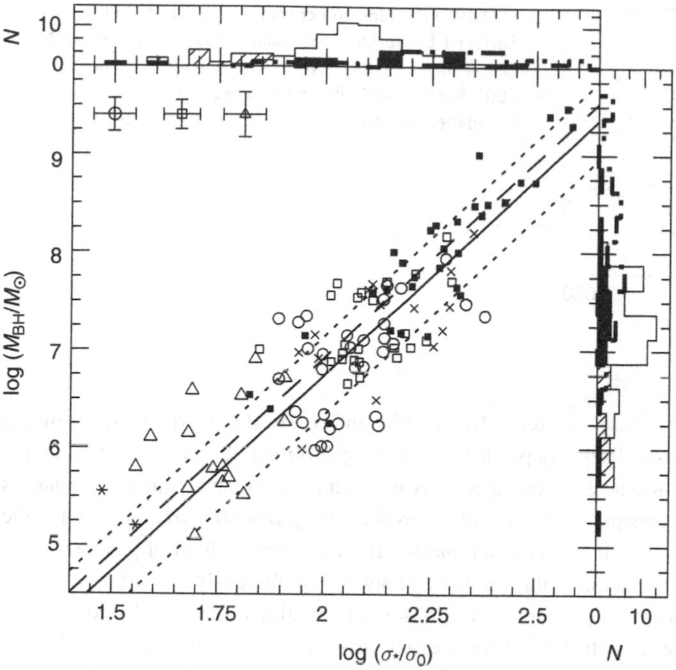

FIGURE 13.2 The $M_\bullet - \sigma$ relation between the central black hole mass and stellar velocity dispersion σ_* (measured in $km\,s^{-1}$) in the bulge, for both active and inactive galaxies [350].

$M_\bullet > 10^6 M_\odot$,[2] followed by stellar formation in the spheroidal bulge. The virial theorem (Eq. 13.9) would then apply between the gravitational potential and the velocity disperson of stars, as observed. As the black hole grows via accretion, it radiatively drives ever more intense winds that would shut off the accretion flow at some point, thereby regulating the activity of the central source. Furthermore, like a thermostat or a hypothetical 'valve', it would also serve to decouple the rest of the galaxy and its evolution. Nevertheless, it is logical to conclude that all galaxies – AGN hosts and non-active galaxies – undergo episodes of periodic central activity driven by the SMBH. Observationally, the $M_\bullet - \sigma$ relation appears to hold for both active and non-active galaxies, implying that galaxy formation may be initiated by the central SMBH for all galaxies in periodic phases of activity. Finally, the exceptions from the $M_\bullet - \sigma$ relation in Fig. 13.2 are the NLSy1s with high L/L_{Edd}, which is also an argument that NLSy1s are young AGN with rapidly growing black hole mass [342].

13.1.5 Size of the emitting region

The Spectra of AGN exhibit variability in the lines as well as the continuum (not always correlated), typically on a scale of days to weeks. This timeframe constrains the size of the emitting region,

$$D = c\Delta t, \tag{13.11}$$

where Δt is the period of observed variability and is interpreted as the *light crossing time*; multipled by c it gives the maximum size of the region responsible for emission. The variablity of BLR lines indicates their size to be less than 0.1 pc, and the variability of the continuum indicates the size of the central source to be even smaller. Figure 13.3 shows the variability of emission lines. The continuum in the Seyfert 1 galaxy 3C 390.3 belongs to an interesting class of broad-line radio galaxies. Over an approximately ten-year timeframe, there is pronounced variation in the continuum flux, as well as line-flux profiles of the Balmer series, Hα, Hβ, etc. While the strengths of the narrow [O III] 4959, 5007 lines remains constant, the continuum and the broad emission lines underwent a strong increase from August 1994 to October 2005. The non-variability of NLR lines points to their origin in a region much farther away than the BLR from the central source. The double-humped shape of the Hα line profile is generally interpreted as the signature of accretion disc emission due to rotational Doppler motion.

13.1.6 States of black hole activity

The SMBH paradigm depends on the release of gravitational energy during accretion of matter on to the central black hole. Since X-ray spectra originate from the closest

[2] The primordial cloud masses need to be large so as not be disrupted by supernova-driven winds [349].

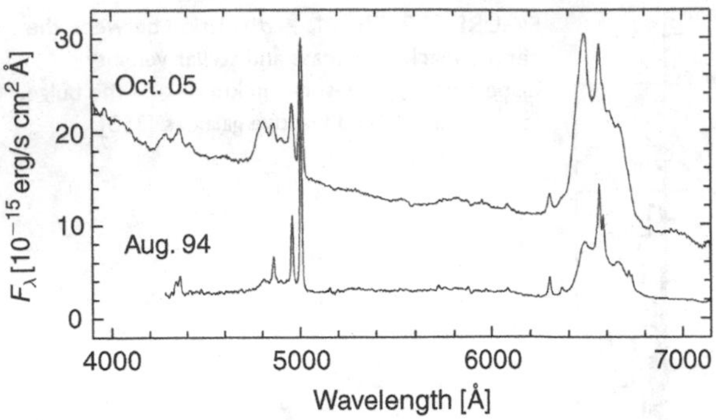

FIGURE 13.3 Line and continuum variability in the Seyfert I broad-line radio galaxy 3C390.3 (Courtesy: M. Dietrich). The unbroadened forbidden [O III] lines at $\lambda \sim$ 5000 Å, and the double-peaked Hα line profile at $\lambda \sim 6400$ broadened due to black hole reverberation, are shown.

observable regions to the central source, including possibly the innermost stable orbits within the accretion disc, the X-ray fluxes are indicative of the level of black hole activity. A dichotomy reveals itself by the two approximate bands of emitted energy, in the *soft* X-ray spectra between 0.5–2 keV, and relatively hard X-ray spectra between 3–10 keV. The two 'states' of emission from an AGN may be characterized either by a high level of soft X-ray flux, or a low level of hard X-ray flux, giving rise to the terminology 'high / soft' or 'low / hard' states, relating black hole activity and the hardness of the spectrum. Figure 13.4 shows the low and high states for a bright AGN Mrk 335 (as named in the Markarian catalogue of bright active galaxies).

These two 'states' of black hole activity may be understood in the sense that a high X-ray flux indicates considerable material interactions in the accretion disc, that would tend to reprocess mechanical energy and 'thermalize' the emitting plasma. The resulting flux is characterized by peak emission towards energies and temperatures in the 1-2 keV ($\sim 10^{6-7}$ K) range, or soft X-rays. L-shell excitation in Fe ions plays a big role in soft X-ray production. Such a situation would correspond to a high accretion rate. Contrariwise, when the accretion rate is low, the observed spectrum is dominated by the non-thermal component characteristic of black hole activity, associated with a relatively quiescent accretion disc and low flux levels of hard X-ray emission.

There are myriad observations of emission lines originating in the hot coronal line region surrounding the accretion disc. In the aforementioned AGN Mrk 335 (Fig. 13.4), was recently observed by X-ray instruments aboard the Gamma-Ray Burst Explorer Mission *Swift*, and the *X-Ray Multi-Mirror Mission – Newton* (XMM) space observatories, and found to be in an historically low X-ray flux state in 2007, in contrast to earlier observations of much higher flux [360]. But an important caveat is in order when interpreting the variability in flux levels from AGN. To a significant extent, the interpretation is model dependent, indicating different levels of activity in the central SMBH environment. We mention two scenarios that are often invoked: the *partial absorber model* and the *reflection model*. If an absorbing cloud of gas intervenes along the line of sight then obviously the flux would drop and lead to observed variations. In the reflection model, the thermal disc emission is complemented by hard emission from the hot corona being reflected from the cooler disc (e.g., [328]), and emission lines due to fluorescence and (e + ion) recombination (the most interesting example is the fluorescent Fe 6.4 keV X-ray line discussed later).

Remarkably, in spite of the greatly diminished flux from Mrk 335, the soft X-ray emission lines of highly ionized Fe ions, N VII, O VII and others are sufficiently bright to enable high-resolution spectroscopy, as also shown in Fig. 13.4. Therefore, the X-ray properties and variability can be studied to constrain spectral models. Spectral analysis of especially the forbidden (f), intercombination (i) and resonance (r) lines of He-like O VII (Fig. 8.8), and line ratios $R \equiv f/i$ and $G \equiv (i+f)/r$ are powerful diagnostics of density, temperature, and ionization equilibria (as discussed in Chapters 10 and 8).

The O VII line ratios in Mrk 335 have been measured with high uncertainties, and reported to be r/i = 0.38±0.25 and G = $(f+i)/r$ = 4.30±2.70 [361]. Neglecting the large error bars (for illustrative purposes), these ratios seem to suggest the intercombination i-line to be nearly three times stronger than the resonance r-line, quite different from normal coronal situation where the reverse is true (cf. Fig. 8.10). The suggested mechanism for this line intensity inversion between the i and f lines could be photo-excitation from the $1s2s(^3S_1)$ level of O VII upwards to the $1s2p(^3P^o_1)$ level, which then decays preferentially into the i-line $1s2p(^3P^o_1) \rightarrow 1s^2(^1S_0)$, thereby making it stronger at the expense of the f-line $1s2s(^3S_1) \rightarrow 1s^2(^1S_0)$. The high value of the observed G ratio ~ 4.3, in contrast to normal coronal value of ~ 0.8,

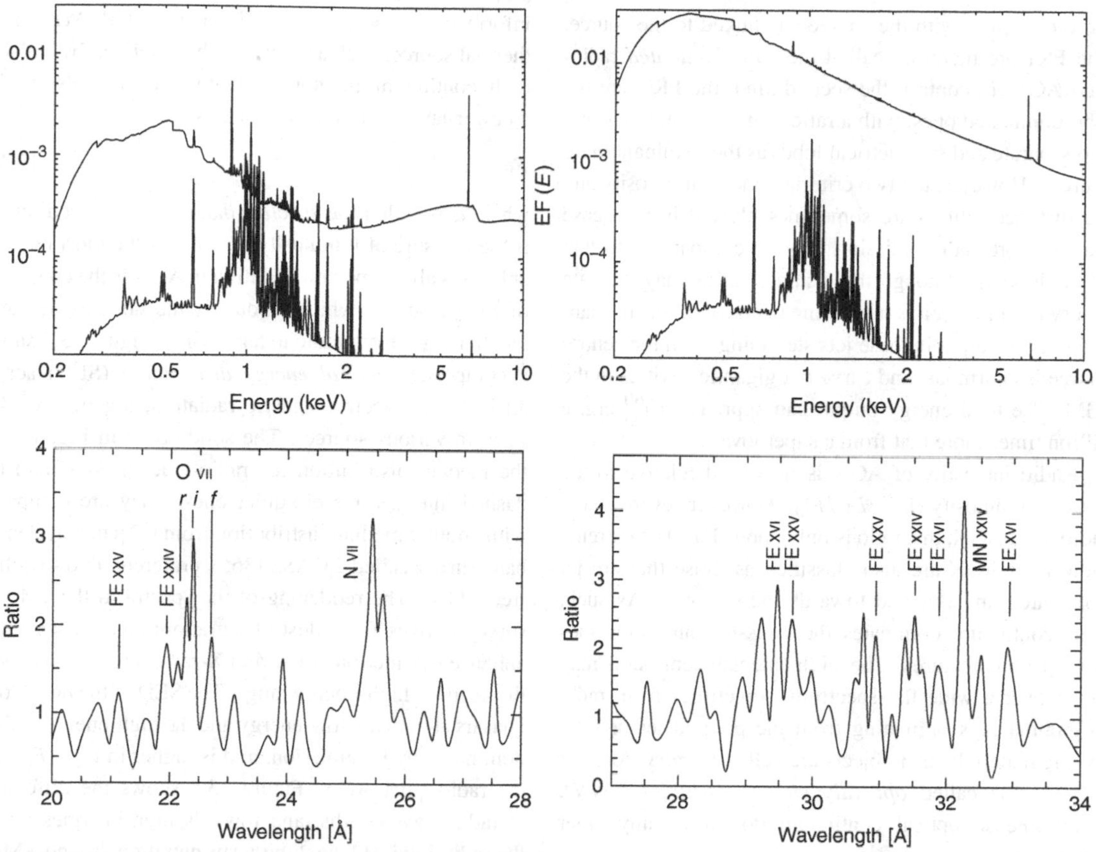

FIGURE 13.4 X-ray fluxes of the narrow-line Seyfert 1 galaxy Mrk 335 [360] in a low state (top curve left panel), compared to a high-state (top curve right panel). The spectral lines included in the models are also shown as the bottom curve in both panels (courtesy: L. Gallo and D. Grupe). The soft X-ray spectra show a number of coronal ions, particularly Fe XVI–Fe XXV (bottom right), in addition to the Heα complex of O VII lines *r, i, f* (bottom left) whose line ratios enable density and temperature diagnostics (Chapters 7 and 9).

indicates a photoionized plasma rather than a collisionally ionized one (Fig. 8.14).

13.1.7 Radio intensity

Historically, radio-loud objects at high-redshift were initially called *quasi-stellar radio sources*, or quasars. Here, 'radio-loud' is defined by comparing the ratio of the radio luminosity[3] in the $\nu \sim 5$ GHz ($\lambda \sim$ 6 cm) range to the total bolometric luminosity, e.g., the flux ratio $F(5\ \text{GHz})/F(B) > 10^3$ [362]. In 1963 Maarten Schmidt discovered that the optical emission lines from the quasar 3C 273 were redshifted, and that the source was at a large distance from the Milky Way ($z = 0.158$). It followed that 3C 273 was an extremly bright object, given its apparent luminosity. A large number of such quasi-stellar objects or QSOs are now known, but it turns out that most of the QSOs are indeed radio-quiet.

Although only about 10% of AGN are characterized by significant radio emission, they were in fact first discovered as radio sources. B. L. Fanaroff and J. M. Riley (hereafter FR [363]) divided the radio-loud sources into two groups, essentially defined by their morphology. Images of radio-loud AGN appear to have two large and radio-bright lobes on either side of a source in the middle or the core. A jet is often clearly visible close to either the core or the lobes, all thereby connected by a relatively straight line. The distinction between the two types is given by the ratio of the distance between the brightest spots on either side to the total size. If this ratio is less than about 0.5, a radio-loud AGN is referred to as FR1. In FR1, the central source is the dominant source of

[3] We have eschewed discussion of radio lines since it is rather specialized spectroscopically, and primarily related to molecular emission, say from CO, SiO, etc.

radio emission, with the jet seen as linked to the source. The FR1 are therefore called the *core-dominated* radio-loud AGN. In contast, the second kind, the FR2, are the lobe-dominated ones, with a ratio of greater than 0.5, and two separate and symmetrical lobes as the dominant radio sources. However, the two criteria – radio luminosity and the distance ratio – are sometimes blurred in the sense that the core radio emission may be comparable to that of the lobe(s), although the distance ratios may indicate otherwise. Finally, it is interesting to note that the mechanical energy output in these jets stemming from the central source is enormous, and carve out gigantic cavities in the IGM. The total energy output can approach 10^{60} erg, a billion times more that from a supernova.

Radio intenstity of AGN is measured relative to the optical luminosity, i.e., L_R/L_O. However, even in the radio-loud AGN, this ratio is only about 1%. The extreme radio-loud AGN are also classified as those that are jet dominated and oriented towards the observer. As such, their continuum dominates the emission and subsumes any spectral features. The high intensity emission may range over a wide flat spectrum, sometimes from radio to gamma rays orginating from the pointed jet or relativistic material; such objects are called *blazars*. Another sub-class is called *optically violent variables* (OVV), if an intense optical continuum dominates any other emission.

At least 10% of luminous radio-loud quasars constitute a sub-class of *broad absorption line* (BAL) objects that exhibit not only lines that are broad but also blueshifted. They imply high-velocity outflows that absorb impinging radiation from the central source along the line of sight to the observer. Recent spectropolarimetric studies show that some quasars seen face-on have non-equatorial outflows whose spectra are polarized parallel to the radio axis by an equatorial scattering region [364]. Polarization studies therefore also reveal information on the morphology and dynamics of such quasars.

13.2 Spectral characteristics

Historically, the AGN classification scheme depends on the observed spectral properties in the optical. Among the most common are the Balmer Hα and Hβ lines, and the forbidden 5007 Å [O III] line. But over the years observations in other wavelength ranges have acquired increasing importance in revealing the substructures and kinematics of AGN. For example, the Fe II lines from AGN are observed from the near-infrared to the near-ultraviolet, and potentially contain a plethora of information, discussed later in more detail. Whereas a thermal source, such as a star, is characterized by a black-body continuum, the non-thermal source in AGN implies a power-law continuum, defined as

$$F_\nu \sim \nu^{-\alpha}, \tag{13.12}$$

where α is called the *spectral index* (the observed flux F is the measure of luminosity L). One of the most remarkable facts about spectral variation in AGN is the constancy of energy output, which is roughly the same in all wavelength ranges from radio to hard X-ray. Figure 13.5 shows a composite *spectral energy distribution* (SED) across all bands of electromagnetic radiation, log νF_ν vs. log ν, from various sources. The solid curve in Fig. 13.5 is the median distribution for radio-loud quasars, and the dashed line is for radio-quiet ones. They are compared with another median distribution from a 2 μm wavelength band survey called 2MASS [365], preferential to so-called 'red-AGN'. The reddening of the spectra in the 2MASS survey derives from dust obscuration, and consequently enhanced extinction at UV and X-ray energies with lower fluxes in the high-energy range. The SED of the radio-loud quasars at both the low-energy and the high-energy ends is dominated by jet emission, and is higher in log νF_ν than the radio-quiet AGN. Figure 13.5 shows the difference at radio wavelengths, and towards high energies where the radio-loud AGN exhibit a strong hump around 1 MeV $\sim 10^{20}$ Hz (somewhat beyond the rising solid curve on the right-hand side in Fig. 13.5).

Although the common feature in all AGN SEDs is the high-energy flux, which remains relatively constant, it is clear from Fig. 13.5 that fits to the power-law continuum in different multi-wavelength bands may differ significantly. Figure 13.6 shows a composite quasar spectrum obtained from the Sloan Digital Sky Survey [367]. It shows the characteristically strong UV and optical lines, superimposed on an underlying continuum fitted with two spectral indices, $\alpha_\nu = -0.46$ in the ultraviolet and $\alpha_\nu = -1.58$ in the optical.

Likewise, the spectral index in the X-ray ranges from +0.5 to −2.0 in the soft X-ray (0.1–3.5 keV), and from about −0.5 to −1.0 in the hard X-ray (2–10 keV). X-ray astronomers often use the parameter $\Gamma = (\alpha + 1)$, called the *photon index*,

$$F_E \sim E^{-\Gamma}. \tag{13.13}$$

The spectral or the photon indices are related to AGN activity, which depends on the accretion rate on to the SMBH through the parameter $L/L_{\rm Edd}$ (e.g., [339, 368]).

Why are the few emission lines in Fig. 13.6 so commonly observed, not only in AGN but also from many

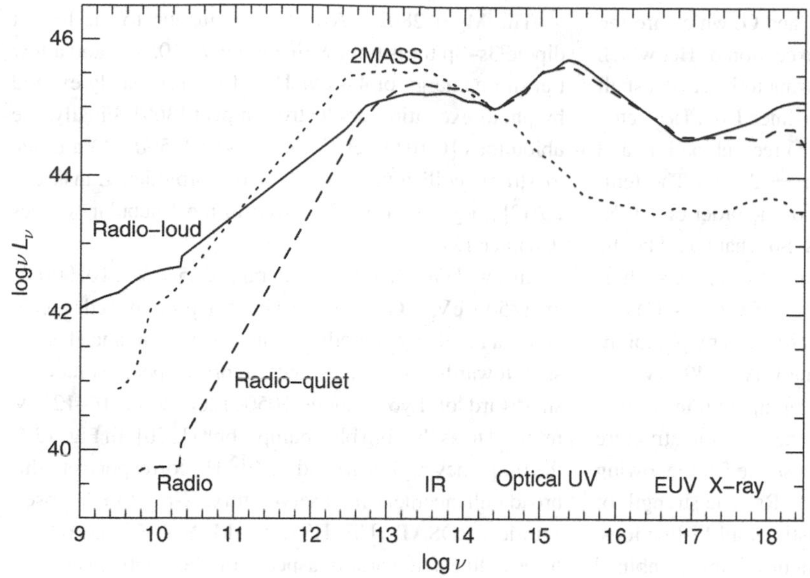

FIGURE 13.5 Spectral energy distribution of active galactice nuclei. The solid line represents the median energy distribution for radio-loud quasars, and the dashed line refers to radio-quiet quasars. The dotted line is a median energy distribution of 'red AGN' from the 2 μm survey called 2MASS [365], which refers to AGN where the central source is obscured (modelled after [366], courtesy: J. Kuraszkiewicz and B. Wilkes). Note that 1 eV = 2.42 $\times 10^{14}$ Hz.

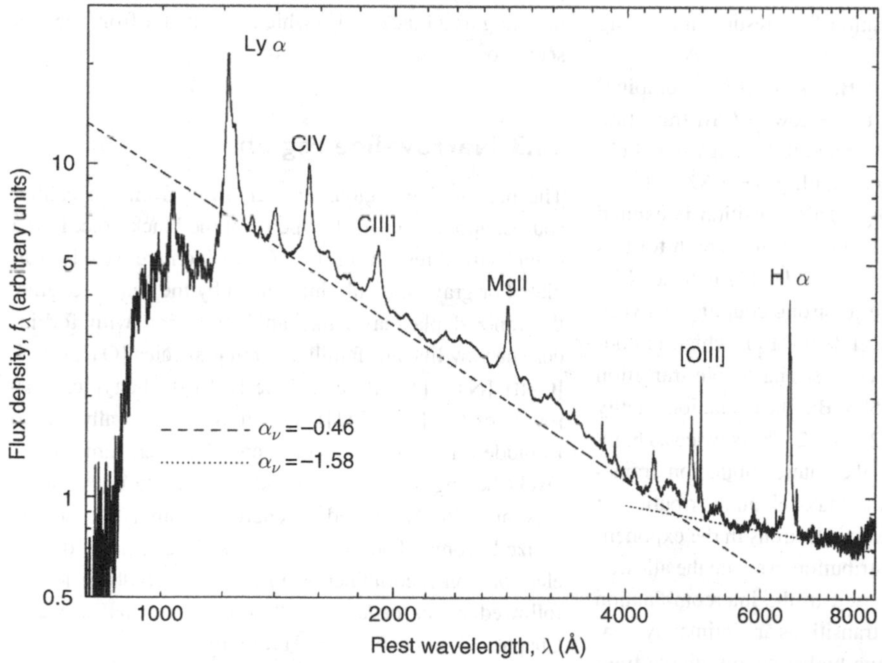

FIGURE 13.6 A composite quasar spectrum from the Sloan Digital Sky Survey [367]. The figure shows some of the prominent lines in the UV and the optical, as well as power-law fits with two different spectral indices α to the underlying non-thermal (non-stellar) continuum.

other astrophysical sources? It is worth examining the underlying atomic physics. There are several reasons (apart from the obvious fact that the lines happen to lie in the wavelength region being considered): (i) abundance of elements, (ii) *locally optimum physical conditions* for spectral formation, such as the excitation temperature and critical densities for a particular transition in an ion, (iii) temperatures and densities where particular ions are abundant and (iv) the atomic parameters that determine the strengths of lines. The sometimes intricate interplay of these factors determines the line intensity.

Most of the lines are from H, C and O, which are the most abundant elements (with the exception of He, which has very high excitation energies owing to its closed shell atomic structure). Recall that the H-lines Lyα, Hα , etc., are formed via $(e + H^+) \rightarrow H^0(n\ell)$ recombination, and cascades to $n = 2$ levels: $n\ell \rightarrow n = 2\ell, 3\ell$. The temperatures in the BLR and NLR are of the order of 10^4 K, which implies low ionization stages. So what could be the reason for the presence of strong lines from ions such as C IV, given that the ionization energy of C III → C IV is 47.9 eV? Multiply ionized ions are produced by photoionization from the central source. The C IV 1549 UV line is due to the strong (lowest) $2s - 2p$ dipole transition(s) $1s^2 2s\ ^2S_{1/2} \rightarrow 1s^2 2p\ ^2P^o_{1/2,3/2}$; the two fine structure components are usually blended as a single feature, owing to Doppler velocity broadening in BLRs. The strength of the $2s - 2p$ transition is characteristic of all Li-like ions, such as C IV.[4] The oscillator strength for the combined transition(s) is relatively large, $f = 0.285$. Although the transition energy is about 8 eV, photo-excitation by the strong UV background radiation field results in a strong line.

The other carbon line, the Be-like C III intercombination line at 1909 Å, is due to the lowest C III transition $1s^2s^2\ ^1S_0 - 1s2p\ ^3P^o_1$ with an excitation energy of 6.4 eV. But here the f-value is very small, $f = 1.87 \times 10^{-7}$. So why is the line so strong? This transition is excited largely by electron impact. The collision strength for this transition $\Omega(2^1S_0 - 2^3P^o_1)$ is considerably enhanced by *resonances* (Chapter 5), due to strong coupling between the $2^3P^o_1$ and the higher level $1s2p(^1P^o_1)$, which is connected to the ground state via a strong dipole transition with $f(2^1S_0 - 2^1P^o_1) = 0.7586$. But the excitation energy of the allowed transition $E(2^1S_0 - 2^1P^o_1)$ is twice as high, about 12.7 eV (977 Å), as the intercombination transition $(2^1S_0 - 2^3P^o_1)$. Thus, the Maxwellian distribution of electrons at 10^4 K has far fewer electrons in the exponentially decaying tail of the distribution to excite the allowed 977 Å line in C III, as opposed to the intercombination line at 1909 Å. But the two transitions are intimately connected by atomic physics: the higher energy dipole transition effectively enhances the intercombination line *indirectly* via resonance phenomena, often central to atomic processes (cf. Chapter 3).

[4] Another Li-like ion of great importance in astronomy is O VI. The two fine-structure lines are at λλ 1032, 1038 Å in the for ultraviolet. Because the O VI lines are from a higher ionization stage, the two components are separated more than C IV, and are often resolved. The O VI lines have great diagnostic value in studies of the ISM, Galactic halo, the solar corona and many other objects. To a lesser extent, Li-like N V lines at ~1240 Å are also commonly observed.

The Mg II 2802.7 Å line is again due to the lowest dipole 3s–3p transition, with a large $f = 0.303$, and a low transition energy of 4.4 eV. Therefore, it is easily excited by photo-excitation or electron impact [369]. Finally, the ubiquitous [O III] lines (λλ 4363, 4959, 5007 Å) are due to strong collisional excitation of forbidden transitions $2p^4(^3P_{0,1,2} \rightarrow ^1D_2, ^1S_0)$, seen in most nebular sources (Chapter 12).

In a wide multi-wavelength range, $25 < \lambda < 10,000$ Å, or 1–500 eV, AGN spectra show a pronounced excess above a relatively smooth continuum. The enhanced emission towards the blue shows a broad peak somewhat shortward of Lyα , about 1050–1200 Å or 10–12 eV, referred to as the 'big blue bump' (bbb) [370]. In Fig. 13.5, the frequency region around $\sim 10^{15}$ Hz corresponds to this broad enhancement in observed flux. X-ray to UV observatories, ROSAT, IUE, HST, XMM, SWIFT and others, have delineated various aspects of the 'bbb'. Theoretically, the bbb around 10 eV is expected for emission predicted by the thin-disc models, assuming the disc to be radiating as a black body (which may be far from the real scenario[5]).

13.3 Narrow-line region

The narrow-line regions are far away from the central source, approximately 100 pc from the black hole. In the outer ionized regions of the AGN, which are weakly irradiated or gravitationally influenced by the central engine, the ionized plasma is like an H II region with forbidden, narrow lines of familiar nebular species [O I], [O II], [O III], [N II], [Ne III], etc. Table 13.1 lists the typical narrow lines (cf. [228, 311]). The lines are generally due to forbidden transitions of low energy between same-parity levels belonging to the ground configuration. As such, they are closely spaced in energy in singly or doubly ionized atoms. The H and He lines are usually due to electron–ion recombination into excited Rydberg levels, followed by cascades via allowed transitions (see also Chapter 12 on nebulae and H II regions).

The background ionizing continuum is, of course, different from the stellar continuum in nebulae, but the high-energy flux is heavily attenuated by the intervening regions of the AGN, so that only a limited amount of UV flux is available and produces low ionization stages of elements. As might be expected, the narrow-line spectra of AGN and their velocity distributions may span an entire

[5] Equation 7.4 in [371] derives an approximate relationship in terms of black-body photon emission from inner regions of an accretion disc around 10 eV.

range up to those found in broad line regions (viz. the spectral sub-divisions between Seyferts 1 and 2). In any case, the essential discriminant is that the narrow line regions are more indirectly subjected to AGN activity than the broad line regions, as determined by the geometry associated with the SMBH paradigm depicted in Fig. 13.1. Therefore, the essential spectroscopic physics of NLRs is typical of the H II regions discussed in Chapter 12.

13.4 Broad-line region

By contrast, the ionized clouds in the BLR have high systemic bulk velocities due to gravitational motions around the central black hole. The resulting spectra show extremely broad lines whose Doppler widths approach a FWHM of up to $0.1c$. While the proximity of the BLR to the black hole, about 0.1 pc, explains the high velocities, the physical mechanisms that result in the observed emission are not well-understood. In particular, it is not entirely clear in what way photoioization or collisional processes determine the ionization balance and excite the observed spectrum, especially in the case of anomalous Fe II flux from a sub-class of quasars (see next section). In addition, radiative transfer effects play an important role in determining the observed spectrum. Whereas the BLR spectra should reveal certain characteristics of the central source, theoretical models of the BLR are often unsatisfactory in being able to account for the coupling between the central engine and the ionized gas. Table 13.2 lists some common BLR emission lines in the UV/optical range, primarily due to allowed or intercombination E1 transitions (Chapter 4).

The relative fluxes in the BLR lines differ greatly among AGN, and are distinct from the NLR fluxes. One of the most significant differences, apparent from the list of lines in Tables 13.1 and 13.2, is the absence of forbidden lines in the BLR. This immediately suggests electron densities higher than the critical density for quenching line emission due to transition $i \rightarrow j$, given by

$$N_c > \frac{A_{ji}}{q_{ij}}. \tag{13.14}$$

Since the A-values for forbidden transitions are often very low, say $A \sim 10^{-2}\ \mathrm{s}^{-1}$, and electron impact excitation rate coefficients are, say $q \sim 10^{-10}\ \mathrm{cm}^3\ \mathrm{s}^{-1}$, the critical densities $N_c \geq 10^8\ \mathrm{cm}^{-3}$. These densities are sufficient to quench most forbidden lines, as happens in BLRs in contrast to NLRs.

The extreme velocities observed in the BLR have interesting consequences, owing to Doppler broadening and excitation of lines. For example, for a given velocity of 3000 $\mathrm{km\,s}^{-1}$, the Doppler width of Lyα at 1215 Å is $\Delta\lambda/\lambda = v/c = 0.01$, or more than 10 Å. This implies that the intense Lyα radiation, subject to significant trapping within the BLR owing to its very small optical depths, is capable of exciting lines with $\lambda \approx 1215 \pm 5$Å. An important example is Fe II, whose closely spaced levels can be strongly excited by Lyα fluorescence. These processes, and iron emission from AGN-BLR, are discussed in the next section.

13.5 Fe II spectral formation

We have discussed the forbidden [Fe II] lines in Chapter 12, as observed from the optically thin nebulae and H II regions. Here, we extend the discussion to the AGN-BLR where the physical conditions and excitation processes are quite different. Moreover, the underlying activity due to the central SMBH source manifests itself in defining the characteristics of BLR spectra. In contrast with the discussion of high energy spectra of highly ionized species emanating from the inner regions of AGN that follows later, low ionization stages of iron provide a useful and equally intriguing view of the outer regions – none more so than singly ionized iron. We already know that Fe II lines are prominent in the spectra of many astrophysical objects, such as the Sun and stars in general, all kinds of nebulae, supernovae, etc. But their presence in AGN and quasar spectra constitutes a special problem, owing to the extent and enormous intensity of the observed Fe II emission. For that reason it is necessary to study Fe II spectra in the AGN context, although the discussion of individual optical and infrared lines and line ratios is applicable generally to other sources as well. The ultraviolet spectra of BLRs of AGN show thousands of blended lines, mainly from Fe II [372]. The Fe II lines are so extensive as to form a *pseudo-continuum* that underlies the rest of the BLR spectra. Figure 13.7 shows the spectra of the prototypical strong Fe II emitter, the QSO I Zw 1, classified as a narrow-line Seyfert 1 galaxy [373]. The observed spectra from other atomic species in the 1500–2200 Å range [374, 375] are overlayed on the Fe II pseudo-continuum obtained from non-LTE calculations with an exact treatment of radiative transfer. These 'theoretical templates' of Fe II are useful in the interpretation of Fe II spectra in general, and have been tabulated with approximately 23 000 line fluxes ranging from 1600 Å to 1.2 μm for different BLR models [373].

Observationally, it is also possible to synthesize many different spectra of typical AGN and derive a representative 'template'. One of the most studied QSOs is the aforementioned super-strong Fe II emitter I Zw 1, the

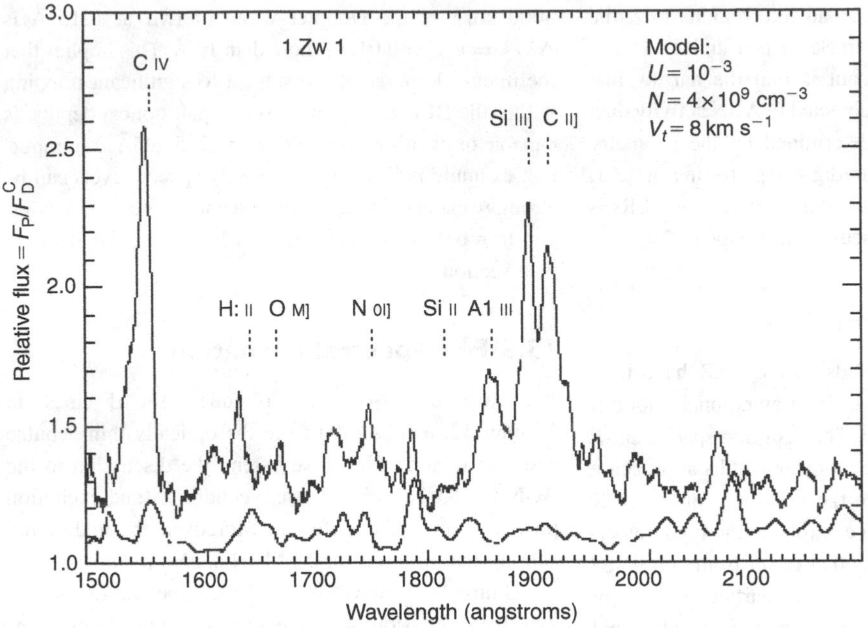

FIGURE 13.7 Ultraviolet spectra of I Zw I with strong Fe II emission that forms a pseudo-continuum (bottom solid line), with flux exceeding that in all other lines (top spectrum) [373].

prototypical NLSy1. Observational templates have been constructed in the optical [338] and in the ultraviolet [352]. The observational templates implicitly assume a 'typical' AGN spectrum, which may be scaled in some way when analyzing individual objects. But differences in physical conditions, such as Doppler blending, raise certain complications (which may of course be ameliorated by combining the analysis with theoretical templates). Often the Fe II spectra present such complications that observational templates are used to *subtract* the Fe II contribution from AGN spectra so as to facilitate the analysis of the remainder of the lines from simpler species, such as C IV and O III.

Numerous Fe II lines have been identified in the optical [338, 376] and in the near-infrared [377, 378]. Since AGN plasmas are essentially photoionized, we expect that photoionization models should be able to reproduce the strengths of Fe II lines. Yet another approach is to invoke mechanical heating of clouds shielded from the central continuum source, with Fe II emission orginating in or close to the outer accretion disc (e.g., [356]). That is also the reason we observe strong Fe II emission from Seyfert 1s, but not from Seyfert 2s. However, in spite of decades of theoretical modelling, combined with many observational programs (e.g., [379]) there remain large discrepancies in reproducing observed intensities. The nature of the Fe II problem becomes clear when one considers the fact that the cumulative Fe II lines are much more intense even than hydrogen lines, typically Fe II (UV/optical)/H$\beta \sim 10$, and ranging from ~ 2–30 for super-strong Fe II emitting QSOs [380, 381]. This, in turn, appears to imply several factors: (i) lack of understanding of the underlying physical mechanisms that form Fe II lines, (ii) incomplete atomic and radiative models, (iii) uncertainty about the nature of line formation regions surrounding the central source of AGN and (iv) abnormal iron abundances.

13.5.1 Fe II *Excitation Mechanisms*

To address these issues, it is first essential to describe the basic spectral physics of Fe II in detail, beginning with the description in Chapter 12 on ionized gaseous nebulae. As we noted in the nebular context, the complexity of the Fe II spectrum arises mainly from the fact that there are a large number of coupled and interacting levels in the low energy region (Fig. 12.5 schematically shows some of the prominent transitions). To recap and extend the earlier discussion, the ground configuration of Fe II and the lowest two *LS* terms are: $3d^64s(a^6D, a^4D)$. Lying close to it is the next excited configuration and its two lowest two terms, $3d^7(a^4F, a^4P)$. All four *LS* terms, and their fine structure levels, are easily excited collisionally and give rise to strong near-infrared forbidden lines, such as the 1.6436 and 1.2567 μm lines (Fig. 12.5). Transitions to higher levels of the same (even) parity result in forbidden optical lines, such as the λλ 4815, 5159 Å. The lowest odd-parity levels are: $3d^64p\ ^{6,4}(D, F, P)^o$. These sextet spin-multiplicty $(2S+1)$ terms are connected to the even-parity ground state ^{6}D via dipole allowed ultraviolet transitions such as at λλ 2599, 2382, 2343 Å. The quartet odd terms, on the other hand, decay to even-parity quartet terms, also via dipole transitions, resulting in allowed optical lines, such as λλ 4233 and 5198 Å.

An interesting variation in the above schematics is introduced by the presence of the level a^{6}S, which decays to the ground level via a forbidden optical transition; it also serves as the lower level for the 5169 Å allowed optical line due to decay from z^{6}P^o. Being an even-parity term, the a^{6}S can only be excited collisionally from the ground term a^{6}D, or other low-lying excited but populated quartet levels; however, it is observed in AGN spectra and is evidence of the occurrence of energetic collisional excitation processes. While the lower levels are collisionally populated at typical $T_e \approx 1$ eV, the energetics become more complicated, owing to photo-excitations from the lower levels up to high-lying levels of Fe II. To elucidate the microphysics of Fe II emission, we discuss four principal excitation mechanisms. Figure 13.8 presents a greatly simplified Fe II Grotrian diagram.

Continuum fluorescence

In Chapter 12, we discussed background UV fluorescent excitation (FLE) in nebulae to explain the anomalous intensity of some lines such as in O III and [Ni II]. The FLE mechanism is driven by continuum UV radiation absorption in strong dipole allowed transitions, followed by cascades into the observed optical emission lines (cf. [382]).

Collisional excitation

At temperatures $T_e \sim 10^4$K and densities $n_e > 10^8$ cm^{-3} it is possible to excite the odd parity levels at ~ 5 eV (Fig. 13.8), followed by UV and optical decays. Thus collisional excitation may be a major contributor to the Fe II emission.

Self-fluorescence

The 'unexpected UV' transitions in Fig. 13.8 emanating from 5 eV levels are due to absorption of the Fe II UV photons by overlapping UV transitions at a large number of coincidental wavelengths.

Lyα fluorescent excitation

The three mechanisms mentioned above are not sufficient to reproduce the observed Fe II emission. But a very effective radiative excitation mechanism is Lyα pumping, schematically shown in Figs. 13.8 and 13.9 (cf. [383, 384, 385]). Also shown is a partial Grotrian diagram of the quartet levels and multiplets that participate in Lyα FLE. In moderately dense plasmas encountered in AGN-BLR, $n_e \approx 10^{9-12}$ cm^{-3}, Lyα photons are trapped ('redistributed'), and excite the low-lying and significantly populated even parity levels of Fe II to much higher levels around 10 eV [383, 384, 386]. One particular case is the excitation of the 3d^{6}5p levels, which subsequently decay into the e(^{4}D,^{6} D) levels giving rise to strong enhancement in the near-IR region 8500–9500 Å, with a strong feature at 9200 Å [378, 384]. Furthermore, cascades give rise to another set of optical and ultraviolet lines (Fig.13.8).

Other examples of Lyα fluorescence are the excitation of a^{4}G – b^{4}G^o multiplet (the b^{4}G^o term lies at about 13 eV), followed by primary cascades into a group of UV lines at λλ 1841, 1845, 1870 and 1873 Å [386], and secondary cascades into a group of near-infrared lines $\sim 1\ \mu$m, as shown in Fig. 13.9 [378]. Many of these Fe II near-infrared lines have been seen from NLSy1 galaxies [378], as well as from other objects such as a Type IIn supernova remnant [387]. The detection of these secondary lines provides reasonably conclusive proof of the efficacy of Lyα FLE not only in AGN but generally in astrophysical sources with conditions similar to AGN-BLR. In addition to Lyα, we also need to consider Lyβ pumping of even higher levels of Fe II within the non-LTE formulation with partial redistribution.

13.5.2 *Fe I–Fe III emission line strengths*

Thus far we have described line formation of strong transitions in Fe II spectra (a fuller discussion, with and without Lyα FLE, is given in [373]). Much of this understanding is based on theoretical templates derived from sophisticated non-LTE Fe II models at a range of temperatures and densities, and including all known processes prevailing in the BLR. Extending the models to include the adjacent ionization stages, Fe I and Fe III including up to 1000 levels (827 from Fe II), theoretical templates have been computed to predict AGN iron emission from Fe I–Fe II–Fe III, as shown in Fig. 13.10 [379]. The theoretical templates can be compared with observational ones [373, 379]. The computed line fluxes show that generally over 90% iron emission is from Fe II at a range of ionization parameters U $\sim 10^{-1.3}$–10^{-3} (see Chapter 12 for a definition of U), and $n_e \sim 10^{9.6}$–$10^{11.6}$ cm^{-3}. The significant discrepancies between theory and observations still point to the fact that the Fe II problem remains at least partially unsolved.[6]

[6] The website www.astronomy.ohio-state.edu/~pradhan gives a list of Fe II fluxes generated from a model with U = 10^{-2}, $\log_{10} N_H = 9.6$, and a fiducial optical line flux $\log_{10}$ F(Hβ) = 5.68 [373, 379]. The model incorporates exact radiative transfer, up to 1000 levels, and Lyα, Lyβ FLE. The line list may also be useful in spectral analysis of sources other than AGN with significant Fe II and Fe III emission.

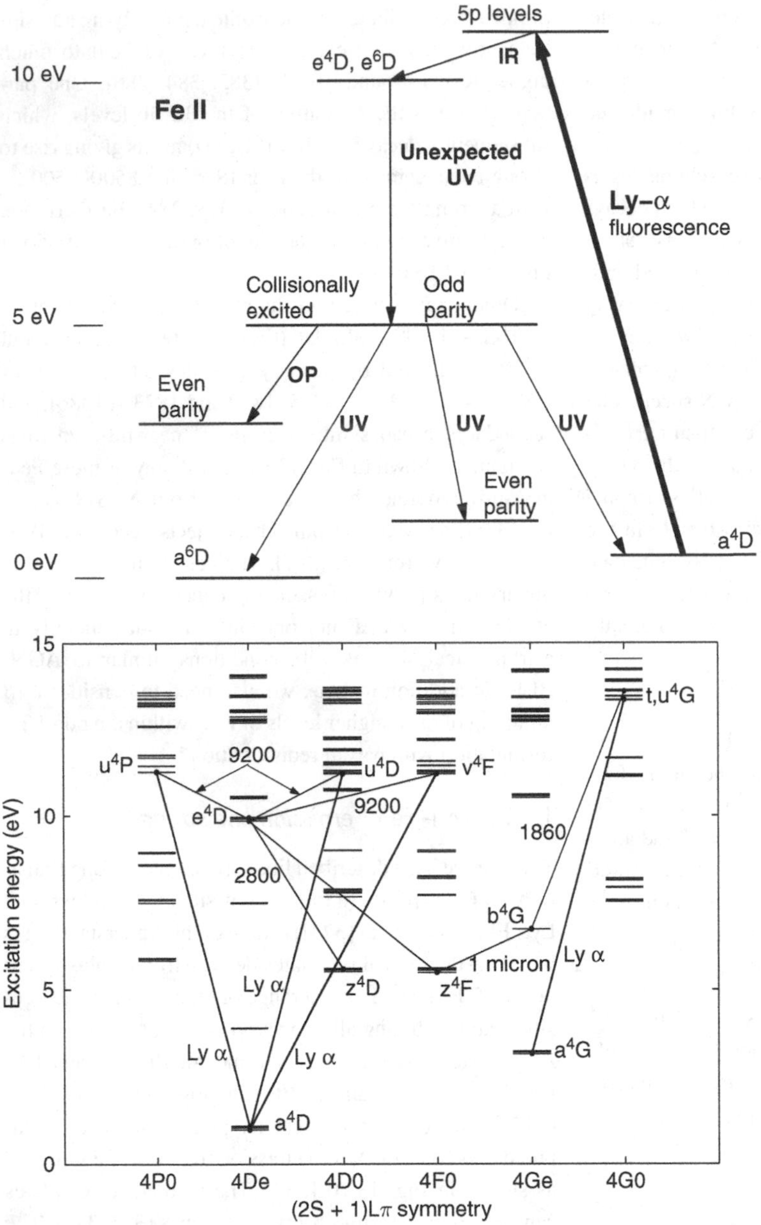

FIGURE 13.8 Simplified energetics of Fe II emission including Lyα fluorescence and related cascades and transitions.

13.5.3 Spectral properties and the central source

The Fe II problem illustrates one aspect of several spectral characteristics of AGN that appear to be driven by the central source. The underlying physical parameter is the relative accretion rate L/L_{Edd}, which in turn is related to the mass of the SMBH. In addition to Fe II, other spectral parameters have been extensively utilized to attempt a spectroscopic unification of AGN from the optical to the X-ray band. These also include the Hβ FWHM, the 1541 Å C IV line equivalent width, the forbidden 5007 Å [O III] line, and the X-ray spectral index α_x. The attempt to correlate some or all of these spectral features in a unified model is sometimes referred to as *Eigenvector 1* or *principal component analysis* [338, 388]. The principal driver of the elements or components of Eigenvector 1 is related to the mass and accretion activity around the SMBH.

The inverse relationship between equivalent width and luminosity observed in AGN is known as the Baldwin effect [389, 390]. The original relationship is for the C IV

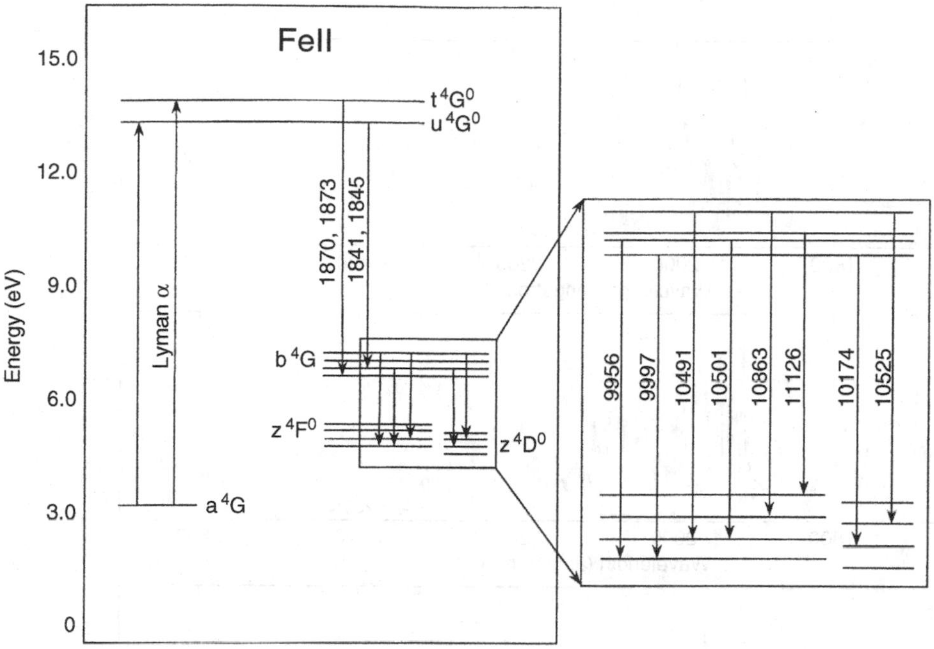

FIGURE 13.9 Near-infrared lines of Fe II excited by Lyα fluorescence [378].

line due the transition(s) 2p $^2P^o_{1/2,3/2} \rightarrow$ 2s $^2S_{1/2}$. Since Li-like ions are relatively easy to ionize, the C IV ion is sensitive to the luminosity of the ionizing source. On the other hand, the 2s–2p transition is easy to excite, with a large oscillator strength, and therefore the corresponding line is observed to be quite strong. Figure 13.11 shows the ionization states of carbon as a function of temperature in coronal equilibrium. It may be seen that C IV remains in the plasma over the *smallest* temperature range of any other ionization state; in other words, it is most sensitive since it gets ionized away most rapidly. Ionization balance curves in photoionized plasmas are similar, though all ionization stages occur at lower electron temperatures, owing to additional photoionization in an external radiation field.

The origin of the Baldwin effect appears to be related to other parameters that characterize the AGN phenomenon, and associated spectral correlations of L vs. Fe II and O III lines. The C IV equivalent width has been *anti-correlated* for a sample of quasars to the ratio L/L_{Edd} and the [O III] 5007 Å line [390]. The Baldwin effect may be generalized to other lines and wavelength ranges where the basic principle applies: an ionization state sufficiently sensitive to the intensity of radiation, and a strong transition. For example, an anti-correlation between the narrow Fe $K\alpha$ line at 6.4 keV (discussed in the next section) and AGN X-ray luminosity has been reported [391]. Basically, the underlying continnum flux in a high-luminosity AGN is more intense, compared with the emission line flux, than in a low-luminosity AGN.

13.5.4 Iron abundance at high redshift

One of the puzzles in QSO research is the consistently high iron abundance even at high redshift, out to $z \sim 6$. The issue is significant as it has cosmological implications. Iron is produced only as the end product of stellar evolution in supernovae Type Ia and Type II. Therefore, detection of Fe emission at the earliest possible epoch in the history of the Universe is a useful indicator not only of chemical evolution but also of chronology. Since this topic is related to others in cosmology – such as reionization following universal re-lighting by the first quasars and stars – we discuss it further in Chapter 12.

13.6 The central engine – X-ray spectroscopy

We now describe the underlying atomic physics up to the innermost regions of AGN. Spectroscopy of the ionic species found in these inner regions reveal the kinematics of each component of the plasma. High-energy X-ray observations probe AGN activity up to the accretion disc

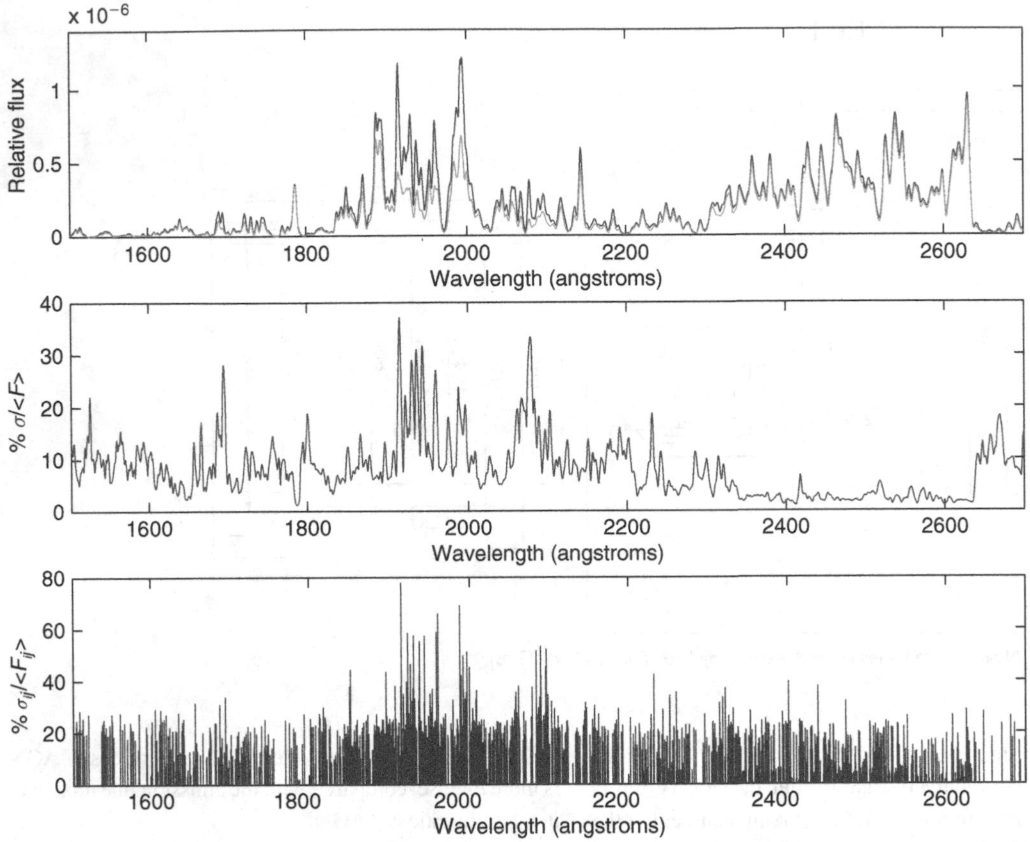

FIGURE 13.10 Theoretical templates of Fe emission from AGN: predicted Fe II–Fe III UV fluxes for a BLR model with ionization parameter $U = 10^{-1.3}$ and $N_e = 10^{11}$ cm^{-3}. The top panel shows minimum and maximum flux F_λ (in erg cm^{-2} s^{-1} Å) obtained by varying the atomic parameters within the range of uncertainties. The middle panel shows the standard deviation of F_λ at each wavelength as a percentage of the average flux $< F_\lambda >$. The bottom panel shows the uncertainty in individual line fluxes for the 2300 strongest Fe II–Fe III lines in the model [379].

and the hot corona surrounding the central source. The discussion in this section covers a range of X-ray spectral properties of AGN in a systematic manner, beginning with the most energetic atomic transitions.

13.6.1 Fe $K\alpha$ X-ray lines and relativistic broadening

The $K\alpha$ complexes are the non-hydrogenic analogues of the simplest atomic transition, the Lyα transition 1s → 2p. At first, that may seem to imply that the spectra should be quite simple as well. But in fact that is not true, as we saw in the detailed analysis of the He $K\alpha$ complex of Fe XXV lines observed in the solar corona (Chapter 9). The spectroscopic analysis of $K\alpha$ complexes is a minor subfield of astrophysical X-ray spectroscopy in itself. This is partly because the $K\alpha$ X-ray lines are usually well-separated (given sufficient resolution of course), unlike the multitude of overlapping L-shell and M-shell spectra (discussed next).

We consider the $K\alpha$ transition in *all* Fe ions. For all ionization stages where the 2p-subshell is filled this transition is affected following ionization of the K-shell, leaving a vacancy which is then filled by a downward 2p → 1s transition, i.e., Fe I–Fe XVII. The two fine-structure components of this fluorescent transition are:

$$1s2s^22p^6(^2S_{1/2}) \rightarrow 1s^22s^22p^5\left(^2P^o_{1/2,3/2}\right)$$

at 6.403 84 and 6.390 84 keV, respectively called the Kα1 and Kα2 transitions.

Higher ionization stages of Fe with an open 2p-subshell, Fe XVIII–Fe XXVI are at higher energies and give rise to a multitude of $K\alpha$ lines due to the transitions given in Table 13.3. It is worth emphasizing that from an atomic physics point of view the so-called '$K\alpha$ lines' are in fact mostly $K\alpha$ resonances, corresponding to

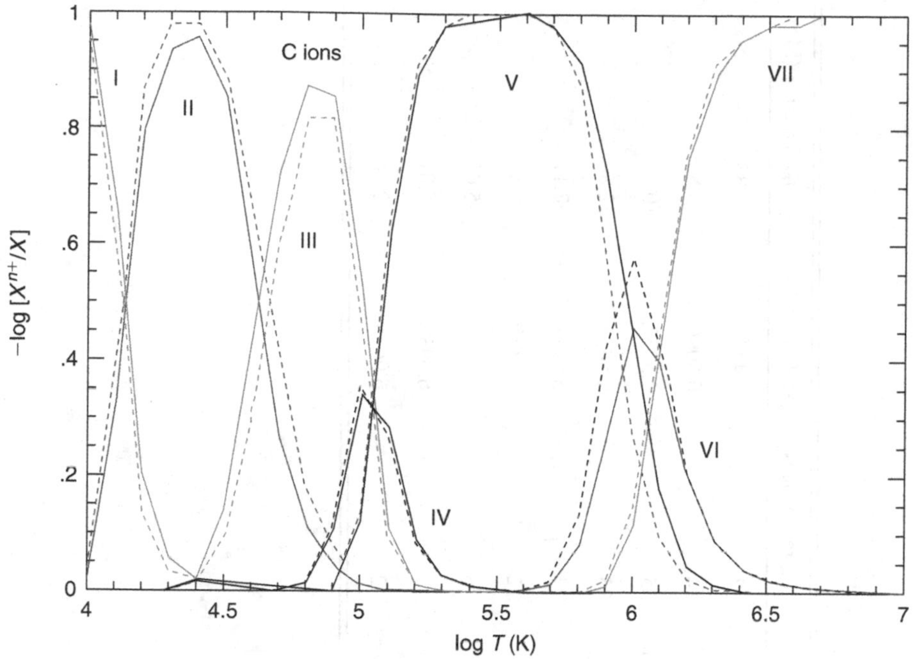

FIGURE 13.11 Ionization fractions of carbon in collisional (coronal) equilibrium.

transitions into or from highly excited states with a K-shell vacancy. The transition energies and the cumulative resonant oscillator strengths for the $K\alpha$ complexes are also given in Table 13.3 [392].

For Fe XVII–Fe XXVI, the transition energy increases from ∼6.4 keV for Fe XVII to up to 6.7 keV for the He-like Fe XXV, and 6.9 keV for the H-like Fe XXVI (recall the detailed discussion in Chapter 7 for He-like ions as the most prominent X-ray diagnostics of high temperature plasmas).

Perhaps the most interesting spectral observations from AGN are the ones showing the effect of gravity of the black hole, and one of the most direct observational evidence of its existence. This signature of the central SMBH is the extremely broad feature centred at 6.4 keV, shown in Fig 13.12, corresponding to the fluorescent $K\alpha$ transition in Fe ions with a filled p-shell. The broad but asymmetric 6.4 keV $K\alpha$ line has been observed from a number of sources. The first reported observation was from a Seyfert 1 galaxy MCG–6-30-15 6 [239]. Figure 13.12 shows a more recent observation from the same galaxy [393]. The line is greatly skewed towards the red (lower) energies all the way down to ∼5 keV. The extreme width of the emission line, and its asymmetry towards the red wing, implies its origin in the innermost stable orbits of the accretion disc and gravitational broadening due to the proximity of the black hole according to the theory of general relativity.

Energies of photons emitted in the vicinity of a massive black hole are lowered by an amount needed to 'climb' out of the gravitational potential well. These photons are therefore *redshifted* with respect to the peak rest-frame energy of the observer (Fig. 13.13). The redshifted photons originate in the innermost regions of AGN, close to the last stable orbits of matter within the accretion disc at $r \sim 3r_s$, or about three times the Schwarzschild radius

$$r_s = \frac{2GM}{c^2}. \tag{13.15}$$

The gravitational redshift is given by

$$1+z = \frac{1}{\sqrt{1-\frac{2GM}{rc^2}}} = \frac{1}{\sqrt{1-\frac{r_s}{r}}}. \tag{13.16}$$

Therefore the estimated gravitational redshift from Eq. 13.16 of a 6.4 keV photon is about 5.2 keV.[7]

Since the 6.4 keV emission line is from L → K transitions in iron ions with a filled 2p-shell, Fe I–Fe XVII, the line emitting environment is relatively 'cold', compared with more highly ionized stages. It would probably correspond to plasma in the accretion disc, moving along magnetic field lines in stable orbits at $T < 10^5$ K. The $K\alpha$

[7] There are several reviews of the spectroscopy of iron $K\alpha$ lines based on the theory of general relativity and related atomic astrophysics, e.g., C. S. Reynolds and M. A. Nowak [394] and D. A. Liedahl and D. F. Torres [395].

TABLE 13.3 Fe $K\alpha$ resonance complexes and absorption cross sections $6.4 < E(keV) \leq 6.9$.

Fe ion	$K\alpha$ transition array	$K\alpha$ resonances	$< E(keV) >$	$< \sigma_{res}(K\alpha) >$ (Mb)
Fe XVIII (F-like)	$1s^2 2s^2 2p^5 \left({}^2P^o_{3/2}\right) \rightarrow 1s2s^2 2p^6 ({}^2S_{1/2})$	2	6.444	1.33
Fe XIX (O-like)	$1s^2 2s^2 2p^4 ({}^3P_2) \rightarrow 1s2s^2 2p^5 \left({}^3P^o_{0,1,2}, {}^1P^o_1\right)$	14	6.5096	5.80
Fe XX (N-like)	$1s^2 2s^2 2p^3 \left({}^4S^o_{3/2}\right) \rightarrow 1s2s^2 2p^4 ({}^4P_{1/2,3/2/,5/2}, {}^2D_{3/2,5/2}, {}^2S_{1/2}, {}^2P_{1/2,3/2})$	35	6.5237	10.12
Fe XXI (C-like)	$1s^2 2s^2 2p^2 ({}^3P_0) \rightarrow 1s2s^2 2p^3 \left({}^5S^o_2, {}^3D^o_{1,2,3}, {}^1D^o_2, {}^3S^o_1, {}^3P^o_{0,1,2}, {}^1P^o_1\right)$	35	6.5633	12.86
Fe XXII (B-like)	$1s^2 2s^2 2p \left({}^2P^o_{1/2}\right) \rightarrow 1s2s^2 2p^2 ({}^4P_{1/2,3/2/,5/2}, {}^2D_{3/2,5/2}, {}^2S_{1/2}, {}^2P_{1/2,3/2})$	14	6.5971	7.11
Fe XXIII (Be-like)	$1s^2 2s^2 ({}^1S_0) \rightarrow 1s2s^2 2p \left({}^3P^o_{0,1,2}, {}^1P^o_1\right)$	2	6.6375	5.36
Fe XXIV (Li-like)	$1s^2 2s ({}^2S_{1/2}) \rightarrow 1s2s2p \left({}^2P^o_{1/2,3/2}, {}^4P^o_{1/2,3/2,5/2}\right)$	6	6.6617	5.02
Fe XXV (He-like)	$1s^2 ({}^1S_0) \rightarrow 1s2p \left({}^3P^o_{0,1,2}, {}^1P^o_1\right)$	2	6.6930	6.01
Fe XXVI (H-like)	$1s ({}^2S_{1/2}) \rightarrow 2p \left({}^2P^o_{1/2,3/2}\right)$	2	6.9655	2.18

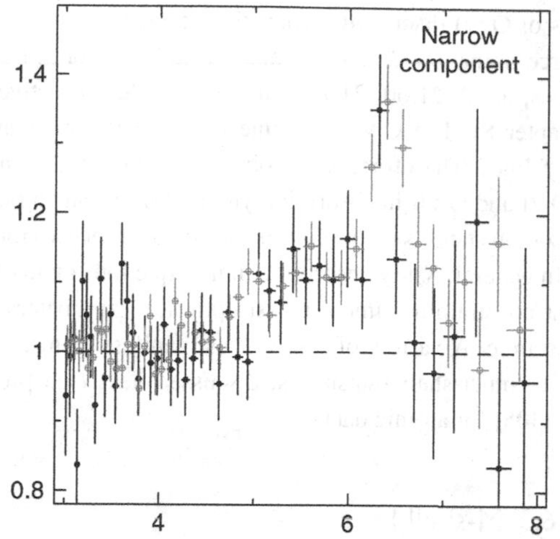

FIGURE 13.12 The relativistically broadened Fe Kα X-ray line from the Seyfert I galaxy MCG–6-30-15 6 [393]. Two sets of data are shown, from the *Advanced Satellite for Cosmology and Astrophysics* (ASCA or Astro-D), and the *Chandra X-ray Observatory* High-Energy Transmission Grating. The asymmetric line profile peaks at 6.4 keV, but is skewed redward down to about 5 keV, owing to X-ray photons originating in close proximity of the central black hole. The line profile also has a narrow component from emission from matter far away from the black hole, and relatively unaffected by gravitational broadening.

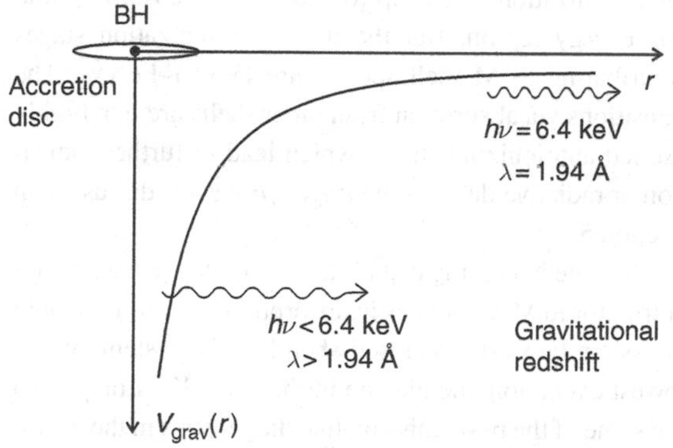

FIGURE 13.13 Gravitational broadening of Fe Kα emission line from the close vicinity of the black hole. Whereas fluorescent emission from iron in matter far away from the black hole is at 6.4 keV, photons originating in the inner region of the accretion disc are highly redshifted to much lower energies and longer wavelengths.

fluorescence is due to K-shell ionization following irradiation by hard X-rays from the disc-corona which is at much higher temperatures $T > 10^6$ K.

Although relativistic gravitational broadening is widely accepted as the cause of the redward, asymmetric line profile in the Seyfert 1 AGN MCG–6-30-15 shown in Fig. 13.12, one should also keep in mind other factors that might be contributing, if not alternative, factors. These include geometrical effects owing to the orientation of the AGN nucleus towards the observer. For instance, the nucleus of MCG-6-30-15 is nearly face-on, with an inclination angle of 60°. Doppler broadening components, both transverse and parallel, would contribute accordingly. One might think of the well-known P-cygni profile (Chapter 10), which is suppressed at the blue end but enhanced towards the red; however, the sharp drop-off in flux immediately after 6.4 keV appears to rule that out. Attempts have also been made to model *reflection components* and resultant reddening of the line owing to absorption or scattering [396].

In addition to the broad Fe $K\alpha$ 6.4 keV line, many AGN spectra also show a narrow line profile at 6.4 keV, which is normally symmetrical. Moreover, the higher energy He-$K\alpha$ and H-$K\alpha$ components from Fe XXV and Fe XXVI, respectively, are also prominently observed in AGN spectra, arising from the hot plasma associated with the disc-corona farther away from the nucleus. Determination of the exact location of the emitting plasma is further complicated by observations that show that the Fe $K\alpha$ line varies little compared with the X-ray continuum variability, whereas it would be expected to vary in phase with it (cf. [397]).

13.6.2 Warm absorber (WA)

Farther out from the central source, X-ray observations reveal the presence of hot gas up to or more than about a million degrees, with strong absorption features and blueshifts indicative of outflows. The WA may represent winds from the accretion disc, carrying away angular momentum from the accreting material towards the SMBH; an extreme example of which is a bipolar jet from the spinning black hole. The outflow appears to be related to the hot corona around the central source, and is at temperatures exceeding those found in the WA. Spectroscopic diagnostics reveals that (i) the WA contains significant amounts of metals ranging from O to Fe, and is therefore often referred to as 'dusty WA' and (ii) the WA spans several orders of magnitude in temperature and density, as revealed by arrays of transitions in several ionic species of each element. Modelling the dusty WA, therefore, requires a range of ionization parameters and electron densities (e.g., [398, 399]). The primary transitions in the WA are due to absorption from inner shells, as described next.[8]

[8] A comprehensive review of atomic data for X-ray astrophysics is given in [400].

However, it is worth noting that the WA lines can also be seen in emission when the central source is occulted, as in Seyfert 2 galaxies [339].

Other outstanding features seen in the WA spectra include K-shell absorption lines from several elements. Figure 13.14 shows the soft X-ray spectrum of the Seyfert 1 galaxy MCG–6-30-15, the same one with the broad relativistic Fe $K\alpha$ 6.4 keV fluorescent line in Fig. 13.12. The soft X-ray spectrum in Fig. 13.14 was obtained by the *Chandra X-ray Observatory* and is worth discussing in detail. It reveals a dusty WA with metals up to iron. The spectrum shows the strong signature lines due to dipole allowed transitions $1s^2(^1S_0) \rightarrow 1s2p\left(^1P^o_1\right)$ in He-like ions, Ne IX around 13.5 Å or 0.94 keV, and O VII around 0.7 keV, together with absorption lines from the whole Rydberg series of transitions $1s^2(^1S_0) - \text{lsnp}\left(^1P^o_1\right)$ in O VII. The spectral 'turnover' at ~17.7 Å, or 0.7 keV, is especially attributed to strong absorption by neutral iron L-shell 2p → 3d transition array, and the O VII Rydberg series limit 1s → np [398]. The astrophysical interpretation is that the supermassive black hole system that illuminates the dust in its path length is itself surrounded by highly ionized gas of temperatures 10^5–10^7 K, imprinting features in the X-ray spectra. The best model for the spectrum is with an iron + oxygen combination, placing the dust in a region close to the black hole, where most of the oxygen has probably been sputtered away. The observed absorption features are due to fine-structure resonances close to excitation – ionization edges of iron and other metals that make up the dust composition (the spectroscopic physics is exemplified below with the relatively simple example of oxygen). Figure 13.14 provides an example of the state-of-the-art observations in high-resolution X-ray astronomy, as well as the emerging field of *condensed matter astrophysics* [401].

Of particular interest was the predicted detection of strong O VI resonance absorption at 22.05 Å due to the KLL transition $1s^22s(^2S_{1/2}) \rightarrow 1s2s2p\left(^2P^o_{3/2}\right)$ [402]. It occurs from the Li-like O VI ground state $1s^22s(^2S_{1/2})$, into a resonant level $1s2s2p\left(^2P^o_{3/2}\right)$ lying *above* the ground state $1s^2(^1S_0)$ of the He-like residual ion O VII. The photoabsorption cross section is shown in Fig. 13.15 [402]. In fact, there are two closely spaced peaks in the cross section due to the two fine structure components $1s2s2p\left(^2P^o_{3/2,1/2}\right)$ of the resonance at λλ 22.05, 21.86, with the former being much stronger than the latter. The O VI KLL line is especially useful since it is at 22.05 Å, and lies *within* the $K\alpha$-complex of the three prominent lines of O VII discussed earlier: the 'resonance' (r or w), intercombination (i or x +y) and forbidden (f or z) transitions, at λλ 21.60, 21.790, and 22.101 Å, respectively (Chapter 8). The O VI KLL line lies between the i and the f lines. Therefore, X-ray observations of the K-lines of O VI and O VII in absorption yield information on both ionization states, say, the column densities of the two ions.

In general, X-ray absorption in mulitple ions of an element constrain a number of astrophysical parameters in the source. A variety of X-ray lines manifest themselves due to inner-shell transitions, discussed next (see [404] and [405] for atomic data).

13.6.3 M-shell lines

The strongest features in the WA are a multitude of lines due to absorption by Fe ions with open M-shells ($n = 3$) but filled L-shells ($n = 2$). As these lines are often unresolved in low-resolution X-ray spectra, they are sometimes labelled *unidentified transition arrays*, or 'UTAs'. However, 'UTA' is a misnomer, since the lines are in fact quite well-known and may all be identified. The Fe M-shell absorption lines occur primarily from strong inner-shell dipole allowed $2p \rightarrow 3d$ transitions clustered around 0.7–0.8 keV or 16–17Å [406]. In principle, all Fe ions with a 2p filled shells, Fe XVII (Ne-like), and lower ionization stages up to neutral Fe I, absorb around this energy region. But the dominant ionization stages contributing to M-shell spectra are Fe VIII–Fe XVI. The transitions via absorption from inner shells are into highly excited autoionizing states, which lead to further ionization or radiative decay – the *Auger processes* discussed in Section 5.9.

We note in passing that although Fe XVII is not a major cotributor to M-shell lines in absorption, it is prominent in emission. Fe XVII is a closed-shell Ne-like system, whose lowest excitation energies lie in the ~15–17 Å range and are some of the best emission line diagnostics in the X-ray, as described in Chapter 8.

13.6.4 L-shell lines

More highly ionized ions with open L-shells, from Fe XVIII (F-like) up to Fe XXIV (Li-like), have much less cumulative absorption than those with filled L-shells. They are well-resolved, since the energy separation among levels of the same n ($\Delta n = 0$) increases with ion charge as $\Delta E_n \sim Z$, and among those with different n as $\Delta E_{n,n'} \sim Z^2$. The L-shell lines may be seen strongly in absorption since the allowed oscillator strengths are large

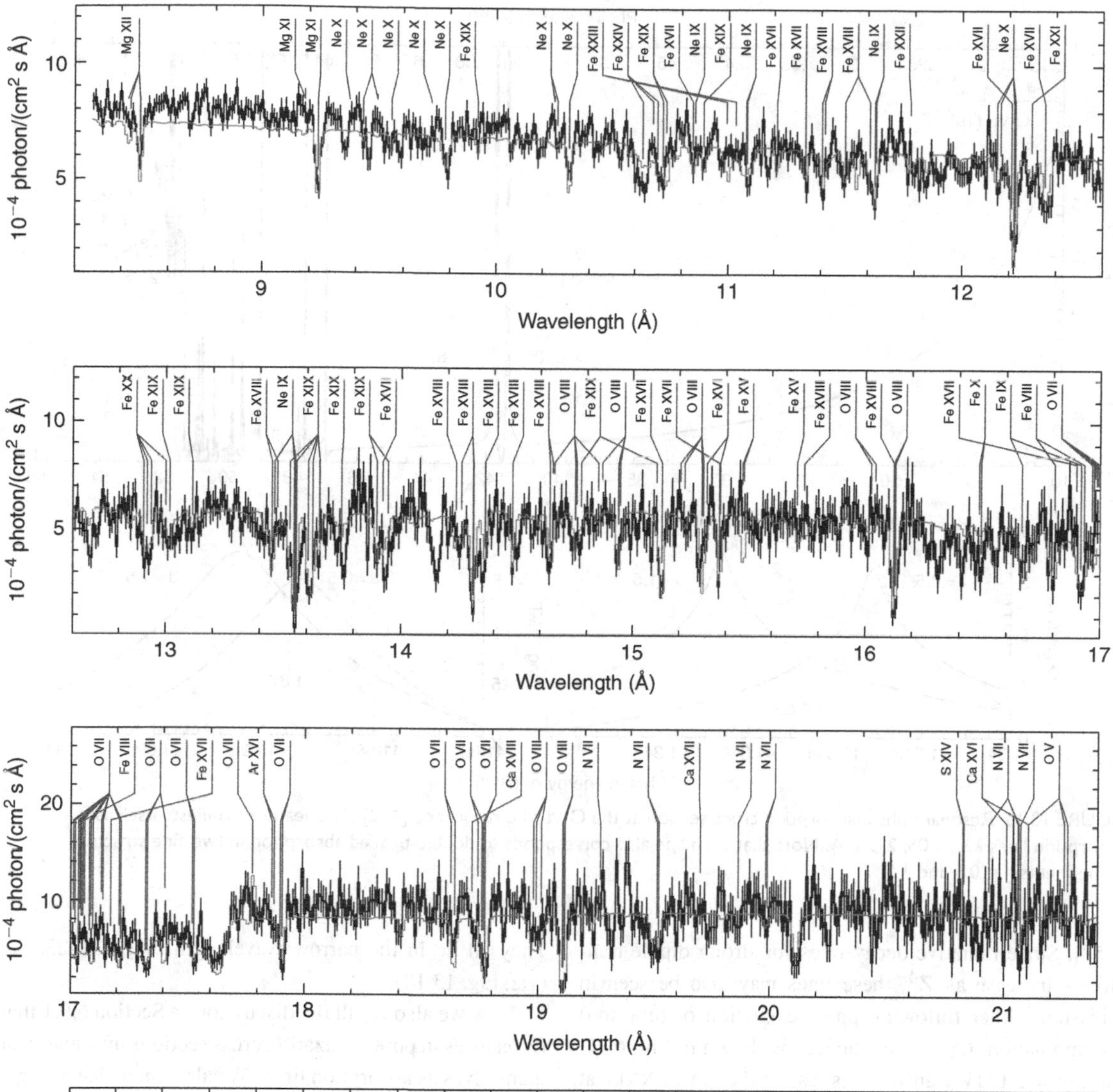

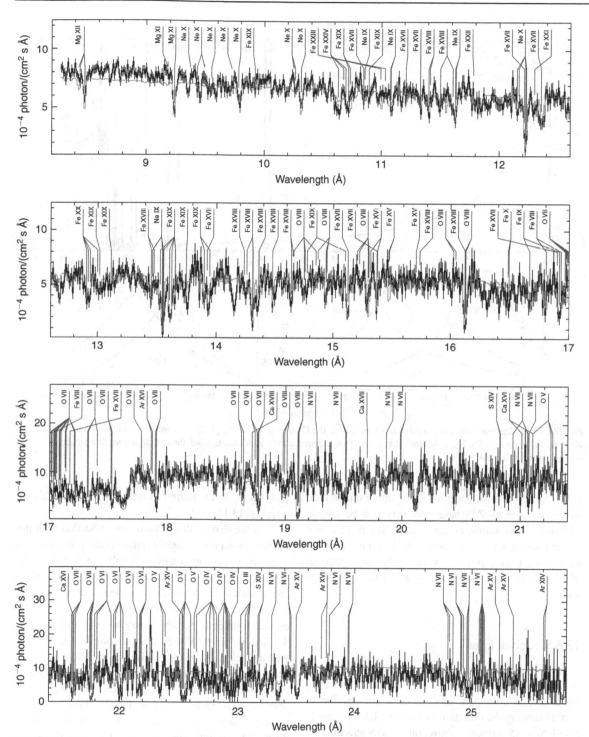

FIGURE 13.14 High-resolution Chandra X-ray Observatory spectrum of the Seyfert 1 galaxy MCG–6-30-15 in the soft X-ray region ([398], Courtesy: J. Lee). The X-ray spectrum at $E < 1\,keV$ ($\lambda > 12\,\text{Å}$) reveals a *dusty warm absorber*, ionized gas with many absorption features from highly ionized ions, such as He-like N, O, Ne, and Fe. The KLL resonance absorption feature due to Li-like O VI at 22.05 Å (see Fig. 13.15 [402]), as well as similar $K\alpha$ features of other O-ions are observed [403].

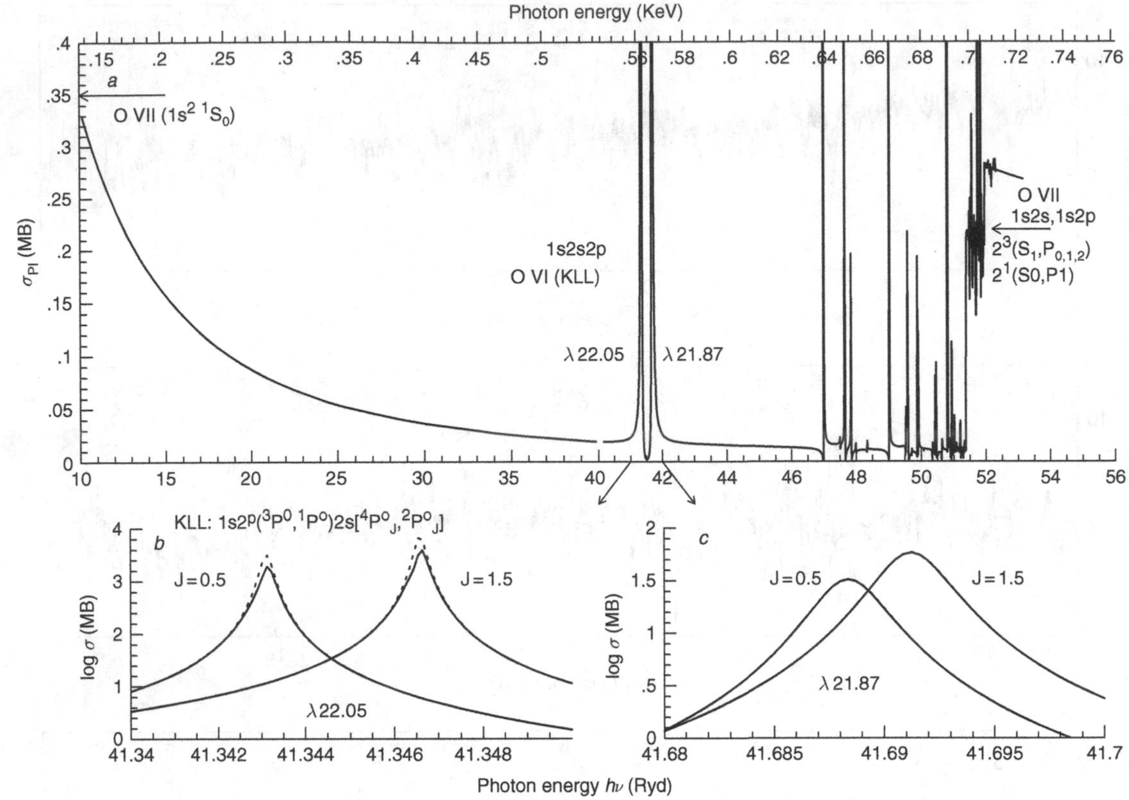

FIGURE 13.15 Resonant photoabsorption cross section at the O VI KLL resonance [402]. This feature manifests itself as two absorption lines λλ 22.05, 21.86 Å. Note that each line also corresponds to double-peaked absorption in two fine structure components, J = 0.5 and 1.5.

[158]. Since radiative decay rates for strong dipole transitions increase as Z^4, these lines may also be seen in emission, either following photo-excitation or (e + ion) recombination. One such example is shown in Fig. 6.17: the $\Delta n = 0$ EUV transition $1s^22s - 1s^22p$ in Fe XXIV at 209 Å; higher transitions $n > 2 \rightarrow n = 2$ lie in the X-ray region [158].

13.6.5 K-shell lines

As we have seen, in addition to bound–bound transitions, resonant transitions with excitation from inner shells play a prominent role in X-ray spectral formation. Following the detection of the O VI KLL feature in MCG–6-30-15 discussed above, other AGN spectra were found to have $K\alpha$ absorption lines from *all* ionization states of oxygen, O I–O VI. They are due to $K\alpha$ inner-shell resonances in ions from O VI ($1s^22s + h\nu \rightarrow 1s2s2p$) to O I ($1s^22s^22p^4 + h\nu \rightarrow 1s2s^22p^5$). Figure 13.16 shows the photoionization cross sections that produce $K\alpha$ resonances in each ion. Note the logarithmic scale, and hence the magnitude of these strong resonaces (see also [399]). They all lie in the narrow wavelength range 22–23.5 Å (viz. Fig. 13.14).

Now we also recall the discussion in Section 6.9.1 that resonances in photoionization cross sections may manifest themselves as absorption lines. We also know that absorption lines are most useful in determining column densities and abundances of elements, provided their oscillator strengths are known. Equation 6.69 defines the resonance oscillator strength f_r, which may be evaluated from the detailed σ_{PI} provided the resonance profile is sufficiently well-delineated. In practice, this is often difficult and elaborate methods need to be employed to obtain accurate positions and profiles (the background and the peaks) of resonances. As in Table 13.3 for Fe irons Table 13.4 gives these oscillator strengths, f_r and other quantities for the inner-shell $K\alpha$ transitions in O ions. The calculated equivalent width W_a, obtained from the autoionization profiles of resonances, and the peak value of the resonance cross section σ_{max} are also given.

Exercise 13.1 *A measurement of X-ray photoabsorption in the 'warm absorber' region of an AGN via the KLL*

TABLE 13.4 $K\alpha$ resonance oscillator strengths f_r (Eq. 6.69) for oxygen ions.

Ion	E_r (Ryd)	E_r (keV)	λ (Å)	f_r	W_a(meV)	σ_{max} (MB)
O I	38.8848	0.5288	23.45	0.113	31.88	48.05
O II	39.1845	0.5329	23.27	0.184	23.21	107.7
O III	39.5000	0.5372	23.08	0.119	27.00	59.92
O III	39.6029	0.5386	23.02	0.102	12.08	114.5
O III	39.7574	0.5407	22.93	0.067	24.27	37.48
O IV	40.1324	0.5458	22.73	0.132	27.11	66.15
O IV	40.2184	0.5470	22.67	0.252	14.32	239.2
O IV	40.5991	0.5521	22.46	0.027	21.91	17.00
O V	40.7826	0.5546	22.35	0.565	14.01	549.0
O VI	41.3456	0.5623	22.05	0.576	1.090	7142.0
O VI	41.6912	0.5670	21.87	0.061	12.16	67.36

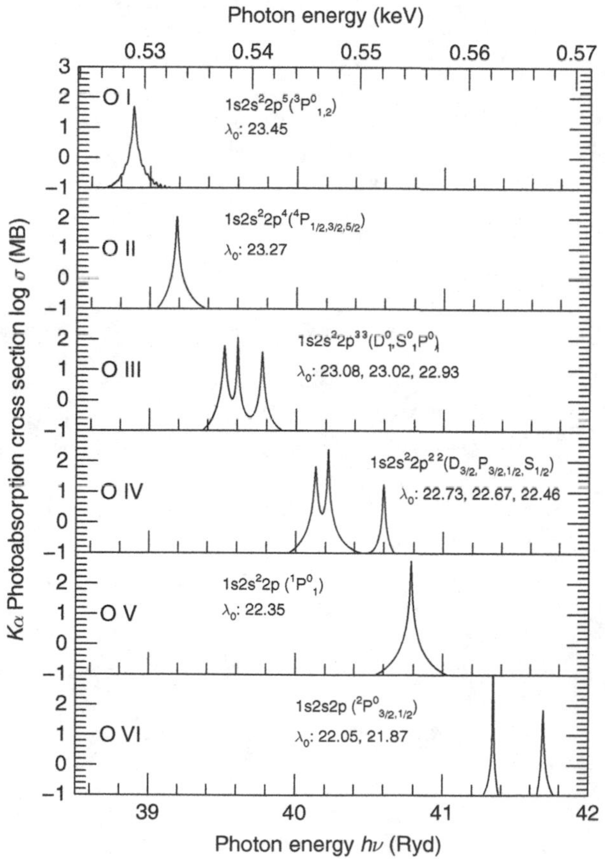

FIGURE 13.16 $K\alpha$ resonances in oxygen ions [403]. These resonances in photoionization cross sections appear as absorption lines in observed spectra.

resonance in O VI *gives the equivalent width of 29 mÅ at 22.09 Å. Identify the atomic transition; calculate the column density of O* VI*, and the line centre optical depth* τ_0 *(compare with results given in [402, 398]).*

As we have seen, multi-wavelength spectroscopy is required to study the AGN phenomena, since the observed spectral energy distribution covers all bands of electromagnetic radiation. The widely accepted SMBH paradigm appears to explain the overall features of AGN, geometrically and spectroscopically. Nevertheless, the precise kinematics and coupling between the central black hole engine and the disparate regions of AGN are far from being entirely understood. Some of the major problems may be traced to inadequate and inaccurate atomic physics in AGN models, such as the anomalous Fe II emission in the infrared, optical and ultraviolet, and Fe $K\alpha$ line(s) in the X-ray. Some relevant reviews are atomic X-ray: spectroscopy of accreting black holes [407] and fluorescent iron lines [408]. To address the spectroscopic needs the authors' websites list atomic data sources relevant to AGN studies: www.astronomy.ohio-state.edu/~pradhan (or /~nahar).

14 Cosmology

Which atoms were formed first, in what proportion and when? The relationship between atomic spectroscopy and cosmology rests on the answer to these questions. According to *big bang nucleosynthesis* (BBN), before the creation of the first atoms, the Universe would have been filled with a highly dense ensemble of nuclei, free electrons, and radiation. The standard model from high-energy particle physics implies that most observable matter is made of baryons, such as protons and neutrons; electrons are leptons and much less massive. The baryons are themselves made of more exotic fundamental particles, such as quarks, gluons and so forth. According to the BBN theory, given a fixed baryon-to-photon ratio in the first three minutes of origin, a few *primordial nuclear species* made of baryons appeared. The atomic nuclei created during the BBN were predominantly protons and helium nuclei ($^2He_3, ^2He_4$), with very small trace amounts of deuterium (heavy hydrogen $^2H_1$) and lithium ($^3Li_6, ^3Li_7$). Atomic physics then determines that singly ionized helium He II (*not* hydrogen!) would have been the first atoms(ions) formed.

The process of formation is (e + ion) recombination: He III + e $\rightarrow$ He II + $h\nu$. This temporal marker in the history of the Universe is referred to as the *recombination epoch.*[1] The reason that He II was the first atomic species is not difficult to see, given the extremely hot plasma that preceded the recombination epoch when nuclei and electrons were free in the fully ionized state. The atomic species that would form first is the one with the highest ionization potential, as detailed balance dictates.[2] Consider that E_{IP}(HeII) $= Z^2 = 4$ Ry = 54 eV, as opposed to E_{IP} (He I) = 1.8 Ry = 24.6 eV, and E_{IP} (H I) = 1 Ry $\equiv$ 13.6 eV. It follows that He II can exist at much higher temperatures, i.e., at earlier (hotter) times, than either He I or H I. The study of the recombination epoch is also important to ascertain the primordial abundances, with a percentage abundance ratio for H:He of ~93:7 by number, and ~76:24 by mass. The determination of the precise ratio is a crucial test of BBN cosmology and the baryonic matter in the early Universe.

At the earliest times, radiation and matter were coupled in the sense that photons scatter from free matter particles via Thomson or Compton scattering, and have short mean free paths [409]. Since all radiation energy was thus 'trapped', the Universe was in a radiation dominated state and essentially opaque. The conditions would have been as in an ideal black body characterized by a radiation temperature and a Planck distribution. That would correspond to an extemely hot radiation background, the forerunner of the much cooler present-day *cosmic microwave background* (CMB). Having cooled due to cosmological expansion, the radiation temperature at the present epoch corresponds to a black body at a characteristic Planck temperature of 2.725 K, predominantly in the microwave range. As the Universe expanded and cooled, radiation and matter decoupled and the Universe became matter-dominated, as radiation began to escape interactions with matter. Radiation–matter decoupling was followed by the recombination epoch, when electrons and primordial nuclei recombined. When matter made the transition from fully ionized to neutral state, the mean free paths of photons increased and the Universe became increasingly transparent as radiation escaped away, eventually at the speed of light. However, Compton scattering of particles with photons prior to this epoch would distort the otherwise isotropic black-body radiation to a small extent – but potentially detectable – an effect known as the *Sunyaev–Zeldovich effect* [409]. While the Universe remains generally isotropic, observations

[1] It may seem somewhat illogical to refer to the first-ever combination of electrons and nuclei as *recombination.* It in fact refers to conditions when electrons and nuclei would *remain* combined without immediate break-up.

[2] Note that througout the text we have continually emphasized the detailed balance inverse relationship between photoionization and photorecombination, elaborated in Chapter 6.

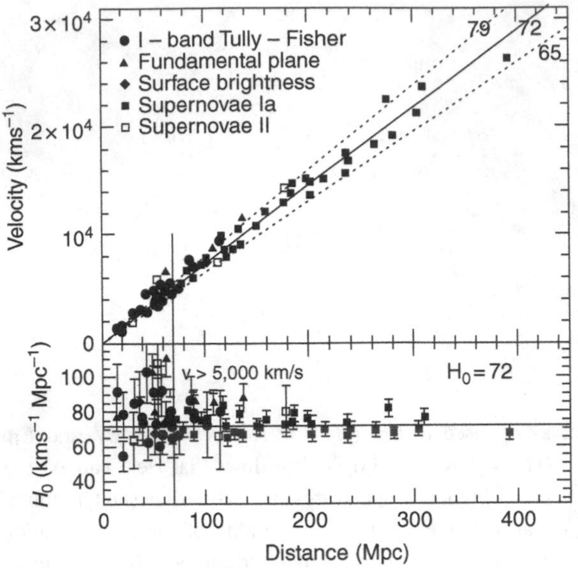

FIGURE 14.1 Hubble's law and the Hubble constant (http://nedwww.ipac.caltech.edu/level5/Sept01/Freedman/Freedman7.html). Straight lines corresponding to three values of H_0 are shown. The bottom rectangle shows the residuals of data points with respect to the $H_0 = 72$ line. Observations of Type Ia supernovae provide data out to the farthest distances.

of the early Universe should contain information on any anisotropies in the radiation background, manifest in the CMB even until the present epoch. Such anisotropies have indeed been detected in recent observations by the satellite *Wilkinson Microwave Anisotropy Probe* (WMAP) of the CMB power spectrum. A new space probe PLANCK was launched in 2009, and is now set to measure anisotropies over the entire sky with higher precision.

14.1 Hubble expansion

Uniform expansion of the isotropic and homogeneous cosmos implies that all objects recede from any observer anywhere in the Universe with a constant velocity v. Empirical observations lead to *Hubble's law*

$$v(t) = H_0 \times D(t), \tag{14.1}$$

where v is the velocity and D is the distance of an object (e.g., a far-away galaxy) at an epoch t, sometimes referred to as the 'look-back' time. H_0 is the *Hubble constant* with the commonly accepted value of $\sim 72\,\text{km s}^{-1}\,\text{Mpc}^{-1}$ (note the units in terms of velocity and distance). Figure 14.1 shows the linear fit to observed data for H_0 values of 65 and 79. The limits of the range for H_0, and have direct implications for the derived age of the Universe. Since the linear Hubble relation has the slope H_0, its inverse $1/H_0$ has the dimension of time and, in fact, directly yields the age of the Universe. A value of $67\,\text{km s}^{-1}\,\text{Mpc}^{-1}$ yields an age of approximately 13.7 gigayears, about the age of the oldest stars according to stellar models. H_0 of $50\,\text{km s}^{-1}\,\text{Mpc}^{-1}$ implies about 20 billion years and 100 $\text{km s}^{-1}\,\text{Mpc}^{-1}$ about 10 billion years; the latter value is obviously in conflict with the estimated ages of the oldest stars. While there are several methods of determining the Hubble constant, the latest measurements of the Hubble constant are based on observations of Type Ia supernovae and yield a value $H_0 = 74.2 \pm 3.6\,\text{km s}^{-1}\,\text{Mpc}^{-1}$ with less than 5% uncertainty [410]. Finally, it should be mentioned that estimates also depend on the cosmology adopted, i.e., cosmological models that take account of all the known or unknown matter and energy in the Universe.

The interpretation of Hubble's law as a linear velocity–distance relation, implying uniform-velocity expansion, is to be reconsidered later based on two facts: (i) the amount and distribution of matter in the Universe would determine the velocity of expansion owing to gravitational interaction among masses and (ii) recent observations that show a net *acceleration* of present day galaxies as opposed to those in the past. Both of these factors depend on cosmological parameters, particularly the matter and energy density in the Universe.

The recessional velocity of all objects from one another (assuming isotropy) implies a wavelength shift in radiation from any object observed by any observer, relative to the rest-frame wavelength, in analogy with the Doppler effect. The observed wavelength appears longer or *redder* than the rest wavelength as

$$z \equiv \frac{\lambda(\text{obs}) - \lambda(\text{rest})}{c}. \tag{14.2}$$

The wavelength redshift z (Eq. 14.2) is expressed in terms of the relativistic Doppler effect and velocities as

$$(1+z) = \gamma(1+v/c) = \sqrt{\frac{1+v/c}{1-v/c}}, \tag{14.3}$$

where γ is the Lorentz factor $1/\sqrt{1-v^2/c^2}$. Since $\gamma \approx 1$ for v $\ll$ c, $v \sim cz$, i.e. for small z.

The effective temperature of the Universe at $z > 0$ is given by

$$T(z) = T_0(1+z). \tag{14.4}$$

$T_0 = 2.725$ K is the background temperature at the present epoch $z = 0$. The radiation associated with a black body at this cold temperature is in the microwave region, as discovered by Penzias and Wilson [411], and constitutes the aforementioned CMB. As defined, the CMB temperature increases linearly with z, i.e., looking back in time towards a hotter Universe.[3] The transition from radiation to matter dominated Universe is thought to occur at $z \sim 35\,000$, or $T(z) \sim 10^5$ K. Cosmological models yield a corresponding timeframe of about 3000 years after the big bang. But the temperature was still too hot for electrons to recombine permanently with nuclei. Eventually, when the temperatures decreased to <3000 K after about 200 000 years, the first atoms appeared and remained without being immediately reionized.

In the sections below we describe the spectral signatures and atomic models used to understand the early Universe.

14.2 Recombination epoch

The recombination epoch is the earliest possible time accessible to atomic spectroscopy. Since the CMB temperature[4] increases with z (Eq. 14.4), it follows that the ionization energies of He II, He I, and H I can be used in an equation-of-state for the early Universe at high z to estimate the temperature range for recombination [412]. It is then found that during the recombination epoch of primordial atoms He II (e + He III) formed at $5500 \leq z \leq 7000$, He I (e + He II) at $1500 \leq z \leq 3500$, and H I (e + H II) at $500 \leq z \leq 2000$ [413].

The Saha equation-of-state yields the temperatures at which recombination occurs for each atomic species. For hydrogen recombination (e + p)$\rightarrow$H^o, we have

$$\frac{n_p n_e}{n_H} = \frac{1}{h^3}\left[\frac{(2\pi kT)(m_p m_e)}{m_H}\right]^{3/2} e^{-I_H/kT}, \tag{14.5}$$

where $I_H = 1$ Ry $= 13.6$ eV. The atomic physics of H I and He II lines is similiar, since both have Rydberg spectra (Chapter 2). The wavelengths of the same Rydberg transitions in H I are four times longer than in He II,

$$\lambda_{n,n'}(\text{HeII}) = \left[\frac{\lambda_{n,n'}(\text{HI})}{4}\right]. \tag{14.6}$$

For example, the Lyα transition in H I at 1216 Å corresponds to the 304 Å line in He II. It follows that some Rydberg lines of H I and He II would overlap in observed spectra. That could complicate the analysis of direct measurements of line intensities, which depends on the respective abundances.

Spectral formation of He I lines is, of course, quite different, as has been discussed extensively in the context of He-like ions (Chapters 4, 8, and 13 and Fig. 4.3). It is worth re-examining the Grotrian diagram of neutral helium in a slightly different way than before, as shown in Fig. 14.2, with the low energy levels and prominent lines divided according to spin multiplicity. The diagram excludes fine structure, and we know from Chapter 4 that radiative rates differ by orders of magnitude among the variety of multipole transitions characteristic of He-like

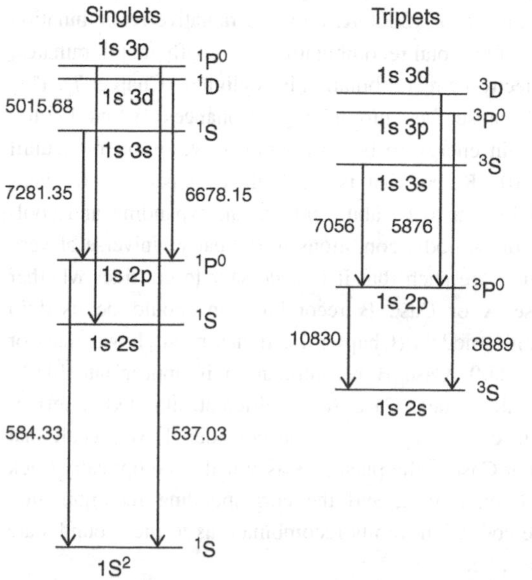

FIGURE 14.2 Energy levels of neutral helium. The triplet transition wavelengths on the right have been averaged over the fine structure (cf. Fig. 4.3).

[3] In reality, of course, it is the radiation from the earlier epoch that is finally reaching us today.

[4] In general, the temperature would correspond to the appropriate radiation background, which would be at shorter wavelengths and higher z.

transitions (Fig. 4.3). However, relativistic effects per se are small for He I, at least in the sense that fine-structure energy separations are small. One of the consequences is that the singlet and the triplet *LS* term structure is collisionally uncoupled. Singlet–triplet cross excitations are far less probable than those within each multiplicity. However, fine structure effects could be quite important for high-n levels where singlet–triplet mixing of levels occurs. Since the outer electron is farther removed and more weakly interacting with the core than the inner orbitals, the spin–orbit interaction of the outer electron, relative to the Coulomb interaction, is stronger than that for inner electrons. That requires the use of intermediate coupling or jj-coupling (Chapters 2 and 4) for high-nl orbitals.

The primary atomic processes that need to be taken into account are *level-specific collisional excitation* (Chapter 5), radiative transitions (Chapter 4) and (e + ion) recombination (Chapter 7). A number of these atomic parameters are available in literature, although not comprehensively. Radiative transition probabilities have been computed with high accuracy for all fine-structure levels of He I up to $n \leq 10$, $\ell \leq 7$ [73, 414]. Collision strengths were computed using the R-matrix method in *LS* coupling up to $n \leq 5$ [415]. Free–bound (e + ion) recombination rate coefficients and emissivities have been tabulated up to $n \leq 25$, $el \leq 5$ [216, 416]. It is noteworthy that these calculations are almost all in *LS* coupling and relativistic fine structure is not considered.

At the prevailing temperatures during the recombination era, $T < 3000$ K, only the radiative recombination part of the total recombination rate coefficient dominates; dielectronic recombination is negligible (Chapter 7). This is because the autoionizing resonances 1s2pnl lie too high in energy to be accessible for recombination until $T \sim 10^5$ K, but that is too high a temperature for neutral helium to be abundant. As an expanding and cooling black body, conditions in the early Universe at very high z are such that it is necessary to consider whether Case A or Case B recombination should be used in atomic models (Chapter 7). It has been shown that for $z > 800$, Case B recombination is appropriate [412]. At later times, Case A is sufficient, although intermediate cases may need to be considered. We recall that under Case B the plasma is assumed to be optically thick in Lyman lines, and the corresponding recombination rate coefficient omits recombinations to the ground state $n = 1$, i.e.,

$$\alpha_B(T) = \sum_{n>1} \alpha_R(T, n). \qquad (14.7)$$

The presence of a strong radiation field at very high-z implies that non-LTE collisional–radiative models are necessary for high precision (e.g. [412, 417, 418]).

The spectral signatures of He III → He II recombination will not be readily visible as recombination lines since that occured before hydrogen recombination. Therefore, those photons would probably have been scattered by the predominant form of ionized matter, electron–proton plasma, in the early Universe. But *any* detection of these lines would constitute evidence of the earliest epoch of atomic formation.

14.3 Reionization and Lyα forests

As the Univerese cooled, the recombination era associated with very high redshift eventually led to neutral matter, predominantly hydrogen. Primordial hydrogen clouds coalesced under self-gravity into the first large-scale structures, the precursor of latter-day galaxies. Under extremely dense conditions, and given sufficient masss, the first black holes and QSOs would have been formed. We saw in the previous chapter the connection that is now known between quasars and AGN on the one hand, and most galaxies on the other hand. The 'active' phases of galaxies are thought to be governed by accretion onto supermassive black holes. Quasi-stellar objects, in particular, are the source of stupendous amounts of energy. Therefore, the extreme luminosity of the first QSOs would have reionized the hydrogen clouds at some period, called the *reionization epoch*. Spectral signatures of reionization, and in general absorption by neutral hydrogen clouds, can be found in observations towards high-z quasars. Any hydrogen clouds lying in between a high-z quasar and the Earth would reveal absorption features resulting in H I excitation and ionization. We detect the QSO by its Lyα emission due to electron–proton recombination feature originating in the ionized plasma from the source itself. Then the Lyα absorption of the QSO signature by the intervening H I clouds at all redshifts $z \leq z$ (QSO) constitutes its entire emission–absorption spectra.

Such is indeed the case. Figure 14.3 shows the spectra of two quasars, 3C 273, which is relatively nearby at $z = 0.158$, and Q1422 + 2309 at medium redshift $z = 3.62$. The two spectra are qualitatively quite different. 3C 273 shows the large Lyα emission peak, together with some significant absorption towards the blue, indicating absorption or ionization due to H I Lyman lines. The Lyα absorption feature is subsumed by the large amount of emission, but does manifest iteself on the blue side of the Lyα emission peak. The spectrum of Q1422 + 2309, on the other hand, is full of a multitude of absorption

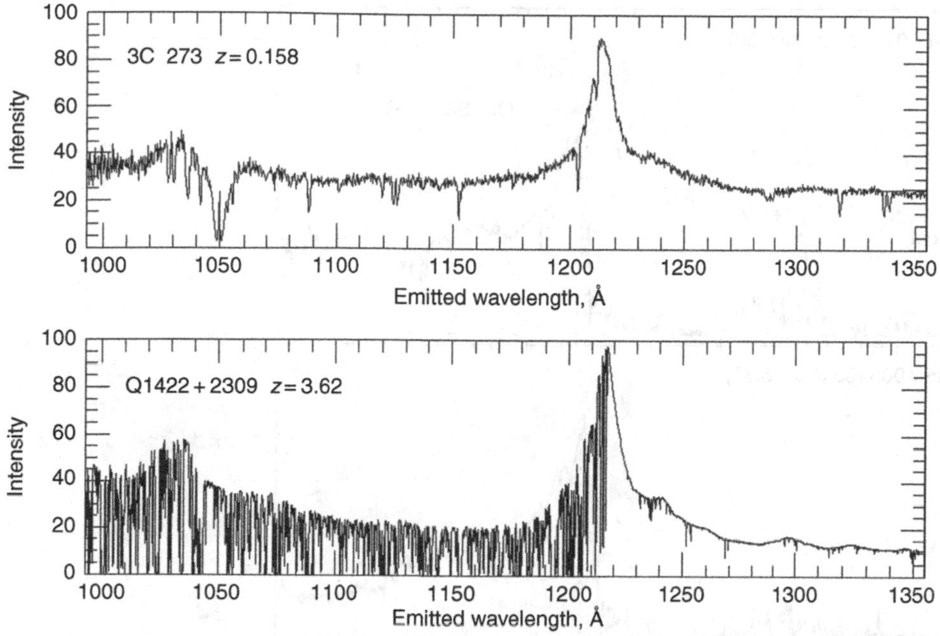

FIGURE 14.3 Lyα forests towards low- and high-redshift quasars. The observations show Lyα emission intrinsic to the quasar, and absorption lines blueward of the Lyα peak produced by Lyα clouds in the intervening IGM. The higher-z Q1422+2309 shows the 'forest' prominently, with mainly Lyα but also other Lyman lines (reproduced by permission from W. Keel (http://www.astr.ua.edu/keel/agn/forest.html).

lines at lower z than the quasar itself, i.e., blueward of the Lyα peak. This is referred to as the *Lyα forest* – heavy attenuation by Lyman absorption features by H I clouds at different redshifts in the intervening intergalactic medium towards the quasar.

It is then logical to ask at what point in time did the first QSOs light up and reionize the Universe? The answer is revealed in the spectroscopy of quasars up to the highest possible redshift observable. Figure 14.4 shows moderate-resolution spectra obtained from the Keck spectrographs [419] of four quasars in the range close to $z \sim 6$. The top panel in Fig. 14.4 identifies some of the most prominent emission features in the spectra. This includes the Lyβ and O VI blend at rest wavelength $\lambda\lambda$ 1026 and 1036 Å, respectively, and the closely spaced Lyα and N V features at $\lambda\lambda$ 1216 and 1240 Å, in addition to other unresolved sets of lines. At the extreme left, and the red-shifted feature at the shortest wavelength, is the Lyman limit $\lambda_0 = 912$ Å. Comparison of Fig. 14.4 with the much lower z quasars in Fig. 14.3 reveals a remarkable fact: the Lyα forest seems to be thinning with increasing redshift. This is due to the increasing density of neutral hydrogen towards higher z, which at some critical z absorbs nearly *all* of the light, giving rise to a large 'trough'-like absorption structure. The highest-z quasar at $z = 6.28$ exhibits nearly complete absorption up to wavelengths $(1+z) \times \lambda_0(1216) = \lambda_z < 8850$ (the bottom-most panel). The existence of this phenomenon was predicted, and is known as the *Gunn–Peterson trough* [420]. Its detection corresponding to quasars at $z \sim 6$ presents the strongest evidence to date of the critical time when the Universe approched the reionization epoch [419].

14.3.1 Damped Lyman alpha systems

Spectral analysis of Lyα clouds entails a particular type of line profile that we studied in connection with the curve of growth, shown in Fig. 9.6. We refer to the 'damped' part of the curve of growth corresponding to high absorber densities. The Lyα systems are vast regions of neutral H I with large column densities. As such, the line profiles have a square-well shape, characteristic of extremely high densities, beyond saturation of the central line profile, when absorption occurs and is observed mainly in line wings. The resulting shape is an increasingly larger flat bottom at zero flux, growing farther apart horizontally, and with vertical boundaries bracketing the line profile. Extensive neutral H I regions therefore give rise to the observed *damped Lyman alpha* (DLA) systems. Figure 14.5 shows the spectra of four sub-DLA and DLA

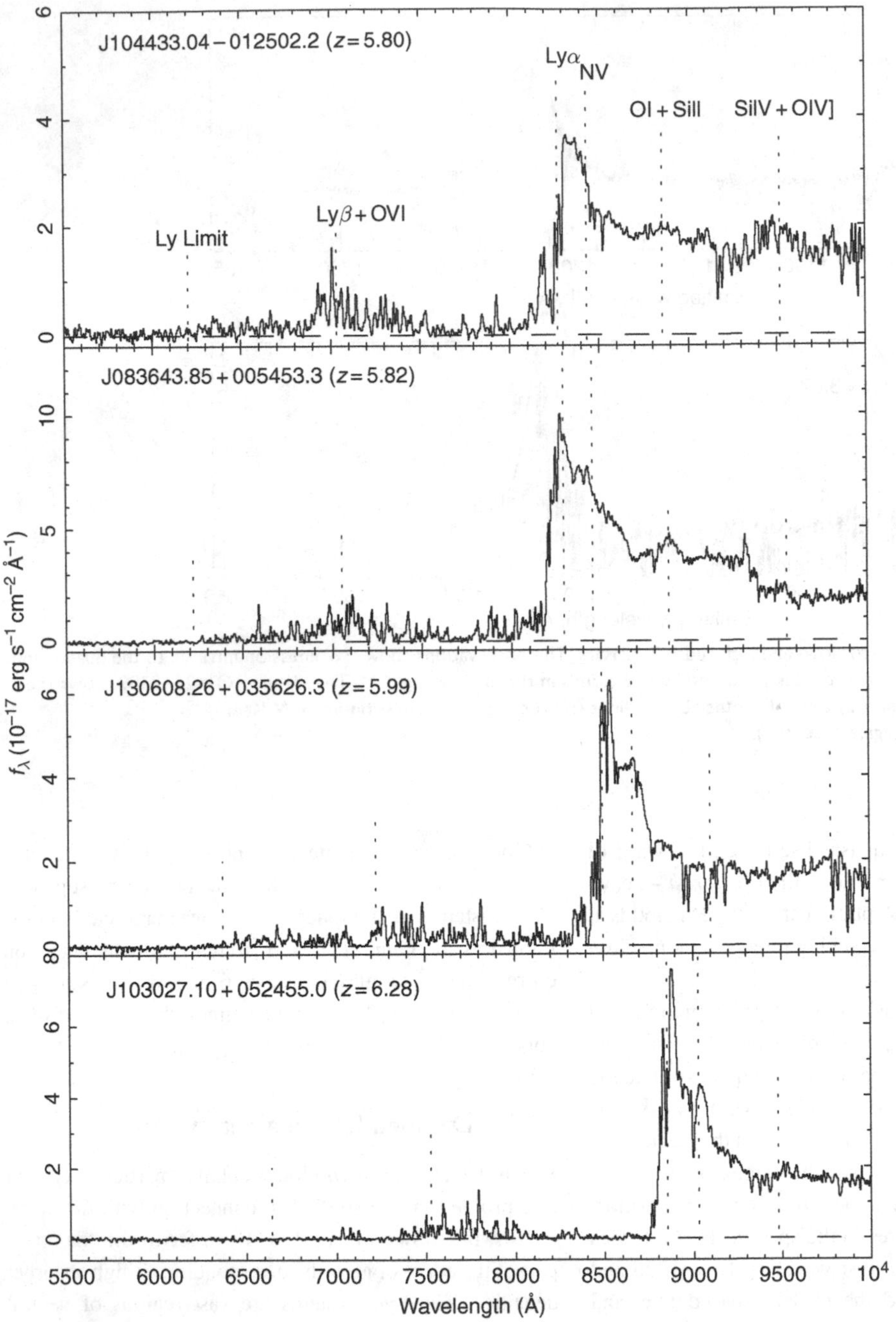

FIGURE 14.4 Signature of the reionization epoch: evolution of Lyα absorption with redshift and the Gunn–Peterson trough in the spectra high-z quasars around $z \sim 6$ [419] (reproduced with permission from R. Becker).

systems [421] viewed towards four quasars observed with the *Ultraviolet Echelle Spectrograph* (UVES) on the Very Large Telescope (VLT). The redshifts and the large H I column densities are marked individually; the higher-z systems clearly display the characteristic DLA profile (the observed spectra are fitted to different profiles, as shown).

14.4 CMB anisotropy

During the recombination era, matter and radiation are sufficiently coupled to nearly wipe out directly observable and identifiable spectral signatures. But the imprint of the recombination era could still manifest itself on large

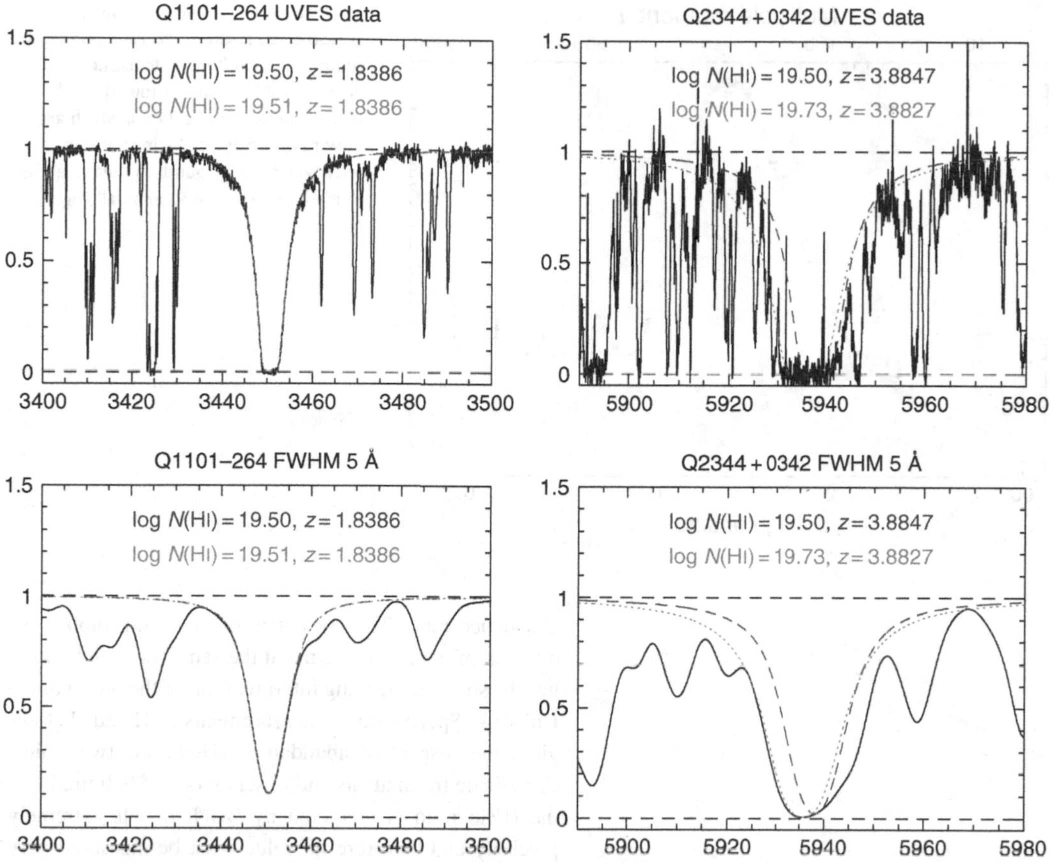

FIGURE 14.5 Absorption spectra of damped Lyα (DLA) systems. The panels on the left correspond to lower-z QSOs than the ones on the right, and display sub-DLA line profiles [421] (reproduced with permission from S. Ellison). Another colour image of more pronounced DLA absorption is provided on our website www.astronomy.ohio-state.edu/~pradhan.

scales in the radiation background as observable even today. In particular we refer here to *CMB anisotropy*. Owing to expansion, the early Universe cooled, but any non-uniformity in the distribution of primordial matter implied the presence of regions denser than others. Their gravitational fingerprints were then imprinted on the otherwise smooth Planckian radiation background. However, though measurable, the effect is indirect and reveals itself in the power spectrum on small angular scales over all space. The analysis is carried out in terms of a decomposition of the radiation background in spherical harmonics, quantified by the multipole moments C_l (in analogy with the standard Y_{lm} defined in Chapter 2; the azimuthal number m is not physically relevant, since it depends only on orientation). The measurements are made over all sky, on varying angular scales. Figure 14.6 shows the WMAP measurements going up to multipole moments $\ell = 1000$ (upper horizontal axis). Temperature fluctuations in the CMB resulting from anisotropic distribution of matter is in units of μm K^2. The wiggles at small angular scales represent the anisotropies imposed by inhomogeneities in the distribution of matter, and include those arising in the recombination era. The curve in Fig. 14.6 is a measure of departure from an otherwise Planck radiation distribution characteristic of a perfect radiation-only blackbody. It would have been flat if there were no matter-induced distortions in the early Universe, that persist even today, frozen into or imprinted upon the CMB. Given that the measured quantities involved are so small, it follows that the physics of recombination needs to be computed as precisely as possible.

Atomic physics is helpful in discerning the contributions to these 'distortions' imposed on the CMB at very high z during the recombination epoch. The three forms of atomic recombination, He III $\rightarrow$ He II, He II $\rightarrow$ He I and H II $\rightarrow$ H I, would each have a different spectrum. As such, and given sufficient sensitivity in future probes (viz. PLANCK), they could be distinguished. Generally,

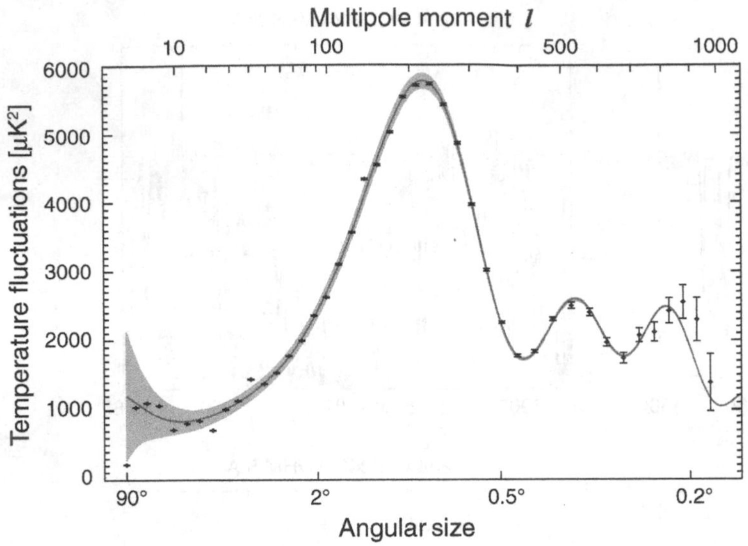

FIGURE 14.6 WMAP observations of anisotropies in the CMB radiation field (courtesy: http://map.gsfc.nasa.gov). The curve would be flat without these matter-induced distortions, which are now revealed as oscillations in the microwave background, as measured by WMAP all-sky surveys at small angular scales.

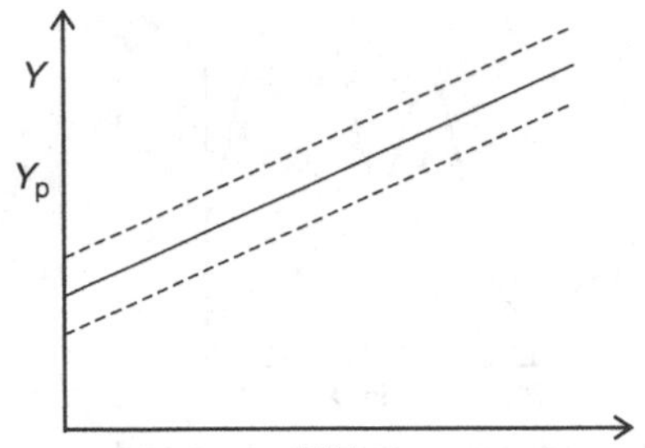

FIGURE 14.7 Schematic representation of increasing helium abundance Y vs. metal abundance Z/H . The (linear?) increase is due to secondary production of He in stars, as opposed to the primordial value Y_p. According to present-day measurements and BBN models, the uncertainties denoted by dashed lines put Y_p in a narrow range between 0.23 and 0.25.

(e + ion) recombination is seen as emission lines owing to radiative transitions downwards following recombination into high-*n* levels. But numerical simulations also indicate some interesting *negative* features due to absorption in the emission spectrum expected from He II → He I recombination lines [417]. This radiative transfer effect occurs if some lower states are fully occupied and radiative decays into those does not proceed faster than upward absorption to higher levels, resulting in some net absorption features, in addition to the mostly emission spectrum.

14.5 Helium abundance

Atomic physics and spectroscopy also provide the supporting evidence for one of the main pillars of standard cosmology – the primordial helium-to-hydrogen abundance ratio. The constancy of this ratio implies the creation of the two elements at the same time, and underlies the singular big bang interpretation of the origin of the Universe. Spectroscopic measurements of H and He lines yield the respective abundances. However, two points complicate the analysis and observations of helium. First, the H/He ratio is predicted by BBN models extremely precisely and therefore its value must be measured very accurately. Second, helium is also produced in stars and therefore it must be ascertained that the observations refer to primordial BBN, and not the helium produced by stellar nucleosynthesis. Before proceeding further, it is useful to remind ourselves of the notation employed in stellar astrophysics (Chapter 11): X → H, Y → He and metals → Z. The primordial value of He is denoted Y_p. Owing to stellar production of He and heavier elements, it follows that there should be a direct correlation between Y and Z, such that primordial Y_p is a limiting value of Y as Z → 0. In other words, Y_p depends on the precise value of the baryon-to-photon ratio η that underpins standard BBN models, and also on measurements differentiating it from the helium produced by stars.

Figure 14.7 is a schematic diagram of the linear correlation between the observed helium abundance as a function of Z or η. The aforementioned WMAP observations have placed a very tight constraint on the value of η, and thereby the baryonic density, usually denoted as Ω_b (not to be confused with the collision strength Ω). The BBN model then yields $Y_p = 0.2483 \pm 0.0004$. A contemporary challenge for atomic spectroscopy is then to determine the helium abundance observationally to similar precision. However, recent determinations using Z

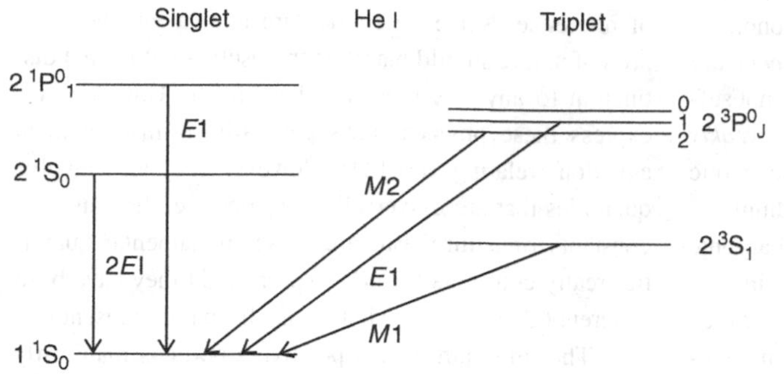

FIGURE 14.8 Grotrian diagram and selected fine-structure transitions in the helium atom (cf. Figs 4.3 and 14.2).

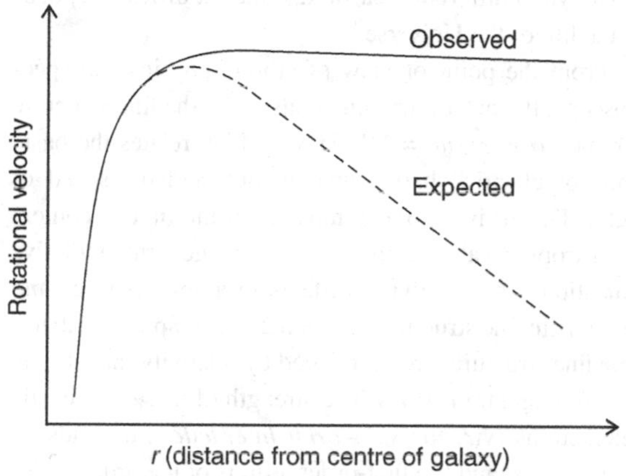

FIGURE 14.9 Schematic representation of rotational velocities of galaxies with respect to the centre.

abundances derived from H II regions (e.g., nitrogen and oxygen) range from about 0.23 to 0.25 [423, 424], with uncertainties that are much higher than those allowed by the WMAP data [425]. As is evident from Fig. 14.7, the linear relationship between Y and Z involving Y_p may be expressed as

$$Y = Y_p + \frac{dY}{dZ/dH}\left(\frac{Z}{H}\right). \tag{14.8}$$

To reduce the uncertainties in the atomic physics it is necessary to carry out calculations for a variety of atomic processes to a heretofore unprecedented accuracy of better than 1% or, in other words, a several-fold increase in precision from contemporary state-of-the-art results. As we have amply shown, the deduction of abundances from observed line intensities is subject to uncertainties in the atomic physics and parameters included in the spectroscopic models. Line emissivities of He must be computed taking account of all relevant atomic processes, such as electron impact excitation and ionization, (e + ion) recombination, photo-excitation, etc. Figure 14.8 underlies the complexity of the helium atom (which we have already encountered for He-like ions in previous chapters, albeit to requirements where lesser accuracy is adequate). It seems clear that further atomic calculations should consider the hitherto neglected relativistic effects and fine structure explicitly.

14.6 Dark matter: warm–hot intergalactic medium

There is now considerable evidence that if we consider the total amount of matter and energy in the Universe, then only about 4% of it is directly 'visible'. That is, all the observable galaxies and other objects made of known material whose existence we can ascertain. It is thought to be baryonic matter – protons and neutrons in the nuclei of atoms. So how do we (i) know that there is more matter and energy than we ordinarily 'see' – the other 96% – and (ii) how do we measure the presumbly predominant but invisible 'dark' component of the Universe? It is common to separate the 96% of the unknown entities into *dark matter* and *dark energy*, since physical phenomena related to each manifest themselves in different astrophysical situations. We discuss dark energy in a later section.

Matter would remain invisible (or 'dark' in common parlance) if it does not interact with – neither emits nor absorbs – electromagnetic radiation that we can observe. However, the gravitational influence of dark matter should be 'felt' in its effect on other observable objects. Such is the case when we attempt to determine the *rotation curves* of galaxies. Since they rotate around a central massive concentration, we would expect rotational velocities to decrease with distance from the centre. However, the observed situation is quite different. The rotational speeds remain roughly constant with distance from the centre of the galaxy; the measured rotation curves of galaxies are almost flat (Fig. 14.9). The explanation is that there is unseen matter, which exerts gravitational pull on the stars and gas that exist beyond the visible component of the haloes of galaxies. But the exact nature of such dark matter is unknown; although astronomers have long looked for candidates such as *massive compact halo objects* (MACHOs) – possibly brown dwarfs not sufficiently massive to turn into normal stars, and *weakly interacting massive particles* (WIMPs) – some exotic particles pervading the galactic–intergalactic medium.

There is, however, another component of baryonic dark matter, which appears to be present in the intergalactic medium or galactic haloes. These so-called *missing baryons* have been predicted by cosmological models [426, 427]. Atomic spectroscopy provides the needed strands of evidence: recent observations with *Chandra X-ray Observtory* and *X-Ray Multi-Mirror Mission – Newton* yield statistically significant levels of detection of absorption lines of Li-like, He-like and H-like oxygen ions O VI, O VII and O VIII at their respective Lyα or $K\alpha$ wavelengths 22.019, 21.6019 and 18.9689 Å (for further discussion see, for example, [428, 429, 430]).[5] In addition to X-ray observations, ultraviolet measurements, with the erstwhile space observatory *Far Ultraviolet Spectroscopic Explorer (FUSE)*, also reveal the existence of highly ionized O VI absorber clouds at wavelengths of the $1s^22s\ ^2S_{1/2} - 1s^22p\ ^2P^o_{3/2,1/2}$ doublet at 1032 and 1038 Å (e.g. [431, 432]).

This potential reservoir of missing baryons (about 20–40% of the total number of baryons), as high-temperature plasma at 10^{5-7} K, has been labelled the *warm–hot intergalactic medium* (WHIM or WHIGM). The highly ionized atomic species at extremely low WHIM densities ($\sim 10^{-5}\ n_e$) are almost 'invisible' unless (i) observed against the background of a very bright source, such as a bright AGN or quasar (even a blazar [428]) and (ii) along a line of sight that passes through a substantial cloud of otherwise diffuse ionized gas. In that case the absorption lines may offer sufficient contrast with the continuum to enable determination of column densities in the range of 10^{15} cm^2 or higher, and thereby an estimate of the total baryonic dark matter in the Universe.

14.7 Time variation of fundamental constants

The principle of relativity as postulated by Galileo and Einstein is essentially that the laws of nature are the same for all observers – there is no absolute or preferred frame of reference. Since all frames are equivalent, the basic laws of nature should manifest themselves with equal distinction to any observer anywhere in the Universe. We express these physical laws generally as mathematical equations relating variables. However, they also involve quantities that are universally the same, i.e., fundamental constants of nature. But are these fundamental quantities really constants for *all* time, or might they have been different (albeit slightly) in the past from their present values? The importance of a positive answer is manifestly obvious, for it would imply that the same laws of nature would yield different measured values at different epochs in the life of the Universe.[6]

From the point of view of atomic physics and spectroscopy, the most intriguing constant is the fine-structure constant $\alpha = e^2/\hbar c = 1/137\,036$, which relates the basic units of electric charge, quantization and the speed of light. The α is also the most relevant in the context of astrophysical spectroscopy as it governs radiative transitions and relativistic effects in atomic physics and relativistic fine structure (Section 2.13.2). Spin, and therefore fine structure, are introduced by relativity into atomic structure and transitions. The strength of these relativistic interactions (viz. the *spin–orbit interaction*) depends on powers of α, and the first-order term is of the order α^2.

Thus the variation of the fine-structure constant $\alpha = e^2/\hbar c$ is of fundamental interest in cosmology. However, if there is a variation of the fundamental constants over time, the effect must be very small, since otherwise it would manifest itself in observed phenomena rather easily. A variety of efforts are under way to measure any such possible deviation from the canonical (and by definition present-day) values. Recent laboratory measurements using $^{171}Yb^+$ transition frequency [433] give an upper limit of 2×10^{-15}year^{-1}, resulting in a change of α of the order of 10^{-5} in 10 gigayears. To measure this effect using astronomical observations, therefore, we need long look-back timescales. They have to be performed with high-redshift objects, and the lines to be studied need to be sufficiently strong features so as to be detectable from high-z, as well as to enable the analysis of any deviations from their 'normal' appearance in the spectrum.

The most common methods are measurements of high-redshift Lyα forest *metal absorption lines*. Among the first ones was the *alkali-doublet method* [434], with transitions from the singlet ground level to the two fine-structure doublets in alkali-like systems, such as the C IV and Si IV systems in QSO spectra [435]. These ions, as well as other

[5] Recall from the discussion in Chapter 13 that the O VI absorption 'line' is in fact a resonance – resonant absorption due to the inner-shell $K\alpha$ transition $1s^22s \rightarrow 1s2s2p$. There is somewhat larger uncertainty in the precision of the resonance wavelength as opposed to the lines O VII Heα and O VIII Lyα due to dipole allowed $E1$ transitions.

[6] Or different *cosmologies*. The word 'cosmogony' is often used to describe theories of the origin of the Universe.

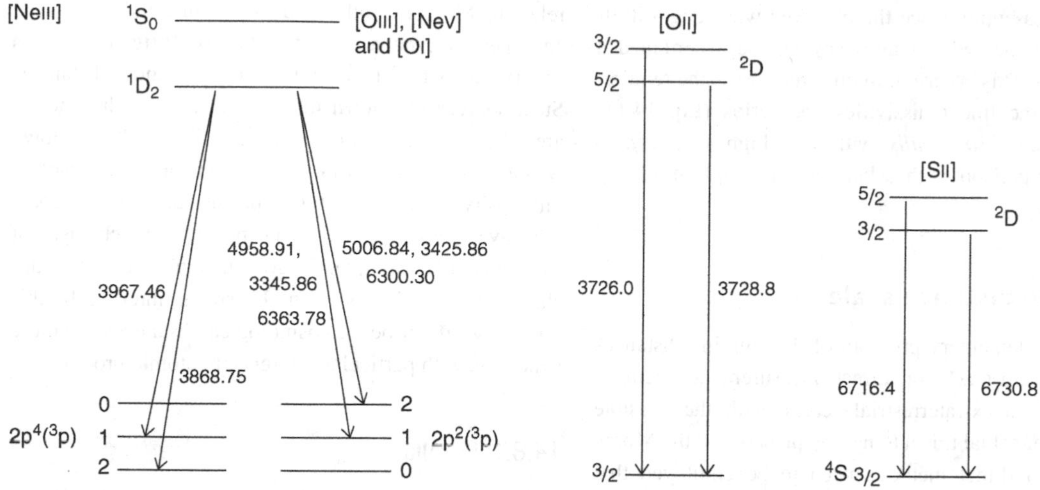

FIGURE 14.10 Fine-structure level structure of forbidden emission lines of nebular ions as potential candidates for time variation of $\alpha(t)$ [438].

alkali-like ions such as Mg II, form a pair of absorption lines due to the transitions $ns(^2S_{1/2}) \rightarrow ^2 P^o_{1/2,3/2}$.[7] The wavelength separation between the two lines depends on α^2 [15], and hence a direct probe of any variation in α.

Absorption line studies have been extended to more complex multiplets in heavy atomic systems, including the Fe group elements employing the so-called *many-multiplet method* using Lyα forests of quasar lines [436]. Based on high-resolution Keck spectra of several different samples of quasars [436], a value of $\Delta\alpha/\alpha_0 = -0.57 \pm 0.10 \times 10^{-5}$ has been reported, where α_0 is the standard value at the present epoch.

Emission lines, as discussed in Chapter 8, may also be used to study variations of α. Forbidden lines arise from higher-order multipole terms involving relativistic effects (Chapter 4). Therefore, their wavelength separation would depend on $\alpha(t)$ in a time-dependent manner, and could potentially serve as a chronometer of the age of the Universe. As we know from nebular studies (Chapter 12), forbidden fine structure lines of the [O III] doublet at 5006.84 and 4958.91 Å are extremely bright, and nearly ubiquitous in the optical spectra of H II regions in many sources (viz. Fig. 8.3), including high-redshift AGN and QSOs (Chapter 13). Generally, for any two lines in a forbidden multiplet one may define a *time-dependent* ratio [437]

$$R(t) = \frac{\lambda_2 - \lambda_1}{\lambda_2 + \lambda_1}, \tag{14.9}$$

at cosmological time t related to the ratio at the present epoch $t = 0$ by

$$\frac{R(t)}{R(0)} = \frac{\alpha^2(t)}{\alpha^2(0)}. \tag{14.10}$$

This is then a measure of the variation in α as a function of the cosmological look-back time t. An analysis of [O III] multiplet line ratios of the spectra of quasars from the Sloan Digital Sky Survey [367] has produced a statistically invariant result $\Delta\alpha/\alpha_0 = 0.7 \pm 1.4 \times 10^{-4}$ [437].

The [O III] line ratio technique may also be generalized into a 'many-multiplet' method for emission lines [438], as opposed to the one based on absorption lines [436]. We first note that the line ratios of lines originating from the same upper level depend only on intrinsic atomic properties, the energy differences between the fine-structure levels and corresponding spontaneous decay Einstein A coefficients, and are independent of external physical conditions, such as the density, temperature and velocities. For a three-level system, we may write the line ratio as

$$\mathrm{R} = \frac{N_3 A_{31} h\nu_{31}}{N_3 A_{32} h\nu_{32}}, \tag{14.11}$$

where level 3 is the common upper level. Thus, the expression on the right is simply $A_{31}h\nu_{31}/A_{32}h\nu_{32}$. If the A-values and energies are known to high accuracy, then we may use the observed ratio to identify pairs of forbidden emission lines in many atomic systems (cf. Fig. 8.3), such as the forbidden multiplets of [Ne III], [Ne V], [O II] and [S II], shown in Fig. 14.10. Note that an identification

[7] It is useful in many instances in astrophysics to note the analogous energy level structures of H-like and alkali-type ions, both forming 'doublet' line pairs of the type np(1/2, 3/2)→ns(1/2)), e.g., the Lyα 1,Lyα 2 fine structure components in H-like ions, or 3p→3s doublets in C IV, Si IV, Mg II, etc. A further generalization is to $K\alpha$ 1,$K\alpha$ 2 components of X-ray transitions, discussed in Chapter 13.

scheme is essential, since the observed wavelength itself depends on the redshift and may not be ascertained a priori [438]. This criterion, in turn, reinforces the requirement that the line emissivities and ratios (Eq. 14.11) be computed *theoretically* with very high accuracy, to enable comparison with what must be high-resolution spectroscopy.

14.8 The distance scale

The rather elementary problem of determining distances using a 'yardstick' for direct measurement becomes impossible on extraterrestrial scales, with the possible exception of actual travel to nearby planets and the Moon. Therefore, indirect methods need to be employed that might enable accurate determination of distances not only to nearby stars but also to far away galaxies, and to establish the extragalactic *distance scale*. It is necessary to do so for a variety of practical reasons in astronomy, such as to ascertain masses, sizes, luminosity and proper motions of astrophysical objects. In addition, just as mapping out distances between locations gives us a global image of the Earth, the cosomological distance scale yields valuable information on the spatial–temporal history, and the future of the Universe itself.

Astronomers often refer to the various practical methods as the *cosmological distance ladder* [439]. This is because each step or method, beginning with the one that helps ascertain the distances to nearby objects, provides the calibration for the next step, as in climbing a ladder. As we shall see, with the exception of the trigonometric parallax method, all other methods of determining astronomical distances must rely on some fundamental *intrinsic property* of the observed object. The most common such property is the *absolute luminosity*. But the problem is that we can only measure the *apparent luminosity* at a distance that is unknown a priori. However, knowing both allows us to calculate the distance from the distance modulus $(m - M)$ (Eq.10.6)

$$D\,(\text{pc}) = 10^{\frac{m-M+5}{5}} = 10 \times 10^{0.2(m-M)}. \qquad (14.12)$$

But using this relation obviously requires us to know the absolute luminosity irrespective of distance.[8] We mentioned earlier that observations of bright Cepheid variable stars (Chapter 10) do provide a means of measuring absolute luminosity via the period–luminosity relation. More recently, and as we discuss later, Type 1a supernovae have proven to be even brighter sources observable out to much farther cosmological distances. Such sources of known intrinsic or absolute luminosity are called *standard candles*, just as a light bulb of known wattage. In crucial ways, the determination of absolute luminosity depends on atomic physics and spectroscopy. We have already seen that the pulsation mechanism of Cepheids, for example, is governed by radiative transitions in specific zones in the stellar interior. In this section, we describe the cosmological distance scale more generally, with particular reference to atomic processes.

14.8.1 Parallax

The standard technique of using trigonometry to determine distances constitutes the first observational step in the cosmological distance ladder. Using the mean radius of the Earth's orbit around the Sun (1 AU) as the baseline, one can triangulate and determine the distances to nearby stars. But the method depends on measurement of at least one angle of a triangle, or the *parallax angle*, as shown in Fig. 14.11. The basic relation, which also defines the unit *parsec* (pc) (Chapter 10), is

$$d\,(\text{pc}) = \frac{1}{p\,(\text{arcsec})}, \qquad (14.13)$$

where p is the angle of parallax measured in arcseconds; correspondingly the distance is in pc. The fact that Eq. 14.13 requires a direct measurement of an extremely small angle p (Fig. 14.11) imposes a serious observational constraint. This is particularly so, since angles much

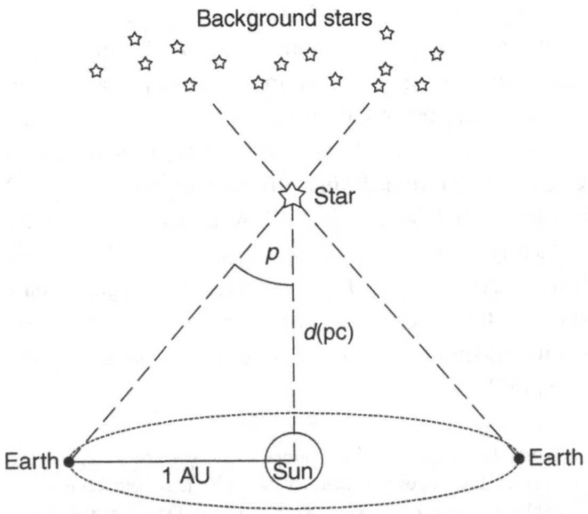

FIGURE 14.11 The parallax angle p of a star can be measured from the extremities of the Earth's orbit around the Sun, against the far away background stars that have negligible parallax.

[8] Distances are often calibrated in terms of the distance modulus. For instance the $(m - M)$ for our nearest neighbour, the galaxy Large Magellanic Cloud (LMC), is about 18.5, or at a distance of about 50 kpc.

smaller than an arcsecond need to be measured, because even the nearest star system, α Centauri, is 4.3 light years away. Since 3.26 light years = 1 pc, the parallax of αCen is $0.76''$. Therefore, accurate trigonometric parallax measurements from the Earth limit the distance scale only to nearby stars 10 – 50 pc away (<100 pc for the brightest visible stars).

In 1990, the launch of the *High Precision Parallax Collecting Satellite (Hipparcos)* dramatically improved the situation. Hipparcos measured high-precision parallaxes with milliarcsecond resolution for about 100 000 stars. In addition, the *Tycho star catalog* based on lower-precision Hipparcos data was compiled with a list of parallaxes of over a million stars out to about 1000 pc (1 kpc).

14.8.2 Spectroscopic parallax

The role of atomic spectroscopy is crucial in nearly all other methods for building the cosmic distance ladder – essentially, the relationship(s) between absolute luminosity and a related property. The expression *spectroscopic parallax* is referred to a combination of photometry and spectroscopy (Chapter 1), but has nothing to do with the geometrical parallax discussed above (it is a misnomer really). The basic idea is to observe the stellar spectral type with spectroscopy, together with its colour index in some range, say blue to violet, using photometry. That enables placing the star on the HR diagram (Fig. 10.2) to estimate its absolute magnitude, and hence the distance.

A method widely employed for clusters of stars is called *main sequence fitting*. Since the stars of a cluster are approximately the same age and distance, we may construct a partial HR diagram for the cluster stars based on their apparent luminosity magnitudes and colours or spectral types. Plotting and comparing the cluster diagram based on apparent magnitudes by overlaying on the full HR diagram, based on absolute magnitudes, practically allows us to 'read-off' the absolute luminosities. The cluster distance then follows according to the distance modulus relation Eq. 14.12.

Another useful spectroscopic relationship is known as the *Wilson–Bappu effect*: the absolute visual magnitude M_v is related linearly to the *width* or broadening of the Ca II K line produced by the transition $3p^6 4s\ ^2S_{1/2} \rightarrow 3p^6 4p\ ^2P^o_{1/2}$ at 3968 Å. The Wilson–Bappu effect [440] was discovered for late-type stars, such as the solar-type G stars or cooler K and M stars. It is applicable to stars with $M_v > 15$, or absolute luminosities less than 15th magnitude stars (recall the inverse relation between increasing magnitude number and decreasing luminosity – Chapter 10). The effect is remarkably independent of stellar spectral type; it is mainly a manifestation of Doppler broadening in the line core due to chromospheric activity driven by magnetic fields in nearly all cool stars, particularly K and M giants and supergiants. Surface gravity, effective temperature, radiative transfer and metallicity also bear on Ca II line formation. Calibration of the Wilson–Bappu effect (e.g., [441]) generally enables distance determinations even up to a few hundred kpc.

14.8.3 Cepheid distance scale

The next and perhaps the most reliable step in the cosmic distance ladder is the Cepheid period-luminosity (PL) relation discovered by H. Leavitt [442]. As mentioned in Chapter 10, Cepheids act as standard candles because their intrinsic luminosities vary periodically: *the longer the period, the greater the luminosity*. Recapping the discussion in Chapter 10, the pulsation periods depend critically on the opacity in the interior via the so-called κ-mechanism [443]. As explained in Chapter 11, the opacity κ does *not* decrease monotonically with temperature towards the stellar core, as one might expect, because of increasing ionization of atomic electrons that absorb radiation, and hence less opacity. Rather, there are distinct zones, mainly the H and He zones, where the opacity has large enhancements or bumps. While the helium opacity is most important, the metallicity is also crucial, since there is another bump due to metal opacity – the Z-bump primarily due to iron (Fig. 11.3). These enhanced opacity zones dampen the flow of radiation, periodically heating and cooling these layers, which, in turn, make the star expand and contract, or pulsate as observed. Figure 14.12 shows the period–luminosity variation, or PL curves, for four Cepheids from the *Harvard Variables (HV)* catalogue.

The RR Lyrae stars, which are metal-poor galactic halo stars, also pulsate, but are much less luminous than the Cepheids and therefore not useful as cosmological distance indicators. Based on PL curves, such as the ones shown in Fig. 14.12, one can determine absolute luminosities of Cepheids. It has been found that there are two distinct populations of Cepheid, the *classical Cepheids*.[9] which have high metallicity (Type I), and another class (Type II) with significantly lower metallicity, such as the prototypical Cepheid W Virginis. Figure 14.13 shows

[9] The prototypical Cepheid is the star Delta Cephei. The most well-known Cepheid is the North star or Polaris with a period of four days, but its luminosity variation of only about 1% and hence not discernible by eye.

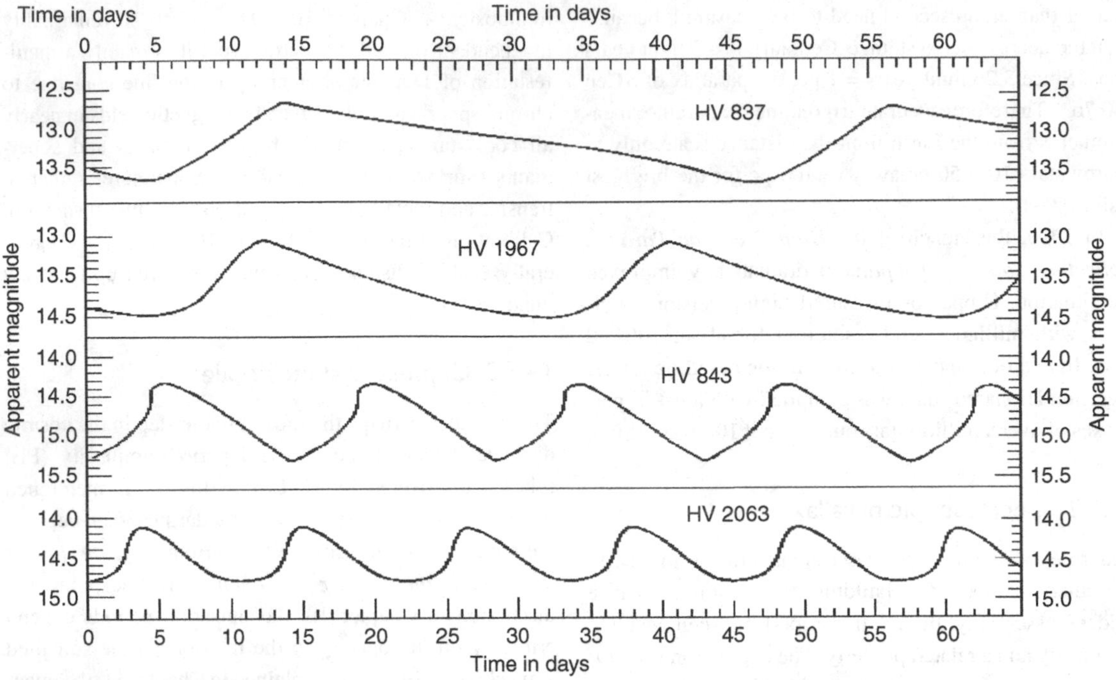

FIGURE 14.12 Cepheid pulsation periods and apparent luminosity of four Cepheids. The topmost panel has the longest period and the highest luminosity.

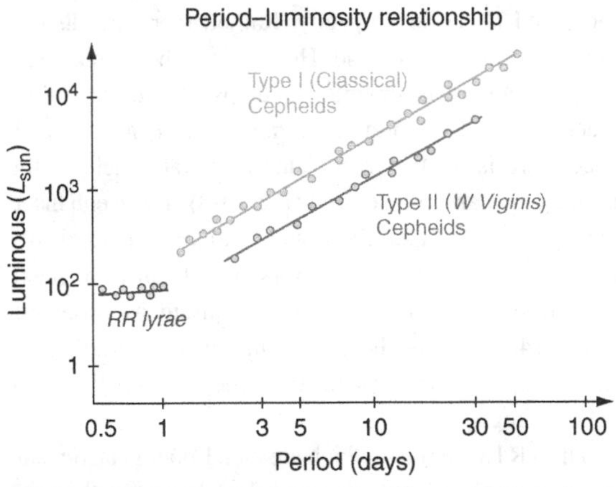

FIGURE 14.13 The period–luminosity relations for Types I and II Cepheids, and the metal-poor RR Lyrae stars (http://outreach.atnf.csiro.au/education/senior/astrophysics).

three sets of PL data, for the metal-poor RR Lyrae stars, as well as the two groups of Cepheids: the higher metallicity Type I Cepheids and the lower metallicity Type II Cepheids. The Type II Cepheids yield a significantly lower PL relation than the Type I Cepheids, which are also more luminous. The metallicity and type needs to be determined spectroscopically with sufficient accuracy for calibration. That is often difficult since metallacity is usually the Fe/H ratio, and photospheric Fe lines are difficult to resolve and analyze. The standard Population I Cepheid PL relation can be expressed in terms of the absolute visual magnitude M_v and the period (days) as

$$M_v = -2.81 \log_{10}(P) - 1.43. \tag{14.14}$$

More recent calibration of the classical Cepheids is now available, based on the 2 μm survey (2MASS) photometry, including reddening effects [444]. The Cepheid distance scale allows measurements out to approximately 10 kpc, although Cepheids have been observed up to Mpc distances.

14.8.4 Rotation velocity and luminosity

According to the standard mass–luminosity (M/L) correlation noted for stars in Chapter 10, the mass of a galaxy is also proportional to its intrinsic luminosity. Moreover, as pointed out earlier in connection with the presence of dark matter, the rotational velocity is proportional to the mass. Since the total mass is predominantly hydrogen, an extremely useful spectroscpic observation may be made: the 21 cm H I hyperfine structure line in the radio waveband. The 21 cm line is due to the transition between the two coupled-spin states of the electron and the proton, parallel or anti-parallel, triplet or singlet. Owing to the rotation of the galaxy the 21 cm line exhibits a double Doppler peak, redshifted and blueshifted. The total width of the Doppler-broadened line is then related to the luminosity by the *Tully–Fisher relation* [445]

$$L \propto W^{\alpha} \propto R^3, \qquad (14.15)$$

where W is the width of the line. The virial theorem also provides a relationship with the radius R in Eq. 14.15. The line width is directly related to the Doppler-broadened twin peaks. If $\Delta\lambda$ is the shift of red or the blue peak then we have

$$W = \frac{2\Delta\lambda}{\lambda} = \frac{2v \sin i}{c}. \qquad (14.16)$$

In practice, one needs to account for the orientation at angle i of the galaxy to our line of sight, and hence the relation of W to the (maximum) rotational velocity. The Tully–Fisher relation Eq. 14.15 is found to be strikingly linear, with little velocity dispersion, and a good estimate of absolute luminosities in selected photometric bands out to distances of galaxies >100 Mpc away.

14.8.5 Supernovae

As we noted in Chapter 10, Type II supernovae are the end-products of a huge variety of massive stars (e.g., $M \gtrsim 8M_{\odot}$) due to gravitational core collapse. But Type Ia supernovae all have similar origin in terms of the progenitor mass. If *all* Type Ia supernovae arise from nuclear fusion of the same mass, given by the Chandrasekhar limit, the energy generated in the corresponding supernova should be the same, and hence the same intrinsic luminosity or absolute magnitude. In addition, the temporal evolution of their light curves following the supernova explosion are also found to be similar. Since these explosions are tremendously powerful, they are observed out to much farther distances than the Cepheids. The key question, of course, is to ascertain that all Type Ias are indeed the same. The answer again lies in spectroscopic calibration of detailed spectral features.

We begin with spectral identification of the different SN types. Figure 14.14 shows the spectra of Type II and Types Ia, b, c at the same time after explosion, ~1 week [446]. The basic observational difference between Types I and II is that *the Type II supernovae spectra contain hydrogen lines*, whereas the Type Ia, b, c do not. In a Type II SN, the H I lines are formed in the outermost ejecta of the exploding star, which consists mainly of hydrogen. Types Ib and Ic are physically similar to Type II in that their progenitors are also sufficiently massive stars, but evolved to a stage where they have lost their hydrogen envelopes before the onset of core collapse. Most SNe Ib, c progenitors are thought to lose their hydrogen via binary interactions in symbiotic star systems. Massive stars, such as the Wolf–Rayet stars also undergo immense amounts

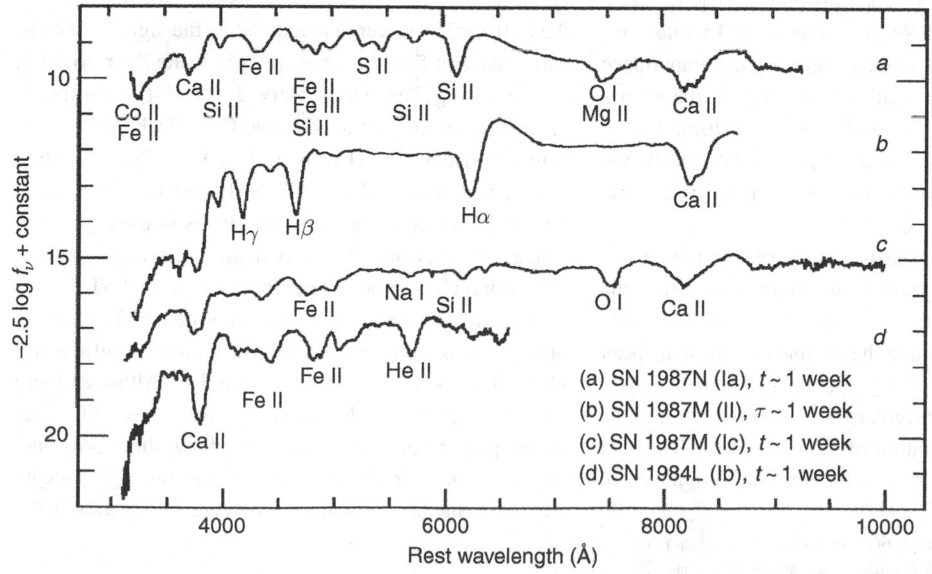

FIGURE 14.14 Optical spectra of supernovae Ia, II, Ib and Ic (top to bottom) at an early time, ~1 week after explosion [446]. Note the predominance of hydrogen lines in Type II, but their absence in the others. The suffix capital letter refers to the alphabetical order of detection in the particular year; e.g., the most widely observed supernova, the Type II SN 1987A in the nearby galaxy the Large Magellanic Cloud, was the first supernova detected in 1987.

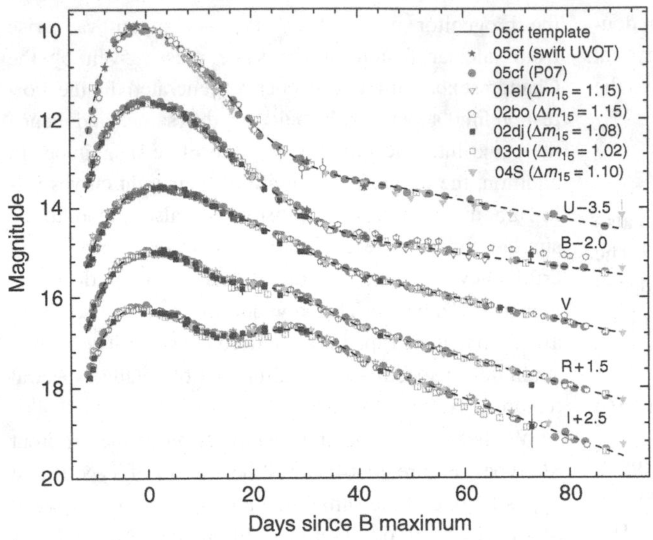

FIGURE 14.15 Light curves of several Type Ia supernovae in different colour magnitudes [448].

of mass loss through radiatively driven winds.[10] Types Ib and Ic are similar, except that the former still have their helium envelopes, as evidenced by the He I features (the lowest curve (d) in Fig. 14.14), whereas the former (curve (c)) do not show helium lines. The physical mechanisms of Types II, Ib and Ic are similar, but their spectra span a wide range depending on the progenitor mass and composition.

On the other hand, the observational features of Type Ia supernovae are remarkably similar. Figure 14.15 is a compilation of light curves in U, B, V, R, I colour bands (Chapter 10). Each curve contains photometric observations of a number of SNe Ia.[11] Figure 14.15 illustrates dramatically the whole point underlying the importance of SNe Ia as standard candles: not only are the shapes similar, *the light (energy) emitted in each distinct band appears to be the same for all Type Ias*. Obviously, the *sum* over all colour bands, i.e., the total or bolometric luminosity, should also be the same.

Another feature of SNe Ia light curves is the uniformity of their decline from peak luminosity during the early phase. A correlation has been established between the absolute luminosity and the decline rate from the peak value to its value after 15 days, parametrized by the quantity Δm_{15} [447]. The brighter the SN Ia, the broader the width Δm_{15} or the duration of the peak phase. This fact is crucial to the calibration of the absolute luminosity of SNe Ia. In addition to the photometric light curves, the spectroscopic homogeneity of SNe Ia has also been confirmed. Figure 14.16 compares detailed spectral features of three SNe Ia in galaxies with different Hubble velocities (cf. Eq. 14.3), but at the same early temporal epoch (as in Fig. 14.14) of about ~1 week past peak luminosity [446]. The most outstanding feature is the *blueshifted* Si II λλ 6347.10, 6371.36 (blended at ∼ λ 6355) due to the lowest dipole allowed transitions ('resonance' lines): $3s^2\ 4s(^2S_{1/2}) \rightarrow 4p(^2P^o{}_{3/2,1/2})$. Nucleosynthesis of silicon is understood to be due to the fusion of carbon and oxygen in the progenitor C-O white dwarf. Therefore the Si II λ 6355 feature is regarded as the signature of the early phase of SNe Ia. Other lines indicate their origin in the expanding (hence the observed blueshift) photosphere of the progenitor star, such as the Ca II H&K 3934, 3968 (Fig. 10.8) and the CaT near-IR lines λλ 8498, 8542, 8662 (Fig. 10.9) discussed in Chapter 10. At later times, the Fe group elements manifest themselves in the expanding ejecta, which eventually overwhelms the receding photosphere and photospheric features, such as the CaT lines.

Despite the commonality exemplified in Fig. 14.16, how do we ascertain that all SNe Ia are indeed sufficiently identical to be *precise* standard candles? Although there must be significant deviations in the masses of white dwarf progenitors, as characterized by their *zero-age-main-sequence* (ZAMS) masses,[12] the range of ejected masses in Sne Ia do not appear to range widely away from

[10] The most spectacular example of observed mass-loss is again the luminous blue variable Eta Carinae (cover jacket), which has undergone periodic episodes of huge mass loss in what may be a frantic effort to prevent core collapse by an extremely massive star.

[11] It is customary to denote the plural 'supernovae' as SNe, and the singular as SN.

[12] This is when stars first begin producing thermonuclear energy and arrive at the main sequence in the HR diagram (Fig. 10.2), following gravitational contraction of the protostellar masses to the temperature–density regime to initiate nuclear p–p ignition.

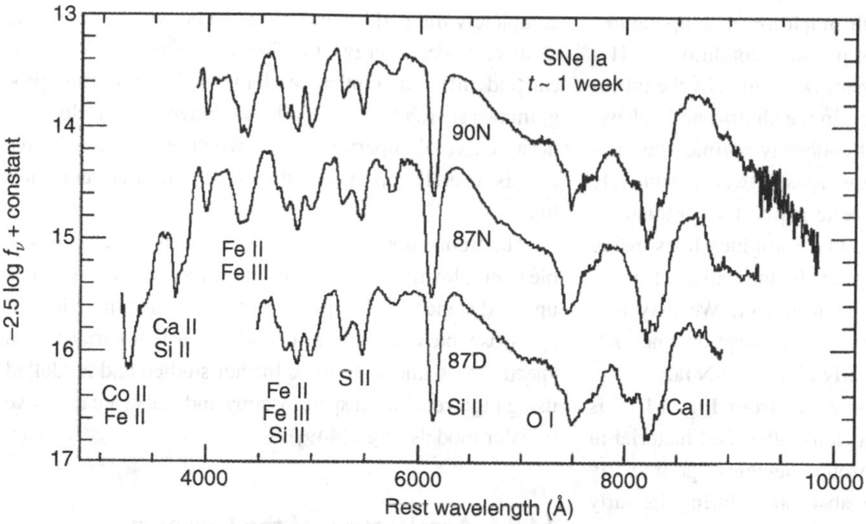

FIGURE 14.16 Spectral homogeneity of SNe Ia at different redshifts. Measured in velocities, they range from 970 kms $^{-1}$ (SN 1990N), 2171 kms $^{-1}$ (SN 1987N) and 2227 kms $^{-1}$ (SN 1987D) [446].

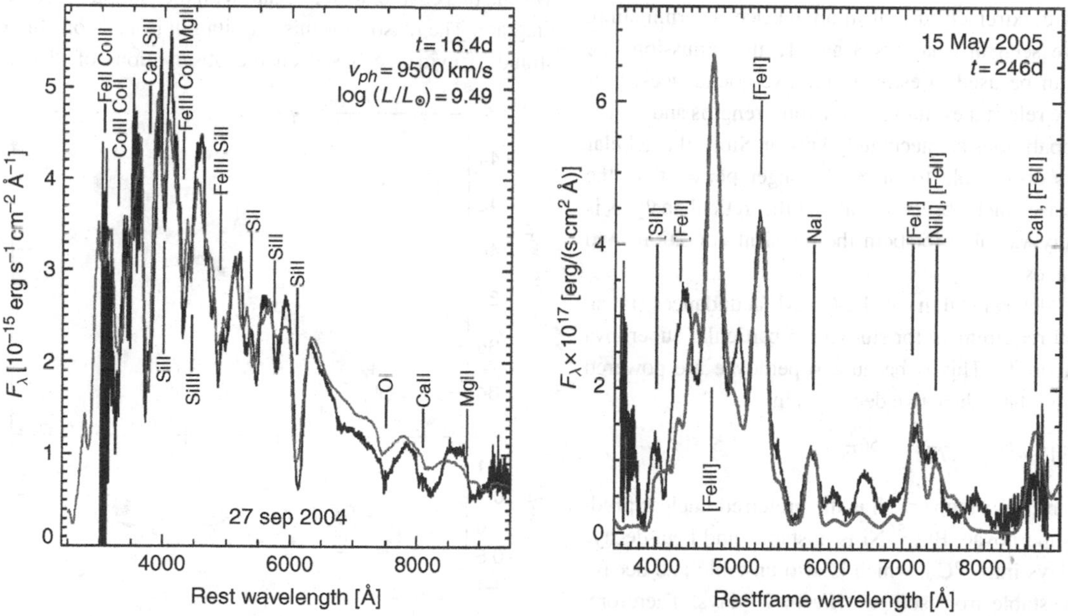

FIGURE 14.17 Temporal evolution of SN Ia spectrum in two different stages: the photospheric epoch (left) and the nebular epoch (right). During the early photospheric epoch spectral formation occurs in allowed lines, whereas in the late nebular epoch it is predominantly forbidden lines. The figures refer to observations of SN 2004eo, superimposed by radiative transfer models [449].

the canonical Chandrasekhar limit of 1.4 $M_\odot$. But the physical environments of SNe Ia span a wide range of host galaxies over much of the history of the Universe. The nucleosynthetic yields of common elements differ, and are observed to show variations. Furthermore, the detonation mechanism that pervades throughout the white-dwarf progenitor is not entirely explained by models. This is where it becomes important to determine the precise elemental composition, physical conditions, and kinematics of supernovae ejecta using detailed spectroscopy. It is necessary to carry out high-resolution spectroscopic observations at different epochs since various spectral features manifest themselves at particular times, as the physical conditions in the expanding ejecta evolve.

Figure 14.17 illustrates two sets of spectra representing an early phase with prominent photospheric features, and a late nebular phase with much lower temperatures and densities characteristic of physical conditions in H II regions. The basic atomic physics is revealed by the intrinsic nature of the observed lines. In the photospheric phase, and relatively high temperature–density regime, the lines are due to dipole allowed transitions betweeen relatively far apart atomic levels of opposite parity. By contrast, the late-time spectra show the familiar forbidden lines owing to the excitation of low-lying levels, indicative of a less energetic and optically thin environment. We may also note the aforementioned large Si II absorption line(s) at λ 6355 that characterize the early phase of SN Ia.

Another important fact apparent from Fig. 14.17 is the relative *abundances* of nucleosynthesized material in supernovae. Lines of iron-group elements, particularly Fe-Co-Ni, are seen mainly in absorption during the early phase. But as the high density SN ejecta expands into a thinner (i.e., optically thin) nebula, forbidden emission lines of these elements become the strongest features. These are extremely useful in abundance determination. We have seen in Chapters 8 and 12 that emission line ratios can be used to estimate relative abundances, provided the relevant excitation collision strengths and transition probabilities are accurately known. Since the nebular phase is observable for a much longer period than the early phase, such observational and theoretical analysis is especially valuable for both the physical conditions and abundances.

The determination of Fe-Co-Ni abundances is an essential requirement for studying a particular supernova (Type I or II). This is because supernovae are powered largely by the radioactive decay chain

$$^{56}\text{Ni (6d)} \rightarrow {}^{56}\text{Co (77d)} \rightarrow {}^{56}\text{Fe}.$$

The nickel isotope ^{56}Ni is the preferred nuclear product in supernovae. But ^{56}Ni is unstable, and beta-decays in six days into ^{56}Co, which is also unstable and decays into the stable iron isotope ^{56}Fe in 77 days. Therefore, during most of the decay phase shown in Fig. 14.15 the energetics of the supernova is driven by the radioactive decay of cobalt to iron. Nearly two-thirds of all iron in the Universe is believed to be nucleosynthesized and released in SNe Type Ia, and the remainder in SNe Type II. The γ-rays emitted in these nuclear reactions eventually degrade into lower energy X-rays, ultraviolet, optical and infrared radiation. The higher energy photons are reprocessed within the ejecta, which also fuels its expansion.[13] Initially, the γ-radiation is completely trapped by the dense ejecta, which thermalize to drive lower energy emission in the ultraviolet, optical and infrared. Optical and infrared monitoring programmes search for this sudden brightening to follow the decay curve of supernova light, which eventually decays and is usually observed in nebular infrared emission lines.

The tremendous luminosity of SNe Ia makes it possible to employ them as standard, or standardizable, candles up to distances >1 Gpc. However, an examination of spectroscopic calibration makes it clear that variations in spectra exist, and need to be further studied and modelled using high-resolution spectroscopy and elaborate radiative transfer models (e.g., [449]).

14.8.6 Acceleration of the Universe and dark energy

We end this chapter as we began, with the 'simple' Hubble diagram. The reason for this revisitation is an astonishing strand of evidence based on the observations of distant

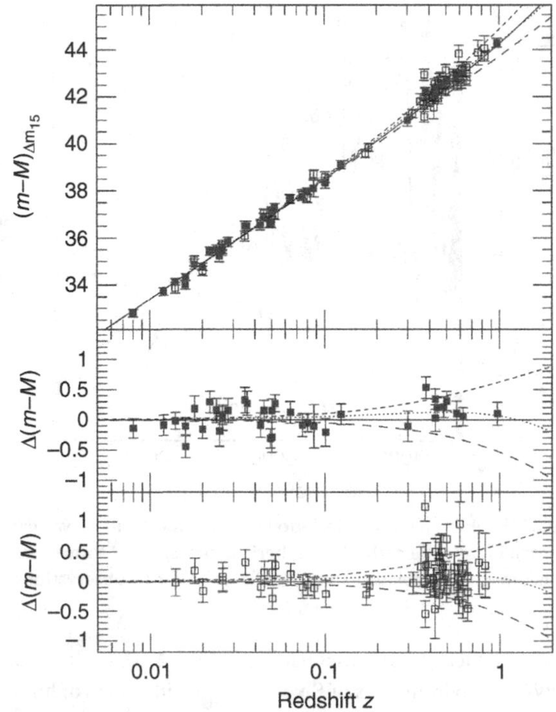

FIGURE 14.18 Evidence for accelerating expansion of the Universe from SNe Ia (from [452]). The points in the top panel in the upper right corner correspond to high-z SNe and deviate systematically, albeit slightly, from the linear Hubble relation at low z. The middle panel are the results from the High-z Supernova Search Team [451] and the bottom panel from the Supernova Cosmology Project [450].

[13] In core-collapse Type II SNe, the dominant release of energy is via neutrinos produced during nucleosynthesis.

SNe Ia. Recent work in the past decade or so appears to indicate that the expansion of the Universe is *accelerating*. The usual Hubble diagram, in Fig. 14.1, shows uniformly linear expansion characterized by its slope H_0. But using SNe Ia as standard distance indicators, two independent groups, the Supernova Cosmology Project [450] and the High-z Supernova Search Team [451], have obtained data that deviates from the linear Hubble law, as shown in Fig. 14.18. The deviations occur towards higher z, and although slight they are regarded as a statistically significant signature of accelerating expansion. According to the distance modulus (m-M) vs. z plotted in Fig. 14.18, the more distant SNe are fainter than the ones that are closer. That means that the nearby objects are moving with faster velocities than in those in the more distant past, i.e., moving farther away faster than the linear Hubble law would predict.

What could be causing a more rapid expansion of the Universe? As far as we know, gravity plays the ultimate role. So one possible answer is that some form of *dark energy* is causing a negative pressure or repulsion to counterbalance the attraction due to matter, which would otherwise make the Universe collapse gravitationally, or at least slow down the expansion. The idea is reminiscent of Einstein's famous *cosmological constant* Λ, which he introduced to explain a similar conundrum in theoretical models of a static Universe based on the general theory of relativity. Einstein abandoned the apparently unphysical term in the field equations when Hubble discovered the expansion of the Universe, and the ensuing big bang scenario that has since been amply verified.[14] Thus the notion of 'dark energy' seems to be harking back to the future, and is perhaps the most intriguing area of current research in cosmology.

14 Einstein called it his 'greatest blunder', apparently unhappy at having had to introduce an ad-hoc expression into the otherwise elegant mathematical framework of the general theory of relativity.

Appendix A Periodic table

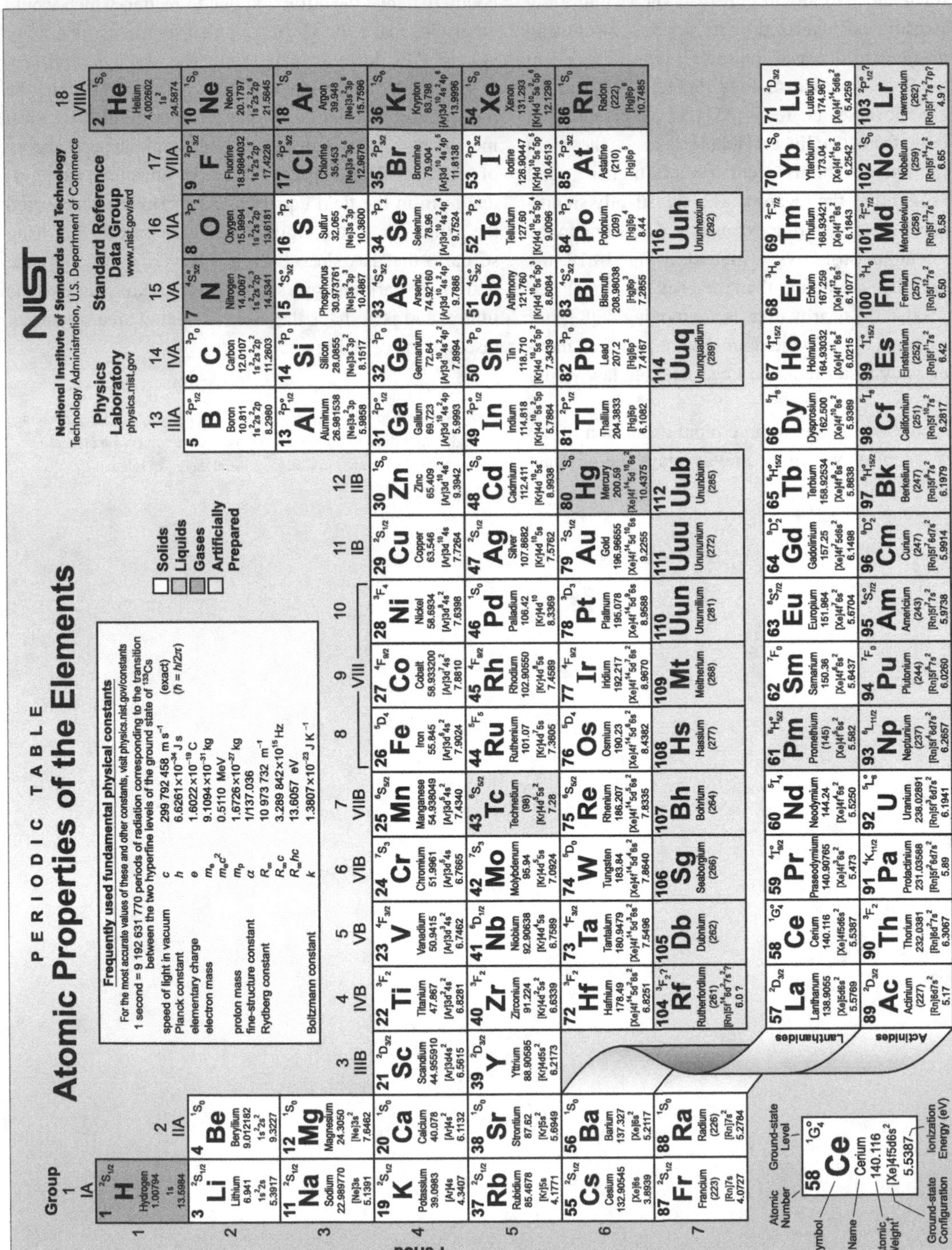

Appendix B Physical constants

Notation: cv = conventional value, vc = vacuum, BM = Bohr magneton

Quantity	Symbol	Value
Angstrom	Å	1.00×10^{-10} m
Astronomical distance unit	AU	1.496×10^{11} m
Atomic mass unit (^{12}C = 12 scale)	$m_u = 1\,u$	$1.660\,538\,86(28) \times 10^{-27}$ kg
Atomic time unit	$\tau_0 = \frac{h^3}{8\pi^3 m_e e^4}$	2.4189×10^{-17} s
Avogadro's number	N_A, L	$6.022\,141\,5(10) \times 10^{23}$ mol^{-1}
Bohr magneton	μ_B	$9.2740154 \times 10^{-24}$ J T^{-1}
	in eV	$5.78838263 \times 10^{-5}$ eV T^{-1}
	in Hz	$1.39962418 \times 10^{10}$ s^{-1} T^{-1}
	in wavenumber	46.686437 m^{-1} T^{-1}
	in K	0.6717099 K T^{-1}
Bohr radius	$a_0 = \alpha/4\pi R_\infty$	$5.291\,772\,108E18 \times 10^{-11}$ m
Boltzmann constant	$k = k_B = \frac{R}{N_A}$	$1.3806504(24) \times 10^{-23}$ J K^{-1}
	in eV	8.617385×10^{-5} eV K^{-1}
Characteristic impedance of vc	$Z_0 = \mu_0 c$	376.730 313 461
Compton wavelength	$\frac{h}{m_e c}$	2.426310×10^{-10} cm
	$\frac{h}{2\pi m_e c}$	3.861592×10^{-11} cm
Conductance quantum	$G_0 = 2e^2/h$	$7.748\,091\,7004(53) \times 10^{-5}$ S
Coulomb constant	$k_e = 1/4\pi\epsilon_0$	$8.987551\,787 \times 10^{9}$ Nm^2 C^{-2}
Earth's radius	R_E	6.37×10^{6} m
Electron–alpha-particle mass ratio	m_e/m_α	0.000137093354
Eletron charge	e	$1.602\,176\,487 \times 10^{-19}$ C
Electron compton wavelength	$\lambda_{c,e}$	2.42631×10^{-12} m
Electric constant (vc permittivity)	$\varepsilon_0 = 1/(\mu_0 c^2)$	$8.854\,187\,817 \times 10^{-12}$ F m^{-1}
Electron–deuteron mass ratio	m_e/m_d	0.000272443707
Electron g-factor	g_e	2.002319304386
Electron magnetic moment		1.001159652193 BM
Electron mass	m_e	$9.109\,382\,15 \times 10^{-31}$ kg
		$5.4857990943(23) \times 10^{-4}$ u
Electron molar mass		$5.48579903 \times 10^{-7}$ kg mol^{-1}
Electron–muon mass ratio	m_e/μ	0.00483633218
Electron–proton mass ratio	m_e/p_e	0.000544617013

Quantity	Symbol	Value
Electron radius	$r_e = \frac{e^2}{4\pi\epsilon_0 m_e c^2}$	$2.817\,940\,2894(58) \times 10^{-15}$ m
Electron rest mass energy		0.510998910(13) MeV
Electron specific charge	q	$1.75881962 \times 10^{11}$ C kg^{-1}
Electron speed in first Bohr orbit	$\frac{a_0}{\tau_0}$	2.18769×10^{8} cm s^{-1}
Electron volt	eV	$1.60217733 \times 10^{-19}$ J
Faraday constant	$F = N_A e$	96 485.3383(83) C mol^{-1}
Fermi coupling constant	$G_F/(\hbar c)^3$	$1.166\,39 \times 10^{-5}$ GeV^{-2}
Fine-structure constant	$\alpha = \frac{e^2 \mu_0 c}{2h}$	$7.297\,352\,537\,6 \times 10^{-3}$
	$\frac{1}{\alpha}$	137.0360
First radiation constant	$c^1 = 2\pi h c^2$	$3.741\,771\,18 \times 10^{-16}$ Wm^2
For spectral radiance	c_{1L}	$1.191\,042\,82 \times 10^{-16}$ Wm^2 sr^{-1}
Frequency of first Bohr orbit		6.5797×10^{15} s^{-1}
Gas constant	R	8.314 472(15) J K^{-1} mol^{-1}
		1.9858775(34) cal K^{-1} mol^{-1}
		8.205746×10^{-5} m^{3}atm K^{-1} mol^{-1}
		1.9872 cal g^{-1} mol^{-1} K^{-1}
Gravitational acceleration	g	9.80665 m s^{-2}
Gravitational constant	G	$6.67428(67) \times 10^{-11}$ m^3 kg^{-1} s^{-2}
Hartree energy (atomic unit)	$E_h = 2R_\infty hc$	$4.359\,744\,17 \times 10^{-18}$ J
		27.21165 eV
Ice point	$T = 0\,°C$	273.15 K
Inverse conductance quantum	$G_0^{-1} = h/2e^2$	12 906.403 7787(88)
Josephson constant	$K_J = 2e/h$	$1.835\,978\,91(12) \times 10^{14}$ Hz V^{-1}
Josephson constant (cv)	K_{J-90}	$4.835\,979 \times 10^{14}$ Hz V^{-1}
Loschmidt constant at	$n_0 = N_A/V_m$	
$T = 273.15$ K, $p = 101.325$ kPa		$2.686\,777\,3 \times 10^{25}$ m^{-3}
Magnetic constant (vc permeability)	μ_0	$4\pi \times 10^{-7}$ N/A^2
Magnetic flux quantum	$\phi_0 = h/2e$	$2.067\,833\,667 \times 10^{-15}$ Wb
Molar mass constant	$M_u = \frac{M(^{12}C)}{12}$	1×10^{-3} kg/mol
Molar Planck constant	$N_A h$	$3.990\,312\,716 \times 10^{-10}$ Js mol^{-1}
Molar volume of an ideal gas at	$V_m = RT/p$	
$T = 273.15$ K, $p = 100$ kPa		$2.2710\,981(40) \times 10^{-2}$ m^3 mol^{-1}
$T = 273.15$ K, $p = 101.325$ kPa		$2.2413\,996(39) \times 10^{-2}$ m^3 mol^{-1}
Nuclear magneton	$\mu_N = e\hbar/2m_p$	$5.050\,783\,43(43) \times 10^{-27}$ J T^{-1}
Planck charge	$q_P = \sqrt{4\pi\varepsilon_0 \hbar c}$	$1.875545870(47) \times 10^{-18}$ C
Planck constant	h	$6.626\,068\,96 \times 10^{-34}$ Js
	$\hbar = h/(2\pi)$	$1.054\,571\,628 \times 10^{-34}$ Js
Planck length	$l_P = \sqrt{\frac{\hbar G}{c^3}}$	1.616252×10^{-35} m
Planck mass	$m_P = \sqrt{\frac{\hbar c}{G}}$	2.17644×10^{-8} kg
Planck temperature	$T_P = \sqrt{\frac{\hbar c^5}{G k^2}}$	1.416785×10^{32} K
Planck time	$t_P = \sqrt{\frac{\hbar G}{c^5}}$	5.39124×10^{-44} s
Proton mass	m_p	$1.672\,621\,637 \times 10^{-27}$ kg
	in au	1.00727647 u
	in eV	9.3827231×10^{8} eV

Quantity	Symbol	Value
Proton Compton wavelength	$\lambda_{C,p}$	$1.32141002 \times 10^{-15}$ m
Proton magnetic moment	$\lambda_{C,p}$	$1.41060761 \times 10^{-26}$ J T^{-1}
Quantum of circulation	$\frac{h}{m_e}$	3.63695×10^{-4} J s kg^{-1}
		7.27389 erg s g^{-1}
Radiation constant	c^1	3.7418×10^{-16} W m^2
Radiation constant	c^2	1.43879×10^{-2} m K
Rydberg constant	$R_\infty = \frac{\alpha^2 m_e c}{2h}$	10 973 731.568 525 m^{-1}
	$\frac{1}{R_\infty}$	911.26708 I.A.
Rydberg energy (atomic unit)	$\mathrm{Ry} = hcR_\infty$	2.17992×10^{-11} erg
		13.605 6923(12) eV
Second radiation constant	$c_2 = hc/k$	$1.438\,775\,2 \times 10^{-2}$ mK
Solar constant		1367 W/m2
		433.3 Btu ft^2 h^{-1}
Speed of light in vacuum	c	299 792 458 m s^{-1}
Standard atmosphere	atm	101 325 Pa
Standard temperature and pressure	STP	$T = 273.15$ K, $p = 101.325$ kPa
Stefan–Boltzmann constant	$\sigma = \frac{\pi^2}{60}\frac{k^4}{\hbar^3 c^2}$	$5.670\,400 \times 10^{-8}$ W m^{-2} K^{-4}
Sun's radius		6.96×10^8 m
Sun's mean subtended full angle		0.532 degrees
		31.99 arc min
Thomson cross section	$(8\pi/3)r_e^2$	$6.652\,458\,73 \times 10^{-29}$ m
Triple point	H_2O	273.16 K
Wien displacement law constant for	$b = \frac{hc}{4.965114231k}$	
power max		$2.897\,768\,5 \times 10^{-3}$ mK
photons max		3669 m K

The Earth's atmospheric composition (%volume)

N_2	78.08%
O_2	20.95%
CO_2	0.033%
Ar	0.934%
Ne	1.82×10^{-3}%
He	5.24×10^{-4}%
Kr	1.14×10^{-4}%
Xe	8.7×10^{-6}%
H_2	5.0×10^{-5}%
CH_4	2.0×10^{-4}%
N_2O	5.0×10^{-5}%

Appendix C Angular algebra and generalized radiative transitions

The angular algebra for radiative processes is not straightforward to solve. However, we will describe only some basics of angular momenta algebra relevant to solutions of problems with three angular momentum functions. These are needed to derive probabilities for radiative transitions and relevant atomic parameters. A common integral in particle physics is

$$\langle l'm'|\mathrm{Y}_{LM}|lm\rangle = \int_0^{2\pi} \mathrm{d}\phi \int_0^{\pi} \mathrm{Y}^*_{l'm'}\mathrm{Y}_{LM}Y_{lm}\sin\theta \mathrm{d}\theta. \tag{C.1}$$

These integrals are sometime called Gaunt's coefficients.

C.1 3-j symbols

An integral of three related angular functions j_1, j_2 and j_3, such that they satisfy the triangular conditions

$$j_1 + j_2 - j_3 \geq 0, \qquad j_1 - j_2 + j_3 \geq 0, \qquad -j_1 + j_2 + j_3 \geq 0, \tag{C.2}$$

and $j_1 + j_2 + j_3$ is an integer, can be expressed conveniently by a 3-j symbol as

$$\frac{(-1)^{j_1-j_2-m_3}}{(2j_3+1)^{1/2}}(j_1m_1j_2m_2|j_1j_2j_3-m_3) = \begin{pmatrix} j_1 & j_2 & j_3 \\ m_1 & m_2 & m_3 \end{pmatrix}, \tag{C.3}$$

where the right-hand side is the 3-j symbol. Numerical computaion of a 3-j symbol is given by

$$\begin{pmatrix} j_1 & j_2 & j_3 \\ m_1 & m_2 & m_3 \end{pmatrix} = (-1)^{j_1-j_2-m_3} \times \left[\frac{(j_1+j_2-j_3)!(j_1-j_2+j_3)!(-j_1+j_2+j_3)!(j_1+m_1)!(j_1-m_1)!(j_2+m_2)!(j_2-m_2)!(j_3+m_3)!(j_3-m_3)!}{(j_1+j_2+j_3+1)!}\right]^{1/2}$$
$$\times \sum_k \frac{(-1)^k}{k!(j_1+j_2-j_3-k)!(j_1-m_1-k)!(j_2+m_2-k)!(j_3-j_2+m_1+k)!(j_2-j_1-m_2+k)!}. \tag{C.4}$$

Two frequently encountered 3-j symbols are

$$\begin{pmatrix} j_1 & 0 & j_3 \\ 0 & 0 & 0 \end{pmatrix} = (-1)^{j_1}\frac{1}{\sqrt{2j_3+1}}\delta j_1, j_3$$
$$\begin{pmatrix} j_1 & j_2 & j_3 \\ 0 & 0 & 0 \end{pmatrix} = (-1)^{J/2}\left[\frac{(j-2j_1)!(j-2j_2)!(j-2j_3)!}{(j+1)!}\right]^{1/2} \times \frac{(j/2)!}{(j/2-j_1)!(j/2-j_2)!(j/2-j_3)!}, \tag{C.5}$$

where $j = j_1 + j_2 + j_3$ is even; otherwise the expression becomes zero.

For even permutations of columns, the numerical value remains unchanged while for odd permutations, the value changes by the factor $(-1)^{j_1+j_2+j_3}$. The orthogonality properties are

$$\sum_{j_3m_3}(2j_3+1)\begin{pmatrix} j_1 & j_2 & j_3 \\ m_1 & m_2 & m_3 \end{pmatrix}\begin{pmatrix} j_1 & j_2 & j_3 \\ m_1' & m_2' & m_3 \end{pmatrix} = \delta\left(m_1m_1'\right)\delta\left(m_2m_2'\right),$$
$$\sum_{m_1m_2}\begin{pmatrix} j_1 & j_2 & j_3 \\ m_1 & m_2 & m_3 \end{pmatrix}\begin{pmatrix} j_1 & j_2 & j_3' \\ m_1 & m_2 & m_3' \end{pmatrix} = \frac{\delta\left(j_3j_3'\right)\delta\left(m_3m_3'\right)}{(2j_3+1)}\delta(j_1j_2j_3), \tag{C.6}$$

where $\delta(j_1j_2j_3) = 1$ if j_1, j_2, j_3 satisfy the triangular conditions, and zero otherwise.

The 3-j symbols are most commonly used for physical systems where two momenta vector-couple to form a resultant to give another good quantum number.

C.2 6-j symbols

While 3-j symbols are involved in coupling two angular momenta, 6-j symbols appear in problems concerned with couplings of three angular momenta. Consider a system of total angular momentum j composed of three subsystems of angular momenta j_1, j_2 and j_3. There is no unique way that j can form from the three angular momenta. For example, we can have $j_1 + j_2 = j'$ and $j = j_3 + j'$. The wavefunction in this representation is then $|(j_1 j_2)j', j_3 j\rangle$. We can also have $j_2 + j_3 = j'$ and $j = j_1 + j''$. The wavefunction in this coupling scheme is then $|j_1(j_2 j_3)j'', j\rangle$. The overlap between the two representations is proportional to a 6-j symbol:

$$\begin{aligned}&\langle (j_1 j_2)j', j_3 j | j_1(j_2 j_3)j'', j\rangle \\ &= (-1)^{j_1+j_2+j_3+j}\sqrt{(2j'+1)(2j''+1)} \\ &\times \begin{Bmatrix} j_1 & j_2 & j' \\ j_3 & j & j'' \end{Bmatrix},\end{aligned} \tag{C.7}$$

where the quantity in curly brackets on the right-hand side is a 6-j symbol. It differs from the Racah-coefficient W in a sign factor,

$$\begin{Bmatrix} j_1 & j_2 & j_3 \\ l_1 & l_2 & l_3 \end{Bmatrix} = (-1)^{j_1+j_2+l_1+l_2}\mathrm{W}(j_1 j_2 l_2 l_1;\, j_3 l_3); \tag{C.8}$$

similarly for Jahn coefficients, U:

$$\begin{aligned}&\mathrm{U}(j_1 j_2 l_2 l_1;\, j_3 l_3) = (-1)^{j_1+j_2+l_1+l_2} \\ &\times \sqrt{(2j_3+1)(2l_3+1)}\begin{Bmatrix} j_1 & j_2 & j_3 \\ l_1 & l_2 & l_3 \end{Bmatrix}.\end{aligned} \tag{C.9}$$

Numerical computation of a 6-j symbol is obtained using the formula

$$\begin{aligned}\begin{Bmatrix} j_1 & j_2 & j_3 \\ l_1 & l_2 & l_3 \end{Bmatrix} &= (-1)^{j_1+j_2+l_1+l_2}\Delta(j_1 j_2 j_3)\Delta(j_1 l_2 l_3)\Delta(l_1 j_2 l_3)\Delta(l_1 l_2 j_3) \\ &\times \sum_k \frac{(-1)^k (j_1+j_2+l_1+l_2+1-k)!}{k!(j_1+j_2-j_3-k))!(l_1+l_2-j_3-k)!(j_1+l_2-l_3-k)!(l_1+j_2-j_3-k)!} \\ &\times \frac{1}{(-j_1-l_1+j_3+l_3+k)!(-j_2-l_2+j_3+l_3+k)!}\end{aligned} \tag{C.10}$$

where

$$\Delta(abc) = \left[\frac{(a+b-c)!(a-b+c)!(-a+b+c)!}{(a+b+c+1)!}\right]^{1/2}. \tag{C.11}$$

The 6-j symbol is invariant under interchange of columns, and the interchange of any two numbers in the bottom row with the corresponding two numbers in the top row. A 6-j symbol is automatically zero unless each of the four triads (j_1, j_2, j_3), (j_1, l_2, l_3), (l_1, j_2, l_3), (l_1, l_2, j_3) satisfies the triangular conditions and the elements of each triad sum up to an integer. The orthogonality condition is

$$\begin{aligned}&\sum_l (2l+1)(2j''+1)\begin{Bmatrix} j_1 & j_2 & j' \\ l_1 & l_2 & l \end{Bmatrix}\begin{Bmatrix} j_1 & j_2 & j'' \\ l_1 & l_2 & l \end{Bmatrix} \\ &= \delta(j', j'').\end{aligned} \tag{C.12}$$

The 6-j symbol is extensively used in the computation of reduced matrix elements of tensor operators.

C.3 Vector and tensor components

A tensor $\mathbf{T_k}$ of order k is a quantity with $2k+1$ components, T_{kq} with $q = k, k-1, \ldots 0, \ldots, -k$. The spherical components of a vector $\mathbf{A}$ can be expressed in the form of components of a tensor of order 1 as follows:

$$\begin{aligned}A_0 &= A_z \\ &= |A|\cos\theta = |A|\sqrt{\frac{4\pi}{3}}\mathrm{Y}_{1,0}, \\ A_{+1} &= -\frac{1}{\sqrt{2}}(A_x + \mathrm{i}A_y) = -\frac{|A|}{\sqrt{2}}\mathrm{e}^{\mathrm{i}\phi}\sin\theta \\ &= |A|\sqrt{\frac{4\pi}{3}}\mathrm{Y}_{1,+1}, \\ A_{-1} &= \frac{1}{\sqrt{2}}(A_x - \mathrm{i}A_y) = \frac{|A|}{\sqrt{2}}\mathrm{e}^{-\mathrm{i}\phi}\sin\theta \\ &= |A|\sqrt{\frac{4\pi}{3}}\mathrm{Y}_{1,-1}.\end{aligned} \tag{C.13}$$

An irreducible tensor, also known as a spherical tensor $\mathbf{T_k}$, whose components under rotation of the coordinate system transform as the spherical harmonics Y_{lm}, transform and obey the same commutation rules of the angular momentum $\mathbf{J}$ of the system with Y_{kq}, that is

$$\begin{aligned}&[(J_x \pm \mathrm{i}J_y), T_{kq}] \\ &= \sqrt{(k \mp q)(k \pm q + 1)}T_{k,q+1},\ [J_z, T_{kq}] = qT_{kq}.\end{aligned} \tag{C.14}$$

When k=1, the above commutation rules coincide with those for the spherical components of a vector $\mathbf{A}$.

C.4 Generalized radiative transitions

With reference to the discussion in Chapter 4, evaluation of the transition matrix element $\langle j|\mathbf{D}|i\rangle$ for complex atoms is rather more involved, owing to angular and spin dependences, on the one hand, and the inherent complexity of computing wavefunctions for a many-electron system, on the other hand. The latter problem is the more difficult one, and hitherto we have described various methods to solve it. Now we sketch the basic expressions for carrying out the spin-angular algebra of vectorial (tensorial) additions of spin and orbital angular momenta. The simplification inherent in the division between the two parts is made possible by the *Wigner–Eckart theorem*, which enables the exact and a-priori computation of the spin-angular problem exactly, and the radial matrix element approximately.

At first glance, transition matrix elements $\iiint \psi^* O \psi' d\tau$ look rather complex if the operator O involves products of several spherical harmonics $\mathrm{Y}_{lm}(\vartheta,\varphi) = \langle \hat{r}|lm\rangle$, especially if spin is also added. As demonstrated in the case of magnetic two-body magnetic integrals (Section 2.13.5), it would be impractical to perform the integrations over the direction space, that is the angles (ϑ,φ). It is here that the spherical calculus enables such operations to be performed in the space of the state vectors $|lm\rangle$: instead of computing integrals over 4π, one exploits the algebraic properties of rotations in three-dimensional space, starting with the basics of angular momentum algebra, as developed by E. U. Condon and G. H. Shortley [3], but later using the more powerful identities derived by G. Racah (e.g., [13]). In spherical or 'rotational', rather than Cartesian, coordinates the components of the radius vector $\boldsymbol{r}$ read

$$r_\pm = \mp\frac{x \pm \mathrm{i}\,y}{\sqrt{2}}, \qquad r_0 = z\,. \tag{C.15}$$

The components of J (integer like L or half-integer like s) act as step-up and step-down operators on the quantum number m of Slater states:

$$J_\pm\,|j\;m\rangle = \hbar\sqrt{(j\mp m)(j\pm m+1)}\;|j\;m\pm 1\rangle \tag{C.16}$$
$$J_0\,|j\;m\rangle = \hbar\,m\,|j\;m\rangle\,,$$

reproducing the eigenvalues of the angular operator L^2, Eq. 2.10, without resort to spherical harmonics Y. Similarly, the tensor operator $\mathsf{C}^{[k]}_\kappa$, like P^κ_k of Eq. 2.17, representing the components of the unit radius vector $\boldsymbol{r}/r$ or its dyadic products of order k but acting on Slater states (Eq. C.16), can raise or lower the value of l as seen in its matrix elements

$$\begin{aligned}
c^k(lm,l'm') &\equiv \left\langle lm\left|\mathsf{C}^{[k]}_\kappa\right|l'm'\right\rangle \\
&= (-1)^m\,\langle l'm'\,k\kappa|lm\rangle\,\langle l0\,k0|l'0\rangle \\
&= (-1)^k\sqrt{(2l{+}1)(2l'{+}1)}\begin{pmatrix} l & k & l' \\ -m & \kappa & m'\end{pmatrix} \\
&\quad\times\begin{pmatrix} l & k & l' \\ 0 & 0 & 0\end{pmatrix}.
\end{aligned} \tag{C.17}$$

In Eq. C.17 we have introduced the basic quantities related to coupling of *two* angular momentum quantum numbers: the *Clebsch–Gordon coefficients* or the *vector coupling coefficient* (VCC) in the bra-ket notation in the first equation, and the equivalent 3-j symbol in the second equation. Their relationship to spherical harmonics is as outlined above.

The magnetic components $\kappa = k, k-1, \ldots, -k$ of any spherical operator $\mathsf{T}^{[k]}$ can be related to a single quantity, the *reduced matrix element* of this operator according to the *Wigner–Eckart theorem*

$$\begin{aligned}
\left\langle JM\left|\mathsf{T}^{[k]}_\kappa\right|J'M'\right\rangle &= \frac{\langle JM\,J'M'|k\kappa\rangle}{\sqrt{2J+1}}\langle J\|\mathsf{T}^{[k]}\|J'\rangle \\
&= (-1)^{J-M}\begin{pmatrix} J' & k & J \\ -M' & \kappa & M\end{pmatrix}\langle J\|\mathsf{T}^{[k]}\|J'\rangle\,,
\end{aligned} \tag{C.18}$$

incidentally relating the transparent vector coupling or Clebsch–Gordan coefficients to the highly symmetric 3-j symbols. Moreover, the reduced matrix element of C is readily derived. Here are more examples of reduced matrix elements, beginning with the identity operator I:

$$\begin{aligned}
&\langle J\|\mathsf{I}\|J'\rangle = \sqrt{2J+1}\,\delta_{J\,J'} \\
&\langle J\|\boldsymbol{J}\|J\rangle = \sqrt{J(J+1)(2J+1)} \\
&\langle slj\|\mathrm{Y}_k\|sl'j'\rangle = (-1)^{j-1/2}\sqrt{\frac{(2j+1)(2k+1)(2j'+1)}{4\pi}} \\
&\quad\times\begin{pmatrix} j & k & j' \\ -1/2 & 0 & 1/2\end{pmatrix}\left\{1+(-1)^{l+k+l'}\right\}/2.
\end{aligned} \tag{C.19}$$

ending with the first example of a coupled state (and without the need to formulate spin angular functions[1]).

Without any presumption of the physical nature of states, the inital i and final j levels may be desginated with

[1] Simply to retain the letter P for C would miss the decisive difference in phase factors imposed on spherical operators and embodied in Eq. C.15. Summation over pairs of identical magnetic quantum numbers (adding up to angular quantum numbers inside 6-j or Racah coefficients) would otherwise not work when embedded in such a complex context as two-body magnetic couplings (Eq. 2.196). The respective other-orbit terms (the ones with twice s) give a flavour of tensor operations involving *two* systems i and j. It shows that simple tools like the addition theorem of spherical harmonics (Section 2.1.1) had to give way to spherical tensor operations. But those detailed evaluations are well beyond the scope of this text.

their relevant quantum numbers as $l^n(\alpha_1 L_1 S_1)n_i l_i L_i S_i$ and $l^n(\alpha_1 L_1 S_1)n_j l_j L_j S_j$, where $l^n(\alpha_1 L_1 S_1)$ is the total angular momentum of the electrons staying inert during the transition (here the symbol α denotes a generic expression for all other characteristic parameters, such as configuration and principal quantum numbers of the 'parent' ion core). We generalize the transition probability with respect to the degeneracies of the initial and final states. If L_i and S_i are the total oribital and spin angular momenta of the initial state, and degeneracy $g_i = (2S_i + 1)(2L_i + 1)$, then the line strength S is expressed as (in the length formulation)

$$S_{ij} = \sum_{M_{S_i}, M_{S_j}} \sum_{M_{L_i}, M_{L_j}} |\langle l^n(\alpha_1 L_1 S_1)n_j l_j L_j S_j M_{L_j} M_{S_J}|\mathbf{D}|l^n \times (\alpha_1 L_1 S_1)n_i l_i L_i S_i M_{L_i} M_{S_i}\rangle|^2, \tag{C.20}$$

where the sum is over all initial and final degenerate levels. In the radiative perturbation operator $e/(mc)(\mathbf{p}.\mathbf{A})$ there is no spin dependence, that is, the spin cannot change during the transition. This leads to the selection rule for dipole E1 transitions,

$$S_j = S_i, \text{ i.e.}, \Delta S = 0. \tag{C.21}$$

However, the spin rule can be violated, owing to departure from LS coupling via the spin–orbit interaction. In that case, if S is no longer a good quantum number, then we need to further consider the $\mathbf{J} = \mathbf{L} + \mathbf{S}$ intermediate coupling scheme, discussed later.

The dipole moment operator $\mathbf{D}$ is equivalent to an irreducible tensor of order 1 with three spherical components D_q,

$$D_0 = D_z = |D|\sqrt{\frac{4\pi}{3}}\mathrm{Y}_{10},$$

$$D_{+1} = -\frac{1}{\sqrt{2}}(D_x + \mathrm{i}D_y) = |D|\sqrt{\frac{4\pi}{3}}\mathrm{Y}_{1,+1}, \tag{C.22}$$

$$D_{-1} = \frac{1}{\sqrt{2}}(D_x - \mathrm{i}D_y) = |D|\sqrt{\frac{4\pi}{3}}\mathrm{Y}_{1,-1},$$

which can be expressed in short as $D_q = |D|\sqrt{\frac{4\pi}{3}}\mathrm{Y}_{1q}$. Using the Wigner–Eckart theorem (Eq. C.18) the dipole transition matrix element is

$$\langle(\alpha_1 L_1)n_j l_j L_j M_{L_j}|D_q|(\alpha_1 L_1)n_i l_i L_i M_{L_i}\rangle = (-1)^{L_j - M_j}\begin{pmatrix} L_j & 1 & L_i \\ -M_{L_j} & q & M_{L_i}\end{pmatrix} \times \langle(\alpha_1 L_1)n_j l_j L_j||\mathbf{D}||(\alpha_1 L_1)n_i l_i L_i, \tag{C.23}$$

where we have dropped the parent configuration l^n as the quantum states are specified. The properties of 3-j symbols dictate that the matrix element is zero unless

$$\Delta L = L_j - L_i = 0, \pm 1, \ \Delta M = M_{L_j} - M_{L_i} = 0, \pm 1. \tag{C.24}$$

However, for each of the three possible transitions $\Delta M = 0, \pm 1$, only one term in D_q is non-zero, i.e.,

$$\langle(\alpha_1 L_1)n_j l_j L_j M_{L_j}|D_0|(\alpha_1 L_1)n_i l_i L_i M_{L_i}\rangle \times \text{ for } \Delta M = 0,$$
$$\langle(\alpha_1 L_1)n_j l_j L_j M_{L_j}|D_1|(\alpha_1 L_1)n_i l_i L_i M_{L_i}\rangle \times \text{ for } \Delta M = 1,$$
$$\langle(\alpha_1 L_1)n_j l_j L_j M_{L_j}|D_{-1}|(\alpha_1 L_1)n_i l_i L_i M_{L_i}\rangle \times \text{ for } \Delta M = -1. \tag{C.25}$$

Again, using the Wigner–Eckart theorem (Eq. C.18), the summation over M_s reduces the matrix element to

$$S_{ij} = \left|\langle(\alpha_1 L_1)n_j l_j L_j||\mathbf{D}||(\alpha_1 L_1)n_i l_i L_i\rangle\right|^2. \tag{C.26}$$

Since the angular momenta are coupled, the matrix element can be expressed in terms of a 6-j symbol (which describes the coupling of three angular momentum quantum numbers, as opposed to two) by the 3-j symbol as

$$\langle(\alpha_1 L_1)n_j l_j L_j|\mathbf{D}|(\alpha_1 L_1)n_i l_i L_i\rangle = (-1)^{l_j + L_1 + L_i + 1}\sqrt{(2L_i + 1)(2L_j + 1)} \times \begin{Bmatrix} L_j & 1 & L_i \\ l_i & L_1 & l_j \end{Bmatrix}\langle n_j l_j||\mathbf{D}||n_i l_i\rangle. \tag{C.27}$$

From the algebraic properties of the 6-j symbol this equation is zero, unless

$$\Delta L = L_j - L_i = 0, \pm 1, \ \Delta l = l_j - l_i = 0, \pm 1. \tag{C.28}$$

The permutation properties of the 6-j symbol also dictate that $\Delta l = 0$. The reduced matrix element can be written in a scalar form a

$$\langle n_j l_j||\mathbf{D}||n_i l_i\rangle = \left\langle n_j l_j|||D|\sqrt{\frac{4\pi}{3}}\mathrm{Y}_1||n_i l_i\right\rangle = (-1)^{l_j + g}\sqrt{l_{\max}}\langle n_j l_j|D|n_i l_i\rangle, \tag{C.29}$$

where

$$\langle n_j l_j|D|n_i l_i\rangle \neq \int_0^\infty R^*_{n_j l_j} er R_{n_i l_i} r^2 \mathrm{d}r. \tag{C.30}$$

Here $D = er$ and $l_{\max}$ is the larger of l_i and l_j. So the reduced matrix element Eq. C.29 vanishes for all l_j except $l_j = l_i + 1$, which implies that $l_{\max} = (l_i + l_j + 1)/2$.

The square of matrix element in S_{ij} is expressed conveniently using the Racah coefficient W as

$$|\langle(\alpha_1 L_1)n_j l_j L_j||\mathbf{D}||(\alpha_1 L_1)n_i l_i L_i\rangle|^2$$
$$= (2L_i+1)(2L_j+1)\begin{Bmatrix} l_i & L_i & L_1 \\ l_j & L_j & 1 \end{Bmatrix}^2$$
$$\times \left|\langle n_j l_j||\mathbf{D}||n_i l_i\rangle\right|^2$$
$$= (2L_i+1)(2L_j+1)\mathrm{W}^2\,(l_i L_i l_j L_j;\, L_1 1)$$
$$\left|\langle n_j l_j||\mathbf{D}||n_i l_i\rangle\right|^2. \quad \text{(C.31)}$$

Combining Eq. C.31 with the reduced matrix element Eq. C.29, we obtain the line strength

$$S_{ij} = (2L_i+1)(2L_j+1)\frac{(l_i+l_j+1)}{2}$$
$$\times\, \mathrm{W}^2(l_i L_i l_j L_j;\, L_1 1)|\langle n_j l_j|D|n_i l_i\rangle|^2. \quad \text{(C.32)}$$

The oscillator strength f_{ij} and the radiative decay rate A_{Ji} can now be obtained from S_{ij} as before. Since $S_{i,j}$ does not depend on M_{S_i}, $\sum_{M_{S_i}} = 2L_i + 1 = g_i$. Hence, the corresponding f-value is

$$f_{ij} = \frac{E_{ji}}{3g_i e^2}\, S_{ij} \quad \text{(C.33)}$$
$$= \frac{E_{ij}}{3}(2L_j+1)l_{\max}\mathrm{W}^2(l_i L_i l_j L_j;\, L_1 1)$$
$$\times \left|\int_0^\infty R_{n_j l_j}(r) R_{n_i l_i}(r) r^3 \mathrm{d}r\right|^2,$$

and the A-coefficient may be obtained from Eq. 4.113.

Appendix D Coefficients of the fine structure components of an *LS* multiplet

The numerical values of the coefficients, $C(L_i, L_j; J_i, J_j)$ obtained by Allen [184], for the fine structure components of a *LS* multiplet are given in this table[1]

Numerical values of the coefficient $C(L_i, L_j; J_i, J_j)$ for relative strengths of fine-structure components of the *LS* multiplets, $L_i \rightarrow L_j$; $g = (2S_i + 1(2L_i + 1)(2L_j + 1)$

$2S_i$+1 =	1	2	3	4	5	6	7	8	9	10	11
						LS multiplet: SP					
g =	3	6	9	12	15	18	21	24	27	30	33
x_1	3	4	5	6	7	8	9	10	11	12	13
y_1		2	3	4	5	6	7	8	9	10	11
z_1			1	2	3	4	5	6	7	8	9
						LS multiplet: PP					
g =	9	18	27	36	45	54	63	72	81	90	99
x_1	9	10	11.25	12.6	14	15.4	16.9	18.3	19.8	21.3	22.8
x_2		4	2.25	1.6	1.25	1.04	0.88	0.75	0.68	0.61	0.55
x_3			0	1	2.25	3.6	5	6.4	7.9	9.3	10.8
y_1		2	3.75	5.4	7	8.6	10.1	11.65	13.2	14.7	16.2
y_2			3	5	6.75	8.4	10	11.6	13.1	14.7	16.2
						LS multiplet: PD					
g =	15	30	45	60	75	90	105	120	135	150	165
x_1	15	18	21	24	27	30	33	36	39	42	45
x_2		10	11.25	12.6	14	15.4	16.9	18.3	19.8	21.3	22.8
x_3			5	5	5.25	5.6	6	6.4	6.9	7.3	7.8
y_1		2	3.75	5.4	7	8.6	10.1	11.65	13.2	14.7	16.2
y_2			3.75	6.4	8.75	11	13.1	15.2	17.3	19.3	21.4
y_3				5	6.75	8.4	10	11.6	13.1	14.7	16.2
z_1			0.25	0.6	1	1.43	1.88	2.33	2.8	3.27	3.75
z_2				1	2.25	3.6	5	6.4	7.86	9.3	10.8
z_3					3	6	9	12	15	18	21

[1] Ref: NORAD: www.astronomy.ohio-state.edu/ nahar/nahar_radiativeatomicdata/index.html

$2S_i+1 =$	1	2	3	4	5	6	7	8	9	10	11
						LS multiplet: DD					
$g =$	25	50	75	100	125	150	175	200	225	250	275
x_1	25	28	31.1	34.3	37.5	40.7	44	47.3	50.6	53.8	57.2
x_2		18	17.4	17.2	17.5	17.9	18.3	19	19.6	20.1	20.9
x_3			11.25	8	6.25	5.14	4.37	3.81	3.37	3.03	2.75
x_4				5	1.25	0.22	0.0	0.14	0.48	0.95	1.5
x_5					0	2.23	5	8	11.1	14.3	17.5
y_1		2	3.9	5.7	7.5	9.25	11	12.75	14.4	16.1	17.8
y_2			3.75	7	10	12.85	15.6	18.4	21	23.6	26.3
y_3				5	8.75	12	15	17.8	20.6	23.4	26
y_4					5	7.8	10	12	13.9	15.7	17.5
						LS multiplet: DF					
$g =$	35	70	105	140	175	210	245	280	315	350	385
x_1	35	40	45	50	55	60	65	70	75	80	85
x_2		28	31.1	34.3	37.5	40.7	44	47.3	50.6	53.8	57.2
x_3			21	22.5	24	25.8	27.5	29.4	31.2	33.1	35
x_4				14	14	14.4	15	15.7	16.5	17.4	18.2
x_5					7	6.2	6	6	6.1	6.3	6.5
y_1		2	3.9	5.7	7.5	9.2	11	12.8	14.4	16.1	17.8
y_2			3.9	7.3	10.5	13.6	16.5	19.4	22.2	25.1	27.8
y_3				5.6	10	13.9	17.5	21	24.4	27.5	30.8
y_4					7	11.4	15	18.3	21.4	24.4	27.3
y_5						7.8	10	12	13.9	15.7	17.5
z_1			0.11	0.29	0.5	0.74	1	1.28	1.56	1.84	2.14
z_2				0.4	1	1.71	2.5	3.33	4.2	5.1	6
z_3					1	2.4	4	5.7	7.5	9.3	11.2
z_4						2.22	5	8	11.1	14.3	17.5
z_5							5	10	15	20	25
						LS multiplet: FF					
$g =$	49	98	147	196	245	294	343	392	441	490	539
x_1	49	54	59	64.1	69.3	74.4	79.6	84.8	90	95.2	100.4
x_2		40.4	41.2	42.7	44.5	46.2	48.2	50.1	52.2	54.2	56.4
x_3			31.1	28.9	27.6	26.7	26.3	25.9	25.9	25.9	26.
x_4				22.4	17.5	14.4	12.3	10.7	9.5	8.5	7.7
x_5					14	7.6	4.38	2.5	1.36	0.67	0.26
x_6						6.2	0.89	0	0.49	1.6	3.06
x_7							0	3.5	7.89	12.6	17.5
y_1		2	3.94	5.8	7.7	9.5	11.4	13.2	15	16.8	18.6
y_2			3.88	7.5	11	14.2	17.5	20.7	23.9	26.9	30
y_3				5.6	10.5	15	19.2	23.3	27.4	31.1	35
y_4					7	12.6	17.5	22	26.3	30.3	34.2
y_5						7.7	13.1	17.5	21.4	25	28.5
y_6							7	10.5	13.1	15.4	17.3

$2S_i+1 =$	1	2	3	4	5	6	7	8	9	10	11
						LS multiplet: FG					
$g =$	63	126	189	252	315	378	441	504	567	630	693
x_1	63	70	77	84	91	98	105	112	119	126	133
x_2		54	59	64.1	69.3	74.4	79.6	84.8	90	95.2	100.4
x_3			45	48.2	51.5	55	58.4	62	65.8	69.3	73
x_4				36	37.5	39.3	41.2	43.4	45.5	47.8	49.9
x_5					27	27	27.5	28.3	29.3	30.4	31.5
x_6						18	16.9	16.5	16.5	16.8	17
x_7							9	7.5	6.9	6.6	6.5
y_1		2	3.94	5.8	7.7	9.5	11.4	13.2	15	16.8	18.6
y_2			3.94	7.6	11.2	14.5	17.9	21.2	24.4	27.6	30.7
y_3				5.8	11	15.7	20.2	24.6	28.9	33	37.1
y_4					7.5	13.7	19.2	24.5	29.3	34	38.5
y_5						9	15.6	21.2	26.3	31	35.5
y_6							10.1	16	20.6	24.6	28.5
y_7								10.5	13.2	15.4	17.5
						LS multiplet: FG					
$g =$	63	126	189	252	315	378	441	504	567	630	693
z_1			0.06	0.17	0.3	0.46	0.62	0.81	1	1.2	1.41
z_2				0.21	0.56	1	1.5	2.05	2.63	3.23	3.85
z_3					0.5	1.29	2.25	3.34	4.51	5.7	7
z_4						1	2.5	4.29	6.25	8.3	10.5
z_5							1.88	4.5	7.5	10.7	14
z_6								3.5	7.9	12.6	17.5
z_7									7	14	21
						LS multiplet: GG					
$g =$	81	162	243	324	405	486	567	648	729		
x_1	81	88	95	102.1	109.2	116.4	123.4	130.6	137.7		
x_2		70	73	76.1	79.9	83.5	87	90.9	94.2		
x_3			59	58.4	58.4	59	59.6	60.4	61.3		
x_4				48.2	44.5	41.8	39.7	38.2	37.1		
x_5					37.5	30.9	26.3	22.8	20.2		
x_6						27	18.4	13	9.4		
x_7							16.9	7.7	3.36		
x_8								7.5	0.67		
x_9									0		
y_1		2	3.96	5.9	7.8	9.7	11.6	13.4	15.3		
y_2			3.94	7.7	11.4	14.9	18.4	21.8	25.1		
y_3				5.8	11.2	16.2	21.1	25.7	30.4		
y_4					7.5	14.2	20.2	26	31.5		
y_5						9	16.5	23.2	29.2		
y_6							10.1	17.8	24.3		
y_7								10.5	17.3		
y_8									9		

$2S_i+1 =$	1	2	3	4	5	6	7	8	9	10	11	
						LS multiplet: GH						
$g =$	99	198	297	396	495	594	693					
x_1	99	108	117	126	135	144	153					
x_2		88	95	102.1	109.2	116.3	123.4					
x_3			77	82	87.4	92.6	97.9					
x_4				66	69.1	72.9	76.5					
x_5					55	56.6	58.4					
x_5						44	44					
x_6							33					
y_1		2	3.96	5.9	7.8	9.7	11.6					
y_2			3.96	7.8	11.4	15.1	18.6					
y_3				5.9	11.3	16.5	21.6					
y_4					7.7	14.6	21.1					
y_5						9.4	17.5					
y_6							11					
z_1			0.04	0.11	0.2	0.31	0.43					
z_2				0.13	0.36	0.65	1					
z_3					0.3	0.8	1.44					
z_4						0.57	1.5					
z_5							1					
				LS multiplet: HH						*LS* multiplet: HI		
$g =$	121	242	363	484	605	726	857	143	286	429	572	715
x_1	121	130	139	148.1	157.2	166.2	175.3	143	154	165	176	187
x_2		108	113	118	123.7	129.2	134.7		130	139	148.1	157.2
x_3			95	96.3	98.1	100	102			117	124	131
x_4				82.3	79.9	78.3	77.3				104	109
x_5					69.2	63.7	59.5					91
x_6						56.6	48.2					
x_7							44.1					
y_1		2	3.97	5.9	7.8	9.8	11.7		2	3.97	5.9	7.8
y_2			3.97	7.8	11.6	15.2	19.8			3.97	7.8	11.6
y_3				5.9	11.4	16.8	22				5.9	11.6
y_4					7.7	14.8	21.6					7.8
y_5						9.4	17.8					
y_6							11					
z_1										0.03	0.08	0.14
z_2											0.09	0.25
z_3												0.2

Appendix E Effective collision strengths and *A*-values

In this table all data pertain to fine-structure transitions; however, in cases where the fine-structure collision strengths are not available, the total *LS* multiplet value is listed under the *first* fine-structure transition within the multiplet, followed by blanks for the other transitions in the multiplet.

Ion	Transition	λ (Å)	$A(s^{-1})$	$\Upsilon(T \times 10^4$ K)			
				$T = 0.5$	1.0	1.5	2.0
H I	1s − 2s	1215.67	8.23 + 0	2.55 − 1	2.74 − 1	2.81 − 1	2.84 − 1
	1s − 2p	1215.66	6.265 + 8	4.16 − 1	4.72 − 1	5.28 − 1	5.85 − 1
He I	$1^1S - 2^3S$	625.48	1.13 − 4	6.50 − 2	6.87 − 2	6.81 − 2	6.72 − 2
	$1^1S - 2^1S$	601.30	5.13 + 1	3.11 − 2	3.61 − 2	3.84 − 2	4.01 − 2
	$1^1S - 2^3P^o$	591.29	1.76 + 2	1.60 − 2	2.27 − 2	2.71 − 2	3.07 − 2
	$1^1S - 2^1P^o$	584.21	1.80 + 9	9.92 − 3	1.54 − 2	1.98 − 2	2.40 − 2
	$2^3S - 2^1S$	15553.7	1.51 − 7	2.24 + 0	2.40 + 0	2.32 + 0	2.20 + 0
	$2^3S - 2^3P^o$	10817.0	1.02 + 7	1.50 + 1	2.69 + 1	3.74 + 1	4.66 + 1
	$2^3S - 2^1P^o$	8854.5	1.29 + 0	7.70 − 1	9.75 − 1	1.05 + 0	1.08 + 0
	$2^1S - 2^3P^o$	35519.5	2.70 − 2	1.50 + 0	1.70 + 0	1.74 + 0	1.72 + 0
	$2^1S - 2^1P^o$	20557.7	1.98 + 6	9.73 + 0	1.86 + 1	2.58 + 1	3.32 + 1
	$2^3P^o - 2^1P^o$	48804.3	—	1.45 + 0	2.07 + 0	2.40 + 0	2.60 + 0
He II	1s − 2s	303.92	5.66 + 2	1.60 − 1	1.59 − 1	1.57 − 1	1.56 − 1
	1s − 2p	303.92	1.0 + 10	3.40 − 1	3.53 − 1	3.63 − 1	3.73 − 1
Li II	$1^1S - 2^3S$	210.11	2.039 − 2	5.54 − 2	5.49 − 2	5.43 − 2	5.38 − 2
	$1^1S - 2^1S$	—	1.95 + 3	3.81 − 2	3.83 − 2	3.85 − 2	3.86 − 2
	$1^1S - 2^3P^o$	202.55	3.289 − 7	9.07 − 2	9.17 − 2	9.26 − 2	9.34 − 2
	$1^1S - 2^1P^o$	199.30	2.56 + 2	3.82 − 2	4.05 − 2	4.28 − 2	4.50 − 2
C I	$^1D_2 - {}^3P_0$	9811.03	7.77 − 8	6.03 − 1	1.14 + 0	1.60 + 0	1.96 + 0
	$^1D_2 - {}^3P_1$	9824.12	8.21 − 5	⇓	⇓	⇓	⇓
	$^1D_2 - {}^3P_2$	9850.28	2.44 − 4	⇓	⇓	⇓	⇓
	$^1S_0 - {}^3P_1$	4621.57	2.71 − 3	1.49 − 1	2.52 − 1	3.20 − 1	3.65 − 1
	$^1S_0 - {}^3P_2$	4628.64	2.00 − 5				
	$^1S_0 - {}^1D_2$	8727.18	5.28 − 1	1.96 − 1	2.77 − 1	3.40 − 1	3.92 − 1
	$^3P_1 - {}^3P_0$	6.094 + 6	7.95 − 8	2.43 − 1	3.71 − 1	—	—
	$^3P_2 - {}^3P_0$	2304147	1.71 − 14	1.82 − 1	2.46 − 1	—	—
	$^3P_2 - {}^3P_1$	3704140	2.65 − 7	7.14 − 1	1.02 + 0	—	—

Ion	Transition	λ (Å)	$A(s^{-1})$	$\Upsilon(T \times 10^4\ K)$			
				$T = 0.5$	1.0	1.5	2.0
	$^5S^o_2 - {}^3P_1$	2965.70	6.94 + 0	4.75 − 1	6.71 − 1	8.22 − 1	9.50 − 1
	$^5S^o_2 - {}^3P_2$	2968.08	1.56 + 1	⇓	⇓	⇓	⇓
C II	$^2P^o_{3/2} - {}^2P^o_{1/2}$	1.5774 + 5	2.29 − 6	1.89 + 0	2.15 + 0	2.26 + 0	2.28 + 0
	$^4P_{1/2} - {}^2P^o_{1/2}$	2325	7.0 + 1	2.43 − 1	2.42 − 1	2.46 − 1	2.48 − 1
	$^4P_{1/2} - {}^2P^o_{3/2}$	2329	6.3 + 1	1.74 − 1	1.77 − 1	1.82 − 1	1.84 − 1
	$^4P_{3/2} - {}^2P^o_{1/2}$	2324	1.4 + 0	3.61 − 1	3.62 − 1	3.68 − 1	3.70 − 1
	$^4P_{3/2} - {}^2P^o_{3/2}$	2328	9.4 + 0	4.72 − 1	4.77 − 1	4.88 − 1	4.93 − 1
	$^4P_{3/2} - {}^4P_{1/2}$	4.55 + 6	2.39 − 7	6.60 − 1	8.24 − 1	9.64 − 1	1.06 + 0
	$^4P_{5/2} - {}^2P^o_{1/2}$	2323	–	2.29 − 1	2.34 − 1	2.42 − 1	2.45 − 1
	$^4P_{5/2} - {}^2P^o_{3/2}$	2326	5.1 + 1	1.02 + 0	1.02 + 0	1.04 + 0	1.05 + 0
	$^4P_{5/2} - {}^4P_{1/2}$	1.99 + 6	3.49 − 14	7.30 − 1	8.53 − 1	9.32 − 1	9.71 − 1
	$^4P_{5/2} - {}^4P_{3/2}$	3.53 + 6	3.67 − 7	1.65 + 0	1.98 + 0	2.23 + 0	2.39 + 0
C III	$^3P^o_2 - {}^1S_0$	1907	5.19 − 3	1.12 + 0	1.01 + 0	9.90 − 1	9.96 − 1
	$^3P^o_1 - {}^1S_0$	1909	1.21 + 2	⇓	⇓	⇓	⇓
	$^3P^o_0 - {}^1S_0$	1909.6	–	⇓	⇓	⇓	⇓
	$^1P^o_1 - {}^1S_0$	977.02	1.79 + 9	3.85 + 0	4.34 + 0	4.56 + 0	4.69 + 0
	$^3P^o_1 - {}^3P^o_0$	4.22 + 6	3.00 − 7	8.48 − 1	9.11 − 1	9.75 − 1	1.03 + 0
	$^3P^o_2 - {}^3P^o_0$	1.25 + 6	–	5.79 − 1	6.77 − 1	7.76 − 1	8.67 − 1
	$^3P^o_2 - {}^3P^o_1$	1.774 + 6	2.10 − 6	2.36 + 0	2.66 + 0	2.97 + 0	3.23 + 0
C IV	$^2P^o_{3/2} - {}^2S_{1/2}$	1548.2	2.65 + 8	–	8.88 + 0	–	8.95 + 0
	$^2P^o_{1/2} - {}^2S_{1/2}$	1550.8	2.63 + 8	–	⇓	–	⇓
N I	$^2D^o_{5/2} - {}^4S^o_{3/2}$	5200.4	6.13 − 6	1.55 − 1	2.90 − 1	–	4.76 − 1
	$^2D^o_{3/2} - {}^4S^o_{3/2}$	5197.9	2.28 − 5	1.03 − 1	1.94 − 1	–	3.18 − 1
	$^2P^o_{3/2} - {}^4S^o_{3/2}$	3466.5	6.60 − 3	5.97 − 2	1.13 − 1	–	1.89 − 1
	$^2P^o_{1/2} - {}^4S^o_{3/2}$	3466.5	2.72 − 3	2.98 − 2	5.67 − 2	–	9.47 − 2
	$^2D^o_{5/2} - {}^2D^o_{3/2}$	1.148 + 7	1.24 − 8	1.28 − 1	2.69 − 1	–	4.65 − 1
	$^2P^o_{3/2} - {}^2P^o_{1/2}$	2.59 + 8	5.17 − 13	3.29 − 2	7.10 − 2	–	1.53 − 1
	$^2P^o_{3/2} - {}^2D^o_{5/2}$	10397.7	5.59 − 2	1.62 − 1	2.66 − 1	–	4.38 − 1
	$^2P^o_{3/2} - {}^2D^o_{3/2}$	10407.2	2.52 − 2	8.56 − 2	1.47 − 1	–	2.52 − 1
	$^2P^o_{1/2} - {}^2D^o_{5/2}$	1040.1	3.14 − 2	6.26 − 2	1.09 − 1	–	1.90 − 1
	$^2P^o_{1/2} - {}^2D^o_{3/2}$	10407.6	4.80 − 2	6.01 − 2	9.70 − 2	–	1.57 − 1
N II	$^1D_2 - {}^3P_0$	6529.0	5.35 − 7	2.57 + 0	2.64 + 0	2.70 + 0	2.73 + 0
	$^1D_2 - {}^3P_1$	6548.1	1.01 − 3	⇓	⇓	⇓	⇓
	$^1D_2 - {}^3P_2$	6583.4	2.99 − 3	⇓	⇓	⇓	⇓
	$^1S_0 - {}^3P_1$	3062.9	3.38 − 2	2.87 − 1	2.93 − 1	3.00 − 1	3.05 − 1
	$^1S_0 - {}^3P_2$	3071.4	1.51 − 4				
	$^1S_0 - {}^1D_2$	5754.6	1.12 + 0	9.59 − 1	8.34 − 1	7.61 − 1	7.34 − 1
	$^3P_1 - {}^3P_0$	2.055 + 6	2.08 − 6	3.71 − 1	4.08 − 1	4.29 − 1	4.43 − 1
	$^3P_2 - {}^3P_0$	7.65 + 5	1.16 − 12	2.43 − 1	2.72 − 1	3.01 − 1	3.16 − 1
	$^3P_2 - {}^3P_1$	1.22 + 6	7.46 − 6	1.01 + 0	1.12 + 0	1.21 + 0	1.26 + 0

Ion	Transition	λ (Å)	$A(\text{s}^{-1})$	$\Upsilon(T \times 10^4\ \text{K})$			
				$T = 0.5$	1.0	1.5	2.0
	$^5S^o_2 - {}^3P_1$	2144	4.80 + 1	1.19 + 0	1.19 + 0	1.21 + 0	1.21 + 0
	$^5S^o_2 - {}^3P_2$	2140	1.07 + 2				
N III	$^2P^o_{3/2} - {}^2P^o_{1/2}$	5.73 + 5	4.77 − 5	1.32 + 0	1.45 + 0	1.55 + 0	1.64 + 0
	$^4P_{1/2} - {}^2P^o_{1/2}$	1748	3.39 + 2	1.89 − 1	1.98 − 1	2.04 − 1	2.07 − 1
	$^4P_{1/2} - {}^2P^o_{3/2}$	1754	3.64 + 2	1.35 − 1	1.51 − 1	1.62 − 1	1.68 − 1
	$^4P_{3/2} - {}^2P^o_{1/2}$	1747	8.95 + 2	2.81 − 1	2.98 − 1	3.09 − 1	3.16 − 1
	$^4P_{3/2} - {}^2P^o_{3/2}$	1752	5.90 + 1	3.67 − 1	3.99 − 1	4.23 − 1	4.35 − 1
	$^4P_{3/2} - {}^4P_{1/2}$	1.68 + 6	–	1.01 + 0	1.10 + 0	1.14 + 0	1.16 + 0
	$^4P_{5/2} - {}^2P^o_{1/2}$	1744.4	–	1.78 − 1	2.01 − 1	2.19 − 1	2.29 − 1
	$^4P_{5/2} - {}^2P^o_{3/2}$	1747	3.08 + 2	7.93 − 1	8.44 − 1	8.80 − 1	8.98 − 1
	$^4P_{5/2} - {}^4P_{1/2}$	7.10 + 5	–	6.12 − 1	6.67 − 1	6.95 − 1	7.11 − 1
	$^4P_{5/2} - {}^4P_{3/2}$	1.23 + 6	–	1.88 + 0	2.04 + 0	2.12 + 0	2.16 + 0
N IV	$^3P^o_2 - {}^1S_0$	1483.3	1.15 − 2	9.37 − 1	9.05 − 1	8.79 − 1	8.58 − 1
	$^3P^o_1 - {}^1S_0$	1486.4	5.77 + 2	⇓	⇓	⇓	⇓
	$^3P^o_0 - {}^1S_0$	1487.9	–	⇓	⇓	⇓	⇓
	$^1P^o_1 - {}^1S_0$	765.15	2.40 + 9	3.84 + 0	3.53 + 0	3.41 + 0	3.36 + 0
	$^3P^o_1 - {}^3P^o_0$	1.585 + 6	6.00 − 6	–	–	–	–
	$^3P^o_2 - {}^3P^o_0$	4.83 + 5	–	–	–	–	–
	$^3P^o_2 - {}^3P^o_1$	6.94 + 5	3.63 − 5	–	–	–	–
N V	$^2P^o_{3/2} - {}^2S_{1/2}$	1238.8	3.41 + 8	6.61 + 0	6.65 + 0	6.69 + 0	6.72 + 0
	$^2P^o_{1/2} - {}^2S_{1/2}$	1242.8	3.38 + 8	–	⇓	–	⇓

Ion	Transition	λ(Å)	$A(\text{s}^{-1})$	Υ(0.5)	Υ(1.0)	Υ(1.5)	Υ(2.0)
O I	$^1D_2 - {}^3P_0$	6393.5	7.23 − 7	1.24 − 1	2.66 − 1	–	5.01 − 1
	$^1D_2 - {}^3P_1$	6363.8	2.11 − 3	⇓	⇓		⇓
	$^1D_2 - {}^3P_2$	6300.3	6.34 − 3	⇓	⇓		⇓
	$^1S_0 - {}^3P_1$	2972.3	7.32 − 2	1.53 − 2	3.24 − 2	–	6.07 − 2
	$^1S_0 - {}^3P_2$	2959.2	2.88 − 4	⇓	⇓		⇓
	$^1S_0 - {}^1D_2$	5577.3	1.22 + 0	7.32 − 2	1.05 − 1	–	1.48 − 1
	$^3P_0 - {}^3P_1$	1.46 + 6	1.74 − 5	1.12 − 2	2.65 − 2	–	6.93 − 2
	$^3P_0 - {}^3P_2$	4.41 + 5	1.00 − 10	1.48 − 2	2.92 − 2	–	5.36 − 2
	$^3P_1 - {}^3P_2$	6.32 + 5	8.92 − 5	4.74 − 2	9.87 − 2	–	2.07 − 1
O II	$^2D^o_{5/2} - {}^4S^o_{3/2}$	3728.8	3.50 − 5	7.95 − 1	8.01 − 1	8.10 − 1	8.18 − 1
	$^2D^o_{3/2} - {}^4S^o_{3/2}$	3726.0	1.79 − 4	5.30 − 1	5.34 − 1	5.41 − 1	5.45 − 1
	$^2P^o_{3/2} - {}^4S^o_{3/2}$	2470.3	5.70 − 2	2.65 − 1	2.70 − 1	2.75 − 1	2.80 − 1
	$^2P^o_{1/2} - {}^4S^o_{3/2}$	2470.2	2.34 − 2	1.33 − 1	1.35 − 1	1.37 − 1	1.40 − 1
	$^2D^o_{5/2} - {}^2D^o_{3/2}$	4.97 + 6	1.30 − 7	1.22 + 0	1.17 + 0	1.14 + 0	1.11 + 0
	$^2P^o_{3/2} - {}^2P^o_{1/2}$	5.00 + 7	2.08 − 11	2.80 − 1	2.87 − 1	2.93 − 1	3.00 − 1

Ion	Transition	λ(Å)	$A(s^{-1})$	Υ(0.5)	Υ(1.0)	Υ(1.5)	Υ(2.0)
	$^2P^o_{3/2} - ^2D^o_{5/2}$	7319.9	1.07 − 1	7.18 − 1	7.30 − 1	7.41 − 1	7.55 − 1
	$^2P^o_{3/2} - ^2D^o_{3/2}$	7330.7	5.78 − 2	4.01 − 1	4.08 − 1	4.14 − 1	4.22 − 1
	$^2P^o_{1/2} - ^2D^o_{5/2}$	7321.8	6.15 − 2	2.90 − 1	2.95 − 1	3.00 − 1	3.05 − 1
	$^2P^o_{1/2} - ^2D^o_{3/2}$	7329.6	1.02 − 1	2.70 − 1	2.75 − 1	2.81 − 1	2.84 − 1
O III	$^1D_2 - ^3P_0$	4932.6	2.74 − 6	2.13 + 0	2.29 + 0	2.45 + 0	2.52 + 0
	$^1D_2 - ^3P_1$	4958.9	6.74 − 3	⇓	⇓	⇓	⇓
	$^1D_2 - ^3P_2$	5006.7	1.96 − 2	⇓	⇓	⇓	⇓
	$^1S_0 - ^3P_1$	2321.0	2.23 − 1	2.72 − 1	2.93 − 1	3.17 − 1	3.29 − 1
	$^1S_0 - ^3P_2$	2332.1	7.85 − 4	⇓	⇓	⇓	⇓
	$^1S_0 - ^1D_2$	4363.2	1.78 + 0	4.94 − 1	5.82 − 1	6.10 − 1	6.10 − 1
	$^3P_1 - ^3P_0$	883562	2.62 − 5	5.24 − 1	5.45 − 1	5.59 − 1	5.63 − 1
	$^3P_2 - ^3P_0$	326611	3.02 − 11	2.58 − 1	2.71 − 1	2.83 − 1	2.89 − 1
	$^3P_2 - ^3P_1$	518145	9.76 − 5	1.23 + 0	1.29 + 0	1.34 + 0	1.35 + 0
	$^5S^o_2 - ^3P_1$	1660.8	2.12 + 2	1.07 + 0	1.21 + 0	1.25 + 0	1.26 + 0
	$^5S^o_2 - ^3P_2$	1666.1	5.22 + 2	⇓	⇓	⇓	⇓
O IV	$^2P^o_{3/2} - ^2P^o_{1/2}$	2.587 + 5	5.18 − 4	2.02 + 0	2.40 + 0	2.53 + 0	2.57 + 0
	$^4P_{1/2} - ^2P^o_{1/2}$	1426.46	1.81 + 3	1.21 − 1	1.33 − 1	1.42 − 1	1.48 − 1
	$^4P_{1/2} - ^2P^o_{3/2}$	1434.07	1.77 + 3	8.67 − 2	1.02 − 1	1.15 − 1	1.24 − 1
	$^4P_{3/2} - ^2P^o_{1/2}$	1423.84	2.28 + 1	1.80 − 1	2.00 − 1	2.16 − 1	2.28 − 1
	$^4P_{3/2} - ^2P^o_{3/2}$	1431.42	3.28 + 2	2.36 − 1	2.68 − 1	2.98 − 1	3.18 − 1
	$^4P_{3/2} - ^4P_{1/2}$	1.68 + 6	–	1.04 + 0	1.09 + 0	1.13 + 0	1.16 + 0
	$^4P_{5/2} - ^2P^o_{1/2}$	1420.19	–	1.15 − 1	1.36 − 1	1.55 − 1	1.69 − 1
	$^4P_{5/2} - ^2P^o_{3/2}$	1427.78	1.04 + 3	5.08 − 1	5.67 − 1	6.15 − 1	6.48 − 1
	$^4P_{5/2} - ^4P_{1/2}$	3.26 + 5	–	7.14 − 1	6.88 − 1	7.06 − 1	7.36 − 1
	$^4P_{5/2} - ^4P_{3/2}$	5.62 + 5	1.02 − 4	2.04 + 0	2.05 + 0	2.12 + 0	2.20 + 0
O V	$^3P^o_2 - ^1S_0$	1213.8	2.16 − 2	7.33 − 1	7.21 − 1	6.74 − 1	6.39 − 1
	$^3P^o_1 - ^1S_0$	1218.3	2.25 + 3	⇓	⇓	⇓	⇓
	$^3P^o_0 - ^1S_0$	1220.4	–	⇓	⇓	⇓	⇓
	$^1P^o_1 - ^1S_0$	629.7	2.80 + 9	2.66 + 0	2.76 + 0	2.82 + 0	2.85 + 0
	$^3P^o_1 - ^3P^o_0$	7.35 + 5	5.81 − 5	7.26 − 1	8.39 − 1	8.65 − 1	8.66 − 1
	$^3P^o_2 - ^3P^o_0$	2.26 + 5	–	2.74 − 1	6.02 − 1	7.51 − 1	8.16 − 1
	$^3P^o_2 - ^3P^o_1$	3.26 + 5	3.55 − 4	3.19 + 0	2.86 + 0	2.80 + 0	2.77 + 0
O VI	$^2P^o_{3/2} - ^2S_{1/2}$	1031.9	4.15 + 8	4.98 + 0	5.00 + 0	5.03 + 0	5.05 + 0
	$^2P^o_{1/2} - ^2S_{1/2}$	1037.6	4.08 + 8	⇓	⇓	⇓	⇓
Ne II	$^2P^o_{1/2} - ^2P^o_{3/2}$	1.28 + 5	8.55 − 3	2.96 − 1	3.03 − 1	3.10 − 1	3.17 − 1
Ne III	$^1D_2 - ^3P_0$	4012.8	8.51 − 6	1.63 + 0	1.65 + 0	1.65 + 0	1.64 + 0
	$^1D_2 - ^3P_1$	3967.5	5.42 − 2	⇓	⇓	⇓	⇓
	$^1D_2 - ^3P_2$	3868.8	1.71 − 1	⇓	⇓	⇓	⇓
	$^1S_0 - ^3P_1$	1814.6	2.00 + 0	1.51 − 1	1.69 − 1	1.75 − 1	1.79 − 1
	$^1S_0 - ^3P_2$	1793.7	3.94 − 3	⇓	⇓	⇓	⇓
	$^1S_0 - ^1D_2$	3342.5	2.71 + 0	2.00 − 1	2.26 − 1	2.43 − 1	2.60 − 1

Ion	Transition	λ(Å)	$A(s^{-1})$	Υ(0.5)	Υ(1.0)	Υ(1.5)	Υ(2.0)
	$^3P_0 - {}^3P_1$	3.60 + 5	1.15 − 3	3.31 − 1	3.50 − 1	3.51 − 1	3.50 − 1
	$^3P_0 - {}^3P_2$	1.07 + 5	2.18 − 8	3.00 − 1	3.07 − 1	3.03 − 1	2.98 − 1
	$^3P_1 - {}^3P_2$	1.56 + 5	5.97 − 3	1.09 + 0	1.65 + 0	1.65 + 0	1.64 + 0
Ne IV	$^2D^o_{5/2} - {}^4S^o_{3/2}$	2420.9	4.58 − 4	8.45 − 1	8.43 − 1	8.32 − 1	8.24 − 1
	$^2D^o_{3/2} - {}^4S^o_{3/2}$	2418.2	5.77 − 3	5.63 − 1	5.59 − 1	5.55 − 1	5.50 − 1
	$^2P^o_{3/2} - {}^4S^o_{3/2}$	1601.5	1.27 + 0	3.07 − 1	3.13 − 1	3.12 − 1	3.09 − 1
	$^2P^o_{1/2} - {}^4S^o_{3/2}$	1601.7	5.21 − 1	1.53 − 1	1.56 − 1	1.56 − 1	1.55 − 1
	$^2D^o_{5/2} - {}^2D^o_{3/2}$	2.237 + 6	1.48 − 6	1.37 + 0	1.36 + 0	1.35 + 0	1.33 + 0
	$^2P^o_{3/2} - {}^2P^o_{1/2}$	1.56 + 7	2.82 − 9	3.17 − 1	3.43 − 1	3.58 − 1	3.70 − 1
	$^2P^o_{3/2} - {}^2D^o_{5/2}$	4714.3	3.88 − 1	8.56 − 1	9.00 − 1	9.08 − 1	9.09 − 1
	$^2P^o_{3/2} - {}^2D^o_{3/2}$	4724.2	4.37 − 1	4.73 − 1	5.09 − 1	5.15 − 1	5.16 − 1
	$^2P^o_{1/2} - {}^2D^o_{5/2}$	4717.0	1.15 − 2	3.40 − 1	3.68 − 1	3.73 − 1	3.74 − 1
	$^2P^o_{1/2} - {}^2D^o_{3/2}$	4725.6	3.93 − 1	3.24 − 1	3.36 − 1	3.39 − 1	3.39 − 1
Ne V	$^1D_2 - {}^3P_0$	3301.3	2.37 − 5	2.13 + 0	2.09 + 0	2.11 + 0	2.14 + 0
	$^1D_2 - {}^3P_1$	3345.8	1.31 − 1	⇓	⇓	⇓	⇓
	$^1D_2 - {}^3P_2$	3425.9	3.65 − 1	⇓	⇓	⇓	⇓
	$^1S_0 - {}^3P_1$	1574.8	4.21 + 0	2.54 − 1	2.46 − 1	2.49 − 1	2.51 − 1
	$^1S_0 - {}^3P_2$	1592.3	6.69 − 3	⇓	⇓	⇓	⇓
	$^1S_0 - {}^1D_2$	2972.8	2.85 + 0	6.63 − 1	5.77 − 1	6.10 − 1	6.49 − 1
	$^3P_1 - {}^3P_0$	2.428 + 5	1.28 − 3	1.68 + 0	1.41 + 0	1.19 + 0	1.10 + 0
	$^3P_2 - {}^3P_0$	90082	5.08 − 9	2.44 + 0	1.81 + 0	1.42 + 0	1.26 + 0
	$^3P_2 - {}^3P_1$	1.432 + 5	4.59 − 3	7.59 + 0	5.82 + 0	4.68 + 0	4.20 + 0
	$^5S^o_2 - {}^3P_1$	1137.0	2.37 + 3	1.11 + 0	1.43 + 0	1.39 + 0	1.34 + 0
	$^5S^o_2 - {}^3P_2$	1146.1	6.06 + 3	⇓	⇓	⇓	⇓
Ne VI	$^2P^o_{3/2} - {}^2P^o_{1/2}$	7.642 + 4	2.02 − 2	3.22 + 0	2.72 + 0	2.37 + 0	2.15 + 0
	$^4P_{1/2} - {}^2P^o_{1/2}$	1003.6	1.59 + 4	1.54 − 1	1.37 − 1	1.26 − 1	1.18 − 1
	$^4P_{1/2} - {}^2P^o_{3/2}$	1016.6	1.43 + 4	1.85 − 1	1.53 − 1	1.36 − 1	1.27 − 1
	$^4P_{3/2} - {}^2P^o_{1/2}$	999.13	3.20 + 2	2.69 − 1	2.32 − 1	2.10 − 1	1.96 − 1
	$^4P_{3/2} - {}^2P^o_{3/2}$	1012.0	3.33 + 3	4.51 − 1	3.73 − 1	3.32 − 1	3.08 − 1
	$^4P_{3/2} - {}^4P_{1/2}$	2.24 + 5	–	5.34 − 1	5.73 − 1	5.95 − 1	6.22 − 1
	$^4P_{5/2} - {}^2P^o_{1/2}$	992.76	–	2.56 − 1	2.11 − 1	1.89 − 1	1.76 − 1
	$^4P_{5/2} - {}^2P^o_{3/2}$	1005.5	1.14 + 4	7.88 + 0	6.75 − 1	6.09 − 1	5.68 − 1
	$^4P_{5/2} - {}^4P_{1/2}$	92166	–	4.12 − 1	4.23 − 1	4.36 − 2	4.56 − 1
	$^4P_{5/2} - {}^4P_{3/2}$	1.56 + 5	–	1.10 + 0	1.16 + 0	1.20 + 0	1.26 + 0
Ne VII	$^3P^o_2 - {}^1S_0$	887.22	5.78 − 2	1.29 − 1	1.72 − 1	2.05 − 1	2.28 − 1
	$^3P^o_1 - {}^1S_0$	895.12	1.98 + 4	⇓	⇓	⇓	⇓
	$^3P^o_0 - {}^1S_0$	898.76	–	⇓	⇓	⇓	⇓
	$^1P^o_1 - {}^1S_0$	465.22	4.09 + 9	1.39 + 0	1.56 + 0	1.63 + 0	1.66 + 0
	$^3P^o_1 - {}^3P^o_0$	2.20 + 5	1.99 − 3	–	–	–	–
	$^3P^o_2 - {}^3P^o_0$	69127.6	–	–	–	–	–
	$^3P^o_2 - {}^3P^o_1$	1.01 + 5	1.25 − 2	–	–	–	–

Ion	Transition	λ(Å)	$A(s^{-1})$	ϒ(0.5)	ϒ(1.0)	ϒ(1.5)	ϒ(2.0)
Na III	$^2P^o_{1/2} - {}^2P^o_{3/2}$	7.319 + 4	4.59 − 2		3.00 − 1		
Na IV	$^1D_2 - {}^3P_0$	3417.2	2.24 − 5		1.17 + 0		
	$^1D_2 - {}^3P_1$	3362.2	1.86 − 1		⇓		
	$^1D_2 - {}^3P_2$	3241.7	6.10 − 1		⇓		
	$^1S_0 - {}^3P_1$	1529.3	7.10 + 0		1.63 − 1		
	$^1S_0 - {}^3P_2$	1503.8	1.05 − 2		⇓		
	$^1S_0 - {}^1D_2$	2803.7	3.46 + 0		1.57 − 1		
	$^3P_0 - {}^3P_1$	2.129 + 5	5.57 − 3		1.77 − 1		
	$^3P_0 - {}^3P_2$	62467.9	1.67 − 7		1.11 − 1		
	$^3P_1 - {}^3P_2$	90391.4	3.04 − 2		4.71 − 1		
Na V	$^2D^o_{5/2} - {}^4S^o_{3/2}$	2068.4	1.39 − 3		5.51 − 1		
	$^2D^o_{3/2} - {}^4S^o_{3/2}$	2066.9	2.70 − 2		3.68 − 1		
	$^2P^o_{3/2} - {}^4S^o_{3/2}$	1365.1	4.23 + 0		2.39 − 1		
	$^2P^o_{1/2} - {}^4S^o_{3/2}$	1365.8	1.76 + 0		1.20 − 1		
	$^2D^o_{5/2} - {}^2D^o_{3/2}$	2.78 + 6	1.56 − 6		6.96 − 1		
	$^2P^o_{3/2} - {}^2P^o_{1/2}$	2.70 + 6	3.66 − 7		4.38 − 1		
	$^2P^o_{3/2} - {}^2D^o_{5/2}$	4010.9	9.07 − 1		5.02 − 1		
	$^2P^o_{3/2} - {}^2D^o_{3/2}$	4016.7	1.28 + 0		2.79 − 1		
	$^2P^o_{1/2} - {}^2D^o_{5/2}$	4017.9	1.35 − 1		2.01 − 1		
	$^2P^o_{1/2} - {}^2D^o_{3/2}$	4022.7	9.75 − 1		1.90 − 1		
Na VI	$^1D_2 - {}^3P_0$	2816.1	–	1.55 + 0	1.45 + 0	1.39 + 0	1.38 + 0
	$^1D_2 - {}^3P_1$	2872.7	4.06 − 1	⇓	⇓	⇓	⇓
	$^1D_2 - {}^3P_2$	2971.9	1.27 + 0	⇓	⇓	⇓	⇓
	$^1S_0 - {}^3P_1$	1356.6	1.69 + 1	1.73 − 1	1.72 − 1	1.72 − 1	1.73 − 1
	$^1S_0 - {}^3P_2$	1343.9	–	⇓	⇓	⇓	⇓
	$^1S_0 - {}^1D_2$	2568.9	5.27 + 0	1.07 − 1	1.16 − 1	1.28 − 1	1.39 − 1
	$^3P_1 - {}^3P_0$	1.43 + 5	6.14 − 3	7.24 − 1	7.70 − 1	7.73 − 1	7.58 − 1
	$^3P_2 - {}^3P_0$	5.37 + 4	–	5.02 − 1	5.21 − 1	5.08 − 1	4.94 − 1
	$^3P_2 - {}^3P_1$	8.61 + 4	2.11 − 2	2.03 + 0	2.13 + 0	2.10 + 0	2.05 + 0
Mg II	$^2P^o_{3/2} - {}^2S_{1/2}$	2795.5	2.6 + 8	1.59 + 1	1.69 + 1	1.78 + 1	1.86 + 1
	$^2P^o_{1/2} - {}^2S_{1/2}$	2802.7	2.6 + 8	⇓	⇓	⇓	⇓
Mg IV	$^2P^o_{1/2} - {}^2P^o_{3/2}$	4.487 + 4	1.99 − 1	3.44 − 1	3.46 − 1	3.49 − 1	3.51 − 1
Mg V	$^1D_2 - {}^3P_0$	2993.1	5.20 − 5	1.31 + 0	1.33 + 0	1.32 + 0	1.30 + 0
	$^1D_2 - {}^3P_1$	2928.0	5.41 − 1	⇓	⇓	⇓	⇓
	$^1D_2 - {}^3P_2$	2782.7	1.85 + 0	⇓	⇓	⇓	⇓
	$^1S_0 - {}^3P_1$	1324.4	2.14 + 1	1.42 − 1	1.48 − 1	1.46 − 1	1.44 − 1
	$^1S_0 - {}^3P_2$	1293.9	2.45 − 2	⇓	⇓	⇓	⇓
	$^1S_0 - {}^1D_2$	2417.5	4.23 + 0	1.91 − 1	1.97 − 1	2.02 − 1	2.08 − 1
	$^3P_0 - {}^3P_1$	1.354 + 5	2.17 − 2	2.48 − 1	3.00 − 1	3.18 − 1	3.18 − 1
	$^3P_0 - {}^3P_2$	39654.2	1.01 − 6	2.31 − 1	2.92 − 1	3.04 − 1	2.99 − 1
	$^3P_1 - {}^3P_2$	5.608 + 4	1.27 − 1	8.30 − 1	1.03 + 0	1.08 + 0	1.07 + 0

Ion	Transition	λ(Å)	$A(s^{-1})$	Υ(0.5)	Υ(1.0)	Υ(1.5)	Υ(2.0)
Mg VII	$^1D_2 - {}^3P_0$	2441.4	–	7.96 − 1	8.57 − 1	9.11 − 1	9.42 − 1
	$^1D_2 - {}^3P_1$	2509.2	1.17 + 0	⇓	⇓	⇓	⇓
	$^1D_2 - {}^3P_2$	2629.1	3.36 + 0	⇓	⇓	⇓	⇓
	$^1S_0 - {}^3P_1$	1189.8	4.58 + 1	2.08 − 1	1.85 − 1	1.75 − 1	1.73 − 1
	$^1S_0 - {}^3P_2$	1174.3	–	⇓	⇓	⇓	⇓
	$^1S_0 - {}^1D_2$	2261.5	6.16 + 0	5.25 − 1	4.46 − 1	3.90 − 1	3.82 − 1
	$^3P_1 - {}^3P_0$	9.03 + 4	2.44 − 2	2.75 − 1	3.37 − 1	3.95 − 1	4.14 − 1
	$^3P_2 - {}^3P_0$	3.42 + 4	–	1.90 − 1	3.01 − 1	3.88 − 1	4.09 − 1
	$^3P_2 - {}^3P_1$	5.50 + 4	8.09 − 2	7.69 + 0	1.08 + 0	1.32 + 0	1.39 + 0
Al II	$^3P^o_2 - {}^1S_0$	2661.1	–	3.062 + 0	3.564 + 0	3.612 + 0	3.54 + 0
	$^3P^o_1 - {}^1S_0$	2669.9	–	⇓	⇓	⇓	⇓
	$^3P^o_0 - {}^1S_0$	2674.3	–	⇓	⇓	⇓	⇓
	$^1P^o_1 - {}^1S_0$	1670.8	1.46 + 9	2.045 + 0	3.251 + 0	4.096 + 0	4.717 + 0
	$^3P^o_1 - {}^3P^o_0$	1.6426 + 6	4.10 − 6	–	–	–	–
	$^3P^o_2 - {}^3P^o_0$	5.4124 + 5	–	–	–	–	–
	$^3P^o_2 - {}^3P^o_1$	8.072 + 5	2.45 − 5	–	–	–	–
Si II	$^2P^o_{3/2} - {}^2P^o_{1/2}$	3.48 + 5	2.17 − 4	5.59 + 0	5.70 + 0	5.78 + 0	5.77 + 0
	$^4P_{1/2} - {}^2P^o_{1/2}$	2335	4.55 + 3	5.50 − 1	5.16 − 1	4.88 − 1	4.67 − 1
	$^4P_{1/2} - {}^2P^o_{3/2}$	2350	4.41 + 3	4.33 − 1	4.02 − 1	3.81 − 1	3.65 − 1
	$^4P_{3/2} - {}^2P^o_{1/2}$	2329	1.32 + 1	8.32 − 1	7.80 − 1	7.37 − 1	7.06 − 1
	$^4P_{3/2} - {}^2P^o_{3/2}$	2344	1.22 + 3	1.13 + 0	1.05 + 0	9.97 − 1	9.56 − 1
	$^4P_{3/2} - {}^4P_{1/2}$	9.23 + 5	–	4.92 + 0	4.51 + 0	4.18 + 0	3.94 + 0
	$^4P_{5/2} - {}^2P^o_{1/2}$	2319.8	–	5.71 − 1	5.34 − 1	5.08 − 1	4.88 − 1
	$^4P_{5/2} - {}^2P^o_{3/2}$	2335	2.46 + 3	2.33 + 0	2.19 + 0	2.08 + 0	1.99 + 0
	$^4P_{5/2} - {}^4P_{1/2}$	3.53 + 5	–	1.68 + 0	1.67 + 0	1.63 + 0	1.57 + 0
	$^4P_{5/2} - {}^4P_{3/2}$	5.70 + 5	–	7.36 + 0	6.94 + 0	6.58 + 0	6.32 + 0
Si III	$^3P^o_2 - {}^1S_0$	1882.7	1.20 − 2	6.96 + 0	5.46 + 0	4.82 + 0	4.41 + 0
	$^3P^o_1 - {}^1S_0$	1892.0	1.67 + 4	⇓	⇓	⇓	⇓
	$^3P^o_0 - {}^1S_0$	1896.6	–	⇓	⇓	⇓	⇓
	$^1P^o_1 - {}^1S_0$	1206.5	2.59 + 9	5.30 + 0	5.60 + 0	5.93 + 0	6.22 + 0
	$^3P^o_1 - {}^3P^o_0$	7.78 + 5	3.86 − 5	1.78 + 0	1.81 + 0	1.83 + 0	1.83 + 0
	$^3P^o_2 - {}^3P^o_0$	2.56 + 5	3.20 − 9	3.66 + 0	3.62 + 0	3.53 + 0	3.43 + 0
	$^3P^o_2 - {}^3P^o_1$	3.82 + 5	2.42 − 4	1.04 + 1	1.04 + 1	1.02 + 1	1.00 + 1
Si IV	$^2P^o_{3/2} - {}^2S_{1/2}$	1393.8	7.73 + 8	1.69 + 1	1.60 + 1	1.61 + 1	1.62 + 1
	$^2P^o_{1/2} - {}^2S_{1/2}$	1402.8	7.58 + 8	⇓	⇓	⇓	⇓
Si VI	$^2P^o_{1/2} - {}^2P^o_{3/2}$	1.964 + 4	2.38 + 0	2.42 − 1			
S II	$^2D^o_{5/2} - {}^4S^o_{3/2}$	6716.5	2.60 − 4	4.90 + 0	4.66 + 0	4.44 + 0	4.26 + 0
	$^2D^o_{3/2} - {}^4S^o_{3/2}$	6730.8	8.82 − 4	3.27 + 0	3.11 + 0	2.97 + 0	2.84 + 0
	$^2P^o_{3/2} - {}^4S^o_{3/2}$	4068.6	2.25 − 1	1.67 + 0	2.07 + 0	1.98 + 0	2.07 + 0
	$^2P^o_{1/2} - {}^4S^o_{3/2}$	4076.4	9.06 − 2	8.31 − 1	8.97 − 1	9.87 − 1	1.03 + 0
	$^2D^o_{5/2} - {}^2D^o_{3/2}$	3.145 + 6	3.35 − 7	7.90 + 0	7.46 + 0	7.11 + 0	8.65 + 0

Ion	Transition	λ(Å)	$A(s^{-1})$	Υ(0.5)	Υ(1.0)	Υ(1.5)	Υ(2.0)
	$^2P^o_{3/2} - ^2P^o_{1/2}$	2.14 + 6	1.03 − 6	2.02 + 0	2.54 + 0	2.13 + 0	2.22 + 0
	$^2P^o_{3/2} - ^2D^o_{5/2}$	10320.4	1.79 − 1	5.93 + 0	4.77 + 0	4.75 + 0	4.68 + 0
	$^2P^o_{3/2} - ^2D^o_{3/2}$	10286.7	1.33 − 1	3.41 + 0	2.74 + 0	2.74 + 0	2.71 + 0
	$^2P^o_{1/2} - ^2D^o_{5/2}$	10373.3	7.79 − 2	2.47 + 0	1.99 + 0	1.99 + 0	1.97 + 0
	$^2P^o_{1/2} - ^2D^o_{3/2}$	10336.3	1.63 − 1	2.20 + 0	1.76 + 0	1.76 + 0	1.73 + 0
S III	$^1D_2 - ^3P_0$	8833.9	5.82 − 6	9.07 + 0	8.39 + 0	8.29 + 0	8.20 + 0
	$^1D_2 - ^3P_1$	9068.9	2.21 − 2	⇓	⇓	⇓	⇓
	$^1D_2 - ^3P_2$	9531.0	5.76 − 2	⇓	⇓	⇓	⇓
	$^1S_0 - ^3P_1$	3721.7	7.96 − 1	1.16 + 0	1.19 + 0	1.21 + 0	1.24 + 0
	$^1S_0 - ^3P_2$	3797.8	1.05 − 2	⇓	⇓	⇓	⇓
	$^1S_0 - ^1D_2$	6312.1	2.22 + 0	1.42 + 0	1.88 + 0	2.02 + 0	2.08 + 0
	$^3P_1 - ^3P_0$	3.347 + 5	4.72 − 4	2.64 + 0	2.59 + 0	2.38 + 0	2.20 + 0
	$^3P_2 - ^3P_0$	1.20 + 5	4.61 − 8	1.11 + 0	1.15 + 0	1.15 + 0	1.14 + 0
	$^3P_2 - ^3P_1$	187129	2.07 − 3	5.79 + 0	5.81 + 0	5.56 + 0	5.32 + 0
	$^5S^o_2 - ^3P_1$	1683.5	6.22 + 3	–	3.8 + 0	3.7 + 0	3.6 + 0
	$^5S^o_2 - ^3P_2$	1698.86	1.70 + 4	⇓	⇓	⇓	⇓
S IV	$^2P^o_{3/2} - ^2P^o_{1/2}$	1.05 + 5	7.73 − 3	–	6.42 + 0	6.41 + 0	6.40 + 0
	$^4P_{1/2} - ^2P^o_{1/2}$	1404.9	5.50 + 4	–	5.50 − 1	4.80 − 1	4.60 − 1
	$^4P_{1/2} - ^2P^o_{3/2}$	1423.9	3.39 + 4	–	6.60 − 1	6.30 − 1	6.10 − 1
	$^4P_{3/2} - ^2P^o_{1/2}$	1398.1	1.40 + 2	–	8.70 − 1	8.30 − 1	8.00 − 1
	$^4P_{3/2} - ^2P^o_{3/2}$	1017.0	1.95 + 4	–	1.47 + 0	1.40 + 0	1.34 + 0
	$^4P_{3/2} - ^4P_{1/2}$	2.91 + 5	–	–	3.04 + 0	2.85 + 0	2.72 + 0
	$^4P_{5/2} - ^2P^o_{1/2}$	1387.5	–	–	9.5 − 1	9.1 − 1	8.8 − 1
	$^4P_{5/2} - ^2P^o_{3/2}$	1406.1	3.95 + 4	–	2.53 + 0	2.41 + 0	2.33 + 0
	$^4P_{5/2} - ^4P_{1/2}$	1.12 + 5	–	–	2.92 + 0	2.71 + 0	2.56 + 0
	$^4P_{5/2} - ^4P_{3/2}$	1.85 + 5	–	–	7.01 + 0	6.57 + 0	6.20 + 0
S V	$^3P^o_2 - ^1S_0$	1188.3	6.59 − 2	9.11 − 1	9.10 − 1	9.14 − 1	9.05 − 1
	$^3P^o_1 - ^1S_0$	1199.1	1.26 + 5	⇓	⇓	⇓	⇓
	$^3P^o_0 - ^1S_0$	1204.5	–	⇓	⇓	⇓	
	$^1P^o_1 - ^1S_0$	786.48	5.25 + 9	7.30 + 0	7.30 + 0	7.29 + 0	7.27 + 0
	$^3P^o_1 - ^3P^o_0$	2.71 + 5	9.16 − 4		2.72 − 1		
	$^3P^o_2 - ^3P^o_0$	88401.7	–		4.00 − 1		
	$^3P^o_2 - ^3P^o_1$	1.312 + 5	5.49 − 3		1.24 + 0		
S VI	$^2P^o_{3/2} - ^2S_{1/2}$	933.38	1.7 + 9	1.18 + 1	1.19 + 1	1.19 + 1	1.19 + 1
	$^2P^o_{1/2} - ^2S_{1/2}$	944.52	1.6 + 9	⇓	⇓	⇓	⇓
	$^2P^o_{1/2} - ^2P^o_{3/2}$	1.0846 + 5	7.75 − 3	5.85 + 0	6.67 + 0	7.10 + 0	7.27 + 0
Cl II	$^1D_2 - ^3P_0$	9383.4	9.82 − 6		3.86 + 0		
	$^1D_2 - ^3P_1$	9123.6	2.92 − 2		⇓		
	$^1D_2 - ^3P_2$	8578.7	1.04 − 1		⇓		
	$^1S_0 - ^3P_1$	3677.9	1.31 + 0		4.56 − 1		
	$^1S_0 - ^3P_2$	3587.1	1.97 − 2		⇓		
	$^1S_0 - ^1D_2$	6161.8	2.06 + 0		1.15 + 0		

Ion	Transition	λ(Å)	$A(s^{-1})$	Υ(0.5)	Υ(1.0)	Υ(1.5)	Υ(2.0)
	$^3P_0 - {}^3P_1$	3.328 + 5	1.46 − 3		9.33 − 1		
	$^3P_0 - {}^3P_2$	1.004 + 5	4.57 − 7		4.43 − 1		
	$^3P_1 - {}^3P_2$	1.437 + 5	7.57 − 3		2.17 + 0		
Cl III	$^2D^o_{5/2} - {}^4S^o_{3/2}$	5517.7	7.04 − 4	1.94 + 0	2.05 + 0	2.04 + 0	2.04 + 0
	$^2D^o_{3/2} - {}^4S^o_{3/2}$	5537.9	4.83 − 3	1.29 + 0	1.36 + 0	1.36 + 0	1.35 + 0
	$^2P^o_{3/2} - {}^4S^o_{3/2}$	3342.9	7.54 − 1	7.69 − 1	8.37 − 1	8.88 − 1	9.20 − 1
	$^2P^o_{1/2} - {}^4S^o_{3/2}$	3353.3	3.05 − 1	3.85 − 1	4.18 − 1	4.44 − 1	4.61 − 1
	$^2D^o_{5/2} - {}^2D^o_{3/2}$	1.516 + 6	3.22 − 6	4.45 + 0	4.52 + 0	4.51 + 0	4.48 + 0
	$^2P^o_{3/2} - {}^2P^o_{1/2}$	1.081 + 6	7.65 − 6	1.73 + 0	1.76 + 0	1.81 + 0	1.86 + 0
	$^2P^o_{3/2} - {}^2D^o_{5/2}$	8480.9	3.16 − 1	3.75 + 0	4.20 + 0	4.33 + 0	4.32 + 0
	$^2P^o_{3/2} - {}^2D^o_{3/2}$	8433.7	3.23 − 1	2.01 + 0	2.19 + 0	2.34 + 0	2.25 + 0
	$^2P^o_{1/2} - {}^2D^o_{5/2}$	8552.1	1.00 − 1	1.44 + 0	1.56 + 0	1.60 + 0	1.60 + 0
	$^2P^o_{1/2} - {}^2D^o_{3/2}$	8500.0	3.03 − 1	1.45 + 0	1.65 + 0	1.71 + 0	1.72 + 0
Cl IV	$^1D_2 - {}^3P_0$	7263.4	1.54 − 5	5.10 + 0	5.42 + 0	5.88 + 0	6.19 + 0
	$^1D_2 - {}^3P_1$	7529.9	5.57 − 2	⇓	⇓	⇓	⇓
	$^1D_2 - {}^3P_2$	8045.6	2.08 − 1	⇓	⇓	⇓	⇓
	$^1S_0 - {}^3P_1$	3118.6	2.19 + 0	2.04 + 0	2.27 + 0	2.32 + 0	2.30 + 0
	$^1S_0 - {}^3P_2$	3204.5	2.62 − 2	⇓	⇓	⇓	⇓
	$^1S_0 - {}^1D_2$	5323.3	4.14 + 0	9.35 − 1	1.39 + 0	1.73 + 0	1.92 + 0
	$^3P_1 - {}^3P_0$	2.035 + 5	2.13 − 3		4.75 − 1		
	$^3P_2 - {}^3P_0$	74521	2.70 − 7		4.00 − 1		
	$^3P_2 - {}^3P_1$	1.1741 + 5	8.32 − 3		1.50 + 0		
Cl V	$^2P^o_{3/2} - {}^2P^o_{1/2}$	67049	2.98 − 2		1.05 + 0		
Ar II	$^2P^o_{1/2} - {}^2P^o_{3/2}$	69851.9	5.27 − 2		6.35 − 1		
Ar III	$^1D_2 - {}^3P_0$	8038.7	2.21 − 5		4.74 + 0		
	$^1D_2 - {}^3P_1$	7751.1	8.23 − 2	⇓	⇓	⇓	⇓
	$^1D_2 - {}^3P_2$	7135.8	3.14 − 1	⇓	⇓	⇓	⇓
	$^1S_0 - {}^3P_1$	3109.1	3.91 + 0		6.80 − 1		
	$^1S_0 - {}^3P_2$	3006.1	4.17 − 2	⇓	⇓	⇓	⇓
	$^1S_0 - {}^1D_2$	5191.8	2.59 + 0		8.23 − 1		
	$^3P_0 - {}^3P_1$	2.184 + 5	5.17 − 3		1.18 + 0		
	$^3P_0 - {}^3P_2$	63686.2	2.37 − 6		5.31 − 1		
	$^3P_1 - {}^3P_2$	89910	3.08 − 2		2.24 + 0		
Ar IV	$^2D^o_{5/2} - {}^4S^o_{3/2}$	4711.3	1.77 − 3	2.56 + 0	6.13 + 0	1.64 + 0	1.46 + 1
	$^2D^o_{3/2} - {}^4S^o_{3/2}$	4740.2	2.23 − 2	1.71 + 0	1.30 + 0	1.14 + 0	9.70 − 1
	$^2P^o_{3/2} - {}^4S^o_{3/2}$	2853.7	2.11 + 0	3.01 − 1	2.93 − 1	3.06 − 1	3.25 − 1
	$^2P^o_{1/2} - {}^4S^o_{3/2}$	2868.2	8.62 − 1	1.49 − 1	1.46 − 1	1.53 − 1	1.63 − 1
	$^2D^o_{5/2} - {}^2D^o_{3/2}$	7.741 + 5	2.30 − 5	6.35 + 0	6.13 + 0	6.03 + 0	5.93 + 0
	$^2P^o_{3/2} - {}^2P^o_{1/2}$	564721	4.94 − 5	2.24 + 0	2.33 + 0	2.53 + 0	2.72 + 0
	$^2P^o_{3/2} - {}^2D^o_{5/2}$	7237.3	5.98 − 1	4.29 + 0	4.44 + 0	4.40 + 0	4.34 + 0
	$^2P^o_{3/2} - {}^2D^o_{3/2}$	7170.6	7.89 − 1	2.45 + 0	2.47 + 0	2.44 + 0	2.39 + 0

Ion	Transition	λ(Å)	$A(s^{-1})$	Υ(0.5)	Υ(1.0)	Υ(1.5)	Υ(2.0)
	$^2P^o_{1/2} - {}^2D^o_{5/2}$	7333.4	1.19 − 1	1.78 + 0	1.79 + 0	1.76 + 0	1.72 + 0
	$^2P^o_{1/2} - {}^2D^o_{3/2}$	7262.8	6.03 − 1	1.61 + 0	1.69 + 0	1.68 + 0	1.66 + 0
Ar V	$^1D_2 - {}^3P_0$	6135.2	3.50 − 5	4.37 + 0	3.72 + 0	3.52 + 0	3.42 + 0
	$^1D_2 - {}^3P_1$	6435.1	1.61 − 1	⇓	⇓	⇓	⇓
	$^1D_2 - {}^3P_2$	7005.7	4.70 − 1	4.37 + 0	3.72 + 0	3.52 + 0	3.42 + 0
	$^1S_0 - {}^3P_1$	2691.0	5.89 + 0	1.17 + 0	1.18 + 0	1.11 + 0	1.03 + 0
	$^1S_0 - {}^3P_2$	2686.8	5.69 − 2	⇓	⇓	⇓	⇓
	$^1S_0 - {}^1D_2$	4625.5	5.18 + 0	1.26 + 0	1.25 + 0	1.24 + 0	1.23 + 0
	$^3P_1 - {}^3P_0$	1.307 + 5	8.03 − 3		2.57 − 1		
	$^3P_2 - {}^3P_0$	49280.5	1.24 − 6		3.20 − 1		
	$^3P_2 - {}^3P_1$	79040	2.72 − 2		1.04 + 0		
Ar VI	$^2P^o_{3/2} - {}^2P^o_{1/2}$	45275	9.69 − 2		7.98 − 2		
K III	$^2P^o_{1/2} - {}^2P^o_{3/2}$	46153.2	1.83 − 1		1.78 + 0		
K IV	$^1D_2 - {}^3P_0$	7110.9	4.54 − 5		1.90 + 0		
	$^1D_2 - {}^3P_1$	6795.0	1.98 − 1		⇓		
	$^1D_2 - {}^3P_2$	6101.8	8.14 − 1		⇓		
	$^1S_0 - {}^3P_1$	2711.1	1.00 + 1		2.92 − 1		
	$^1S_0 - {}^3P_2$	2594.3	8.17 − 2		⇓		
	$^1S_0 - {}^1D_2$	4510.9	3.18 + 0		7.98 − 1		
	$^3P_0 - {}^3P_1$	1.539 + 5	1.48 − 2		4.21 − 1		
	$^3P_0 - {}^3P_2$	43081.2	1.01 − 5		2.90 − 1		
	$^3P_1 - {}^3P_2$	59830.0	1.04 − 1		1.16 + 0		
K V	$^2D^o_{5/2} - {}^4S^o_{3/2}$	4122.6	4.59 − 3	9.25 − 1	8.51 − 1	8.24 − 1	8.18 − 1
	$^2D^o_{3/2} - {}^4S^o_{3/2}$	4163.3	8.84 − 2	6.17 − 1	5.67 − 1	5.50 − 1	5.45 − 1
	$^2P^o_{3/2} - {}^4S^o_{3/2}$	2494.2	5.19 + 0	1.49 − 1	3.68 − 1	4.94 − 1	5.47 − 1
	$^2P^o_{1/2} - {}^4S^o_{3/2}$	2514.5	2.14 + 0	7.40 − 2	1.84 − 1	2.47 − 1	2.73 − 1
	$^2D^o_{5/2} - {}^2D^o_{3/2}$	4.22 + 5	1.42 − 4	5.24 + 0	5.31 + 0	5.13 + 0	4.96 + 0
	$^2P^o_{3/2} - {}^2P^o_{1/2}$	3.11 + 5	2.96 − 4	4.43 − 1	6.27 − 1	7.83 − 1	9.02 − 1
	$^2P^o_{3/2} - {}^2D^o_{5/2}$	6315.1	1.21 + 0	2.56 + 0	3.07 + 0	3.31 + 0	3.40 + 0
	$^2P^o_{3/2} - {}^2D^o_{3/2}$	6221.9	1.86 + 0	1.39 + 0	1.76 + 0	1.93 + 0	2.00 + 0
	$^2P^o_{1/2} - {}^2D^o_{5/2}$	6448.1	1.41 − 1	9.92 − 1	1.28 + 0	1.41 + 0	1.46 + 0
	$^2P^o_{1/2} - {}^2D^o_{3/2}$	6349.2	1.25 + 0	9.83 − 1	1.14 + 0	1.21 + 0	1.24 + 0
Ca II	$^2P^o_{3/2} - {}^2S_{1/2}$	3933.7	1.47 + 8	1.56 + 1	1.75 + 1	1.92 + 1	2.08 + 1
	$^2P^o_{1/2} - {}^2S_{1/2}$	3968.5	1.4 + 8	⇓	⇓	⇓	⇓
Ca IV	$^2P^o_{1/2} - {}^2P^o_{3/2}$	32061.9	5.45 − 1		1.06 + 0		
Ca V	$^1D_2 - {}^3P_0$	6428.9	8.42 − 5		9.04 − 1		
	$^1D_2 - {}^3P_1$	6086.4	4.26 − 1		⇓		
	$^1D_2 - {}^3P_2$	5309.2	1.90 + 0		⇓		
	$^1S_0 - {}^3P_1$	2412.9	2.31 + 1		1.16 − 1		
	$^1S_0 - {}^3P_2$	2281.2	1.45 − 1		⇓		
	$^1S_0 - {}^1D_2$	3997.9	3.73 + 0		7.93 − 1		
	$^3P_0 - {}^3P_1$	1.1482 + 5	3.54 − 2		2.02 − 1		

Ion	Transition	λ(Å)	$A(s^{-1})$	ϒ(0.5)	ϒ(1.0)	ϒ(1.5)	ϒ(2.0)
	$^3P_0 - ^3P_2$	30528.8	3.67 − 5		2.24 − 1		
	$^3P_1 - ^3P_2$	41574.2	3.10 − 1		7.60 − 1		
Fe III	$^5D_4 - ^5D_3$	229146		2.38 + 0	2.87 + 0	3.02 + 0	3.01 + 0
	$^5D_4 - ^5D_2$	135513		9.70 − 1	1.23 + 0	1.31 + 0	1.32 + 0
	$^5D_4 - ^5D_1$	107294		4.75 − 1	5.91 − 1	6.29 − 1	6.36 − 1
	$^5D_4 - ^5D_0$	97436		1.43 − 1	1.78 − 1	1.90 − 1	1.94 − 1
	$^5D_3 - ^5D_2$	331636		1.65 + 0	2.03 + 0	2.16 + 0	2.18 + 0
	$^5D_3 - ^5D_1$	201769		6.12 − 1	7.94 − 1	8.45 − 1	8.46 − 1
	$^5D_3 - ^5D_0$	169516		1.70 − 1	2.23 − 1	2.36 − 1	2.35 − 1
	$^5D_2 - ^5D_1$	515254		1.04 + 0	1.28 + 0	1.36 + 0	1.36 + 0
	$^5D_2 - ^5D_0$	346768		2.35 − 1	3.09 − 1	3.31 − 1	3.33 − 1
	$^5D_1 - ^5D_0$	1060465		4.00 − 1	4.85 − 1	5.15 − 1	5.20 − 1
	$^3H_6 - ^3G_5$	22189.8		2.80 + 0	2.72 + 0	2.67 + 0	2.60 + 0
	$^3H_6 - ^3G_4$	20448.4		1.18 + 0	1.20 + 0	1.19 + 0	1.16 + 0
	$^3H_6 - ^3G_3$	19655.2		2.77 − 1	2.90 − 1	2.97 − 1	2.96 − 1
	$^3H_5 - ^3G_5$	23505.2		1.26 + 0	1.28 + 0	1.26 + 0	1.23 + 0
	$^3H_5 - ^3G_4$	21560.3		1.60 + 0	1.69 + 0	1.70 + 0	1.66 + 0
	$^3H_5 - ^3G_3$	20680.3		1.07 + 0	1.12 + 0	1.12 + 0	1.10 + 0
	$^3H_4 - ^3G_5$	24516.1		3.43 − 1	3.75 − 1	3.88 − 1	3.88 − 1
	$^3H_4 - ^3G_4$	22407.9		1.18 + 0	1.23 + 0	1.23 + 0	1.21 + 0
	$^3H_4 - ^3G_3$	21458.8		1.80 + 0	1.94 + 0	1.96 + 0	1.92 + 0

References

[1] E. T. Whittaker and G. N. Watson. *A Course of Modern Analysis* (Cambridge University Press, 1973).

[2] H. N. Russell and F. A. Saunders. *Astrophys. J.*, **61** (1925), 38.

[3] E. U. Condon and G. H. Shortley. *The Theory of Atomic Spectra* (Cambridge University Press, 1970).

[4] L. I. Schiff. *Quantum Mechanics* 3rd edn (McGraw-Hill, 1978).

[5] G. K. Woodgate. *Elementary Atomic Structure* (McGraw-Hill, 1970).

[6] B. H. Bransden and C. J. Joachain. *Physics of Atoms and Molecules* (Longman Scientific and Technical, co-published with John Wiley and Sons Inc., New York, 1990)

[7] V. A. Fock. *Z. Physik*, **98** (1935), 145.

[8] C. Froese Fischer. *The Hartree–Fock Method For Atoms* (John Wiley & Sons, 1977).

[9] D. Layzer. *Ann. Phys. (N.Y.)*, **8** (1959), 257.

[10] W. Eissner, M. Jones and N. Nussbaumer. *Comput. Phys. Commun.*, **8** (1974), 270.

[11] F. A. Parpia, C. Froese-Fischer and I. P. Grant. *Comput. Phys. Commun.*, **94** (1996), 249.

[12] W. Eissner and H. Nussbaumer. *J. Phys. B*, **2** (1969), 1028 (and references therein).

[13] M. Weissbluth. *Atoms and Molecules* (Academic Press, 1978).

[14] C. G. Darwin. *Proc. Roy. Soc. London,* **A117** (1928), 654.

[15] H. A. Bethe and E. E. Salpeter. *Quantum Mechanics of the One- and Two-Electron Atoms* (Plenum/Rosetta, New York, 1977).

[16] S. N. Nahar and J. M. Wadehra. *Phys. Rev. A*, **43** (1991), 1275.

[17] G. Breit. *Phys. Rev.*, **29**, (1929), 553; **36**, (1930), 383; **39** (1930), 616.

[18] J. J. Sakurai. *Advanced Quantum Mechanics* (Pearson Education, 2006).

[19] M. Blume and R. E. Watson. *Proc. Roy. Soc.*, **A270** (1962), 127.

[20] A. Dalgarno. In *The Analysis of Emission Lines*, Space Telescope Science Institute Symposium Series, 8 (Cambridge University Press, 1995), 158.

[21] E. Kallne, J. Kallne, A. Dalgarno *et al. Phys. Rev. Lett.*, **52** (1984), 2245.

[22] P. Beiersdorfer, M. Bitter, M. Marion and R. E. Olson. *Phys. Rev. A*, **72** (2005), 032725.

[23] M. J. Seaton. *Proc. Phys. Soc.*, **68** (1955), 457.

[24] M. J. Seaton. *Planet. Space Sci.*, **12** (1964), 55.

[25] A. Dalgarno. *Atoms in Astrophysics* (eds. P. G. Burke, W. B. Eissner, D. G. Hummer and I. C. Percival, Plenum Publishing, 1983), 103.

[26] R. S. Walling and J. C. Weisheit. *Phys. Repts.*, **162** (1988), 1.

[27] National Institute for Fusion Science (NIFS) of Japan. Research Report Series: NIFS-DATA-95.

[28] P. G. Burke and K. A. Berrington. *Atomic and Molecular Processes: An R-matrix Approach* (Institute of Physics Publishing, Bristol, 1993).

[29] J. R. Taylor. *Scattering Theory: The Quantum Mechanics of Nonrelativistic Collisions* (John Wiley & Sons Inc., 1972).

[30] R. G. Newton. *Scattering Theory of Waves and Particles* (McGraw-Hill Book Co., New York, 1966).

[31] U. Fano and A. R. P. Rau. *Atomic Collisions and Spectra* (Academic Press, Inc. 1986).

[32] N. F. Mott and H. S. W. Massey. *Theory of Atomic Collisions* (Oxford University Press, London, 1965).

[33] M. J. Seaton. *J. Phys. B*, **18** (1985), 2111.

[34] P. G. Burke and M. J. Seaton. *Meth. Comp. Phys.*, **10** (1971), 1.

[35] M. J. Seaton. *Phil. Trans. Roy. Soc.* **A245** (1953), 469.

[36] M. J. Seaton, Y. Yu, D. Mihalas and A. K. Pradhan. *Mon. Not. R. Astr. Soc.*, **266** (1994), 805.

[37] The Opacity Project Team, *The Opacity Project*, 2 vols (Institute of Physics Publishing, Bristol, UK, 1995–1996).

[38] D. G. Hummer, K. A. Berrington, W. Eissner *et al. Astron. Astrophys.*, **279** (1993), 298.

[39] P. G. Burke, A. Hibbert and W. D. Robb. *J. Phys. B*, **4** (1971), 153.

[40] P. L. Kapur and R. E. Peierls. *Proc. Roy. Soc. (Lond.)*, **A166** (1938), 277.

[41] E. P. Wigner and L. Eisenbud. *Phys. Rev.*, **72** (1947), 29.

[42] A. M. Lane and R. G. Thomas. *Rev. Mod. Phys.*, **30** (1958), 257.

[43] P. Descouvemont and D. Baye. *Rep. Prog. Phys.*, **73** (2010), 036301.

[44] P. G. Burke and D. Robb. *Advances in Atomic and Molecular Physics* (Academic Press, 1975), 142.

[45] A. Hibbert. *Comput. Phys. Commun.*, **9** (1975), 141.

[46] K. A. Berrington, W. Eissner and P. H. Norrington. *Comput. Phys. Commun.*, **92** (1995), 290.

[47] P. J. Buttle. *Phys. Rev.*, **160** (1967), 719.

[48] P. G. Burke and M. J. Seaton. *J. Phys. B*, **17** (1984), L683.

[49] M. S. Crees, M. J. Seaton and P. M. H. Wilson. *Comput. Phys. Commun.*, **15** (1978), 23.

[50] C. P. Ballance and D. C. Griffin. *J. Phys. B*, **39** (2006), 3617.

[51] A. D. Whiteford, N. R. Badnell, C. P. Ballance *et al. J. Phys. B*, **34** (2001), 3179.

[52] W. Eissner and G.-X. Chen. *J. Phys. B* (2010, in preparation).

[53] P. H. Norrington and I. P. Grant. *J. Phys. B*, **20** (1987), 4869.

[54] I. P. Grant, B. J. McKenzie, P. H. Norrington, D. F. Mayers and N. C. Pyper. *Comput. Phys. Commun.*, **21** (1980), 207.

[55] N. S. Scott and K. T. Taylor. *Comput. Phys. Commun.*, **25** (1982), 347.

[56] W. Eissner. *The Effect of Relativity in Atoms, Molecules and the Solid State* (ed. S. Wilson *et al.*, Plenum Press, New York, 1991).

[57] A. G. Sunderland, C. J. Noble, V. M. Burke and P. G. Burke. *Comput. Phys. Commun.*, **145** (2002), 311.

[58] C. A. Ramsbottom, C. E. Hudson, P. H. Norrington and M. P. Scott. *Astron. Astrophys.*, **465** (2007), 765.

[59] J.-J. Chang. *J. Phys. B*, **14** (1975), 2327.

[60] J.-J. Chang. *J. Phys. B*, **10** (1977), 3195.

[61] S. Ait-Tahar, I. P. Grant and P. H. Norrington. *Phys. Rev. A*, **54** (1996), 3984.

[62] V. M. Burke and I. P. Grant. *Proc. Phys. Soc.*, **90** (1967), 297.

[63] R. D. Cowan. *Theory of Atomic Structure and Spectra*, Los Alamos Series in Basic and Applied Sciences, 3 (University of California Press, 1981).

[64] W. Eissner and M. J. Seaton. *J. Phys. B*, **7** (1974), 2533.

[65] M. Hershkowitz and M. J. Seaton. *J. Phys. B*, **6** (1973), 1176.

[66] A. K. Pradhan, D. W. Norcross and D. G. Hummer. *Phys. Rev. A*, **23** (1981), 619.

[67] W. B. Eissner. *Physics of Electronic and Atomic Collisions*, VII ICPEAC, 1971 (North Holland, 1972), 460.

[68] D. H. Sampson, H. L. Zhang and C. J. Fontes. *Physics Reports*, **477** (2009), 111.

[69] H. L. Zhang. *Phys. Rev. A*, **57** (1998), 2640.

[70] M. J. Seaton. *Adv. At. Molec. Phys.*, **11** (1976), 83

[71] V. M. Burke and M. J. Seaton. *J. Phys. B*, **19** (1986), L527.

[72] D. Luo and A. K. Pradhan. *Phys. Rev. A*, **41** (1990), 165.

[73] G. W. F. Drake. *Atomic, Molecular and Optical Physics Handbook*, (ed. G. W. F. Drake, American Institute of Physics, Woodbury, NY, 1996).

[74] A. Burgess, M. C. Chidichimo and J. A. Tully. *Astron. Astrophys. Suppl. Ser.*, **121** (1997), 187.

[75] H. Bethe. *Ann. Physik.*, **2** (1930), 325.

[76] V. Krainov, H. R. Reiss, B. M. Smirnov. *Radiative Processes in Atomic Physics* (John Wiley and Sons Inc., New York, 1997).

[77] S. Chandrasekhar. *Astrophys. J.*, **102** (1945), 223.

[78] D. H. Menzel and C. L. Pekeris. *Mon. Not. R. Astr. Soc.*, **96** (1935), 77.

[79] R. M. Pengelly and M. J. Seaton. *Mon. Not. R. Astr. Soc.*, **127** (1964), 165.

[80] R. M. More and K. H. Warren. *J. Physiq. France*, **50** (1989), 35.

[81] C. W. Allen. *Astrophysical Quantities*, 3rd edn, (The Athlone Press, University of London, 1976).

[82] G. B. Rybicki and A. P. Lightman. *Radiative Processes in Astrophysics* (John Wiley and Sons, New York 1979).

[83] M. W. Smith and W. L. Wiese. *Astrophys. J. Supp. Ser.*, **196** (1971), 103.

[84] A. K. Pradhan. *Phys. Rev. A*, **28** (1983), 2113.

[85] A. K. Pradhan. *Phys. Rev. A*, **28** (1983), 2128.

[86] A. Burgess and J. A. Tully. *Astron. Astrophys.*, **254** (1992), 436.

[87] M. J. Seaton. *J. Phys. B*, **19** (1986), 2601.

[88] S. N. Nahar. *Atom. Data and Nucl. Data Tables*, **96** (2009), 26.

[89] W. Eissner and C. J. Zeippen. *J. Phys. B*, **14** (1981), 2125.

[90] G. W. F. Drake. *Phys. Rev. A*, **3** (1971), 908.

[91] I. M. Savukov, W. R. Johnson and U. I. Safronova. *Atom. Data and Nucl. Data Tables*, **85** (2003), 83.

[92] S. N. Nahar. *Phys. Scr.*, **48** (1993), 297.

[93] I. Bray and A. T. Stelbovics. *Phys. Rev. A*, **46** (1992), 6995.

[94] I. Bray and A. T. Stelbovics. *Comput. Phys. Commun.*, **85** (1995), 1.

[95] I. Bray, D. V. Fursa, A. S. Kheifets and A. T. Stelbovics. *J. Phys. B*, **35** (2002), R117.

[96] M. Gailitis. *Sov. Phys.–JETP*, **17** (1963), 1328.

[97] M. J. Seaton. *J. Phys. B*, **2** (1969), 5.

[98] M. A. Bautista and A. K. Pradhan. *Astrophys. J.*, **492** (1998), 650.

[99] G.-X. Chen, A. K. Pradhan and W. Eissner. *J. Phys. B*, **36** (2003), 453.

[100] National Institute for Standards and Technology, *www.nist.gov*.

[101] G.-X. Chen and A. K. Pradhan. *Phys. Rev. Lett.*, **89** (2002), 013202.

[102] R. J. W. Henry. *Physics Reports*, **68** (1981), 1.

[103] A. K. Pradhan and J. W. Gallagher. *Atom. Data and Nucl. Data Tables*, **52** (1992), 227.

[104] H. L. Zhang and A. K. Pradhan. *J. Phys. B*, **28** (1995), L285.

[105] S. Ait-Tahar, I. P. Grant and P. H. Norrington. *Phys. Rev. A*, **54** (1996), 3984.

[106] J. Pelan and K. A. Berrington. *Astron. Astrophys. Suppl. Ser.*, **122** (1997), 177.

[107] S. J. Smith, M. Zuo, A. Chutjian, S. S. Tayal and I. D. Williams. *Astrophys. J.*, **463** (1996), 808.

[108] M. Montenegro, W. Eissner, S. N. Nahar and A. K. Pradhan. *J. Phys. B*, **39** (2006), 1863.

[109] T. A. A. Sigut and A. K. Pradhan. *J. Phys. B*, **28** (1995), 4879.

[110] S. J. Smith, A. Chutjian, J. Mitroy *et al. Phys. Rev. A*, **48** (1993), 292.

[111] V. I. Lengyel, V. T. Navrotsky, E. P. Sabad and O. I. Zatsarinny. *J. Phys. B*, **23** (1990), 2847.

[112] M. Zuo, S. J. Smith, A. Chutjian *et al. Astrophys. J.*, **440** (1995), 421.

[113] M. E. Bannister, X. Q. Guo, T. M. Kojima and G. H. Dunn. *Phys. Rev. Lett.*, **72** (1994), 3336.

[114] I. P. Zapesochnyi, A. I. Dascehenko, V. I. Frontov *et al. Sov. Phys-JETP Lett*, **39** (1984), 51.

[115] V. I. Lengyl, V. T. Navrotsky, E. P. Sabad. *J. Phys. B*, **23** (1990), 2847.

[116] A. K. Pradhan and J. F. Peng. *The Analysis of Emission Lines*, Space Telescope Science Institute Symposium Series 8 (Cambridge University Press, 1995).

[117] A. K. Pradhan and H. L. Zhang. *Electron collisions with atomic ions*. In *Photon and Electron Interactions with Atoms, Molecules, and Ions*, Landolt-Börnstein, **17 B** (Springer, 1992).

[118] C. D. Caldwell and M. O. Krause. *Atomic, Molecular and Optical Physics Handbook*, Ch. 59 (ed. G. W. F. Drake, American Institute of Physics, 1996).

[119] H. Jakubowicz and D. L. Moores. *J. Phys. B*, **14** (1981), 3733.

[120] P. L. Bartlett and A. T. Stelbovics. *Phys. Rev. A*, **66** (2002), 012707.

[121] D. C. Gregory, L. J. Wang, F. W. Meyer and K. Rinn. *Phys. Rev. A*, **35** (1987), 3256.

[122] D. H. Sampson. *J. Phys. B*, **15** (1982), 2087.

[123] W. Lotz. *Astrophys. J. Supp. Ser.*, **14** (1968), 207; Ibid. Z. *Phys.*, **216**, (1968), 241.

[124] A. Burgess and M. C. Chidichimo. *Mon. Not. R. Astr. Soc.*, **203** (1983), 1269.

[125] K. J. LaGattuta and Y. Hahn. *Phys. Rev. A*, **24** (1981), 2273.

[126] G. S. Voronov. *Atom. Data and Nucl. Data Tables.*, **65** (1997), 1.

[127] V. P. Shevelko. *Atoms and Their Spectroscopic Properties* (Springer, 1997).

[128] W. Bambynek *et al.*, *Rev. Mod. Phys.*, **44** (1972), 716; Erratum: **46** (1976) 853.

[129] H. Kjeldsen, F. Folkmann, J. E. Hensen, *et al. Astrophys. J.* **524** (1999), L143.

[130] H. Kjeldsen, B. Kristensen, F. Folkmann, and T. Andersen, *J. Phys. B*, **35** (2002), 3655.

[131] J.-P. Champeaux, *et al. Astrophys. J. Supp. Ser.*, **148** (2003), 583.

[132] A. M. Covingtion *et al. Phys. Rev. Lett.*, **87** (2001), 243002.

[133] S. N. Nahar. *Phys. Rev. A*, **53** (1996), 1545.

[134] S. N. Nahar. In *New Quests in Stellar Astrophysics. II. The Ultraviolet Properties of Evolved Stellar Populations*, (ed. M. Chavez, E. Bertone, D. Rosa-Gonzalez and L. H. Rodriguez-Merino, Springer, 2009), 245

[135] J. W. Cooper. *Physical Review*, **128** (1962), 681.

[136] S. Sahoo and Y. K. Ho. *Phys. Plasma*, **13** (2006), 06301.

[137] D. R. Bates. *Proc. Roy. Soc.*, **188** (1947), 350.

[138] M. J. Seaton. *Proc. Roy. Soc.*, **208** (1951), 418.

[139] R. F. Reilman and S. T. Manson. *Astrophys. J. Supp. Ser.*, **40** (1979), 815.

[140] D. A. Verner, D. G. Yakovlev, I. M. Band and M. B. Trzhaskovskaya. *Atom. Data and Nucl. Data Tables*, **55** (1993), 233.

[141] M. J. Seaton. *Mon. Not. R. Astr. Soc.*, **119** (1959), 81.

[142] NORAD (Nahar-OSU-Radiative Database) www.astron omy.ohio-state.edu/~nahar.

[143] P. J. Storey and D. G. Hummer. *Comput. Phys. Commun.*, **66** (1992), 129.

[144] M. A. Bautista and A. K. Pradhan. *J. Phys. B*, **28** (1995), L173.

[145] S. N. Nahar. *Phys. Rev. A*, **58** (1998), 3766.

[146] S. N. Nahar. *J. Quant. Spectrosc. Radiat. Transfer*, **109** (2008), 2417.

[147] S. N. Nahar. *Phys. Rev. A*, **65** (2002), 052702.

[148] S. N. Nahar and A. K. Pradhan. *J. Phys. B*, **26** (1993), 1109.

[149] Yu Yan and M. J. Seaton, *J. Phys. B*, **20** (1987), 6409.

[150] S. N. Nahar and A. K. Pradhan. *Astrophys. J.*, **397** (1992), 729.

[151] S. N. Nahar and A. K. Pradhan. *Radiation Phys. Chem.*, **70** (2004), 323.

[152] S. N. Nahar. *Phys. Rev. A*, **69** (2004), 042714.

[153] M. J. Seaton. *Rep. Prog. Phys.*, **46** (1983), 167.

[154] J. Dubau and M. J. Seaton. *J. Phys. B*, **17** (1984), 381.

[155] S. N. Nahar and A. K. Pradhan. *Phys. Rev. A*, **44** (1991), 2935.

[156] K. Sakimoto, M. Terao and K. A. Berrington. *Phys. Rev. A*, **42**, (1990), 291.

[157] R. H. Bell and M. J. Seaton. *J. Phys. B*, **18** (1985), 1589.

[158] S. N. Nahar, A. K. Pradhan and H. L. Zhang. *Phys. Rev. A*, **63** (2001), 060701.

[159] U. Fano. *Phys. Rev.*, **124** (1961), 1866.

[160] A. Wishart. *Mon. Not. R. Astr. Soc.*, **187** (1979), 59.

[161] A. Wishart. *J. Phys. B*, **12** (1979) 3511.

[162] J. T. Broad and W. P. Reihnardt. *Phys. Rev. A*, **14** (1976), 2159.

[163] H. C. Bryant, B. D. Dieterle, J. Donahue *et al. Phys. Rev. Lett.*, **38** (1977), 228.

[164] D. J. Pegg. *Rad. Phys. Chem.*, **70** (2004), 371.

[165] J. Macek and P. G. Burke. *Proc. Phys. Soc. Lond.*, **92** (1967), 351.

[166] C. D. Lin. *Phys. Rev. A*, **14** (1976), 30.

[167] J. Cooper and R. N. Zare. *Lectures on Theoretical Physics*, Vol 11-C (ed. S. Geltman, K. T. Mahanthappa, W. E. Brittin, Gordon and Breach Publishers, New York, 1969), 317.

[168] U. Fano and D. Dill. *Phys. Rev. A*, **6** (1972), 185.

[169] D. Dill. *Phys. Rev. A*, **7** (1973), 1976.

[170] Y. Hahn and K. L. Lagattuta. *Phys. Repts.*, **166** (1988), 195.

[171] S. N. Nahar and A. K. Pradhan. *Phys. Rev. Lett.*, **68** (1992), 1488.

[172] S. N. Nahar and A. K. Pradhan. *Phys. Rev. A*, **49** (1994), 1816.

[173] E. A. Milne. *Phil. Mag. Series. 6.*, **47** (1924), 209.

[174] S. N. Nahar, A. K. Pradhan, H. L. Zhang. *Astrophys. J. Supp. Ser.*, **133** (2001), 255.

[175] D. A. Verner and G. J. Ferland. *Astrophys. J. Supp. Ser.*, **103** (1996), 467.

[176] H. S. W. Massey and D. R. Bates. *Rep. Prog. Phys.*, **9** (1942), 62.

[177] D. R. Bates and H. S. W. Massey. *Phil. Trans. Roy. Soc.*, **A239** (1943), 269.

[178] M. J. Seaton and P. J. Storey. *Atomic Processes and Applications*, (ed. P. G. Burke and B. L. Moiseiwitch, North Holland, 1976), 133.

[179] A. Burgess. *Astrophys. J.*, **139** (1964), 776.

[180] A. Burgess and H. P. Summers. *Astrophys. J.*, **157** (1969), 1007.

[181] P. C. W. Davies and M. J. Seaton. *J. Phys. B*, **2** (1969), 747.

[182] D. R. Bates and A. Dalgarno. *Atomic and Molecular Processes* (ed. D. R. Bates, Academic Press, 1962), 245.

[183] A. Burgess, *Astrophys. J.*, **141** (1965), 1588.

[184] C. W. Allen. *Space Sci. Rev.*, **4** (1965), 91.

[185] L. P. Presnyakov and A. M. Urnov. *J. Phys. B*, **8** (1975), 1280.

[186] A. K. Pradhan. *Phys. Rev. Lett.*, **47** (1981), 79.

[187] H. Nussbaumer and P. J. Storey. *Astron. Astrophys.*, **126** (1983), 75.

[188] S. N. Nahar and A. K. Pradhan. *Astrophys. J. Supp. Ser.*, **111** (1997), 339.

[189] S. N. Nahar and A. K. Pradhan. *Astrophys. J.*, **447** (1995), 966.

[190] D. G. Hummer. *Mon. Not. R. Astr. Soc.*, **268** (1994), 109.

[191] S. N. Nahar. *New Astronomy*, **13** (2008), 619.

[192] S. N. Nahar. *Astrophys. J. Supp. Ser.*, **156** (2004), 93.

[193] S. Schippers, T. Bartsch, C. Brandau *et al. Phys. Rev. A*, **59** (1999), 3092

[194] S. Mannervik, D. R. DeWitt, L. Engstrom *et al. Phys. Rev. Lett.*, **81** (1998), 313.

[195] S. M. V. Aldrovandi and D. Pequignot. *Revista Brasileira de Fisica*, **4** (1974), 491.

[196] J. M. Shull and M. van Steenberg. *Astrophys. J. Supp. Ser.*, **48** (1982), 95.

[197] C. Romanik. *Astrophys. J.*, **330** (1988), 1022.

[198] Z. Altun, A. Yumak, N. R. Badnell, J. Colgun and M. S. Pindzola. *Astron. Astrophys.*, **420** (2004), 779.

[199] D. W. Savin, *et al. Astrophys. J. Supp. Ser.*, **123** (1999), 687.

[200] A. K. Pradhan, S. N. Nahar, H. L. Zhang. *Astrophys. J. Lett.*, **549** (2001), L265.

[201] H. L. Zhang, S. N. Nahar and A. K. Pradhan, *Phys. Rev. A*, **64** (2001), 032719.

[202] S. Schippers, A. Müller, G. Gwinner, J. Linkemann, A. Saghiri and A. Wolf. *Astrophys. J.*, **555** (2001), 1027.

[203] A. K. Pradhan, G. X. Chen, S. N. Nahar and H. L. Zhang. *Phys. Rev. Lett.*, **87** (2001), 183201.

[204] H. L. Zhang, S. N. Nahar and A. K. Pradhan. *J. Phys. B*, **32** (1999), 1459.

[205] A. H. Gabriel and C. Jordan. *Mon. Not. R. Astr. Soc.*, **145** (1969), 241.

[206] A. H. Gabriel. *Mon. Not. R. Astr. Soc.*, **160** (1972), 99.

[207] J. Oelgoetz and A. K. Pradhan. *Mon. Not. R. Astr. Soc.*, **327** (2001), L42.

[208] S. N. Nahar, J. Oelgoetz and A. K. Pradhan. *Phys. Scripta*, **79** (2009), 055301.

[209] P. Beiersdorfer, T. W. Philips, K. L. Wong, R. E. Marrs and D. A. Vogel. *Phys. Rev. A*, **46** (1992), 3812.

[210] L. A. Vainshtein and U. I. Safronova. *Atom. Data and Nucl. Data Tables*, **25**, (1978), 49.

[211] F. Bely-Dubau, J. Dubau, P. Faucher and A. H. Gabriel, *Mon. Not. R. Astr. Soc.*, **198** (1982), 239.

[212] S. N. Nahar and A. K. Pradhan. *Astrophys. J. Supp. Ser.*, **162** (2006), 417.

[213] S. N. Nahar and A. K. Pradhan. *Phys. Rev. A*, **73** (2006), 062718.

[214] C. P. Bhalla, A. H. Gabriel and L. P. Presnyakov. *Mon. Not. R. Astr. Soc.*, **172** (1975), 359.

[215] G. J. Ferland, B. M. Peterson, K. Horne, W. F. Walsh, S. N. Nahar. *Astrophys. J.*, **387** (1992), 95.

[216] D. G. Hummer and P. J. Storey. *Mon. Not. R. Astr. Soc.*, **297** (1998) 1073.

[217] P. J. Storey and G. G. Hummer. *Mon. Not. R. Astr. Soc.*, **272** (1995), 41.

[218] G. J. Ferland. CLOUDY: University of Kentucky, Department of Physics and Astronomy Internal Report (1993).

[219] S. N. Nahar and M. A. Bautista. *Astrophys. J.*, **120** (1999), 327.

[220] D. T. Woods, J. M. Shull, C. L. Sarazin. *Astrophys. J.*, **249** (1981), 399.

[221] S. N. Nahar. *Astrophys. J.*, **120** (1999), 131.

[222] R. S. Sutherland and M. A. Dopita. *Astrophys. J. Supp. Ser.*, **88** (1993), 253.

[223] D. G. Hummer and P. J. Stoery. *Mon. Not. R. Astr. Soc.*, **224** (1987), 801.

[224] R. Kisielius, P. J. Storey, A. R. Davey, L. T. Neale. *Astron. Astrophys. Suppl. Ser.*, **133** (1998), 257.

[225] N. Smith. *Mon. Not. R. Astr. Soc.*, **346** (2003), 885.

[226] M. J. Seaton and D. E. Osterbrock. *Astrophys. J.*, **125** (1957), 66.

[227] C. Esteban, M. Peimbert, J. Garcia-Rojas *et al. Mon. Not. R. Astr. Soc.*, **355** (2004), 229.

[228] D. E. Osterbrock. *Astrophysics of Gaseous Nebulae and Active Galactic Nuclei*, 2nd edn (University Science Books, 1989).

[229] J.-U. Ness, R. Mewe, J. H. M. M. Schmitt *et al. Astron. Astrophys.*, **367** (2001), 282.

[230] G. R. Blumenthal, G. W. Drake and W. H. Tucker. *Astrophys. J.*, **172** (1972), 205.

[231] R. Mewe and J. Schrijver. *Astron. Astrophys.*, **65** (1978), 114.

[232] A. K. Pradhan and J. M. Shull. *Astrophys. J.*, **249** (1981), 821.

[233] A. K. Pradhan. *Astrophys. J.*, **262** (1982), 477.

[234] F. Delahaye and A. K. Pradhan. *J. Phys. B*, **35** (2002), 3377.

[235] R. S. Walling and J. C. Weisheit. *Phys. Repts.*, **162** (1988), 1.

[236] J. Oelgoetz and A. K. Pradhan. *Mon. Not. R. Astr. Soc.*, **354** (2004), 1093

[237] J. Oelgoetz, C. J. Fontes, H. L. Zhang *et al. Mon. Not. R. Astr. Soc.*, **382** (2007), 761.

[238] J. Oelgoetz, C. J. Fontes, H. L. Zhang and A. K. Pradhan. *Phys. Rev. A*, **76** (2007), 062504.

[239] Y. Tanaka *et al. Nature*, **375** (1995), 659.

[240] Y-D. Xu, R. Narayan, E. Quataert, F. Yuan and F. K. Baganoff. *Astrophys. J.*, **640** (2006), 319.

[241] R. S. Sutherland and M. A. Dopita. *Astrophys. J. Supp. Ser.*, **88** (1993), 253.

[242] T. Kato, T. Fujiwara and Y. Hanaoka. *Astrophys. J.*, **492** (1998), 822.

[243] R. D. Blum and A. K. Pradhan. *Phys. Rev. A*, **44** (1991), 6123.

[244] D. Mihalas. *Stellar Atmospheres*, 2nd edn (W. H. Freeman and Company, 1978).

[245] R. J. Rutten. *Radiative Transfer in Stellar Atmospheres*, Utrecht University Lecture Notes, 8th edn 2003 (http://esmn.astro.uu.nl).

[246] E. H. Avrett. *Lecture Notes: Introduction to Non-LTE Radiative Transfer and Atmospheric Modeling*, 2008 (http://www.cfa.harvard.edu/~avrett/).

[247] N. N. Naumova and V. N. Khokhlov. *J. Opt. Technol.*, **73** (2006), 509.

[248] M. Baranger. *Phys. Rev.*, **111** (1958), 494; **112** (1958), 855.

[249] M. J. Seaton. *J. Phys. B*, **20** (1987), 6363 (ADOC I).

[250] H. Griem. *Phys. Rev. A*, **165** (1968), 258.

[251] G. Alecian, G. Michaud and J. Tully. *Astrophys. J.*, **411** (1993), 882.

[252] M. S. Dimitrijevic and N. Konjevic. *Astron. Astrophys.*, **172** (1987), 345.

[253] S. Sahal-Brechot and E. R. A. Segre. *Astron. Astrophys.*, **13** (1971), 161.

[254] A.-M. Dumont, S. Collin, F. Paletou *et al. Astron. Astrophys.*, **407** (2003), 13.

[255] H. Zanstra. *Pub. Dominion Astrophys. Obs.*, **4** (1931), 209.

[256] M. A. Dopita and A. S. Sutherland. *Astrophysics of the Diffuse Universe* (Springer, 2003).

[257] L. Spitzer. *Physical Processes in the Interstellar Medium* (John Wiley & Sons, 1978).

[258] D. Luo, A. K. Pradhan and J. M. Shull. *Astrophys. J.*, **335** (1988), 498.

[259] D. F. Gray. *The Observation and Analysis of Stellar Photospheres*, 2nd edn (Cambridge University Press, 1992).

[260] G. B. Rybicki and D. G. Hummer. *Astrophys. J.*, **290** (1994), 553.

[261] T. Lanz and I. Hubeny. *Astrophys. J. Supp. Ser.*, **169** (2007), 83.

[262] J. P. Philips. *New Astronomy*, **12** (2007), 378.

[263] J. C. Martin, K. Davidson, R. M. Humphreys *et al. Bulletin of the American Astronomical Society #211*, **39** (2007), 841.

[264] J. J. Cowan, C. Sneden, J. W. Truran and D. L. Burris. *Stellar Abundance Observations*, Proceedings of the Oak Ridge Symposium on Atomic and Nuclear Astrophysics (Institute of Physics Publishing, 1998), 621.

[265] J. E. Lawler, C. Sneden, J. J. Cowan, I. I. Evans and E. A. Den Hartog. *Astrophys. J. Sup. Ser.*, **182** (2009), 51.

[266] C. Sneden, J. E. Lawler, J. J. Cowan, I. I. Evans and E. A. Den Hartog. *Astrophys. J. Sup. Ser.*, **182** (2009), 80.

[267] J. Sinmerer, C. Sneden, J. J. Cowan *et al. Astrophys. J.*, **617** (2004), 1091.

[268] A. McWilliam. *Ann. Rev. Astron. Astrophys.*, **35** (1997), 503.

[269] M. Asplund. *Ann. Rev. Astron. Astrophys.*, **43** (2005), 481.

[270] J. R. Kuhn, R. I. Bush, M. Emilio and P. H. Scherrer. *Astrophys. J.*, **613** (2004), 1241.

[271] R. L. Kurucz (2005), http://kurucz.harvard.edu/papers/TRANSMISSION.

[272] J. Canaro, N. Cardiel, J. Gorgas *et al. Mon. Not. R. Astr. Soc.*, **326** (2001), 959.

[273] M. Melendez, M. A. Bautista and N. R. Badnell. *Astron. Astrophys.*, **469** (2007), 1203.

[274] J. L. Kohl, G. Noci, S. R. Cranmer and J. C. Raymond. *Astron. Astrophys. Rev.*, **13** (2006), 31.

[275] K. J. H. Phillips, R. Mewe, L. K. Harra-Murnion *et al. Astrophys. J. Supp. Ser.*, **138** (1999), 381.

[276] *The Eleventh Cambridge Workshop on Cool Stars, Stellar Systems and the Sun*. ASP Conference Series Vol. 223 (Ed. R. J. Garcia Lopez, R. Bebolo and M. R. Zapatero Osorio, 2001).

[277] D. J. Hillier and D. L. Miller. *Astrophys. J.*, **496** (1998), 407.

[278] K. Davidson and R. M. Humphreys. *Annu. Rev. Astron. Astrophys.*, **35** (1997), 1.

[279] V. Kilmov, S. Johansson and V. S. Letokhov. *Astron. Astrophys.*, **385** (2002), 313.

[280] F. J. Rogers and C. A. Iglesias. *Science*, **263** (1994), 50.

[281] F. Delahaye and M. H. Pinsonneault. *Astrophys. J.*, **625**, (2005), 563.

[282] D. G. Hummer and D. Mihalas. *Astrophys. J.*, **331** (1988), 794.

[283] D. Mihalas, W. Dappen, G. G. Hummer. *Astrophys. J.*, **331** (1988), 815.

[284] W. Dappen, D. Mihalas, D. G. Hummer, B. W. Mihalas. *Astrophys. J.*, **332** (1988), 261.

[285] D. Mihalas, D. G. Hummer, B. W. Mihalas, W. Dappen. *Astrophys. J.*, **350** (1990), 300.

[286] W. Däppen, L. Anderson and D. Mihalas. *Astrophys. J.*, **319** (1987), 195.

[287] A. Nayfonov, W. Dappen, D. G. Hummer and D. Mihalas. *Astrophys. J.*, **526** (1999), 451.

[288] D. G. Hummer. *American Institute of Physics Conference Proceedings*, **168** (1988).

[289] I. Hubeny, D. G. Hummer and T. Lanz. *Astron. Astrophys.*, **282** (1994), 151.

[290] A. Kawka and S. Vennes. *Astrophys. J.*, **643** (2006), 402.

[291] W. L. Wiese, D. E. Kelleher and D. R. Paquette. *Phys. Rev. A*, **6** (1972), 1132.

[292] M. J. Seaton. *J. Phys. B*, **23** (1990), 3255 (ADOC XIII).

[293] M. J. Seaton *J. Phys. B*, **20** (1987), 6363.

[294] M. Asplund, N. Grevesse, A. J. Sauval and P. Scott. *Ann. Rev. Astron. Astrophys.*, **47** (2009), 481

[295] J. N. Bahcall, A. M. Serenelli and S. Basu. *Astrophys. J.*, **621** (2005), L85.

[296] F. Delahaye and M. H. Pinsonneault. *Astrophys. J.*, **649** (2006), 529.

[297] A. K. Pradhan and S. N. Nahar. Recent directions astrophysical quantitative spectroscopy and radiation hydrodynamics, *AIP Conf. Proc.* **1171** (2009), 52.

[298] M. J. Seaton and N. R. Badnell. *Mon. Not. R. Astr. Soc.*, **354** (2004), 457.

[299] S. N. Nahar and A. K. Pradhan. *J. Phys. B*, **27** (1994), 429.

[300] C. Mendoza, M. J. Seaton, P. Buerger *et al. Mon. Not. R. Astr. Soc.*, **378** (2007), 1031.

[301] G. Michaud and J. Richer. *Contrib. Astron. Obs. Skalnate Pleso*, **35** (2005), 1 (astro-ph/0802.1707).

[302] S. Hubrig. *Contrib. Astron. Obs. Skalnate Pleso*, **27** (1998), 296.

[303] D. S. Leckrone, C. R. Proffitt, G. M. Wahlgren, S. G. Johansson and T. Brage. *Astron. J.*, **117** (1999), 1454.

[304] S. Bowyer and R. F. Malina, eds., *Astrophysics in the Extreme Ultraviolet.* (Kluwer Academic Publishers, 1996).

[305] J. Richer, G. Michaud and C. Profitt. *Astrophys. J. Supp. Ser.*, **82** (1992), 329.

[306] G. Alecian, G. Michaud and J. A. Tully. *Astrophys. J.*, **411** (1993), 882.

[307] G. Alecian and M. J. Stift. *Astron. Astrophys.*, **454** (2006), 571.

[308] M. J. Seaton. *Mon. Not. Roy. Astr. Soc.*, **382** (2007), 245.

[309] J. Bailey *et al. Phys. Rev. Lett.*, **99** (2007), 265002.

[310] J. Bailey *et al. Rev. Sci. Instrum.*, **79** (2008), 113140.

[311] D. E. Osterbrock and G. J. Ferland. *Astrophysics of Gaseous Nebulae and Active Galactic Nuclei*, 3rd edn (University Science Books, 2006).

[312] Y. G. Tsamis *et al.*, *Mon. Not. R. Astr. Soc.*, **338** (2004), 687.

[313] M. Peimbert, P. J. Storey and S. Torres-Peimbert, *Astrophys. J.*, **414** (1993), 626.

[314] J. A. Baldwin, G. J. Ferland, P. G. Martin *Astrophys. J.*, **374** (1991), 580.

[315] R. H. Rubin, J. P. Simpson, M. R. Haas and E. F. Erickson. *PASP*, **103** (1991), 834.

[316] K. P. M. Blagrave, P. G. Martin, R. H. Rubin *et al. Astrophys. J.*, **655** (2007) 299.

[317] A. K. Pradhan and H. L. Zhang. *Astrophys. J. Lett.*, **409** (1993), L77.

[318] S. Torre-Peimbert and A. Arieta. *Bulletin of American Astronomical Society*, **189** (1996), 9711.

[319] I. S. Bowen. *PASP*, **59** (1947), 196.

[320] L. B. Lucy. *Astron. Astrophys.*, **294** (1995), 555.

[321] M. A. Bautista, J.-F. Peng, A. K. Pradhan. *Astrophys. J.*, **460** (1996), 372.

[322] M. Barlow, J. E. Drew, J. Meaburn, R. M. Massey. *Mon. Not. R. Astr. Soc.*, **268** (1994), L29.

[323] R. H. Rubin, R. J. Dufour, G. J. Ferland *et al. Astrophys. J. Lett.*, **474** (1997), L131.

[324] A. K. Pradhan and J. W. Gallagher. *Atom. Data and Nucl. Data Tables*, **52** (1992), 227.

[325] A. K. Pradhan and H. L. Zhang. *Electron Collisions with Atomic Ions*. In *Photon and Electron Interactions with Atoms, Molecules, Ions*, Landolt-Börnstein, **17** (Springer-Verlag, 2001), 1–102.

[326] A. K. Pradhan, M. Montenegro, S. N. Nahar, W. Eissner. **366** (2006), L6.

[327] N. C. Sterling, H. L. Dinerstein, C. W. Bowers and S. Redfield. *Astrophys. J.*, **625** (2005), 368. Ch

[328] R. C. Reis, A. C. Fabian, R. R. Ross *et al. Mon. Not. R. Astr. Soc.*, **387** (2008), 1489.

[329] A. Konigl and J. F. Kartje. *Astrophys. J.*, **434** (1994), 446.

[330] M. Elvis. *Astrophys. J.*, **545** (2000), 63.

[331] C. M. Urry and P. Padovani. *PASP*, **107** (1995), 803. The version of the unified model cartoon used is from: http://www.cv.nrao.edu/course/astr534/ExtraGalactic.html.

[332] M. Modjaz, J. M. Moran, P. T. Kondratco and L. J. Greenhill. *Astrophys. J.*, **626** (2005), 104.

[333] C. K. Seyfert. *Astrophys. J.*, **97** (1943), 28.

[334] D. E. Osterbrock and R. W. Pogge. *Astrophys. J.*, **297** (1985), 166.

[335] T. Boller, W. N. Brandt and H. H. Fink. *Astron. Astrophys.*, **305** (1996), 53.

[336] K. Leighly. *Astrophys. J. Supp. Ser.*, **125** (1999), 317.

[337] D. Grupe, H.-C. Thomas and K. Beurmann. *Astron. Astrophys.*, **367** (2001), 470.

[338] T. A. Boroson and R. F. Green. *Astrophys. J. Supp. Ser.*, **80** (1992), 109.

[339] D. Grupe. *Astron. J.*, **127** (2004), 1799.

[340] J. W. Sulentic, P. Marziani, D. Dultzin-Hacyan. *Annual Reviews of Astron. Astrophys.*, **38** (2000), 521.

[341] D. Grupe and S. Mathur. *Astrophys. J.*, **606** (2004), 41.

[342] D. Grupe and S. Mathur. *Astrophys. J.*, **633** (2005), 688; Ibid. *Astron. Astrophys.*, **432** (2005), 462

[343] R. D. Blandford and C. F. McKee. *Astrophys. J.*, **255** (1982), 419.

[344] A. Wandel, B. M. Peterson and M. A. Malkan. *Astrophys. J.*, **526** (1999), 579.

[345] S. Kaspi, P. S. Smith, H. Netzer *et al. Astrophys. J.*, **533** (2000), 631.

[346] B. M. Peterson *et al. Astrophys. J.*, **542** (2000), 161.

[347] M. Vestergaard and B. M. Peterson. *Astrophys. J.*, **641** (2006), 689.

[348] L. Ferrarese and D. Merritt. *Astrophys. J.*, **539** (2000), L9.

[349] J. Silk and M. Rees. *Astron. Astrophys.*, **331** (1998), L1.

[350] J. E. Greene and L. Ho. *Astrophys. J.*, **641** (2006), L21.

[351] M. Nikolajuk, B. Czerny, J. Ziokowski and M. Gierlinski. *Mon. Not. R. Astr. Soc.*, **370** (2006), 1534.

[352] M. Vestergaard. *Astrophys. J.*, **571** (2002), 733.

[353] J. Magorrian, S. Tremaine and D. Richstone. *Astron. J.*, **115** (1998), 2285.

[354] N. Häring and H. Rix. *Astrophys. J.*, **684** (2004), L89.

[355] T. M. Tripp, J. Bechtold and R. F. Green. *Astrophys. J.*, **433** (1994), 533.

[356] S. Collin-Souffrin, M. Joly, D. Pequignot and A.-M. Dumont. *Astron. Astrophys.*, **166** (1986), 27.

[357] J. E. Greene and L. Ho. *Astrophys. J.*, **630** (2005), 122.

[358] T. E. Papadakis. *Astron. Astrophys.*, **425** (2004), 1133.

[359] I. M. McHardy, F. K. Gunn, P. Uttley and M. R. Goad. *Mon. Not. Roy. Astr. Soc.*, **359** (2005), 1469.

[360] D. Grupe, S. Komossa, L. C. Gallo *et al. Astrophys. J.*, **681** (2008), 982.

[361] A. L. Longinotti, A. Nucita, M. Santos-Lleo and M. Guainazzi. *Astron. Astrophys.*, **484** (2008), 311.

[362] K. I. Kellerman, R. Sramek, M. Schmidt, D. B. Shaffer and R. F. Grenn. *Astron. J.*, **98** (1989) 1195.

[363] B. L. Fanaroff and J. M. Riley. *Mon. Not. R. Astr. Soc.*, **167** (1974), 31.

[364] M. S. Brotherton, C. De Breuck and J. J. Schaefer. *Mon. Not. R. Astr. Soc.*, **372** (2006), L58.

[365] J. Kuraszkiewicz *et. al. Astrophys. J.*, **590** (2003), 128.

[366] M. Elvis *et. al. Astrophys. J. Supp. Ser.*, **95** (1994), 1.

[367] D. Vanden Berk, *et al. Astron. J.*, **122** (2001), 549.

[368] R. D. Williams, S. Mathur, R. W. Pogge. *Astrophys. J.*, **610**, (2004), 737.

[369] T. A. A. Sigut and A. K. Pradhan. *J. Phys. B*, **28** (1995), 4879.

[370] S. Huang, Y. Zhou and T. Wang. *Chinese Astron. Astrophys.*, **24** (2000), 405.

[371] J. H. Krolik. *Active Galactic Nuclei* (Princeton University Press, 1999), 136, 189.

[372] B. J. Wills, H. Netzer and D. Wills. *Astrophys. J.*, **288** (1985), 94.

[373] T. A. A. Sigut and A. K. Pradhan. *Astrophys. J. Supp. Ser.*, **145** (2003), 15.

[374] P. Marziani, J. W. Sulenic, D. Dultzin-Hacyan, M. Calvani and M. Moles. *Astrophys. J. Supp. Ser.*, **104** (1996), 37.

[375] A. Laor, J. N. Bahcall, B. T. Jannuzi *et al. Astrophys. J.*, **489** (1997), 656.

[376] M.-P. Veron-Cetty, M. Joly and P. Veron. *Astron. Astrophys.*, **417** (2004), 515.

[377] R. J. Rudy, S. Mazuk, C. C. Venturini, C. Puetter and F. Hamann. *PASP*, **113** (2001), 916.

[378] A. Rodriguez-Ardila, S. M. Viegas, M. G. Pastoriza and L. Prato. *Astrophys. J.*, **565** (2002), 140.

[379] T. A. A. Sigut, A. K. Pradhan and S. N. Nahar. *Astrophys. J.*, **61** (2004), 611.

[380] M. Joly. *Ann. Phys. Fr.*, **18** (1993), 241.

[381] M. J. Graham, R. G. Clowes and L. E. Campusano. *Mon. Not. R. Astr. Soc.*, **279** (1996), 1349.

[382] H. Netzer. In *Physics of Formation of Fe* II *Lines Outside of LTE*, (ed. R Viotti, A. Vittone and M. Friedjung, Dordrecht:Reidel, 1988), 247.

[383] M. V. Penston. *Mon. Not. R. Astr. Soc.*, **229** (1987), 1.

[384] T. A. A. Sigut and A. K. Pradhan. *Astrophys. J. Lett.*, **499** (1998), L139.

[385] A. Rodriguez-Ardila, M. G. Pastoriza, S. M. Viegas, T. A. A. Sigut, A. K. Pradhan. *Astron. Astrophys.*, **425** (2004), 457.

[386] S. Johansson and C. Jordan. *Mon. Not. R. Astr. Soc.*, **210** (1984), 239.

[387] C. Fransson *et al. Astrophys. J.*, **572** (2002), 350.

[388] J. W. Sulentic, D. Dultzin-Hacyan, P. Marziani, *RevMexAA*, **28** (2007), 83.

[389] J. A. Baldwin. *Astrophys. J.*, **214** (1997), 679.

[390] A. Baskin and A. Laor. *Mon. Not. R. Astr. Soc.*, **350** (2004), L31.

[391] P. Jiang, J. X. Wang and T. G. Wang. *Astrophys. J.*, **644** (2006), 725.

[392] A. K. Pradhan, S. N. Nahar, M. Montenegro *et al. J. Phys. Chem. A*, **113** (2009), 12356.

[393] J. C. Lee, K. Iwasawa, J. C. Houck, A. C. Fabian, H. L. Marshall and C. R. Canizares. *Astrophys. J. Lett.*, **570** (2002), L47.

[394] C. S. Reynolds and M. A. Nowak. *Phys. Repts.*, **377** (2003), 389.

[395] D. A. Liedahl and D. F. Torres. *Can. J. Phys.*, **83** (2005), 1177.

[396] L. Miller, T. J. Turner and J. N. Reeves. *Astron. Astrophys.*, **483** (2008), 437.

[397] C. S. Reynolds, J. Wilms, M. C. Begelman, R. Staubert and E. Kendziorra. *Mon. Not. R. Astr. Soc.*, **349** (2004), 1153.

[398] J. Lee, P. M. Ogle, C. R. Canizares *et al. Astrophys. J. Lett.*, **554** (2001), L13.

[399] J. Garcia, C. Mendoza, M. A. Bautista *et al. Astrophys. J. Supp. Ser.*, **158** (2005), 68.

[400] T. R. Kallman and P. Palmeri. *Rev. Mod. Phys.*, **79** (2007), 1.

[401] J. C. Lee, J. Xiang, B. Ravel, J. Kortright, K. Flanagan. *Astrophys. J.*, **702** (2009), 970.

[402] A. K. Pradhan. *Astrophys. J. Lett.*, **545** (2000), L165.

[403] A. K. Pradhan, G.-X. Chen, F. Delahaye, S. N. Nahar, J. Oelgoetz. *Mon. Not. R. Astr. Soc.*, **341** (2003), 1268.

[404] P. Palmeri, P. Quinet, C. Mendoza *et al. Astrophys. J. Supp. Ser.*, **177** (2008), 408.

[405] J. Dubau, D. Porquet, O. Z. Zabaydullin. arXiv:astro-ph/0303435.

[406] S. Kaspi *et al. Astrophys. J. Supp. Ser.*, **574** (2002), 643.

[407] D. A. Liedahl and D. H. Torres. *Can. J. Phys.*, **83** (2005), 1177.

[408] C. S. Reynolds and M. A. Nowak. *Phys. Rept.*, **377** (2003), 389.

[409] R. A. Sunyaev and Ya B. Zeldovich. *Astrophys. Space Sci.*, **7** (1970), 3; Ibid. *Ann. Rev. Astron. Astrophys.*, **18** (1980), 537.

[410] A. G. Reiss *et al. Astrophys. J.*, **699** (2009), 539.

[411] A. A. Penzias and R. W. Wilson. *Astrophys. J.*, **142** (1965), 419.

[412] S. Seager, D. D. Sasselov and D. Scott. *Astrophys. J.*, **128** (2000), 407.

[413] R. A. Sunyaev and J. Chluba. *ASP Conf. Ser.*, 2008 (astro-ph/0710.2879).

[414] G. W. F. Drake and D. C. Morton. *Astrophys. J. Supp. Ser.*, **170** (2007), 251.

[415] I. Bray, A. Burgess, D. V. Fursa and J. A. Tully. *Astron. Astrophys. Suppl. Ser.*, **146** (2000), 481.

[416] B. Ercolano and P. J. Storey. *Mon. Not. Roy. Astr. Soc.*, **372** (2006), 1875.

[417] J. A. Rubino-Martin, J. Chluba and R. A. Sunyaev. *Mon. Not. R. Astr. Soc.*, **485** (2008), 377.

[418] W. Y. Wong, A. Moss and D. Scott. *Mon. Not. R. Astr. Soc.*, **386** (2008), 1023.

[419] R. H. Becker *et al. Astron. J.*, **122** (2001), 2850.

[420] J. E. Gunn and B. A. Peterson. *Astrophys. J.*, **142** (1965), 1633.

[421] S. L. Ellison, M. T. Murphy and M. Dessauges-Zavadsky. *Mon. Not. R. Astr. Soc.*, **392** (2009), 998.

[422] D. N. Spergel *et al. Astrophys. J. Supp. Ser.*, **148** (2003), 175.

[423] Y. I. Izutov and T. X. Thuan. *Astrophys. J.*, **602** (2004) 200.

[424] M. Fukugita and M. Kawasaki. *Astrophys. J.*, **646** (2006), 691.

[425] R. L. Porter, G. J. Ferland, K. B. MacAdam and P. J. Storey. *Mon. Not. R. astr. Soc.*, **393** (2009) L36.

[426] R. Cen and J. P. Ostriker. *Astrophys. J.*, **514** (1999), 1.

[427] R. Dave *et al. Astrophys. J.*, **552** (2001), 473.

[428] D. A. Buote, L. Zappacosta, T. Fang *et al. Astrophys. J.*, **695** (2009), 1351.

[429] F. Nicastro *et al. Astrophys. J.*, **629** (2005), 700.

[430] P. Richter, F. B. S. Paerels and J. S. Kaastra. *Space Sci. Rev.*, **134** (2008), 25.

[431] S. V. Penton, J. M. Shull and J. T. Stocke. *Astrophys. J.*, **544** (2000) 150.

[432] T. M. Tripp *et al. Astrophysics in the Far Ultraviolet: Five Years of Discovery with FUSE*, ASP Conf. Ser. 348, (ed. G. Sonneborn, H. W. Moos and B.-G. Andersson, 2006), 341.

[433] E. Peik, B. Lipphardt, H. Schnatz *et al. Phys. Rev. Lett.*, **93** (2004), 170801.

[434] J. N. Bahcall, W. L. Sargent and M. Schmidt. *Astrophys. J.*, **149** (1967) L11/

[435] A. F. Martinez Fiorenzano, G. Vladilo and P. Boniffacio. *Mem. Soc. Astron. Italiana Suppl.*, **3** (2003), 252.

[436] M. T. Murphy, J. K. Webb, V. V. Flambaum. *Mon. Not. R. Astr. Soc.*, **345** (2003), 609.

[437] J. N. Bahcall, C. L. Steinhardt and D. Schlegel. *Astrophys. J.*, **600** (2004), 520.

[438] D. G. Grupe, A. K. Pradhan and S. Frank. *Astron. J.*, **130** (2005), 355.

[439] M. Rowan-Robinson. *The Cosmological Distance Ladder*, W. H. Freeman and Company, New York, 1985).

[440] O. C. Wilson and M. K. V. Bappu. *Astrophys. J.*, **125** (1957), 661.

[441] G. Pace, L. Pasquini and S. Ortolani. *Astron. Astrophys.*, **401** (2003), 997.

[442] S. Henrietta Levitt. *Harvard College Observatory*, **LX(IV)** (1908), 87.

[443] J. P. Cox. *Theory of Stellar Pulsations* (Princeton, 1980).

[444] D. J. Majaess, D. G. Turner, D. J. Lane. *Mon. Not. R. Astr. Soc.*, **390** (2008), 1539.

[445] R. B. Tully and J. R. Fisher. *Astron. Astrophys.*, **54** (1977), 661.

[446] A. Filippenko. *Ann. Rev. Astron. Astrophys.*, **35** (1997), 309.

[447] M. M. Phillips. *Astrophys. J.*, **413** (1993), L105.

[448] X. Wang *et al. Astrophys. J.*, **697** (2009), 380.

[449] D. N. Sauer and P. A. Mazzali. *New Astronomy Reviews*, **52** (2008), 370.

[450] S. Perlmutter *et al. Astrophys. J.*, **517** (1999), 565.

[451] A. G. Riess *et al. Astron. J.*, **116** (1998), 1009.

[452] B. Leibundgut. *Ann. Rev. Astron. Astrophys.*, **39** (2001), 67.

Index